ASHRAE

Humidity Control Design Guide

For Commercial and Institutional Buildings

Lewis G. Harriman III
Member, ASHRAE

G. W. Brundrett, Ph.D., CEng.
Fellow, ASHRAE

R. Kittler, PEng.
Fellow, ASHRAE

Project Contractor - 1047-RP

Lew Harriman, Principal Investigator
Mason-Grant Consulting
P.O. Box 6547
Portsmouth, NH 03802 USA
www.MasonGrant.com

Project Monitoring Committee

Chair: Craig Wray, Lawrence Berkeley Laboratory
Rodney Lewis
Lynn Werman
TC 3.5: Jaroslav Wurm
TC 8.7: Christopher Morton
Sponsors: Douglas Kosar, GTI
Phillip Fairchild, DOE/ORNL

Project Guided & Funded by:

- ASHRAE Technical Committees
  TC 3.5 Desiccant & sorption technology
  TC 7.5 Mechanical dehumidifiers
  TC 8.7 Humidifying equipment
  TC 9.12 Tall buildings
- The Gas Technology Institute
- U.S. Department of Energy
  (Through Oak Ridge National Laboratory)

ISBN 1-883413-98-2

American Society of Heating, Refrigerating and Air Conditioning Engineers, Inc.
1791 Tullie Circle, N.E., Atlanta, GA 30329
(404) 636-8400 www.ashrae.org

DEDICATED TO THE ADVANCEMENT OF THE PROFESSION AND ITS ALLIED INDUSTRIES

DISCLAIMER

Humidity Control Design Guide

For Commercial and Institutional Buildings

ISBN 1-883413-98-2

Authors

Lew Harriman
Mason-Grant Consulting
P.O. Box 6547
Portsmouth, NH 03802 USA
LewHarriman@MasonGrant.com

Geoffrey W. Brundrett
Cooks Hill, Kingswood
Fordsham, WA6 6JT
United Kingdom
GeoffBrundrett@Waitrose.com

Reinhold Kittler
Dectron Internationale
2316 Mariner's Mark Way, # 304
Virginia Beach, VA 23451
rkittler@aol.com

Chapter Co-Authors

Larry Berglund
Tohoku University, Japan
larry@sabine.pln.archi.tohoku.ac.jp

Donald Gatley
D.P Gatley & Associates
Atlanta, Georgia
dpgatley@earthlink.net

Joseph Lstiburek
Building Science, Inc.
Westford, Massachusetts
joe@buildingscience.com

William Lull
Garrison-Lull
Princeton Junction, New Jersey
wplull@alum.mit.edu

Bernard Morton
Naples, Florida

Chris Morton
Morton Engineering
Waconia, Minnesota
mortonenginc@aol.com

Berlin Nelson
North Dakota State University
Fargo, North Dakota
berlin_nelson@ndsu.nodak.edu

J. David Odom III
CH2M/Hill
Orlando, Florida
jodom@ch2m.com

Preface

This project resulted from an awakening of interest in humidity control among building owners, Architects and HVAC Engineers.

During the 1990's, it became apparent that many commercial buildings have problems related to moisture and humidity control. The causes of these problems are complex, and they touch all the disciplines involved in a building project. This particular book deals with only one aspect of preventing moisture problems—the HVAC system. The book was produced as a response to requests from ASHRAE members for more and better information about designing for humidity control. It appears to be the first time in the 101-year history of the Society that ASHRAE has focused a book on this topic.

We are very pleased to have had the opportunity to collect and present this information. All of the authors share the opinion that this attention to humidity control is long overdue, and that if designers have access to better information, the environment for building occupants all over the world will be better in many important ways. Human comfort, energy economics, equipment design and construction costs can all be improved by a better understanding of humidity and how it can be controlled.

At the same time, we are humbled by the magnitude of the project we have undertaken. Imagine how difficult it would be to produce a book titled: "Temperature Control Design Guide." The possibilities for books on that topic are endless, and the opinions of experienced professionals are often contradictory and strongly-held. We faced a similarly daunting task in writing this book.

With that in mind, we trust that the reader will consider this book simply a good beginning rather than the last word on humidity control. Much remains to be done. We sincerely hope that the reader will share his or her experiences, observations, questions, corrections and considered judgements in the form of additional written work, expanding the range and depth of information available to our profession and to the public at large.

Acknowledgments

We are especially grateful to the U.S. Department of Energy, its Offices of Distributed Energy Resources and Power Technologies, and also to the Gas Technology Institute for their technical and financial support of this project. We also appreciate the three years of effort by the dedicated professionals who gave their time to serve on the project monitoring committee and on the ASHRAE technical committees that guided our work. Further, the authors are indebted to the writer Cynthia Harriman, for her tireless weed-cutting of our dense prose and her hunting down of errors in usage, grammar and typography. Wherever the book is clear, accurate and complete, it is largely because of the oversight of these generous volunteers.

Finally, the information in this book is the product of our personal experiences and those we have collected from hundreds of other professionals through conversations, technical and trade papers and our day-to-day professional and commercial activities. Consequently, this work is dedicated to our families, teachers, clients and colleagues, without whom this book could not exist.

Lew Harriman
Portsmouth, New Hampshire
September, 2001

Humidity Control Design Guide

For Commercial and Institutional Buildings

Major Sections & Chapter Titles

Chapter 5 - Corrosion & Relative Humidity (Continued)

Chapter 20 - Retail Buildings (Continued)

Chapter 23 - Museums, Libraries & Archives (Continued)

Chapter 1

Purpose, Assumptions & Terminology

Introduction

Controlling humidity in commercial buildings is simple. Just add or subtract enough moisture to or from the air to compensate for any surplus or deficit of humidity inside the building.

The difficulty comes in changing the amount of that addition and subtraction in response to changing loads, so that humidity stays within an acceptable range. Also, the designer must achieve control within an acceptable budget for both construction and operation. To do this, the designer must understand the owner's purpose for the project. That understanding allows the HVAC designer to make more effective assumptions as the project proceeds.

The authors have taken the same approach to writing this book. We begin our work by explaining our understanding of the book's purpose. Based on that understanding, we proceed to make assumptions about our readership and its needs. These assumptions have governed our decisions about what information to include, and they influenced both the structure and the depth of that information. To help *you* decide whether this book will meet *your* needs, you may wish to consider our purpose and assumptions.

Purpose

The purpose of this book is to help technical professionals design humidity control systems for commercial buildings. If you are not a technical professional, or are interested mostly in residential or industrial buildings, this book is likely to meet fewer of your needs.

Assumptions

We assume the reader:

1. Is a technical professional, although not necessarily one with an HVAC engineering degree or professional engineer's registration.
2. Has a basic understanding of HVAC terminology, elementary algebra and the common variables of equations used in designing HVAC systems.

As an overall focus, we write primarily for the system designer who "has to get it done by Friday." We expect that the text may also be useful to others such as architectural designers, building developers and owners, maintenance professionals, students, teachers and researchers. However, where decisions have been made about technical depth and detail, we have tried to meet the needs of the HVAC designer before meeting the needs of others.

Strengths And Limitations Of This Book

We expect that this information will be most useful for designing humidity control systems for commercial and institutional buildings, rather than for industrial, agricultural or residential buildings.

Emphasis on low-rise construction

Our book is focused on the low-rise construction typical of 90% of the commercial space built each year in North America. The book contains less information that is specific to high-rise buildings alone.

Includes information to help establish the humidity control level

After the owner clearly defines the purpose of controlling humidity, the next step is to establish the humidity control level. The book includes information about the effects of different humidity levels on comfort, corrosion, electrostatic discharge, mold and mildew, dust mites and insects as well as bacteria and viruses. We hope this information will help the owner and designer establish a control level that will achieve the purpose of the project.

Most useful for controlling humidity between 30 and 60% rh

We assume the designer's system will be intended to control humidity within a defined range, and that its range will fall somewhere between 30 and 60% rh. There are other reference books that would be more helpful than this text for designing systems to maintain a control level above or below that mid-range humidity.

This book emphasizes dehumidification

This book was created by ASHRAE research project 1047-RP. That project was originally generated in response to a strong demand from the ASHRAE membership for better information concerning *dehumidification* systems for hot and humid climates.

After the project began, the Authors and the Project Monitoring Committee both agreed there is also a pressing need for information that helps the designer of *humidification* systems in dry climates. So we have included information on that topic as well. But the principal emphasis of this book is on dehumidification.

This book is frequently redundant

The Authors and the Project Monitoring Committee believe that most readers will first read the specific sections that are most related to their immediate projects. So, we have tried to make each chapter as complete as possible *by itself.* Consequently some of the text and graphics appear in more than one chapter. Repetition reduces the number of annoying references to other chapters, but it will be redundant for those intrepid souls who read the book cover-to-cover.

For specifics, you'll have to contact equipment suppliers

The worldwide range of equipment is far too broad to include here. Also, critical equipment details become obsolete as manufacturers change their designs. Finally, enough detail to be useful would violate the commercial impartiality required for ASHRAE publications. So this book only helps begin the design process.

This information should help you define the basic layout and general requirements of a humidity control system, including general types of dehumidifiers and humidifiers, their controls and the loads they must remove. After that, you will need to have more extensive discussions with equipment suppliers to complete a full set of plans and specifications.

Terminology

The heating and air conditioning industry is constantly changing, so its terminology is sometimes unclear and not always consistent. Beyond the psychrometric variables described in Chapter 3, we use a few other words and phrases which should be clear if you are to understand what we say in this book:

Humidity moderation

By moderation, we mean that the HVAC system helps the building avoid extremes of humidity, but that humidity can still swing, uncontrolled, throughout a broad range over 24 hours. Humidity *moderation* is easier and less expensive than humidity *control.*

Humidity control

By control, we mean that the indoor humidity is held within a defined range at all times. That range may be wide or narrow, and it may only have a high or low limit rather than both. But when a building is said to require humidity control, we assume the system must not allow the indoor humidity to rise or fall beyond the limits specified by the owner. *Control* is more difficult and costs more to achieve than *moderation*. This book is focused primarily on humidity control rather than just moderation.

Mechanical dehumidifier

A mechanical dehumidifier removes moisture by cooling the air so that its excess moisture condenses and is drained away as liquid water. The process uses mechanical means—compressors, condensers, pumps and motors—to achieve the cooling of air, and therefore its dehumidification. Such equipment is also called a "cool-reheat" dehumidifier, or a "cooling-based" dehumidifier.

Any cooling coil will condense water and dehumidify air if the airs dew point is high enough. But what we mean by a mechanical dehumidifier is a piece of equipment that has been designed specifically for removing moisture rather than for cooling, and one which operates independently of the building's cooling equipment.

Desiccant dehumidifier

A desiccant dehumidifier removes moisture by attracting water vapor out of the air to the surface of a desiccant. The desiccant adsorbs (or absorbs) water in its vapor phase. No condensation of liquid water is involved. Water vapor moves out of the humid air to the dry desiccant surface in response to the difference in vapor pressures. After a great deal of water has been adsorbed by the desiccant, hot air is used to dry out the saturated desiccant for reuse.

Desiccant dehumidifiers are sometimes called "active desiccant wheels", or "chemical dehumidifiers." Like mechanical dehumidifiers, they operate independently of cooling or heating equipment as they dry the air.

Load reducer

By "load reducer," we mean a passive desiccant wheel or other heat exchanger. These are also called enthalpy wheels, heat recovery wheels, total heat wheels, total enthalpy wheels or air-to-air heat exchangers or heat pipes.

Chapter 15 contains detailed information about these devices. They are placed between the exhaust air and the incoming ventilation air, so they can transfer energy between those two air streams. Some of them can recover the dehumidification effect from exhaust air during the summer, and others simply recover the heating or cooling effect of that exhaust air.

We call these heat exchangers "load reducers" to distinguish them from dehumidifiers and humidifiers, which by themselves, can remove or add moisture to the air. A load reducer cannot, by itself, control humidity. They only work when they have exhaust air flowing through them, and that exhaust air must be dry enough or humid enough to reduce the load from the incoming ventilation air. A system with a load reducer still requires a humidifier or dehumidifier if it is to control humidity.

Ventilation air

By ventilation air, we mean *outdoor* air. In other words, air that enters the system from outside the building, as opposed to air that has been recirculated from someplace inside the building. Ventilation air is also sometimes called either "fresh air" or "outdoor air."

Ventilation pretreatment system

By this we mean a separate system which dehumidifies or humidifies ventilation air *before* that incoming outdoor air mixes with the return air from the rest of the building.

We use this term to describe a system which is principally concerned with pretreating the moisture content of the incoming air. But beyond those functions, such a pretreatment system might *also* clean the air by removing particulate with filters, and adjust the temperature of the incoming air with heaters or coolers.

Corrections & Suggestions, Please

Finally, we assume that as the first ASHRAE book ever dedicated to the subject of humidity control, the material contained here is simply a good beginning. It can certainly be improved as readers use the information and learn from that experience. We encourage you to contribute to the ongoing improvement of this book. We welcome all constructive comments and suggestions on any aspect of this text. Please address your remarks to:

Lew Harriman
Mason-Grant Consulting
P.O. Box 6547
Portsmouth, NH 03802 USA
LewHarriman@MasonGrant.com
Tel: (603) 431-0635
Fax: (603) 427-0015

Chapter 2
Humidity Control Basics

By Lew Harriman

Humidity Control

The importance of humidity control depends on your personal perspective. To the nursing home patient, it means a comfortable environment year-round. To the museum curator, it means slower deterioration of artwork and historical artifacts. To the maintenance professional, it means a reliable system with no complaints from occupants. Each participant has a different perspective.

This chapter is intended to establish a common level of understanding about key concepts for the major groups involved in a humidity control project. Beginning with common understanding of the basics, the project team can save time and money by focusing on the purpose of humidity control as it affects the specific and unique circumstances of the project in question.

Fig. 2.1 The Owner

The owner can reduce costs and speed the project by clearly defining the reasons that humidity must be controlled, and by explaining what happens when humidity is not controlled. This information will guide dozens of cost/benefit decisions that must be made by other members of the team.

Owner's Role - Define the Purpose of the Project

Unlike temperature control, humidity control is not automatically included in commercial buildings. Many owners are quite tolerant of wide variations in humidity from hour-to-hour and season-to-season. So a designer seldom provides humidity control unless the owner specifically asks for it.

Assuming the owner needs humidity control enough to specify it, the owner's most important responsibility is to explain the purpose of humidity control for the building, and the consequences to the building's function if humidity control is not adequate. Without a clear understanding of the project purpose, the construction team must guess, and their guesses will not always meet the owner's expectations. When the purpose is clear, the building team can make effective professional judgements and more accurately estimate the cost of technical alternatives.

Taking the time to define the project purpose is much less expensive for the owner before the project begins. Postponing clear definition of the owner's expectations wastes time and effort, and frequently requires the assistance of the legal profession. [1.]

Good definitions of project purpose

To be useful, the definition of the project purpose should include the reasons for humidity control, along with the consequences of failing to provide it. A common, but nonfunctional instruction from an irate owner might be: "I don't want to spend a lot of money, but make sure the humidity control is done right. The occupants in our other buildings are always complaining."

The owner has a responsibility to go further, because the construction team needs better instructions. What purpose did the last building serve? Why were the occupants complaining? Which ones are complaining? How often? At what time of year? Were they feeling too dry or too humid? How much money is "too much" to spend to avoid the problem? The list of questions goes on indefinitely. The process can be accomplished more quickly with a better definition of project purpose. Here are some examples.

Poor Definition: "Control the humidity as usual for a hotel"

Better Definition: "Keep temperature and humidity in the hotel within the ASHRAE-defined summer and winter human comfort zones"

Comprehensive Definition: "Control the temperature and humidity in the hotel such that:

a. Neither condensation nor humidity over 60% rh can occur either inside the building or inside building cavities to cause mold, mildew, or other moisture damage to building materials or furnishings.

b. Guest rooms, corridors, meeting rooms, dining rooms and offices are maintained above 30% rh, and below 60% rh during all occupied hours.

c. The system has adequate capacity and modulation to achieve control for 98% of the hours in a typical year."

The *poor definition* is at least a starting point for a further discussion. The owner presumably has some impression of what "normal-for-a-hotel" means. It also signals the construction team that the subject has some degree of importance simply because it appears in the owner's instructions. But the definition leaves a lot to the imagination, and left alone, it virtually guarantees misunderstandings and problems.

The *better definition* says much more. It defines the owner's expectations and implies the reasons for controlling humidity. Apparently the concern is comfort, and comfort is defined in a meaningful, quantified manner. But in a hotel, of course, there are special-purpose areas like kitchens and swimming pools. Presumably these do not carry the same requirement as meeting rooms and guest rooms. Also, guest rooms are unoccupied nearly all day, so one might consider relaxing the specification during unoccupied hours.

The *comprehensive definition* clarifies the owner's intentions much further. Now we know the owner is concerned about mold and mildew damage, not simply comfort. Also, air pressures and flows must be controlled to avoid mold problems behind walls, not just inside guest rooms. Equally important, the owner has told the team what shortcomings are acceptable—the system need not be designed for the 30-year extreme. It can fail to achieve the goals for up to 2% of the hours in a typical year without causing the owner an excessive problem. The comprehensive definition tells the team ***what result is expected, what may happen if the result is not achieved, and what shortcoming is acceptable.***

The comprehensive definition results in a more expensive HVAC system. On the other hand, the poor definition can lead to far greater expense because of the risk of miscommunication and moisture damage. Each owner must decide the value of humidity control, and communicate his expectations to the building team clearly if he wants predictable results.

Owner and occupant expectations: moderation vs. control

In many commercial buildings, there is no particular need for a specified control level. Or at least no need that justifies additional expense. On the other hand, occupants and owners expect that with modern HVAC systems, they will always be "comfortable." Expectations become implied specifications, which are a frequent source of conflict on building projects.

The owner can reduce the potential for conflicts by controlling expectations when there is no specified humidity control level. Simply stated, if the owner does not define the humidity *control* requirement, he should expect that a typical HVAC system will only "*moderate*" humidity. Moderation means that the humidity will drift from hour to hour and season to season, but will probably not be above 70% rh or below 10% rh except during extreme weather. The key word is "probably". Unless the owner specifies the purpose and consequences of humidity control to the construction team, he would be well-advised to control the expectations of the occupants, promising nothing other than *moderation* of humidity for most of the time, rather than implying *comfort or control* all of the time.

Owner's Checklist

1. What is the **owner's purpose** in controlling humidity?

2. **What happens** when humidity is not controlled?

---

3. Which parameter most controls success of the project? ❑ Relative Humidity ❑ Dew Point ❑ Both

4. What range is tolerable, given the purpose of the project? Summer: From ______ to _____ Winter: From ______ to ______

5. Which extreme condition shall be used to size equipment?
   Summer: ❑ 35 hours beyond design ❑ 87 hours ❑ 175 hours
   Winter: ❑ 35 hours beyond design ❑ 87 hours

---

6. What will be the typical occupancy of the building?

Weekday

# of People	Hour	# of People	Hour
_______	0100	_______	1300
_______	0200	_______	1400
_______	0300	_______	1500
_______	0400	_______	1600
_______	0500	_______	1700
_______	0600	_______	1800
_______	0700	_______	1900
_______	0800	_______	2000
_______	0900	_______	2100
_______	1000	_______	2200
_______	1100	_______	2300
_______	1200	_______	2400

Saturday/ Sunday

# of People	Hour	# of People	Hour
_______	0100	_______	1300
_______	0200	_______	1400
_______	0300	_______	1500
_______	0400	_______	1600
_______	0500	_______	1700
_______	0600	_______	1800
_______	0700	_______	1900
_______	0800	_______	2000
_______	0900	_______	2100
_______	1000	_______	2200
_______	1100	_______	2300
_______	1200	_______	2400

Holiday

# of People	Hour	# of People	Hour
_______	0100	_______	1300
_______	0200	_______	1400
_______	0300	_______	1500
_______	0400	_______	1600
_______	0500	_______	1700
_______	0600	_______	1800
_______	0700	_______	1900
_______	0800	_______	2000
_______	0900	_______	2100
_______	1000	_______	2200
_______	1100	_______	2300
_______	1200	_______	2400

Architectural Designer - A Low-Leakage Building

Commercial buildings leak air. The amount of leakage determines the cost of humidity control, and the degree of risk of damage from condensation. The Architect has tremendous influence over both the cost and effectiveness of humidity control through design of the building envelope and supervision of its construction.

A humidity-controlled building in some ways resembles a refrigerator. If a refrigerator had holes drilled in its door, or leaks at the seams of panels, humid air would migrate into the panels, where moisture would condense and grow mold. And if the air seal around a refrigerator door were cracked, cut or missing, humid air would enter the box, condensing inside and making it difficult to control temperature and impossible to control humidity. Different problems would occur in the winter. The refrigerator would pull in dry air, so the food would dry out.

Commercial buildings are actually more similar to *very leaky* refrigerators. Even the tightest buildings leak enough to provide a complete change of air every three hours! Most buildings are much worse. Outside air leaks in to provide a complete air change every one or two hours, and even as often as every 30 minutes.[2] Unlike ventilation air, the temperature and moisture of infiltration air cannot be adjusted to avoid problems.

Humid air in the summer leaks through the wall and condenses inside the wall cavities where it contacts surfaces kept cold by the cooling system and on cold pipes. In the winter, cold leakage air cools the surfaces inside the walls, and then humid air from inside the building condenses moisture inside the wall cavities.

If the entire building really behaved like a refrigerator, leakage could be eliminated easily by simply providing an excess of conditioned ventilation air, so that the wall cavities would be filled with conditioned air instead of untreated leakage air. But buildings are much more complex than refrigerators. A positive internal air pressure in one part of the building does not mean that all parts of the building are under positive air pressure with respect to the outside. Studies have shown this fact time and time again.[2]

So when the owner requests humidity control, the Architect should pay more-than-usual attention to the design of the building envelope, with the goal of minimizing air leakage. Of course that's easy to say, but very difficult to do. No Architect intends for the building to leak at all. But based on frequent field problems, building scientists have developed useful advice for the architectural designer who is beginning a new project.

Air infiltration vs. vapor diffusion

Much attention has been lavished on the problem of vapor diffusion through building materials. To be sure, water vapor does diffuse through materials to equalize the water vapor pressure on both sides of the material or wall. Diffusion results in condensation when the vapor hits a cold surface, one which is below the dew point of the air

Fig. 2.2 The Architectural Designer
The humidity-controlled building must not leak large amounts of air, and materials must be selected so that interior finishes do not trap moisture in walls, and so that thermal bridges do not create condensing surfaces.

Fig. 2.3
Take Steps To Minimize Air Leaks... Vapor Diffusion Is Seldom A Problem
In a typical commercial building, the moisture load from infiltration is 10 to 200 times larger than the load from permeation. Hence the importance of minimizing air leaks.

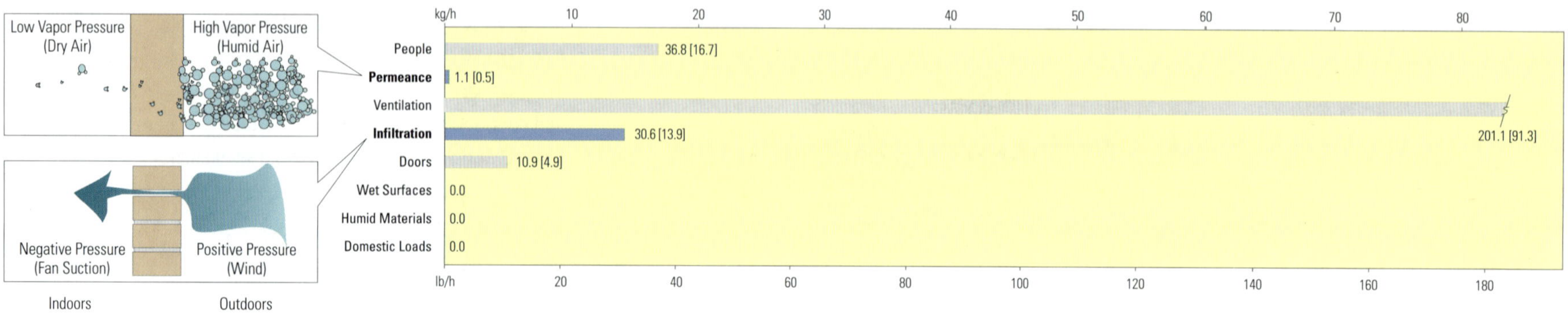

contacting that surface. The Architect is certainly responsible for designing the wall, ceiling and floor so that vapor diffusion does not cause condensation. To avoid that potential problem, many useful calculations and computer programs are readily available to assist the Architect and the HVAC designer.[3]

However, the much larger problem is air infiltration through gaps in the exterior building envelope. Air infiltration produces 10 to 200 times more water vapor leakage than vapor diffusion, as shown graphically in figure 2.3. Unfortunately, engineering tools for predicting and quantifying that huge problem are essentially nonexistent, because the problem depends on the number and size of construction gaps, and the relative air pressures on each side of the wall. No construction team intends to have any gaps in the wall, and the wind pressure varies minute-to-minute. So in the absence of credible values to fill those terms, equations to calculate leakage are educated guesses. But that does not mean that the architectural and HVAC designers are without resources to minimize this key problem. Suggestions include:

Install a continuous air barrier

Regardless of climate, air leaking into the walls will cause problems. In cold climates, humid *indoor* air can condense its moisture if it gets into cold exterior walls. In hot and humid climates, humid *outdoor* air leaks in and condenses moisture on cooler surfaces inside walls. Installing an air barrier reduces these problems.

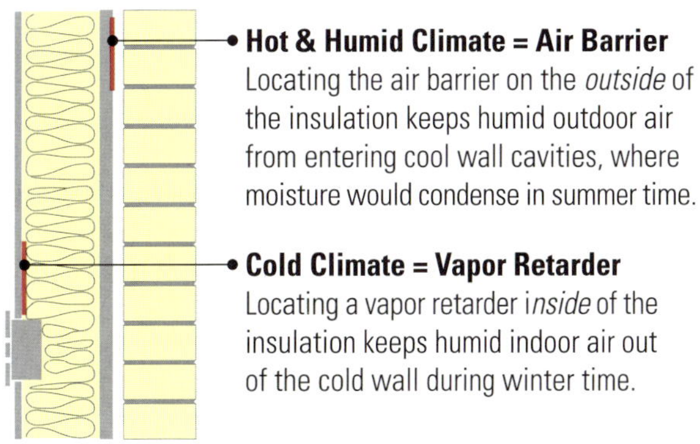

Fig. 2.4
Locating Air Barriers And Vapor Retarders

The placement of the air barrier is critical. In hot and humid climates, it should be located on the *outdoor* side of the insulation. In very cold climates, the air barrier should be located on the *indoor* side of the insulation. In climates such as Kansas City, Missouri or Seoul, Korea, both very hot summers and very cold winters are typical. In those climates the air barrier should be towards the outside—unless the building is humidified during the winter, as in a swimming pool enclosure or a museum. Then the air barrier should be placed towards the indoors with respect to the insulation. The logic is not immediately obvious, but it is consistent: an air barrier is best located where it can keep the most humid air *out of most of the wall* for most of the year.

In addition, for any air barrier to be effective it must be continuous. All seams and joints must be sealed with tape, not just overlapped. This includes the joints between the air barrier film and all doors and window frames, and the whole length of the joints where walls meet the roof and foundation.

A good HVAC design and installation is also necessary, but the Architect and Owner should recognize that the foundation of humidity control is a tight building. Without that foundation, humidity control will be very difficult and costly to achieve, no matter how well-designed the HVAC system might be.

Avoid suction inside the exterior wall

One reason that commercial buildings leak so much air is that HVAC systems often create suction near the exterior wall by accident. Then air outdoor leaks in continuously. Although the HVAC designer and contractor are responsible for avoiding this problem, the architectural designer can also help avoid such suction.

When the HVAC equipment does not have ducted returns, the plenum above the ceiling is under negative pressure constantly, which leads to very high outdoor air infiltration rates.[2] Figure 2.14 shows this situation, along with what each member of the team can do to minimize the problems it creates.

The HVAC designer can reduce the problem by specifying a return duct system which is hard-connected to the rooftop HVAC unit. The architectural designer can also reduce the potential problem by specifying that the interior wallboard must continue all the way to the roof, and must be sealed to the underside of the roof decking.

Specify a whole-building leak test

After construction is complete, the Owner, Architect, Engineer and maintenance staff assumes it has been built to specifications, and will not leak an excessive amount of air. Where humidity levels are especially high or low, as in museums, swimming pools or ice arenas, quantifying the air leakage after construction is useful. This can be done by specifying a whole-building leak test.

Two leak test procedures are often used by indoor air quality consultants and by energy auditors. Smaller buildings can be tested at low cost with a blower door. In that procedure, the door is replaced with a moveable partition containing an air blower. The building is "inflated", and the blower air flow is adjusted until the pressure inside the building is either 4 or 50 Pascals higher than the pressure outside the building [either 0.016 or 0.2 in. wc. higher inside the building] . The air flowing into the building at that pressure difference is defined as the leakage.[5]

Larger buildings are usually tested by measuring the dilution of a tracer gas. A harmless tracer gas, usually sulfur hexafluoride, is released inside the static, closed building with the HVAC system off. The concentration of the gas is measured periodically and compared to the original concentration. The dilution of the gas over time allows calculation of the air infiltration rate.[5]

After the leakage volume is quantified, it can be compared to the amount of excess air provided by the ventilation system. In hot and humid climates, the HVAC designer and contractor can adjust the ventilation air to maintain a positive internal air pressure, which ensures that dry air will be leaking *out*—rather than humid air leaking *into* the building. In cold weather, ventilation and exhaust air (beyond the code-required minimums) should be adjusted to maintain a neutral or slightly negative internal air pressure, to keep humid indoor air from leaking outwards into cold wall cavities.

Vapor retarders

In most commercial buildings, air barriers are more useful than vapor retarders. Humid air infiltration into cold walls carries far more damaging moisture than what vapor permeation can push through solid materials. Also, many walls will gain excess moisture at some point through air or water leakage. Placing a vapor retarder in the wall makes it more fragile, because the water may become trapped. A vapor retarder may prevent the seasonal drying that avoids a major problem. On the other hand, vapor retarders are quite important in cold climates, and for any building with a high indoor moisture level, such as museums and swimming pool enclosures. The logic for placing vapor retarders is the same as for air barriers: keep high humidity out of the walls. Here are some guidelines which may be useful:

Locate vapor retarders only on the warm side of the wall

In cold climates and in humidified buildings, the vapor retarder should be located inside from the insulation. Vapor will be held inside the occupied space, rather than in the cooler wall cavities, where it could condense.

Avoid doubled vapor retarders

Some of the most widespread and costly building failures have been caused when vapor retarders are doubled by accident, creating a vapor tight "pressure cooker" by sealing both sides of an exterior wall. Then any moisture that gets into that wall never dries out, and major damage follows. The classic example is when the architectural designer and owner decide on a face-sealed external glass fiber insulation wall for the exterior (EIFS), while at the same time the interior designer places an impermeable vinyl wall covering on the inside of a hotel room in a humid climate. Tens of millions of dollars

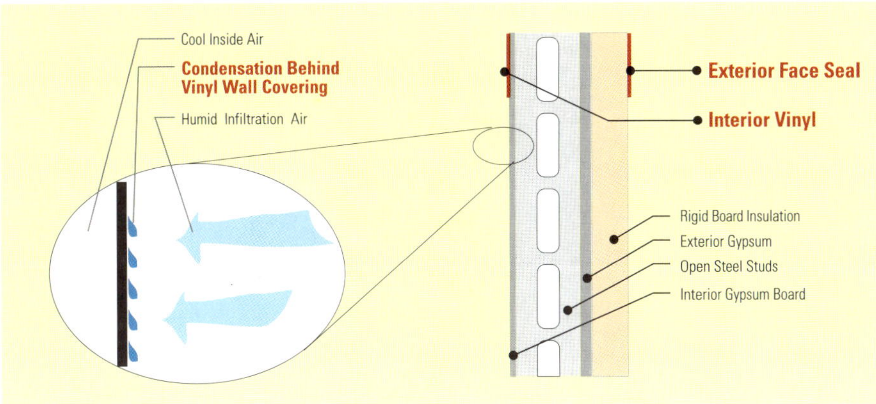

Fig. 2.5
Avoid Trapping Water Inside A Wall
Except in swimming pools, an inside surface vapor retarder such as vinyl wall covering is a very poor practice. It traps moisture inside the wall, promoting the growth of mold and mildew. Avoid vinyl wall covering on indoor surfaces of exterior walls.

are lost in North American hotels and condominiums every year because of this poor practice.[6] Sooner or later some humid air and/or rainwater gets into those walls, and moisture condenses behind the vinyl. The condensation and high humidity supports mold and mildew, destroying both the wall covering and the wall board, corroding steel studs, and generating the musty odors typical of moisture-damaged buildings.

To prevent such problems, the exterior wall should be able to dry towards the inside (except when the inside is highly humid, as in the case of museums and swimming pools). So a highly-permeable interior wall finish can reduce the potential for damage.

Avoid thermal bridges

In cold climates, adding humidity inside the building can increase the risk of condensation on cold surfaces. Classic problems include condensation on single glazing, on metal casings of windows and doors and on roofing fasteners. All of these elements can either be safe from condensation, or can cause significant condensation problems depending on whether they create a "thermal bridge" between the cold weather outdoors and the humid air inside the building.

If a roofing nail penetrates all the way through the roof membrane, the roof deck, and the insulation, the inside tip of that nail will be cold. Moisture will condense on the nail tip and drip inside the building. Saturated roofing insulation is often caused by internal condensation rather than by rain leaks. Also, frames for doors and windows that are formed from a single piece of steel or aluminum can carry heat out of the building, leaving the inside surface of the frame very cold. This also causes internal condensation and puddling.

To avoid these problems, the architectural designer should detail the exterior walls so there are "thermal breaks" in all metal components. A thermal break places some insulating material—usually an engineering plastic—between metal components. With that insulating joint, there is no "through-path" for heat to move through the exterior wall. In extremely cold climates, energy and building codes will alert the designer to the importance of such thermal breaks. But in more moderate climates, their use may not be so common. Thermal breaks become especially critical in any climate when the building must be humidified during cold months.

The bigger challenge may be in renovation. Buildings that have no difficulty with condensation before humidification may develop severe problems when extra humidity is suddenly added to meet changes in building use. For example, rooms or floors in antique administration buildings of colleges might be converted to museums or libraries. These may develop great difficulties when humidity is added during cold months because there are no thermal breaks in the original construction. In those situations, the architectural designer may find it necessary to replace building components and add vapor retarders near the inside surface of exterior walls to avoid condensation when humidity must remain high during the winter.

Architectural Designer's Checklist

Air Barriers, Vapor Retarders and Thermal Breaks

1. Is the building dehumidified in summer time?	❑ **Yes**. Specify a continuous air barrier outside from the insulation. ❑ **No**. Air barrier is useful, but not essential.
2. Is the building humidified during winter, or is there a swimming pool or spa indoors, or is the winter design temperature below -15°F [-26.2°C]?	❑ **Yes**. Specify a vapor retarder inside from the insulation. Also ensure that windows, doors and exterior components have thermal breaks ❑ **No**. Check local codes to determine if thermal breaks are required.

Quantifying Air Leakage Through the Building Envelope

3. Humidity control is important, but not mission-critical.	❑ No need to quantify the air leakage after construction. The HVAC designer should calculate air infiltration load based on the assumption that the **exterior envelope will not leak more than:** ❑ Very tight buildings.................... 0.10 cfm•ft² [0.5 l/s•m²] ❑ Average commercial buildings:... 0.30 cfm•ft² [0.5 l/s•m²] ❑ Rather loose buildings............... 0.60 cfm•ft² [0.5 l/s•m²] ❑ Other leakage rate..................... ____ cfm ft² [0.5 l/s•m²]
4. Functions performed inside the building ***will suffer*** unless humidity is controlled within the range defined by the owner.	❑ **Large** building - Specify a tracer gas test to validate the fact that air leakage is below _____ cfm•ft² @ 0.016 in.wc [________ l/s•m² @ 4 Pa] ❑ **Small** building - Specify a fan pressurization test to ensure that air leakage is below _____ cfm•ft² @ 0.016 in.wc [________ l/s•m² @ 4 Pa]

HVAC Designer - Control the Outdoor Air

The HVAC designer is the team member most responsible for successful control of humidity. Dozens of critical details compete for attention, and these will be described in depth in later chapters. To gain a useful overview and avoid distractions of details, the designer might consider these basic suggestions.

Pretreat the ventilation air

Willie Sutton, a famous bank robber during the 1930's, was once asked why he robbed banks. He replied: "Because that's where the money is." For a similar reason, the HVAC designer should focus on the ventilation air—because that's where the problem is.

Outdoor air brings large amounts of excessively dry or humid air into the building continuously. That air accounts for 50 to 80% of the dehumidification and humidification loads in a typical commercial building. Consequently, *in most commercial buildings, pretreating the ventilation air usually provides adequate humidity control without the need to change the rest of the system.*

By drying the incoming air below the dew point inside the building, the ventilation air becomes a "sponge", absorbing moisture loads from the building and controlling humidity during humid months. By humidifying ventilation air above the building's moisture level set point during winter, the air makes up any humidity deficit in the building. Pretreating ventilation air is not the only way to control humidity in commercial buildings, but it is often the most cost-effective method.

When there is no dedicated system to dehumidify and humidify the ventilation air, the rest of the HVAC system must be able to remove the load created by that untreated air—because it remains the largest humidity-related load on the system. Failing to quantify and remove the ventilation load at design dew point conditions is the most common reason that commercial buildings have difficulty controlling humidity. Some examples follow.

Fig. 2.6 The HVAC Designer

The HVAC designer leads the team with respect to humidity control, and is principally responsible for its success. In most cases, controlling humidity inside the building is very simple as long as the HVAC designer controls the humidity of all incoming ventilation air, and ensures that the system does not create suction at the exterior wall.

Temperature-controlled unitary equipment removes very little moisture

Room cooling and ventilating units such as unit ventilators, packaged terminal air conditioners (PTACs), fan-coil units and packaged terminal heat pumps (PTHPs) are designed primarily for cooling. The equipment is often used in school classrooms, hotel rooms and dormitories. In some designs, the units pull a small amount of untreated outdoor air into the room whenever the fan is operating, which meets the requirement for local ventilation. However, the unit will not remove excess humidity when the room is not calling for cooling. So humidity builds up and causes problems.

Fan-coil units, which cool air using chilled water, have additional problems when the chilled water temperature changes. During periods of peak sensible heat loads, the chilled water temperature can rise above the intended set point. Also, during periods of low heat load, some automation systems save energy by allowing the chilled water temperature to rise. In both situations, the fan-coil unit loses its dehumidification capacity, and moisture rises above the desired set point.

For better results, provide pretreated air to each unit from

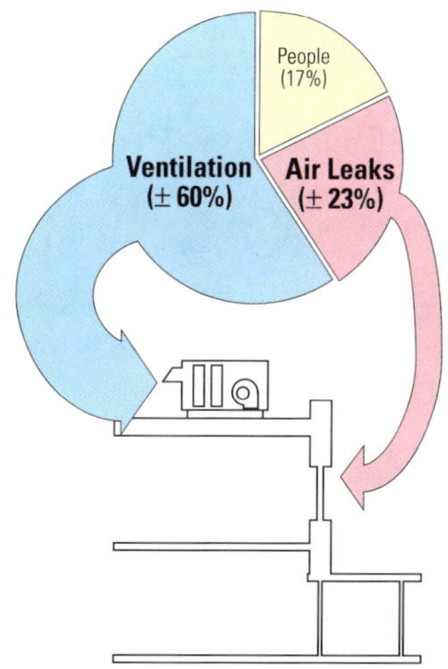

Fig. 2.7 Outdoor Air Load

Moisture load comes primarily from outdoor air ventilation and infiltration. Controlling the quality and quantity of outdoor air is fundamental to humidity control.

a dedicated ventilation system which dehumidifies and humidifies outdoor air. This solution costs more, but provides the humidity control lacking in rooms served by unitary equipment.

"Temperature-Tempered Air" Is Not Dry
When ventilation air is only temperature-tempered by cooling to 80°F [26.7°C], that air stays wet. Outdoor air dew points are always below that temperature.

Inside Air Dew Point
In many climates, ventilation air requires dehumidification all year long. In Tampa, Florida, the outdoor dew point is frequently above 50°F [10°C] even during "winter" months.

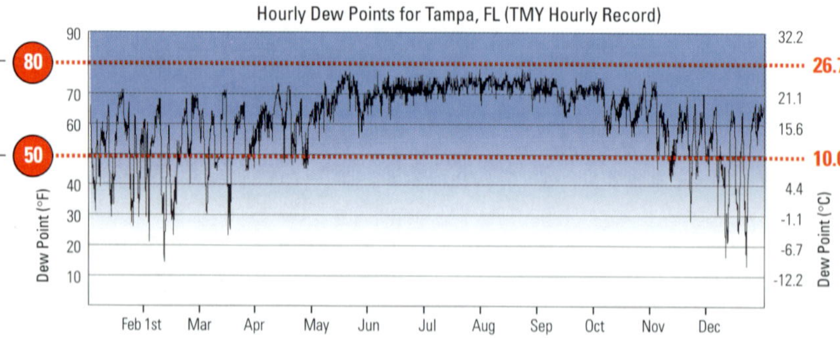

Fig. 2.6 "Tempering" Caution
If only the temperature of outdoor air is controlled, that air may be carrying excess moisture into the building.

Temperature-tempering alone is inadequate for outdoor air

In some hotels, schools, dormitories and elder care facilities, ventilation air is provided to corridors, and then drawn into rooms by toilet exhaust fans. Since corridors are not occupied for long periods, the ventilation air is often just "tempered" instead of fully-conditioned. Generally, this means it is cooled to 78 or 80°F [25.5 or 26.6°C]. Unfortunately, if the outdoor air is never cooled below that high dew point, the air will never be dehumidified at all. The ventilation air then creates a massive, constant moisture load. This fact is displayed graphically in figure 2.8.

The better solution is to control the dew point of the entering air, not just its dry bulb temperature. When using cooling-type dehumidification, chill the ventilation air to a low dew point, then reheat the air to avoid overcooling the corridors. Alternatively, use an active desiccant system to remove moisture to the appropriate dew point and post-cool the air if corridors are above the desired temperature.

Be cautious with temperature-controlled packaged rooftop units

Packaged rooftop cooling units are designed to cool and dehumidify some outdoor air along with a larger amount of recirculated air. Most such units, however, do not contain cooling coils that can dehumidify the air adequately when the outdoor air fraction rises above 15% of the total air flow. Also, while these units may provide continuous ventilation, they probably will not provide continuous dehumidification, because the compressors are controlled by the temperature in the space.

A classic example of problems with such units would be that of a retail store with a need for more ventilation air than 15% of the total flow, combined with the unfortunate practice of oversizing the unit for the sensible cooling load. Well-meaning designers often oversize equipment to avoid insufficient cooling capacity. These designers also assume—erroneously—that extra capacity will remove moisture. Unfortunately, oversizing for the cooling load has the opposite effect. When the unit is oversized, the compressors will remove the cooling load with very little run-time. Then the compressors shut down, and the moisture on the coil re-evaporates into the air. In the meantime, the ventilation air pumps a continuous stream of highly humid air into the building. Very little dehumidification occurs with oversized cooling units when they are controlled by the space temperature.

The better solution is to dedicate a separate system to providing humidity-controlled ventilation air, with a separate distribution system to carry the treated outdoor air throughout the building.

Consider dedicated equipment for dehumidification

Combining cooling and dehumidification in the same device sets up an inherent "conflict of interest" for the controls. During many hours, the building may not need cooling, but might be in great need of dehumidification. If the cooling coil chills all the air in the building it will overcool the building, using excess energy and making occupants uncomfortable. The only remedy would be to add heat back into the air after cooling, which is often prohibited by energy codes unless waste heat is used, such as that from condensers.

When moisture loads are high, and when the owner needs control rather than simply moderation, the most effective strategy is to install separate equipment dedicated to dehumidification. The dehumidifier can be controlled with a humidistat, and the air conditioner by a thermostat. Both parameters are controlled independently, gaining tight control of both temperature and humidity without wasting energy. Dedicated dehumidifiers are based on either active desiccant or cooling technology as shown in Figure 2.9. These equipment alternatives are discussed in detail in Chapter 13.

Fig. 2.9 Dedicated Dehumidifiers

When control rather than simply moderation is the goal, the designer needs to consider dedicating a specific piece of equipment to respond to a humidistat signal. Two types of dedicated dehumidifiers are used in commercial buildings.

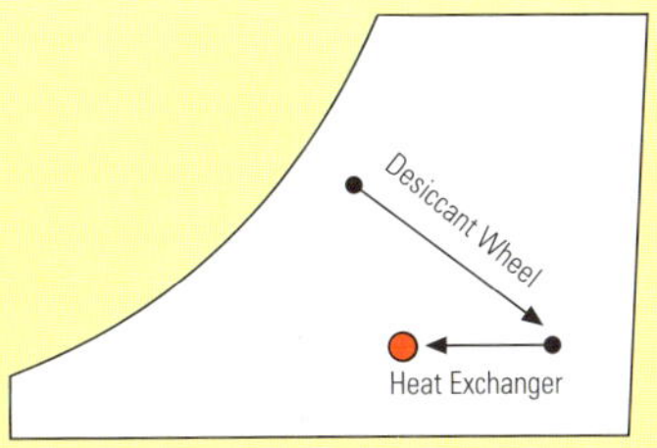

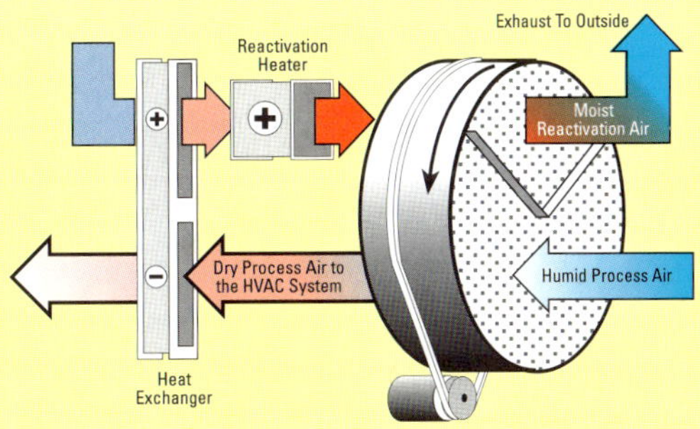

Active Desiccant Dehumidifier

Air is dried by desiccant material in the rotating wheel. The heat from sorption is removed by the post-cooling heat exchanger.

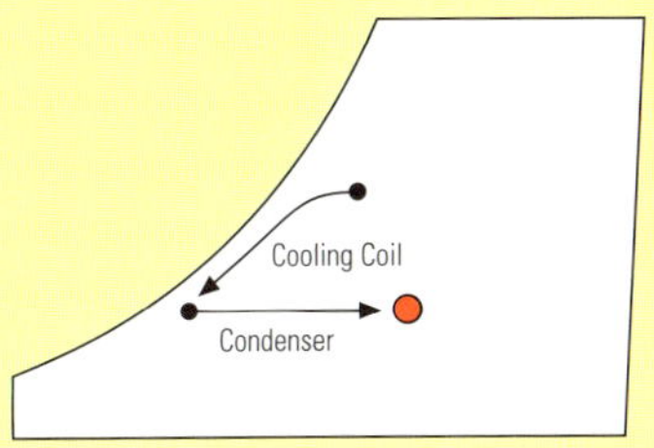

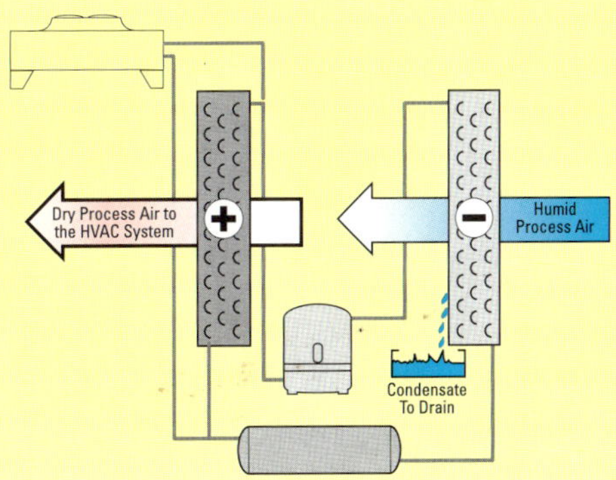

Mechanical Dehumidifier

Moisture is condensed from the air by a cooling coil. The unit's condenser adds heat back to the air to reduce its relative humidity.

Reduce the load

The owner controls the moisture load, because the building use determines that load. But the HVAC designer can sometimes use mechanical equipment to reduce the load on the dehumidification and humidification equipment by using energy recovery devices. Exhaust air from the building can be run through a passive desiccant heat exchanger. The recovered energy can be used to pretreat incoming outdoor air, reducing the dehumidification and humidification loads.

When the exhaust air can leave the building at the same point where the outdoor air enters the system, adding a heat exchanger actually reduces the net installed cost. A passive desiccant heat exchanger allows the size of cooling and heating equipment to be reduced, more than offsetting the cost of the heat exchanger.

A passive device does not control humidity by itself. It simply reduces the load on the equipment that controls humidity. But load reduction can often help equipment with less dehumidification capacity to satisfy the owner and reduce installed cost . Load reduction is useful at all times, and is especially cost effective when humidity moderation—not control at a defined level—is the only need.

Specify tight ductwork

Duct leaks damage buildings. The HVAC designer should specify that ductwork be sealed along all transverse and longitudinal joints (Seal Class B).8 When air pressure in the duct is above or below 4" wc. [more than 1000 Pa], all wall penetrations should be sealed as well (Seal Class A). As the Sheet Metal and Air Conditioning Contractors Association (SMACNA) explains: "The designer should assume that unsealed ducts may leak at the following rates :"

Pressure Difference Inches w.c. [Pa]	Leakage cfm/100 ft² [l/s •m²]
0.1 [25]	11 [0.57]
0.25 [62]	20 [1.04]
0.5 [125]	31 [1.67]
1.0 [249]	48 [2.50]

Consider what that means for a simple, low pressure system operating with supply and return ducts at average pressure differences of 1" w.g. [249 Pa] and a few short runs of 14 " x 16 " rectangular duct with a total length of 200 ft. SMACNA guidelines suggest that system should be expected to leak about 385 cfm continuously if not sealed. [182 l/s •m^2] Here's why ducts must be sealed to avoid this leakage:

Return ducts need to be sealed because otherwise they create suction in the building cavities. All cavities eventually connect to the exterior wall. If the cavities are under suction, they will pull untreated air through the building wall, and cause condensation in both humid and cold climates. Cold air entering will chill surfaces inside the walls, and internal humidity will condense. In hot climates, humid infiltration air will condense on cold surfaces inside the building, beginning with uninsulated cold supply air ductwork and continuing further to condense on cool walls and ceiling grids, and supply air grills, registers and diffusers.

Exhaust air ducts from laboratories are usually well-sealed because it's obviously important to prevent toxic vapors from leaking. But exhaust ductwork from toilets and kitchen hoods is not always sealed, because problems caused by those leaks are less intuitively obvious. When exhaust ducts leak, they pull air from building cavities. Often, the cavities are connected to other cavities near the exterior wall, so the exhaust fan effectively pulls untreated air into the building through that wall, causing condensation. Sealing exhaust ductwork avoids these problems, and is especially important when the exhaust fan operates continuously.

Supply air ducts also need to be sealed because otherwise they will leak treated air into building cavities causing condensation problems. In hot and humid climates, cold supply air leaking into humid building cavities chills surfaces low enough to cause condensation. This is a familiar problem to plumbers and roofers, who are often called on to "fix the leaks" that are not water leaks at all, but simply cool surfaces dripping condensate caused by cold supply air leaking into humid spaces. Similar problems happen during cold months, when warm, humidified supply air leaks into cold building cavities, causing condensation and dripping stains on ceilings and in walls. Sealing the supply ducts (including the joints between the ducts and diffusers) helps avoid these problems.

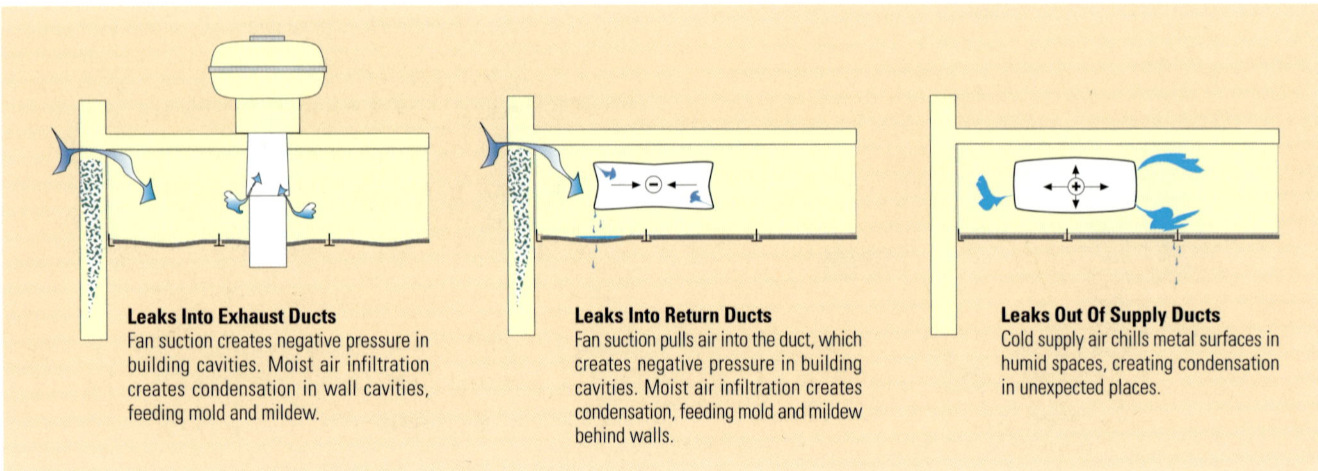

Fig. 2.10 Seal All Ductwork
Any air leakage into or out of ductwork reduces the system's ability to moderate or control humidity.

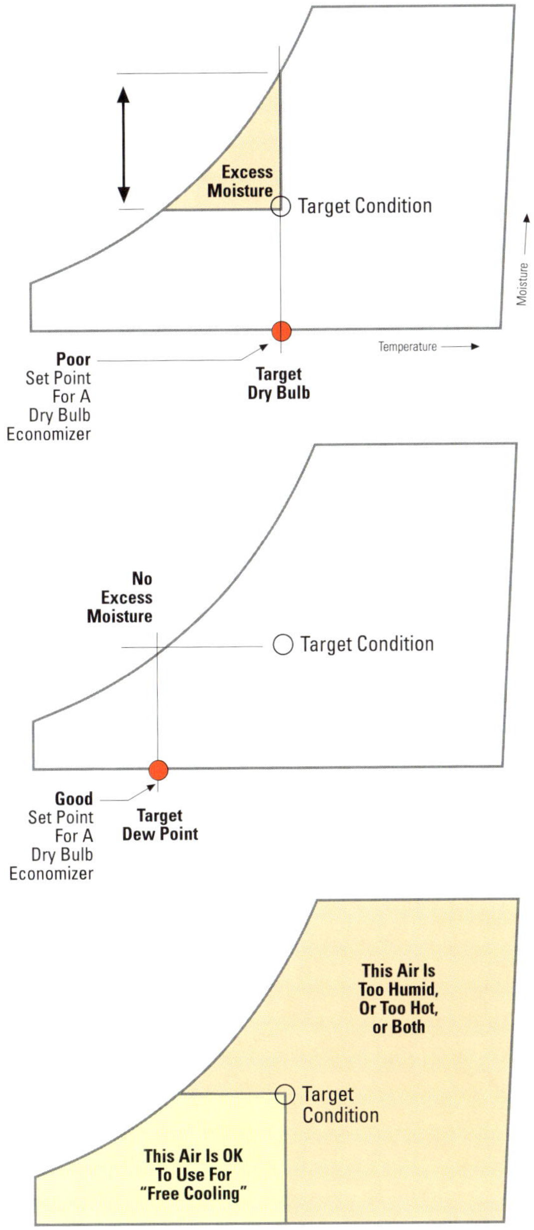

Fig. 2.11 Economizer Cautions

"Free cooling" may accomplish cooling, but load the building with moisture. Setting the economizer with the humidity in mind can avoid creating serious problems.

Set cooling economizers at the control dew point

A temperature-controlled, air-side economizer uses outdoor air to cool the building when the outdoor temperature is lower than the temperature of air inside the building. This is sometimes called "free cooling." Unfortunately, such temperature-controlled economizers may prevent humidity control during most of the year. Although the outdoor air might be usefully cool, it may also be highly humid or much too dry. Consider the summer extreme. The peak moisture hours occur at moderate temperatures, usually between 65 and 85°F [18.3 and 29.4°C]. So just when the economizer decides to bring in outdoor air for cooling, it brings in a huge moisture load by accident. Since the cooling system is shut off, no dehumidification occurs, and the building loads up with moisture.

Enthalpy economizers do not always solve this problem. An enthalpy economizer uses outdoor air when the total heat outdoors (the enthalpy) is less than the total heat inside. In theory that decision includes the moisture levels. But in fact, a decision based only on enthalpy does not consider dew point differences between inside and outdoor air. Air with a low enthalpy can carry a higher moisture content than what is desirable in the space. This happens for thousands of hours each year during the evening, night and morning hours when outside temperature is low but moisture remains high.

The better solution is to set any economizer so it *never uses outdoor air with a higher dew point than the level desired in the space*. For example, if the inside control condition is 75°F, 50% rh, the specific humidity is 65 gr/lb, which is a dew point of 55°F. [23.9°C, 50% rh = 9.3 g/kg, a dew point of 12.8°C] So if the temperature-based economizer is set so it never uses air with a higher dry bulb temperature than 55°F [12.8°C], then it will not overload the space with moisture while using the outdoor air for cooling.

The reader should also be aware that operating economizers in buildings with winter humidification is more complex. Those issues are addressed in the Chapter 14.

Avoid oversizing cooling equipment

When the humidity only needs moderation rather than control, HVAC designers often rely on cooling equipment to dehumidify the building. Although this strategy seldom works well, its dehumidification performance is slightly better when the equipment is *undersized*, rather than oversized for the peak cooling load.

A cooling system can dehumidify *whenever the coil is actually cooling the air*. But if the unit is controlled on temperature and not on dew point, and if the cooling capacity is oversized for the heat load, the system may not provide any dehumidification at all.

The classic example is a two-story office building, built to meet an energy code that reduces the sensible heat load from the building envelope to very low levels. The HVAC designer, mindful of the tendency of occupants to add office equipment and add heat load, and mindful of the tendency of building owners to add employees, decides to leave extra cooling capacity in the design. In fact, he might find himself leaving a "safety factor" of 20 to 30% in some of the cooling units for that reason and because of available equipment sizes. That safety factor, combined with already-conservative calculations based on the extreme dry bulb weather condition, means the system will have 50 to 75% excess cooling capacity during most of the cooling hours in the year. The system will not need to cool for more than 15 to 30 minutes in every hour. When compressors shut off, any moisture on the coil re-evaporates into the air, and the ventilation air continues to add more moisture to the building. The building feels cold and damp in the morning and evening during summer as well as during spring and fall—when sensible heat loads are low.

To improve moisture removal, the much better alternative is to slightly "undersize" the system with respect to the sensible cooling load. This will ensure that the system runs for a larger percentage of every hour, which will improve its dehumidification performance.

One way to check the dehumidification performance of any cooling-only system is to calculate how it will perform when the building sees its peak moisture load. The *ASHRAE Handbook—Fundamentals* contains both peak dew point and peak dry bulb conditions. (This is only true for the 1997 and later editions of the *Handbook*. Previous editions do not contain that information) Peak moisture loads occur at moderate temperatures in summer mornings, in contrast to the more familiar peak temperature and peak solar loads, which occur during summer afternoons.

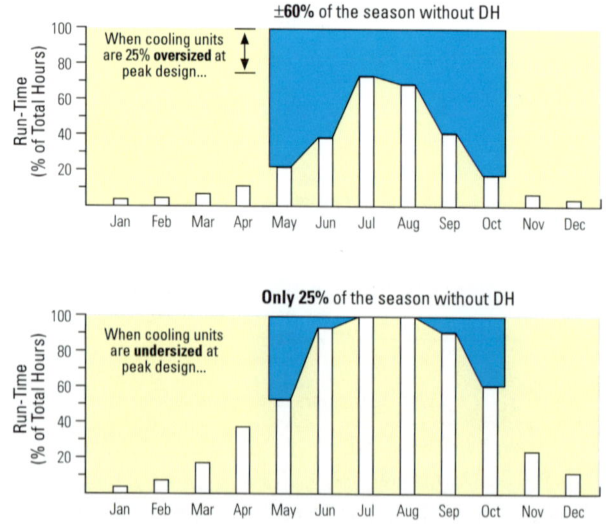

Fig. 2.12

Oversizing Cooling Units Means Less Dehumidification

The all-too-understandable tendency is to oversize cooling equipment to be certain of maintaining temperature at peak load. But that means units will not run for most of the hours of the year, so dehumidification cannot occur.

Slightly undersizing equipment at the peak cooling load is a better strategy if low-cost humidity moderation is a goal. Units will run longer, removing more moisture.

The HVAC designer and the owner should remain aware, however, that no cooling-only system will control humidity—it can only moderate its extremes. When humidity control (rather than humidity moderation) is the goal, the system must include equipment that responds directly to a humidistat.

Eliminate any suction at the exterior wall

The HVAC designer has a pivotal role in avoiding infiltration of unconditioned air through the building envelope. Avoiding such leakage is essential to preventing moisture damage to the building, and essential to maintaining control of humidity in occupied spaces.

In humid climates, maintain a slight positive air pressure inside the building. Provide a slight *excess* of conditioned ventilation air compared to the air exhausted from the building.

While that strategy is certainly fundamental to success, it is not sufficient to prevent problems. More care is needed, because buildings are not simple, single-cell spaces. They have complex internal pressure relationships between rooms and walls, corridors and doors, elevators and lobbies. Air moves in response to all differences in pressures. Uncontrolled air infiltration through the exterior wall is at the root of most moisture problems in humid climates.

So the HVAC designer must take care to maintain the exterior wall—and any internal cavities that connect to that wall—under a constant, slight positive air pressure. Any net positive air pressure will be sufficient. In fact, "less is more" because too much internal pressure wastes energy. Field studies have shown that even as little as 0.008" w.g. [2 Pa] is sufficient to prevent problems, provided that such pressure is constant, and reflects the *difference between the wall cavities and the outside of the building*, as opposed to the difference between inside the rooms and outside the building.[9]

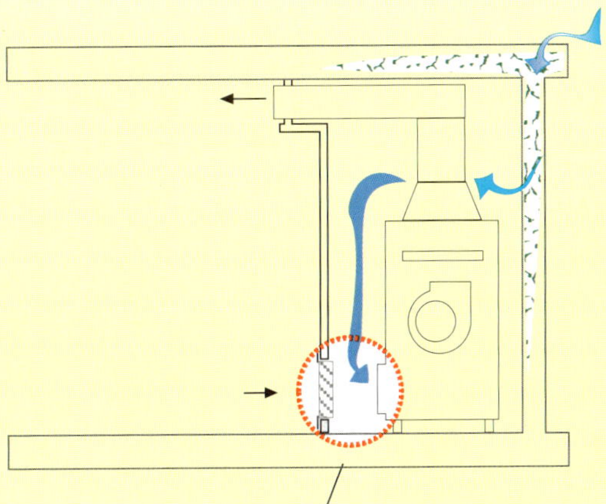

Missing Return Duct
This unit pulls air from the room...but also pulls air from building cavities, increasing infiltration from outside. Hard-connecting the return grill to the units inlet with ductwork avoids wall suction, reducing condensation and mold growth in the walls.

Fig. 2.13 Avoid Local Suction
The building as a whole might be under positive pressure, but that does not prevent certain spots on the wall being under local negative pressure. Local negative pressure is just as damaging as general depressurization.

Caution for cold weather

Note that in cold weather, the designer should take care to *avoid* positive pressure inside the building, because that would force the more humid indoor air into cold building walls, where moisture could condense and damage the building. So during cold weather, keep the building under neutral or very slightly negative pressure with respect to the outdoors to avoid this potential problem.

Avoid suction in wall cavities

One classic problem in hotel and apartment construction involves wall-mounted cooling units which accidentally create suction near one small section of the exterior wall. Although the building as a whole may be under positive pressure, cooling units in walls can create local suction in the nearby wall cavity. For example, not all such units have tightly sealed fan compartments. And not all have ducted return connections. In some cases, the fan pulls air from a plenum that connects to the wall cavity in addition to the room. When that happens, the fan creates suction in the wall cavity, which pulls air through any cracks in the exterior wall. Figure 2.13 describes such problems. The solution is to seal up the internal fan compartments of these units, and also to make sure a return air duct is sealed to both the unit and to the surface of the interior wall.

Enough conditioned makeup air

To keep the building under positive pressure, begin by quantifying the exhaust air. Then add a slight excess of ventilation air to ensure positive internal pressure. This seems simple, but it may be difficult to quantify all exhausts, particularly until all equipment has been installed and the system has been balanced. Toilet exhaust and kitchen hoods are the obvious items to locate and quantify, but there are many other less-obvious exhausts. Machinery and equipment may include exhaust air and fume hoods need makeup air. The sources vary by application. In all buildings, the designer needs a quantified list of exhausts before he can properly size the ventilation system.

Also, some spaces in commercial buildings need to be kept under negative pressure to control odors. This is true in restaurant kitchens, swimming pools and spas. In those spaces, the designer must take special care not to create too much negative pressure, which would pull air through walls from the outdoors. Pull treated air from other parts of the building but take care to supply an adequate quantity of conditioned air to those other spaces, making up for any air exhausted from kitchens, pools and toilets.

Specify a building air balance report

Unless the HVAC designer specifies an air balance report, maintaining a positive building pressure is largely a matter of chance. When humidity control is important, the specification should include air testing and balancing not only for each system, but for the building pressure as well. Consulting the test and balance contractor for suggestions before and during the design phase usually reduces the cost and increases the effectiveness of this key function.

Consider pressure monitoring

The designer might consider specifying continuous monitoring of wall pressure vs. outside pressure when there is reason to be especially concerned about humidity control or moisture damage. Meaningful monitoring is not simple to accomplish, because the differences will be small (between 0.004 and 0.016" wc. [1 and 4 Pa]). At that level, the slightest breeze outside can reverse the desired pressure difference several times every few seconds. To ensure meaningful data, pressure monitoring systems must be able to measure:

- Pressure differences of 0 to 0.04" w.g. [0 to 10 Pa].
- Average outdoor pressure (without distortions from instantaneous wind velocity).
- Many readings per second, recording some average over time rather than any instantaneous pressure difference.
- Enough locations inside the exterior walls to provide a meaningful representation of the average pressure differences for all sides of the building.

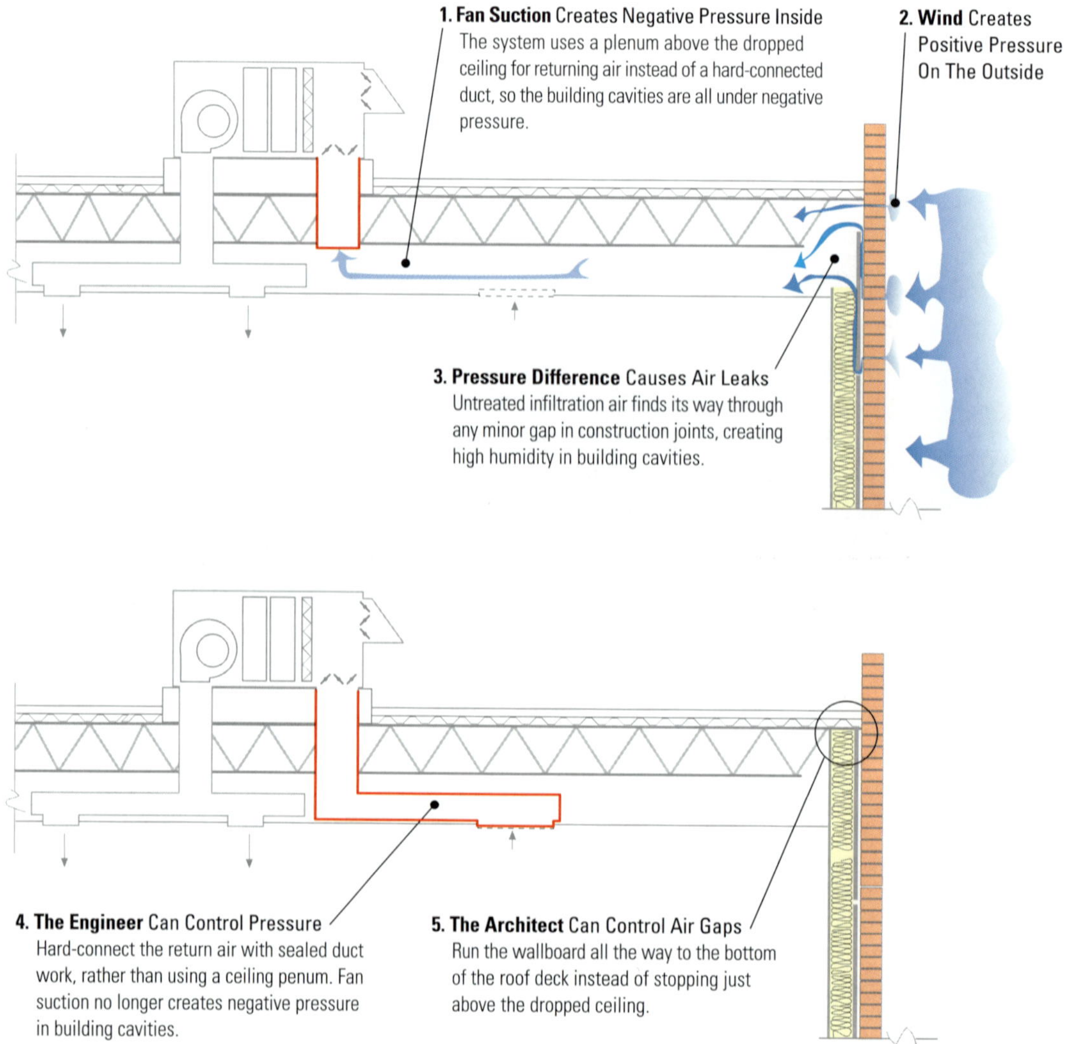

Fig. 2.14 In Humid Climates, Ensure Positive Pressure

The Architect, HVAC designer and the contractors each have specific tasks that will achieve—or fail to achieve—positive internal air pressure, a key element in controlling humidity and preventing moisture damage in hot and humid climates.

Above 65 % rh

Obtain advice from manufacturers of humidifiers, and from manufacturers who make environmental rooms for high-humidity requirements.

Below 35 % rh or 35°F [1.7°C] Dew Point

Obtain advice from manufacturers of desiccant dehumidifiers and from dryroom manufacturers.

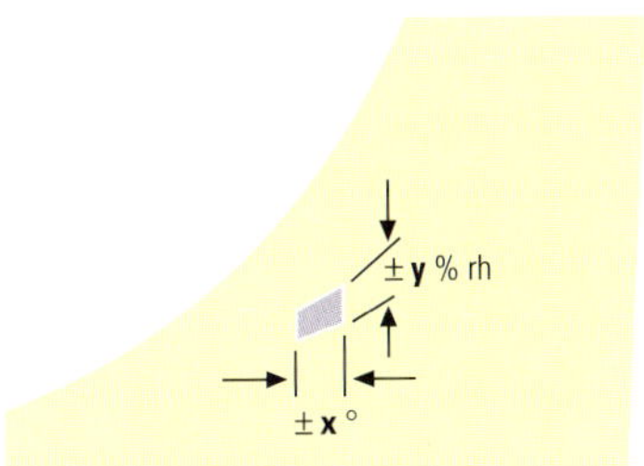

Tight Tolerances All Year

Obtain advice from manufacturers of dehumidifiers, humidifiers and controls, or from environmental room manufacturers.

Fig. 2.15 Special Humidity Levels

Outside normal humidity ranges, the HVAC designer should consider seeking help from specialists to avoid excessive costs and needless problems.

Cautions for special humidity requirements

The HVAC designer should recognize that the situation calls for more attention than usual when humidity is outside the human comfort range, or when it must be kept within a narrow range year-round. Often, the most practical and directly useful advice can be obtained from the manufacturers of the specialized equipment needed in those situations, and from manufacturers of environmental rooms. Several circumstances can alert the designer to the fact that the situation calls for "special" humidity control requirements. These include:

Narrow-range tolerances

When the owner needs to keep relative humidity in a narrow range year-round, the equipment cost goes up substantially, along with the cost of controls, and the cost of the building envelope. The narrower the range, the higher the costs. Also, costs increase as loads go up and as the floor plan becomes more complex. Some examples will help illustrate this point.

One extreme would be a multistory museum with "international style" glass walls on both northern and southern exposures, in a climate with extremes in both seasons, such as Chicago, IL. Assume the owner has called for controlling relative humidity to ± 2% rh all year long. In such a building, the thermal loads vary widely every day as well as through the year. One part of the building needs cooling and dehumidification, another requires heating and humidification at the same moment. And the ± 2% rh tolerance is at the limit of the accuracy of humidity sensors, and beyond the ability of a large system to respond quickly. So even if the specification were achievable technically, it would be economically impractical.

A less extreme, but still very challenging situation would be a similar-size museum in the same city, but with a late 19th century thick stone exterior wall, few windows, and a requirement to keep relative humidity constant year-round with a tolerance of ± 5% rh. The HVAC system remains expensive and the controls complex. Dedicated dehumidifiers and humidifiers are installed in several zones, and sensor calibration and system response remains a problem. But the specification is probably achievable, provided the building will not suffer from condensation damage in old walls during the winter.

Much less challenging installation would be a room used as an archive, which has no external walls or windows and is located inside a temperature-controlled museum. Such a room has little or no ventilation requirement and very little exposure to wide swings in outside thermal and moisture loads. The space itself is relatively simple compared to a multistory building, so air temperature differences are not going to be large across that space. Maintaining such a room within a ± 2% rh tolerance year-round will not be especially complex or difficult.

The key point to remember is that tight tolerances are likely to carry extra costs, but those costs can vary from a small percentage increase to doubling or tripling the entire cost of the system, depending on the size and complexity of the space, and the amount of variation in the loads over the day and year.

Relative humidity above 35% in cold climates

When the relative humidity must be kept above 35% rh in cold climates, there is a considerable risk of internal condensation. Condensation can occur in ductwork, because the need to add moisture to the building is so great that it is difficult to carry that much moisture in air at moderate temperatures. Also, any uninsulated ducts carrying highly-humid air run the risk of condensation if the duct passes through cool areas of the building envelope. Finally, the risk of condensation is very high inside the building wall. Any leak of cold air into the building can lead to condensation from highly humid indoor air.

When faced with this challenge, the architectural designer and owner should be aware that the building envelope will have to be specially-designed to resist air leaks, and designed to avoid condensation caused by water vapor diffusion.

Relative humidity over 65%

If the humidity must be kept above 65% rh constantly, as in produce storage areas, floral displays, biomedical research rooms and cigar humidor rooms, the materials chosen for the space must resist fungus and be easy to clean. Also, the walls, ceiling and floor must be heavily insulated to avoid any cold spots where condensation could occur. Keeping small spaces with low thermal loads above 65% rh is difficult, but even more challenging is keeping larger spaces with wide swings in thermal loads at that level. Seeking the advice of manufacturers of humidifiers and environmental rooms saves money and design time.[10]

Relative humidity below 35% or dew points below 35°F [1.7°C]

Keeping humidity very low also requires special considerations in building construction and in HVAC design. In general, such requirements are not found in commercial buildings, but they can occur in research buildings, or office buildings which contain pilot manufacturing of pharmaceuticals, confectionery or electronics.

Industrial buildings regularly have low-humidity requirements, and keeping large rooms as low as a -40°C/F dew point is quite common in battery manufacturing and in processing of hygroscopic medical products. In those facilities, metal walls, airtight joints, double doors and welded ductwork are basic necessities.[12.]

Such ultra-low humidity requirements can be met most economically by consulting the manufacturers of desiccant dehumidification equipment or dryroom manufacturers, who can provide complete installations.

HVAC Designer's Checklist

Convert Owner's Requirements to Design Values

1. Based on the owner's definition of the project purpose, what humidity value will be controlled?	❑ Relative Humidity ❑ Dew Point ❑ Both
2. What ranges are tolerable to the owner? (Convert rh or dew point values to humidity ratios)	Summer: From ______ to _______ gr/lb [g/kg] (*min.* _____ in.hg [Pa]) Winter: From ______ to _______ gr/lb [g/kg] (*max.* _____ in.hg [Pa])
3. Based on the owner's tolerance for operating hours beyond design, what outdoor air values define the peak moisture loads? (See Chapter 31 for data)	Summer Dew Point:: ❑ 0.4% (35 hr) ❑ 1.0% (87 hr) ❑ 2.0% (175 hr) _______ gr/lb [g/kg] with ______ °F [°C] Vapor Pressure: _______ in.hg. [Pa] Winter Dry Bulb: ❑ 99.6%(35 hrs) ❑ 99.0% (87 hrs) _______ gr/lb [g/kg] with ______ °F [°C] Vapor Pressure: _______ in.hg. [Pa]
4. Based on the owner's occupancy estimates and/or usable floor area, how much outdoor ventilation air needed to meet building codes?	Peak occupancy __________ x ___cfm/person [l/s/person] = _________ Usable area __________ ft^2 [m^2] x ____ cfm/ft^2 [l/s/m^2] = _________
5. Based on the total exhaust air, the ventilation air and excess air needed for maintaining positive internal air pressure, what is the total outdoor air requirement?	Total exhaust air ________ cfm [l/s] + 10% = _____________ cfm [l/s] Required ventilation air from item 4..................= _____________ cfm [l/s] Larger of these air flows ....= _____________ cfm [l/s] Outdoor air requirement

Minimizing Loads From Outdoor Air

6. Can the outdoor air moisture load be reduced by using a heat exchanger to recover waste energy from exhaust air?	❑ **Yes**. By using a heat exchanger, the installed system cost is reduced. ❑ **No**. It is not practical to bring the exhaust air back to the point where outdoor air enters the building.
7. Can the outdoor air moisture load be removed or reduced by installing a dedicated ventilation air pretreatment unit?	❑ **Yes**. By removing the ventilation air load, internal cooling and heating equipment capacity can be reduced. ❑ **No**. It is not practical to treat ventilation air separately. The moisture removal capacity of the cooling equipment will be checked at the peak outdoor moisture (Item 2 above).

HVAC Designer's Checklist

Equipment Selection & Sizing

8. Dedicated dehumidification equipment ***will be installed*** to control humidity independently from temperature.

- ❑ **Yes**. The dehumidification equipment will respond to a humidistat or dew point sensor in the space. Cooling equipment capacity can be reduced since the dehumidifier will remove all or most of the moisture load.
- ❑ ***No***. Humidity can be allowed to rise and fall with weather changes. To moderate these swings, the following precautions will be taken:
  - ❑ None. The owner expects high humidity in spring, summer and fall, and has no need for humidity control that justifies design effort or equipment expense.
  - ❑ Cooling equipment performance will be calculated under peak dew point operating conditions, to be sure it can remove the maximum moisture load.
  - ❑ DX Cooling equipment will be slightly undersized for the sensible load to ensure frequent operation and therefore more frequent dehumidification.
  - ❑ Cooling units will be staged to come on at different temperatures, so that at least one unit operates during periods of high latent (moisture) loads even when sensible loads are low.
  - ❑ Chilled water temperature will not be allowed to rise higher than the temperature needed to keep humidity under control in the space.
  - ❑ Chilled water coils will be split horizontally. The water flow through the lower half of each coil will be controlled by a humidistat.

Duct Work & Air Pressure Management

9. All supply and return air ***will be ducted*** to and from the HVAC units and the conditioned spaces. No open plenum returns shall be allowed.

- ❑ **Yes**. Fan suction cannot pull unconditioned air through the envelope.
- ❑ **No**. It is not economically practical to eliminate all open plenums. The owner and achitectural designer are not sufficiently concerned about the risks of wall condensation, corrosion or mold growth to incur the cost of ducted return and supply air distribution systems.

10. All duct work is sealed to SMACNA standards of either Seal Class A or B.

- ❑ **Yes**. The building team expects that total air **leakage from ducts will be less** than the amount calculated below. Duct work shall be sealed to SMACNA standards of either:
  - ❑ Seal Class A: All transverse joints, longitudinal seams and duct wall penetrations.
  - ❑ Seal Class B: All transverse joints and longitudinal seams.
- ❑ **No**. The building team **expects the duct work to leak more than** 0.48 cfm/ft^2 [2.5 l/s • m^2] of duct surface. Total leakage will exceed:
  Duct surface area _________ x 0.48 [2.5 l/s • m^2] = ___________

HVAC Designer's Checklist — Page 3 of 3

Humidity Sensors & Controls

11. Humidity sensors are located so they will not encounter air above 90% rh, and so they will control humidity at the locations that deliver the greatest benefits with respect to the purpose of the project.

- ❑ **Yes.** Locations are defined in three dimensions in relation to physical reference points from which sensor locations can be checked.
- ❑ **No**. Humidity does not need to be controlled—only moderated. No humidity sensors or controllers are specified.

12. Contract documents call for all humidity sensors to be tested for calibration after the system begins operation. Those calibration results shall be recorded in the balancing report and any needed adjustments entered into the building management system.

- ❑ **Yes**. Contract documents clearly define the procedure and instrument to be used for calibration, and also prohibit reliance on pre-installation calibration that has not been checked after the system is operational.
- ❑ **No**. Humidity does not need to be controlled—only moderated. No humidity sensors or controllers are specified.

13. Air and water flow-setting devices and their locations are included in the contract documents, so the commissioning authority can ensure the system operates as the designer intended.

- ❑ **Yes**. Contract documents clearly define the location and type of flow measurement and adjustment devices needed to achieve equipment performance specified by the manufacturer of the dehumidifiers and/or humidifiers.
- ❑ **No**. Humidity does not need to be controlled—only moderated. No dedicated dehumidifiers or humidifers are specified for the building.

Commissioning & System Balance

14. The organization with commissioning authority is defined in the contract documents.

- ❑ **Yes**. Contract documents define who is responsible for commissioning, and that organization has reviewed and commented on the design documents.
- ❑ **No**. Humidity does not need to be controlled—only moderated. No commissioning is specified with respect to humidity control.

15. Flow values for all humidity control devices are defined in the contract documents, which also require the contractor to measure and record performance in accordance with manufacturer's instructions and to supply the results to the owner.

- ❑ **Yes**. Contract documents clearly define those values as well as the form of the field report expected after commissioning is complete.
- ❑ **No**. Humidity does not need to be controlled—only moderated. No commissioning is specified with respect to humidity control.

16. The contract documents call for all installed air filters to be clean before final air balance, and before the system is turned over to the owner.

- ❑ **Yes**. Contract documents clearly require that all filters must be less than 20% loaded before balancing and before the system is turned over to the owner.
- ❑ **No**. The owner does not expect clean filters before accepting the system.

Contractors - Low-Leakage Building & Ductwork

During construction, buildings either improve or lose their ability to control moisture. The determining factor is the amount of accidental air leakage. The Contractor can help ensure proper humidity control through attention to a few critical items during construction.

Seal the exterior wall

Air leaks in the exterior wall cause condensation damage. Job site supervisors are certainly aware of the problems caused by water leaks, but they may not fully appreciate the problems caused by air leaks. Those consequences have been discussed in the previous sections to guide architectural and HVAC designers, which the contractor can review at his leisure. To avoid those problems, the air barrier at the exterior wall must be continuous, which is to say that all the individual sheets must be taped together. The entire sheet must also be taped to all window and door frames, foundations and roofing.

Dry all material before applying interior finish

Catastrophic problems can be avoided by ensuring that the structure of the building is dry before interior finish is applied.

In cold climates, the classic problem occurs in a new school, built on a tight schedule requiring completion before the academic year begins in September. That schedule means the building must be "closed-in" during early spring, when rain and snow have prevented effective drying of wet concrete slabs and block walls. Heaters are brought to the problem building, drying the surfaces quickly. Surface finishes and cabinets are installed. Then spring heat brings moisture from deep inside the materials to the finished surface. High humidity also generates condensation on cold floors and walls. The wall coverings then peel off, and cabinets, doors and ceiling tiles warp and sag, and must be replaced.

Similar problems occur in hot and humid climates, but they are less seasonal. Hotels, built on a fiercely tight schedule are common victims. Rain soaks the site, saturating block walls and concrete slabs, and humid weather prevents drying. Interior vinyl wall covering is applied to the wallboard. The wallboard absorbs moisture given off by saturated concrete block walls as they dry in the hot sun. The rooms begin to smell, and the wall covering, originally applied to dry wallboard, develops stains caused by mildew and peels off the now-saturated gypsum board.

Both humid and cold climate problems can be avoided easily by drying the building thoroughly before finishes are applied. As recently as 1990, this advice would have been difficult to follow. But over the last 10 years, a worldwide infrastructure of drying contractors has developed to serve the insurance industry as it seeks to reduce the cost of claims after floods, fires and disasters.[7]

Seal all ductwork and its connections

If pipes leak, the problem is immediate and obvious. When air ducts leak, problems take time to develop and are not obviously connected to the leaking air. But air leaks interfere severely with maintaining control of humidity. When leaks are large, humidity control may be impossible, and moisture damage is probable.

The problem mechanisms were explained in detail in the section providing advice to HVAC designers. From the contractors' perspective, it is important to ensure that ducts are sealed in accordance with the specifications. If the designer follows suggestions provided in this book, this will mean sealing all transverse and longitudinal joints on ducts operating at less than 4.0" wc. pressure difference [995 Pa], and also sealing all duct wall penetrations on systems operating at higher pressures.

Fig. 2.16 General Contractor

The builder can allow—or prevent—humidity control, depending on how tightly the building is built. Without air leaks, controlling humidity is relatively simple.

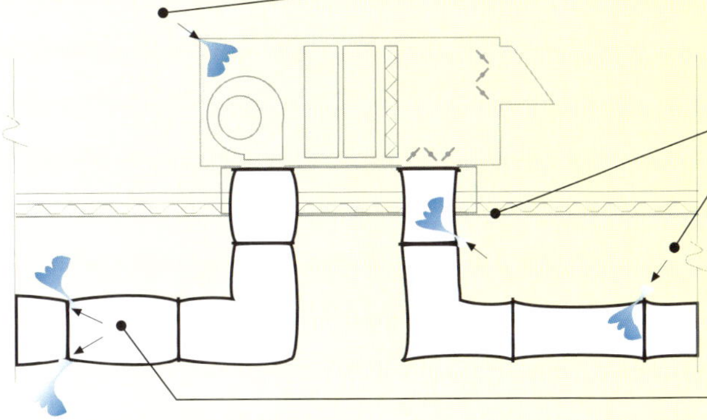

1. Eliminate Air Handler Leaks

Most commercial-grade air handlers leak a great deal of air. Seal open seams with tape and mastic, and make sure all access panels are gasketed, with all screws in place.

2. Seal Return Ducts

If return ductwork leaks, it creates low pressure in building cavities, which sucks untreated air into the wall from the outdoors. Sealing return ductwork helps avoid condensation from untreated air in these cavities.

3. Seal Supply Ducts

When supply ductwork leaks in summer, cold air leaks into building cavities, cooling surfaces, which condense moisture. In winter, warm supply air carries moisture into cold building cavities, where it condenses once again. Sealing supply air ducts prevents these problems.

Fig. 2.17 HVAC Contractor

The sheet metal installation greatly affects the success of any humidity control project. Ductwork and air handlers must not leak air.

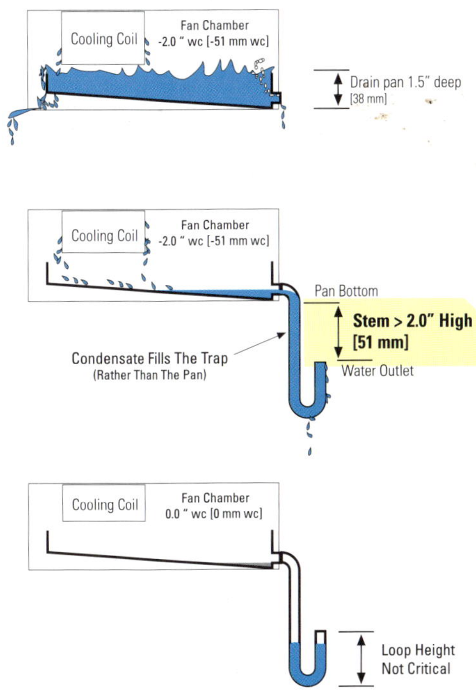

Without a Trap

The fan lifts up the condensate, which spills over if the fan suction exceeds the depth of the pan. Standing water allows microbial growth. Water spilling out of the pan can leak out of the unit and drip into the building.

With a Trap

The condensate drains out of the unit because the fan cannot lift the water out of the stem of the trap. The stem is measured from the *bottom of the pan* to the *top of the water outlet* from the trap.

Note... The Stem Height Is Critical

The height of the stem must be greater than the suction of the fan. eg: A fan with -2.0" wc pressure needs a trap with a stem height *over* 2.0" [A fan with suction of 51 mm of water column needs a drain stem higher than 51 mm].

Loop Height Not Critical - But Deeper Is Better

The distance from the water outlet to the bottom of the loop is not critical. But deeper loops do have a benefit. They take longer to dry out, which prevents the fan pulling in untreated air if the coil is not condensing water.

Fig. 2.18 Condensate Traps

The condensate drain line must be equipped with a trap, and the stem height of that trap must equal the pressure of the fan. Otherwise the condensate will stay in the air handler and eventually leak into the building.

Contractor's Checklist

Building Envelope

1. An air barrier is installed to prevent outdoor air from flowing into building cavities. That barrier is continuous, and is sealed at its edges and at the joints between building components.

❑ **Yes**. Building air leakage is expected to be less than tight buildings. Specifically, at pressure differences of 4 Pa or less, it will leak less than:
Total exterior surface ________ x 0.10 cfm•ft² [0.5 l/s•m²] = ________

❑ **No.** There is no specification to install an air barrier. The owner and achitectural designer are not sufficiently concerned about the risks of wall condensation, corrosion or mold growth to incur the cost of such a barrier. Air leakage is expected to be greater than the figure calculated above.

2. A vapor retarder is installed to limit water vapor diffusion from inside the conditioned space into cold walls. The retarder is continuous. It is sealed at its seams and edges, and at all penetrations to unconditioned areas such as electrical outlets, conduit and duct penetrations and above dropped ceilings.

❑ **Yes.** Water vapor diffusion from the space into unconditioned spaces and cavities will be less than:
Total interior surface ________ x 0.45 gr/h • ft² = __________ gr/hr
[Total interior surface ________ x 0.24 g/h • m² = __________ g/hr]

❑ **No.** There is no specification to install a vapor retarder. The owner and achitectural designer are not sufficiently concerned about the risks of wall condensation, corrosion or mold growth to incur the cost of installing a retarder. Vapor diffusion into unconditioned spaces and cavities is expected to be greater than the figure calculated above. Such an omission is safe for the structure when ALL of the conditions below are TRUE of the building:

- ❑ There is no swimming pool or spa in the building
- ❑ There is no forced humidification during cold months
- ❑ The winter design temperature is above -15°F [-26°C]

Duct Work & Commissioning

3. All supply and return air is ducted to and from the HVAC units and the conditioned spaces. There are no loose or missing connections. All duct work is sealed to SMACNA standards of either Seal Class A or B.

❑ **Yes**. The contractor has visually inspected the duct system to be certain it has no gaps, and that sealing meets the specified SMACNA seal class.

❑ **No**. It is not economically practical to eliminate open plenums and air leaks. The owner and achitectural designer are not sufficiently concerned about the risks of wall condensation, corrosion or mold growth to incur the cost of ducted and sealed return and supply air distribution systems.

4. The HVAC system has been balanced so that it matches the expected air flows and pressures within 5% of the values specified by the HVAC designer.

❑ **Yes**. The system has been balanced and supporting documentation provided to the owner as specified.

❑ **No**. Neither the owner nor the HVAC designer requires validation of correct system operation.

5. Humidity sensors have been calibrated after the system became operational, and that documentation has been submitted to the owner.

❑ **Yes**. The sensor calibration has been adjusted after installation, and supporting documentation provided to the owner as specified.

❑ **No**. Neither the owner nor the HVAC designer requires accurate measurement or control of humidity.

Fig. 2.19 Building Operations Staff

The owner's operations staff inherits the work of the rest of the building team (both the good and the bad). Most important is to remember that—more than any other element—success depends on controlling the humidity of incoming makeup air, and avoiding suction that pulls untreated air through joints in the exterior wall.

Building Staff - Maintain Air Balance

The principal role of the building operation staff is to maintain the system and adapt it to changes in building use. While doing that, it is useful to remember the importance of keeping the building under a net positive air pressure in humid climates, and under a neutral or slight negative air pressure during cold weather.

Correcting shortcomings

Unfortunately, the world is seldom perfect. Often, the first task for the building staff is to correct construction and design problems, and to adapt the system to changes in building use that occurred before the building was ever put to productive use! Correcting initial shortcomings often includes the following items.

Establishing positive air pressure indoors

If the building is not under positive air pressure in a humid climate it will be difficult to mitigate, let alone control, humidity. The first step is to establish whether the building is in fact under positive or negative pressure relative to the outside of the structure. Hold a smoke pencil near the frame of a slightly open exterior doorway. If smoke blows inward consistently, the building as a whole is probably under a net negative pressure. That may be correctable by installing clean filters on the inlets to the building's ventilation air. If clean filters do not bring the building to positive air pressure, increase the volume of ventilation air or reduce the exhaust air, or both.

If the ventilation system is already supplying its maximum, or if the exhaust flow cannot be reduced, the problem is more costly to solve, and will probably require expenditure for equipment to add more conditioned ventilation air.

Sealing-up ductwork and equipment

Humidity control is very difficult to achieve when ductwork or equipment leaks air. So the entire system must be sealed. If the designer or contractor has not accomplished this task, the building operation staff can enhance humidity performance of any system by taking the relatively inexpensive step of sealing all joints.

Even when all ducts are sealed, designers and contractors often neglect to seal the joints in the air handling equipment itself. After the building has been used for a while, screws go missing and panels come loose on air handlers and access doors. Any air leak on air handlers is very detrimental to humidity control, because that's where the pressure differences are greatest. Large pressure differences mean large air leaks.

Finding and sealing exterior wall leaks

The standard commercial building leaks large amounts of air through the building envelope. These air leaks can cause great damage to the building as well as prevent humidity control. Finding and

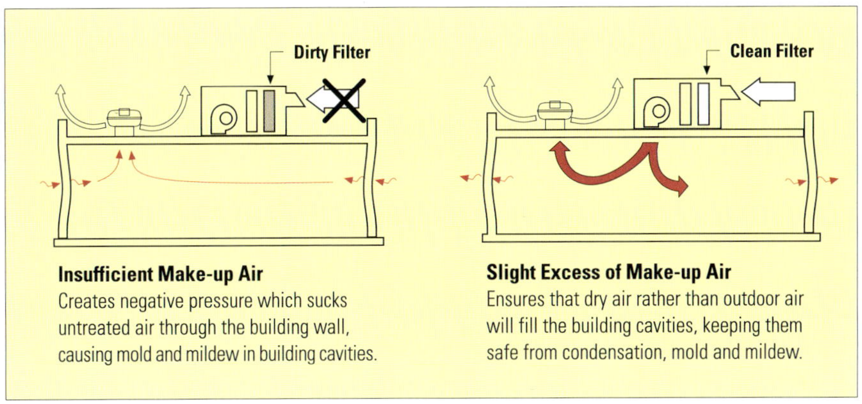

Fig. 2.20 Improving Dehumidification

The building must stay under positive air pressure, which means ventilation air must not be shut off, and outside air filters must be clean, to allow full air flow.

sealing these air leaks greatly improves the system's ability to mitigate or control moisture.

Typical leak locations include the cracks where the exterior wall meets the roofing system and around windows and door frames, particularly under window frames where cracks are in shadow and are therefore difficult to see.

There are many cost-effective techniques available to find air leaks. These range from simply holding a "smoke pencil" inside the building near the exterior wall on a windy day, to viewing the exterior of the building through infrared imaging equipment on a cold day. [13.]

Keep in mind that the important leaks are the ones in the exterior wall, and that such leaks are not likely to be obvious by inspecting the inside of the conditioned spaces. Climb above dropped ceilings, and closely inspect the outside of the building to locate leaks.

Connect condensate drains and install traps

Every cooling coil will condense moisture sooner or later. But not all contractors or designers take the time to ensure that all coils have condensate drains and traps. Without the right trap, water will either be re-evaporated into the air stream, or the drain pan will overflow, dripping water into the cooling unit or into the ductwork. In a new system, ensure that all cooling coil drain pans are connected to condensate drains, and that all the drains have traps with stems high enough to prevent air leakage as shown in figure 2.18.

Maintain make-up air systems

Ventilation air creates the largest dehumidification and humidification loads. If the makeup air system is not removing those loads, they are passed into the building, where the rest of the mechanical system may lack the capacity to remove it. Several items deserve continuing attention:

Keep filters clean

If the filters on the outdoor air intake are dirty, they restrict the air flow. If the restriction is severe, it may create a negative air pressure inside the building. When that happens, suction pulls unconditioned air into the building through leaks in the building walls, which results in moisture damage. Changing outdoor air inlet filters at least once a month avoids choking off the incoming conditioned air.

Do not shut off ventilation air to "Save Energy" or "Improve Control"

Any building occupied by people requires ventilation air for health and for comfort. Treating that outdoor air costs money. A common error is to reduce or eliminate ventilation air to restore cooling or heating capacity, or to save operating costs.

While those are important goals, shutting off the makeup air is not the way to achieve them. Shutting off ventilation air violates all building codes. Secondly, lack of ventilation air leads to indoor air quality problems, and sometimes to costly health problems and lawsuits that greatly exceed any energy savings achieved. Finally, shut-

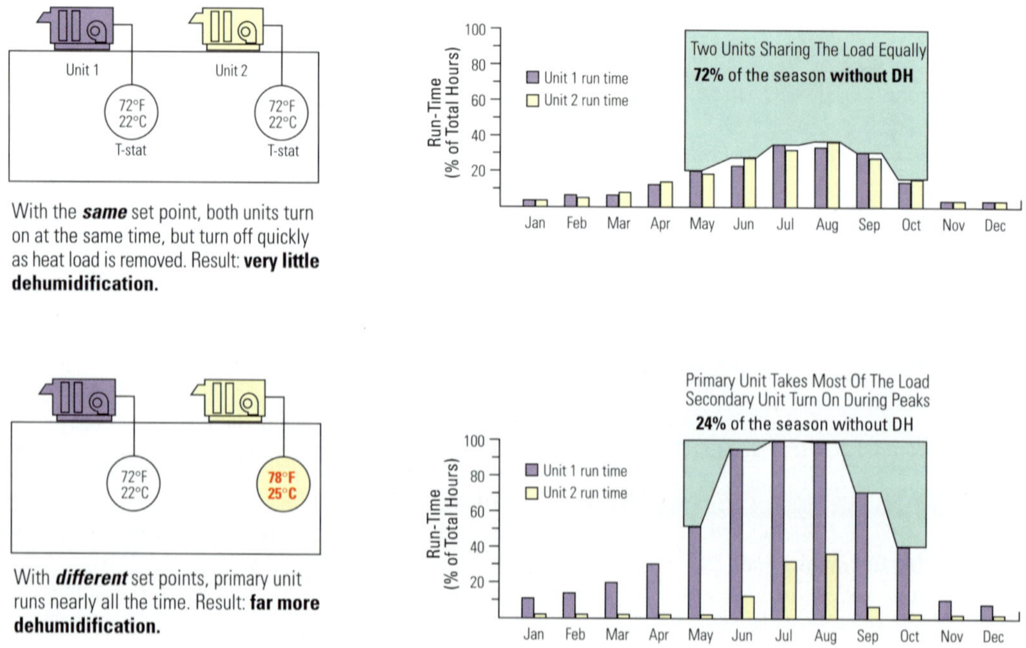

Fig. 2.21 Increase Run-Time

When a zone is served by more than one cooling unit, humidity can be more effectively moderated if at least one of the units runs all the time. Setting the other units to switch on their compressors at a higher temperature means that the first unit will run longer, improving dehumidification.

ting off makeup air without shutting off exhaust air shifts the makeup load from the pretreatment system to the building envelope. Nature does not tolerate a vacuum, so exhaust air is made up by untreated infiltration air when the ventilation system is shut off.

On the other hand, *minimizing* the amount of ventilation air is always a good idea (provided that the exhaust air is reduced at the same time, to be certain that the system does not suck air through the exterior walls). The operations staff has an obligation to consult with local code officials to determine the minimum ventilation air required for the current use of the building before reducing that flow to save energy or to regain capacity.

When raising temperatures to save energy, provide continuous dehumidification

Resetting temperature during unoccupied periods is an excellent way to reduce operating costs. But the operations staff should be aware that humidity is affected by decisions about temperature if there is no dedicated system to control humidity.

For example, a large building using a chiller/boiler HVAC system might raise the chilled water temperature to save costs when the cooling loads are low. However, if that cooling system is also responsible for dehumidification, raising its chilled water temperature prevents the coils from cooling air deeply enough to dehumidify the air. So when chilled water is reset to higher temperatures, it is wise to have a separate system which is dedicated to dehumidification. Such systems are usually located on the ventilation air, because that's where the excess moisture is greatest. If no dedicated dehumidification system is available, then the chilled water temperature *must* remain low enough to remove the moisture load, even when the cooling loads on the building are low.

Raising the building's temperature set point is also an excellent way to save cooling energy, but it can interfere with dehumidification. One typical example would be a large retail store in a hot climate. Such a building might use packaged rooftop heating/cooling units for temperature control and ventilation, with no specific humidity level required. When the store closes, the temperature set point might be raised from 73 to 81°F [from 22.8 to 27.2°C] to reduce operating costs. However, ventilation continues by code, because of the need to dilute volatile organic vapors from the merchandise. Since the cooling loads are low at night and the temperature set point has been raised, the cooling compressors hardly ever operate, which saves energy. On the other hand, the building slowly fills up with humid ventilation air.

In the morning when the staff returns, the building is hot as well as humid. It takes quite a while to recover temperature, because the cooling units must remove excess moisture as they cool the building. Also, high humidity overnight allows the merchandise to adsorb moisture from the air, and the system struggles to dry out that large mass of material.

The best solution is to have a separate system to control humidity in the store. While other units control temperature, one unit can be controlled by a room humidistat or a supply air dew point temperature sensor. That unit would dehumidify all ventilation air without being concerned with the store's temperature. In open-plan buildings with many rooftop units, one conventional cooling unit could be dedicated to this purpose.

Tune the cooling systems for dehumidification

When mitigation rather than humidity control is the goal, and after the building has positive internal air pressure, the operations staff may be able to tune existing cooling systems for better dehumidification without installing additional equipment.

Reduce chilled water temperature for outdoor air coils

The moisture load comes primarily from ventilation air. Chilling that air deeply removes moisture before it can become a problem inside the building. If the system uses chilled water, the staff can reduce its temperature and gain dehumidification capacity. If the coldest chilled water can be sent *first* to the coils that chill the incoming ventilation air, this strategy can make major improvements without high capital costs. The only construction expense might be to re-route piping to ensure that the cold water leaving the chiller goes first to the coils that dehumidify the outdoor air.

Reduce fan speed

Moisture loads are highest when sensible heat loads are low. The staff can take advantage of this fact of nature. Reduce the fan speed rather than shutting off compressors when the thermostat is satisfied. With lower fan speed, the air takes longer to move through the coil, so air is cooled more deeply, which drops out more moisture.

This strategy can be used for all types of cooling systems. However, when reducing air flow through direct expansion (DX) coils, one must take care not to reduce the flow below the minimum established by the manufacturer. At very low flow rates, air might become overcooled to the point where its condensate freezes on coil surfaces, blocking air flow entirely.

On some room cooling units, the fan speed is separate from the temperature setting. Such units can be set at low fan speed, while leaving the temperature set to "High Cooling" or "Dehumidification" mode. That combination will provide constant dehumidification, although the room may become quite chilly. Other similar units have only a single control for "Maximum Cooling", which brings the fan to high speed as well as turning on the compressor. The strategy will not work with such units.

Packaged rooftop units and even residential units often have variable-speed, or multispeed options for the supply fan. In many cases, installed equipment may be capable of this feature, but the designer or the contractor may have originally chosen not to take advantage of it. In such units, reducing the default fan speed costs very little, but greatly improves dehumidification capacity.

Raise temperature set point for secondary cooling units

In open-plan commercial buildings like retail stores and office buildings, the HVAC system may be packaged rooftop heating/cooling units in pairs for partial redundancy. If both units are controlled by the same thermostat setting, they switch on together, and cool the zone very quickly, shutting off before they can dehumidify.

A better strategy controls both units with a single, multistage thermostat, setting the units to switch on at different temperatures. For example, the primary unit could be set to switch on at 72°F [22°C], and the secondary unit could be set for 78°F [25.5°C]. That way, the primary unit will run nearly all the time, providing constant dehumidification. The second unit will only turn on when the sensible heat loads are very high.

This strategy is best employed with a single multistage thermostat rather than separate thermostats. There is enough variation in thermostat hardware and in air currents in small spaces to create problems when separate sensors control separate units. In very large spaces, multiple sensors may be needed, because zone loads may also be quite different. But in most buildings, a single multistage thermostat is likely to give less trouble than separate, independent controls for each unit.

Set economizers to low dry bulb temperatures

Air-side economizers use outdoor air for "free cooling" of the building when the temperature or the enthalpy outside is lower than the condition of air inside the building. But cool air can still be saturated with moisture. To avoid high humidity when the economizer brings in outdoor air, set the maximum outdoor air temperature to the dew point temperature desired inside the building. For example, if the building should be kept at 75°F, 50% rh, then the maximum dew point desired is 55°F. Setting the economizer so it never takes air if its temperature is above 55°F will ensure that the "free cooling" air does not overload the building with moisture while it removes sensible heat.

[For example, if the building should be kept at 23.9°C, 50% rh, then the maximum dew point desired is 12.8°C. Setting the economizer so it never takes air if its temperature is above 12.8°C will ensure that the "free cooling" air does not overload the building with moisture while it removes sensible heat.]

Building Operations Staff's Checklist

Building Pressurization For Hot & Humid Climates

1. Is the building operating under a slight positive air pressure?	❑ **Yes.** When the wind outside is essentially still, and a lighted cigarette or smoke pencil is held in a doorway or window that is cracked slightly open, smoke is pushed away from rather than pulled into the building. ❑ **No.** Smoke is pulled into the building through one or more cracked-open windows or doors.
2. Are the building cavities under positive or negative air pressure?	❑ **Positive**. When a lighted cigarette or smoke pencil is held next to an uncovered electrical outlet or cracked-open dropped-ceiling tile, smoke is not pulled into the building cavity. ❑ **Negative**. Smoke is pulled into building cavities through electrical outlets and dropped-ceiling tiles.
3. If the building as a whole or building cavities are under negative pressure, can the problem be corrected by installing clean make-up air filters or adjusting fan speeds?	❑ **Yes**. Installing clean air filters for the make-up air, or ensuring the make-up air system fan is operating at the correct speed solves the problem. ❑ **No**. Outdoor air filters are clean, the make-up air system is operating properly and the building is still negative. Contact an HVAC system troubleshooter for a more thorough on-site inpection and analysis.

Keep condensate drains clear, and clean the condensate pans

Clogged condensate drains are a common problem in HVAC systems. After the cooling coils remove water from the air, the condensate must be drained away rapidly. This prevents standing water from being reabsorbed by the air flowing through the coil after the compressors shut off. Running a cleaning brush through the drain before the cooling season will prevent the buildup of fibers and slime that typically block such drains.

Also, slime and microbial growth can build up in drain pans that do not drain completely. Taking a few moments to scrub the scum and dirt out of the drain pan will also help prevent drain blockage, and will avoid the microbial growth that leads to indoor air quality problems. After cleaning, fix the drain pan so that it slopes and drains properly, even if that means replacing the original pan. Then test the new arrangement with a large, full bucket of water to be certain that no standing water remains in the pan to support microbial growth.

Building Operations Staff's Checklist

Proper Operation of Equipment, Sensors & Controls

4. Are humidity sensors reporting signals consistent with higher-accuracy handheld hygrometers used for spot-checks?

❏ **Yes**. Humidity sensors report values similar to higher-accuracy instruments.

❏ **No**. Sensors need calibration or repair because they do not report humidity values that reflect actual conditions.

5. Are all dehumidifiers and humidifiers receiving control signals consistent with the humidity needs of the controlled spaces?

❏ **Yes**. Equipment is receiving signals from the control system that are consistent with the needs of the space.

❏ **No**. Controls need repair because equipment is receiving signals that do not reflect the needs of the space.

6. Are all humidifiers and dehumidifiers operating mechanically?

❏ **Yes**. Equipment seems to be operating properly from a mechanical perspective

❏ **Yes—But!** Equipment is operating but the space is out of control. The device needs complete performance testing.

❏ **No**. Some component is not operating in response to control signals and therefore needs repair.

Buildings Without Summertime Dehumidification

7. Are dedicated dehumidifers installed for summertime humidity control?

❏ **Yes**. The dehumidification equipment should respond to a humidistat or dew point sensor in the space. If that is not the case, troubleshoot the dehumidification equipment and humidity sensors.

❏ **No**. Dedicated equipment is not insalled because the designer believed that the owner wanted to save money by allowing humidity to rise and fall with weather changes. To moderate these swings, the staff can take the following measures:

❏ None. The owner expects high humidity in spring, summer and fall, and has no need for humidity control that justifies design effort or equipment expense.

❏ Cooling units will be staged to come on at different temperatures, so that at least one unit operates during periods of high latent (moisture) loads even when sensible loads are low.

❏ Chilled water temperature will not be allowed to rise higher than the temperature needed to keep humidity under control in the space.

❏ Chilled water coils will be replaced and split horizontally so the lower half the coil is controlled by a humidistat and can operate even when no cooling is called for in the space.

❏ An HVAC designer will be engaged to design a dedicated dehumidification system, with particular attention paid to dehumidifying the make-up air.

References

1. Moisture, humidity and construction malpractice

More than 30% of construction malpractice claims against Architects and Engineers are prompted by damage caused by moisture and humidity. This startling fact is described by a 5-year study of more than 5,000 construction claims performed by the Design Professional Insurance Company. The study is summarized in an article that appeared in the July 15th, 1991 issue of *Engineering News-Record* Magazine: *Engineering News-Record*, Two Penn Plaza, 9th Floor, New York, NY 10121. (212) 904-3249. www.enr.com

2. Measured infiltration

Air infiltration through building envelopes is difficult to measure. Ventilation air flow must be shut off, and measurements must extend over time, gathering long-term averages rather than simply recording what may be transient extremes of a few moment. One excellent compilation of field studies is *Measured Air leakage of Buildings*, edited by Heinz Treschel and Peter Lagus. This 438-page book was published by the American Society of Mechanical Engineers in 1986 (ASTM STP 904, 1916 Race St, Philadelphia, PA 19103 www.astm.org. ISBN 0-8031-0469-3) It contains chapters with measurements from different types of buildings, including aircraft hangers, office buildings and many varieties of residential buildings. It also includes chapters which discuss measurement techniques.

Another useful reference is *Air Change Rate and Air Tightness in Buildings, ASTM STP 1067*, published in 1989 (ISBN 0-8031-1451-6). It deals with similar issues, but is more concentrated on residential construction, including multistory army barracks, manufactured housing, high-rise apartments and single-family dwellings. It also addresses the complex problem of interzonal air movement between rooms and floors inside such "cut-up" buildings—which is the key to understanding ventilation effectiveness. Simply putting outdoor air into a building at a single point does not mean the fresh air will find its way through the HVAC system to all spaces.

To find highly practical advice based on field experience, the reader might wish to consult *Indoor Air Quality and HVAC Systems*, written by David W. Bearg. (Lewis Publishers, CRC Press, 2000 Corporate Blvd, Boca Raton, FL 33431. 1993 ISBN 0-87371-574-8) Through a wide variety of case histories of field investigations, Mr. Bearg explains the many challenges as well as successful techniques for quantifying air leakage into and out of buildings.

The references above were written quite some time ago. But the leakage characteristics of commercial buildings built in North America have not changed substantially since those references were published. This assertion is supported by an article that appeared in the March, 1999 issue of the *ASHRAE Journal*, titled "Myths About Building Envelopes" by Andrew K. Persily. Using measured results of field tests, he found that buildings constructed in the 1980's and 1990's are not "tighter" than those built in the 1950's, 60's and 70's.

For an even more detailed and enlightening survey of existing buildings, the reader should consult the final report of a survey of the leakage characteristics of 70 commercial buildings in Florida. The field research was conducted through the Florida Solar Energy Center by James Cummings and his co-workers Withers, Moyer, Fairey and McKendry, and is titled *Uncontrolled air flow in non-residential buildings.* The report shows how the HVAC system itself is responsible for most of the unwanted leakage into typical commercial buildings. This infiltration is caused largely by inadequate amounts of conditioned makeup air and by unsealed ductwork that creates suction near the exterior walls. The report can be obtained from FSEC: Final Report of FSEC project number FSEC-CR-878-96, April 15th, 1996, Florida Solar Energy Center, 1679 Clearlake Rd, Cocoa, FL 32922.

3. Condensation in exterior walls

To find techniques for predicting condensation in exterior walls, the designer can consult Chapter 22 of the 1997 edition of the *ASHRAE Handbook—Fundamentals*. (Thermal and moisture control in insulated assemblies). That chapter shows how to calculate the temperature of each layer in the wall, which can then be compared to the dew point temperature at the same points. If the dew point exceeds the surface temperature, condensation is certain.

Computer programs are also available to reduce the time needed for these dew point analysis. One of these is "MOIST", written by Douglas Burch and J. Chi at the US National Institute of Standards and Technology. Developed with public funds, it can be obtained at no cost by sending an e-mail message to moist@nist.gov. (MOIST, release 3.0, NIST special publication 917 by Douglas M. Burch and Joseph Chi, September 1997)

For more comprehensive analyses of wall moisture behavior including wetting and drying rates and profiles, a program called "WUFI" is available from both the US Department of Energy's Oak Ridge National Laboratory and from the Fraunhofer Institute of Building Science in Germany. WUFI allows the designer to predict not only internal condensation, but also drying behavior of walls that have experienced condensation or wetting from wind-driven rain, based on a complete annual simulation of hourly weather at the site. In most cases, the key issue is not whether the wall will condense moisture for a few hours at extreme design conditions, but rather how much moisture will condense, causing what moisture content in materials, over what amount of time and with what degree of automatic drying through normal operation of the building and normal climate variations. This program provides answers to such complex questions in both quantitative detail and compelling graphic display.

Künzel, Hartwig M. and Karagiozis, Achilles N. and Holm A., 2000. *WUFI ORNL/IBP; A Hygrothermal Design Tool for Architects and Engineers*, ASTM Manual 40, Heinz Treschel, Editor. American Society for Testing And Materials, Philadelphia, PA. Note: The actual software is developed and published jointly by the Fraunhofer Institute of Building Science, Germany, Dr. Hartwig Künzel, 011-49-8024-64345 and by the US Department of Energy, Oak Ridge National Laboratory, Oak Ridge, TN, Dr. Achilles Karagiozis (865) 576-3924.

4. Building leakage tests using blower doors

The American Society of Mechanical Engineers (1916 Race St, Philadelphia, PA 19103 www.astm.org.) publishes standard techniques for testing the leakage rates of buildings with blower doors:

1. *ASTM E 779 Method for determining air leakage rate by fan pressurization*
2. *ASTM E 1258 Test Method for Airflow Calibration of Fan Pressurization Devices*
3. *ASTM E 1186 Practice for Air Leakage Site Detection in Building Envelopes*

For field experience using these standard methods, consult the publications described by reference 2. Examples of actual hardware used for these tests can be seen at: www.energyconservatory.com.

A brief course instructing owners and designers in visualizing and quantifying air flow in buildings was created by Dean Heerwagen, Judith Heerwagen, David Hudacek & Lisa Kirkendall of the Department of Architecture, University of Washington. The three-part course material is available for downloading in the form of PDF files at: www-archfp.ced.berkeley.edu/vitalsigns/res/rps.html

5. Building leakage tests using tracer gas

The American Society of Mechanical Engineers (1916 Race St, Philadelphia, PA 19103 www.astm.org.) publishes *ASTM E 741 Test Method for Determining Air Leakage by Tracer Dilution*. Tracer gas measurements are especially useful for quantifying leakage between zones inside a building, and also for establishing the actual sources of leakage air. Tracer gases can uniquely identify air from different locations—which is not possible with blower doors. For practical examples of using tracer gases in the field, consult the publications described by reference 2.

6. Mold & mildew inside walls

In an often-cited survey of 9,000 properties, the Executive Engineer's Committee of the American Hotel & Motel Association estimated that their members spend over $68 million each year to remediate problems caused by excess moisture and humidity. A major cause of that damage is humid air pulled into the building by exhaust systems. The survey is titled "Mold and Mildew in Hotels and Motels". (AH&MA 1201 New York Ave. NW, Suite. 600 Washington, DC 20005-3931 (202) 289-3100 www.ahma.com)

Chapter 21 of this book (Hotels) describes the work of building scientists over the 10 years since the publication of the original AH&MA report. The mechanisms of the mold and mildew problem are now understood in much more detail. That's why this book emphasizes the importance of sealing all exhaust and return air ductwork to eliminate the suction in wall cavities that pulls outdoor air into the building.

7. Construction drying

Moisture trapped in building materials such as concrete slabs, ceiling tile, drywall and wet-sprayed fire proofing causes problems like mold and mildew inside walls, warped wooden floors and recurring condensation in unexpected places. There is a large industry that focuses on removing excess moisture from buildings after floods, fires and disasters. Some additional references include:

1. "Drying Concrete" by L.G. Harriman III. This article appeared in the March 1995 issue of *Construction Specifier* magazine. It describes how moisture is measured, and how excess moisture is removed from concrete. (Construction Specification Institute, 601 Madison St., Alexandria, VA 22314-1791 (703) 684-0300. www.csinet.org)

2. "Construction Dehumidifiers" by Major J. Hanley. This two-part article describes techniques used in both concrete and steel building construction to remove excess moisture. It appeared in the October/November and December issues of *Military Engineer* magazine, published by the Society of American Military Engineers (SAME) 607 Prince St, Alexandria, VA 22314 (703) 549-3800 www.same.org.

3. "Specifications for Humidity Control During Installation" describes how humidity is controlled by portable equipment during installation of wooden flooring. These specifications are published by the Maple Flooring Manufacturers' Association (MFMA) 60 Revere Drive, Northbrook, IL 60062 (847) 480-9138 www.maplefloor.com

8. Quantifying duct leakage

The Sheet Metal Manufacturers and Air Conditioning Contractors National Association (SMACNA) publishes guides and standards for the construction of air ductwork. Their "HVAC Duct Systems Inspection Guide" (15D, 1989) describes how much air should be expected to leak into or out of ductwork built to typical standards. A brief but comprehensive commentary on this topic is contained on pages C-1 through C3 of that publication. (SMACNA 8224 Old Courthouse Rd., Tyson's Corner, Vienna, VA. 22182 (703) 790-9890 www.smacna.org)

9. Positive internal air pressure in hot & humid climates

Very little positive air pressure is needed to prevent moisture damage from infiltrating air. The amount of pressure is not especially important. As little as 0.004 in. wc. [1.0 Pa] has been shown to be adequate. But it is critical to maintain that positive pressure continuously, 24-hours, 365-days per year, and maintain it across the entire external surface of the building. Simply measuring pressure difference at a single location can create a very misleading impression. Some internal zones may be positive with respect to the outdoors, and others may be negative. In a multistory building, for example, the lower floors are likely to be negative, and the higher floors positive because of pressure differences caused by the stack effect. That situation may not be apparent if a single pressure measurement is taken on an upper floor. These issues are discussed in useful detail by an article written by George DuBose, J.David Odom and Philip Fairey titled "Why HVAC Commissioning Procedures Do Not Work in Humid Climates" (*ASHRAE Journal*, December, 1993. p. 25-31)

10. Requirements for humidity over 65% rh.

When humidity must be maintained above 65%, it becomes difficult to avoid condensation in ductwork, and inside the building. A great deal of moisture must be added at one point—nearly saturating the air—in order to maintain the desired lower level at other locations. Special precautions are necessary in both ductwork design and building envelope design to avoid condensation. For these situations, the HVAC designer will find the best advice will come from manufacturers of humidification equipment, and from industry references such as *The Humidification Handbook*, Second Edition, 1998. (Dri-Steem Inc, 14949 Technology Drive. Eden Prairie, MN 55344 (612) 949-2415 www.dristeem.com)

11. Requirements for humidity below 35 %rh.

When humidity must be held below 35%, moisture load calculations become more critical, ductwork and walls *must not* leak air, and dehumidification equipment must be able to produce dew points well-below the freezing point of water. For these situations, the HVAC designer will find the best advice is obtained from manufacturers of desiccant dehumidification equipment, and industry references such as *The Dehumidification Handbook*, Second Edition, 1990. (Munters Cargocaire, 79 Monroe St, Amesbury, MA 01913 (978) 241-1100 www.muntersamerica.com)

12. Finding air leaks in exterior walls

Finding leaks in exterior walls is quite a challenge. Infrared imaging has been used for this purpose, and can be quite successful when there is a significant temperature difference between the inside of the building, and the air outdoors. The larger the temperature difference, the more obvious the leak area will be when the wall is viewed through an infrared imaging device. Information about this technique is contained in a brief course created by Dean & Judith Heerwagen of the Department of Architecture, University of Washington. The course describes how to visualize and quantify air flow in buildings. It is available for downloading in the form of PDF files at: www-archfp.ced.berkeley.edu/vitalsigns/res/rps.html

Chapter 3

Psychrometrics of Humidity Control

By Lew Harriman

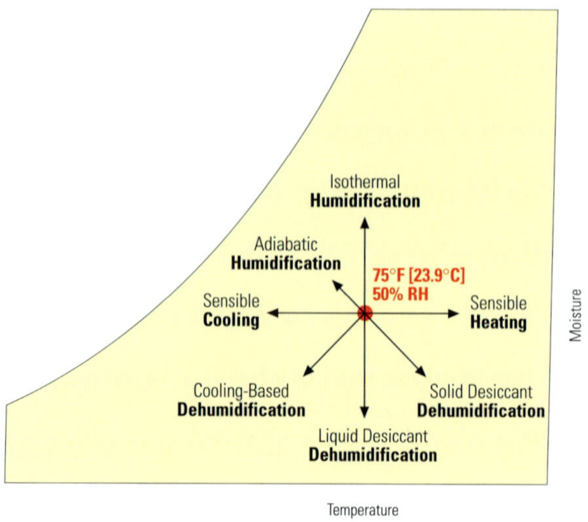

Fig. 3.1 Psychrometric Chart

The chart displays the key thermodynamic characteristics of air, and it lets HVAC designers quickly estimate the energy required to change the temperature or humidity of air. Each direction of change can be accomplished by using a different type of mechanical equipment. These processes are described and quantified by the text and graphics in this chapter.

Psychrometrics & Humidity Control

To control humidity, the designer must understand how the moisture content of air varies, and understand the equipment and energy implications of changing the moisture content of air.

Psychrometry is the science concerned with measuring the thermal characteristics of air. It provides simple tools for quantifying the work that must be accomplished when humidity is controlled. These tools are displayed in the form of the psychrometric chart—which can be used for *understanding* the various processes of air conditioning, as well as for quantifying the energy they require.

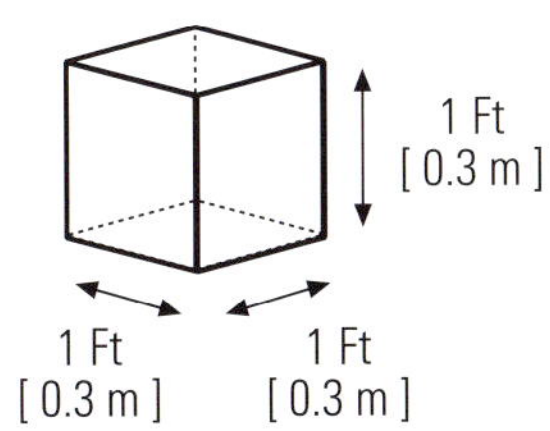

Fig. 3.2 Weight And Volume Of Air

Weight at sea level and 72°F [22.2°C]:

- *One cubic foot weighs 0.075 lbs [28.2 Liters weigh 0.43 kg]*

Volume at sea level and 72°F [22.2°C]:

- *One pound of air occupies 13.3 ft³ [0.4536 kg of air occupies 375 liters]*

Psychrometric Variables

To understand the psychrometric chart and the variables it displays, begin by considering an air sample weighing one pound [0.45 kg] at a comfortable temperature of 75°F [23.9°C] with a similarly comfortable moisture level of 50% relative humidity. Such a condition would be typical of a well-controlled commercial building in the summertime. We'll use this air sample to describe psychrometric variables, to show how they are displayed on a psychrometric chart and to explain how each variable relates to humidity control.

Dry bulb temperature

This is the familiar "air temperature" measured by an ordinary thermometer which has a dry sensing bulb. Our air sample has a dry bulb temperature of 75°F [23.9°C]. On a psychrometric chart, the dry bulb temperature is displayed on the horizontal axis at the base of the chart, as shown in figure 3.4.

Fig. 3.3 Dry Bulb Thermometer

Fig. 3.4 Dry Bulb Temperature

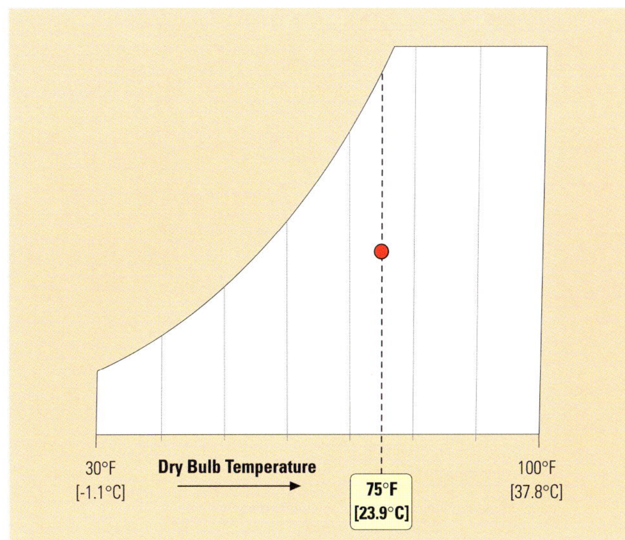

Knowing the current vs. desired dry bulb temperatures of the air is essential to controlling humidity, because the maximum amount of moisture that air can contain depends on its temperature. The warmer the air, the more moisture it can hold.

Relative humidity

The air sample has a relative humidity of 50%. In other words, it contains 50% of the maximum amount of water vapor that it could contain at its current dry bulb temperature.[1]

On a psychrometric chart, the relative humidity appears as a series of curves. As shown in figure 3.5, the 50% curve "connects the dots" that represent 1/2 of the maximum moisture that air can contain at each dry bulb temperature. The 100% rh curve forms the upper boundary of the chart, and is also called the "saturation curve." It represents the maximum moisture content of air at each dry bulb temperature.

Relative humidity is important to humidity control in commercial buildings because it affects human comfort. Occupants and owners have personal experience that associates mid-range relative humidity with comfort, and associates high or low relative humidity with discomfort. So relative humidity is well known to the public. Also, the lowest-cost humidity controls for HVAC equipment measure relative humidity directly, and often display only that variable.

Actually, human comfort is governed more strongly by water vapor pressure than by relative humidity. But relative humidity is more familiar and more easily measured. So conversations between owner, Architect, HVAC designer and contractors about humidity control are likely to include relative humidity more often than other humidity-related parameters.

The popular focus on relative humidity can both assist and bedevil the designer of humidity control systems. The term provides a means of communicating with the nontechnical professional. On the other hand, it leads to confusion. When occupants report comfort problems, they may know *either* the relative humidity *or* the temperature, but neglect to report the other variable, making the information nearly useless for troubleshooting. Also, small changes in temperature cause large changes in relative humidity. As air moves through a building, different heat loads make the relative humidity quite different in each space, which often confuses the nontechnical professional. In short, the designer of humidity control systems will be communicating with others by discussing relative humidity, but some education will usually be needed during the conversation.

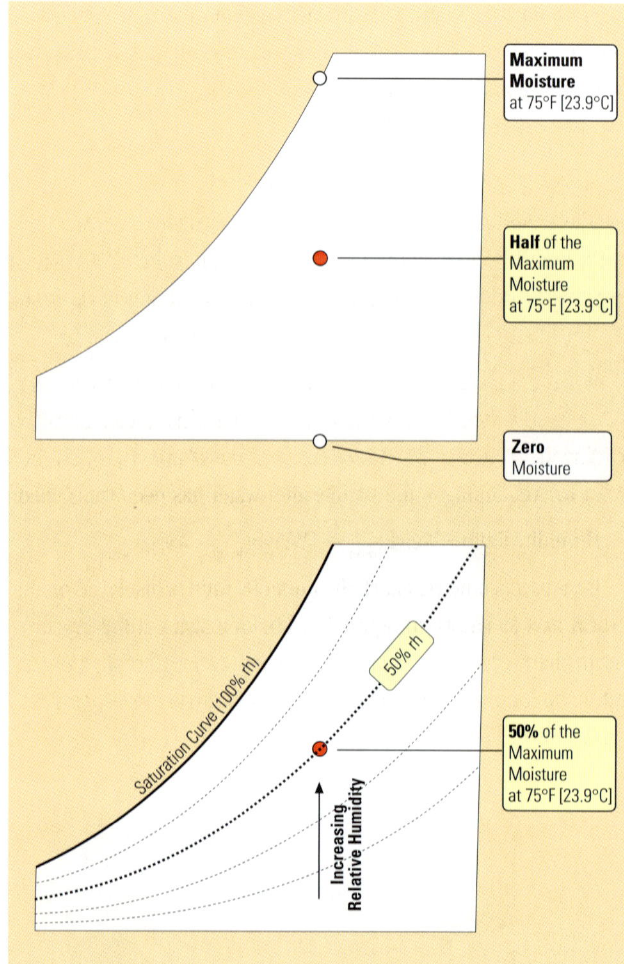

Fig. 3.5 Relative Humidity

The maximum moisture content of air changes with temperature. Relative humidity expresses the air's moisture content as a percent of its maximum capacity at its current dry bulb temperature.

Fig. 3.6 Electronic RH Instrument

Many electronic sensors measure the absorption of water vapor into a capacitor. The change in capacitance is displayed as relative humidity.

In contrast, during the design, operation and maintenance of commercial HVAC systems, the primary focus is on the humidity ratio or the dew point temperature. These two variables represent "absolute" amounts of water in the air.[2]

Humidity ratio (also known as mixing ratio)

The humidity ratio is completely essential to the designer of humidity control systems. The value clearly defines the total mass of water vapor in the air, and therefore allows definition of the work that must be done to change the moisture condition of that air.

When the air temperature is 75°F [23.9°C] and its relative humidity is 50%, the humidity ratio is 0.0093. This is calculated by dividing the *weight of the water vapor* in the sample by the *weight of dry air* remaining in the sample after water has been subtracted:

$$\text{Humidity Ratio} = \text{Weight}_{water} \div (\text{Weight}_{dry\ air} - \text{Weight}_{water})$$

On a psychrometric chart, the humidity ratio is displayed on the vertical axis as shown in figure 3.7. Looking again at the psychrometric chart, one can see the chart is simply an x-y graph, with dry bulb temperature as the x-axis, and the moisture content (the humidity ratio) as the y-axis.

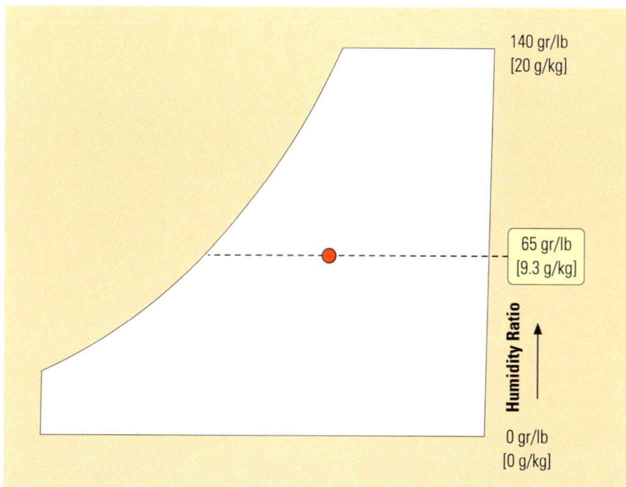

Fig. 3.7 The Humidity Ratio

The moisture content of air—the principal value used to design dehumidification and humidification systems.

Manufacturers of dehumidifiers and humidifiers often convert the decimal fraction to whole numbers by multiplying the humidity ratio by the number of smaller units in a pound or kilogram. In I-P measurement, 7,000 grains weigh one pound. Multiplying the humidity ratio by the number of grains in a pound yields a whole number. (0.0093 x 7000 = 65 gr/lb). In the SI system of units, multiplying the humidity ratio by the number of grams in a kilogram also yields a number with few decimal places (0.0093 x 1000 = 9.3 g/kg) For many, the number "65 gr/lb" seems to have more immediate meaning than "0.0093 lb/lb" [9.3 g/kg rather than 0.0093 kg/kg].

Specific humidity

The specific humidity is often confused with the humidity ratio, because the calculation is very similar, and it yields numbers that are nearly identical. But the specific humidity is different. It is the ratio of the weight of water in the sample to the *weight of the whole sample*:

$$\text{Specific Humidity} = \text{Weight}_{water} \div (\text{Weight}_{dry\ air} + \text{Weight}_{water})$$

The value is seldom used in air conditioning or heating, and does not appear on most psychrometric charts. As noted above, the humidity ratio is the more useful parameter. It is the weight of the water, divided by the *weight of the remaining dry air,* rather than divided by the weight of *entire* sample.

Dew point temperature

The dew point temperature is the temperature at which moisture will begin to condense from the air. Our air sample's dry bulb temperature is 75°F [23.9°C], and its relative humidity is 50%. Given those conditions, moisture will begin to condense out of the air if it is cooled below 55°F [12.8°C]. When the dry bulb temperature of the air cools to the same value as the dew point temperature, the air is said to be "saturated"... it has reached 100% relative humidity.

Finding the dew point temperature on the psychrometric chart is straightforward as shown in figure 3.8. From the point defined by the dry bulb temperature and relative humidity, move horizontally to the left to intersect the 100% rh curve (the saturation curve). Then proceed straight down to the bottom of the chart and read the temperature. That value is the dew point temperature of the air.

The complete and precise definition of dew point temperature is "the temperature of moist air which is saturated at the same pressure, with the same humidity ratio of the given sample of moist air." The complete definition highlights the fact that the dew point depends on the air pressure, which becomes important when working with precise humidity control at high altitudes. But for most work in commercial buildings, it is useful to think of the dew point as simply being the temperature at which water begins to condense from the air. That description is less comprehensive but also correct, and it focuses the designer and owner on the process of condensation. Understanding when condensation does and does not occur is critical to understanding humidity-related behavior of both cooling equipment and the building envelope.

Like the humidity ratio, the dew point temperature clearly defines the amount of water in the air in non-relative terms. So dew point is often used to clearly communicate moisture content, and is often used in contract documents to avoid the potential confusion caused by specifying the relative humidity alone.

Fig. 3.8 Dew Point Temperature

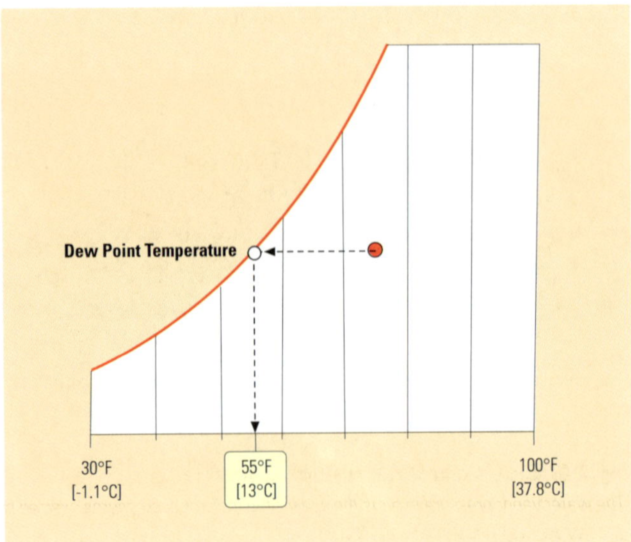

Furthermore, an HVAC system can keep a building in a narrow range of dew point at far less expense and complexity than one which controls relative humidity in a narrow range. Unlike relative humidity, the dew point is not affected by changes in dry bulb temperature as the air moves through equipment and through the building. Relative humidity changes widely and constantly with temperature. Dew point only changes with additions or subtractions of water vapor, which are less frequent and less extreme than temperature variations in commercial buildings.

Dew point temperature is somewhat familiar to the general public, since television weather forecasters often note the current value. Dew point can be useful in predicting the probable lower limit of the nighttime dry bulb temperature. Unless a weather front brings a new air mass into the area, the dry bulb temperature seldom falls lower than the dew point, since heat released by condensing water vapor tends to keep the dry bulb temperature near the dew point.

Wet bulb temperature

The temperature at which evaporation takes place is called the wet bulb temperature.

This parameter is measured by placing a wetted wick around the sensing bulb of a conventional thermometer. As water evaporates from the wick, it draws heat for evaporation from the bulb and from the surrounding air. Evaporation cools the bulb and the surrounding air, because a portion of their sensible heat is lost to the evaporative process. Within a few minutes the evaporation rate becomes constant (assuming the wick stays wet, and air continues to flow across the wick). At that point the thermometer reading will have fallen to the wet bulb temperature—the temperature at which water can evaporate at a constant rate into that particular air mass.

Fig. 3.9
Moisture Condenses When Air Cools Below Its Dew Point
Air in the boundary layer near a cold surface gives up some of its heat to that surface. If the air cools sufficiently, it may fall below its dew point temperature, so it can no longer retain all the water vapor it carried originally. Then dew forms on the nearest surface, as shown here. In this photo, the surface temperature is 32°F [0.0°C], and the dew point of the surrounding air is 60°F [15.5°C]. The volume of liquid condensate increases when there is a greater difference between the temperature of the cold surface and the dew point of the surrounding air.

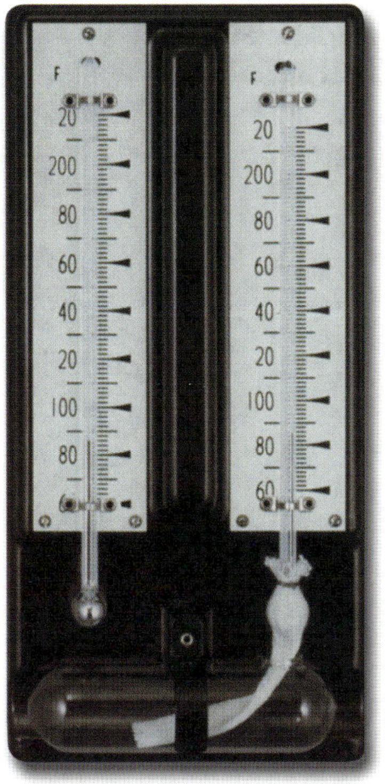

Fig. 3.10 Wet Bulb Thermometer

The thermometer with a wet wick in this photo is the wet bulb thermometer. It measures the temperature at which evaporation occurs in the air sample, provided that the wick is fully-wetted, and that air flows across that wick at a rate of at least 500 fpm [2.54 m/s].

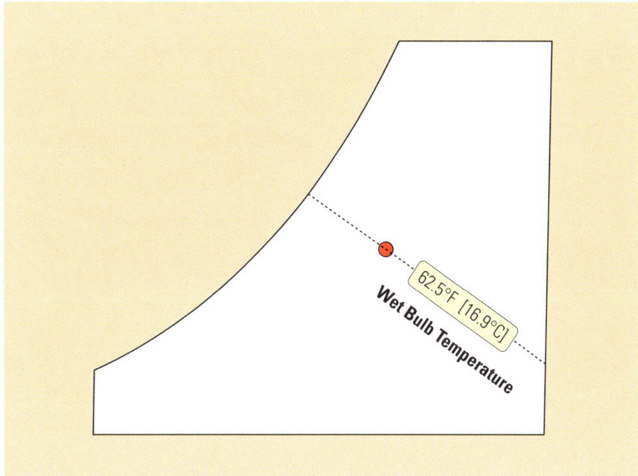

Fig. 3.11 Wet Bulb Temperature

Air in our sample is at 75°F db [23.9°C] and 50% rh. At that condition, its wet bulb temperature is 62.5°F [16.9°C]. The wet bulb temperature lines are diagonals, as shown on figure 3.10.

One can also relate the chart to the physical process of evaporation in another way: When air is fully saturated, it cannot absorb more moisture, so no evaporation (and no evaporative cooling) can take place. When the air is saturated, both wet bulb and dry bulb temperatures are the same. Indeed, when one follows the wet bulb line upwards, it is apparent that the line intersects the saturation curve at the same dry bulb temperature as the wet bulb value of that line.

Wet bulb temperature is important to humidity control in several ways. Since the wet bulb value is the temperature of evaporation, it defines the coolest possible temperature that can be produced by evaporative equipment like cooling towers, which get rid of the excess heat moved out of a building by cooling systems.

Further, humans use evaporative cooling to keep their bodies at constant temperature when the dry bulb temperature rises. The maximum cooling power available through evaporation is limited by the difference between the dry bulb and wet bulb temperature of the air. When wet bulb temperatures are very high, people are uncomfortably hot when they are active and therefore producing extra body heat. Given such effects on systems and people, wet bulb temperature values have been used to design HVAC systems, defining equipment operating parameters and comfort limits.

Also in the past, the wet bulb temperature was frequently used to analyze and control an HVAC system. Measuring the wet bulb temperature was the least expensive way to obtain the value of a second psychrometric variable (after the dry bulb temperature). With the value of any two parameters known, the value of other parameters can be calculated, or simply read from the psychrometric chart. In recent years, however, the cost of instruments to measure relative humidity electronically has fallen substantially. As a result, relative humidity is used much more commonly than wet bulb temperature to diagnose and control humidity-related system operation.

Vapor pressure

One can visualize the vapor pressure variable by considering a tank of air with a mercury manometer attached to measure the pressure of air inside that tank as shown in figure 3.12.

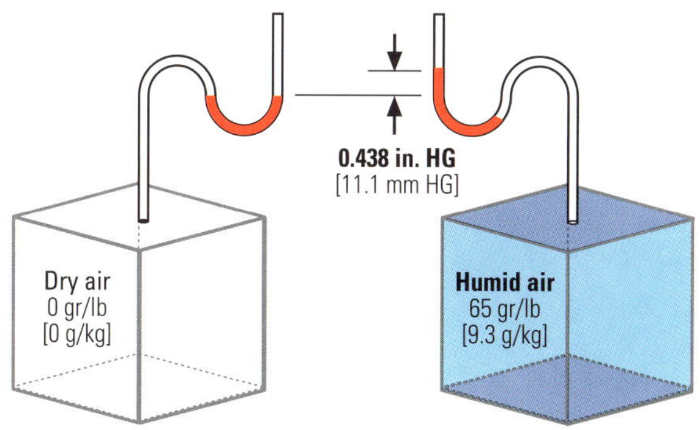

Fig. 3.12 Each Water Molecule Exerts Pressure

The water vapor pressure in air is the sum of the individual pressures exerted by each of the water molecules in the sample.

Begin with the assumptions that the air is perfectly dry, at a temperature of 75 °F [23.9°C] and located at sea level. Take the manometer reading for that circumstance as zero. By adding enough water vapor to increase the relative humidity to 50% at the same temperature, the vapor will increase the pressure in the tank enough to push the column of mercury up the tube by 0.438 inches [11.1 mm].

On the psychrometric chart, the vapor pressure value appears at the right, on a second scale to the right of the humidity ratio as shown in figure 3.13.

Vapor pressure is important to design and operation of all HVAC systems, because differences in vapor pressure cause water vapor to slowly diffuse from one part of the building to another. Also, the force created by differences in vapor pressure is substantial, as one can easily visualize by considering the work done by pushing a column of heavy metal like mercury up as high as 0.438" [11.1 mm]. This force can cause significant problems. For example, the difference in vapor pressure between saturated wood siding and dry winter air helps force paint to peel, flake and fall off buildings. Also, differences in vapor pressure slowly force vapor to diffuse through solid materials like brick, wall board and insulation. Vapor movement can cause condensation, corrosion and mold growth when water vapor meets cold surfaces inside building cavities.

Quantifying vapor pressure differences has only modest importance for the designer of humidity control systems. The vapor pressure difference between inside and outside helps quantify the moisture load caused by diffusion through the building walls, floor and roof. But in commercial buildings, the exterior wall area is usually small in comparison to the enclosed volume, so the permeation load is quite insignificant compared to loads such as air infiltration and ventilation.

On the other hand, vapor pressure differences gain importance for buildings that are humidified during cold weather, such as a museum, or any building that has a high internal moisture load year-round, such as one which contains a swimming pool. In these situations, there is a considerable outward migration of water vapor into cold walls. The rate is influenced by differences in vapor pressure. The designer must quantify the vapor pressures inside the space compared to outside the building to be certain of how to minimize moisture permeation, which could otherwise result in condensation inside the walls.

Please note that for clarity, we have somewhat simplified the concept of vapor pressure. A more detailed and thermodynamically-complete explanation of vapor pressure is described by note 3 at the end of this Chapter.

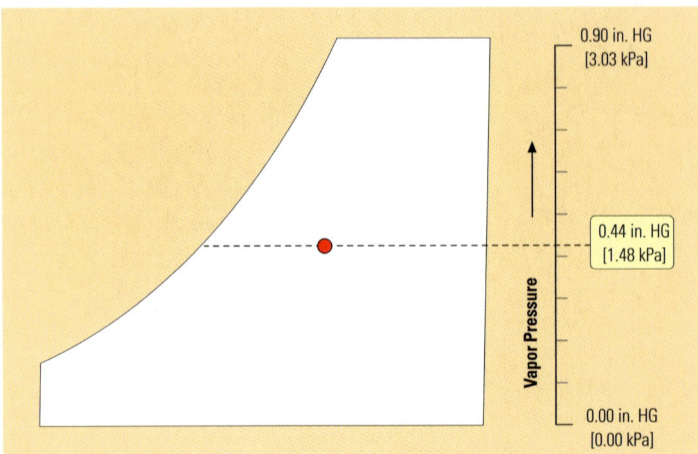

Fig. 3.13 Vapor Pressure

Enthalpy

The enthalpy of the sample quantifies the total heat content of the air. The total heat is the sum of two components: sensible heat plus latent heat. One can visualize this value by thinking of sensible heat as the energy measured by a dry bulb thermometer, and latent heat as the energy required to evaporate the water that is now contained in the air.

At 75°F [23.9°C] and 50% rh, our air sample has an enthalpy of 28.1 Btu/lb [65.4 kJ/kg]. On the psychrometric chart, the enthalpy lines are diagonals, nearly (but not quite) parallel to the wet bulb lines as shown in figure 3.14.

Reading enthalpy values from most psychrometric charts requires care. A straight edge must be placed through the state point of the air such that the same enthalpy value is read on the scale at the top left as well as either the bottom or right edge of the chart. It is easy to misalign the straight edge, and read an incorrect value for enthalpy from one or the other of those scales.

Accurate enthalpy values are important, because the total heat content of the air determines the total energy needed to change the condition of the air from its current condition to the desired condition. Small differences in enthalpy can make large differences in equipment capacity, especially in commercial buildings where high air flow rates are commonly needed for heating and cooling.

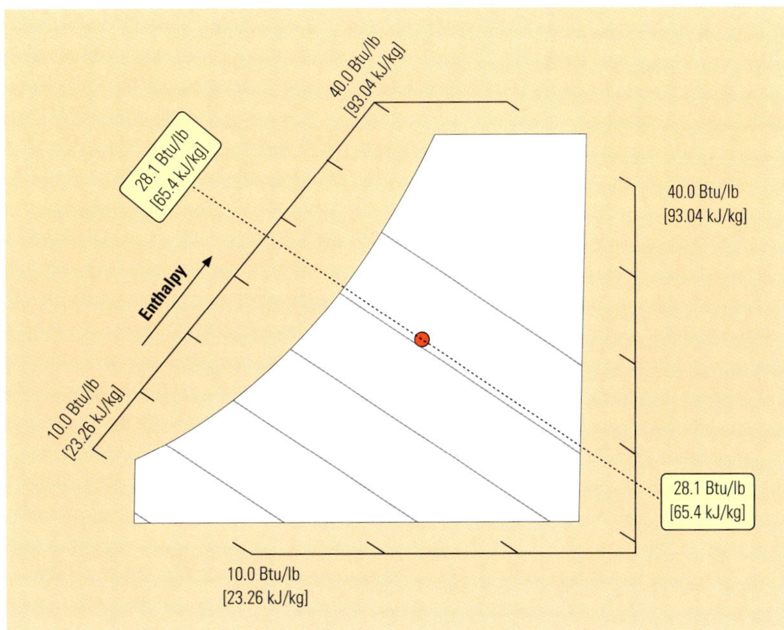

Fig. 3.14 Enthalpy

Enthalpy describes the total heat content of the air—the sum of the sensible heat plus the heat needed to evaporate the moisture it contains (its latent heat).

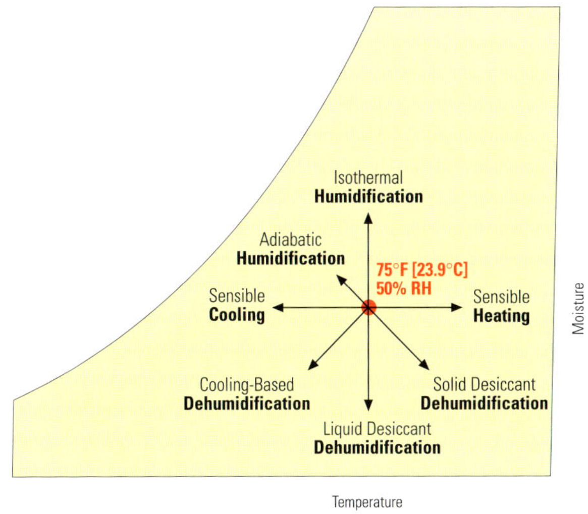

Fig. 3.15
Air Conditioning Processes Displayed on the Psychrometric Chart

Relating Equipment to Air Conditioning Processes

The processes required to change the temperature and moisture of air can be plotted on a psychrometric chart. Each process is characterized by the direction of the line drawn between the initial and final conditions of that air, as shown on figure 3.15.

Also, each process is accomplished by a different type of mechanical equipment. The descriptions below are intended to relate the abstract processes on the psychrometric chart to one of the several possible devices that can actually effect each type of change.

As these processes are described, the reader should keep in mind that air moving through actual equipment does not follow paths identical to the theoretical constructs shown here. The real world is more complex. No physical process can be as perfectly efficient as implied by lines on a chart. But these descriptions help the reader understand the basic functions of common system components.

Sensible heating

In purely sensible heating, a device takes entering air at one condition, and simply adds dry heat to raise its dry bulb temperature without changing the moisture content of the air. Examples include an indirect-fired gas furnace or electrical resistance heater.

This process is shown on figure 3.16. Air enters the heater at 75°F [23.9°C] and 50% rh. It leaves the device at 95°F [35°C]. Because the air is warmer, its capacity to hold moisture is much greater than when it entered the heater. So the amount of moisture it contains now represents a lower percent of its maximum moisture content. The relative humidity is now only 26% of the air's maximum capacity instead of 50%.

No water has been removed, so the dew point of the air has not changed. Water will still condense from the air when it contacts a surface with a temperature lower than 55°F [12.8°C]. But the airs *relative humidity* is lower than it was before it entered the heater.

Fig. 3.16 Sensible Heating
Increasing the sensible heat in the air—without changing its moisture content.

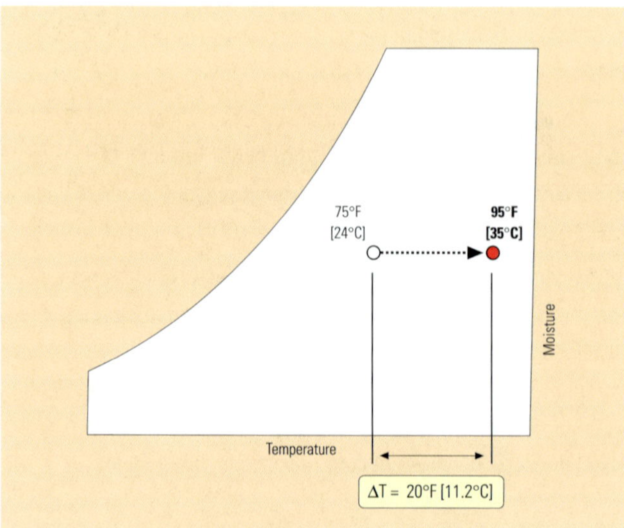

The energy needed to affect that change continuously for one hour depends on the flow rate of air moving through the heater, and on the magnitude of the temperature change:

To affect this particular temperature change in an air stream flowing at rate of 1,000 cfm [472 L/s], the heater must add 21,600 Btu [6.3 kW] to the flowing air every hour.

$$\textbf{Btu/h} = 1{,}000 \times (95 - 75) \times 1.08$$

$$\textbf{Btu/h} = 21{,}600$$

$$[\textbf{kW} = 6.33]$$

Sensible Heating — Equation 3.1 I-P

$$\textbf{Btu/h} = \text{cfm} \times (t_2 - t_1) \times 1.08$$

- **Btu/h**: **Sensible heating energy** (British Thermal Units per Hour)
- **cfm**: Air flow rate (Cubic feet per minute)
- t_2: Air temperature *leaving* the coil (Degrees Fahrenheit)
- t_1: Air temperature *entering* the coil (Degrees Fahrenheit)
- **1.08**: Heat required every hour to heat air by one degree Fahrenheit when it flows at a rate of one cfm (Btu/h • °F • cfm)

Sensible Heating — Equation 3.1 SI

$$\textbf{kW} = \text{L/s} \times (t_2 - t_1) \times 0.0012$$

- **kW**: **Sensible heating energy** (Kilowatts)
- **L/s**: Air flow rate (Cubic liters per second)
- t_2: Air temperature *leaving* the coil (Degrees Celsius)
- t_1: Air temperature *entering* the coil (Degrees Celsius)
- **0.0012**: Heat required every hour to heat air by one degree Celsius when it flows at a rate of one liter per second (kW/L/s • °C)

Fig. 3.17 Indirect-fired Gas Heaters
Gas burners with heat exhangers are used to heat supply air without changing its moisture content. Electric resistance heaters are also used for this purpose.

Fig. 3.18 Sensible Cooling Coils

Cooling coils that only cool the air without dehumidifying it are usually "shallow". That is, they have few rows of cooling tubes since they do not need to chill air deeply to remove moisture.

Sensible cooling

When the air only needs cooling, some device removes part of the air stream's sensible heat, reducing the dry bulb temperature of the air without changing its moisture content. Typically, sensible cooling is accomplished by a cooling coil connected to a system which moves the first air stream's sensible heat to a second air stream indirectly, using a heat transfer fluid.

An example is shown in figure 3.19. Air enters the cooling coil at 75°F [23.9°C] and 50% rh. It leaves the cooler at 55°F [12.8°C]. At the lower temperature, its moisture capacity is also lower. In fact, the current moisture content (its humidity ratio) represents 100% of the moisture that the air can contain at this new, lower temperature. The relative humidity is now 100%. But the moisture content has not changed, and therefore the dew point is the same. In fact, the air temperature has fallen exactly to its dew point—55°F [12.8°C]—and if its temperature falls any lower, the air will lose moisture as water vapor condenses onto the nearest cold surface.

Fig. 3.19 Sensible Cooling

Reducing the temperature of air without changing its moisture content.

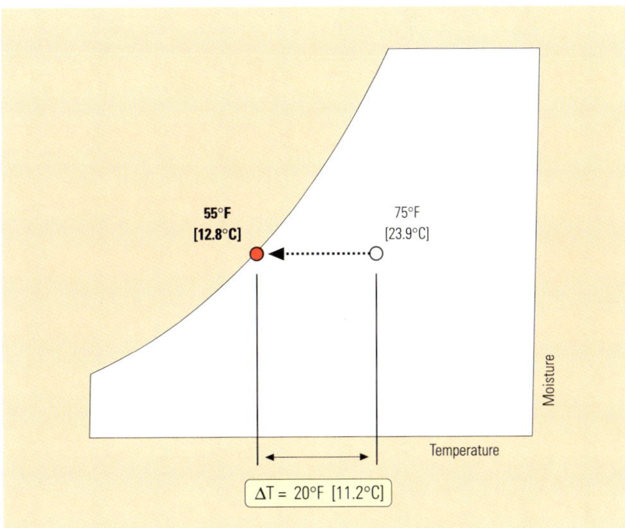

The energy requirement for sensible cooling is calculated with the same equation used for calculating a sensible heating requirement. The air flow is multiplied by the temperature difference and by a constant that converts the air flow into mass flow.

To affect this temperature change in an air stream flowing at a rate of 1,000 cfm [472 L/s], the cooling coil must subtract 21,600 Btu [6.3 kW] from the flowing air every hour:

Btu/h = 1,000 x (75 - 55) x 1.08

Btu/h = 21,600

[**kW** = 6.33]

Sensible Cooling — **Equation 3.2 I-P**

$$\text{Btu/h} = \text{cfm} \times (t_2 - t_1) \times 1.08$$

- **Sensible cooling energy** (British Thermal Units per Hour)
- Air flow rate (Cubic feet per minute)
- Air temperature *leaving* the coil (Degrees Fahrenheit)
- Air temperature *entering* the coil (Degrees Fahrenheit)
- Energy required every hour to cool air by one degree Fahrenheit when it flows at a rate of one cfm (Btu/h • °F • cfm)

Sensible Cooling — **Equation 3.2 SI**

$$\text{kW} = \text{L/s} \times (t_2 - t_1) \times 0.0012$$

- **Sensible cooling energy** (Kilowatts)
- Air flow rate (Cubic Liters per second)
- Air temperature *leaving* the coil (Degrees Celsius)
- Air temperature *entering* the coil (Degrees Celsius)
- Energy required every hour to cool air by one degree Celsius when it flows at a rate of one liter per second (kW/L/s • °C)

Dehumidification using cooling coils

Most dehumidification in commercial buildings is accomplished by cooling coils. Air is chilled low enough to drop its dry bulb temperature below its initial dew point temperature. As the air drops below its dew point temperature, water vapor condenses on the cold surfaces of the cooling coil, and drains away from the air as liquid condensate.

Fig. 3.20 Dehumidification by Cooling

Cooling air below its dew point temperature forces some of the water vapor it contains to condense out—the air is dehumidified.

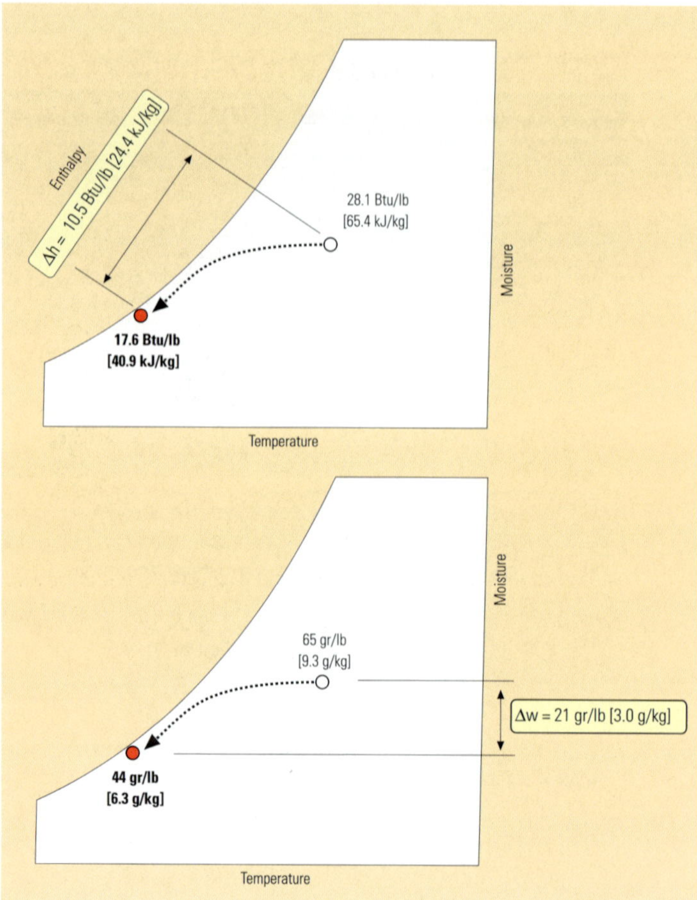

An example is illustrated by figure 3.21. Air enters the cooling coil at 75°F and 65 gr/lb [23.9°C, 9.3 g/kg]. It leaves the coil at 45°F and 44 gr/lb [7.2°C, 6.3 g/kg]. The air has been cooled—and also dehumidified.

Since both sensible and latent heat have been removed from the air, designers find it most convenient to use the enthalpy difference (Δh) between the entering and leaving conditions to calculate the energy required to accomplish the change. The enthalpy of the entering air is 28.1 Btu/lb [65.4 kJ/kg]. The enthalpy of the air as it leaves the coil is 17.6 Btu/lb [40.9 kJ/kg]. The energy required depends on the mass of the air flow and the enthalpy difference.

To remove this much sensible heat and moisture, the cooler must subtract 47,250 Btu [13.8 kW] from the flowing air every hour.

Btu/h = 1,000 x (28.1 - 17.6) x 4.5

Btu/h = 47,250

[**kW** = 13.8]

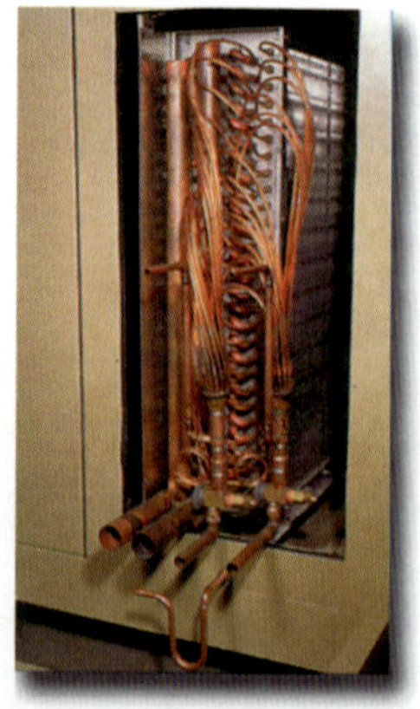

Fig. 3.21 Deep Cooling Coil

A dehumidification coil uses many rows of cooling tubes to chill air well below its dew point temperature, which forces moisture to condense out of the air stream.

Cooling with Dehumidification — Equation 3.3 I-P

$$\textbf{Btu/h} = \text{cfm} \times (h_1 - h_2) \times 4.5$$

- **Dehumidification cooling energy** (British thermal units per hour)
- Air flow rate (Cubic feet per minute)
- Air enthalpy *leaving* the coil (Btu per lb)
- Air enthalpy *entering* the coil (Btu per lb)
- Energy required every hour to remove one Btu per lb of heat from air when it flows at a rate of one cubic foot per minute (Btu/h/cfm • Btu/lb)

Cooling with Dehumidification — Equation 3.3 SI

$$\textbf{kW} = \text{L/s} \times (h_1 - h_2) \times 0.0012$$

- **Dehumidification cooling energy** (Kilowatts)
- Air flow rate (Cubic Liters per second)
- Air enthalpy *leaving* the coil (Kilojoules per kilogram)
- Air enthalpy *entering* the coil (Kilojoules per kilogram)
- Energy required every hour to remove one kilojoule per kilogram of heat from air when it flows at a rate of one liter per second. (kW/L/s • kJ/kg)

The dehumidification effect—the mass of moisture removed—can also be calculated through the difference in humidity ratios. Air entering the coil has a humidity ratio of 65 gr/lb, and leaving the coil its humidity ratio is reduced to 44 gr/lb [9.3 g/kg vs. 6.3 g/kg]:

$$\textbf{lb/h} = \frac{1{,}000 \times (65 - 44) \times 4.5}{7000}$$

$$\textbf{lb/h} = \frac{94{,}500}{7000}$$

$$\textbf{lb/h} = 13.5$$

$$[\textbf{kg/h} = 6.1]$$

Moisture Removal — Equation 3.4 I-P

$$\textbf{lbs/h} = \frac{\text{cfm} \times (w_1 - w_2) \times 4.5}{7000}$$

- **Moisture removal** (Pounds of water per hour)
- Air flow rate (Cubic feet per minute)
- Air moisture content *leaving* the coil (Grains per lb)
- Air moisture content *entering* the coil (Grains per lb)
- Pounds per hour of dry air when delivered at one cubic foot per minute
- Grains in one pound

Moisture Removal — Equation 3.4 SI

$$\textbf{kg/h} = \frac{\text{L/s} \times (w_1 - w_2) \times 4.32}{1000}$$

- **Moisture removal** (Kilograms of water per hour)
- Air flow rate (Liters per second)
- Air moisture content *leaving* the coil (Grams per kilogram)
- Air moisture content *entering* the coil (Grams per kilogram)
- Kilograms per hour of dry air when delivered at one cubic liter per second
- Grams in one kilogram

Visualizing this dehumidification work as a physical result, the cooling coil removed 13.5 pounds (pints) of water from that 1,000 cfm over the course of an hour. [In SI units, 6.4 liters of water from 472 l/s of flowing air over the course of an hour].

Sensible Heat Ratio

For a process that combines cooling with dehumidification, the sensible heat ratio (SHR) is a useful psychrometric tool. The ratio is the amount of *sensible* heat removed, compared to the *total heat* removed. In other words, what percent of the work removed sensible heat rather than moisture? A sensible heat ratio of 1.0 would mean that all the work involves removing sensible heat. A value of zero would mean that all the work involved removing moisture, and that no sensible heat need be removed. In this case, nearly 70% of the work concerned removal of sensible heat.

The sensible heat ratio is calculated by dividing the energy associated with the change in sensible temperature by the total energy change (the enthalpy change):

$$\textbf{SHR} = \frac{(75 - 55) \times 1.08}{(28.1 - 17.6) \times 4.5}$$

$$\textbf{SHR} = 0.69$$

Sensible Heat Ratio — Equation 3.5 I-P

$$\textbf{SHR} = \frac{(t_2 - t_1) \times 1.08}{(h_2 - h_1) \times 4.5}$$

- **Sensible Heat Ratio** (Dimensionless)
- Sensible temperature change (Degrees Fahrenheit)
- Enthalpy (Total heat) change (Btu/lb)

Sensible Heat Ratio — Equation 3.5 SI

$$\textbf{SHR} = \frac{(t_2 - t_1)}{(h_2 - h_1)}$$

- **Sensible Heat Ratio** (Dimensionless)
- Sensible temperature change (Degrees Celsius)
- Enthalpy (Total heat) change (kJ/kg)

The value of the sensible heat ratio of the required change has implications for the configuration of the cooling coil. When the SHR is low (when most of the cooling removes moisture rather than sensible heat), the coil needs more surface area on which to condense moisture. Cooling coils optimized for dehumidification tend to be deep in the direction of air flow, with fins spaced widely to allow water to drain freely from the coil.

Conversely, when the SHR is high (meaning most of the work involves removing sensible heat), there is no need to provide wide passages between fins to drain condensate. Coils for air conditioning processes with high sensible heat ratios tend to have closely-spaced fins, and fewer rows of tubes in the direction of air flow. The useful point to keep in mind is that equipment configured for a process with *one* sensible heat ratio is not as useful for a process with a quite *different* sensible heat ratio. Coils optimized for removing only sensible heat will not be equally effective in removing moisture.

Cooling equipment capacity is usually rated based on an assumed sensible heat ratio for the process results it is intended to achieve. This "equipment SHR" will tell the designer quite a bit about the machine's capacity to dehumidify air. High sensible heat ratios imply low capacity for removing moisture, and vice-versa.

Dehumidification via active desiccants

Another way to dehumidify air is to pass it through a honeycomb wheel impregnated with a desiccant. The desiccant attracts moisture because its surface vapor pressure is lower than the vapor pressure in the humid air. The air takes the path shown on figure 3.22.

Air enters at 75°F and 65 gr/lb [24°C, 9.3 g/kg]. It leaves the desiccant media considerably warmer and dryer at 99°F and 30 gr/lb [37°C, 4.3 g/kg].

The dehumidification effect—the mass of moisture removed—can be calculated though the difference in humidity ratios. Air *entering* the desiccant had a humidity ratio of 65 gr/lb [9.3 g/kg], and *leaving* the desiccant the humidity ratio is 30 gr/lb [4.3 g/kg].

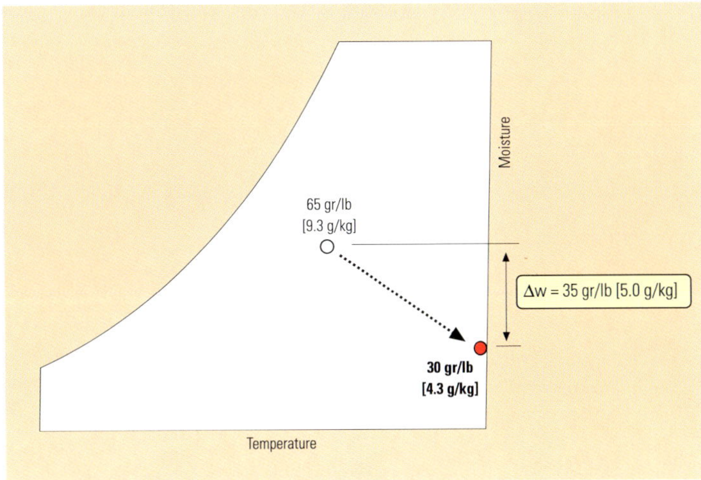

Fig. 3.22 Dehumidification Using Dry Desiccants

$$\textbf{lb/h} = \frac{1{,}000 \times (65 - 30) \times 4.5}{7000}$$

$$\textbf{lb/h} = \frac{157{,}500}{7000}$$

$$\textbf{lb/h} = 22.5$$

$$[\textbf{kg/h} = 10.2]$$

Fig. 3.23 Active Desiccant Dehumidifier

These units dry the air by using differences in vapor pressure. A dry desiccant surface has a low vapor pressure, while the vapor pressure of humid air is high. Moisture is attracted out of the air to the desiccant by this pressure difference. The word "active" describes the fact that heated air is used to dry off the desiccant for re-use after it becomes saturated with moisture.

Visualizing this dehumidification work from a physical perspective, the desiccant removed 22.5 pounds (pints) of water from that 1,000 cfm over the course of an hour. [In SI units, 10.6 liters of water from 472 l/s of air over the course of an hour].

To fully understand this process, note that the air temperature rises as the desiccant removes its moisture. This phenomenon may

Moisture Removal — **Equation 3.6 I-P**

$$\textbf{lbs/h} = \frac{\text{cfm} \times (W_1 - W_2) \times 4.5}{7000}$$

- **Moisture removal** (Pounds of water per hour)
- Air flow rate (Cubic feet per minute)
- Air moisture content *leaving* the unit (Grains per lb)
- Air moisture content *entering* the unit (Grains per lb)
- Pounds per hour of dry air when delivered at one cubic foot per minute
- Grains in one pound

Moisture Removal — **Equation 3.6 SI**

$$\textbf{kg/h} = \frac{\text{L/s} \times (W_1 - W_2) \times 4.32}{1000}$$

- **Moisture removal** (Kilograms of water per hour)
- Air flow rate (Liters per second)
- Air moisture content *leaving* the unit (Grams per kilogram)
- Air moisture content *entering* the unit (Grams per kilogram)
- Kilograms per hour of dry air when delivered at one cubic liter per second
- Grams in one kilogram

be easier to understand as the reverse of evaporative cooling. *Adding* water to air lowers the temperature of the air because part of its sensible heat is used for evaporation. Conversely, *subtracting* water from air raises its temperature, because that latent heat is reconverted into sensible heat. The total heat of the air (its enthalpy) stays the same. Air entered the desiccant wheel with an enthalpy of 28.4 Btu/lb [66 kJ/kg], and it leaves the desiccant containing the same enthalpy. Air is drier, but warmer.

Since there is no energy change (no enthalpy change) in this process, one might assume there is no cost of energy to accomplish the drying. That would be true if an infinite supply of dry desiccant were available. But desiccant eventually becomes saturated with moisture, and must be reactivated (dried for re-use). Removing water from desiccant consumes a considerable amount of energy—between 1.5 and 3 times the heat required to vaporize the water adsorbed by the desiccant. Drier air requires drier desiccant, and therefore more heat for reactivation.

The total energy required to reactivate the material depends on the drying task, the type of desiccant used and the equipment design details. Consequently, energy requirements for an active desiccant drying process cannot be estimated with a psychrometric chart.

Further, hardware realities also affect the temperature of air leaving the desiccant. In most equipment designs, the desiccant retains heat from when it was heated during reactivation. Some portion of that residual heat is transferred from the warm desiccant to the air stream being dried. So the actual line on the psychrometric chart is not a line of constant enthalpy. It slants further to the right, reflecting the fact that any residual heat from reactivation slightly increases the enthalpy of the air being dried. Consequently, active desiccant equipment is generally followed by a heat exchanger or a cooling coil which removes the excess sensible heat produced by removing moisture from air.

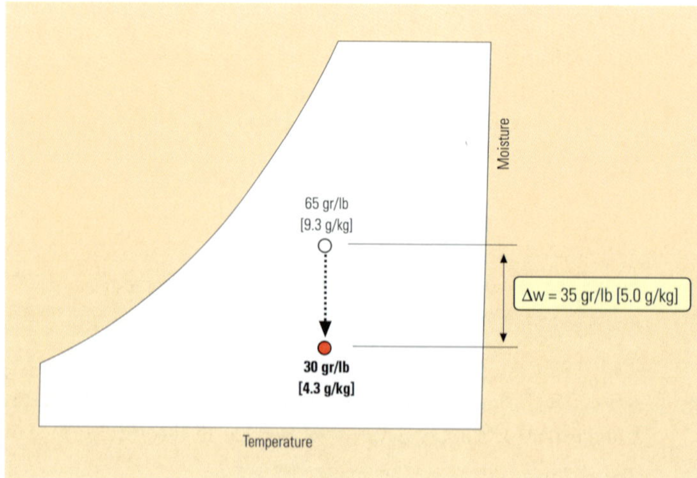

Fig. 3.24
Dehumidification Using Liquid Desiccants

Dehumidification via liquid desiccants

Another option for dehumidifying air is to pass it through media saturated with liquid desiccant. Such devices usually contain internal cooling , so the path taken by the air can be straight down, as shown in figure 3.24.

Air enters at 75°F and 65 gr/lb [24°C, 9.3 g/kg]. It leaves the media at the same temperature, but considerably drier at 30 gr/lb [4.3 g/kg]. Pure dehumidification has taken place without any change to the dry bulb temperature.

As with active solid desiccant processes, the energy required to drive an active liquid desiccant process depends on the drying task, the type of liquid desiccant used and the details of the equipment design. There is no simple way to estimate energy consumption of a liquid desiccant process from a psychrometric chart alone.

Fig. 3.25 Liquid Desiccant Systems
These units dry the air as it flows across a contact media that has been saturated with a concentrated liquid desiccant salt solution. When that solution has been precooled, a liquid desiccant system can dehumidify the air without changing its temperature—the cool solution absorbs the latent heat while removing moisture.

Adiabatic humidification

Adding moisture to air can be accomplished *adiabatically*—with no addition of energy from external sources. In this process, air is passed through media saturated with water. The air evaporates the water by itself, using a part of its sensible heat to accomplish the task. The process takes the path shown in figure 3.26.

Air enters at 75°F and 65 gr/lb [23.9°C, 9.3 g/kg]. It leaves the media considerably cooler, and containing more moisture at 65°F and 82 gr/lb [18°C, 11.7 g/kg]. Although the temperature and moisture content of the air changes, its enthalpy stays constant at 28.4 Btu/lb [66 kJ/kg]. There is no energy added to the process apart from a small amount of power used to force air through the media and to pump water over the air contact surface.

This process is used by three types of equipment used frequently in commercial buildings: evaporative humidifiers, evaporative coolers, and cooling towers. Each type of equipment uses this same process to achieve a different purpose.

Evaporative humidifiers add moisture to air in very simple equipment at very low energy cost. Energy used for evaporation is taken from the air itself. If that air contains excess sensible heat, there is no cost of operating the equipment other than pumps, fans and the cost of feedwater.

Evaporative coolers chill air through the same process. Evaporation uses up part of the sensible heat of the air, cooling that air as it absorbs moisture. The same hardware can serve two different functions: cooling the air and humidifying it.

Cooling towers make use of evaporative cooling to produce cool water rather than cool air. Warm water carrying heat from an air conditioning process flows across evaporative media. Air from outside the building contacts the warm water, and part of the water is evaporated into the air. The remaining water gave up part of its heat during the evaporative process, so that water is cooler than when it began to flow across the media. The cooler water is returned to the air conditioning equipment to absorb more heat from the building.

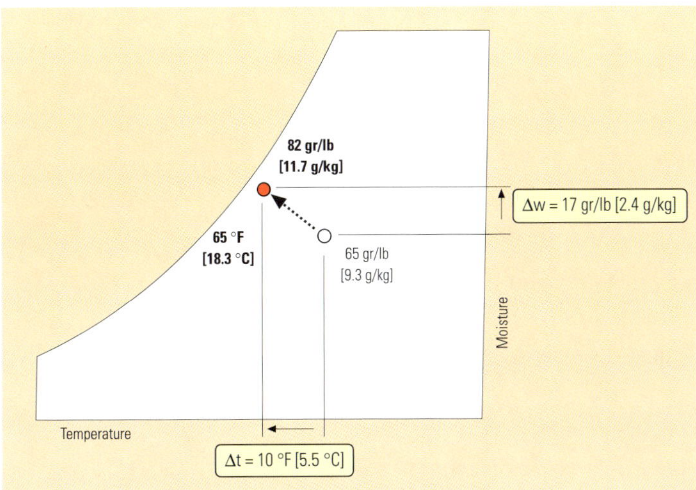

Fig. 3.26
Adiabatic Humidification

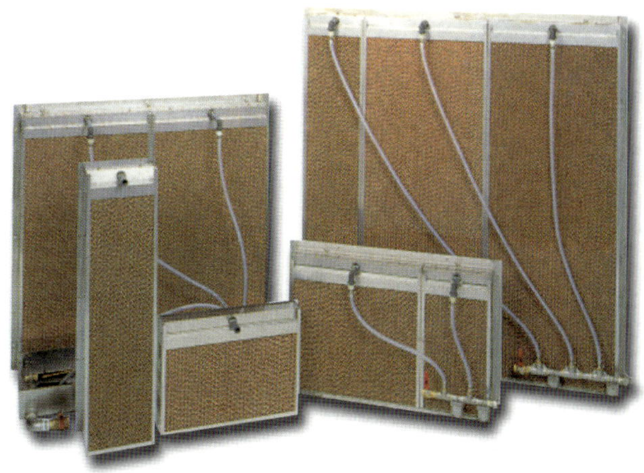

Fig. 3.27 Evaporative Pad-Type Humidifiers

These devices add moisture to air by using the air's own sensible heat for evaporation. So as the air absorbs moisture, it cools down.

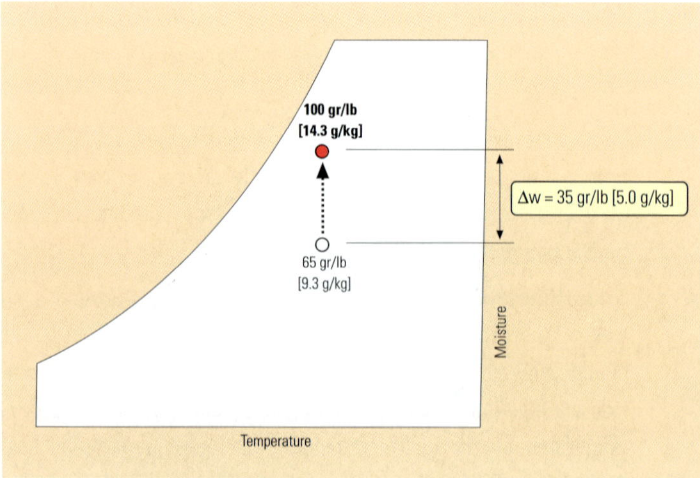

Fig. 3.28 Isothermal Humidification

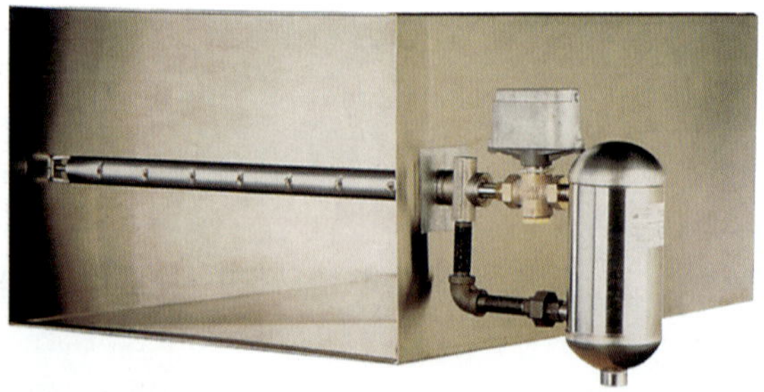

Fig. 3.29 Steam Humidifier

Steam adds moisture to air almost isothermally. The temperature of the supply air stays nearly constant.

Isothermal humidification

When moisture is added to the air without changing its dry bulb temperature, the process is called *isothermal* humidification—the humidification process takes place at constant temperature. Warm water vapor or saturated steam is absorbed by the air in a process that takes the path shown in figure 3.28.

Air enters at 75°F and 65 gr/lb [23.9°C, 9.3 g/kg]. It leaves the device at the same temperature, but containing considerably more moisture at 100 gr/lb [14.3 g/kg]. The dew point has risen from 55°F to 67°F [from 12.8°C to 19.4°C]. The total heat of the air has risen from an enthalpy of 28.4 Btu/lb to 33.7 Btu/lb [from 66 kJ/kg to 78.4 kJ/kg].

The amount of moisture absorbed by the air stream depends on the air flow rate, and the difference in the humidity ratio of the air before and after the humidifier:

$$\textbf{lb/h} = \frac{1{,}000 \times (100 - 65) \times 4.5}{7000}$$

$$\textbf{lb/h} = \frac{157{,}500}{7000}$$

$$\textbf{lb/h} = 22.5$$

$$[\textbf{kg/h} = 10.2]$$

Moisture Addition — Equation 3.7 I-P

$$\textbf{lbs/h} = \frac{\text{cfm} \times (W_2 - W_1) \times 4.5}{7000}$$

- **Moisture addition** (Pounds of water per hour)
- Air flow rate (Cubic feet per minute)
- Air moisture content *leaving* the unit (Grains per lb)
- Air moisture content *entering* the unit (Grains per lb)
- Pounds per hour of dry air when delivered at one cubic foot per minute
- Grains in one pound

Moisture Addition — Equation 3.7 SI

$$\textbf{kg/h} = \frac{\text{L/s} \times (W_2 - W_1) \times 4.32}{1000}$$

- **Moisture addition** (Kilograms of water per hour)
- Air flow rate (Liters per second)
- Air moisture content *leaving* the unit (Grams per kilogram)
- Air moisture content *entering* the unit (Grams per kilogram)
- Kilograms per hour of dry air when delivered at one liter per second
- Grams in one kilogram

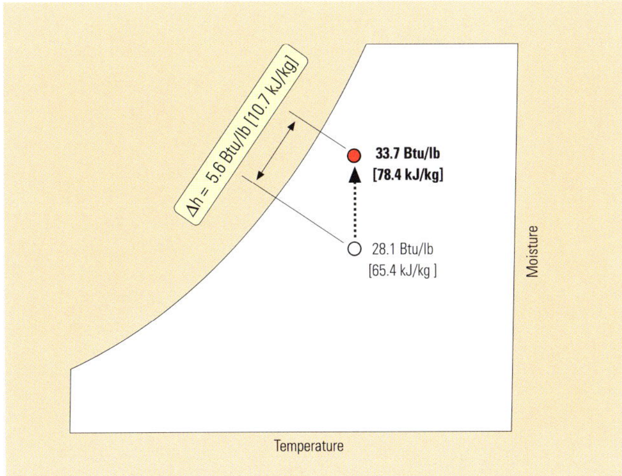

Fig. 3.30
Energy Requirement for Isothermal Humidification

The amount of energy needed to accomplish this change in 1000 cfm [472 L/s] can be easily calculated by using the enthalpy difference between air entering and leaving the device (Δh):

Btu/h = 1,000 x (33.7 - 28.1) x 4.5

Btu/h = 25,200

[**kW** = 7.6]

Additional Resources

Psychrometric variables and processes form the essential foundation of any effective design for a humidity control system. In most cases, the system designer will find that information contained in this chapter will suffice for that foundation. But occasionally, the requirement becomes more complex, such as when dealing with very precise humidity control, or humidity control at high altitudes. In those cases, the designer is encouraged to consult the psychrometrics chapter of the *ASHRAE Handbook—Fundamentals*. That text contains the definitive set of psychrometric values produced for the society by the US National Institute of Standards and Technology, and also a much more detailed discussion of the science of psychrometrics.

Isothermal Humidification Energy — **Equation 3.8 I-P**

$$\textbf{Btu/h} = cfm \times (h_2 - h_1) \times 4.5$$

- **Btu/h** — **Humidification energy** (British thermal units per hour)
- **cfm** — Air flow rate (Cubic feet per minute)
- **h_2** — Air enthalpy *leaving* the unit (Btu per lb)
- **h_1** — Air enthalpy *entering* the unit (Btu per lb)
- **4.5** — Energy required every hour to add one Btu per lb of heat from air when it flows at a rate of one cubic foot per minute (Btu/h/cfm • Btu/lb)

Isothermal Humidification Energy — **Equation 3.8 SI**

$$\textbf{kW} = L/s \times (h_2 - h_1) \times 0.0012$$

- **kW** — **Humidification energy** (Kilowatts)
- **L/s** — Air flow rate (Cubic Liters per second)
- **h_2** — Air enthalpy *leaving* the unit (Kilojoules per kilogram)
- **h_1** — Air enthalpy *entering* the unit (Kilojoules per kilogram)
- **0.0012** — Energy required every hour to add one kilojoule per kilogram of heat from air when it flows at a rate of one liter per second (kW/L/s • kJ/kg)

Notes

1. To be completely accurate, the relative humidity is the ratio of water vapor pressure in the air sample to the water vapor pressure of air at that same dry bulb temperature if it were completely saturated. ($P_w \div P_s$) But when working near sea level in commercial buildings, one can think of relative humidity as simply a percent of maximum humidity of air at the stated dry bulb temperature.

2. The correct definition of absolute humidity quantifies the density of water vapor in the sample. It is the ratio of the weight of water vapor to the volume of the sample. This value is seldom

used in designing HVAC systems since water mass and relative humidity affect comfort and economics more strongly than does vapor density. But the description "absolute humidity" is seductive, and is often erroneously used in place of humidity ratio to describe the non-relative water content of air. That description is quite common, but not correct.

3. In a gas, the total pressure of the sample is the sum of the pressures exerted by all the molecules in the sample. Water vapor is one of those groups of molecules, and the pressure exerted by water vapor is directly proportional to the number of its molecules in the sample. More water = higher water vapor pressure. Further, the relationship is linear; there is a one-to-one correspondence between water content and vapor pressure. As such, it is another "non-relative" measurement of water content.

In HVAC design, the vapor of greatest concern is usually water. So the generic term: "vapor pressure", generally means "water vapor pressure." Also, in the SI system of units, pressure is measured in kilopascals rather than millimetres of mercury, so the correct value of vapor pressure in our sample is either 0.415" HG in the I-P system, or 1.4 kPa in the SI system.

Photo Credits

We appreciate the support of the following companies, which provided the photographs used in Chapter 3. Contact these companies if you need to reproduce their copyrighted images.

Fig. 3.4 *© Cole-Parmer Instruments, Niles, Illinois*

Fig. 3.6 *© Cole-Parmer Instruments, Niles, Illinois*

Fig. 3.9 *© Crate & Barrel, Chicago, Illinois*

Fig. 3.10 *© Davis Instruments, Hayward, California*

Fig. 3.17 *© Munters, Amesbury, Massachusetts*

Fig. 3.18 *© Munters, Amesbury, Massachusetts*

Fig. 3.20 *© Munters, Amesbury, Massachusetts*

Fig. 3.23 *© Munters, Amesbury, Massachusetts*

Fig. 3.25 *© Kathabar, New Brunswick, New Jersey*

Fig. 3.27 *© Munters, Amesbury, Massachusetts*

Fig. 3.29 *© Pure, Chaska, Minnesota*

Psychrometric Chart

Sea Level (Barometric Pressure = 29.921 in.hg.)

This chart was donated by James Judge, P.E. of Linric Company, Bedford, NH. The chart was constructed using the algorithms contained in the ASHRAE Handbook of Fundamentals 1997.

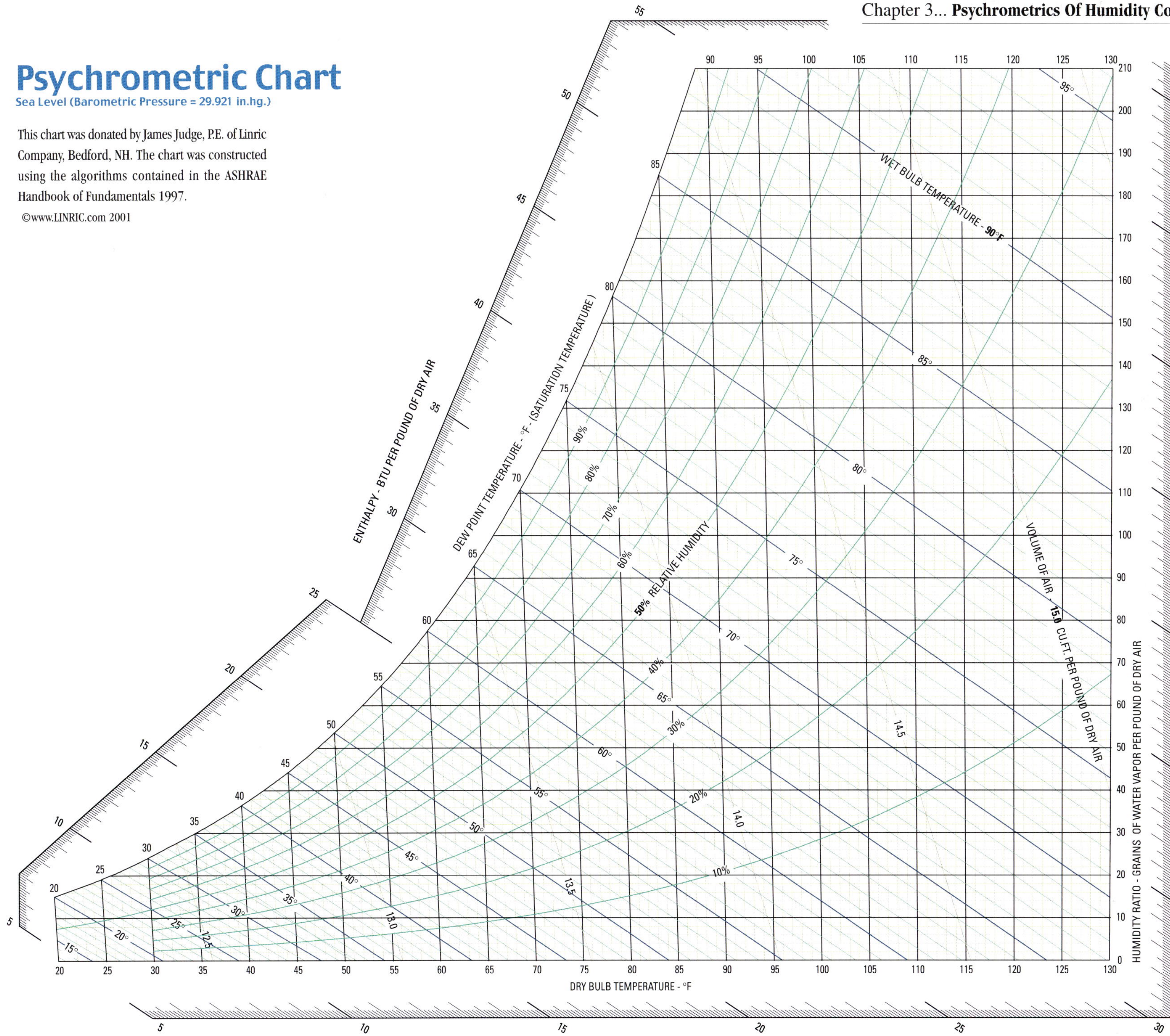

Chapter 4
Human Comfort and Humidity

By Larry Berglund with Lew Harriman

Fig. 4.1 Ending The Thermostat Wars
The HVAC designer can help end the thermostat wars by controlling humidity independently from temperature. Keep the dew point ***down*** *in the summer and* ***up*** *in the winter to increase the number and variety of people who can achieve thermal comfort in the same space.*

Comfort in Commercial Buildings

Human comfort is complex, and 100% satisfaction is difficult to achieve. The cartoon in figure 4.1 illustrates the familiar case of people with different thermal needs sharing the same space. Experience in commercial and institutional buildings provides similar anecdotal evidence of discomfort in some groups and not others. The school building's maintenance staff says: "Women teachers are never satisfied..." (more often when that staff is male). The shivering wife in the restaurant says: "Men always want to live in an icebox..."

But in spite of complaints, most HVAC systems succeed in providing comfortable conditions most of the time. Generally that comfort is achieved by controlling air temperature alone. Of all the variables that affect human comfort, air temperature is the most important. On the other hand, the cartoon illustrates the problem with focusing exclusively on temperature. Some of the occupants will be in the odd position of feeling too cold in the summer or too hot in the winter.

*The HVAC designer can reduce these problems by controlling humidity in addition to temperature, and by controlling humidity **separately** from temperature. By controlling two variables instead of just one, the HVAC system can adjust conditions to suit a broader range of different body types, activities and clothing preferences. This may explain why independent humidity control is gaining popularity in commercial and institutional settings. Such buildings are usually occupied by a very diverse collection of people doing different things in the same space at the same time.*

This chapter focuses on humidity-related comfort issues found in commercial buildings, as opposed to the full range of human circumstances. We encourage the reader to keep this in mind, since our relatively narrow focus might otherwise create an oversimplified impression of comfort science.

Introduction to Basic Variables of Human Comfort

Human comfort requires a body temperature of 98.6°F [37°C]. When the body deviates from that temperature, comfort is reduced. To understand how the body keeps its temperature constant, it is useful to understand each of the variables in the human comfort equation, and to understand what happens to comfort when these variables rise and fall. Key variables include:

The surrounding air

- Dry bulb temperature
- Vapor Pressure (Dew point)
- Relative humidity
- Velocity at the skin surface

The building, its equipment & furnishings

- Surface temperatures of each object surrounding the occupants

The individual

- Metabolic Rate (Activity level)
- Amount of clothing coverage
- Heat transmission of clothing
- Moisture transmission of clothing
- Body mass vs. surface area

Dry bulb temperature

In commercial buildings, the air temperature is the key determinant of comfort. If it climbs much above 80°F or falls much below 68 °F [above 26.6 or below 20.0°C], most occupants are likely to be uncomfortable regardless of other factors. This is self-evident from our common human experience, and needs no further explanation.

Fig. 4.2
Comfort Depends Primarily On Dry Bulb Temperature

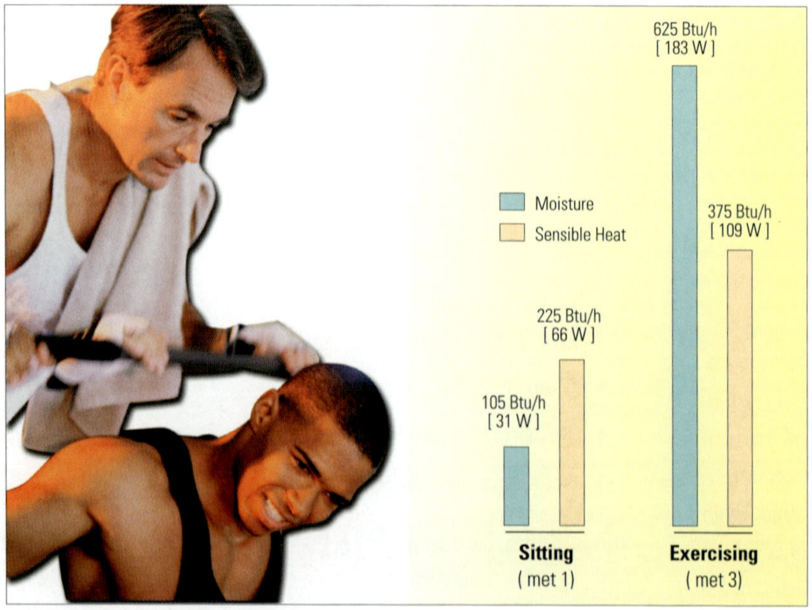

Fig. 4.3 Importance of Dew Point

The graph shows how, as activity increases, more heat is released by evaporation. That's why the comfort of active people is enhanced when the dew point is kept low—dry air allows faster release of excess heat through evaporation.

Dew point and vapor pressure

As long as people are at rest, the moisture content of air does not strongly influence comfort. Moisture is not usually extreme in commercial buildings, and people can adapt to changes by adding or subtracting clothing layers. But when people are more active, controlling the dew point becomes more important because it dictates the vapor pressure of the air, which in turn governs evaporation.

As the body's activity level increases, it reaches the limit of its ability to release heat by convection and radiation. So it releases more heat by evaporation. The rate of evaporation—therefore the rate of cooling—depends on the difference between the vapor pressure of the air in the room and the vapor pressure in the saturated air at the skin surface. As that difference increases, the body can be cooled more rapidly. In other words, the body can release heat quickly and easily when the dew point is kept low, but more slowly and with more difficulty when the dew point is allowed to rise.

Compounding this effect, vapor diffusion through clothing layers also depends on the difference in vapor pressure between the body and the air. This explains why surgeons working in operating rooms held at 65°F [18.3°C] can be uncomfortably hot and may sweat profusely. The garments that protect doctors from infectious fluids are barriers designed to keep fluids out, and therefore they pass water vapor very slowly. Heat normally released by evaporation is trapped inside such garments unless the operating room is held at a low dew point (low vapor pressure). Surgeons may perceive the problem as too high a temperature—but the real problem is the high dew point, which retards normal evaporation and cooling via perspiration.

In cold regions during winter months, the problem is reversed. Without humidification, the dew point in the building is likely to be very low, which means drying and evaporative heat release is much faster than is comfortable for the human body. Also, membranes in the eyes and nose dry too quickly, causing eye irritation and breathing discomfort. In summary, human comfort is optimal when the dew point is in the middle range.

In addition to the dew point (the "absolute" moisture level) the relative humidity can also affect comfort. But the effect of relative humidity levels are indirect, and are not instant.

Relative humidity

Materials absorb and release moisture according to the relative humidity of air at their surface. Consequently, relative humidity can also affect human comfort. As the relative humidity rises, clothes pick up moisture. When that happens, the friction between skin and fibers increases, which eventually creates the clingy, uncomfortable sensation so common in hotel sheets in humid climates.

High relative humidity also allows insects and fungus to obtain water from moist solid food sources. Both insects and fungus have negative effects on human comfort, but the effects are not instant consequences of the current humidity—fungus and insects require time to grow.

Air velocity at the skin surface

The faster air moves across the skin, the faster heat is added to or removed from the body. This explains why in a restaurant during the summer, the hard-working server likes to stand under the air distribution diffuser, while the restaurant parton prefers to be seated well-away from that cold air jet. The server needs to release heat, and therefore seeks fast-flowing cold air to stay comfortable. But the patron is seated and inactive, generating little heat. Such sedentary people can be quickly overcooled by fast-moving cool air.

The designer must remain aware of these effects when deciding where to locate supply air diffusers. Active people need more air flow. On the other hand, cold air jets must be directed *away* from less-active people such as patrons in restaurants or residents in nursing homes and hospitals.

Fig. 4.4 Relative Humidity
Friction increases with moisture absorption in fabrics, which is a function of relative humidity and skin wetting.

Fig. 4.5 Air Velocity
As air speed increases, more heat is transferred from the body by convection. This explains the popularity of fans in hot climates, and the discomfort of high winds in cold regions.

Fig. 4.6 Surface Temperature vs. Radiation
The body loses heat by radiation to colder objects, and gains heat from hotter surfaces. In this scene, the older person with slower blood circulation needs an extra clothing layer to improve comfort if seated near a cold window in winter.

Surface temperatures

Through radiation, every object sends heat *to* and receives heat *from* every other object in the universe. On a smaller scale, the same is true in a room. The human body radiates heat to the walls, ceiling, floor, windows and furnishings, and these objects radiate heat to the human body in return. The net gain or loss of heat by the human body depends on the temperature difference between the surface of the body, and the surfaces in the room.

This explains why in the spring and fall, people sitting in a restaurant near a large glass window may be too cold, while at the same moment, people seated across the room near an inside wall may feel too hot. The cold glass absorbs heat from the patrons near the window. But the patrons seated away from the windows, perhaps near a wall washed by hot air from the heating system, may feel too hot.

Radiation also explains why cooks working in a commercial kitchen during winter may be especially uncomfortable. They receive heat on their chest and back from the grill and fryers, but lose heat from their head, neck and legs to the cold makeup air entering the narrow aisles. As the old joke says, they should be comfortable "on average" because half their body is too hot, and the other half is too cold—but of course nobody is comfortable with such simultaneous extremes.

However, with smaller differences in nearby surface temperatures, the human body can compensate by increasing blood flow. Faster blood flow moves heat from one part of the body to another. In this way, the body "averages" the heat received from and lost to surrounding surfaces. Consequently, the mean radiant temperature—the average of the surrounding surface temperatures—is a quantifiable factor in human comfort. That value is derived by multiplying each surface temperature by its radial degrees of surface and then dividing the sum of these separate surfaces by 360° of total exposure.

Radiant heating and cooling systems use this principle to keep people comfortable without heating or cooling the entire mass of the air and the building. A radiant cooling panel mounted on the ceiling in a conference room absorbs heat from the heads of the conferees, so they feel comfortable even though the air temperature otherwise be too high. During winter, radiant heaters in warehouses keep people warm without the need to heat all the air in the huge building.

Activity level (metabolic rate)

The more active the person, the more fuel is burned to maintain that activity, therefore more heat is released. Active people must release more heat than sedentary people to keep their body temperatures constant. That's why the restaurant patron might feel quite cold, while the server may be uncomfortably hot at the same dry bulb temperature.

The influence of metabolic rate on comfort is most obvious when both active and sedentary people occupy the same space. Eldercare facilities provide another good example. The older, less-active resident wants to keep the air temperature high, because their low metabolic rate makes it difficult to keep their body temperature up to a comfortable level. In sharp contrast, the hard-working staff is moving rapidly, lifting and carrying and generating quite a bit of excess heat—so they prefer a lower temperature.

Metabolic rates are quantified in "met" units. A value of met 1 refers to a normal adult, seated at rest, such as a person reading a book. A sleeping adult generates less heat at approximately met 0.7. A value of met 3 describes an adult walking briskly. A basketball player might generate heat at a met rate between 5 and 8. Connecting this variable to our previous example, an eldercare resident probably operates at met 0.8 most of the day, while the active staff operates at met 2.0 to 3.5 at the same time.

Fig. 4.7
Metabolic Rate
Sedentary people generate less heat and are more comfortable at higher temperatures. Active people must release more heat by evaporation when the temperature is high.

Such mixed-occupancy buildings are ideal candidates for HVAC systems that have independent control of humidity. The dry bulb temperature can be kept high to keep residents comfortable. The dew point can be kept low to help active staff release larger amounts of metabolic heat through evaporation.

Clothing

Clothing insulates the body. More clothing prevents heat loss. Less clothing allows more cooling by the surrounding air. The relationship is obvious. The insulating value is expressed in units called "clo." A clo value of 1 describes an ensemble including trousers, underpants, long-sleeve shirt and suit jacket. Adding a vest and t-shirt brings the clo value to 1.14.

What may be less obvious is that the percent coverage is just as important as the insulating value of the fabric. Consider a women's clothing ensemble consisting of knee-length skirt, short sleeve shirt, sandals and panty hose. That covering sums to a clo value of 0.54. To achieve a clo value of 1.1 with that same skirt, the woman must add a long-sleeve shirt, half slip and long-sleeve sweater. Her relatively uncovered legs lose quite a bit of heat, and the long-sleeve sweater and other garments are needed to compensate for that heat loss.

Fig. 4.8
Clothing Coverage & Comfort
Although the young man may be comfortable, the young lady will feel too cold at the same air temperature. This photo shows a situation where social factors rather than insulation are the primary influence for clothing selection. Controlling humidity allows the HVAC designer to accommodate a wider variety of activity levels and clothing preferences—raising the dry bulb for those who are lightly-dressed, while keeping the dew point low for those who are more heavily clothed, or those who are more active.

In many cases, occupants adjust their clothing layers to the circumstances. One brings a sweater to a movie theater in summer, because the building might be "precooled" in advance of peak heat loads. The patrons are too cold initially, so the sweater is necessary for comfort.

But in other situations, the occupants may not have complete flexibility regarding their clothing. Returning to the restaurant example, servers often wear uniforms that cover more surface than is comfortable in summertime—given their high met rate. Or the material might be too much of a barrier to water vapor. Cotton transmits water vapor more freely than synthetics. Also, that same uniform may also be less than ideal in winter, when more coverage would be comfortable.

Once again, controlling humidity can help the HVAC system to accommodate a wider variety of clothing preferences or business-generated clothing mandates. ***Adding moisture*** in winter reduces the overcooling effect of evaporation. ***Subtracting humidity*** in summer allows the body to cool itself by evaporation in spite of clothing that might be less than optimal for the dry bulb temperature and met rate of the occupant.

Body mass and surface area

Other, more subtle influences on comfort include body weight compared to total skin surface area. The met rate depends on motion and load. Moving a heavier body creates a higher load, which requires more energy consumption. Higher consumption means more heat must be released. Also, consider two people with the same body rate but different body shapes: one is short and heavy and the other is tall and thin. The tall, thin person can release heat more easily because he has more skin surface, and because he has a thinner layer of insulating fat.

The comfort effects of body forms become apparent in elementary schools. In warm weather, slightly-built students may be quite comfortable at temperatures that make their more full-figured teachers hot and highly uncomfortable. In the winter, the heavier teachers are comfortable at cool temperatures that leave the lightweight students quite blue-lipped.

Again, the comfort mismatch of different body types can be partly offset through humidity control. When other factors are constant, dehumidification in humid months allows larger people to be comfortable at higher dry bulb temperatures. Humidification in winter months allows smaller people to be comfortable at lower dry bulb temperatures.

Fig. 4.9 Body Mass vs. Surface Area
Smaller people have more surface area per unit of body mass than large people, so smaller people loose heat more rapidly when other factors are equal. When temperatures are cool, smaller people need more clothing coverage for equivalent comfort. This sometimes explains different preferences for thermostat set points. Controlling humidity as well as temperature can help provide similar comfort to people with different body sizes.

Summarizing the Basics

Human comfort depends on the complex interaction of many variables, of which humidity is just one. But controlling the moisture content of air allows the HVAC designer to accommodate a wider variety of people and activity than would be possible by controlling temperature alone.

The "optimal" moisture level for comfort varies by application and by occupant population. For commercial and institutional buildings, dew points between 35°F and 57°F [1.7°C and 13.9°C] will provide the greatest comfort for the widest variety of people and circumstances. The logic of this range is supported by the deeper detail of the next section.

Quantifying The Relationship Between Comfort and Humidity

By Larry Berglund

This material in the balance of this chapter was first published as an article titled "Comfort and Humidity" which appeared in the August 1998 edition of the *ASHRAE Journal.*

Humidity affects our comfort in numerous ways both directly and indirectly. It is a factor in our energy balance, thermal sensation, skin moisture, discomfort, tactile sensation of fabrics, health and perception of air quality. In 1966, ASHRAE Standard 55-1966 (Thermal Environmental Conditions for Human Occupancy) introduced a definition for thermal comfort which has become widely used and quoted: "Thermal comfort is that condition of mind that expresses satisfaction with the thermal environment." This definition implies that comfort is a cognitive judgement that involves much input and is the result of physical, physiological and psychological processes.

In regulating body temperature the brain continuously compares body temperatures to desired levels and makes physiological adjustments. Comfort seems to occur when body temperatures are maintained with the minimum of physiological regulatory effort.

Fig. 4.10
Correspondence Between Thermal Sensation & Thermal Comfort Scales

Thermal Sensation Scale		Thermal Comfort Scale	
Very hot	+4	(No corresponding value)	
Hot	+3	Intolerable	4
Warm	+2	Very Uncomfortable	3
Slightly Warm	+1	Uncomfortable	2
Neutral	0	Slightly Uncomfortable	1
Slightly Cool	-1	Comfortable	0
Cool	-2	(No corresponding value)	
Cold	-3	(No corresponding value)	
Very Cold	-4	(No corresponding value)	

The role of regulatory effort and body temperatures in comfort is highlighted by experiments of Chatonnet and Cabanac[2] and observations of Kuno et al. (in ASHRAE Transactions 93, Volume 2). In Chatonnet's experiments, the sensation of placing the hand for 30 seconds in relatively hot or cold water (100 to 86°F [37.8 to 30°C]) were compared while the person experienced different thermal states.

When the person was overheated or hyperthermic, the cold water was pleasant, but the hot water was very unpleasant. When in a cold or hypothermic state, the hand felt pleasant in hot water and unpleasant in cold water. Kuno et al. describe similar observations during transient whole body exposures to hot and cold environments. In a state of thermal discomfort any move away from the thermal stress of the uncomfortable environment is perceived as pleasant during the transition.

Thermal sensation

A good correlation to thermal comfort is thermal sensation.[3] The word and numerical scale commonly used to categorize or label thermal sensation is shown in Fig. 4.11. Thermal comfort is generally associated with a neutral or near neutral whole body thermal sensation. Thermal sensation depends on body temperature, which in turn depends on thermal balance and the effects of environmental factors (temperature, radiation, air motion and humidity), as well as personal factors (metabolism and clothing).[4, 5]

Thermal balance

Humidity affects the evaporation of water from mucous membranes and sweating surfaces and its diffusion through the skin. In turn, evaporation affects the energy balance and thereby body temperatures and thermal sensations. When evaporation processes of the skin are compromised or enhanced, skin temperatures change, which is directly sensed by the temperature sensors of the skin. Although sedentary persons depend much less on perspiration for thermal balance than when operating at higher activity levels, humidity still has a significant direct effect.

The rate of water loss depends on vapor pressure differences between the body and its surrounding air. An average adult resting in a 75°F [23.9°C], 50% relative humidity environment while wearing trousers and a long-sleeved shirt (0.6 clo) loses about 1.1 ounce per hour [31 g/h] of water to the environment.[6] Of these, 0.42 oz/h [12 g/h] are from the nose and respiratory surfaces and the remaining 0.7 oz/h [20 g/h] is lost from water diffusing through dry non-sweating skin.

In terms of energy, these evaporative losses represent 73 Btu/h [21 W] or 20% of the resting person's total heat loss of 360 Btu/h [105 W]. Since the resting person is not performing thermodynamically useful work (lifting a weight etc.), all of the metabolic energy ends up as heat that must be dissipated to the environment. For this example with 73 Btu/h [21 W] of waste heat carried to the environment by water vapor flow, the remaining 287 Btu/h [84 W] is transferred by the dry heat transfer mechanisms of conduction, convection and radiation.

Decreasing the relative humidity of the 75°F [23.9°C] environment from 50 to 20% increases the total evaporation rate to 1.3 oz/h [38 g/h] and the associated energy loss to 89 Btu/h [26 W] or to 25% of the example person's total 360 Btu/h [105 W] energy loss. This additional passive evaporation decreases the needed dry heat transfer for energy balance to 270 Btu/h [79 W]. Further, it results in a slight lowering of the skin's temperature by 0.5°F (0.3°C) to about 91.3°F [32.9°C]. And as a result the person feels a little cooler in the drier 20% rh environment than in the 50% rh environment at the same temperature. To maintain the same skin temperature and feeling of warmth with the 50% to 20% humidity decrease, the temperature would need to increase about 1.8 °F [1°C].

In warmer conditions or with increased activity and metabolism, active perspiration is required for thermal balance. In these cases, the effect of humidity is greater. Metabolism is often characterized with the dimensionless met unit: the ratio of actual to resting metabolism. A resting person with this unit system has a metabolism of 1 met. In the earlier example, if the person walked continuously in the 75°F [23.9°C] 50% rh environment the person's metabolism would be about three times higher (3 met) or 1075 Btu/h [315 W]. The convective and radiative losses would change little. But the respiratory heat loss would increase three times (1.3 oz/h [37 g/h]) in proportion to metabolism. Also, perspiration that was absent in the resting case would now be 8.5 oz/h [241 g/h] and its evaporation would carry away body heat at the 550 Btu/h [161 W] rate.

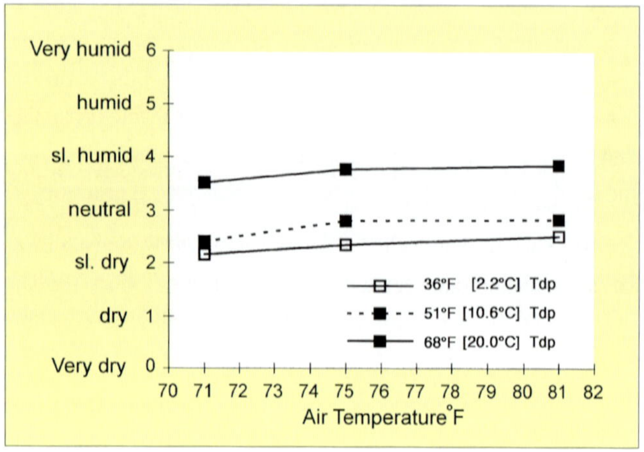

Fig. 4.11
Humidity Perceived By Sedentary Subjects
Sedentary people release relatively small amounts of heat by evaporation, so they are less sensitive to dew point than active people.

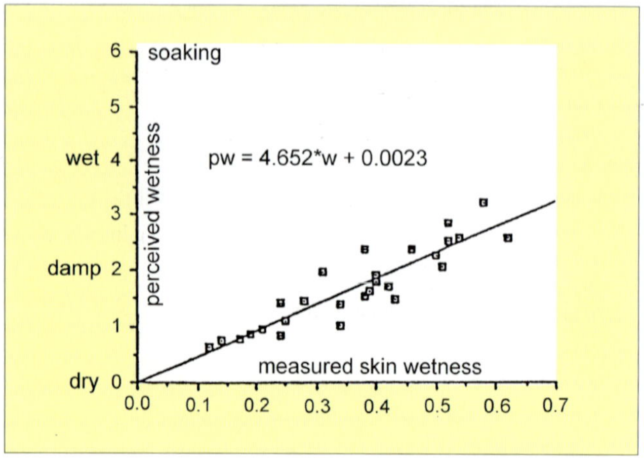

Fig. 4.12
Skin Moisture (Perception vs Measured Values)
There is generally good agreement between perception and measurement of skin wetness.

Acceptability

The ASHRAE comfort standard specifies the environmental conditions necessary for a neutral thermal sensation and also gives a range of parameter values that are expected to provide an environment that is thermally acceptable to at least 80% of the occupants. In terms of temperature, Standard 55-1992 specifies a band of temperatures 6°F [about 3.3°C] wide.

As shown by Fishman[7] and others,[8] the clothing worn by occupants is influenced by the season and outside weather. Clothing worn indoors in summer is generally of lighter weight with a lower insulating value than that worn in the winter. The clothing worn in North America is about 0.5 clo in summer (thin trousers and short sleeved shirt) and about 0.9 clo in winter (heavier trousers with long-sleeved shirt and sweater or business suit). The clo unit is a widely used and convenient measure of thermal resistance: 1 clo = 0.88 ft^2 h °F/Btu [0.155 m^2K/W]. A heavy two-piece business suit with accessories has an insulation value of about 1 clo while a pair of shorts is about 0.05 clo. The thermally acceptable temperature range for this level of indoor winter clothing lies between effective temperatures (ET) of 68 and 74°F [20 and 23.5°C]. For summer clothing the range falls between 73 and 79°F ET [23 and 26°C].

At the temperature boundaries of the comfort zone, an average person may have thermal sensations of approximately +0.5 at the warm side and -0.5 at the cooler ET border. By definition ET is the temperature at 50% rh that will transfer the heat to the environment at the same rate with the same skin temperature and skin wetness as in the actual temperature and humidity. Skin wetness is a measure of skin moisture and is the fraction of skin covered with perspiration necessary to account for the observed rate of water loss. Further, people at the same ET value would be expected to have the same thermal sensation as demonstrated by Gonzalez.[9]

Humidity

Though the temperature boundaries of the comfort zone are well defined and supported by laboratory and field observations, the humidity limits are less certain, particularly at upper humidity levels. The physiological and energy balance considerations that led to ET would indicate there is neither an upper or lower humidity limit in terms of thermal sensation. But laboratory, field and personal experiences suggest that there are humidity limits for comfort and acceptability.

Other aspects of humidity discomfort may not be energy related. The perception of skin moisture and the interactions of clothing fabrics with the skin may be due to the moisture itself. The skin's outer layer of dead squamous cells of the stratum corneum can readily absorb or lose water. With moisture addition, the cells swell and soften. With drying, they shrink and become hard. In this setting, the skin's moisture may be better indicated or characterized by the relative humidity of the skin (RH_{sk}) rather than skin wetness: [10]

$$RH_{sk} = P_m/P_{s,sk} \quad \text{(Equation 4.1)}$$

where P_m is the average vapor pressure of the skin and $P_{s,sk}$ is the saturated vapor pressure of water at skin temperature. Typically, the water content (water/dry skin) of the stratum corneum is about 10%, but it can absorb much more.

Skin moisture may be detected by mechanoreceptors of the skin and hair follicles or some other neural mechanism that senses the skin's swelling and shrinking. At high levels of skin moisture the swelling is sufficient to close or reduce the lumen of sweat glands and reduce sweating (called hydromeiosis).[11] Hydromeiosis occurs at $RH_{sk} > 0.9$. Conversely, under good drying conditions, the skin can shrink to the extent that lesions form.

As previously mentioned, the other term for characterizing skin moisture is skin wetness (w) or the size of the water film as a fraction of total skin area that is necessary to account for the observed evaporative heat loss from the skin (E_{sk}):

$$E_{sk} = w \bullet A_{du} \bullet T_{he} \bullet (P_{s,sk} - P_a) \qquad \text{(Equation 4.2)}$$

where A_{du} is total skin area, T_{he} is evaporative heat transfer coefficient and Pa is the ambient vapor pressure. Skin wetness correlates well with warm discomfort,[12] and people rarely report feeling comfortable at skin wetness levels near or above 25%. With intense sweating, sweat normally begins to drip from some body surfaces when the average skin wetness for the whole body is about 80%.[13] Also, skin wetness and skin relative humidity are related.

$$RH_{sk} = w + (1 - w)\ P_a/P_{s,sk} \qquad \text{(Equation 4.3)}$$

From Equation 4.3 it is clear that RH_{sk} will be greater than w except when w = 1. It is also evident that with a constant w, RH_{sk} increases with ambient absolute humidity. Though the ET temperature boundaries have constant skin wetness levels, the RH_{sk} (governing swelling & softening) increases with absolute humidity.

Humans are sensitive to moisture and can reliably describe the humidity of the environment using word scales as is demonstrated in Figure 4.15 The points in the figure are the average of 20 sedentary subjects. The subjects' humidity judgments appear to be functions of the air's dew point, a measure of absolute humidity, and are relatively unaffected by the ambient temperature. Further, people are also adept at perceiving skin moisture as illustrated in Figure 4.14 where perceived skin wetness is seen to correlate well with measured skin wetness. Each point is the average of five subjects.

Fig.4.13 Skin Wetness vs. Fabric Friction

Friction increases as perspiration wets the fabric of clothing

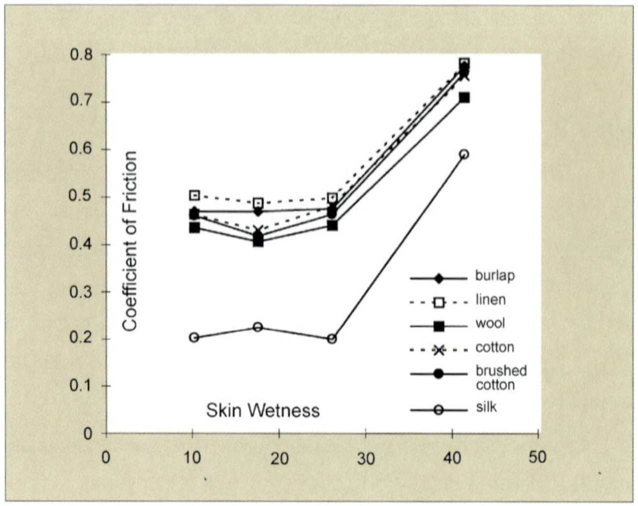

Fig.4.14 Comfort vs. Friction

Comfort declines as fabric friction increases

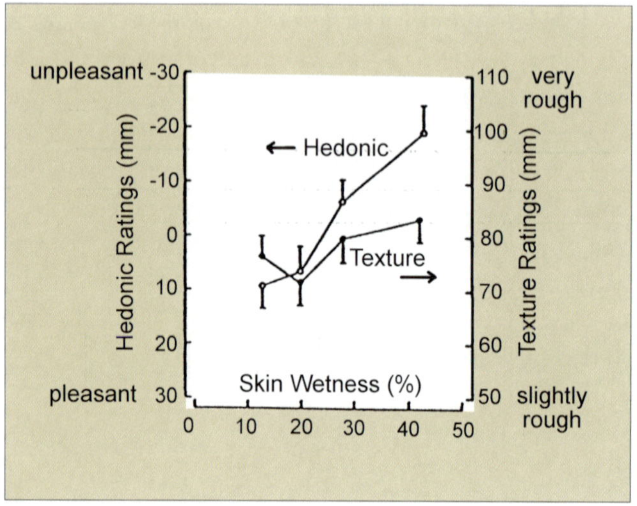

In situations of prolonged sweating, skin wetness slowly increases with time because of accumulating salt on the skin. The increasing salt occurs because the water in perspiration evaporates while the dissolved materials, principally sodium chloride, remain on the surface. The salt lowers the vapor pressure of the sweat film decreasing its rate of evaporation per unit area. The area of the film then naturally increases so that evaporation will equal the rate of sweat secretion.[15] It is thought that part of the relief that bathing brings after a warm day or strenuous activity is that by cleaning the skin, perspiration can evaporate more efficiently with reduced skin wetness.

Clothing can impact acceptability in humid environments. Measurements by Gwosdow[16] reveal that the friction between skin and clothing increases abruptly above skin wetness levels of 25% (Fig. 4.14). Further, fabrics are perceived to have a coarser texture and to be less pleasant with increasing skin moisture (Fig. 4.15). This may be one of the reasons that in the comfort studies cited earlier,[16] people have rarely indicated they were comfortable when their skin wetness levels approached 25% and above.

Following this reasoning, the stickiness and perceived fabric coarseness and unpleasantness with humidity and skin moisture could affect merchandising. Fabrics, clothing and textures will be more pleasant and possibly sell more readily in a dry environment than in a humid warm one. The skin of a potential customer coming in from outside during the summer will become drier. As a result, textiles will feel smoother and more pleasant and be easier to try on than in a neighboring shop that is more humid.

Low humidity

Low humidity affects comfort and health. Comfort complaints about dry nose, throat, eyes and skin occur in low humidity conditions, typically when the dew point is less than 32°F [0°C]. Low humidity can lead to drying of the skin and mucous membrane surfaces. On respiratory surfaces, drying can concentrate mucous to the extent that ciliary clearance and phagocytic activities are reduced, increasing susceptibility to respiratory disease as well as discomfort. Green[17] quantified that respiratory illness and absenteeism increase

Fig. 4.15
Humidity Uncertainty In Comfort Research

The ASHRAE comfort zone chart displays the comfort preferences of sedentary people. From these conflicting editions of the chart, one can see that comfort research for humidity carries more uncertainty than it does for temperature. What does seem consistent is that sedentary people are likely to find comfort in all seasons when the dew point is between 35 and 57°F [between 1.7° and 13.9°C].

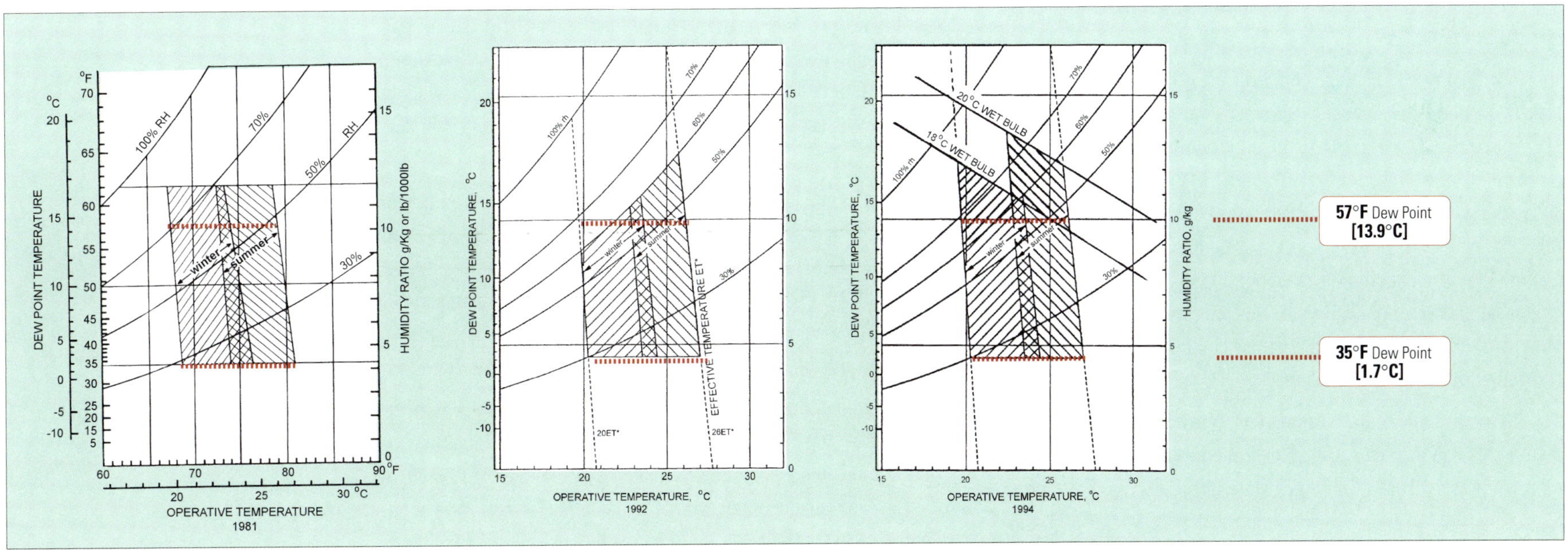

in winter with decreasing humidity. He found that any increase in humidity from the low winter levels decreased absenteeism. Excessive drying of the skin can lead to lesions, skin roughness, discomfort and impair the skin's protective functions. Dusty environments can further exacerbate low humidity dry-skin conditions.[18]

Liviana et al.[19] found drying from low humidity can contribute to eye irritation. Eye discomfort increased with time in low humidity environments $T_{dp} < 36$°F [2°C].

ASHRAE Standard 55 specifies that to decrease the possibility of discomfort due to low humidity, dew point temperature in occupied spaces should not be less than 37°F [2.8°C].

High humidity

Elevated humidity levels reduce comfort. At lower levels of humidity, thermal sensation is a good indicator of overall thermal comfort and acceptability. But at high humidity levels, Tanabe[20] found that thermal sensation alone is not a reliable predictor of thermal comfort. Nevins[21] recommended that on the warm side of the comfort zone, the relative humidity should not exceed 60% to prevent warm discomfort.

That the upper limit of the comfort zone is controversial and not clearly defined is evidenced by the evolution of Standard 55 from 1974 through 1994 (Fig.4.16). The upper humidity limit was a dew point of 63°F [17.2°C] in the 1974 and 1981 standards, based not so much on comfort as on considerations of mold growth and other moisture-related phenomena. In the 1992 edition, Standard 55 specified 60% relative humidity as the upper limit—also based primarily on considerations of mold growth. This limit was challenged for not being based on direct human thermal comfort and for being too restrictive for evaporative coolers.

These air coolers are cost-and-energy-efficient and popular in the hot, dry southwestern part of the United States. In 1994, Addendum 55a was approved with upper humidity limits of 64 °F and 68 °F [18 and 20 °C] wet-bulb temperatures for the winter and summer comfort zones, respectively. Though based on limited comfort data, Addendum 55a was less restrictive to evaporative coolers. The 1999 revision of Standard 55 returns to an upper dew point limit of 60°F [15.6°C] for both comfort zones. Given the constant migration of the upper humidity limit in different editions of Standard 55, it seems clear that human comfort research has been less conclusive with respect to humidity than it has been for temperature.

Overall thermal acceptability

An occupant's overall satisfaction with the thermal environment is probably a cognitive result of temperature, moisture, friction and other sensations. A recent comfort study[14] done at temperatures of 70°F to 81°F [21.1°C to 27°C] with dew points of 36°F to 68°F [2.2°C to 20°C] at three levels of activity (sedentary, intermittent walking and standing, and continuous walking) describes extensive subjective responses of 20 subjects and measures physiological responses from five subjects. The subjective ratings include thermal sensation, perceived skin wetness and thermal acceptability. For thermal acceptability, the participants indicated if the environment was thermally acceptable or not with the instruction that an unacceptable condition would be sufficient to evoke a behavioral action to improve the climate and/or reduce discomfort, e.g., open a window, turn on a fan, change the thermostat setting, alter clothing, complain or leave.

The thermal acceptability (TAC) responses for relative humidities greater than 50% from that study correlated strongly with absolute humidity or humidity ratio (HR) and operative temperature ($r^2 = 0.96$).[22] Figure 4.16 shows lines of constant thermal acceptability (TAC) based on the responses of 20 subjects. The responses for 60%, 70%, 80% and 90% of that population are drawn on the chart.

The shift in thermal acceptability with air temperature is as expected. But figure 4.17 also shows the subtle effect of humidity. Superimposed on the constant TAC lines are the outlines of a psychrometric chart and the boundaries of the summer comfort zone of Addendum 55a (dark lines). The shape of the upper humidity limit of

Addendum 55a conformed reasonably well with the 80% thermal acceptability from this research, reflecting the fact that sedentary subjects are less sensitive to high humidity.

Perceived air quality

Humidity also affects our perceptions of air quality and that aspect of comfort and satisfaction with the environment. Odors can detract from the acceptability of an environment, and humidity may stimulate or hinder odor sources and olfactory sensations. In addition, our judgments about the perceived air quality appear to be influenced by physical factors of temperature, humidity and air motion. Cooler, drier air is somehow perceived as being freer of contaminants and less stale. Even in a clean, non-odorous and well-ventilated space the perceived freshness of the air decreases with increasing humidity and temperature.

Figure 4.18 shows the subjects' judgements of air freshness upon entering a test room.[14] The odorless room was well ventilated, with 30 cfm [15 l/s] per person of clean air. The responses indicate that humidity made the subjects feel that the air was less fresh or staler. A person's olfactory system adapts to odor in a short time so odor intensity decreases with exposure, but in this case the staleness perception did not diminish with time, implying that the chamber air was odorless.

High occupant acceptance of a space's thermal environment and air quality is very important from a designer's or operator's point of view. For this laboratory study, the subjects judged whether the air

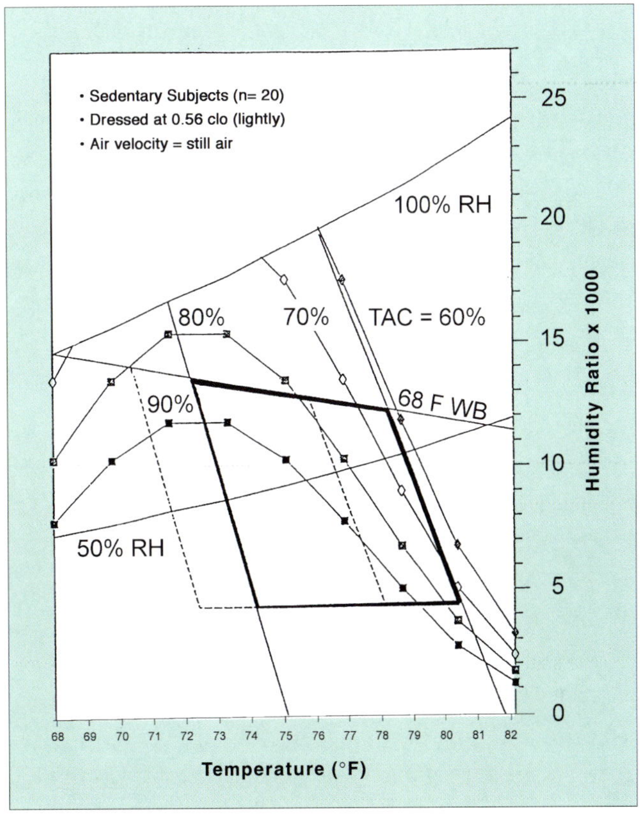

Fig. 4.16
Thermal Acceptability vs. Humidity
The chart shows thermal acceptability for lightly-dressed, sedentary subjects. Note the wide range of humidity tolerance as compared to temperature. In contrast to these charted results, active subjects are more sensitive to high humidity than sedentary subjects, because dew point affects evaporative cooling.

Fig. 4.17 Perceived Air Freshness vs. Time & Dew Point
In this test, 1 indicates a perception of fresh air, 6 indicates stale air. In the absence of other factors, cooler, drier air is perceived to be "fresher."

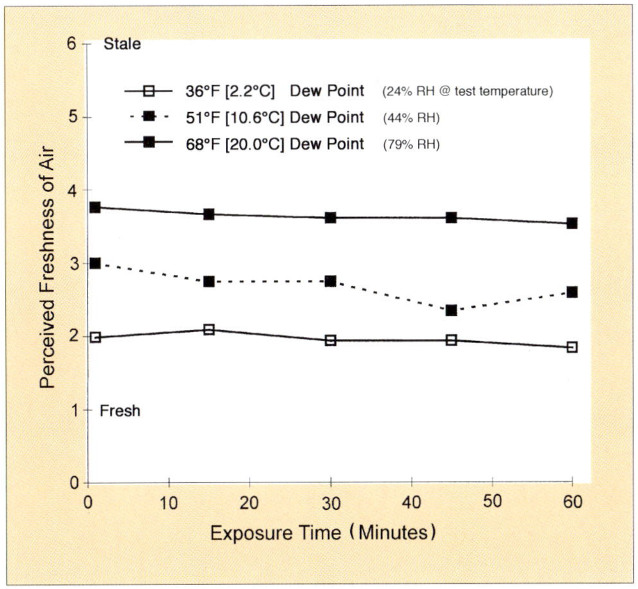

quality was acceptable or not, under the instructions that an unacceptable condition would evoke a behavioral action to improve the environment and/or reduce discomfort. The acceptability of the perceived air quality (Figure 4.19) was affected strongly by humidity in this clean-air environment. The 68 °F [20 °C] dew point condition (RH>65%) was particularly associated with the perception of unacceptable air quality.

Recently, Fang et al.[23] at the Technical University of Denmark found similar results from air contaminated with emissions from common building materials. That is, for a particular pollution concentration, the air quality was perceived to be more acceptable with decreasing temperature and/or humidity. The effect is shown in figure 4.20. Acceptability correlates strongly with enthalpy. Further, the study found that temperature and humidity had minimal or no effect on the emission rate of the materials tested.

Fig. 4.18
Thermal Acceptability at Rising Temperatures vs. Activity & Dew Point
Active subjects, in particular, find higher dew points unacceptable even when temperatures are very low.

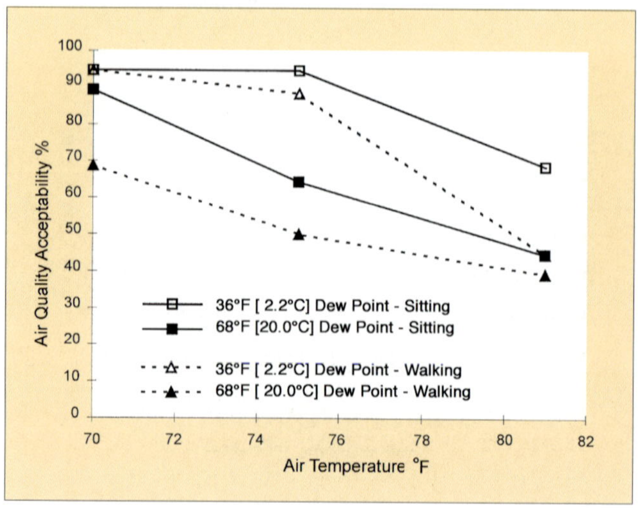

Summary

Humidity affects comfort in a number of ways both directly and indirectly. In hot and humid climates, decreased humidity results in occupants feeling cooler, drier and more comfortable; furthermore, fabrics feel smoother and more pleasant, and the air is perceived to be fresher, less stale and more acceptable.

In warm conditions thermal discomfort increases with humidity. The discomfort appears linked with skin moisture, as persons rarely judge themselves comfortable in situations where skin wetness is above 25%. Discomfort associated with skin moisture could be partly due to friction between skin and clothing. When fabrics ranging from rough burlap to wool, cotton, polyester and smooth silk are pulled across the skin, the measured pull force increases with humidity and perspiration, as does the fabric's perceived texture or roughness.

The effect of humidity on thermal balance, and in turn on skin temperature and thermal sensation, has a clear mechanism. But other effects of humidity on human sensation and health, including humidity's effect on microorganisms, have less clear mechanisms. Further study is warranted to better understand and quantify how humidity affects building occupants and HVAC system users, particularly in the characteristic mixed-occupancy of commercial buildings.

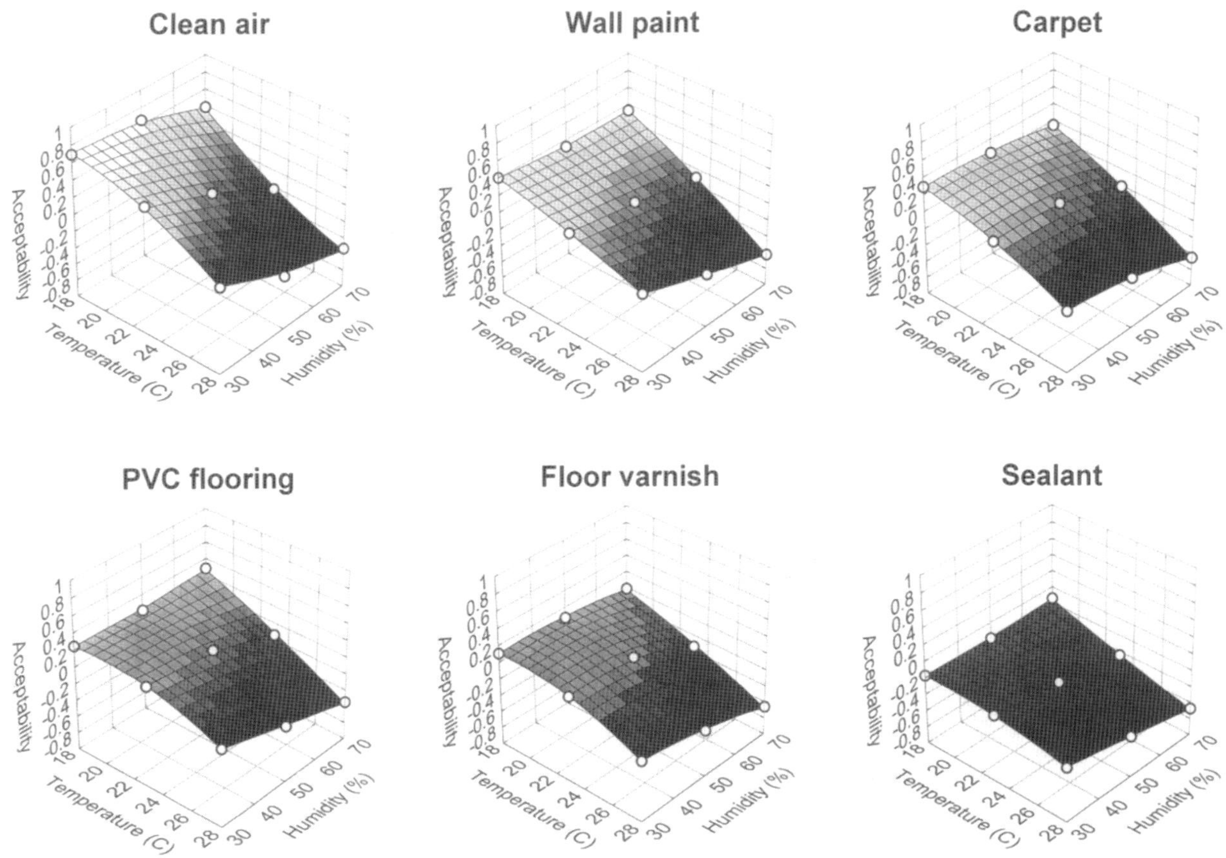

Fig. 4.19

Influence of Perceptable VOCs On Overall Thermal Acceptability

Darker shading on the contour plots show where thermal acceptability falls to its lowest points. When the air carries perceptable volatile organics, thermal acceptability declines for all combinations of temperature and humidity. Note also that in all cases, thermal acceptability is highest when both temperature and moisture are low. (Fang et al - Reference 23)

References

1. Hardy, J. D., J. A. J. Stolwijk and A. P. Gagge. 1971. "Man." Comparative Physiology of Thermoregulation, Chapter 5. Springfield, Ill: Charles C. Thomas.
2. Chatonnet, J. and M. Cabanac. 1965. "The perception of thermal comfort." Int. J. Biometeorology 9:183-193.
3. Gagge, A. P. 1977. "Introduction to thermal comfort." Les Editions de l'INSERM, INSERM 75:11-24.
4. Gagge, A. P., J. Stolwijk and Y. Nishi. 1971. "An effective temperature scale based on a simple model of human physiological regulatory response" ASHRAE Transactions 77(1):247-262.
5. Fanger, P. O. 1972. Thermal Comfort, New York: McGraw-Hill.
6. Gagge, A. P., A. P. Fobelets and L. G. Berglund. 1986. "A standard predictive index of human response to the thermal environment," ASHRAE Transactions 86(2B):709-731.

7. Fishman, D. S. and S. L. Pimbert. 1979. "Survey of subjective responses to the thermal environment in offices." Indoor Climate. Copenhagen: Danish Building Research Institute.

8. Gagge, A. P. and R. G. Nevins. 1976. Effect of Energy Conservation Guidelines on Comfort, Acceptability and Health. Final Report of Contract # CO-04-51891-00 of Federal Energy Administration.

9. Gonzalez, R. R., L. Berglund and A. P. Gagge. 1978. "Indices of thermoregulatory strain for moderate exercise in the heat." Journal of Applied Physiology: Respiration Environment Exercise Physiology 44(6):889-899.

10. Mole, R. H. 1948. "The relative humidity of the skin." Journal of Physiology 107: 399-411.

11. Kerslake, D. McK. 1972. The Stress of Hot Environments. Cambridge: University Press.

12. Berglund, L. G. and D. J. Cunningham. 1986. "Parameters of human discomfort in warm environments." ASHRAE Transactions, vol. 92(2):732-746.

13. Berglund, L. G. and R. R. Gonzalez. 1977. "Evaporation of sweat from sedentary humans in humid environments." Journal of Applied Physiology: Respiration Environment Exercise Physiology 42:767-772.

14. Berglund, L. G. and W. S. Cain. 1989. "Perceived air quality and the thermal environment." The Human Equation: Health and Comfort. Proceedings of ASHRAE/SOEH Conference IAQ '89 Atlanta: ASHRAE, pp. 93-99.

15. Berglund, L. G. and P. E. McNall. 1973. "Human sweat film and composition during prolonged sweating." Journal of Applied Physiology 35:714-718.

16. Gwosdow, A. R., J. C. Stevens, L. G. Berglund and J. A. J. Stolwijk. 1986. "Skin friction and fabric sensations in neutral and warm environments." Textile Research Journal 56:574-580.

17. Green, G. H. 1982. "Positive and negative effects of building humidification." ASHRAE Transactions 88(1):1049-1061.

18. White, I. R. and R. J. G. Rycroft. 1982. "Low humidity occupational dermatosis." Contact Dermatology 8:287-290.

19. Liviana, J. E., F. H. Rohles and O. D. Bullock. 1988. "Humidity, comfort and contact lenses." ASHRAE Transactions 94(1):3-11.

20. Tanabe, S., K. Kimura and T. Hara. 1987. "Thermal comfort requirements during the summer season in Japan," ASHRAE Transactions 93(1):564-577.

21. Nevins, R., et al. 1975. "Effect of changes in ambient temperature and level of humidity on comfort and thermal sensations." ASHRAE Transactions 81(2).

22. Berglund, L. G. 1995. "Comfort criteria-humidity and standards." Proceedings of Pan Pacific Symposium on Buildings and Urban Environment Conditioning in Asia 2:369-382.

23. Fang, L., G. Clausen and P. O. Fanger. 1996. "The impact of temperature and humidity on perception and emissions of indoor air pollutants." Indoor Air '96 4:349-354. Tokyo: Institute of Public Health.

Chapter 5

Corrosion & Relative Humidity

By Geoff Brundrett with Lew Harriman

Fig. 5.1

Corrosion Is A Surface Phenomenon

The relative humidity of air at the center of a room is irrelevant—the only thing that matters is the relative humidity in the first millimetre of air immediately above the metal surface. To avoid corrosion, keep the dew point well below the surface temperature of the coldest metal.

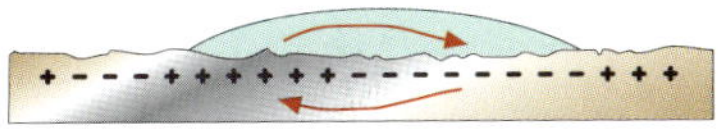

Fig. 5.2 Atmospheric Corrosion
Corrosion begins when water molecules, acting as an electrical conductor (an electrolyte), allow microcurrents to flow between areas of higher and lower electrical potential on a metal surface.

Atmospheric Corrosion & Relative Humidity

In the open air, metal corrodes when an electrical current flows across its surface from areas of high potential to areas of lower potential as shown in figure 5.2. The electrical flow rate, and therefore the corrosion rate, is faster when a conductive film (an electrolyte) is present on the surface to carry the current. Water is an effective electrolyte. And when that water contains conductive molecules such as salts or acids, its electrical resistance is reduced, so more current flows across the metal surface to accelerate the corrosion.

Water vapor molecules in air are constantly moving about. When they collide with a metal surface, they may bounce off if they have sufficient momentum. Or they may stick to the surface if they lose enough of their energy, as when they contact a very cold surface. When the air immediately above the metal surface is close to saturation, more of the water molecules are likely to stick to the surface rather than bouncing off. Consequently, at high rh more layers of water molecules accumulate on a metal surface, as shown in figure 5.4. With a continuous layer of electrolyte, more electrical current can be carried across the metal surface, so the corrosion rate increases.

Figure 5.4 also describes the relationship between corrosion rate and the thickness of the moisture film. At low relative humidities there will be less than a complete layer of water molecules adhering to the surface. There are two basic atmospheric corrosion mechanisms: chemical reaction and electrolytic decomposition. Chemical reactions are very temperature sensitive and are limited, at room temperatures and pressures, to providing a thin oxide coating—usually invisible—on the freshly prepared metal surface. This oxide layer is formed almost immediately and the metal will show little sign of further degradation in a dry atmosphere. This is termed "Stage I" dry corrosion.

As the relative humidity increases, the molecular water layer increases exponentially in thickness. The thicker film behaves progressively more like an electrolyte and establishes corrosion cells.

Fig. 5.3 Molecular Layers
As relative humidity rises, more water molecules lack the energy needed to escape the surface. So layers of water molecules build up to form liquid film on the metal surface. (Tomashov, 1966)

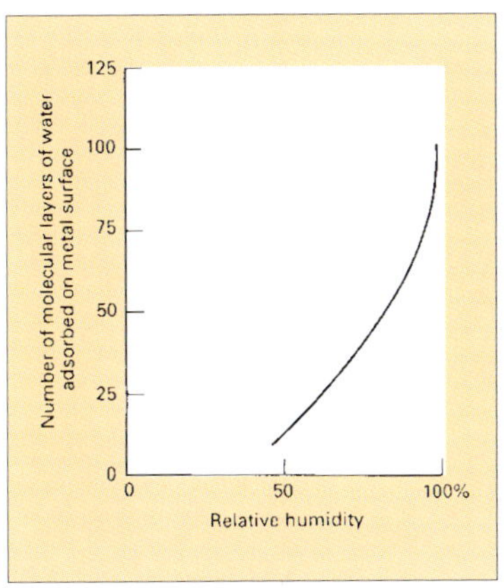

Fig. 5.4 Corrosion Rate Varies With Film Thickness
As water molecules cover more of the surface, they conduct current more easily, and the corrosion rate increases. (Tomashov, 1966)

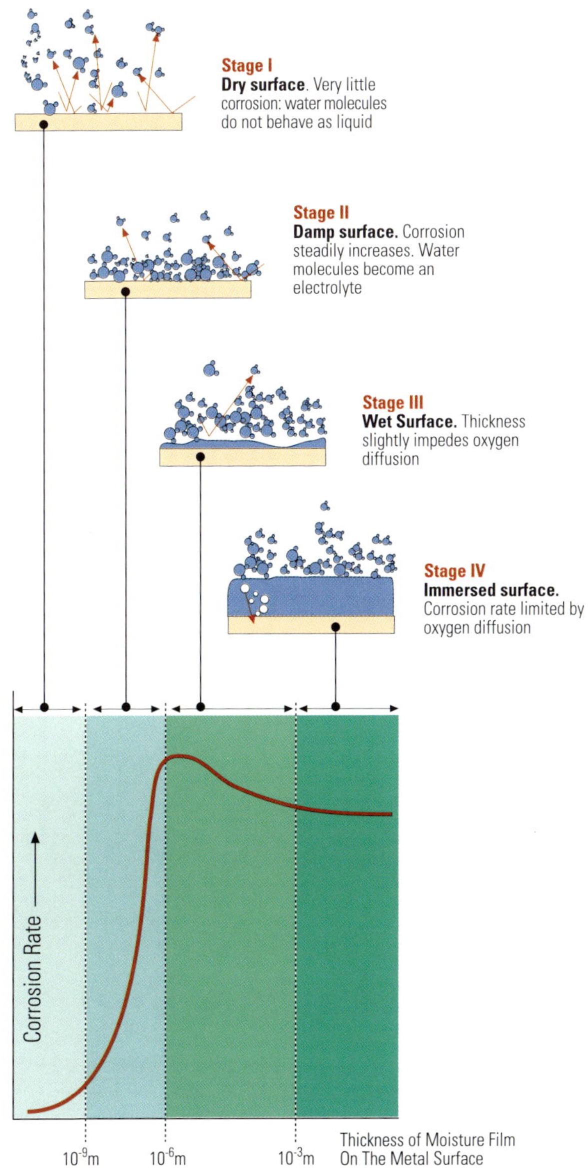

These cells can be created by a wide variety of factors such as differential aeration, presence of different atoms in the metal surface, pores in the protective oxide layer or even the presence of grain boundaries in the metal itself. The corrosion rate increases with the thickening water film. This is termed "Stage II", or moist corrosion, when the corrosion rate accelerates rapidly as the air above the surface comes closer to saturation.

At saturation, liquid condensation occurs and the droplets form a thin film of liquid over the metal surface. The corrosion rate can fall slightly because the thick water film can inhibit the oxygen diffusion rate to the surface. This is termed "Stage III", or wet corrosion.

Finally, the water layer is so thick that the metal can be treated as an immersed object. This is "Stage IV" corrosion, or total immersion. When objects are immersed, their corrosion rate is limited primarily by the rate of oxygen reaching the metal surface.

Fig. 5.5 Capillary Diameter vs. RH Required for Condensation

The vapor pressure in small crevices is lower than on flat surfaces, producing condensation even when the relative humidity is far below 100%.

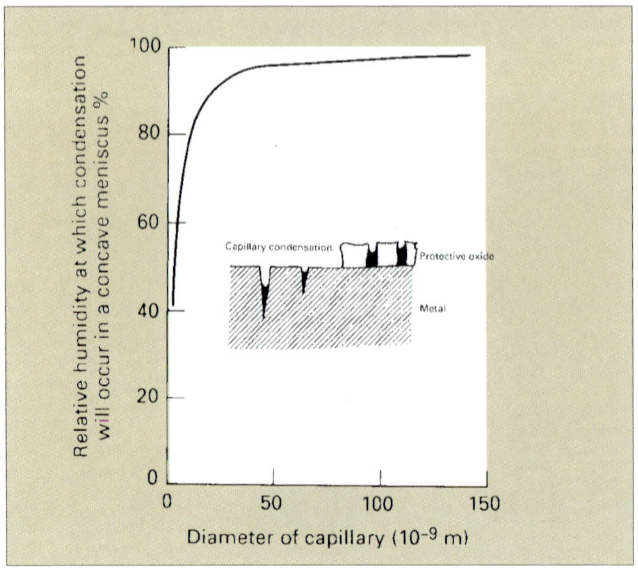

In addition to relative humidity, other factors affect the corrosion rate. Rapid corrosion may take place even in low relative humidity, depending on physical and chemical factors that increase the surface attraction for water molecules, and increase the conductivity of that water once it lands on the surface.

Rough surfaces attract water vapor

The simple electrolyte corrosion mechanism is complicated by the physical nature of the metal surface. As early as 1870, Lord Kelvin demonstrated that the equilibrium vapor pressure of a liquid was a function of both temperature and radius of curvature of the liquid surface. If the surface of the liquid is concave, as in a crevice, then its surface vapor pressure is lower than if the liquid surface were flat. Consequently, condensation can occur at humidities much lower than 100% rh when the surface has narrow microscopic cracks.

This relationship for water at 59°F [15°C] is illustrated in Figure 5.5, which shows how capillary condensation occurs at relative humidities well below saturation. The amount of condensation is determined partly by the size and wettable nature of the metal surface. Minutely fissured metal surfaces or metals partially protected by a porous oxide would be expected to corrode more quickly than smooth, polished surfaces. The fissured surface allows capillary condensation at low ambient relative humidities.

Surface particles attract water vapor

If trace chemical contaminants are present in the air they can form a thin hygroscopic film of surface salt. The film behaves as a desiccant, attracting water to its surface. Soluble salts will enhance condensation of moisture because the vapor pressure above salt solutions is lower than that of pure water.

Experiments on the corrosion rate of steel in contact with particles of sea salt show an increase in corrosion above 60% rh. As shown in Figure 5.6, the corrosion rate becomes very rapid when the relative humidity rises above 79%. That's also the value above which the salt would be expected to dissolve and wet the metal.

The hygroscopic characteristics of salts can be quantified in terms of their "deliquescent relative humidity" in a closed air space above the saturated solution. This value represents the humidity where the salts dissolve in moisture absorbed from the air. Lower equilibrium relative humidity values mean the salt is more hygroscopic. The relative humidity range over which condensation will occur on different salts is also illustrated in Figure 5.6.

Critical relative humidity

The combined effect of the physical and chemical properties of a metal surface means that there is a critical relative humidity above which the corrosion rate "becomes rapid." The actual value depends on the nature and composition of the surface. Clean iron in pure air does not corrode until the air is practically saturated. However, if the air contains even a trace (0.01%) of sulphur dioxide, the critical relative humidity drops to 70%. If the iron has experienced slight corrosion by exposing it to water, then the critical relative humidity is lowered to 65%. More severe prior corrosion, for example by exposing the iron to 3% sodium chloride solution, lowers the critical relative humidity down to 55%.

Fig. 5.6
Hygroscopic Particles Reduce The Critical Relative Humidity

Salts have a lower vapor pressure than metals, so when particles land on the metal surface, they attract moisture to the surface at humidities below 100% rh. Zinc chloride, for example, absorbs water vapor from the air and produces a liquid salt solution on a metal surface at 10% relative humidity. (Dultz, 1950)

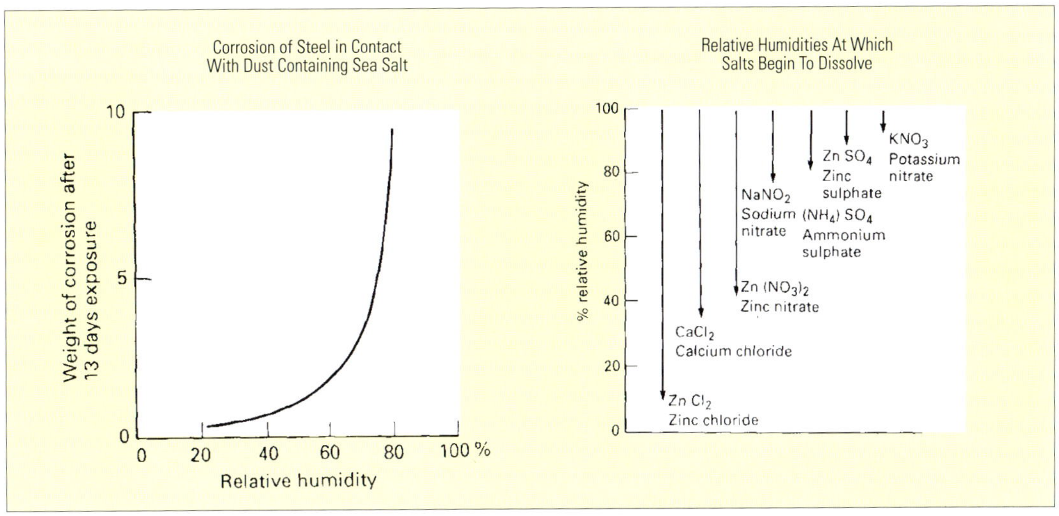

Copper behaves similarly. As shown in figure 5.7, the critical relative humidity is 87% after slight prior corrosion of the copper in a pure atmosphere. This value falls to 80% rh if the copper surface has previously been exposed to gaseous sulphur dioxide, and to much lower values when sulphur dioxide is present. Traces of iodine vapor reduce the critical relative humidity even further to 30-40%.

Fig. 5.7 Gaseous Contaminants Also Increase Corrosion Rates

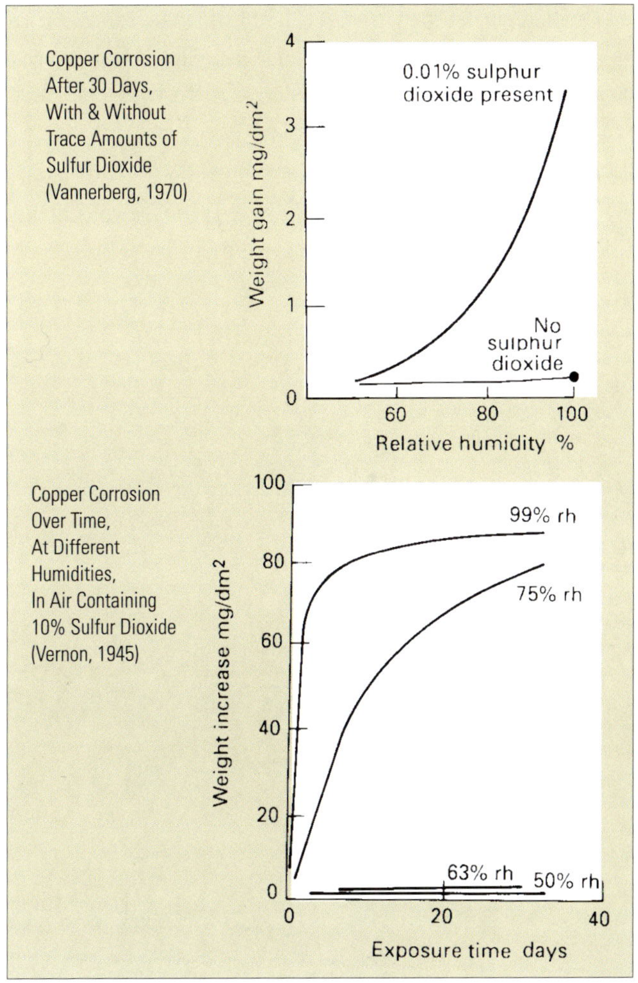

Nickel has a critical relative humidity of 70%. Below that the metal remains bright. Higher relative humidities lead to two chemical processes. The first is the formation of a light haze, easily removed with a polishing cloth. This haze is a mixture of free sulfuric acid and nickel sulphate. After a while this forms the solid basic nickel sulphate which can only be removed by abrasion.

An illustration of the corrosion rates for many metals at different relative humidities is given in Figure 5.8.

Organic acids such as acetic and formic acids are released from most types of wood at high relative humidities and elevated temperatures. These acids can corrode most metals, including steel, lead, cadmium, zinc and possibly magnesium. Acetic acid, for example from wood shavings, can, in combination with the presence of carbon dioxide, attack bronze. No critical relative humidities are known. This potential problem explains why packaging engineers avoid using wood chips to cushion unprotected metal products.

Fig. 5.8 Corrosion Rates Vary by Metal

Each metal corrodes at a different rate. But in general, higher humidity means faster corrosion. (Hudson, 1929)

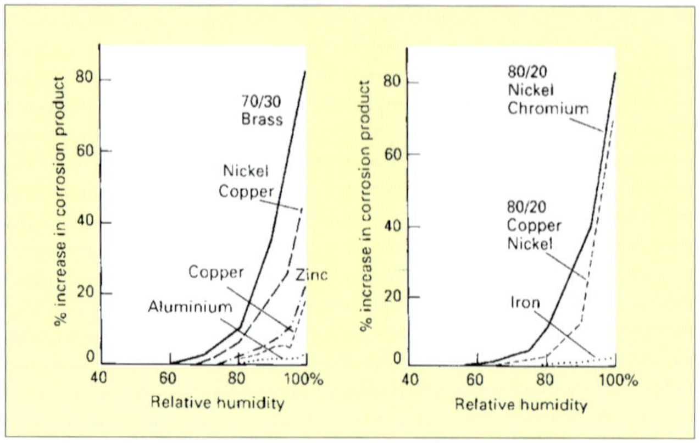

Dust, particularly on steel, has a strong influence on the corrosion rate but less on the critical relative humidity. W.H.J. Vernon pioneered studies on different types of dust, classifying them into groups:

1. Corrosion-free, e.g. an inert solid such as silica.
2. Intrinsically corrosive - produces corrosion wherever it settles, such as ammonium sulphate.
3. Indirectly corrosive - carbonaceous materials can adsorb acidic gases, which in turn can stimulate corrosion.

A summary of some of Vernon's classic and often-referenced findings appears in Figure 5.9.

Fig. 5.9 Dust Types And Their Corrosion Effects

In the real world, a variety of dusts will land on metal surfaces. These combine to produce a different effect on the corrosion rate at each relative humidity. (Vernon, 1945)

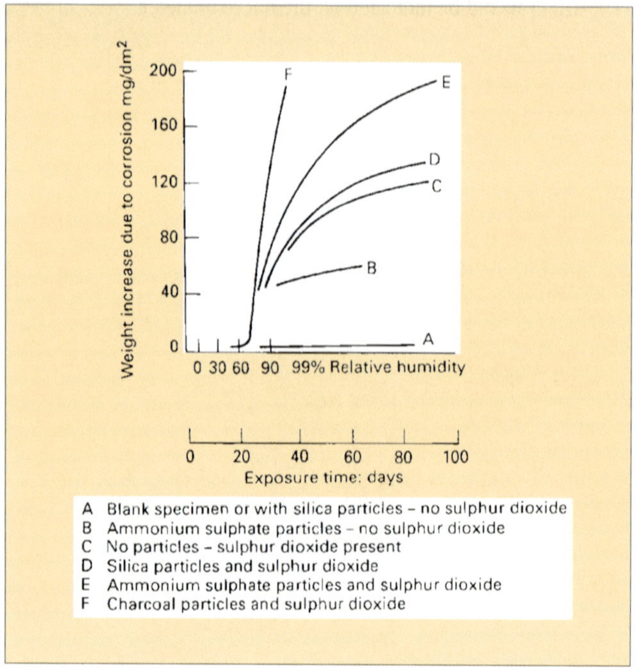

Measuring actual corrosion rates

The general principles above are useful for *understanding* corrosion as it relates to humidity, but they do not describe a specific corrosion *rate* that could aid engineering decisions. To quantify actual rates, one must define both the metal and the environment.

Tomashov has devised a corrosion battery which will measure the corrosion current for a copper-iron pair at any particular site. With that information, an Engineer can better estimate the critical relative humidity in a specific environment.

The construction of this meter and illustrative results are shown in Figure 5.10. The meter is made up of a sandwich of twenty or so plates of copper thinly insulated electrically from an identical number of thin iron plates. All the copper plates are wired in parallel and the iron plates similarly. The two electrical leads, one from the copper plates and the other from the iron plates, are connected to a microammeter. As the electrolytic corrosion process starts, so current will flow and be indicated on the microammeter.

Fig. 5.10
Corrosion Rates Are Site-Specific
Each site has different particulate and gaseous contamination in the air. To predict the corrosion rate at different relative humidities, the best technique is to measure corrosion current at the actual location of the project. (Tomashov, 1966)

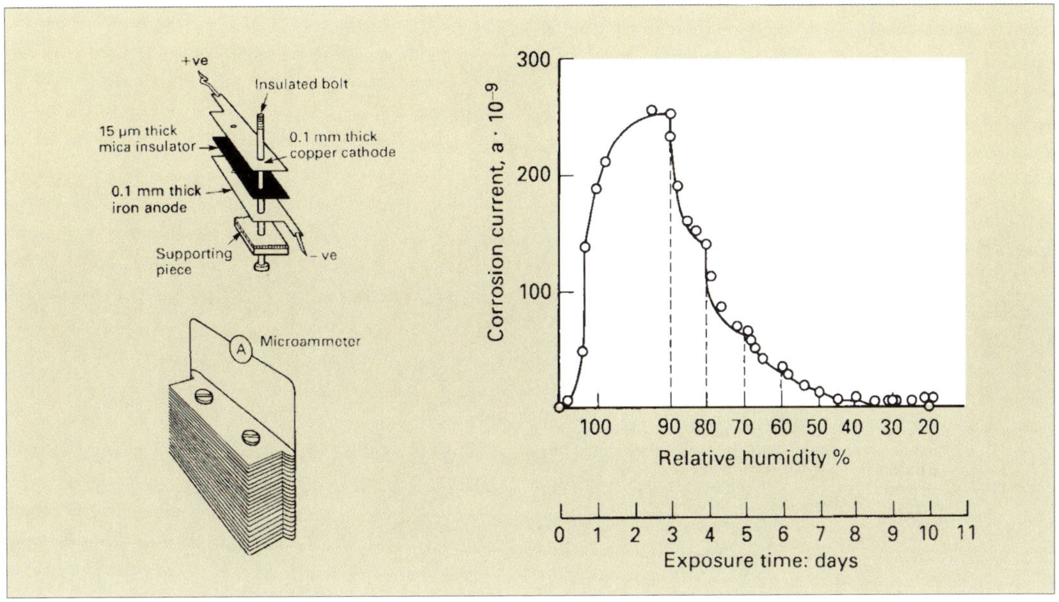

References

Dulz, S.J. J Soc Chem Ind, 60, 304- 315, 1950

Hudson, J.C. *Atmospheric corrosion of metals: Third Report to the Atmospheric Corrosion Research Committee of the British Non Ferrous Metals Research Association.* Trans Faraday Soc, 25, 177-252, 1929

Tomashov, N.D. and Lokotilov, A..A.. Zavodsk Lab, 4, 425, 1958

Tomashov, N.D. *Theory of Corrosion and Protection of Metals* The Science of Corrosion. Translated by B.H. Tytell, I. Geld and H.S. Presser. McMillan Co., New York, 1966

Vannerburg, N.G. and Sydberger, T. Corrosion Sci, 10, 43-48, 1970

Vernon, W.H.J. J Chem Soc, 1857- 1868, 1934

Vernon, W.H.J. J Sci Instr, 22, 228-237, 1945

Further Reading

Evans, U.R. *The Corrosion and Oxidation of Metals.* Edward Arnold, London, 1960 Leidheiser, H. The Corrosion of Copper, Tin and their Alloys. John Wiley. New York, 1971 Tiller, A.K. 'The factors responsible for the atmospheric corrosion of metals'. Report NPK 63/16/151/0238. Electricity Council Research Centre Contract 82/8063. December 1983

Harriman, Lewis G. III, *Preventing metallic corrosion in storage areas.* Proceedings of the International Symposium on Moisture & Humidity, Washington, DC, April 1985. (p. 949-964) Instrument Society of America, Research Triangle Park, NC

Walton, J.R., Johnson, J.B. and Wood, G.C. *Atmospheric corrosion initiation by sulphur dioxide and particulate matter.* Br Corrosion J, 17(2), 59-69, 1982 Wormwell, F. Corrosion of Metals Research, 1924- 1968. National Physical Laboratory, HMSO, 1973

Chapter 6
Electrostatic Discharge & Relative Humidity

By Geoff Brundrett

Fig. 6.1 Mechanisms & Frequency of Electrostatic Shocks

In a dry building, electrostatic shocks occur as he runs to catch the elevator because:

1. Fabric layers repeatedly pull apart, building up the charge.
2. Shoe soles contact, then pull away from flooring, building the charge further.
3. The charge cannot leak away gradually, because the airs low humidity reduces conductivity of surfaces. So charges are retained until they build up to voltages high enough to cause shocks.
4. Perceptable shocks occur more frequently as the relative humidity falls below 35% rh

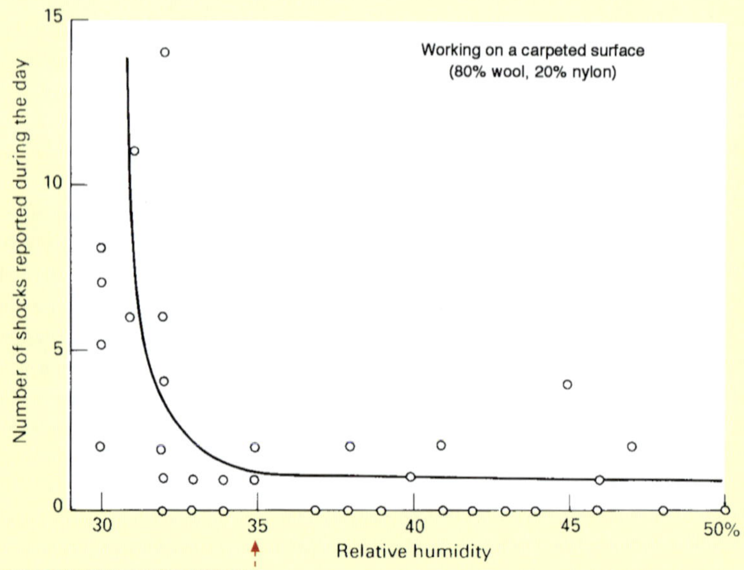

Basics of Electrostatic Discharge

People who live in cold climates are personally familiar with electrostatic discharges in the form of mild shocks that occur in winter. In dry air, sheets pulled from the clothes dryer cling together, and emit audible electrostatic sparks when lofted into the air and smoothed over the bed. The number and magnitude of such "static shocks" increases as the weather becomes drier. Here's why.

Electrostatic charges are created when surfaces separate

When two dissimilar materials are pulled apart, an electrical potential difference is created between the surfaces. The size of the charge depends on the electrical characteristics of the surfaces, the amount of surface, the distance of separation and the frequency of separation. The charge is greater when more surface is separated, or when surfaces are separated frequently, such as when shoes lift off the carpet as one walks across a floor.

Charges build up when they cannot flow to ground

Although electrostatic charges are created any time surfaces are pulled apart, most of the time the charge decays in milliseconds—it leaks away so quickly that the charge has little perceptible effect on either people or objects. But the decay rate depends on the electrical resistance of materials connected to the two newly-separated surfaces. If the resistance is high, the charge cannot flow away quickly to ground. It lingers, awaiting a lower-resistance path to ground. That's when successive separations can build the charge to very high levels, which may be discharged through people or through objects that are connected to ground.

High humidity helps dissipate charges

When water molecules accumulate on a surface, they provide a path of reduced resistance for electrical current to flow, even when the surface is normally a poor conductor. When the relative humidity near a surface is high, more airborne water molecules collide with and stick to the surface. In other words, at high relative humidity, materials are more likely to accumulate a conductive layer of water molecules, and therefore lose electrical resistance on their surfaces.

Using the example of the clinging sheets recently removed from the clothes dryer, we can see why their behavior changes with the seasons. During humid months, sheets rapidly adsorb moisture from the air. Also, materials in the room have adsorbed water molecules from the humid air, so all surfaces are highly conductive. The electrostatic charges decay in microseconds as current leaks across humid surfaces to ground.

On the other hand, in dry weather all surfaces are less conductive, so the charge remains high—perhaps for a minute or even longer. During that period, the high charge can "jump" a narrow air gap to a person or to an object with a lower resistance on the electrical path to ground. When the charge leaps the gap to skin, it can cause a perceptible shock.

High humidity can reduce the number and size of electrical shocks by helping materials dissipate electrostatic charges quickly—before they can build up to high levels, and before they can linger long enough to find a damaging path to ground. In commercial buildings, keeping the relative humidity elevated to at least 35% rh, or keeping the dew point to at least 41 °F [5 °C] will keep the number of electrostatic discharges to similar levels of perception year-round, as shown in Figure 6.1.

Charge Generation & Retention

An electrostatic charge is created by the separation of materials which have previously been in contact. During contact there is some charge transfer between the materials. As the two surfaces separate, the electrical capacitance between them falls inversely as the separation distance. Since the electrical charge is constant and:

Charge Q = capacitance C x voltage V   Equation 3.1

A reduction in capacitance will produce a rise in the voltage of the surface with increasing separation distance. As a walking person lifts one foot from the floor, the increased voltage of the shoe from this effect raises the charge on the total capacitance of his whole body in spite of a partial leak to earth through the shoe still in contact with the ground.

When the foot is lowered again it will have lost some of its charge to the main capacitance of the body and a little through earth leakage. The foot reaches the carpet at a much depleted voltage and recharges from the carpet. Thus at each step the falling and rising local capacitance between foot and carpet pumps charge into the main capacitance between body and ground, giving a ratcheting increase in personal voltage, as shown in figure 6.2. The higher the footsteps, the greater the voltage increments.

This voltage increases with the distance walked up to some maximum value. Charging currents are estimated to be smaller than 10^{-8}A (Potter, 1969). In steady-state conditions the maximum personal voltage will be a direct function of the leakage resistance to ground. This relationship is described by equation 6.2:

Maximum Personal Voltage **Equation 6.2**

V = amps x ohms

- V: **Maximum Personal Voltage (volts)**
- amps: Charging current
- ohms: Leakage resistance

Resistance retains charges

For a given charging current, the maximum personal voltage will depend on the leakage resistance to ground. Higher resistance increases the maximum personal voltage. Assuming a maximum charging current of 10^{-8} A, the personal voltage will be 10V for leakage resistance of 10^9 ohms (Ω). This is illustrated in Figure 6.3. However, for a 1000-fold increase in leakage resistance to 10^{12} Ω, the maximum personal voltage increases also 1000-fold to 10000V. This leakage resistance to earth is the sum of the shoe resistance and the carpet resistance. The carpet resistance depends not only upon the environmental conditions around it, such as moisture content, but also upon the nature of the substrate. In practice it is a complex combination of horizontal and vertical resistance along and through the carpet structure.

The electrical resistance of the shoe component of this circuit varies widely with the material and thickness of the sole of the shoe. Spot-checks on a pair of British leather-soled shoes showed the resistance to be 10^8 Ω and this was unaffected by the wearing of socks. (Quinto, 1953) A spot-check on a pair of rubber-soled sneakers gave an electrical resistance of 5 x 10^{11} Ω (Henry, 1971).

That study measured the electrical resistance from hand to conducting floor for people wearing a range of different kinds of shoes. New dry leather soles could have resistances as high as 10^9W, while for damp, thin soles this went down to 10^3 Ω. Rubber-soled shoes gave resistances of 5 x 10^8 Ω, but with conducting rubber, this could go down to 10^7 or even as low as 10^4. PVC-soled shoes had a resistance of 4 x 10^9 Ω. Microcellular rubber-soled shoes had an electrical resistance greater than 10^{13} Ω. Measurements of the electrical resistance of the footwear of 132 people working in hospital operating theatres showed that rubber soles predominated and that the rest were leather-soled (Guest, 1952). The rubber soles had much higher electrical resistances than the leather ones. Seventy per cent of the rubber-soled shoes had a resistance greater than 10^9 Ω. Only 18‰ of the leather-soled shoes had a resistance greater than 5 x 10^7 Ω.

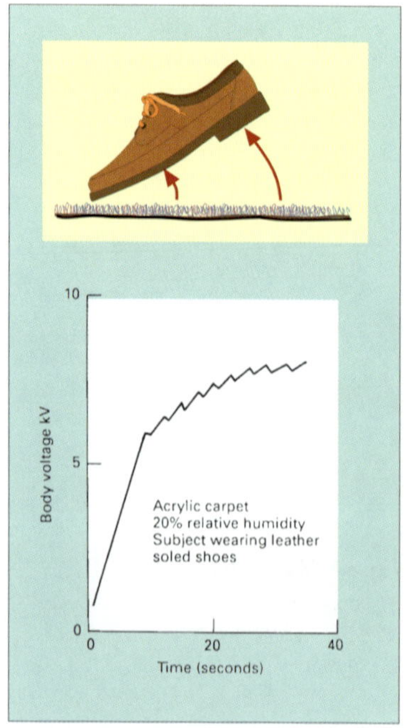

Fig. 6.2 Separating Surfaces

An electrostatic charges is created when surfaces separate. Walking separates surfaces repetitively, which builds the charge higher. (Cusack, 1972)

Fig. 6.3 Charges Increase

As long as the electrical resistance of the path from body to ground is too high to allow the charge to leak away.

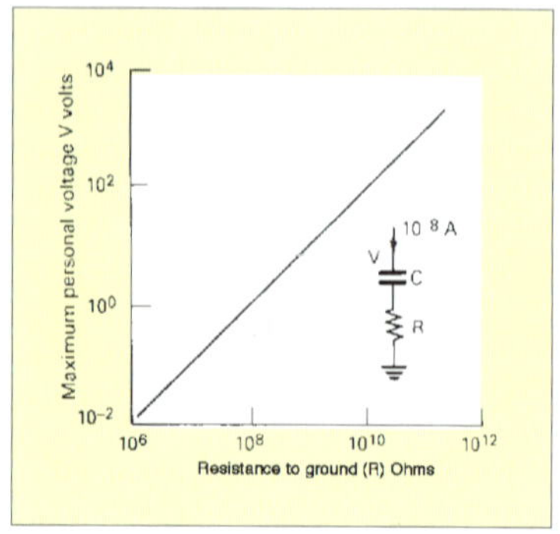

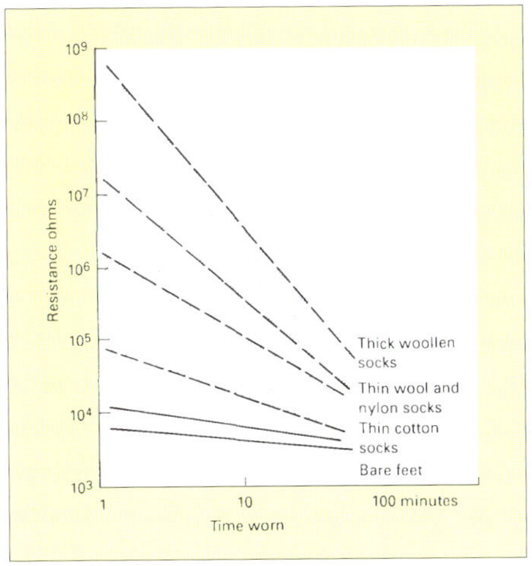

Fig. 6.4 Moisture In Socks

Moisture reduces the resistance of socks, which allows more energy to leak from body to ground.

Fig. 6.5 Resistance of Shoe Soles

In general, natural fibres absorb conductive water molecules more readily than synthetics, reducing electrical resistance and therefore reducing the probablility of shocks.

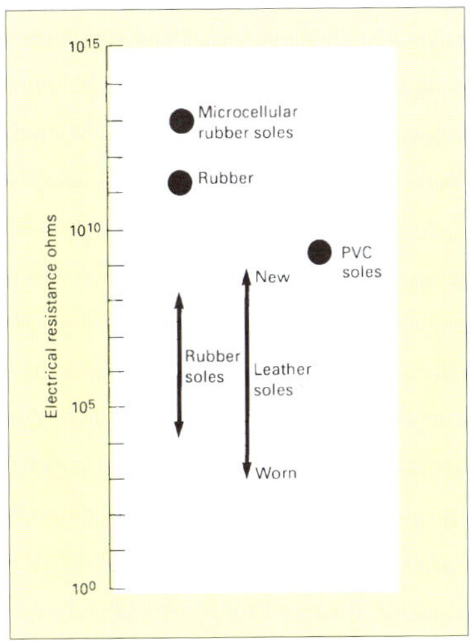

An illustration of the values found by researchers is given in Figure 6.12. The range can vary from 50 to 1100 picofarads (pF), with typical values being 200 pF for men and 100 pF for women. Dry socks, particularly thick woollen ones, have a high resistance value, but this declines quickly because perspiration moistens the sole and lowers the resistance (Figure 6.4 and Bajinski, 1972).

Once a person stops walking he has stopped generating the charging current. The charge which has built up on the individual then leaks away. The time taken for the electrostatic charge to fall to 37% of its initial value is related to the personal capacitance and the leakage resistance:

Resistance values can be used in conjunction with capacitance values from Figure 6.12 to assess how quickly the electrostatic charge can leak away. If the personal capacitance to earth is 200 pF and the total leakage resistance is $10^{12}\ \Omega$, then the time of discharge would be relatively long at 200 seconds. A lower leakage resistance of $10^6\ \Omega$ would result in a discharge time of only 0.00002 seconds (200 microseconds), which is practically instantaneous. In other words, a person would discharge immediately on pausing. One test for antistatic materials is to measure if a 5000-V charge could decay to less than 500V in less than half a second (Ellis, 1974).

The crucial factors for "charging up" a person electrostatically (and for retaining that charge) is the leakage resistance to ground. If resistance is below $10^{10}\ \Omega$ then the personal maximum voltage would be approximately 100V and the discharge time one second. Since shoes are not normally so high in electrical resistance, the dominant factor becomes the electrical resistance of the floor covering.

Time Constant **Equation 6.3**

$$T = \text{farads} \times \text{ohms}$$

- T: **Time Constant (seconds)**
- farads: Capacitance
- ohms: Leakage resistance

Resistance declines as humidity increases

Moisture content is the most important factor in determining the electrical resistance of textile materials. The moisture content is closely related to the ambient relative humidity. This moisture relationship, termed a sorption isotherm, is illustrated in Figure 6.6 for a range of textile fibres. The electrical resistivities of materials decline when relative humidity increases, as the figure shows. Non-absor-

Fig. 6.6 Higher Moisture Content Reduces Resistance

Water molecules improve conductivity of materials. At high humidity, materials adsorb more of those conductive water molecules from the air.

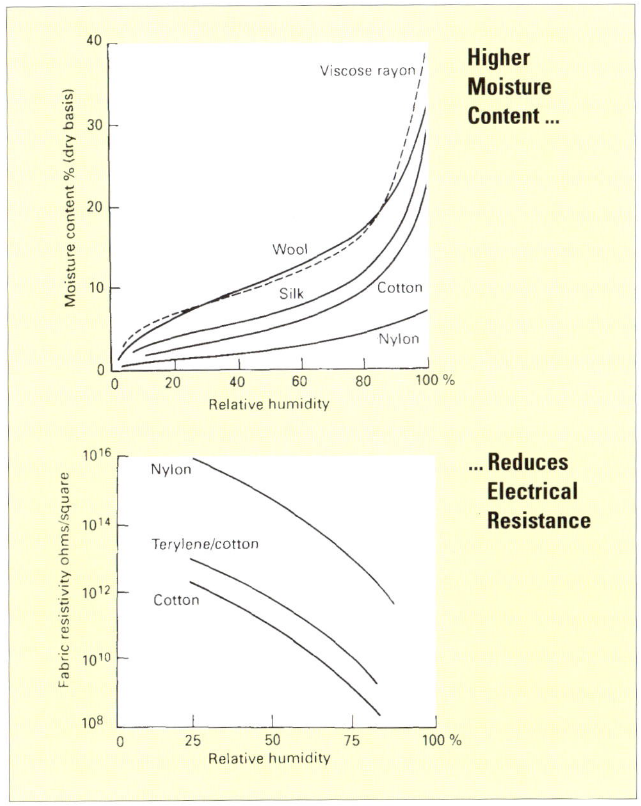

bent fibres such as nylon have a high resistance. There is no simple relationship between water content and resistance, since textiles often contain a mixture of different fibers.

Taking all factors together, adjusting the relative humidity can have a major effect on the frequency and magnitude of electrostatic shocks in a commercial building.

How charges are generated in commercial buildings

In commercial and institutional buildings, electrostatic charges are generated by many activities, but primarily by walking, rising from chairs and from clothing friction.

Walking on carpets & flooring tile

Most electrostatic charges are generated by walking on a floor of high electrical resistance. The body voltage will rise quickly after each step at first and then after many paces reach an alternating high and lower fluctuation in time with the footsteps. An example taken from an electrostatic carpet test was shown in figure 6.2. The voltage generated is influenced by both behavioral and physical factors.

The behavioral factors are distance walked and height of step. The maximum voltage can take many steps to achieve. In small offices this is much less likely to happen than in large open-plan spaces. Higher steps also result in higher voltages, and a 150-mm-high step can induce 50% greater body voltage than that generated by the same person only lifting his feet to a height of 75 mm (Chakravarti, 1976).

The physical factors which affect the body voltage are the type of shoe material and the carpet characteristics. The most important carpet characteristic is its electrical resistance, which always varies with ambient relative humidity. The electrical resistance increases as relative humidity declines. All carpets have higher resistivity in dryer conditions, but the absolute value of resistance is influenced by the weave, the thickness, the mixture of materials used in the carpet pile and backing, and the type of underlay. There are a range of empirical walking tests designed to assess the body voltage which can be produced during a simulated walk. The magnitude and polarity of the voltage are also influenced by the different combinations of shoe sole materials and carpet fibre. This influence is shown in Table 6.1.

Table 6.1
Charges Generated By Shoes On Carpet

	Kilovolts Generated by Shoes on Carpets @ 20 %rh				
Type of Sole	Wool	Nylon	Acrylic	Polyester	Polypropylene
Neolyte (Synthetic)	-17	-16	+3	+7	+8
Oak-tanned Leather	-16	-15	+12	+8	+12
Chrome-tanned Leather	-10	-13	+4	+9	+14

The resistivity of the carpet can be measured in either the vertical or horizontal plane. The unit is the resistance in ohms across a "square" (10 ft x 10 ft, or 100 ft² or 9.3 m2) of the material. In practice the vertical and horizontal resistivities of a square of carpet are numerically similar. The body voltage measured in a walking test is related to the carpet resistivity. This relationship was tested for 27 different carpets, and the results are illustrated in Figure 6.7.

Above relative humidities of 60%, the maximum voltage generated by walking on a carpet tends to be very small—nearly negligible. The voltage rises as relative humidity falls, reaching a maximum when humidity drops to about 10% rh. A few carpets generated maximum voltages at 20-25% rh, and then diminished slightly at lower humidities. Both woollen and nylon carpets could generate between 13 and 17 kV body voltage in dry conditions. In a few cases a nylon carpet can give rise to sufficient voltage to give a threshold electrostatic shock (2000V) at a relative humidity as high as 50%. However, most nylon

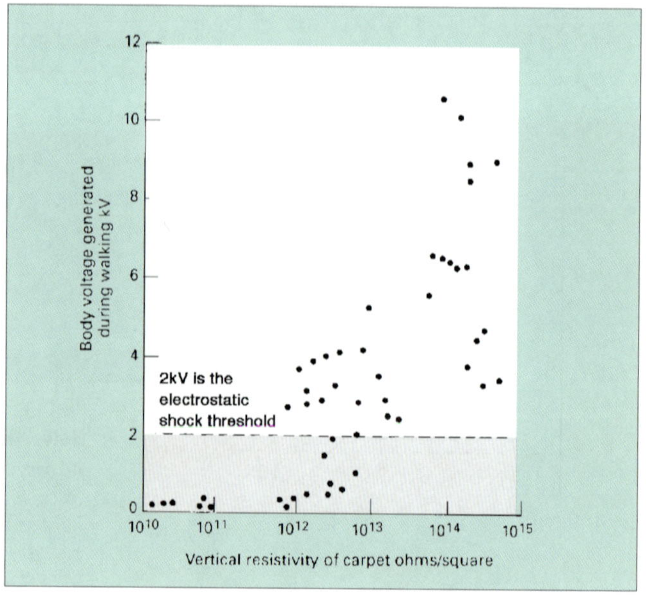

Fig. 6.7 Resistance of Carpets

Varies according to the conductivity of the fibers. When fibers are very conductive, electrostatic discharges are not perceptable. (Wilson, 1972)

carpets behave like those made of wool. Acrylic carpets generate only low voltages until the humidity drops below about 40% rh.

Usually, there is only a small difference between carpets made of artificial fibres and those made of natural fibres. However, if electrostatic shocks are reported at relative humidities around 50%, then the carpet is most likely to be made of nylon. A lower limit of 40% rh will avoid electrostatic shocks when walking on most carpets.

Below 35% rh some number of electrostatic shocks would be perceptible when walking on most carpets. The relationship between body voltage and relative humidity is given in Figure 6.8.

One exceptional case is that of underfloor heating. Then the carpet temperature rises above the temperature of the air. This higher surface temperature means that the relative humidity of air inside the carpet is *lower* than that of the air. Consequently the dew point in the air must be raised higher than usual in order to raise the relative humidity of air trapped in a warm carpet.

Beyond raising the humidity, the electrical properties of carpets can be modified in two ways. The first is addition of a hygroscopic material in the form of a shampoo. This desiccant (called a humectant by the fabric trades) remained moist at lower relative humidities and delayed the onset of electrostatic shocks (Tunmore, Hayek & Valko). Unfortunately, at high relative humidities the carpet could become damp and hence quickly soiled. More recent techniques permit the desiccant to be absorbed within the fibre. This avoids the rapid soiling and slows down the removal of humectant during cleaning. In mild humidity conditions the addition of a particularly hygroscopic fibre has a similar effect. For example, adding 5% of viscose fibre to a nylon carpet significantly lowers the maximum voltage generated at a given relative humidity (Textile Monthly, 1971). Hair underlay has a similar effect down to 35% rh but fails to prevent high voltage charges at lower humidities (Martin, 1971).

The second method uses conductive fibers to reduce resistance. Fine metallic wire, either copper or 8μm-diameter stainless wire, can be woven into the carpet yarn. Only 0.3% by weight is needed to make the carpet antistatic, although there is a tendency for this effect to decline with long use as the wires break and the electrical resistivity of the carpet rises (Chakravarti, Bajinski). Conducting nylon fibres woven into the carpet produce a similar effect. Such fibres, called epitropic, have two components. The outer sheath of the fibre is enriched with sufficient carbon impregnation to conduct current. Only 0.1-0.2% of such fibres are needed in a carpet to reduce the body voltages to less than half of the voltages in an untreated carpet (Ellis, 1974). Conducting latex backing can also reduce carpet resistance.

An electrostatic charge can also occur when walking on ordinary linoleum or tiled floor. The same principles apply, only instead

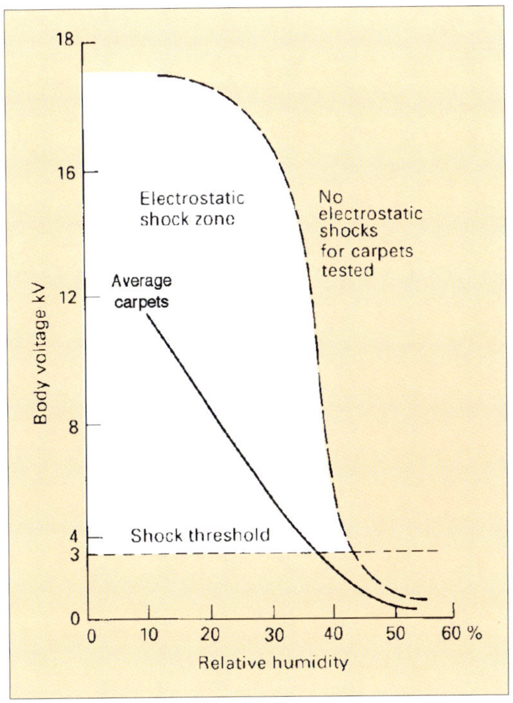

Fig. 6.8

Below 40% rh, Perceptable Shocks Are More Likely

As body voltage increases above 3 kV, shocks become more probable and more perceptable. But shock probability as a function of relative humidity is not uniform for all carpets, because native conductivity and water sorption varies between types of fibers.

Fig. 6.9

Walking On Linoleum

Creates charges just like walking on carpet, although charges tend to be smaller because less surface contact. (Jonassen, 1987)

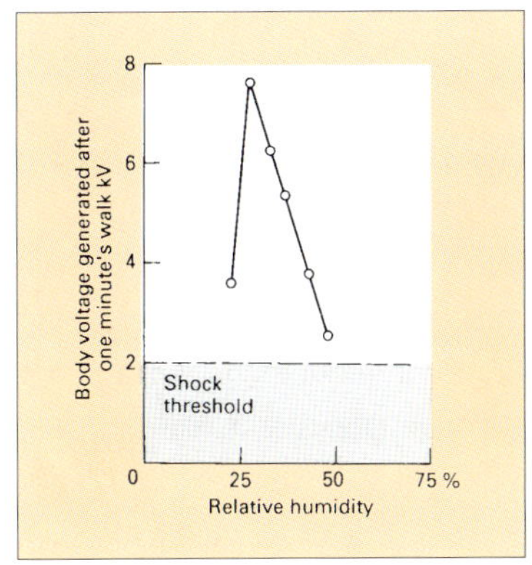

of moisture influencing the electrical resistivity of the carpet, it affects the resistivity of the floor surface. Experiments show that the maximum body voltage generated after one minute's walk at 30% rh can be approximately 8 kV. Interestingly, at lower humidities the voltage declines. This relationship is illustrated in figure 6.9.

Rising from chairs

Any separation of dissimilar materials generates a charge. Such separation occurs when a person rises from a seat or when clothing is removed. The ambient relative humidity determines the leakage resistance of the material. The nature and surface area of the two materials determine the magnitude of the effect. Experiments in hospitals on the body voltage generated when anesthetists rise from their operating theatre stools showed that no charge occurred when the ambient relative humidity was above 45%. However, when the relative humidity fell below 36%, outer overalls of cotton and wool generated body voltages over 1000 V when the seat was upholstered in plastic. With a plain steel seat the voltage was 100 V.

Measurements of the body voltage created by rising out of a chair vary widely with the contact materials. Experiments on chairs covered in plastic are shown in figure 6.10. When the person wore a nylon "boiler suit", the maximum voltage was over 3 kV . The two other materials only generated a very low charge. Voltages up to 5000V were recorded on an operating table when the covering sheet was pulled off. Voltages of 3500V were recorded, even when relative humidity was as high as 53%.

Fig. 6.10. Rising From A Chair

Pulling clothing away from upholstery generates an electrostatic charge. The voltage depends on the type of material in clothing and chair. Less conductive materials create higher charges. (Guest, 1952)

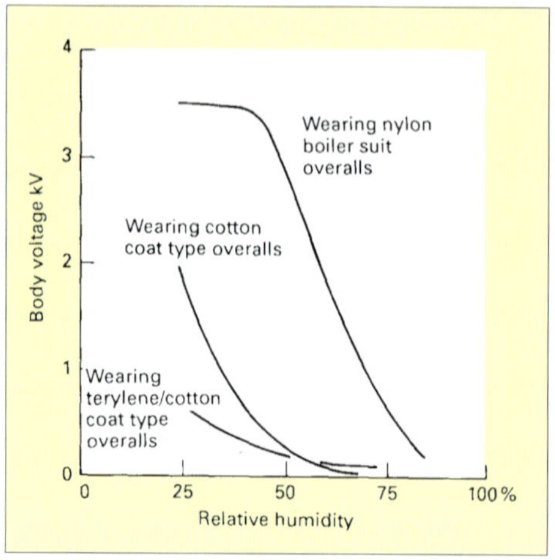

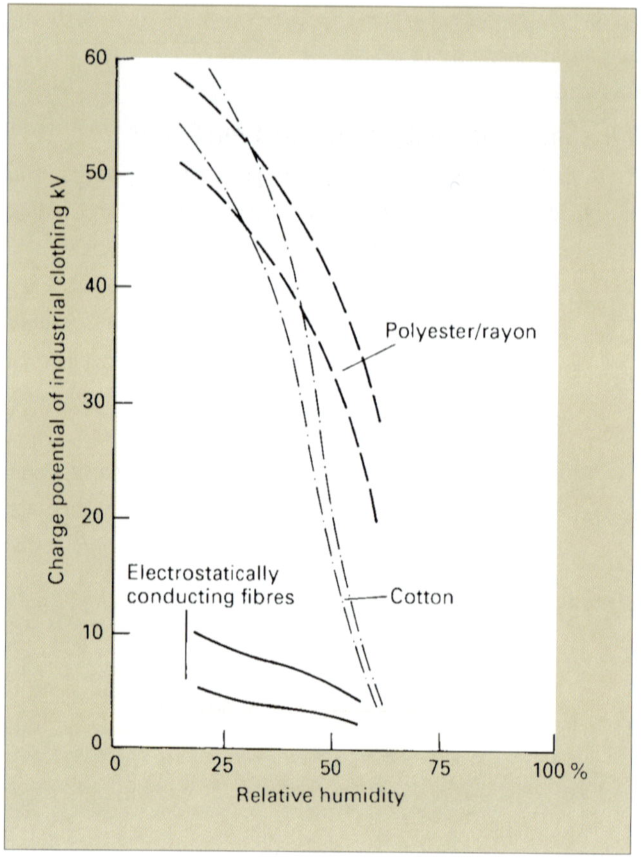

Fig. 6.11. Removing Clothing

Separating clothing layers generates an electrostatic charge. The more hygroscopic materials are conductive at lower humidities, so in general, such materials create less charge and reduce the likelihood of shocks.

Removing clothing

The highest charge potentials are associated with the rapid removal of outer clothing. Voltages over 50 kV were measured for materials made from cotton and for those of polyester/rayon mixtures. The highest voltages were recorded in the driest test conditions (20% rh).

Although cotton clothing can generate such voltages in very dry conditions, the maximum voltage falls rapidly with increasing relative humidity. In contrast, the polyester/nylon mixture is still capable of generating over 20 kV even when humidity rises to 60% rh Clothing made with a proportion of electrically conducting fibres has a much lower voltage generated (Japanese Research Institute, 1971). Results are illustrated in figure 6.11.

Personal Sensitivity to Electrostatic Shock

Not all electrostatic discharges are perceptable by people, and not all people are equally sensitive to such discharges. Three factors influence the magnitude of perceived shock:

1. The total energy stored.
2. The part of the body and the total surface area through which the discharge occurs.
3. The sensitivity of the individual.

The electrical energy stored in a person is the product of his capacitance and voltage. The relationship is:

Energy Stored By The Body — **Equation 6.4**

$$E = \frac{C}{2} \times V$$

Energy Stored (joules)
Capacitance (farads)
Voltage (volts)

Some values of capacitance are shown in figure 6.12. The biggest variable is the surface area and thickness of the sole of the shoe. Large-area, very thin soles can have over 1000 pF capacitance. Sitting down creates a large capacitance. As the contact surface area increases, so does the capacitance. Typical values for a person would be a capacitance of 200 pF from hand to earth. A typical low-level shock occurring at 2 kV requires 0.4 mJ of stored energy. The voltage has a more powerful influence than capacitance. The personal sensation, for different capacitances, is shown in figure 6.13.

This charge is stored on the surface of the skin. The current flows towards the point of discharge. The pain sensation is due to the local burning of skin at this point of discharge. But the discharge will not be noticed if the current density can be reduced by flowing the current across a wide skin area. Discharges through the soles of shoes are not perceptible.

Fig. 6.12. Capacitance

Hgher capacitance allows higher amounts of energy to be stored, increasing the size of the electrostatic discharge when it finally occurs.

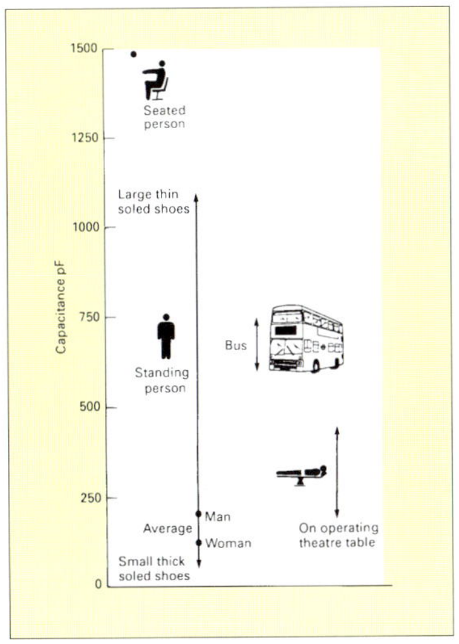

Fig. 6.13. Personal Sensation

The human perception of shocks depends on the total energy released, which is a function of body capacitance and the voltage of the charge. (Brundrett, 1977)

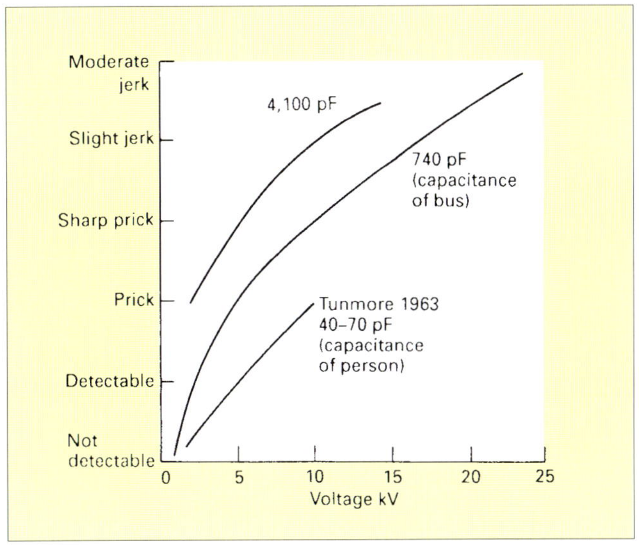

Personal sensitivity to electrostatic shock is a relatively new concept. Early experimenters used themselves and their immediate colleagues. Familiarity and scientific concentration made these subjects relatively insensitive. The average population is more sensitive although there are wide differences between individuals. Assessments by samples of up to 200 people show that there is close agreement between modern authors on the personal sensitivity to the voltage level they can detect when electrically discharged through the hand.

The cumulative distribution of population sensitivity is shown in Figure 6.14. All people will *detect* a spark discharge through the hand when they are charged to a potential between 1 and 2.5 kV. Above that voltage, the shocks are usually perceived as *objectionable*. Finally, above 7 kV, most people find the shocks *unpleasant*.

This more recent and refined analysis of personal sensitivity has led to a lowering of the acceptable body voltage from the 19-kV value used earlier, to the present-day recommendations of 2.4 kV. This is the body voltage limit proposed for testing floor coverings for electrostatic shocks.

In practice there is a large degree of randomness in the reported complaints of electrostatic shocks. It will depend upon the thickness and dampness of the shoe sole, its material, and the distance walked, in addition to the differences between people. One carefully recorded survey shows the kind of complaint record of electrostatic shocks in a large open-plan office as shown in Figure 6.15.

Fig. 6.14 Perceptable vs. Objectionable Shocks

Sensitivity varies between men and women and between studies performed by different researchers. But in general, women perceive electrostatic shocks at lower levels than men.

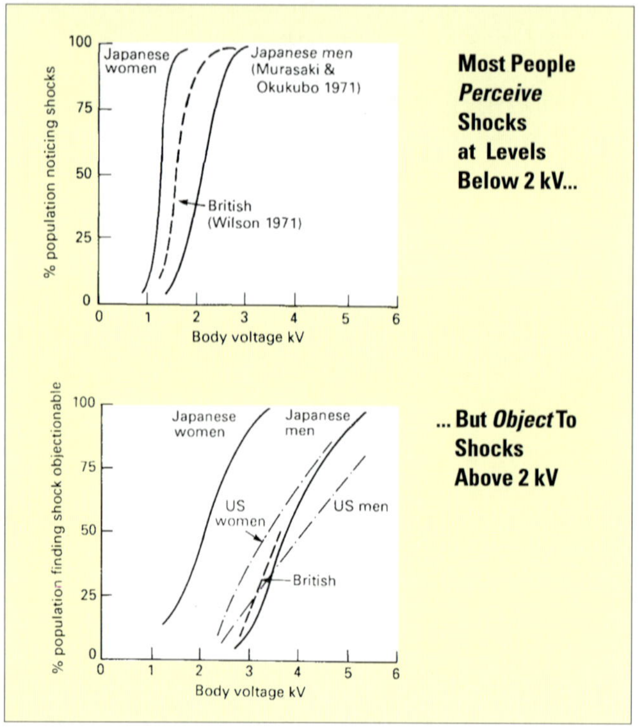

Fig. 6.15 Complaint Frequency

The actual number of complaints depend on a complex mixture of factors, including charge voltage, personal capacitance, clothing and carpet conductivity, personal sensitivity and psychological factors. But in general, shocks become more numerous and more objectionable when the relative humidity drops below 35 %. (Brundrett, 1977)

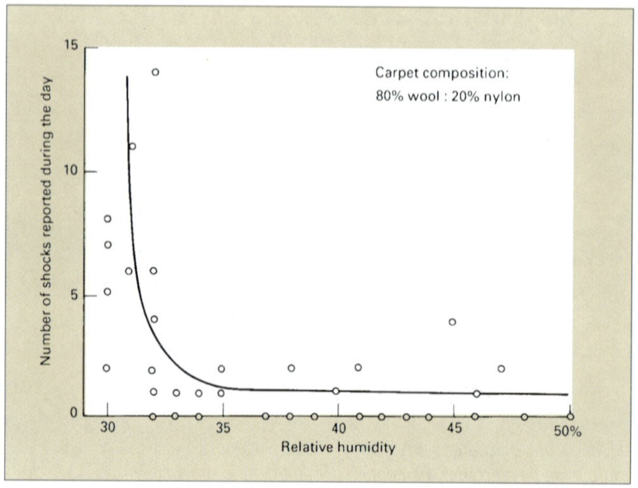

Safety Issues

Beyond the need for comfort, the designer must consider the need for safety. Flammable gases, vapors and dusts can explode. The source of ignition could be an electrostatic discharge. When the air is enriched with extra oxygen, the ignition energy required from the spark could be less than that required to create a detectable shock to a person. This presents extra danger because people in that environment will not be aware of the potential danger. Three conditions are necessary for a flammable gas to ignite (Wilson, 1973):

1. **The gas concentration is within certain limits of flammability**. These limits depend upon the gas itself, the mixture and the physical conditions of pressure and temperature. Cyclopropane, formerly used in hospitals, will only ignite in air at concentrations between 2 and 10% by volume. These limits are considerably widened in an oxygen atmosphere. The gas will ignite when it represents between 2 and 63% of the volume of the mixture.
2. **There is sufficient energy in the spark.** This varies with the gas and is typically 0.03-0.3 mJ in air. Much more energy is needed as the ratio of gas to air moves away from an ideal stoichiometric mixture. However, in an oxygen atmosphere the hazard is much greater. The necessary energy may drop to 1% of the energy needed to ignite gas in normal air.
3. **The spark length is sufficiently long or voltage sufficiently high.** One long spark is much more effective than multiple small sparks discharging the same capacitor. Gas ignition by multiple small discharges took 30 times more energy than required when all energy was concentrated in a single, larger discharge.

Typical values of minimum ignition energy of some gas mixtures are given in Figure 6.16.

Not all the electrostatic energy stored on the person is available for the discharge. During discharge much of the stored energy is dissipated as heat when the current encounters resistance as it flows across the skin. Also, some heat is lost from the spark by electrode quenching. Experiments to measure the body energy needed in practice to ignite an optimum mixture of coal gas with air (and natural gas with air) showed that significantly more stored energy was needed for even an occasional detonation. For the coal gas mixture, the person's stored energy had to be 100 times the minimum spark igni-

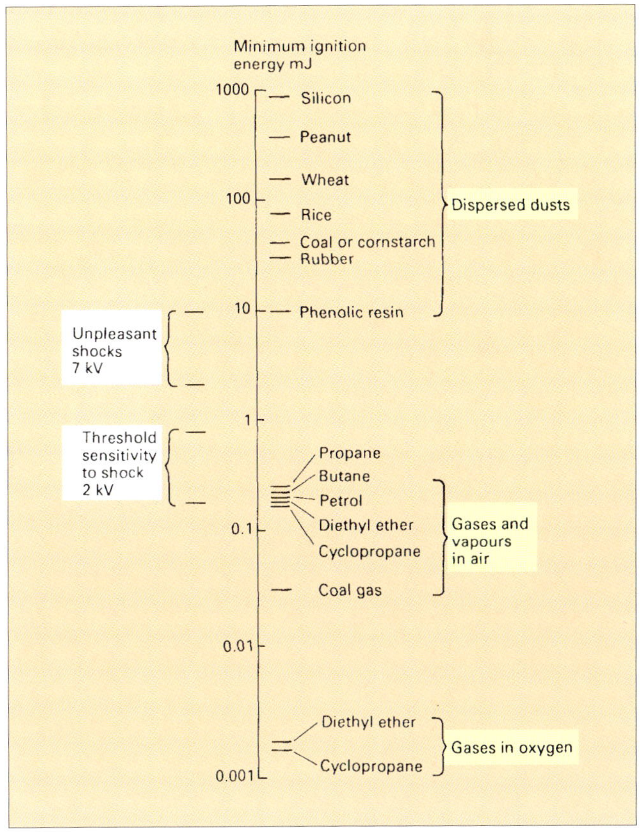

Fig. 6.16 Explosion Hazard vs. Perceptable Shocks

Vapor/air mixtures are especially hazardous. They can ignite with low-energy electrostatic discharges which may not be perceptable to workers, who might otherwise be alerted by perception of electrostatic shocks. (Wilson, 1973) tion energy of the gas mixture to achieve one explosion in 100 tests. The natural gas mixture required 60 times the minimum energy to achieve two explosions in 100 tests.

In normal atmospheres we can conclude that when unpleasant electrostatic shocks occur, there is a serious potential danger to gas/air mixtures. Precautions should be taken to eliminate those shocks.

In oxygen-enriched atmospheres of flammable gases even imperceptible electrostatic discharges could cause explosions, and antistatic precautions are needed at all times.

Air-dust mixtures, while less hazardous than gas mixtures, still demand concern and attention. The three conditions necessary for dust explosion are:

1. The atmosphere must exceed a critical oxygen content. Dust of most common flammable materials will explode in air. But organic dusts such as coal or rice need at least 16% oxygen content to explode. Metallic dusts are more hazardous, and can explode in oxygen concentrations as low as 3%. (Ray, 1975)
2. The ignition energy must be above a critical value. This varies widely from 10 mJ for phenolic resin dust to over 300 mJ for peanut dust.
3. The dust concentration must exceed a minimum concentration. Coal dust must exceed a concentration of 35 g/m^3. Powdered metal dust is more sensitive and need only reach 20 g/m^3 for explosion. These are local concentrations around the ignition source. Dust explosions are unique in self-propagation. A small explosion can lift more dust into the air, creating a larger secondary explosion.

Dust explosions usually require about 100 times the energy that would be needed to initiate a gas explosion. There is little danger of dust explosions from personal sparks. The danger lies in charging up machinery of large electrical capacitance which may then create a powerful discharge.

Guidelines For Controlling Electrostatic Discharge With Humidity

Moisture has a very large influence on the electrical resistance of fibres and surfaces. It can vary over a million times from dry to damp conditions. This makes it the most important single factor in determining the leakage resistance of an electrical charge and therefore the ability of that charge to build up and create a spark.

The energy stored through electrostatic charging is usually small (1 mJ) but at high voltage (~5 kV). The smallest practicable discharge is conventionally believed to be that across a gap of 0.01 mm which requires a voltage difference of 350V. (Eichel, 1967) Voltages up to 100 kV have been reported on heavy machinery. (Bulgin, 1953) In order of *decreasing* hazard, the designer might consider:

Safety

There are certain areas where hazardous materials are handled and the smallest, even imperceptible, discharge could initiate an explosion. Some hospital operating suites, for example, use oxygen-enriched flammable gases. In such areas, the intention is to keep any voltage below 100V and to humidify the atmosphere. There are other tasks which are potentially dangerous but not as sensitive as the first group. An example would be refuelling operations, where pockets of gasoline/air mixtures would be flammable or explosive. But in these circumstances, the ignition energy is normally greater than what people would recognize as electrostatic. Therefore the situation is "self-alarming". Whenever people *do* report static problems, action *must* be taken to eliminate the phenomenon before charges can build to the higher levels need for ignition.

Machinery problems

While all electrostatic problems are unique to each piece of equipment, there is a consensus that problems become more frequent when the relative humidity drops below 40% rh. Humidification is one simple solution, provided the building envelope has been equipped with a vapor retarder on the warm side of all exterior walls.

Personal shocks

This is usually associated with walking across a floor and then touching a metallic contact such as a door handle or filing cabinet. 35% rh eliminates most problems, 40% nearly all. It is rare to have this problem if the relative humidity remains above 40% unless underfloor heating is used. When the floor is heated, the humidity ratio must be raised high enough so that relative humidity at the warm floor surface is above 40%.

Summary

To summarize, for effective elimination of all electrostatic discharge, the relative humidity would have to be kept above 65%. This extraordinary level of humidity may be necessary in areas where oxygen-enriched gases are present. At that level, however, condensation in exterior walls is very difficult to avoid in winter months, even when skilled designers take careful precautions. Structural damage and health hazards from corrosion, mold and mildew are a serious risk for the walls and HVAC systems of such highly humid buildings.

In commercial buildings, most office equipment and most people are not troubled by the relatively infrequent and weak electrostatic discharges that occur at relative humidities down to 40%. A further reduction below 35% rh is needed before the number and size of perceptable discharges increases substantially. Finally, as the relative humidity drops below 20%, materials and office machinery will probably display annoying electrostatic problems unless air ionization or conductive surfaces are installed to dissipate the charge.

References

Bajinski, G. and Lott, S.A. 'Electrostatic hazards from insulated operators'. Report S2 1: Australian Defence Scientific Service, Defence Standard Lab, Victoria, 1972

Bright, A.W. and Hughes, J.F. 'Research on electrostatic hazards associated with tank washings in very large crude carriers'. J Electrostatics, 1, 37-46, 1975

Brundrett, G.W. 'A review of the factors influencing electrostatic shocks in offices'. J Electrostatics, 2, 295 -315, 1977

Bulgin, D. 'Static electricity on rubber tired vehicles'. Br J Appl Phys, Supplement 2, 583-587, 1953

Chakravarti, K. and Pontrelli, G.J. 'The measurement of carpet static'. Text Res J, 129-134, February 1976

Corbett, R.P and Hughes, J. F. 'Electrostatics - benefits and hazards'. Engineering, 1-8, July 1974

Crugnola, A.M. and Robinson, H.M. 'Measuring and predicting the generation of static electricity in military clothing'. US Army Textile Series Report No. 110, September 1959

Cusack, J.A. 'Static control in carpets'. Modern Textiles Part I, 53, 66-72, January 1972;Part 2, 53, 25-29, February 1972

Eden H.F. 'Electrostatic nuisances and hazards', Chapter 18, pp. 424-440 in Electrostatics and its Application (ed. A.D. Moore). John Wiley, New York, 1973

Eichel, F.G. 'Electrostatics'. Chem Eng, 74(6), 153-167, 1967

Ellis, V.S. 'Epitropics - third generation conductive fibres'. Textile Manufacturer, 19-23, July 1974

Fordham-Cooper, W 'Practical estimation of electrostatic hazards'. Br J Phys, Supplement No. 2, 71-77, 1953

Guest, P.G., Sikora, V.W. and Lewis, B. 'Static electricity in hospital operating suites'. US Bureau of Mines Report R14833, January 1952

Greater London Council Bulletin 'Investigation of shocks by static electricity occurring at County Hall Island Block'. No. 87, Item 6, pp. 6/l-6/2. July 1975

Hayek, M. 'Antistatic finishes for textiles'. American Dyestuff Reporter, 43, 368-371 , 1954

Henry, P.S. 'Risk of ignition due to static on outer clothing'. Inst Phys Static Elec Conf Series, 11, 212-221, 1971

Hughes, J.F. 'Electrostatic hazards in supertanker cleaning operations'. Nature, 235(S338), 381-383, 1972

Hughes, J.F., Bright, A.W., Makin, B. and Parker, I.F. 'A study of electrical discharges in a charged water aerosol'. J Phys D. Appl Phys, 6, 966-976, 1973

Hughes, J.F., Corbett, RP., Bright, A.W. and Bailey, A.G. 'Explosion hazards and diagnostic techniques associated with powder handling in large silos'. Inst Phys Conf Series 27, Chapter 4, 264-275, 1975

Jonassen, N. Statisk elektricitet og gulvbelaegninger. Laboratet for Teknisk Fysik, Danmarks Tekniske Hojskole, 1978

Japanese Research Institute of Industrial Safety, 1971. Prevention of electric charge on industrial clothing by means of conducting fibres. Translated by ICI Fibres Division

Jowett, C.E. Electrostatics in the Electronics Environment. MacMillan, London, 1976

Martin, D.H., Radford, R.D. and Lea, K.R. 'A shock free carpet system at 10% relative humidity'. Modern Textiles, 76-79, April 1971

Murasaki, N. and Okukubo, A. 'Some effects of static charges on the human body and possible remedies' 3rd Shirley Institute International Seminar 'Textiles for comfort', Manchester, June 1971

Potter, A.E. and Baker, B.R. Static electricity in the Apollo spacecraft' NASA Tech Note TND 5579, December 1969

Quimby, W.S., Culbertson, T.L. and Mahley, H.S. 'Report of the task force on the potential hazards of synthetic clothing'. Texaco R&D Technical Report to the Static Electricity Committee of the American Petroleum Institute, 1968

Quinton, A. 'Safety procedures in operating theatres'. Br J Phys. Supplement No. 2, 92-94, 1953

Ray, C.T. 'Influence of static electricity on paper printability'. Committee Assignment Report No. 62. Tech Assoc of Pulp and Paper Industry, Atlanta, 1975

Textile Monthly, 'How KI developed a nylon staple for woven carpets'. 81-82, March 1972

Thompson, R.E. 'Electrostatic safety in clothing'. Fire Journal, 15-16, November 1969

Tunmore, B.G. 'Antistatic agents for use on carpets'. ERA Report 5017, September 1963

Veghte and Millard. 'Accumulation of static electricity in arctic clothing'. US Air Force Report AAL-TDR-63- 12, 1963

Valko, E.I., Resoro, G.C. and Ginilewicz, W. 'Elimination of static electricity from textiles by chemical finishing'. Proc Am Assoc Text Cherm Col, 47, 403-409, 1958

Wilson, N. 'The static behavior of carpets'. Text Inst Ind, 10, 235-239, 1972

Wilson, N. 'The risk of fire or explosion due to static charges on textiles'. Shirley Institute Report for the Ministry of Defence, SCRDE 22/286/13/473: MOD Project No. A/78/CL0/47236/CB/CT/4B, 1973

Wilson, N. 'The risk of fire or explosion due to static charges on textiles'. Shirley Institute Report 22/286/13: MOD Contract A/7OIGEN/10004, May 1975

Wilson, N. 'The nature and incendivity of spark discharges from the body'. Shirley Institute Report 21/13/473: MOD Project A/78/CLO/47236/CB (CT) 4B, June 1976

Chapter 7
Mold & Mildew

By Lew Harriman & Berlin Nelson

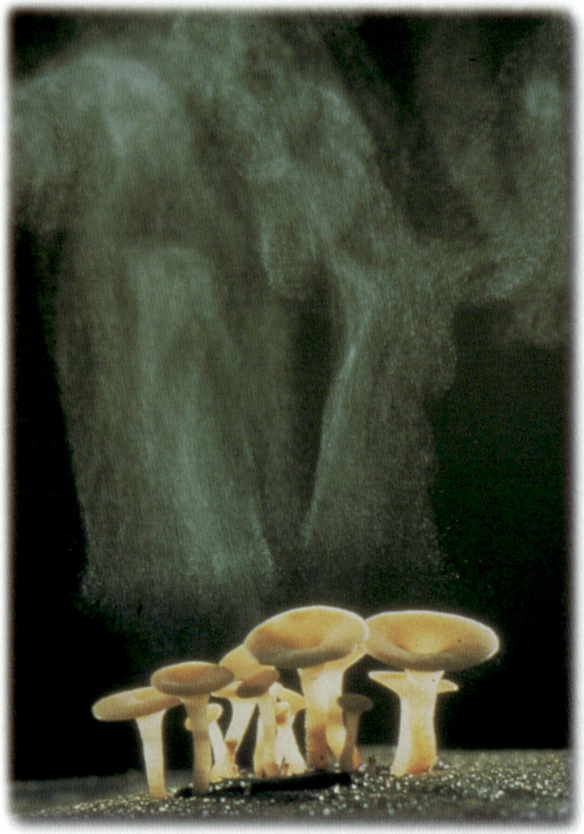

Fig. 7.1 Clouds of Fungal Spores

To reproduce, fungi release millions of spores. A single spore can regenerate the fungus, provided that it lands on a food source which has enough surface moisture to begin the growth cycle.

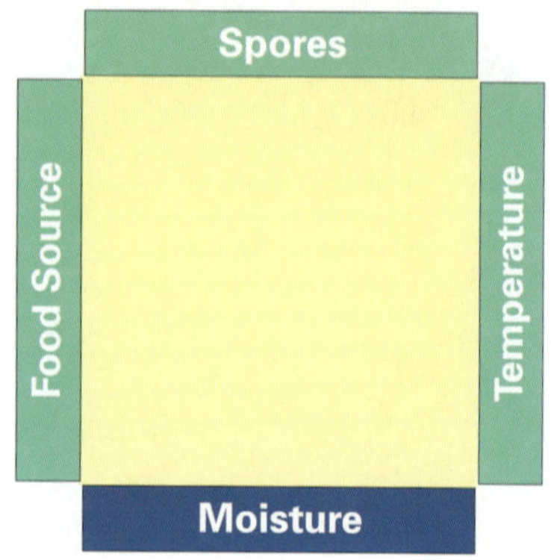

Fig. 7.2 "The Mildew Square"

This shape reminds designers that all fungi need four elements to grow and thrive. Of these, moisture in the food source is the most universal—and most preventable—component. (Peart 1989)

Basics of Mold Growth

When you walk into a building and smell a musty odor, you can be certain that mold and mildew are present in the building envelope, or in the duct work, or both. By the time the odor becomes perceptible, the problem is well-established. By understanding how a fungus grows, the HVAC designer can take steps to help avoid the problems these organisms create in buildings.

Moldis everywhere

Mold and mildew are fungi, which are essential to life on the planet earth. Without fungi, we would quickly be buried under a thick layer of dead tree trunks and leaves. Fungi pull apart all carbon-based debris, reforming the component particles into smaller molecules which can be digested by other life forms. Mycologists—those who study fungi—are still discovering new forms of fungus. At present, they know there are at least 64,000 varieties, and that these thrive on substances as unexpected as jet fuel and oil in fingerprints. In seasons which favor fungal reproduction, air outdoors is likely to contain 3,000 to 5,000 colony-forming units (cfu's) per cubic meter. In scrapings from inside air ducts, counts of 1 to 2 million spores per milligram of dust are common. Consequently, although filters can limit their number, some fungal spores are present in all buildings at all times.

Conditions needed for mold growth

In order to grow, mold needs four elements:

1. One viable spore
2. An acceptable temperature range
3. A usable food source
4. Adequate moisture in the food source

Fig. 7.3 Mildew In Books
When humidity is high, moisture is adsorbed into books. Moisture can build up, providing fungus with the solvent needed to dissolve the surface of the paper and promote growth.

These elements have been described graphically as a "Mildew Square" by Dr. Virginia Peart and Dr. James Kimbrough of the University of Florida. (1989, 1991) That graphic is shown in figure 7.2.

Spores are everywhere. Sooner or later, the right spore for the available food source will find its way to every surface.

Temperature is quite important. Each fungus has an optimal range of temperatures that allow growth. Beyond that range, growth is slow or impossible for that particular fungus, and other fungi will take over the food source. Indeed, when temperature changes kill the first fungi or make them dormant, other fungi will feed on the original fungi as well as on the abandoned food source.

The nature of the **food source** affects the type of fungi that can use it to grow. Thousands of fungi are equipped to digest cellulosic material such as wood, paper and cardboard. A smaller number of fungi are adapted to make use of organic material in oils, plastics or dust that collects on normally sterile surfaces such as metals and glass. For every carbon-based material, there is some form of fungus that can use it as a food source.

Of all essential elements, **moisture** is the simplest and most universal requirement for fungal growth. Without moisture, fungus cannot access the nutrients in its food sources. On dry surfaces, spores will remain dormant until adequate moisture appears.

In moderate temperature ranges and on easily-digestible food sources, many fungi will compete. One way they can compete is through "chemical warfare." Many fungi emit liquids and volatile organic gases that suppress the growth of their competitors. In some cases, the "musty odors" perceptible to humans reflect this fungal competition.

A dramatic example of the importance of moisture and the longevity of spores is seen in the desert tombs of the ancient Egyptians. Pigments that had been safe for many millennia deteriorated in a matter of a few years when moisture from human respiration increased humidity at the painted surfaces.

Mechanics of mold growth

Mold growth proceeds in 4 stages. The first is the most important in relation to humidity control in buildings and HVAC systems.

Stage 1 - Create nutrient broth

First, a spore lands on a food source. That spore remains dormant until the food source absorbs enough moisture at its surface to dissolve the hygroscopic enzymes that cover the surface of the spore. In this first stage of growth, moisture from the food dissolves the enzymes, which then act as corrosive catalysts, dissolving the food. Then the liquid nutrient broth under the spore is pulled back into the spore through a difference in osmotic pressure. The liquid nutrient outside the spore is less concentrated than the dry matter inside the spore. So liquid nutrient diffuses through the spore's exterior wall to equalize concentrations on both sides of that spore wall. *At this first stage of growth, the fungus is easily controlled. As long as the food source lacks sufficient moisture, the enzymes do not dissolve, so nutrients are not available to feed fungal growth.*

Stage 2 - Generate more absorptive surface

When sufficient moisture is available at the surface of the food, growth proceeds to the second stage. The fungus begins to extend its reach for nutrients by growing filaments called hyphae. As more nutrient solution is drawn into the hyphae, more filaments are grown, eventually forming a dense mat called the mycelium. Towards the end of this stage, the organism may become visible to the unaided human eye. Then, as the mycelium thickens it behaves as a vapor retarder, preventing the food surface from drying out, which would slow or halt growth.

Stage 3 - Generate metabolic water

After the mycelium thickens, the fungus can generate and retain enough moisture through its normal metabolism to maintain growth. As food is metabolized, moisture is produced. This water combines with more enzymes to dissolve additional food. At this third stage, lack of moisture from the air will not limit growth. Fungus will grow

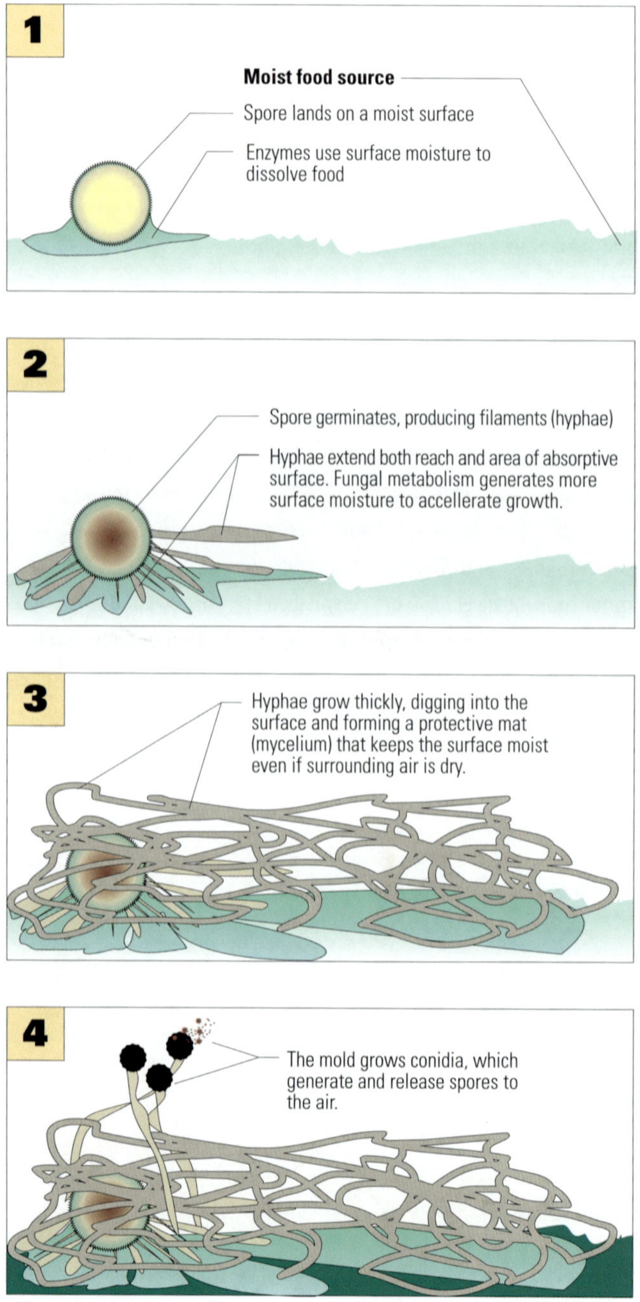

Fig. 7.4 Stages of Mold Growth

In early stages, the food source must be moist to initiate growth. In later stages, fungal metabolism supplies the needed moisture. (Kimbrough 1990)

until it consumes the food, or until the temperature changes, or until it is killed by some other means such as ultraviolet light or a toxic chemical such as bleach. Unfortunately, at this stage, the hyphae extend *below* the surface into the bulk of the food. So even if fungal bodies on the surface are killed, the organism can regenerate from its components buried deep inside the food. This explains why remediation is so difficult after mold growth is firmly established. Surfaces that appear clean can re-grow mold when moisture returns to the food source.

Stage 4 - Reproduce

After the third stage of vigorous growth, the fungus has the resources necessary to reproduce. It forms fruiting bodies which generate spores by the millions. These are released to the air, often accompanied by perceptible vapors—the "musty odors" so common to HVAC systems in humid climates. Spores drift through the air to other food surfaces. The cycle repeats when the new food source has sufficient moisture to dissolve the spores' external enzymes.

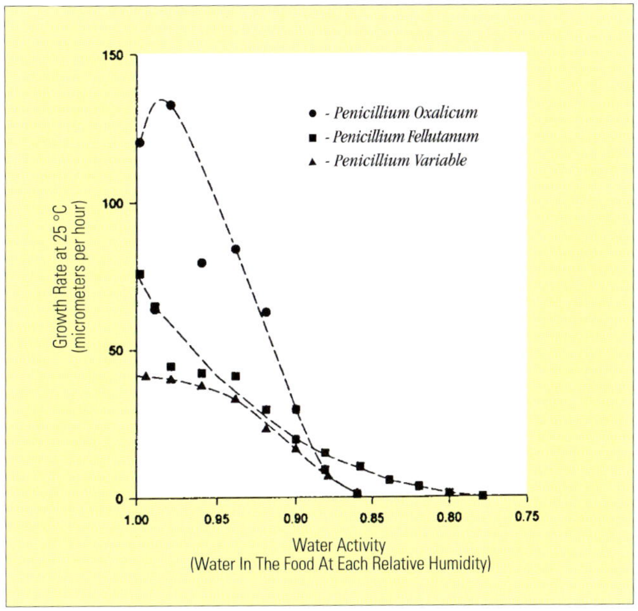

Fig. 7.5 Growth Rates

Vary by type of fungus. Each type thrives in a slightly different range of moisture content and temperature. (Flannigan & Miller 1993)

Controlling humidity to limit fungal growth

Given the fungal growth cycle, it's clear that keeping moisture out of food sources is an excellent way to prevent fungal growth from becoming established. Preventing actual condensation prevents the growth of the most hazardous species of mold, but there are thousands of fungi that do not need liquid water to germinate. No matter how the moisture arrives at the surface, the relative water content of the food, called the "water activity" is what limits mold growth.

Water activity describes the amount of water adsorbed by the specified material when it is in equilibrium with air at a given relative humidity. The term is somewhat confusing, because the mass of water it describes is not absolute. It varies widely, governed by the sorption characteristics of each material. For example, a "water activity" of 0.75 describes the amount of water absorbed by the material when the relative humidity of the air above its surface is 75%. But the mass of water absorbed in cotton at that humidity is 7% of the weight of the dry material, while wool adsorbs 13% of its weight at the same condition. The "water activity" for both materials is still said to be 0.75 when they are in equilibrium with air at 75% relative humidity.

Mycologists think in terms of water activity. HVAC designers think in terms of relative humidity. The two disciplines are very close to using the same concepts. However, the key point for HVAC designers to remember is that fungal growth is not directly governed by the relative humidity as reported by building controls. Air moisture is only relevant as it might influence the amount of water *absorbed into the material* used by the fungus as a food source. Several important implications become clear when this fact is understood.

For example, the HVAC designer will not make the common mistake of assuming that if relative humidity in the controlled space is 50%, the building is safe from mold growth. What really matters is the relative humidity at the *surface of the food*—which will be in the ductwork, the carpet or behind wall coverings and inside building cavities, not in the air at the center of the room where the humi-

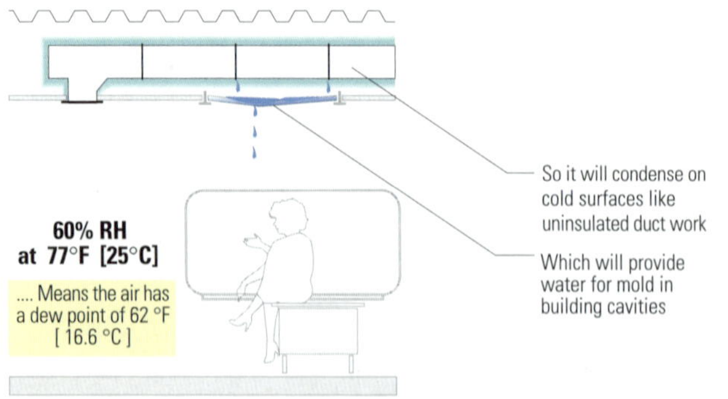

Fig. 7.6 Never Mind The rh... What's The Dew Point?
Cold surfaces and high dew points combine to produce condensation, which create more frequent and serious mold growth than simple high relative humidity.

distat might be located. Also, a central control system reporting a pleasant, safe relative humidity level of 50% tells the HVAC designer nothing useful—unless he is certain that there are no cold surfaces that will bring moisture out of the air and into the food source through condensation.

Further, the HVAC designer will realize that, when humidity rises to high levels every day, the air may be slowly "pumping moisture" into food sources. For example, library books absorb moisture but do not release it easily. When absolute moisture and temperature remain high for extended periods, such as when air conditioning is turned off during school vacations, the extra moisture needed to support fungus is added to the paper. With more moisture and an optimal warm temperature, mold growth appears to suddenly "explode" even though the relative humidity *in the room* may stay below the threshold limit for much of the day. Once started, fungal growth continues because the mycelium prevents moisture from leaving the material even when the humidity in the room is reduced. (Figure 7.3 shows the results of raising a library thermostat set point when there was no dehumidifier installed to keep the rh below 50% at all times.)

Also, the designer will not make the mistake of believing, as household wisdom might suggest, that preventing "stagnant air" will automatically prevent mold growth. If flowing air is humid, it will add moisture to materials, actually encouraging mold growth. But when the air is dry, household wisdom is accurate. Flowing dry air will retard mold growth because it removes moisture from the food.

In any case, the HVAC designer will not go wrong if he or she focuses clearly on the *material moisture content*. The relative humidity and dew point of the air are only relevant as they affect the moisture content of the fungal food source.

Molds Common In Commercial Buildings

Although there are over 645,000 species of fungus known to Mycologists, only a few dozen are thought to be both important and common in buildings. Fungi all have specific temperature, moisture and food source preferences, so it follows that the fungi that thrive in buildings will have preferences which reflect the characteristics of that environment.

Mycologists separate molds into informal categories according to their preferences for temperature and moisture. Xerophilic (dry-loving) molds do well at lower moisture levels than hygrophilic (moisture-loving) molds. Similarly, thermophilic (heat-loving) molds tolerate or thrive at temperatures above the limit of the human comfort range. Mesophilic fungi overlap the comfort range at both ends, and psychrophilic types prefer temperatures just above the freezing point of water. From these characteristics, it follows that as moisture, temperature and food sources change in a building and its HVAC systems, each species will compete more effectively at different times and in different places.

Fig. 7.7 Fungal Demographics Change With Water Availability
As the moisture content of any food source changes, different fungi become more successful competitors. In a building, the moisture content and temperature of each material may change with the seasons, allowing a variety of fungi to thrive at different times. (Flannigan & Miller 1993)

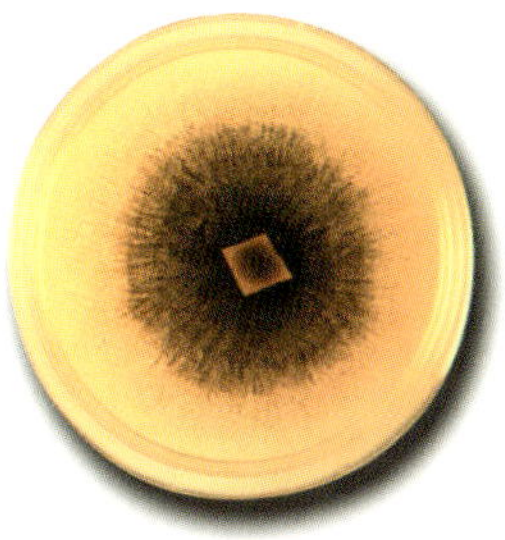

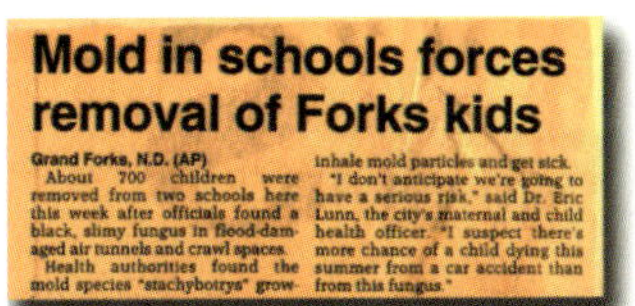

Mold in schools forces removal of Forks kids

Grand Forks, N.D. (AP)

About 700 children were removed from two schools here this week after officials found a black, slimy fungus in flood-damaged air tunnels and crawl spaces.

Health authorities found the mold species "stachybotrys" grow-

inhale mold particles and get sick.

"I don't anticipate we're going to have a serious risk," said Dr. Eric Lunn, the city's maternal and child health officer. "I suspect there's more chance of a child dying this summer from a car accident than from this fungus."

Courtesy of the Fargo Forum

Fig. 7.8 Stachybotrys chartarum

This fungus is toxic to humans and animals, and has been implicated in health problems associated with condensation and water damage in buildings.

Figure 7.7 shows the species of mold which usually colonize plywood and composition board and the paper facing of gypsum board. One quickly observes that the "demographics" of the fungal population change dramatically with relatively small changes in temperature and moisture. This means that a building envelope, which contains not one but dozens of different combinations of temperature and moisture, will provide an ideal environment for dozens of fungi simultaneously. Each species may be dormant during part of the year or in part of the building, but thrive at other times and in other places.

Not all fungi are of equal concern to the designer and the occupants of the building. Many fungi grow slowly, so that ill effects take years or decades to cause problems. Other fungi have negligible effect on either health or structural strength of the building. Also, the effects of some fungi are simply unknown at present. In general, human health is of somewhat more immediate concern than structural problems, because health is affected more quickly than the building's structure.

The heath effects of fungi are currently in debate. Mycologists, physicians, public health professionals and attorneys are in constant conflict about the nature, mechanisms and dimensions of human health effects from fungal infection of buildings and HVAC systems. *Aspergillus*, *Penicillium* and *Stachybotrys* all produce toxic compounds and have all been implicated in health problems in buildings, but conclusive evidence is hard to identify. (Burge 1987, Fink 1976, Miller 1992, Staib 1987)

What is *not* in debate is that mold growth in buildings is neither healthy nor structurally wise, and the HVAC designer and architect should avoid designs that promote fungal growth. Still in debate is the *dimension* of the problems they cause, and to what degree fungal growth can be tolerated. The following section describes one of the fungi most frequently implicated in health problems, and the involvement of one plant pathologist, Dr. Berlin Nelson of the North Dakota State University, with problems associated with that fungus.

Stachybotrys chartarum (formerly Stachybotrys atra)

Berlin Nelson, Ph.D.
Department of Plant Pathology, North Dakota State University at Fargo

This monograph appeared in February 1999 as an feature article on apsnet.org. It is republished here through permission of Dr. Nelson, who retains the copyright.

Stachybotrys chartarum is a fungus that has become notorious as a mycotoxin producer causing animal and human mycotoxicosis. Indeed, over the past 20 years in North America, evidence has accumulated implicating this fungus as a serious problem in homes and buildings and one of the causes of the "sick buildings syndrome." In 1993-1994, there was an unusual outbreak of pulmonary hemorrhage in infants in Cleveland, Ohio, where researchers found *S.chartarum* growing in the homes of the sick infants. This incident has increased the awareness of home/building molds and brought this fungus to the attention of the medical community. In May 1997, for example, the *Journal of the American Medical Association* carried a news article titled "Floods carry potential for toxic mold disease" (Marwick, 1997). Newspaper articles such as "Fungus in "Sick' Building" (*New York Times*, May 5, 1996) or "Mold in schools forces removal of Forks Kids" (*Fargo Forum*, June 1997) and the closure and extensive renovation of several courthouses in Florida, due to molds, are recent testimonials to the importance of indoor molds to human health.

As a plant pathologist-mycologist I became involved with the issue of home and building molds because of the enormous flood in the Red River Valley in 1997. In Grand Forks, ND, over 9,000 homes were flooded and a major part of the town was under water from the Red River. Immediately after the flood I became a member of an advisory group (that included scientists from the Center for Disease Control, National Institute for Occupational Safety and Health, US Public Health Service, and the Environmental Protection Agency), to help city officials deal with microbial problems associated with a major flood. After several years of dealing with this issue, I am struck by the common occurrence, and sometimes the extensive growth of

S. chartarum, and the lack of knowledge by the general public and the health care community about this fungus. This is not to imply that it is a widespread and serious problem, but there is a public need for readily available information on molds. Because *S. chartarum* is toxic, it merits special attention.

The Fungus - *S. chartarum*

Stachybotrys chartarum (synonyms = *S.atra, S.alternans*) was first described as *S.atra* by Corda in 1837 from wallpaper collected in a home in Prague. It is a member of the Deuteromycetes, order Moniliales, family Dematiaceae, and is common on plant debris and in soil. The taxonomic treatment of the genus by Jong and Davis, 1976, is a good reference on identification. The fungus grows well on common mycological media such as potato dextrose (PDA) or cornmeal agar, and sporulates profusely forming dark masses of conidia (figure 7.8).

The fungus is relatively easy to identify because of the unique phialides of the genus and conidial morphology of the species. Conidiophores are determinate, macronematous, solitary or in groups, erect, irregularly branched or simple, septate, dark olivaceous, and often rough walled on the upper part. The phialides are large, 9 to 14 µm in length, in whorls, ellipsoid, olivaceous, and often with conspicuous collarettes. Conidia are ellipsoidal, unicellular, 7-12 x 4-6 µm, dark brown to black and often showing a ridged topography when mature. The ridged nature is readily apparent with scanning electron microscopy, but can also be observed with an oil immersion lens at 1000 x. (figures 7.9 and 7.10) At lower power magnification the spores appear verrucose. Young spores and some mature spores may be smooth.

Fig. 7.9
Conidia of S. chartarum

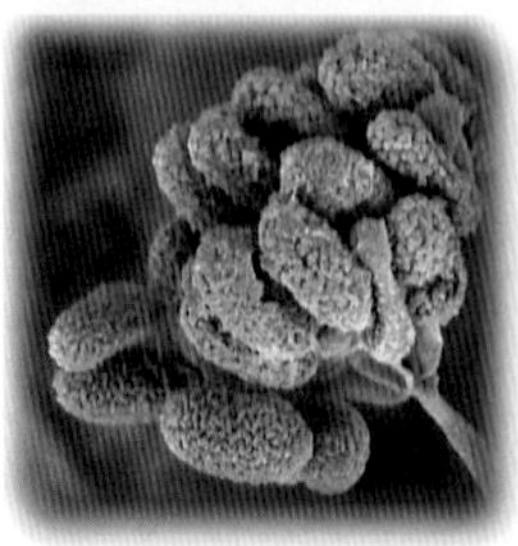

Fig. 7.10
Closeup of Conidia

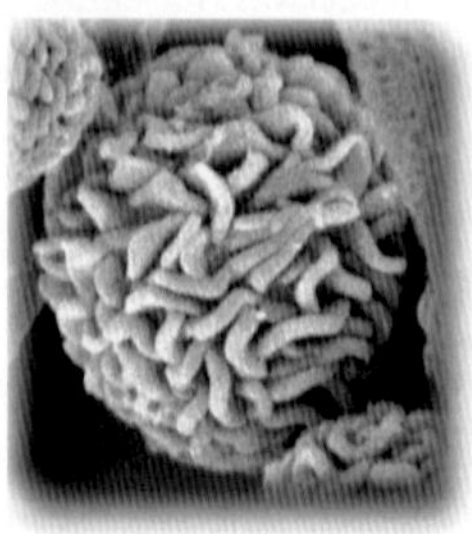

The phialides produce conidia singly and successively into a slime droplet that covers the phialides. Eventually the slime dries and the conidia are covered with the slime residue and remain on the conidiophore as a mass or ball of spores. The spores are, therefore, not readily disseminated in the air compared to other fungi such as Aspergillus. However, when the fungus and substrate dries and is disturbed by agitation or air movement, conidia can become bioaerosols.

S.chartarum growing on natural or man made substrates can often be identified by a person familiar with its growth pattern. However, there are some very dark dematiaceous Hyphomycetes which look similar. Therefore microscopic examination of the fungus is needed to confirm identification. When the fungus is freshly growing, the characteristic phialides and conidia are easy to observe, but when dry, the phialides collapse, are more difficult to observe, and emphasis must be placed on morphology of conidia. The fungus is strongly cellulolytic and will grow under conditions of low nitrogen. A simple way to grow the fungus is to streak some conidia onto wet Whatman filter paper in a petri dish and within a week spores are produced. If you make a small ridge with the paper and place spores on it, the conidiophores will grow at an angle and allow a side view of conidial formation with a stereoscope. This is a convenient method to determine if spores are in chains to distinguish *Stachybotrys* from *Memnoniella*. Also, the filter paper method will allow isolation of *S.chartarum* away from many other fast growing, but non-cellulolytic fungi, that would out-compete *S. chartarum* on rich media.

Mycotoxins produced by *S. chartarum*

Mycotoxins and other biologically active compounds produced by *S.chartarum* are the reason why this fungus is of concern to human health. Mycotoxin poisoning by this fungus is referred to as stachybotryotoxicosis. *Stachybotrys chartarum* produces six types of macrocyclic tricothecenes: roridin E, satratoxins F, G and H, trichoverrols, trichoverrins, verrucarol and verrucarin J. The satratoxins are generally produced in greater amounts than the other trichothecenes, but all compounds are produced in low quantities. They occur in all parts of the fungus. One isolate was reported to contain about 15 ppm trichothecene in the conidia. The difficulty in obtaining, identifying and purifying these toxins has slowed studies on their biological activity. Macrocyclic trichothecenes are highly toxic.

In addition, the fungus produces 9-phenylspirodrimanes (spirolactones and spirolactams) and cyclosporin which are potent immunosuppressive agents. Jarvis et al. (1995) suggested that the combination of trichothecenes and these immunosuppressive agents may be responsible for the observed high toxicity of this fungus.

History of problems with *S. chartarum*

In the Ukraine and other parts of eastern Europe during the 1930's, there were unusual outbreaks of a new disease in horses and other animals that was characterized by symptoms such as shock, dermal necrosis, leukopenia, hemorrhage, nervous disorders and death. In 1938, Russian scientists determined the disease was associated with *S.chartarum* (then known as *S.alternans*) growing on the straw and grain fed to the animals (Fig.7.11)

Fig. 7.11 Stachybotrys on Hay
In addition to wall board facing, the fungus colonizes wet hay easily, which then represents a hazard for animals and for agricultural workers.

Intensive studies were then conducted resulting in the first demonstrated toxicity of *S. chartarum* in animals. Horses were actually fed cultures of the fungus. The contents of one petri plate resulted in sickness and the contents of 30 petri plates resulted in death. Horses seem to be especially susceptible to these toxins; 1 mg of pure toxin is reported to cause death. The Russians coined the term stachybotryotoxicosis for this new disease. Since then, stachybotryotoxicosis has been reported on numerous farm animals from various parts of the world, especially in eastern Europe, but is apparently rare (if even reported) in North America.

In the early 1940's reports of stachybotryotoxicosis in humans appeared in Russia. People affected where those who handled or were in close contact with hay or feed grain infested with *S.chartarum*. Some of these individuals had burned the straw or slept on straw-filled mattresses. Common symptoms in humans were dermatitis, pain and inflammation of the mucous membranes of the mouth and throat, a burning sensation of the nasal passages, tightness of the chest, cough, bloody rhinitis, fever, headache and fatigue. Some members (volunteers?) of the teams investigating this disease rubbed fructifications of the fungus onto their skin to determine its direct toxicity. The fungus induced local and systemic toxic symptoms similar to those observed in naturally occurring cases.

Between the 1950's and the 1980's there were continued publications on stachybotryotoxicosis but few that indicated a potential problem with *S.chartarum* in homes and buildings. In 1986, Croft et al. reported an outbreak of trichothecene toxicosis in a Chicago home. Over a 5-year period, the family complained of headaches, sore throats, flu symptoms, recurring colds, diarrhea, fatigue, dermatitis and general malaise. Air sampling of this home revealed spores of *S.chartarum*. The fungus was found growing on moist organic debris in an uninsulated cold air duct and on some wood fiber ceiling material. *The home had a chronic moisture problem that favored mold growth. Extracts from the duct debris and contaminated building materials were toxic to test animals and several macrocyclic trichothecenes were identified in the extracts. When the mold problem was corrected, these symptoms associated with trichothecene toxicosis disappeared.*

Since the paper by Croft et al (1986), there have been numerous reports of *S.chartarum* in homes/buildings in North America, but few definitive studies implicating the fungus as the cause of mycotoxicosis. One important paper by Johanning et al. (1996), reported on the health of office workers in a flooded New York office building with high concentrations of *S.chartarum* on sheetrock. The study concluded "... self-reported health status indicator changes and lower T-lymphocyte proportions and dysfunction as well as some other immunochemistry alterations were associated with onset, intensity and duration of occupational exposure to toxigenic *S.chartarum* combined with other atypical fungi."

In 1993-1994, a cluster of cases of pulmonary hemorrhage and hemosiderosis in infants occurred in Cleveland, Ohio. Because this is rarely observed in infants an intensive study into the cause of the problem was initiated. There were several factors associated with this outbreak, but an important finding was that all homes of these infants had high levels of total fungi and *S.chartarum* (based on air and surface sampling) (Etzel et al., 1998). Furthermore, isolates of *S.chartarum* from the homes were shown to produce trichothecenes. The homes had previously sustained water damage, which resulted in the mold contamination. This incident brought *S. chartarum* to the attention of the medical community and has highlighted the potential problem with indoor molds. There is controversy, however, about the role of *S.chartarum* in human poisoning in the Cleveland incident and as it relates to indoor molds in general. There have been few toxicological or epidemiological studies; thus some members of the scientific community believe a solid causal relationship has not been firmly established. The Centers for Disease Control recently issued a report critical of the study conducted in Cleveland.

The levels of exposure to *S.chartarum* required to cause human mycotoxicosis, especially from the inhalation of spores, are unknown. We should remember, however, that the accumulation of data over the past 60 years tells us that one should not handle materials contaminated with *S.chartarum* without proper safety procedures. Experience also strongly indicates that indoor environments contaminated with *S.chartarum* are not healthy, especially for children. A useful web site about the Cleveland incident is: http://gcrc.meds.cwru.edu/stachy/default.htm

Where *S. chartarum* occurs indoors

Stachybotrys chartarum is most commonly found in homes or buildings which have sustained flooding or water damage from broken pipes, roof leaks, sewage backup, condensation, etc. Spores are found in the soil and introduced along with floodwaters or dust and dirt entering with water from a roof leak, etc. It is most common on the paper covering of sheetrock (fig. 7.12) but can be found on wallpaper, ceiling tiles, paper products, carpets with natural fibers, paper covering on insulated pipes, in insulation material, on wood, and on general organic debris. Because leaks can occur behind walls and in covered ceiling areas, the fungus may grow profusely, but not be readily visible. It can also be found in ducts if organic debris is present.

Wet conditions are required to initiate growth. The fungus will usually produce large amounts of conidiophores and conidia giving the substrate a black appearance that can be slightly shiny when fresh and powdery when dry. I have observed fungus growing profusely on sheetrock a week after floodwater was drained from a building.

Detection of *S.chartarum* is usually by visual inspection and/or air and surface sampling. Because this fungus is not readily airborne, air sampling in a contaminated indoor environment may show low levels of spores in the air. Inspection of potential sites of contamination, especially in covered/protected places, is necessary to determine where the fungus occurs and the level of contamination. Fungus is often hidden in the walls, ceiling or floor with no visible evidence within the room interior. However, spores can escape through small holes and cracks to contaminate the room air.

If areas contaminated with *S.chartarum* are discovered, do not attempt to solve the problem without following recommended safety procedures for working with toxic molds, especially if contamination is heavy. Workers should use powered air purifying respirators (PAPR) along with proper covering of the skin and eyes. Get advice if there is a serious problem. Disinfecting the surface of contaminated materials, a common reaction to deal with molds, may kill the fungus on the surface, but mycelium within the substrate will often

Fig. 7.12 Stachybotrys In Buildings
Is common after water damage and when condensation drips inside walls. The mold is hazardous. Remediation workers and occupants must take care to avoid inhaling spores, and skin contact should also be avoided.

survive and grow again. Also, mycotoxins may accumulate in contaminated material. Removing contaminated materials is usually the best option. **To prevent further mold development, correct the moisture problem or *S. chartarum* might return!**

Acknowledgments: Thanks to the Electron Microscopy Center at North Dakota State University for the scanning electron micrographs. All other photos are by Berlin Nelson.

Selected references concerning stachybotrys chartarum

Croft, W. A., Jarvis B. C., and Yatawara, C. S. 1986. Airborne outbreak of trichothecene toxicosis. Atmos. Environ. 20:549-552.

Etzel, R., Montana, E., Sorenson, W., Kullman, G., Allan, T., and Dearborn, D. 1998. Acute pulmonary hemorrhage in infants associated with exposure to *Stachybotrys atra* and other fungi. Arch. Pediatr. Adolesc. Med. 152:757-762.

Forgacs, J. 1972. Stachybotryotoxicosis. Pages 95-128 in: Microbial Toxins. S. Kadis, A. Ciegler, and S. Ajl, eds. Volume 8, Fungal Toxins. Academic Press, New York.

Fung, F., Clark, R., and Williams, S. 1998. *Stachybotrys*, a mycotoxin-producing fungus of increasing toxicologic importance. J. Toxicol. Clin. Toxicol. 36:79-86.

Jarvis, B., Salemme, J., and Morais, A. 1995. Stachybotrys toxins. 1. Natural Toxins 3:10-16.

Jarvis, B., Sorenson, W. G., Hintikka, E., Nikulin, M., Zhou, Y., Jiang, J., Wang, S., Hinkley, S., Etzel, R., and Dearborn, D. 1998. Study of toxin production by isolates of *Stachybotrys chartarum* and *Memnoniella echinata* isolated during a study of pulmonary hemosiderosis in infants. Appl. Environ. Microbiol. 64:3620-3625.

Johanning, E., Biagini, R., Hull, D., Morey, P., Jarvis, B., and Landsbergis, P. 1996. Health and immunology study following exposure to toxigenic fungi (*Stachybotrys chartarum*) in a water-damaged office environment. Int Arch. Occup. Env. Health 68:207-218.

Johanning, E.,and Yang, C. S., eds. 1995. Fungi and bacteria in indoor environments. Proceedings of the International Conference, October, 1994, Saratoga Springs, New York. Eastern New York Occupational Health Program, Latham, NY.

Jong, S. C., and Davis. E. E. 1976. Contribution to the knowledge of *Stachybotrys* and *Memnoniella* in culture. Mycotaxon 3:409-485.

Marwick, C. 1997. Floods carry potential for toxic mold disease. JAMA 277:1342.

Sorenson, W. G., Frazer, D. G., Jarvis, B., Simpson, J., and Robinson, V. A. 1987. Trichothecene mycotoxins in aerosolized conidia of *Stachybotrys atra*. Appl. Environ. Microbiol. 53:1370-1375.

Wyllie, T. D., and Morehouse, L. G., eds. 1978. Mycotoxic Fungi, Mycotoxins, Mycotoxicoses - An Encyclopedic Handbook. Volume 2. Marcel Dekker, Inc, New York.

Mold In The Building Envelope & Contents

As the previous section suggests, the most serious problems with fungi in buildings occur when large amounts of liquid water are available along with warm temperatures, as occurs after floods, fires, broken pipes or water leaks in the building envelope. None of these are within the control of the HVAC designer.

However, the HVAC designer often affects mold and mildew problems in the building envelope and its contents. Common mold and mildew symptoms influenced by the HVAC design decisions more than by water leaks can include:

- Purple, pink or black stains on wall covering
- Mold & mildew in attics, behind pictures, in building cavities and on ceiling tile
- Mold & mildew on books, carpets and furnishings

High humidity is a source of water for fungus, but droplets carry more water into the food more quickly. So condensation is a special concern. This is true whether the problem occurs in the ducts themselves, or in the building envelope. Reviewing the mechanism of each problem will help the HVAC designer make better decisions.

Purple, pink or black stains on wall covering

This symptom is visible evidence of the presence of mold and mildew behind the wall covering. Stains appear on the front side because fungal enzymes have dissolved some of the pigments in the material, and the pigmented liquid has soaked through to the front. Stains take the form of either black dots, or more fluid-like marks that look like grape juice or dark coffee had been spilled on the wall.

Staining often occurs in humid climates when the building wall is under negative pressure. The classic example is in hotels. Continuous toilet exhaust creates a very slight suction on the guest room. Even though the corridor may be under positive air pressure, the exterior wall is often negative with respect to the outdoors. So humid air is pulled into the building cavities. Then the humid air is trapped, prevented from entering the room by impermeable vinyl wall covering. When the room air conditioner turns on, it chills the wall surface below the dew point of air trapped in the walls. So the water condenses in the adhesive and paper wallboard facing behind the wall covering, and mold begins to grow, eventually staining the walls and generating musty odors. This mechanism explains why stains are often located near wall-mounted cooling units, or on exterior walls, or on the demising walls that separate rooms but which connect to the exterior wall spaces. (AH&MA 1990 and 1991, Odom 1992, Gatley 1990, Shakun 1990)

A continuous negative air pressure of even as little as 0.008" [2 pa] has been shown repeatedly to cause enough humid air infiltration to cause millions of dollars of water damage in walls of hotels. (AH&MA 1991, Odom & Dubose 1992)

The way to avoid this problem is to *seal all exhaust and return air ductwork*, so that exhaust fans do not accidentally pull suction on the building cavities that the ductwork passes through. Also, each room needs a *direct* feed of *dry* makeup air, so that the exterior wall

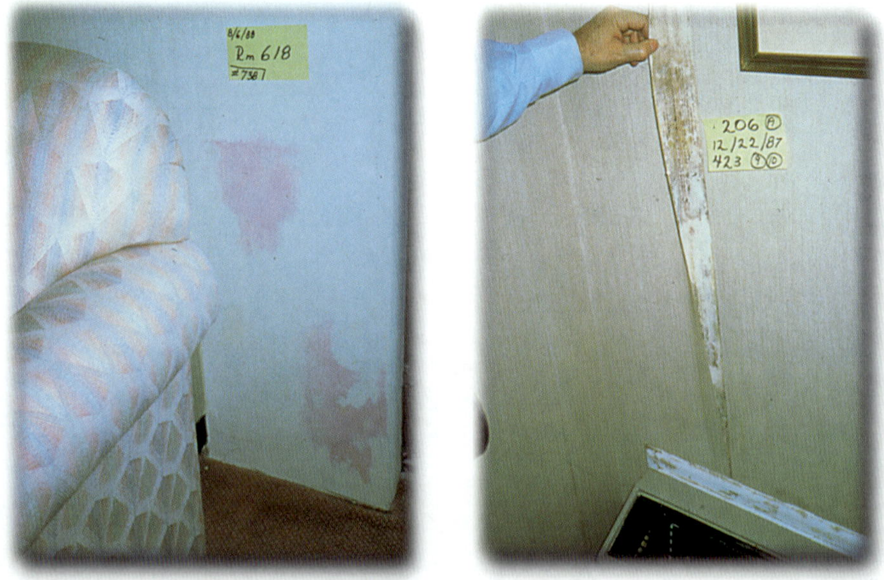

Fig. 7.13
Pink/Purple Stains and Black Spots
Stains are often evidence of mold growth. Humid air in building cavities condenses behind impermeable wall covering. With that moisture, fungi grows in the wallboard paper facing, and in the adhesive that attaches the wall covering. Stains are the result of fungal enzymes dissolving pigments in the wall covering. Black spots are evidence that mold is growing on the exterior surface. After removal by cleaning staff, the stains will reoccur until moisture is removed from the wallboard. (Kimbrough, 1990)

Fig. 7.14 Mold On Furniture

When humid air leaks into buildings, it is often trapped by cool surfaces of furnishings pushed against the wall near an electrical outlet, a cooling unit or near other penetrations like window frames and pinholes that hold nails to support pictures. Mold grows in the moist fabric, on the wall itself or on moist wood or paper surfaces. (Kimbrough, 1990)

of each room is under positive rather than negative air pressure. Then only dry air leaks into building cavities, not humid air.

In theory the problem is reversed in cold climates. Since the air inside the building is more humid, one would not want that air to be pushed into the walls. But in fact, this is seldom a problem in hotels. These buildings are notorious for being under negative air pressure, and achieving positive air pressure in the rooms is a significant challenge. Without special attention, cold-climate hotel rooms are likely to be under negative air pressure, so humid indoor air is less likely to infiltrate into building cavities.

Mold & mildew above ceilings, on walls and ceiling tiles

In both humid and cold climates, mold can grow above ceilings and inside walls. The mechanisms and remedies are slightly different in each climate.

In humid climates, the problem can be caused by the same negative air pressure described above. Some HVAC component causes suction, humid air is pulled into the building, where it eventually provides enough moisture for mold and mildew to grow. A classic example is a simple one-story commercial office building or retail store with a dropped ceiling and no return air duct work. The ceiling plenum is expected to act as a return path for the system.

However, since the HVAC fan pulls a suction on the space above the ceiling, the entire exterior wall is under a negative pressure, which pulls humid air into the building. Then moisture condenses on cold duct work or on cold pipes inside the walls or above the ceiling. The first symptoms may include sagging ceiling tile, dark spots (mold) appearing on those tiles or musty odors that do not seem to come from the HVAC system. If air leaks are large and the suction is great, the first symptom may be condensation dripping from cold supply air diffusers onto occupants. Such problems have been reported in restaurants in humid climates. Exhaust air volumes are very large, and owners sometimes believe they can save energy by shutting off the ventilation preconditioning part of the HVAC system, not realizing that the exhaust will simply pull unconditioned air through the building wall and through the doors.

Another puzzling problem that occurs in humid climates is mold growth behind curtains, behind hanging pictures and on the backs of furniture pushed against a wall, as shown in figure 1.14. These are also caused by humid air leaking into the building because of negative internal air pressure. Humid air leaks through the inside wall at either the nail that hangs the picture, or through an electrical outlet behind a chair pushed against the wall. The picture or chair is relatively cool because it is inside an air conditioned space. So the incoming humid air is very close to saturation as it drifts across the surface of the wall or chair. Since the object is pushed against the wall, and stays put, little or no dry air from inside the room flows against the humid surface to remove moisture. So mold grows on the moist, hidden surfaces near or on the wall.

Curtains may grow mold for similar reasons. While windows are usually quite airtight, the pre-hung units are not always sealed to the exterior of the building. Problems are frequent under the sill, where a crack is difficult to see during installation. Air leaks into the room through that crack, and perhaps through similar cracks in the mounting of the though-wall air conditioner. Whenever the air conditioner cools the room, it blows cold air near the curtains, which then absorb moisture from the incoming humid leakage air. Mold grows on the back of the curtains, where the moisture content of the material is highest.

All of these problems can be avoided by ensuring that the building cavities are under positive air pressure with dry air. This demands sealed exhaust and return ductwork, so that fans do not pull a suction on building cavities. Also, dry ventilation air should be delivered directly to rooms with exterior exposures as opposed to relying on

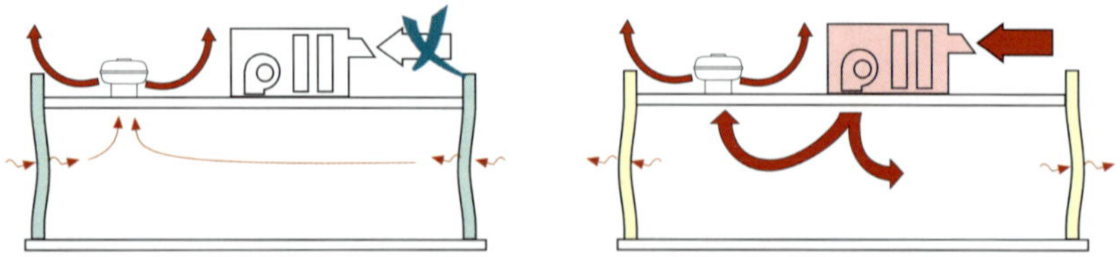

Insufficient Make-up Air
Creates negative pressure which sucks untreated air through the building wall, causing mold and mildew in building cavities.

Slight Excess of Dry Make-up Air
Ensures that treated air rather than outdoor air will fill the building cavities, keeping them safe from condensation, mold and mildew.

Fig. 7.15
Humid Climates Need Positive Indoor Air Pressure
However, in cold climates, a neutral or slight negative internal air pressure helps avoid forcing humid air into cold building cavities where it can condense to feed mold growth.

exhaust fans to pull air into external rooms from interior corridors. That tempting low-cost strategy seldom provides positive pressure at the exterior wall.

In cold climates, mold and mildew problems are caused by humid air from *inside* the building being pushed and pulled into cold building cavities. In low-rise commercial buildings in cold climates, the classic problem begins when there is *no* return duct work. The roof-mounted heating/air conditioning unit pulls a suction on both the building cavities near the exterior wall and on the conditioned space below a dropped ceiling. Cold air comes through cracks in the exterior wall, chilling those wall cavities. Humid air is pulled from the conditioned space. The moisture in the humid air condenses when it contacts cold surfaces above the ceiling, and on cold surfaces in the walls. The more water, the faster the fungus grows.

In moderate climates, condensation is not common. With smaller amounts of condensed water, and with cool temperatures, the number of fungi that can thrive is limited. But the environment may change dramatically in the spring. Rising temperature allows more forms of

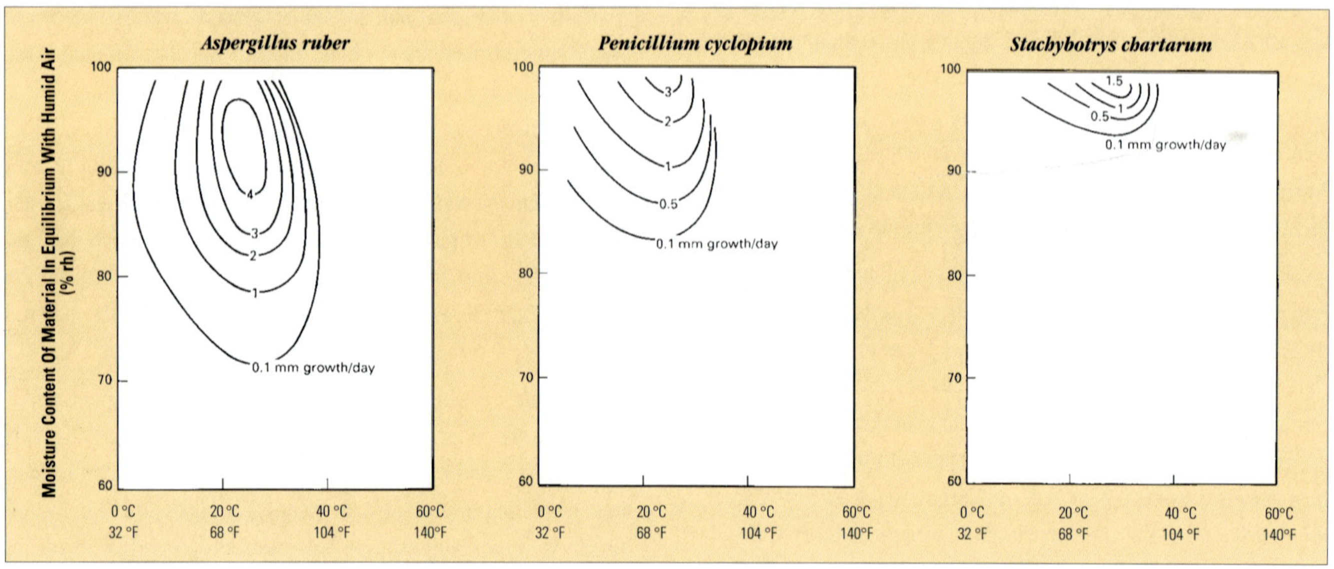

Fig. 7.16 Fungal Growth Rates Vary
Fungi grow at different rates, controlled by their individual characteristics, and by the amount of moisture and heat available to support their metabolism. (Brundrett, 1990)

Fig. 7.17 Seal All Duct Work

Sealing all the ductwork avoids suction in the building envelope, which would pull in humid air or allow it to contact cold surfaces in building cavities.

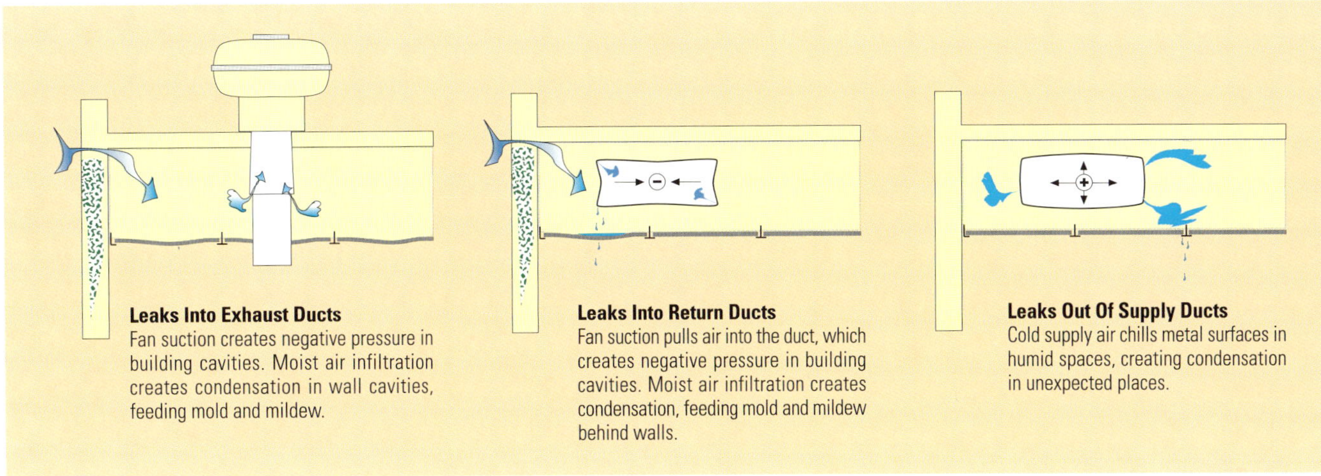

fungi to grow, including those that may thrive on smaller amounts of moisture. Springtime problems may actually be the result of wintertime condensation.

In cold climates, guidance for limiting problems is slightly different. The building exterior should still be as airtight as possible, and as in humid climates all return air ducts and exhaust ducts must be sealed. But the HVAC designer should not allow humid air to be blown into building cavities. The building should be under neutral air pressure with respect to the outdoors for as many hours as possible during the cold season. If anything, the exterior wall could be under a very *slight* negative air pressure with respect to the outdoors. Note this is a very difficult goal to achieve without overdoing it, which would chill the wall cavities to the point where surfaces become cold enough to condense water.

Joseph Lstiburek, Ph.D, P.Eng, the well-known Building Scientist has humorously summarized this advice; "In hot climates the building's *gotta* be a blowhard, but in cold climates it's better to suck."

Fig. 7.18 Moldy Books

This book came from an elementary school in Houston, Texas. To "save energy", the thermostat in the library was reset to a high temperature during summer vacation. But outdoor air ventilation was not reduced, nor was there any dedicated dehumidifier installed to keep the rh below 50%. Since the humid ventilation air was not dried, it "pumped" moisture into the books, allowing prolific mold growth. More than a thousand volumes had to be cleaned or replaced.

Mold in books and carpets

Problems above are usually caused by condensation. But mold can grow without gross water droplets. High relative humidity near the food surface is enough to allow growth.

Libraries

Libraries and used-book stores are often notorious for musty odors, which indicate mold growth. Books are a special case. They absorb moisture at high humidity, but fail to give it up when humidity drops—unless the humidity drops well below 30% rh.

So in a library, the books breath in moisture whenever the humidity is high. This can happen often, for example when the building is closed and the air conditioning system is turned off, or when the thermostat is reset to such a high temperature that no dehumidification is accomplished by the cooling coils. On each cycle of high humidity, the books gather slightly more moisture. Eventually, they accumulate enough to allow mold growth. The problem is accelerated when the HVAC system feeds only "temperature-tempered" outdoor air to the building. For example, if the thermostat is set back to 80°F

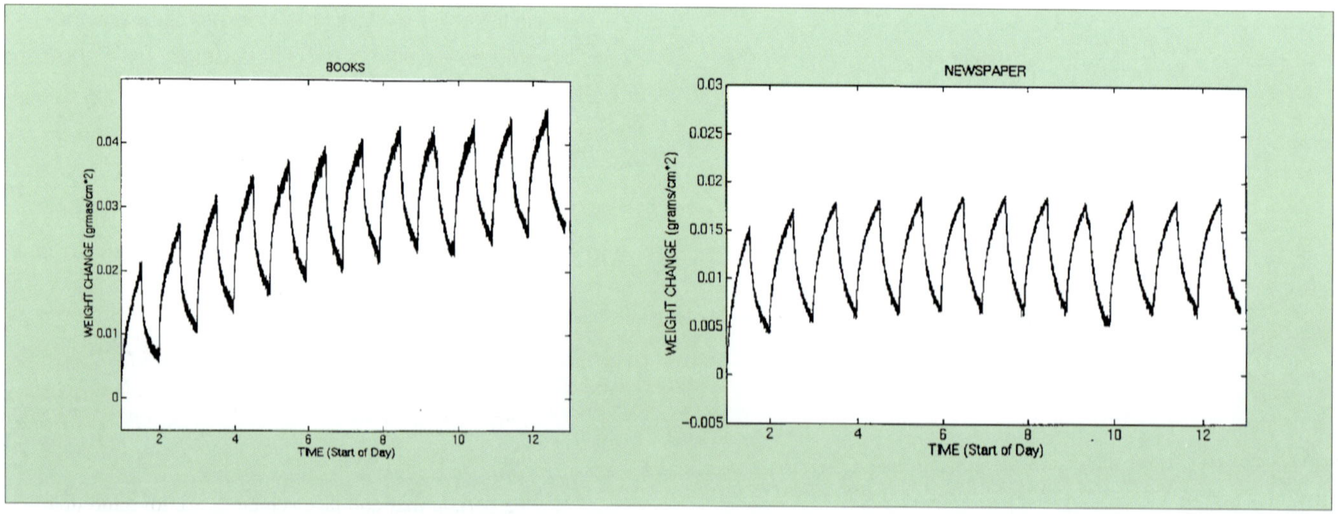

Fig. 7.19 Books Accumulate Moisture

Unlike newspaper and other thin-layer materials, books accumulate moisture and do not give it up easily. In this experiment, both materials were cycled between air at 40% rh and air at 90% rh every hour. Note that the newspaper released its moisture when exposed to 40% rh air. But the books steadily gained water. To dry books effectively, air must be well below 40% rh.
(Bailey et al. 1994)

[26.7°C] during unoccupied periods, and the ventilation air pre-cooling is set back to the same temperature, the incoming ventilation air cannot be dehumidified (at least in North America, Europe and Northern Asia, where there are no design dew points above 78°F [25°C]). Therefore during humid months, the ventilation system would pump immense amounts of water vapor into the building—and then into the books. Thrifty school administrators sometimes return from summer break to find prolific mold growth in books after having "saved energy" by shutting off air conditioning or by raising its set point temperature.

The solution for libraries is to maintain the relative humidity below 50% at *all* times, regardless of the temperature. This strategy generally requires a separate, dedicated dehumidifier, which responds to a humidistat rather than to a thermostat. The incoming ventilation air is usually the best location for that dehumidifier, since the outdoor air is by far the largest moisture load in a library.

Carpets

When carpets are soaked in humid climates—as during periodic cleaning in eldercare facilities—they can grow mold if they are allowed to stay wet. The relative humidity and air velocity at the carpet surface determine the drying rate. If air is stagnant at the floor level, and if the humidity is high (as happens in facilities without dedicated dehumidifiers) then drying time may extend indefinitely, and mold can grow. Preventing this problem is simple, as long as the budget allows for a separate dehumidifier in the system. If that is impractical, then the cooling system can be forced to operate by a humidistat, and heat added back into the air from waste heat such as a standby boiler or other low-cost source.

In rare circumstances, carpets can also grow mold in cold climates. In hotel construction and in military dormitories, there have been examples of concrete floor slabs that extend outboard of the building envelope, creating a thermal bridge to the cold weather. In those buildings, the cold slab just inside the building condenses water from the air, providing moisture for mold in the carpet. Often, cold climate mold problems are caused by such thermal bridges.

Mold In The HVAC System

All HVAC systems have mold and mildew in the duct work. This is true regardless of the system type, its degree of filtration or its geographic location. However, this does not mean that all HVAC systems create mold-related problems for occupants. To create a problem, even one as relatively benign as musty odors, the mold growth in the system must be prolific. "Problem levels" of fungal growth can be avoided through a combination of everyday good design and maintenance practices. Common problems are presented below in order by increasing cost and complexity of their solutions.

Mold on filters

Most filter media is chosen because it is inert. Fibrous glass paper, polyester batting and glass fiber matting are all poor food sources for fungus. ASHRAE-sponsored research and industry studies have repeatedly shown that fungus does not grow well on newly-installed filter media. Therefore, so as long as filters are changed regularly and have fairly clean media in the air stream, they are not likely to provide a good growth environment for mold and mildew. Filters create problems under two circumstances, both of which are easily avoidable.

First, if the filters are not changed frequently, they collect a plentiful supply of food for fungus. Even then, problems are not common unless the clogging is constant for months or years, or the dirty media is exposed to relative humidity over 90% or actually soaked with liquid droplets. (Ahern 1992, Kemp et a.l 1995, Kemp et al. 1995, Pasanen 1991)The solution to mold and mildew on most filters is simple: replace the media. As long as the designer has made sure the media is accessible, the solution to such problems is within the control of the owner's operations staff.

Second, an entirely design-related problem is created when filters are placed immediately downstream of cooling coils or humidifiers. Then the "mold food" as well as the spores collected in the filter are constantly receiving moisture from humid air. Mold will grow on and through the filter media, releasing spores on the other side. This problem is unfortunately quite common. The best solution is to ensure that filters are never installed immediately downstream of cooling coils or humidifiers. But that solution is not always possible in the real world. HVAC designers do not always have the luxury of placing wet components where good design practice suggests.

When there seems no alternative to placing filters downstream of cooling coils, the designer should consider:

1. Placing filters as far downstream as possible
2. Placing filters in all the return air grills, so they can be easily accessed for replacement and so their particulate loading is visually apparent to any room occupant.
3. Making certain that coil face velocities are low, and that fin spacing is wide, to avoid carrying over liquid droplets into the air and onto the filters.
4. Installing a mist eliminator between the coil and the filter to keep water droplets out of the media
5. Warning the owner that downstream filters will need to be changed much more frequently, perhaps with a filter alarm light set at a smaller pressure difference than others, along with a sign on the filter access that makes the point to the maintenance staff.

The same steps would be taken when filters must be installed downstream of humidifiers. But in that circumstance, the designer also has many types of humidifiers to choose from. Some perform better in this position than others. The manufacturers may be able to suggest equipment that will minimize the potential for mold growth in closely-coupled filters.

Mold in and around drain pans

Drain pans under cooling coils are a common source of mold and mildew problems. If the drain pan does not actually drain water—for whatever reason—mold and mildew will certainly grow in the pan and spread through the duct work to the occupied spaces. Also, if water leaks or spills *out* of the drain pan, mold and mildew will grow in the soaked food source. The problems fall into three categories: maintenance, pan design and drain line installation.

If the drain line is clogged—as happens when dirt collects in the line—then the pan can overflow. But this problem is simple to solve with either compressed air or a plumber's snake. Also, keeping the filters changed keeps dirt and fibers from collecting in the drain pan, reducing the potential for a clogged line.

Drain pan design, until recently, has not been the object of much attention by manufacturers or by those who specify equipment. So older equipment designs do not always have drain pans under coils. Often, if a drain pan is in place, it is not long enough to catch droplets blown off cooling coil fins, or not sloped to its drain in two directions, creating standing water in parts of the pan which never drain. Fortunately, correcting such problems is quite simple in new construction. Simply specify that all cooling coils shall have drain pans sloped all the way to their drains, and specify that the manufacturer shall certify that the length of the pan is adequate to capture water droplets coming off the coil when the system is operating at the peak dew point design condition.

In the past, condensate drain lines sometimes received less than adequate attention from designers and installers. The first indication of a problem might be water dripping out of the duct work onto desks and occupants. Common problems have included:

- No drain line or no connection to the line. Either the designer fails to specify a drain line, or the contractor fails to install it. The solution is obvious: there must be a drain line, and it must be connected—through a trap—to the drain pan.
- No trap, or an improperly-designed trap in the drain line. Without a trap, condensation cannot drain from the unit. The fan suction holds water in the pan until it overflows. The solution is to have a trap with a stem height greater than the pressure of the fan. (The stem height is the vertical distance between the *outlet* of the trap and the *bottom* of the drain pan)
- Drain line not connected to a drain. Drain lines that feed water into building cavities are surprisingly common. The contractor must be certain the drain line actually connects to a drain.

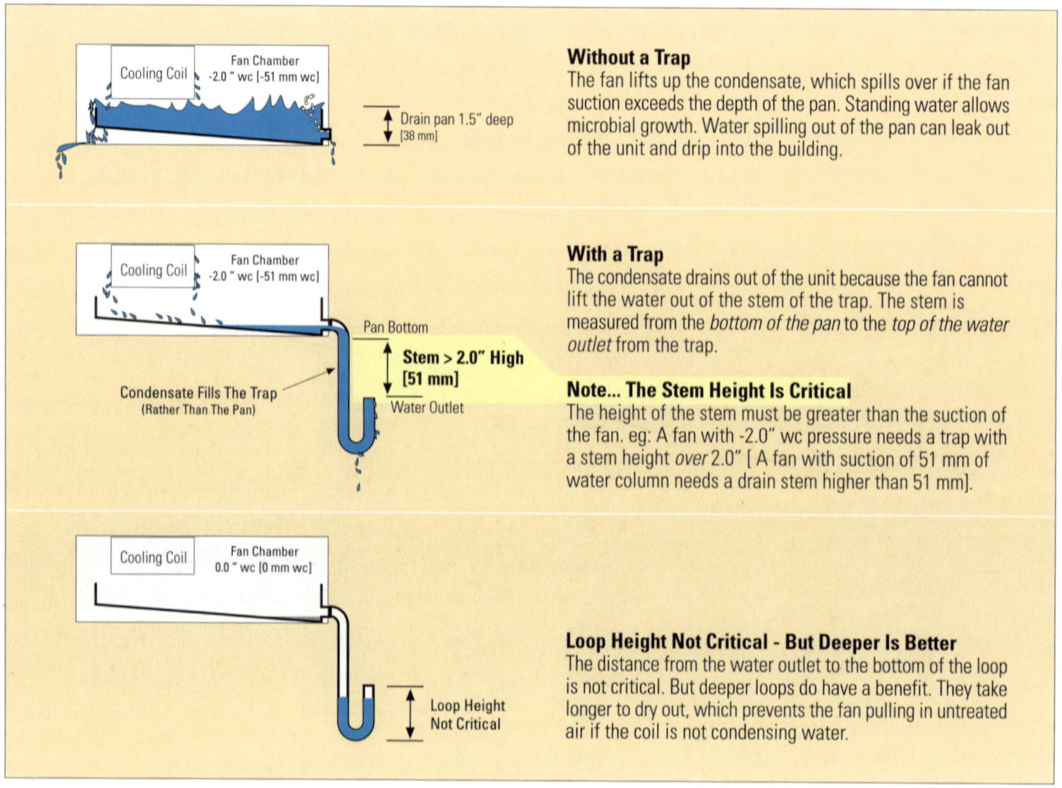

Fig. 7.20
Traps For Coils In Units With A Draw-Through Fan Position
A trap with a high stem is essential to proper drainage of water from condensate pans. Without a trap, the fan pulls water up, out of the pan and into the ductwork, providing ample water for fungal growth.

Mold in unitary equipment

Fan-coil units and other room-mounted unitary equipment do not always have filtration protecting the finned coil surfaces. Dust builds up on the surface over the heating season. When the coil condenses moisture in the cooling season, water becomes available to fungal spores in the dust, and mold begins to grow.

This common problem can be solved by both maintenance and design. The designer can specify equipment with easy-to-replace filters, and which contain coils which the manufacturer can demonstrate are cleanable in-place. The building operations staff can ensure that filters are indeed replaced or cleaned on a monthly or bi-monthly schedule, and ensure the coils are cleaned once a year using the manufacturers' procedures.

In the real world, this simplistic and obvious advice is not always possible to follow. The designer may be forced by other considerations to select equipment that does not have replaceable filters, and/or does not have cleanable coil surfaces. In that case, the designer and owner must simply accept the probability of mold and mildew in the unit. Room air filtration can extend the time until a problem develops, but it will seldom eliminate the potential problem entirely. Designers and owners of buildings where occupant health is compromised—such as hospitals, clinics and eldercare facilities—should carefully weigh the health risks of mold and mildew against any apparent benefit of low-cost unitary equipment that cannot be filtered or cleaned.

Mold in ductwork downstream of cooling coils

Cooling coils produce condensation and high-humidity air. Dust accumulates on the inside of duct work. That combination of dust and humidity eventually leads to mold and mildew growth in duct work downstream of cooling coils (Morey 1990, Thomann 1996). As a practical matter, all duct systems have mold, but very few cause problems. That's evidence that it is not necessary to eliminate mold growth entirely. It is simply important to limit the growth to harmless levels. This can be accomplished easily—or made nearly impossible—by design decisions.

The big problems come with large amounts of water, and large amounts of fungal food. Eliminate both of those, and the problems are small or nonexistent. But both must be eliminated to prevent problems. For example, filtration reduces the amount of fungal food. But if water droplets drift off the coil onto that small amount of food, fungus will flourish. Likewise, if no water drifts off the coil, but dust accumulates in heavy layers, the high humidity air will eventually provide enough moisture to support prolific growth. Finally, if both filtration is good and water droplets eliminated, but the duct surface traps all available dust and humidity, the mold problem is delayed, but still inevitable. Consequently, the best way to avoid problems downstream of cooling coils it to address all of the following cautions in the design:

1. Avoid high air velocity and narrow fin spacings in cooling coils that must remove moisture. When air speeds through narrow fin spaces, or when the face velocity is high, condensed water is blown off the coil into the air. Specify fin spacings at least as wide as 12 fins per inch, and air velocities below 400 fpm. [Wider than 5 fins per cm and velocities below 2 m/s]
2. Avoid tall coils. When the coil is tall, the lower part accumulates a great deal of condensation from above, making it very probable that air will blow water off the fins even when they are spaced far apart. Avoid coils over 3.2 ft high [1 metre high]. When the ductwork demands a tall configuration, split the coil horizontally into two units, and install a drain pan beneath both units.
3. Place filters between particle sources and coils. Locate easy-to-change filters so they are between the source of the particles and the cooling coils. This will keep coil surfaces clean, saving energy and avoiding fungal food accumulation on coils and downstream in humid air. In some cases, locating filters immediately behind the return air grills is the most practical way to keep them dry and to provide easy access for replacement.
4. Immediately downstream of cooling coils, specify smooth metal duct work with access doors for cleaning. Insulate these sections on the *outside* rather than the inside of the duct. Internal insulation and sound lining downstream of

cooling coils is notorious for trapping dust and growing mold at very high rates. Specifying a smooth metal surface in this location, along with access doors for cleaning, will avoid the worst problems that have been recorded with respect to mold and mildew in HVAC systems. The ideal length of the smooth section will depend on the velocity in the duct. Higher velocity carries any water droplets further downstream, so longer smooth sections are appropriate. As a minimum, the smooth and accessible section should extend 8 duct diameters downstream of the coil, allowing the air to return to smooth flow, so that particles are less likely to drop out and deposit on duct walls.

Mold downstream of humidifiers

Designers, contractors and owners do not always pay close attention to guidance from manufacturers of humidifiers. Consequently these devices have sometimes been associated with avoidable mold and mildew problems in duct work. One must keep in mind that humidifiers are designed to add moisture—exactly what mold and mildew need for growth. So the duct work downstream of humidifiers demands careful attention if problems are to be avoided.

The most important design rule is to *seek and follow manufacturer's guidance* regarding duct work downstream of the specified humidifier. Requirements vary between types of equipment. Atomization-type units need smooth, accessible duct work for a longer distance than steam-dispersion units, because large, cold droplets are not absorbed by air as quickly as hot steam vapor. Also, some types of equipment require a drain in the duct work after the humidifier, as some "fallout" water is expected and must be prevented from flowing through the supply duct. Beyond the manufacturer-specific guidance, some clearly good practices are useful to keep in mind when constructing the specification:

1. Avoid obstructions downstream of humidifiers. If turning vanes, for example, are mounted close downstream of humidifiers, moisture is likely to condense as the air pressure changes slightly when it contacts the vanes.

2. Avoid porous insulation and sound lining near humidifiers. Porous surfaces collect dust and moisture, and grow mold. Use external insulation and smooth metal duct downstream of humidifiers.

3. Provide cleaning access downstream of humidifiers. Cleaning is always needed downstream of humidifiers. Be sure to provide that access in the duct layout.

References

Ahearn, D.G. et al. *Colonization studies of various HVAC materials*. Proceedings of the 1992 ASHRAE IAQ Conference, October 1992, San Francisco, CA. ASHRAE. ISBN 0-910110-94-8.

Bailey, D.W. et al. *Investigation of dynamic latent storage effects of building construction and furnishings* (ASHRAE Research Project 455-RP). 1994. Project report with Appendices.

Brundrett, Geoffrey W. , *Moulds, Chapter 4 of Criteria for moisture control*. 1990. Butterworth & Co., Reed International. London. ISBN 0-408-02374-0.

Burge, Harriett A., *Approaches to the control of indoor microbial contamination*. 1987. Proceedings of the ASHRAE IAQ Conference, May 18-20, 1987, Arlington, VA. ASHRAE. Atlanta, GA. ISBN 0-910110-51-4.

deRoos, Jan A. *Summary of mildew problems encountered in hotels in South Texas*. January 9th, 1989. School of Hotel Administration, Cornell University, Ithaca, NY.

Fink, Jordan N. et al. *Interstitial lung disease due to contamination of forced air systems.* Annals of Internal Medicine 84:406-413, 1976.

Flannigan, Brian and Miller, J.David., *Humidity and fungal contaminants.* 1993. Proceedings of the Bugs, Mold & Rot II workshop on control of humidity for health, artifacts and buildings. November, 1993. National Institute for Building Sciences. Washington, DC.

Flannigan, Brian. *Approaches to assessment of the microbial flora of buildings.* Proceedings of the 1992 ASHRAE IAQ Conference, October 1992, San Francisco, CA. ASHRAE. Atlanta, GA. ISBN 0-910110-94-8.

Gatley, Donald P., *Mold and condensation behind vinyl wall covering*. 1990. Gatley & Associates, Atlanta, GA.

Kemp, Stephen J. et al. *Filter collection efficiency and growth of microorganisms on filters loaded with outdoor air.* ASHRAE research project 625-RP. ASHRAE Transactions, 1995. V.101., Pt. 1. Paper #3853. ASHRAE, Atlanta, GA.

Kemp, Stephen J. et al. *Growth of microorganisms on HVAC filters under controlled temperature and humidity conditions.* ASHRAE research project 625-RP. ASHRAE Transactions, 1995. V.101., Pt. 1. Paper # 3860. ASHRAE, Atlanta, GA.

Kimbrough, James W., *Mildew/Moisture Control. Problem avoidance in new and existing hotels/motels.* February 6th, 1990. Koogler & Associates Environmental Services. Gainesville. FL

Miller, J.David. *Fungi and the building engineer*. Proceedings of the 1992 ASHRAE IAQ Conference, October 1992, San Francisco, CA. ASHRAE. Atlanta, GA. ISBN 0-910110-94-8.

Mold & Mildew in Hotels and Motels - Survey Results. 1991.. American Hotel & Motel Association. Washington, DC.

Morey, Phillip. R, Williams, C.M., *Porous insulation in buildings: A potential source of microorganisms*. Proceedings of the 1990 ASHRAE IAQ Conference, Toronto, ONT. ASHRAE. Atlanta, GA.

Odom, J. David, III., *Designing and constructing mildew-free hotels*. 1994. CH2M/Hill Consulting Architects and Engineers. Orlando, FL.

Odom, J. David, III., DuBose, George and Fairey, Phillip. *Moisture problems: Why HVAC commissioning procedures don't work in humid climates*. 1992. Proceedings of the 8th Symposium on improving building design in hot & humid climates. May, 1992. Texas A&M University, College Station, TX.

Pasanen, Pertii. et al., *Criteria for changing ventilation filters*. Proceedings of the 1991 ASHRAE IAQ Conference, Washington, DC. Sept. 4-8 1991. ASHRAE. Atlanta, GA. ISBN 0-910110-83-2

Peart, Virginia. *Mildew and moisture problems in Florida.* February 1989. Institute of food and agricultural sciences. University of Florida, Gainesville, FL.

Robertson, L.D., *Monitoring viable fungi and bacterial concentrations to identify acceptable levels for common indoor environments.* Proceedings of the 11th Symposium on improving building design in hot & humid climates. June, 1998. Texas A&M University, College Station, TX.

Schleibinger, H. et al., *Occurrence of microbially produced aldehydes and ketones (MVOC's) from filter materials of HVAC systems—field and laboratory experiments.* Proceedings of the 1995 ASHRAE IAQ Conference, Denver, CO. October 22-24 1995. ASHRAE. Atlanta, GA. ISBN 1-884313-30-3.

Shakun, Wallace. *A review of water migration at selected Florida hotel/motel sites.* Proceedings of the biennial symposium on improving building practices in hot & humid climates. October 1990. Texas A&M University, College Station, TX.

Staib, Freiderich et al., *Indoor air mycology: aspergillosis, mucormycosis and cryptococcosis caused by fungal spores from indoor air.* Indoor Air '87. Proceedings of the 4th International Conference on Indoor Air & Climate. Berlin, West Germany. 17-21 August 1987 (Volume 1. ISBN 3-89254-033-0.

Thomann, Wayne R., Tulis, Jerry J., *Fungal contamination of HVAC surfaces: The role of the dew point in microbial amplification.* Proceedings of the 1996 ASHRAE IAQ Conference, Baltimore, MD October 6-8 1996. ASHRAE. Atlanta, GA. ISBN 1-883413-41-9

Transcript of the Hotel Mood & Mildew Roundtable Discussion. March 1st, 1990. Executive Engineer's Committee. American Hotel & Motel Association. Washington, DC.

Special Thanks

This chapter could not have been written without the landmark field research accomplished during the 1990's by Dr. Virginia Peart and Dr. James Kimbrough of the University of Florida, and by J. David Odom III and George DuBose of CH2M/Hill of Orlando, Florida, as well as by Donald Gatley and Wallace Shakun of Georgia and Joseph Lstiburek of Building Science Inc in Westford, MA. Also, the Executive Engineers Committee of the American Hotel and Motel Association provided the original impetus to look at mold in North American buildings from a multi-disciplinary perspective. Their focus on the problem was catalyzed by the persistence and generosity of Mr. Jones of MMI Hotel Management of Jackson, Mississippi. Finally, the first broad-based, quantitative research of the problem was performed for the AH&MA through funding provided by the Gas Technology Institute in Chicago and by the Munters Companies in Amesbury, MA.

Photos

Chapter 8
Dust Mites & Insects

By Geoff Brundrett with Lew Harriman

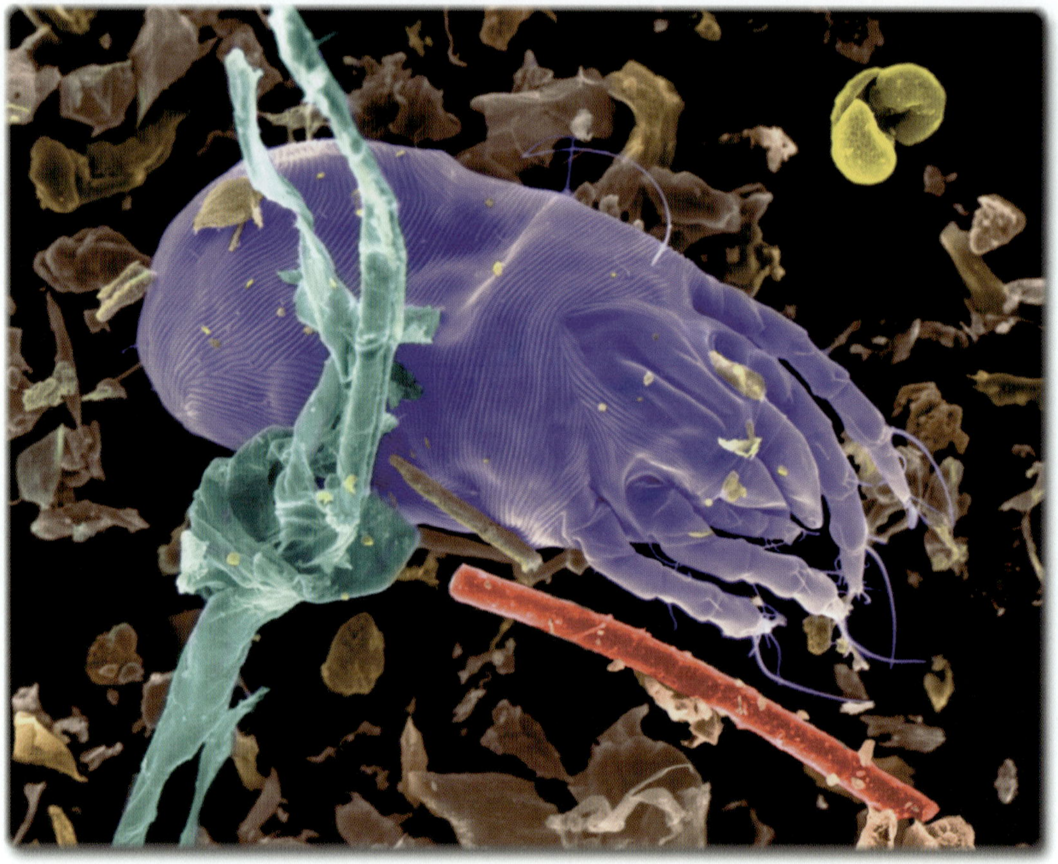

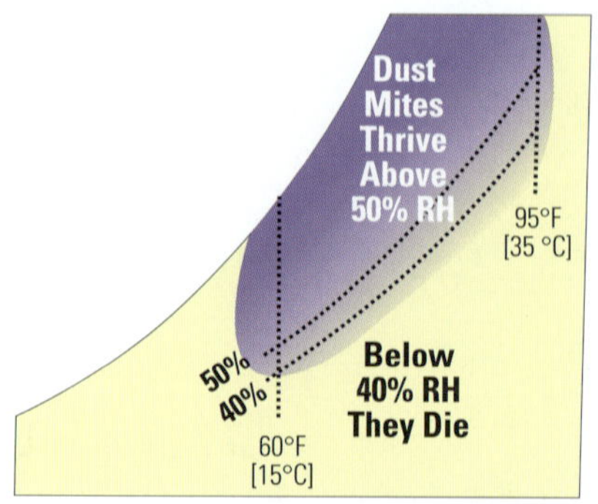

Fig. 8.1 Dust Mite Humidity Requirement
When humidity is over 50% rh, dust mites can thrive by eating skin flakes left behind by people and their pets. Below 40% rh, dust mites quickly desiccate and die.

Dust Mites & Insects in Buildings

In recent years, the news media have focused considerable attention on rising rates of childhood asthma and rhinitis in North America. Asthma is a difficulty in breathing caused by an allergic reaction which inflames and restricts the throat and lungs. Rhinitis is a similar inflammation which blocks the nose and the upper respiratory tract. Health problems have many causes, but these symptoms are observed quite frequently in buildings with excess moisture in the structure and high humidity in the air. In part, the reactions are related to fungus growing in the walls of such buildings. The causes and effects of building-related fungus are discussed in Chapter 7. In addition to structural mold and mildew, allergic reactions are often related to the presence of insects and dust mites.

Dust mites and insects produce fecal pellets that provide a food source for the fungus that can cause allergic reactions. The pellets also contain proteins that can trigger an allergic response. Further, the fungus-laden fecal pellets are small, loose particles. They are easily lofted into the air by the activities of building occupants. Air currents can then carry the pellets into the human breathing zone, where they are pulled into the lungs by normal respiration.

Reducing the humidity in a building can help limit the number of dust mites and insects in a building. Below 50% rh, these pests have difficulty retaining enough moisture for reproduction, and at lower humidity they will desiccate and die.

How insects obtain & retain water

About 50% of the body weight of an insect is water. If this water is lost, the insect will become less active and less able to reproduce. Very few insects can drink liquid water (cockroaches are an exception) Most insects gain the water they need through their food.

Even very dry-looking food sources contain water. Dry grain, for example, can contain up to 10% water by weight and still be safe from fungus and suitable for long-term storage. Moist wood in buildings can contain 26 to 29% water before the fibers are saturated. Human skin flakes and hairs are hygroscopic, and may contain a large percentage of water even when air around them is dry. Dust mites make use of this water as they consume the skin flakes. In addition, an insect's metabolism oxidizes its food, releasing CO_2 and water as well as the energy that allows the organism to grow.

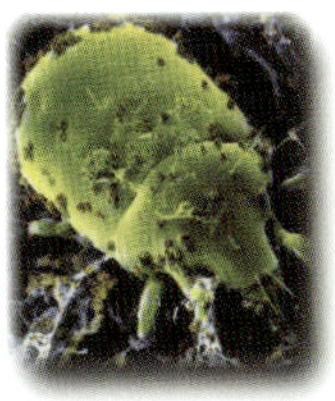

© Meile

Fig. 8.2 Fecal Pellets
Asthma and rhinitis are triggered by allergic reactions to fungi and proteins contained in fecal pellets, such as those shown here covering the dust mite that produced them.

For example, when insects metabolize glucose, 60% of the weight of that carbohydrate is converted to water. When they eat fats, much more water is released. Consuming tripalmitin (a fat) the insect's metabolism generates 1.1 grams of water for every gram consumed. By combining the free water contained in the food and the water released through metabolism, an insect can obtain an amount of water roughly equal to the weight of the food itself (Brundrett, 1990).

For insects, however, *retaining* that essential water can be more difficult than obtaining it. They need oxygen, which many insects obtain

Fig. 8.3 Insect Respiration *(Brundrett, 1990)*

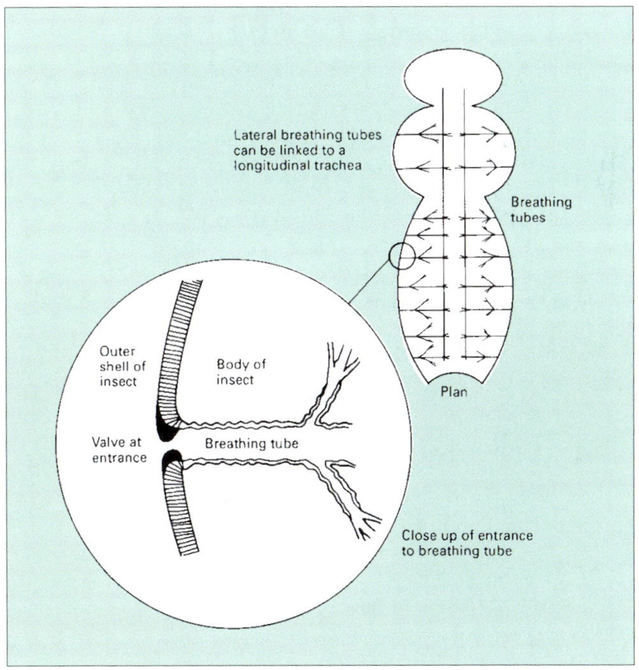

through a series of breathing tubes which penetrate deep into their bodies. (See fig. 8.3) Oxygen in the air enters these tubes and diffuses through the tube walls into the bloodstream. Carbon dioxide leaves the body by the same diffusion process in the opposite direction. Unfortunately for the insect, water vapor also diffuses out of the body along with the carbon dioxide. The water vapor pressure is high inside the body, and much lower in the surrounding air. Water diffuses out through the breathing tubes in response to this difference in vapor pressures. The greater that difference (the drier the air and the warmer the insect), the faster the insect desiccates.

Since oxygen is essential for metabolism, an insect does not have the option of closing up its breathing tubes indefinitely. When the surrounding air stays dry, insects must either obtain moist food, or go into a dormant state during which less oxygen is needed. In dry air, moist food also becomes scarce. The insect is left with no good choices. In an effort to obtain oxygen for metabolism by opening its breathing tubes, it either dies from desiccation, or loses so much water that it cannot reproduce. Dust mites are even more sensitive to low humidity, because they do not have breathing tubes with valves. They gain oxygen and loose water through their entire exterior shell. Dust mites have no ability to limit evaporation by closing up a set of breathing tubes. This is one reason they prefer to inhabit crevices in humid fabrics, where air flow is slow and the dew point is high—so desiccation is less rapid.

Temperature also influences insect metabolism in both positive and negative ways. In warm environments, less energy must be generated internally. Up to some limit, warmer temperatures mean healthier, more active insects. Above that limit, rising internal temperatures raise the water vapor pressure inside the insect, increasing the vapor pressure difference between the body and the surrounding air and increasing the risk of desiccation.

From these relationships, it follows that air with a low dew point is less hospitable for most insects because it has a low vapor pressure. The desiccation effect is magnified when the air is hot as well as dry. Heat raises the insect's internal vapor pressure, and at the same time supplies energy which dries out their food sources.

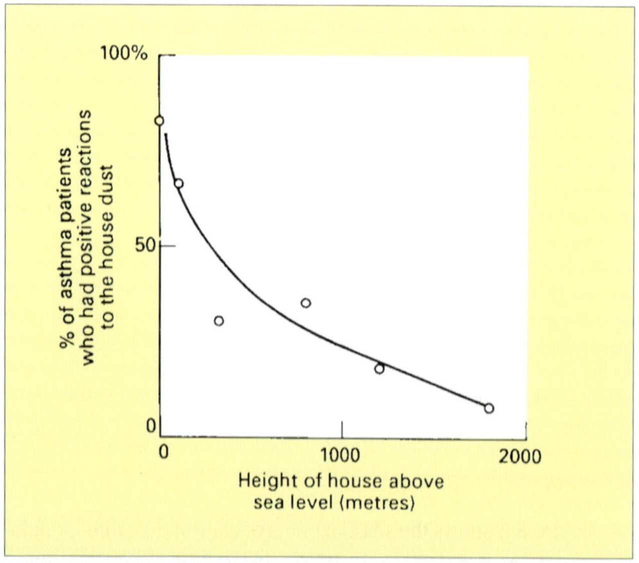

Fig. 8.4

Low Dew Points = Fewer Dust Mites

Higher altitudes have lower average dew points, which make it difficult for dust mites to survive. The graph shows that asthma patients are less likely to be sensitive to dusts from houses at high altitudes. (Verhoorst et al, 1969)

Conversely, moderate temperatures and high humidities are ideal for insects. Warm temperatures mean the insect gains "free" energy from the air. Also, high dew point air has a high water vapor pressure. Food will not dry out, nor will the insect lose much water when opening its breathing tubes to gain oxygen.

The behavior of insects illustrates these principles. To enter a house, termites build protective tubes that start in the ground and proceed up the foundation to a point where they can chew into moist wood. The tubes hide the termites from predators, and also prevent dry air from flowing through their habitat and through their moist food source. Likewise, mosquitoes die in hot, dry and windy environments but thrive in humid air which is warm and still.

Dust mites and humidity

Over time, research has shown that there is a clear relationship between the average dew point in a building and the number of dust mites it contains. Also, research shows that younger and middle-age

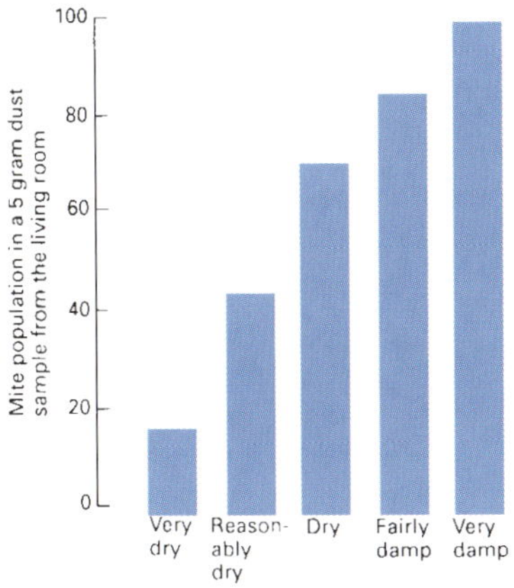

Fig. 8.5
Damp Houses Grow Dust Mites
In higher humidity, dust mites are less likely to desiccate and die. (Speikesma, 1966)

people are more likely to be sensitive to dust mite fecal pellets than are senior citizens.

Figure 8.4 shows the allergic response of asthma patients to house dusts gathered from houses located at different elevations. The higher the elevation, the lower the average annual dew point. The figure shows that dust from houses in lower elevations (more humid environments) caused allergic reactions in more patients.

Figure 8.5 shows the dust mite population in samples of dust from houses with different amounts of moisture in the building. Moisture levels were determined by the number and size of damp spots and/or mildew on walls, and the presence of musty odors. The figure shows that *as the moisture level in the house increases, so does its population of dust mites.*

Figure 8.6 shows the seasonal dependency of dust mite populations in the Netherlands, a northern climate with significant differences in average dew point between winter and summer. The dust mite populations peak in the late fall, after the houses have absorbed humidity all summer, and during the period when rains fall steadily. During winter months the air dries out, and dust mite populations fall accordingly.

Figure 8.7 shows a group of asthmatic patients' sensitivity to dust, and shows the age dependency of their allergic response. Of the total 381 patients, 234 (61%) were sensitive to house dust. But note that as the age of the patients increased beyond 25 years old, they were less likely to be allergic to the dust.

These data show that beyond the fundamental biomechanics of insects, field observations confirm the general effectiveness of a strategy that limits the population of dust mites by keeping the air and building materials dry. But to formulate an effective plan, one must also consider the microgeography of dust mite populations.

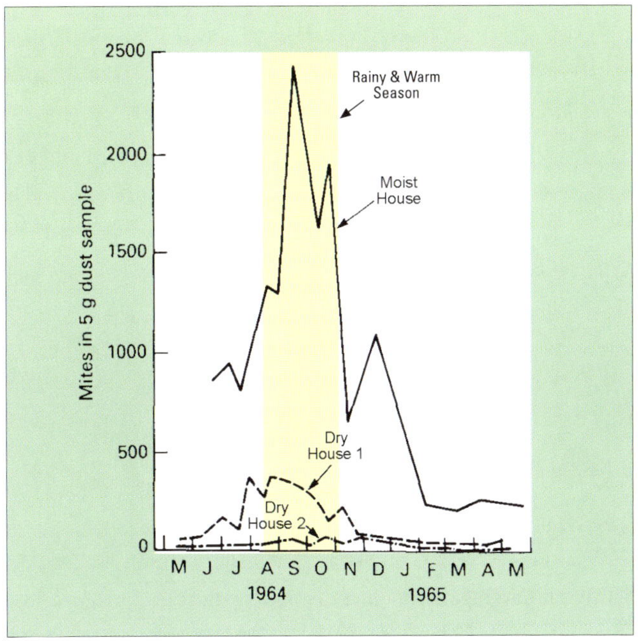

Fig. 8.6
Dust Mites Grow In Humid Weather
During high summer humidity and heavy fall rains, dust mites thrive, particularly in houses which are moist. (Speikesma, 1966)

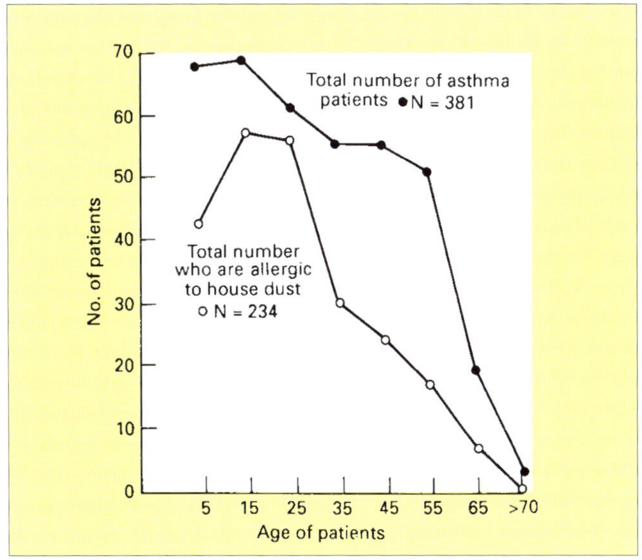

Fig. 8.7 Young & Middle-Age People Are More Sensitive
Senior citizens are less likely to suffer from asthmatic reactions triggered by dust mite fecal pellets. (Voorhorst, 1962)

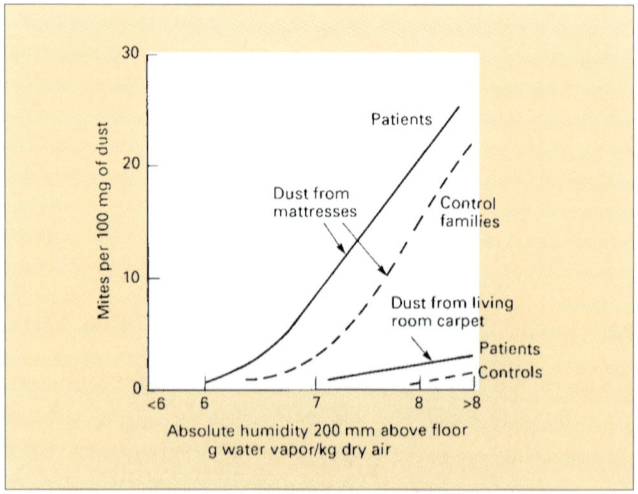

Fig. 8.8 Bedding Contains More Dust Mites Than Carpets

People shed skin flakes and add moisture and heat to bedding, creating the ideal environment for dust mites. They also thrive in carpets when these are not cleaned regularly and when humidity stays over 50% rh. (Korsgaard, 1962)

Dust mite geography

In addition to moisture, dust mites must have food, although being so small, they do not need much. Dust mites eat human and animal skin flakes. People shed skin at a rate of about $^1/_{100}$ of a pound per week [5 g/week]. But most importantly, about 20% of that weight is shed in the exact same spot every week—the bed. Skin flakes build up in the bedding, eventually accumulating in the mattress and pillow under the sheets and pillowcases. In addition to skin flakes, the sleeper releases both moisture and heat to those same locations. Finally, the lack of air flow at the surface of the mattress helps the fabric retain moisture and skin flakes, creating the perfect environment for dust mites to grow, pass fecal pellets and reproduce. Figure 8.8 shows the result. Dust from mattresses contains a much larger number of dust mites than dust from living room carpets. Also, the asthma patients described in figure 8.7 were living in the houses with the highest populations of mites.

Controlling Dust Mites In Buildings

From this information, we can conclude that:

1. Dust mites are most likely to create problems in humid buildings and specifically in bedding, such as that found in military and university dormitories, hotels, apartments and in single family residences. Mites are less common in other types of buildings such as schools and offices, but these can develop high populations of dust mites when the building has infrequent and incomplete cleaning of floors, and when indoor humidity stays above 50% rh.

2. Dust mite populations can be held to very low levels when humidity is kept below 50% and they can be eliminated almost entirely when humidity stays below 40% rh. However, the only relevant air humidity measurement is taken *near the surface of the trapped skin flakes*, not in the middle of the room or high up on a wall.

3. Once dust mites and their fecal pellets have become prolific, keeping humidity below 50% at their food source will prevent expansion of the population, and reducing humidity below 40% will kill existing mites. However, after the fecal material is in place, thorough cleaning—especially in crevices—will be needed to eliminate contamination and avoid allergic reactions.

4. Air filtration does not stop the reproduction of dust mites or their production of fecal contaminants. But filtration can be useful to reduce the concentration of pellets in the air until humidity can be brought below 50% and until the materials containing contaminants can be thoroughly cleaned.

References

Brundrett, G.W, *Criteria for Moisture Control. 1990.* Butterworth & Co. London, UK. pp.146-149

Korsgaard, J. 'Preventive measures in house dust allergy'. Am Rev Respir Dis, 125, 80-84, 1982

Spieksma, F.Th.M. 'Some preliminary remarks on the classification, occurrence and life style of the mite Dermatophagoides pteronyssinus'. Interasma, 5, Utrecht, 37-43, May 1966

Voorhorst, R. *Basic Facts of Allergy* HE Stenfert Kroese NV, Leiden, 1962

Voorhorst, R , Spieksma, F.Th.M. and Varckamp, H. *House Dust Atopy and the House Dust Mite*. Staflen's Scientific Publishing Company, Leiden, 1969

Other Reading

Harsh, G.F. 'The correlation between humidity and house dust allergy', pp. 118-123 in Proc 1st Int Congr Allergy, Zurich, 1951

Ishii, A., Takaoka, M., Ichinoe, M., Kabasawa, Y. and Ouchi, T. 'Mite fauna and fungal flora in house dust from houses of asthmatic children'. Allergy, 34, 379-387, 1979

Kinnaird, C.H. 'Thermal death point of Dermatophagoides pteronyssinus the house dust mite'. Acarologia, 16, 340-342, 1974

Koekkoek, H.H.M. and van Bronswijk, J.E.M.H. 'Temperature requirements of a house dust mite Dermatophagoides pteronyssinus compared with the climate in different habitats of houses'. Entomol Exp Appl, 15,438-445, 1972

Maurya, K.R. and Jamil, Z. 'Factors affecting the distribution of house dust mites under domestic conditions in Lucknow'. Indian J Med Res, 72,284-292, 1980

Mulla, M.S., Harkrider, J.R., Galant, S.P. and Amin. L. 'Some house dust control measures and abundance of Dermatophagoides mites in Southern California'. J Med Entomol, 12, 5-9, 1975

Voorhorst, R., Spieksma-Boezeman, M.I.A. and Spieksma, F.T.M. 'Is the mite the producer of the house dust allergen?' Allergic Asthma, 10, 329, 1964

Chapter 9

Bacteria & Viruses

By Geoff Brundrett

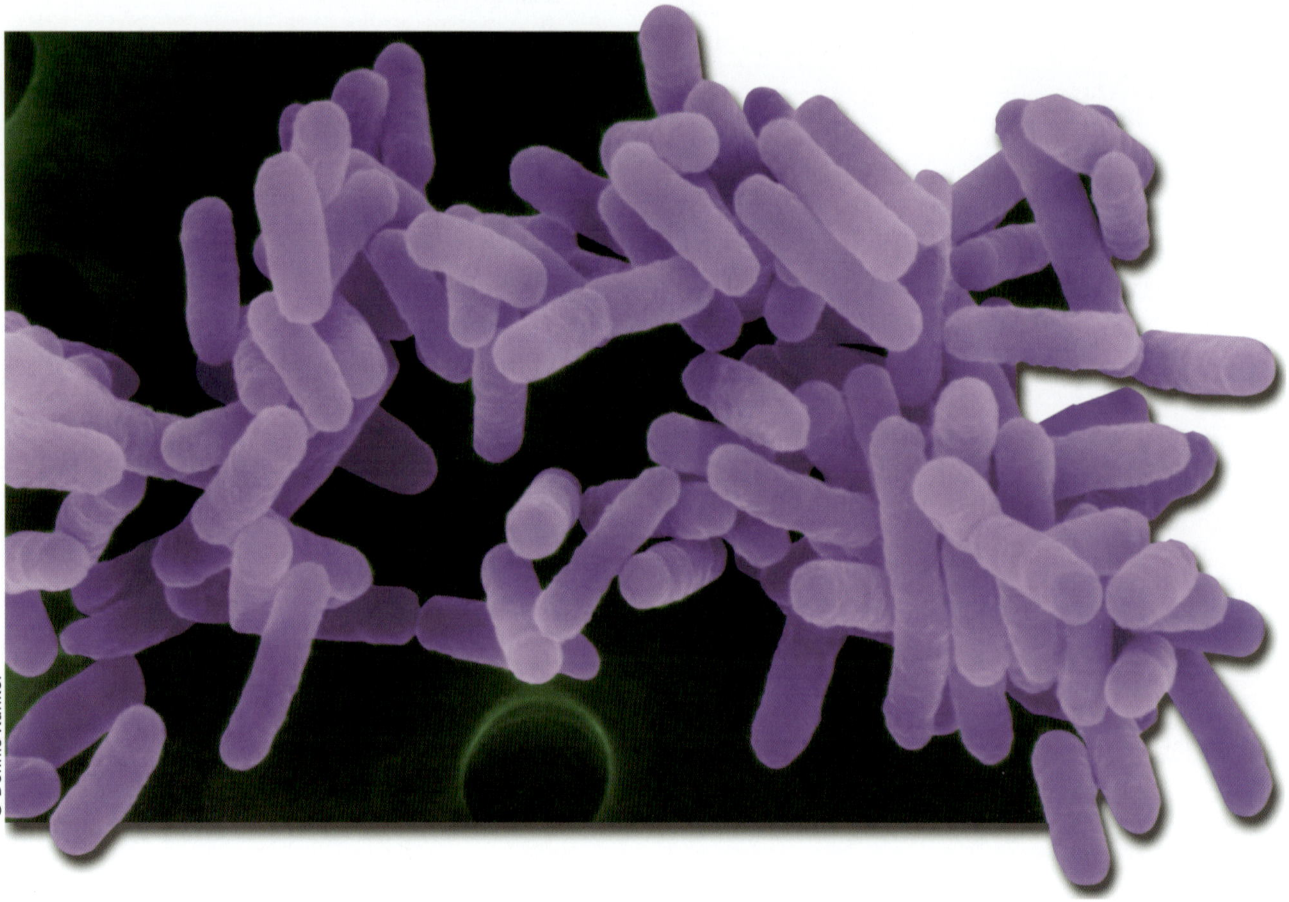

Fig. 9.1 ***Legionella Pneumophila***

Each airborne infectious organism has a range of relative humidity in which it is most vulnerable. The bacterium responsible for Legionnaire's disease dies most quickly when the air is at 30% relative humidity, and 90% rh is almost equally perilous for the organism. It survives longest at 65% rh.

Airborne Infections

Infections are colonizations of the individual by either viruses or bacteria in numbers sufficient to overcome the body's natural defense mechanisms. The infective dose (number of organisms needed to overcome those defenses) is very different for different individuals. It depends on the subject's age, health and susceptibility and the novelty of the microorganism.

Bacteria and viruses infect the body through three paths. Some are transmitted by touch from another infected person. Several of the viruses that cause the common cold are transmitted this way. Others such as the AIDS virus are transmitted by fluid transfer. Finally, diseases can be transferred by organisms that drift through the air. For example, influenza and tuberculosis infect individuals when they inhale microorganisms expelled into the air by someone who suffers from the disease.

Most of the research in this field has been done on a small number of viruses and bacteria, but some patterns emerge from the work. Scientists agree that many factors interact, catalyzing the individual effects of humidity, temperature, irradiance, the nature of the gaseous environment, the solutes and the cell history. Also, fewer microorganisms survive outdoors in daylight than indoors during laboratory tests, probably because of trace contaminants in the atmosphere and because of irradiance damage from sunlight.

The environment affects the individual as well as the organism. Polluted air carries volatile organic vapors and particles that can reduce a given individual's resistance to diseases. Also, human resistance to infection is reduced in humidity extremes. In low humidity, normally protective mucous membranes in the nose and upper respiratory tract dry out, exposing cracked skin to direct contact with infectious organisms. Similarly, when humidity is high, mold and mildew grow, releasing volatile organic gases (musty odors). Some fungal volatiles suppress the human immune response as a side effect of their principal purpose, which is to suppress the growth of competing fungi.

The HVAC designer can reduce the infection rate from airborne viruses and bacteria in three ways: by filtering the air to remove infectious particles; by providing outdoor air to dilute the concentration of such particles; and by ***keeping humidity in the mid-range****, reducing the life span of bacteria and viruses that travel through air.*

Humidity levels that minimize airborne infections

After dispersal into the atmosphere within water droplets, a microorganism goes through three stages:

1. Surface drying of the water droplet. This will result in the cells being coated with a concentrated solution of any solutes present in the water.
2. Desiccation of the cell itself. The speed with which a cell loses its water varies widely from organism to organism, depending on the permeability of its cell wall.
3. Survival in dust once the aerosol has eventually settled.

Not all organisms die through desiccation. In fact, many thrive in dry air and perish in highly humid air. When the building owner wants to reduce the potential for infection, relative humidity in the building should be kept above 30% rh during the dry months and below 60% rh during humid months. Support for this suggestion comes from examining the growth and survival of potentially infectious airborne viruses and bacteria.

Viruses

Viruses are not strictly organisms, because they have no metabolism of their own. They are very small, on the order of 100 nm, (one millionth of a metre) and can only be seen with an electron microscope. They are about one-hundredth of the size of a typical bacterium. To visualize the difference, imagine that a virus is the size of a pea. At that scale a typical bacterium is the size of a basketball.

Viruses consist of a protein coat around a strand of DNA or RNA. They are parasitic. They kill host cells by attaching themselves to the

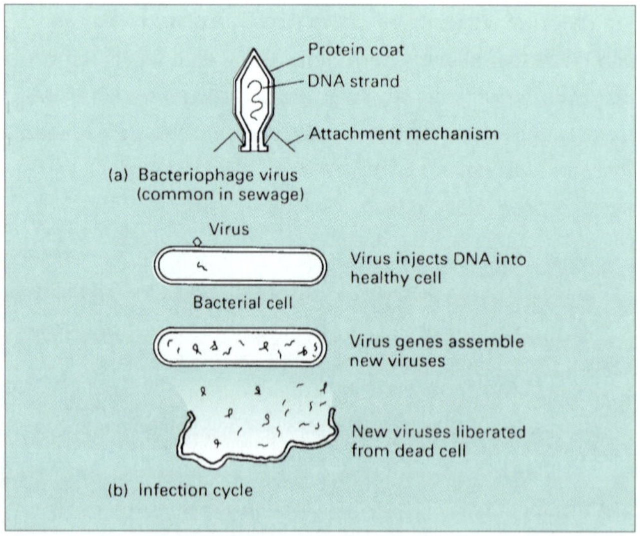

Fig. 9.2 Reproduction of Viruses

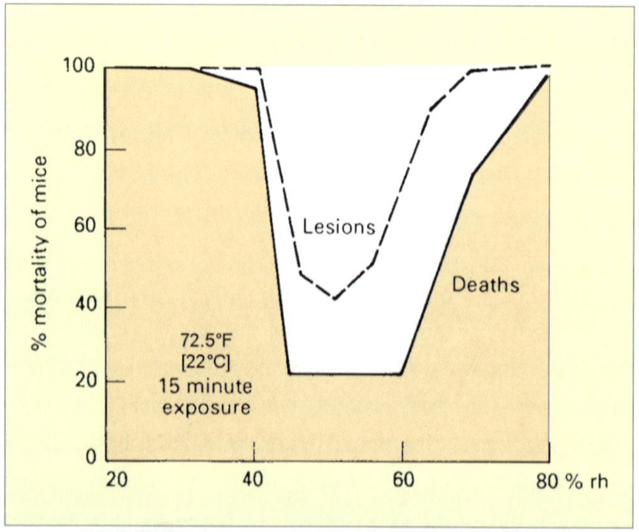

Fig. 9.3 Mortality In Mice Exposed To An Aerosol Of *Infuenza.*

Note that the number of deaths and lung lesions was minimized when humidity was held between 40 and 60% rh.

healthy cell wall and inserting their DNA strand into the living cell. The virus uses the cell's energy and materials, reproducing itself and destroying the cell. When the cell is destroyed, the large number of viruses created repeat the attack on neighboring cells. Viruses infect humans, plants and animals, and also infect bacteria. An illustration of the cycle is given in figure 9.2 (Ford-Robinson, 1981).

Experiments on viral infections have to be made on living tissue, either in animals themselves, in fertile eggs or in special cell cultures. Early work on the *Influenza A* virus showed a clear influence of ambient relative humidity. Mice were exposed for 15 minutes to an aerosol containing the virus. Half of the mass of the aerosol consisted of particles with a diameter less than 1.5 µm. Both lung lesions and mortality were strongly influenced by the ambient relative humidity. Infectivity was highest at relative humidities below 40% and above 80% (figure 9.3 - Lester, 1948) This is consistent with the popular observation that respiratory infections are common at low relative humidities.

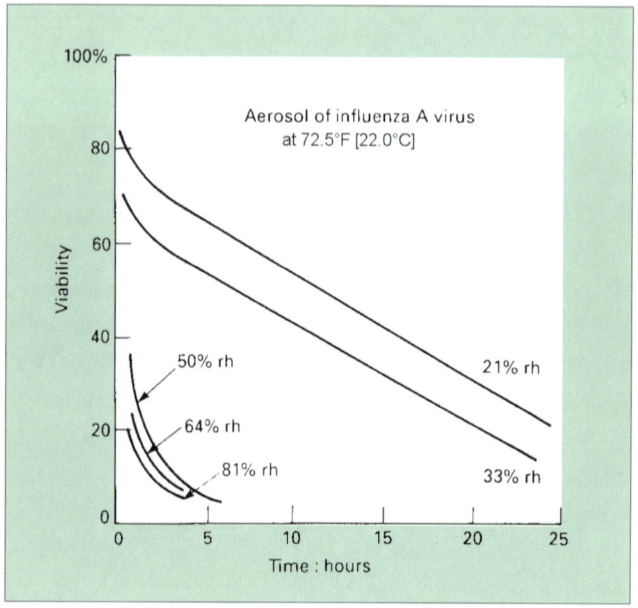

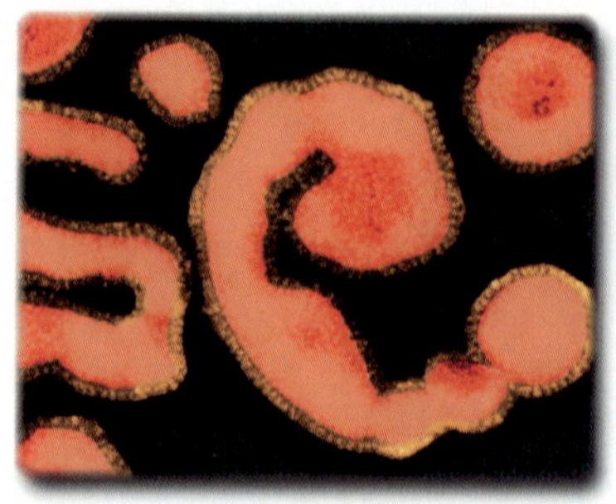

Fig. 9.4 *Influenza A* Virus

Humid air kills this virus more quickly than does dry air.

Fig. 9.5 *Vaccinia* Virus

This virus dies most quickly at higher humidities.

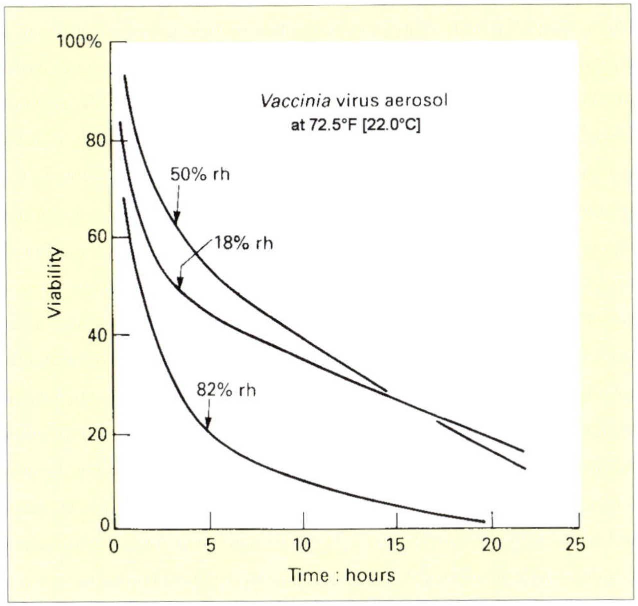

Fig. 9.6 Poliomyelites Virus

In general, dry air kills the virus more quickly, but note the fastest kill-rate occurs when air is held at 50% rh.

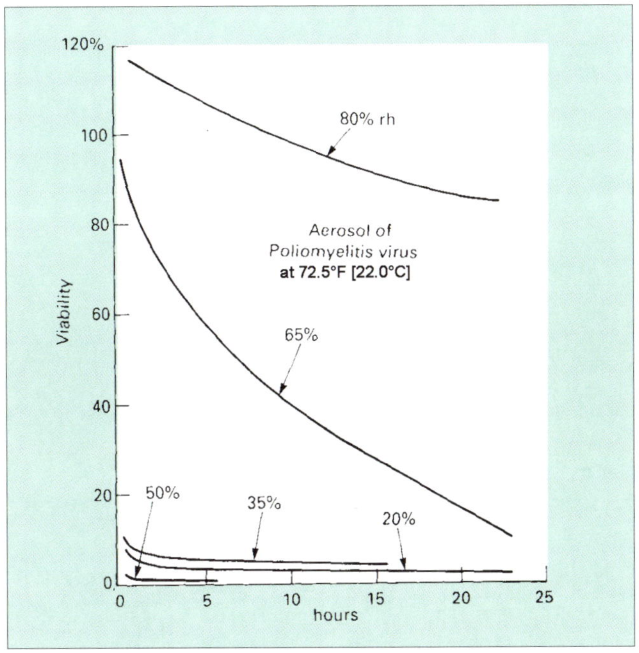

Other experiments with *Influenza A* virus agree, showing enhanced survival at low relative humidities around 20- 33% but also show that cell death increases as relative humidity rises, with maximum mortality at 81% (figure 9.4 - Harper, 1961) The *Vaccinia* virus died quickly at high relative humidities but survived best, at least for 15 hours, at 50% rh (figure 9.5).

Experiments on the survival of the *Poliomyelitis* virus in an aerosol showed the opposite effect of relative humidity to that of *Influenza A* (figure 9.6). The most lethal relative humidity to *Poliomyelitis* was 50% and the viability of the cells was very low at 35% and 20% rh However, survival increased rapidly with relative humidities above 50%. Each virus appears to respond differently to humidity.

Bacteria

Bacteria are the smallest living cells, typically 1 to 4 µm in size. An illustration of a typical cell is shown in figure 9.7. The DNA strand, freely suspended in the jelly-like cytoplasm, instructs the ribosomes to assemble protein chains from the nutrients surrounding them.

Fig. 9.7 Reproduction of Bacteria

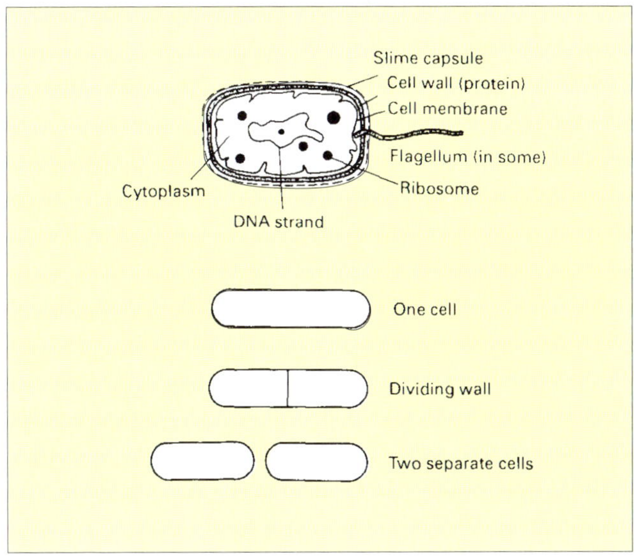

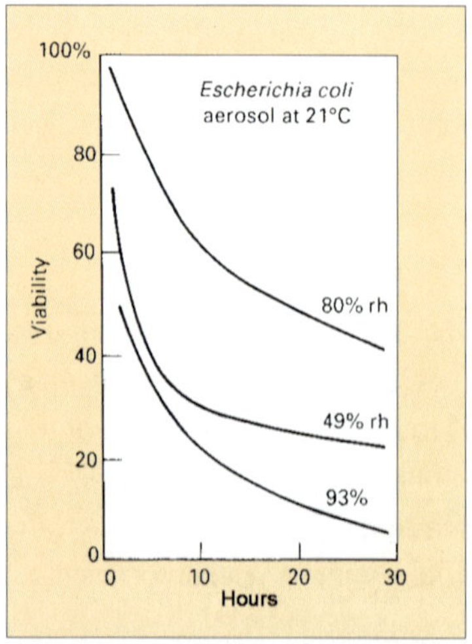

Fig. 9.8
***E. Coli* Bacterium**
The bacterium is most vulnerable at extremely high humidity, but can actually tolerate 80% humidity for longer than 49% rh.

The cell membrane is a very thin layer of protein and lipid (fat) and is freely permeable to water and gases. The cytoplasm itself is typically 80% water and all the metabolic activities occur in solution. Exposure to a drying atmosphere would first remove any surface moisture, and then start to desiccate the cell itself. The speed of desiccation is influenced by both the cell and the atmosphere. The water within the cell is bound with different bonding strengths, and release rate from the cell also depends upon the permeability of the outer wall. The sensitivity of the cell to internal water content is also affected by the metabolic rate. In general, cells with a low metabolic rate survive better in aerosols. Some also believe there is a critical cell moisture content at which the cell is particularly sensitive to damage from external pollutants.

Experiments with bacteria are more readily carried out than those with viruses. Culture techniques are used to sample an aerosol containing the bacteria at frequent intervals. This involves collecting a known volume of the aerosol at frequent intervals and culturing the microorganisms on a nutritional base medium. However, bacteria show the same idiosyncracies as the viruses. Escherichia coli in an aerosol survive well at 80% rh and die quickly at higher relative humidities. This is illustrated in figure 9.8 (Anderson, 1966). An aerosol containing the bacterium *Legionella pneumophila*, the causative agent of Legionnaires' disease, survives best in an atmosphere of 65% rh and dies quickly at 30% rh (figure 9.9 - Hambleton, 1983).

Experiments with the bacterium *pneumococcus* aerosolized from a broth showed two distinct lethal processes. The first was the most rapid and lasted for 5-20 minutes. The second death rate process proceeded more slowly and lasted for two hours or more. Both killing processes gave linear logarithmic survival plots. These are illustrated in figure 9.10 (Dunklin et al, 1948). The slope of these two lines is termed the logarithmic decay constant K. The short-term decay is K_1, the long-term decay K_2. When these survival experiments were carried out over a wide range of relative humidities, they revealed a very narrow range of relative humidity in the vicinity of 50% which is particularly lethal to this relatively delicate microorganism. Similar, somewhat less dramatic effects were observed in tests using *Staphylococcus albus* and *Streptococcus hemolyticus*.

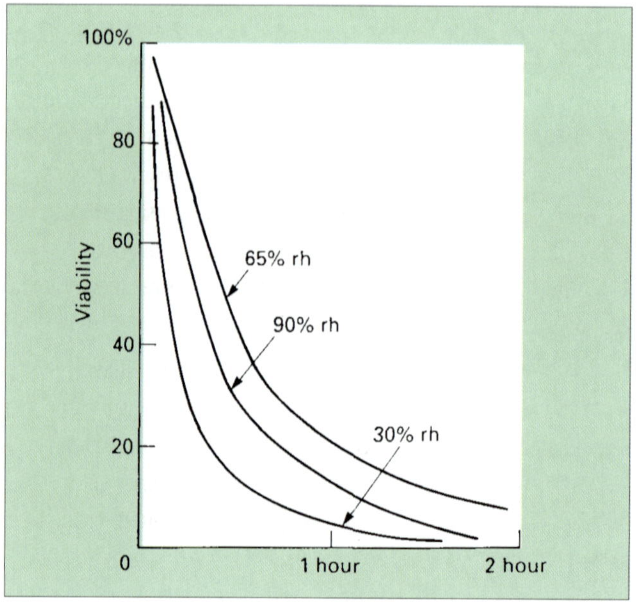

Fig. 9.9
***Legionella Pneumophila* Bacterium**
Low humidity is most effective in limiting the life span of this bacterium.

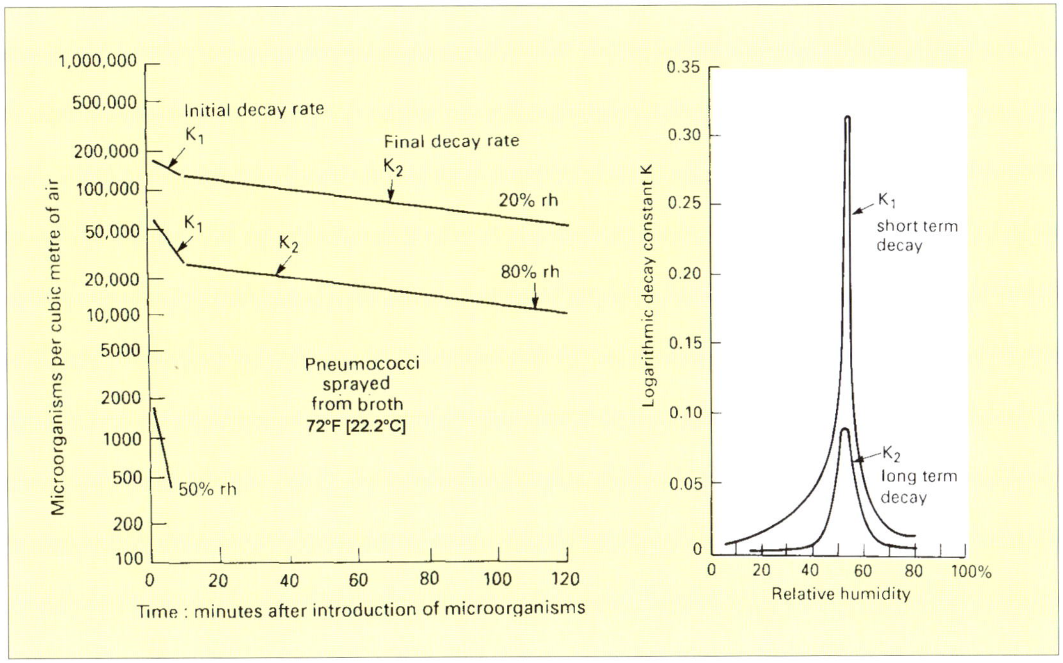

Fig. 9.10 *Pneumococcus* Bacterium

Maximum mortality for airborne Pneumococci comes when the relative humidity is held at 55% rh. This is true for both long and short-term death mechanisms.

When these experiments were repeated using an aerosol of the bacteria generated from distilled water rather than the broth, the sharp lethal effect of relative humidity at 50% disappeared. The pneumococcus was relatively insensitive to the ambient relative humidity except for a slight enhancement in survival for relative humidities below 20%. When the experiments were repeated using *human saliva* as the suspension medium, the survival curves returned to those of the broth, with a sharply defined lethal zone of relative humidity around 50%.

These experiments suggest that salts and other solutes in saliva and broth are concentrated during evaporation, coating the cell and interfering with its natural water balance. The coatings may dehydrate the microorganism to the point where it becomes vulnerable. The lethal action of the salt probably reflects its interference with one or more essential enzyme systems.

These complicating factors are also influenced by aerosol droplet size, which determines the degree of chemical concentration. So it becomes difficult to predict the influence of relative humidity precisely. The complication increases further when irradiation is involved.

The bacteriocidal action of sunlight was recognized in 1877 and the lethal action of radiation on microorganisms is now well established. The relative effectiveness of wavelength in killing bacteria is illustrated in figure 9.11. In a general way this curve resembles the radiation absorption curve of nucleic acid. This mortality relationship appears to be the same for most types of bacteria. Mortality is highest at 0.26 µm wavelength (Koller, 1965).

The lethal effect of the radiation is a function of the total amount of energy received by the microorganisms. The effects of weak radiation over a long time are similar to those of strong radiation over a short time. Time and irradiance are multiplied to give exposure.

Experiments designed to simulate sunlight explored the effect of irradiation and relative humidity on the death rate of the bacterium *Pasteurella tularensis* suspended in both dry and wet aerosols. Sunlamps used a filter to eliminate wavelengths shorter than 0.305 µm (Beebe, 1925).

The results for the *wet* disseminated aerosol showed a strong influence of both irradiance and relative humidity. The kill rate increased linearly with decreasing relative humidity and with increasing irradiance (figure 9.12). Surprisingly, the experiments using a *dry* dispersion of bacteria

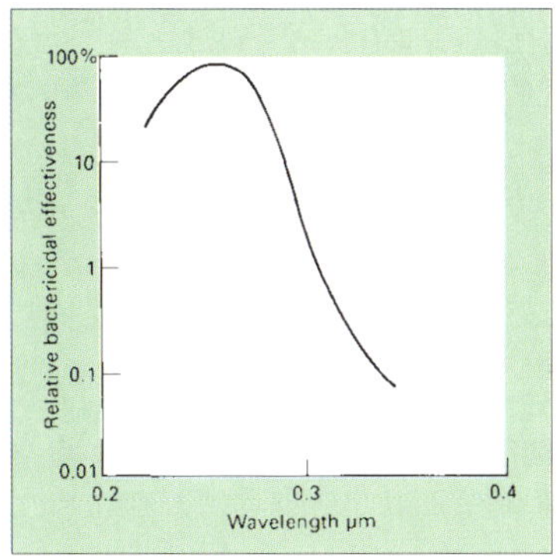

Fig. 9.11 Bacteriocidal Effectiveness of Light

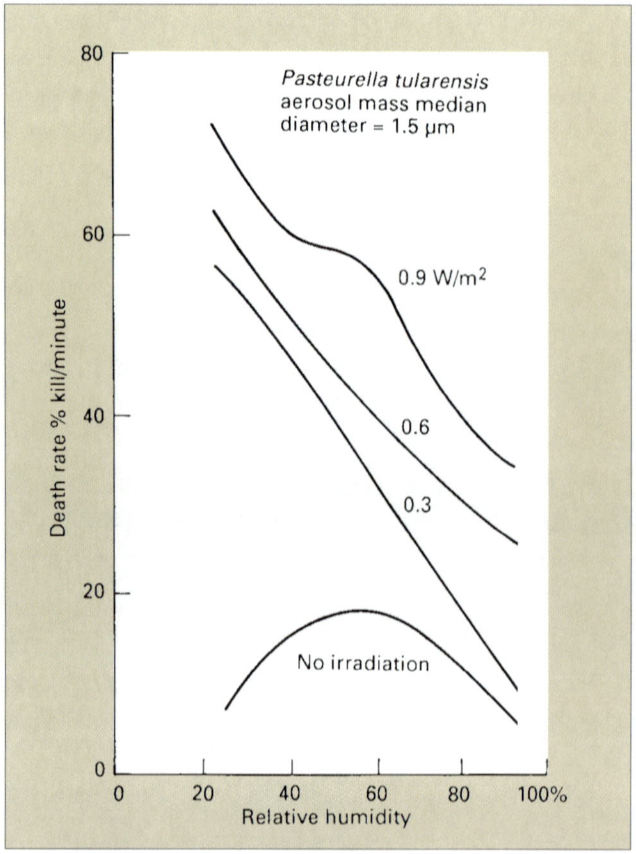

Fig. 9.12 Radiation and Relative Humidity

Death rate of the bacterium pasteurella tularensis as a function of both relative humidity and radiation.

showed a strong effect of irradiance but only a small effect of relative humidity. However, the dry particles consisted of an organic matrix comprising not only the bacterial cells but also milk solids and sugar.

There has been little research on the viability of bacteria in dust. The water adsorption isotherm for the bacterium *Serratia marcescens* is illustrated in figure 9.13 (Bateman et al, 1962). The cells will equilibrate with ambient relative humidity and take up or lose water. In dry conditions (below 20% rh), the cells become desiccated but are relatively immune to damage. At very high humidities the cell is well endowed with water and prospers. At some in-between state there may be enough water present in the cell to permit some metabolic processes to continue but not all. The cell may be particularly susceptible to damage in this stressed state.

There is concern that the incidence of bacterial infections is increasing and that bacteria in dust may be a contributing factor. Outbreaks of infantile gastroenteritis have been reported from hospitals and traced to *Escherichia coli*. Some strains of the organism have been isolated from the air and dust samples from within hospitals. Nosocomial (hospital-specific) pseudomonal infections are becoming more commonly reported. *Pseudomonas aeruginosa* has been isolated from several hospital sites, and bedpans and urine bottles are often heavily contaminated. Cross-infections with *Salmonella* have also been reported.

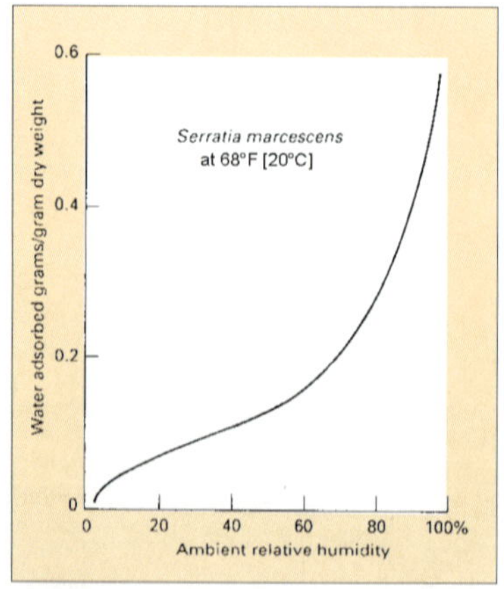

Fig. 9.13

Moisture Adsorption by Bacteria

Adsorption isotherm for the bacterium Serratia marcescens.

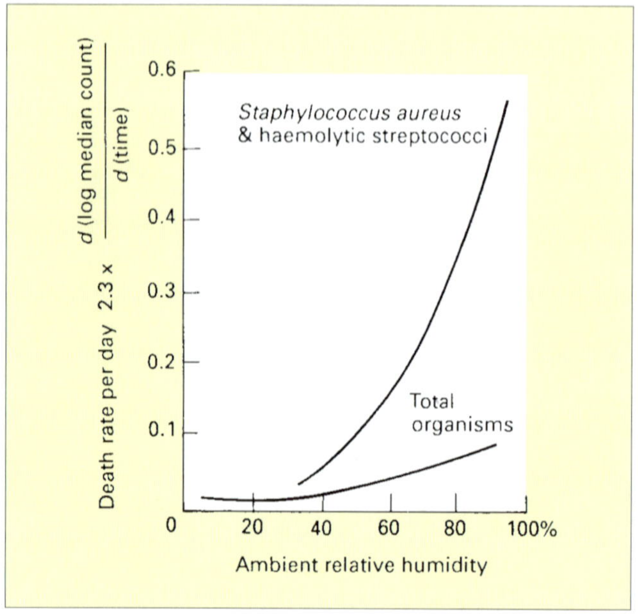

Fig. 9.14 Bacteria in Dust

Death rate for two infectious bacteria in dust, compared with the death rate for all bacteria found in that dust.

Dust from hospitals which had been shown to contain *Streptococci* was exposed to a range of controlled humidities for up to 35 days. The results showed a correlation between atmospheric relative humidity and the death rate of bacteria. (figure 9.14 - Lidwell et al, 1950) Survival was much higher for total bacteria at the lower humidities, and high relative humidities were particularly damaging to the *Staphylococcus aureus* and streptococcal bacteria.

Other experiments have explored the influence on survival of both relative humidity and the kind of surface on which the bacteria rests. Glass, metal and ceramic tile surfaces were used but there was very little difference between them. Relative humidity had a strong effect on both *Pseudomonas vulgaris* and *Proteus morgani*. Survival was very good at 11% rh, 77°F, [25°C], and viable organisms could be cultured after one week. At 53% and 85% rh the surviving bacteria were less than one-thousandth of the initial concentration after 48 hours. These results are illustrated in figure 9.15 (Lidwell et al, 1950).

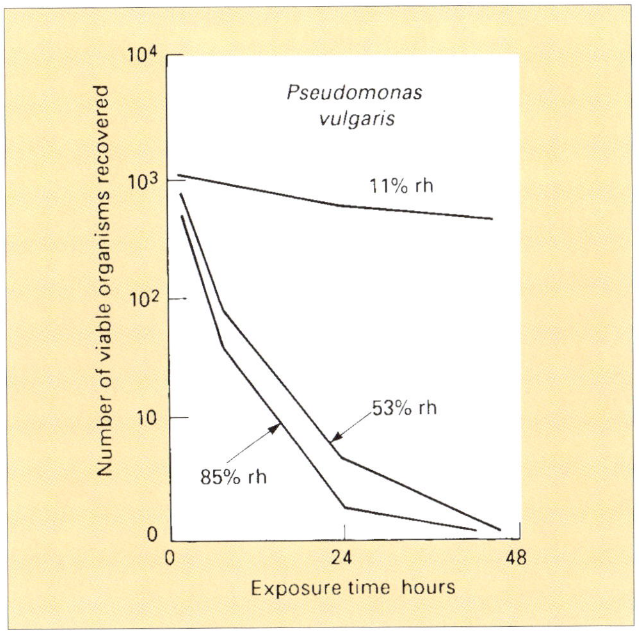

Fig. 9.15
***Pseudonomas Vulgaris* Bacterium**
Survival rates for Pseudonomas vulgaris on metal surfaces at different relative humidities when the temperature is held at 77°F [25°C].

Humidity Influence on Effectiveness of Disinfectants

Gases have long been used as disinfectants. Only in recent years has the importance of the ambient relative humidity become recognized. The effect is still a complex one because it interrelates with the state of hydration of the bacterial cells themselves, the presence of any protecting substances on the surface of the cell and the adsorption of moisture onto the dust or solid surfaces on which the cells rest. (Russell, 1976)

Ethylene oxide gas is widely used, particularly for sterilizing plastic medical supplies. The gas is freely diffusible and penetrates many packaging materials such as paper, cellophane, cardboard and polyvinyl chloride plastic, although it is less readily diffusible through polyethylene. The gas can carry moisture through hydrophobic films which normally have very low moisture penetration. Likewise, moisture can carry gas through polar films which allow water to permeate readily but would normally impede the diffusion of the gas.

Under optimum laboratory conditions the maximum bacteriocidal effect for ethylene oxide is around 33% However, the amount of moisture inside the cell is an equally critical factor. Dehydrated cells are much more resistant. So the recommended operating conditions for disinfection are 30-50% rh and slightly higher if the contamination is heavy. The effectiveness is greater if the cells lie on moisture-absorbing surfaces.

Propylene oxide gas *loses* its effectiveness with increasing relative humidities when the bacterial cells are only air-dried and not fully desiccated. In practice, environmental conditions of 30-50% rh are recommended. Higher relative humidities are recommended if the cells have been desiccated.

Formaldehyde gas is most effective at *higher* relative humidities. The effectiveness is enhanced rapidly with increasing relative humidity up to 50% and thereafter at a lower rate. In practice an ambient of 75% rh is recommended.

The effectiveness of propiolactone *increases with increasing relative humidity* up to an optimum value of 75-85%. This is more directly related to the water content of the bacterial cell than to moisture in the air (Coney, 1961).

Hydrogen peroxide vapor is becoming popular as a disinfectant because unlike other gases, it leaves no toxic or carcinogenic waste products after disinfection. Hydrogen peroxide is water with an extra oxygen atom (H_2O_2). The molecule is unstable, releasing the extra oxygen atom, which oxidizes the cell wall of a bacterium—essentially burning holes in the cell wall. However, to be sure the oxygen atom is released near the organism, the vapor must be adsorbed onto the contaminated surfaces. High humidity before disinfection interferes with that adsorption. To achieve the highest kill rates, surfaces are pre-dried with dry air (a dew point below 20°F [-7°C]). (Jones et al, 1993)

Sulphur dioxide is still used to fumigate agricultural and industrial buildings such as greenhouses. Its effectiveness improves with increasing ambient relative humidity. It is 20 times more effective at 96% rh than at 75% rh in killing the spores of *Botrytis cinerea.*

Summary

Research that quantifies the influence of relative humidity on the infectivity of viruses and bacteria is rare and valuable. Relative humidity is an important factor for most microorganisms, along with radiation and temperature, but each microorganism appears to display its own characteristics. In practice, survival of microorganisms is shorter outdoors than in laboratory tests. This is attributed to an 'open-air' factor, believed to be partly irradiation from sunshine and partly industrial contamination.

Research concerning the survival of pathogens in dust suggests that viability is strongly influenced by the ambient relative humidity. Low relative humidities favored the survival of some of the bacteria responsible for diseases caught in hospitals and nursing homes, namely *E. coli, Salmonella derby, Ps aeruginosa, Pr. vulgaris and Pr. morgani.* Humidification could benefit hospitals which operate in dry climates by reducing the incidence of such nosocomial infections. Dehumidification also benefits hospitals and nursing homes by reducing standing water and condensation that amplifies bacterial growth, and dehumidification benefits all buildings by restricting the growth of fungi which may be toxic or immunosuppressant.

For commercial and institutional buildings other than health care facilities, infection is usually a secondary concern, because most occupants are likely to be healthy most of the time and therefore not emitting airborne infectious agents. *Outside of highly hazardous circumstances, most buildings will be well-served by simple filtration, dilution ventilation and by reducing the potential for infection by shortening the life of airborne microorganisms. The HVAC designer can minimize the life span of the broadest variety of bacteria and viruses by holding humidity above 30% rh during dry months, and below 60% rh during humid months.*

References

Anderson, J.D. 'Biochemical studies of lethal processes in aerosols of E. coli'. J Gen Microbiol, 45, 303-313, 1966

Bateman, J.B., Stevens, C.L., Mercer. W.B. and Carstensen, E.L. 'Relative humidity and the killing of bacteria: the variation of cellular water content with external relative humidity or osmolality'. J Gen Microbiol, 29, 207- 219, 1962

Beebe, J.M. 'Stability of disseminated aerosols of Pasteurella tularensis subjected to simulated solar radiations at various humidities'. J Bacteriol, 78, 18-124, 1959

Coney, H.M. and Uota, M. 'Effect of concentration, exposure time, temperature and relative humidity on the toxicity of sulphur dioxide to the spores of Botrytis cinerea'. Phytopathology SI. 818-819. 1961

Dunklin, E.W. and Puck, T.T. 'The lethal effect of relative humidity on airborne bacteria'. J Exp Med, 87, 87-101, 1948

Ford-Robertson, J. Revise Biology. Charles Letts & Co. Ltd, London, 1981

Green, G.H. 'The positive and negative effects of building humidification'. ASHRAE, 88 (I), 1049-1060, 1982

Guberan, E., Dang, V.B. and Sweetnam, P.M. 'L'humidification de l'air des locaux previentelle les maladies respiratoires pendant l'hiver'. Schweiz med Wschr, 108(22), 827-831, 1978

Hambleton, P., Broster, M.G., Dennis, P.J., Henstridge, R., Fitzgeorge, R. and Conlan, J.W. 'Survival of virulent Legionella pneumophila in aerosols'. J Hyg (Cambridge), 90, 451-460, 1983

Harper, G.J. 'Airborne microorganisms — survival tests with four viruses'. J Hyg (Cambridge), 59, 479-486, 1961

Jones, R., Stuart, D., Large, S., Ghidoni, D., 'Cycle parameters for decontaminating a biological safety cabinet using H_2O_2 vapor.' ACUMEN Series, Vol.1, No.2. The Baker Co., Sanford, ME.,1993.

Koller, L.R. Ultraviolet Radiation, 2nd edition. John Wiley & Sons, New York, 1965

Lester. W. 'The influence of relative humidity on the infectivity of airborne influenza A virus'. J Exp Med, 88, 361-368, 1948

Lidwell, O.M. and Lowbury, E.J. 'The survival of bacteria in dust: the effect of atmospheric humidity on the survival of bacteria in dust'. J Hyg (Cambridge), 48, 21-17, 1950

Lubart, J. 'The common cold and humidity imbalance'. NY State J Med, 816-819, March 1962

McDade, J.J., Hall, L.B. and Street, A.R. 'Survival of Gram-negative bacteria in the environment'. Am J Hyg, 80, 192-204, 1964

Russell, A.D. 'Inactivation of non-sporing bacteria by gases', in Inhibition and Inactivation of Vegetative Microbes (eds F.A. Skinner and W.B. Hugo). Soc of Applied Bacteriology, Symp Series No. 5. Academic Press, London, 1976

Sale, C.S. 'Reducing upper respiratory tract infections by allergy control and humidity control'. Virginia Medical Monthly, 96, 368-371, 1969

Sataloff, J. and Mendule, H. 'Humidity studies and respiratory infections in a public school'. Clin Pediatr, 2(3), 119- 121, March 1963

Wells, W.F. Airborne Contagion and Air Hygiene. An Ecological Study of Droplet Infections. Harvard University Press, Massachusetts, 1955

Chapter 10

Design Procedure

By Lew Harriman & Reinhold Kittler

Peak-Hour Dehumidification Loads For A Medium-Size Retail Store Located In Atlanta, GA

Outdoor Design Dew Point Conditions: 74°F (133 gr/lb), MCDB of 82°F [23.3°C (19.0 g/kg), MCDB of 27.8°C]

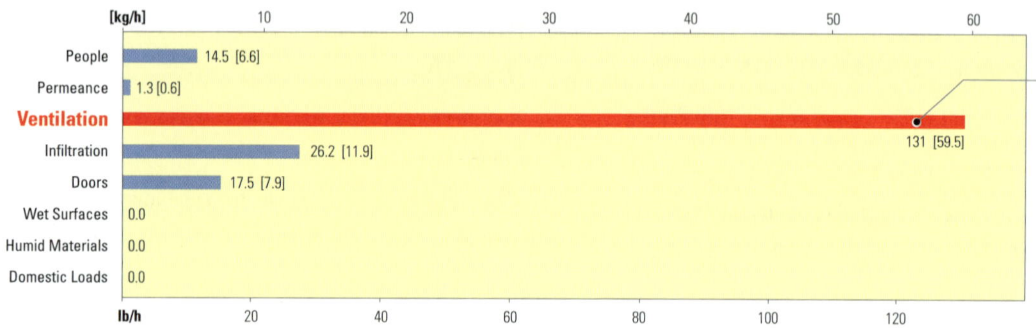

Ventilation Air Carries The Largest Load

That's why it's very helpful to dehumidify and humidify ventilation air with a separate, dedicated preconditioning system when the owner needs true ***control*** *of humidity rather than only moderation of humidity.*

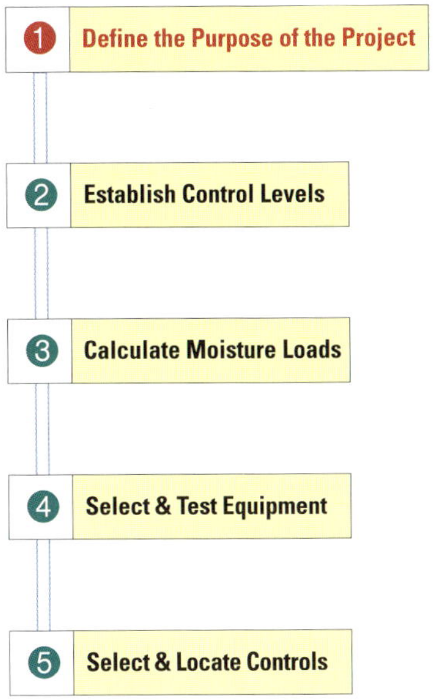

Fig. 10.1 Five-Step Design Process
Designing humidity control projects in this sequence simplifies the process, and avoids costly misunderstandings after the design is complete.

Designing For Humidity Control

When the HVAC designer needs to ***control****, rather than merely* ***moderate*** *humidity, he or she must focus on the moisture level as a separate variable—not merely an artifact of temperature control. Controlling humidity generally demands a separate set of equipment and controls dedicated to that purpose.* This chapter describes a general approach which has proven useful for designing humidity control systems, and moves on to describe how humidity control subsystems can be integrated into common types of HVAC systems that are primarily focused on controlling temperature.

Five-Step Design Process

Any HVAC design can be approached from many starting points. The five-step sequence presented here is based on the approach taken by manufacturers of precision humidity control systems. (Harriman, 1990) Those experts have found that often, clients are not clear on the exact nature of their need for humidity control. By defining those needs up-front, everyone saves considerable time and cost.

Step 1 - Defining the purpose of the project

We assume that you are using this book because an owner has requested control of humidity. Given that request, we can also assume there is a reason or reasons why humidity control is important to that owner. Any project must begin with a clear understanding of the owners' purpose. The HVAC designer needs that understanding to make decisions that balance humidity precision against cost.

For example, if the purpose of the project is to keep workers comfortable in an office, the designer can rely on human comfort research to set control levels and tolerances that most occupants will find acceptable. Human tolerance for humidity variation is quite broad, and occupant expectations change naturally with changing seasons. Such a system can be quite simple and economical. On the other hand, if the purpose of the project is to provide a consistent humidity level for medical research, the owner may have different needs for each research project. The system which provides individual, room-level humidity control is likely to be complex and costly.

The value of humidity control is different between these two cases because the economic consequences of failure to control humidity are very different. In the office environment, humidity variations are merely uncomfortable. In the laboratory, humidity excursions could be catastrophic, losing millions of dollars in research results and costing time to repeat experiments that may not be available at any price. The differing consequences carry different cost implications for the system designer. In the office, the owner neither needs precision, nor expects to pay for it. In the lab, the owner needs precision, and should expect to pay more than the cost of less consistent or less-tailored humidity control.

To avoid expensive and time-consuming misunderstandings and wasted effort, a prudent designer seeks a clear definition of project purpose from the owner. Ideally, the project purpose will be summarized in a few sentences which define:

- **What result is expected**
- **What may happen if the result is not achieved**
- **What shortcoming is acceptable**

Although different clients have different needs, each building type tends to need humidity control for similar reasons. Some good examples of project purpose definitions include:

Office Building: "Control humidity so that:

a. Most of the workers are comfortable all year, minimizing absenteeism.

b. High-speed printers do not jam from excess or inadequate humidity

c. Moisture does not condense inside building cavities.

d. The system must accomplish these goals for all but 2% of the occupied hours in the year."

Library: "Keep humidity *low* enough to be sure that books and carpets will not absorb enough moisture to support mold growth. This must be accomplished 24 hours a day, 365 days a year. Also, keep humidity *high* enough during all occupied hours to minimize discomfort for patrons and staff."

Hotel: "Control the temperature and humidity in the hotel such that:

a. Neither condensation nor high humidity can occur inside the building or inside building cavities to cause mold or other moisture damage to the building or furnishings.

b. Guest rooms, corridors, meeting rooms, dining rooms and offices are maintained within the ASHRAE-defined human comfort zone during all occupied hours.

c. The system should achieve the above goals for all but 2% of the hours in a typical year."

Given a clear definition, the designer and owner can proceed to the next step: agreement on what humidity control levels and tolerances will achieve the purposes of the project.

Step 2 - Setting moisture control levels and tolerances

The HVAC system by itself cannot keep office workers content, nor prevent mold and mildew damage to buildings. Such goals are achieved only when all elements of the building construction, its operation, occupation and use are controlled in harmony. The HVAC system only keeps temperature, humidity and pressure of the air in the building within defined ranges based on defined loads.

So after the project purpose is understood, the owner and HVAC designer must agree on what humidity range and tolerance will best help the owner achieve the overall result.

To help with this task, this book provides information about humidity's effects on people, products and buildings in Chapters 4 through 10. For example, the designer can consult Chapter 4 to find how different ranges of humidity affect human comfort. That chapter does not define a requirement for any specific range, since comfort is rather subjective. But the chapter provides enough information for the designer and owner to discuss and agree on what range is appropriate to achieve the level and consistency of comfort that the owner is willing to pay for. Likewise, Chapter 7 explains how mold and mildew grow, which can help the building team decide what humidity level will minimize that growth.

As an example, consider the project purposes suggested by the fictional owner of an office building: "Control humidity so that:

a. Most of the workers are comfortable all year.

b. High-speed printers do not jam from excess or inadequate humidity.

c. Moisture does not condense inside the building.

d. The system accomplishes the goals above for all but 2% of the occupied hours in the year."

Regarding comfort, the designer consults Chapter 4 and learns that humidity-related comfort is quite subjective, but most people will be comfortable when humidity is kept below 60% rh in the summer and above 30% rh during the winter.

To keep printers operating smoothly, the designer contacts the manufacturer of the high-speed printers in question. He learns that while the equipment will theoretically operate at any humidity between 20% and 75%, normal variations in paper type and quality suggest that there will be fewer jams if humidity is kept between 30 and 60%.

To minimize condensation, the HVAC designer consults the Architect to understand the design details for the building's exterior envelope. After a dew point analysis, the designer concludes that condensation is unlikely as long as the HVAC system does not force cold air into building cavities and as long as the dew point inside the building is kept below 32°F in the winter and below 60°F during the summer [between 0.0°C in winter and 15.5°C during summer]. Keeping the dew point down during the winter reduces diffusion of moisture into cavities from inside the building. During the summer, the designer looks at duct surface temperatures and inside wall temperatures, and concludes that keeping the air dew point below 60°F [15.5°C] provides reasonable assurance that moisture will not condense in building cavities or behind walls.

Finally, the designer needs to decide which extreme weather condition will be used to calculate the moisture loads for sizing the dehumidifiers and humidifiers. That decision should also be made jointly with the owner, because it must balance cost against consistency of control. If a very extreme condition is chosen, the equipment will have capacity to control humidity for all but a few hours each year. On the other hand, larger equipment is expensive and difficult to control during 98% of the year when humidity is *not* extreme. If a less-extreme condition is chosen the equipment will cost less and be easier to control, but there will be more hours every year when the system may lack capacity.

To assist the designer and owner, ASHRAE has calculated summer and winter extremes based on 8 to 30 years of hourly weather observations. The extreme conditions are categorized as percentages of the long-term record. For example, the 0.4 % dew point temperature represents the moisture content that is not exceeded for more than 0.4% of the hours in the long term record. Assuming a typical year, one could also assume that the 0.4% dew point is not likely to be exceeded for more than 35 hours (8760 x 0.004 = 35). ASHRAE publishes extremes of 0.4%, 1%, 2%, 99% and 99.6%. With these values, the owner can decide how many hours each year are acceptable to risk having the system be short of capacity. For dehumidification, the choices are 35 hours, 87 hours or 175 hours. For humidification, the choices are 35 or 87 hours. The ASHRAE dew point design values for 1453 worldwide locations were first printed in Chapter 26 of the *ASHRAE Handbook—Fundamentals* in 1997. (The 1993 and earlier editions of that volume do not include peak dew point data.) The same values that appear in the current *Handbook* are included in Chapter 32 of this book for easy reference.

To continue and summarize our hypothetical office example, we'll assume the owner has decided that minimizing construction cost is more important than year-round control, and that if humidity is less-than-ideal for 87 hours in winter and 175 hours in summer, no great problems will result. Yes, occupants may be less comfortable, and perhaps the printers may be more subject to jams for 262 hours each year—but that degree of risk is acceptable to this owner. Consequently, the specification becomes: "Maintain the moisture content of air in the building between dew points of 30 and 60°F [between -1.1°C and 15.5°C] for all but 262 hours per year."

This example illustrates how an HVAC designer can help the owner convert his needs into a specification that can be accomplished by mechanical equipment. Each building and each owner may have different needs that dictate a different balance between precision, consistency and cost. But this balance should reflect conscious decisions by the owner rather than assumptions by the HVAC designer. This collaborative approach may consume extra time at the beginning of the project, but the owner's expectations will match the designer's intent. Early collaboration saves large amounts of time, money and aggravation by minimizing unpleasant surprises later, when the weather gets humid.

After the control levels and tolerances are decided, and the extreme design conditions are established, the designer can proceed to calculating the peak moisture loads.

Step 3 - Calculating peak moisture loads

Calculating moisture loads is both easier and more difficult than calculating heat loads: easier because the equations are very simple, but more difficult because the designer must make judgements about what numbers to put into those simple equations.

There are two critical aspects of moisture load calculations that are not obvious, even to people who may be expert in other aspects of HVAC design. First, the extreme design condition must be based on moisture, not temperature. Second, the designer must apply much more judgement to moisture load calculations than is the case for heat load calculations. To date, research has been concentrated more on temperature than on moisture, so there are many "unknowns" that affect moisture loads versus far fewer unknowns for heat loads.

Extreme design for moisture

Until late 1997, designers lacked data describing extreme moisture conditions. Before publication of the 1997 edition of the *ASHRAE Handbook—Fundamentals*, the society only published temperature extremes. Worse, the 1993 and earlier handbooks appeared to provide data for moisture extremes, because the average wet bulb temperature was included for hours when the dry bulb temperature is at its peak. Without extreme dew point information, designers combined the extreme dry bulb temperature with its average coincident wet bulb temperature and assumed that the resulting moisture content was the peak moisture. Unfortunately, that is a very inaccurate assumption which greatly underestimates the maximum peak load.

So not only was there no published data for moisture extremes, the ASHRAE handbook *appeared* to contain credible moisture information—showing the engineering community a very misleading value for dehumidification design. Dehumidification systems based on temperature extremes fall far short of capacity when moisture really reaches its peak—which happens at lower temperatures. Figure 10.2 shows how using peak temperature in place of peak moisture can distort the designer's understanding of the moisture load. In summary, the first critical point for the designer to remember is to use peak moisture and not peak temperature values for calculating moisture loads.

The same problem does not occur in calculating humidification loads. While it is also true that the lowest moisture occurs at different times than the lowest temperature, the moisture differences are extremely small—less than 3 grains per lb [Differences are less than 0.4 g/kg]. Such small differences do not affect the sizing of humidifiers in any meaningful way in most commercial buildings.

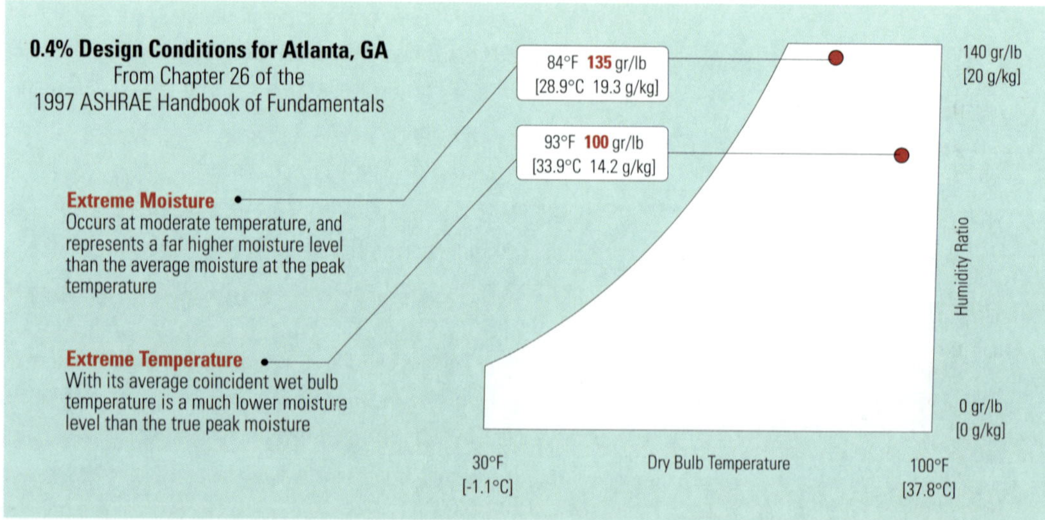

Fig. 10.2
Design Moisture vs. Temperature
For moisture load calculations, it is important to use the peak dew point, not the peak dry bulb temperature as the design extreme. As shown above, the designer would seriously understate the maximum moisture if the peak temperature were used for moisture load calculations. (ASHRAE, 1997)

Engineering judgement in calculations

The second critical point to bear in mind is that a great deal of engineering judgement is needed in calculating moisture loads. Consequently, three equally well-qualified designers may conclude the same building has three different moisture loads.

This fact is discomforting to many technical professionals (and to many clients). One naturally wishes for close agreement on a task which appears to only require arithmetic. But the numbers that go into these simple equations are not always easy to agree on. Once again, the designer is well-advised to consult with the owner and the Architect, to be certain that all parties share the same assumptions.

For example, infiltration air is usually the second-largest element of the moisture load. Infiltration depends on building tightness, which in turn depends on the specifications for the building envelope and on the conformance of the contractor's installation to those specifications. There is no historical evidence to suggest that the infiltration rate can be determined accurately without actual field measurements.

A study of eight medium-size federal office buildings in seven different geographic locations measured leakage rates between 0.15

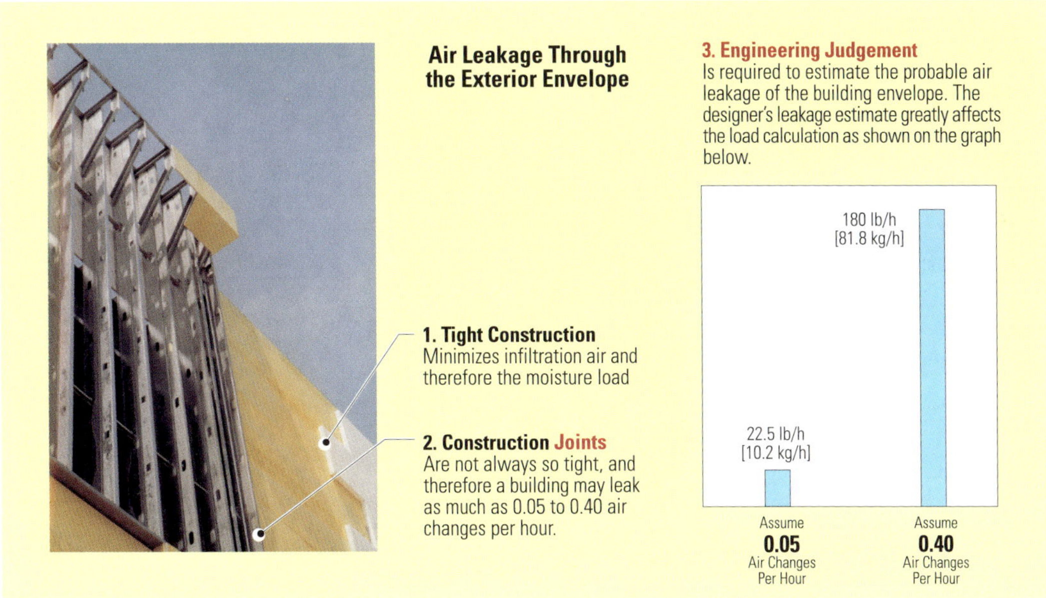

Fig. 10.3 Infiltration Rate Assumption

Infiltration greatly affects the moisture load. Since this and similar assumptions heavily govern load calculation—and therefore the system design—a prudent designer obtains agreement on all critical assumptions from both owner and Architect.

and 0.55 air changes per hour. (Persily & Grot, 1986; Persily, 1999) The best available prediction calculations for those same buildings suggested that the air leakage should have been between 0.01 and 0.25 air changes per hour. One of the buildings predicted to have a leakage rate of 0.01 air changes actually had a measured air leakage of 0.40 air changes per hour. Clearly, if the designer assumed a moisture load from leakage of 0.01 air changes per hour and the finished construction actually created a leakage load 40 times larger, the humidity control system is going to be short of capacity. So engineering judgement is required, and that judgement is best when it is based on input from the owner and architect about building tightness on similar projects in the past.

Frequently, the designer finds it logistically difficult or impossible to obtain such neat, collaborative agreement on key load assumptions. Consequently, the designer makes an estimate based on professional judgement concerning probable building tightness, normal entry and exit traffic, ventilation air rates, likely wind direction and pressure when the doors are open, probable length of time the doors stay open on each transit—and many other difficult-to-quantify variables that affect the load. That's why equally well-qualified professionals may calculate different moisture loads for the same building. To a much greater extent than structural or cooling and heating calculations, experts may disagree on moisture loads. At the very least, documenting critical assumptions for the architect and owner to review can be helpful. Conscious agreement between team members will minimize potential for misunderstanding and costly contention at later, more expensive stages of the project.

In summary, two aspects of moisture load calculation are critical to success: using peak moisture rather than peak temperature for dehumidification calculations, and obtaining agreement on key load assumptions from the owner and Architect—especially the assumptions concerning the amount of ventilation air and likely air leakage through the building envelope. Once the team agrees on the loads, the designer can begin to select, size and place the components of the system.

Step 4 - Selecting, sizing and placing equipment

Regardless of the type of equipment selected, its size or its placement, a prudent designer examines its performance at other than peak load conditions before finalizing any of these decisions. An equipment decision that looks optimal at peak load conditions can be seen as a poor decision when examined at off-peak conditions. By definition, the equipment only sees the peak load for 2% or fewer of the hours of its 15 to 20-year life. For the other 98% of that product's life, it will operate at off-peak conditions. Its behavior under those circumstances determines the cost of its operation and the degree to which it achieves the owner's purposes for the project.

Ideally, the designer uses an hourly building simulation program to examine the building's response to loads in each of the 8760 hours of a typical year. Hourly analysis makes it easier to identify any weaknesses in the design as the building goes through daily and seasonal changes in outdoor loads and internal demands. Also, as time-of-use utility rates become the norm, hourly analysis is essential for accurately estimating the yearly cost of operation for the building. In recent years, this has become more practical as more HVAC designs are computerized and associated electronically to the architectural, site and operational characteristics of the structure.

Even when hourly analysis is not practical, a designer can greatly improve the quality of equipment selection by examining system behavior less than full-load. Unfortunately, there is neither common agreement nor any published part-load conditions for either temperature or humidity control. Each designer determines what part-load conditions are appropriate to "test" the design. For humidity control, many part-load conditions have been used, including:

a. Average annual dry bulb temperature, at saturation. This is obtained from the climatic design conditions in the *ASHRAE Handbook—Fundamentals.*

b. The dew point temperature midway between the peak dew point and the dry bulb temperature used for heating load calculations. This is also obtained through information contained in the climatic design tables in the *ASHRAE Handbook.*

c. The temperature and moisture condition at the approximate center of the area on the psychrometric chart bounded by 50% of the hourly observati?ns in the long-term record. This can be obtained by creating a contour plot from data contained on the ASHRAE CD of joint frequency tables of long-term hourly temperature and moisture observations. (Also see Figure 10.4)

After the equipment selection and placement has been tested at part-load conditions, the designer can consider the selection and placement of sensors and controls.

Step 5 - Selecting and placing sensors and controls

Because of cost limitations, placement of humidity sensors can become quite important in commercial buildings. To minimize cost, sensors usually measure relative humidity rather than dew point, regardless of which parameter actually controls the success of the project. Consequently, temperature changes affect the relative humidity reading, which makes the "geography" of sensor placement important.

For example, consider the office building described earlier in this chapter. The owner wants humidity held between dew points of 30 and 60°F [-1.1 and 15.5°C]. To minimize cost, a humidistat measuring relative humidity will send a signal to the building automation system. That system will also take a dry bulb reading, convert the relative humidity to a dew point value, and control the dehumidifier and humidifier accordingly. But where should that sensor/controller be located? For best results, it will be located near the most relative humidity-sensitive items—the high-speed printers. That location suits the purpose of the project far better than allowing the humidistat to be placed arbitrarily "in the controlled space". Without a specified location, it may be placed at the convenience of the installing technician, perhaps ending up near an exterior door, an open window or the coffee warmers where temperature and moisture can both vary widely, sending misleading and counterproductive signals to the humidifier and dehumidifier.

Part-Load Conditions

One reasonable "part-load" condition is the mid-point for temperature and moisture of the observations that represent 50% of the total hours in the long-term record for that location.

This graphic is available for 900 US and international locations, published in the US Department of Defense's *Engineering Weather Data Handbook* (DOD 32-1163)

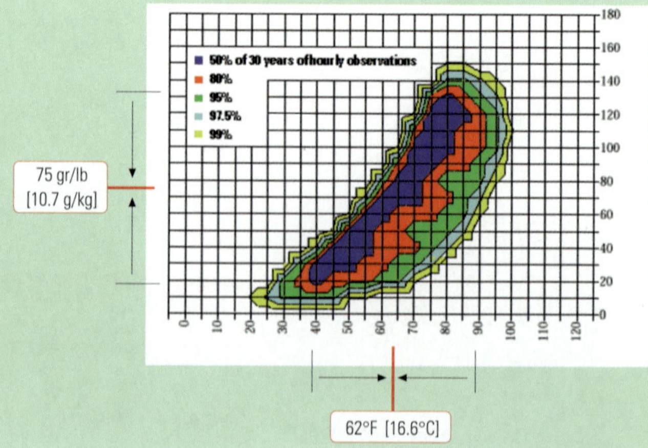

Fig. 10.4 Part Load Conditions

Ideally, the system design will be tested by full-year, 8760-hour simulation. When that is not practical, the designer can still improve confidence in the design by calculating its performance at a "part-load" outdoor air condition—one which is representative of most of the hours of operation at the building's site. This graphic suggests one of several methods for choosing a reasonable part-load condition.

Collaboration Checklist

Defining the Purpose of the Project

1. What is the **owner's purpose** in controlling humidity?
2. **What happens** when humidity is not controlled?

Establishing Moisture Control Levels & Tolerances

3. Which parameter most controls success of the project? ❑ Dew Point ❑ Relative Humidity ❑ Both
4. What range is tolerable, given the purpose of the project? Summer: From _____ to _____ Winter: From _____ to _____
5. Which extreme condition will be used to size equipment? Summer: ❑ 35 hours beyond design ❑ 87 hours ❑ 175 hours; Winter: ❑ 35 hours beyond design ❑ 87 hours

Calculating Moisture Loads

6. What is the extreme moisture and average temperature? Humidity Ratio: _____ gr/lb [g/kg] Dry Bulb _____ °F [°C]
7. What air infiltration rate will be assumed? ❑ **0.10** cfm • ft^2 [0.05 l/s • m^2] ❑ **0.30** cfm • ft^2 [0.15 l/s • m^2] ❑ **0.60** cfm • ft^2 [0.30 l/s • m^2] ❑ _____
8. What is the maximum occupancy of the building? Number of People: _____
9. How much ventilation air should be provided? Min: _____ cfm [l/s] Max: _____ cfm [l/s]

Selecting Equipment

10. What part-load weather conditions are used to check performance of selected equipment? ❑ 8760-hour Analysis ❑ Single-point Analysis At: _____ °F [°C] _____ gr/lb [g/kg]

Placing Sensors

11. Where, specifically, should the moisture be controlled most closely? Room _____ Wall _____ Height _____ Other location constraints:
12. From which other locations is a moisture level signal needed?

Fig. 10.5 Collaboration Checklist

This list contains some of the many decisions that can be improved by collaborative discussions between the HVAC designer, the owner and the architect.

There will be much more discussion of humidity sensor selection, placement and their interface with the control system in later chapters. For now it is enough for the designer to recognize that the issue is quite important to the success of a humidity control project, and to make sure that sensor locations and tolerances are considered and specified clearly in the contract documents.

Integrating Humidity Control into Typical Commercial HVAC Systems

In most commercial buildings, adding humidity control is very simple: add or subtract moisture to the incoming ventilation air, and you will achieve humidity control at reasonable cost with a simple, reliable system. On the other hand, if the outdoor air is allowed to enter the building *without* pretreatment, humidity control will be difficult to achieve at any cost.

This advice is based on field observations, and can be explained by understanding the loads. The outdoor air moisture load is the highest of all load components, and is also the most variable. For example, if a building is located in Chicago, IL is is controlled at 75°F, 50% rh in the summer and 70°F, 30% rh in winter, every 1,000 cfm of ventilation air loads the HVAC system with 41 lbs of excess water every hour at peak design. In winter, that 1,000 cfm pulls 18 lbs of water out of the people, furnishings and the building envelope every hour at peak design. [Every hour at peak design, 472 l/s of outdoor air loads the building with 18.6 kg in summer, and subtracts 8.2 kg in winter.]

Controlling humidity implies maintaining a steady dew point inside the building. Since the outdoor air dew point is never steady, the first and most important task of the humidity control system is to "flatten-out" the range of variation in the incoming ventilation air. In most buildings that need only "mid-range" humidity control, there will be no further equipment needed if the incoming ventilation air is humidified to slightly above the desired dew point in winter, and dehumidified to slightly below the desired dew point in summer.

Fig. 10.6 Outdoor Air Carries The Highest Humidity Load

That's why pretreating the outdoor air before it enters the building usually saves money and simplifies the system.

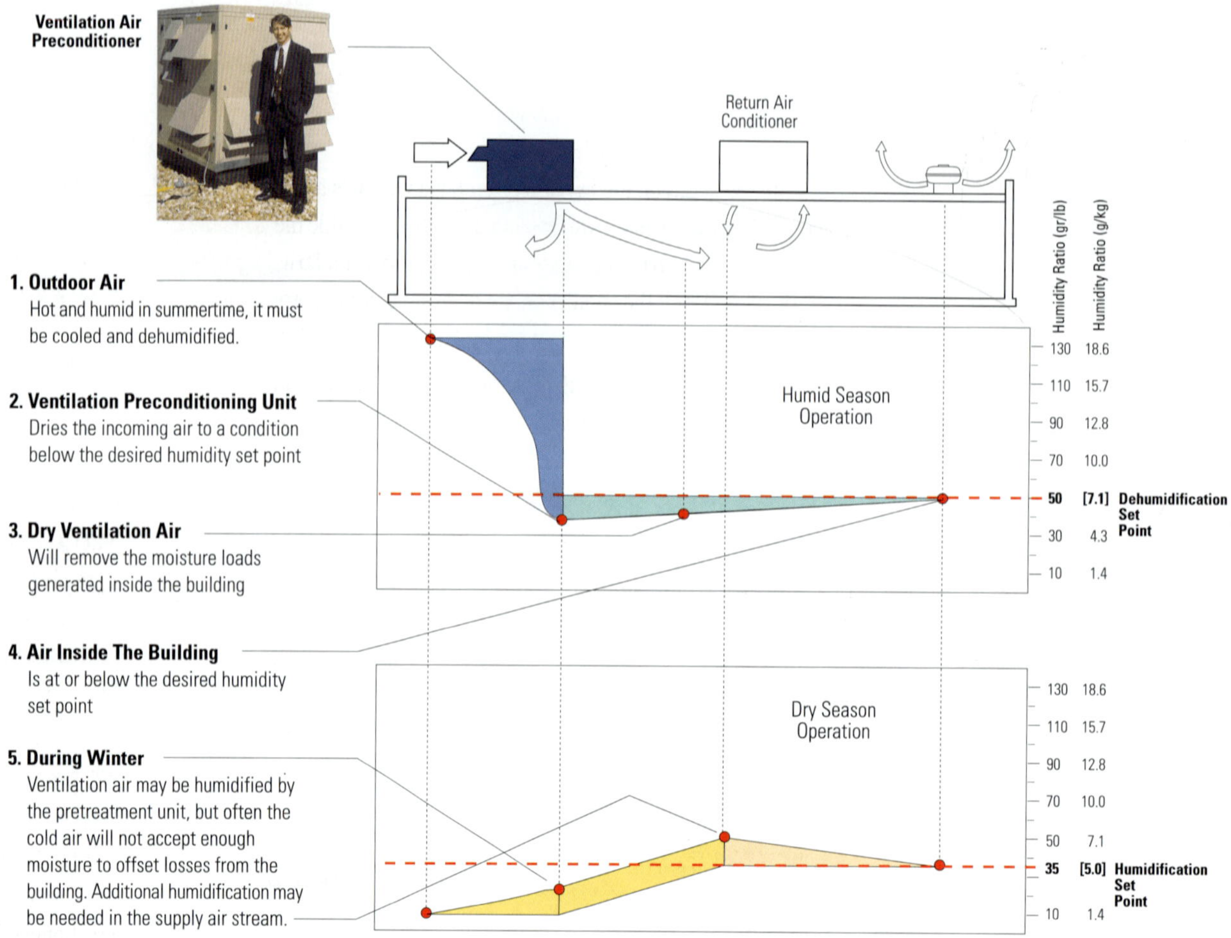

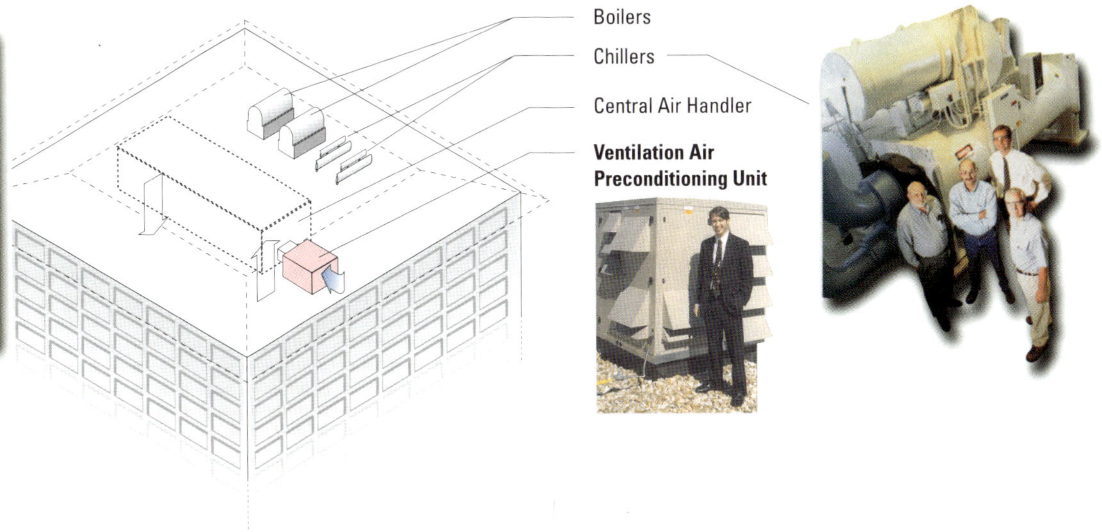

Fig. 10.7 Central Systems

Large or more complex buildings usually have central mechanical systems. Pretreating the ventilation air saves money and considerably simplifies the control system.

In the summer, the overdried ventilation air acts as a "sponge", absorbing excess moisture from loads generated inside the space by people, door activity and infiltration of humid air. Similarly, in winter over-humidifying the ventilation air adds enough moisture to make up for losses from door activity and infiltration of dry air.

By placing the humidifier and dehumidifier in the ventilation air pretreatment system instead of into each recirculating air system, the equipment is smaller, and fewer pieces are needed. So the overall system is simpler, more reliable and less expensive.

Fig. 10.8
The Moisture Content Of Outdoor Air Is Highly Variable
(DOD Engineering Weather Data)

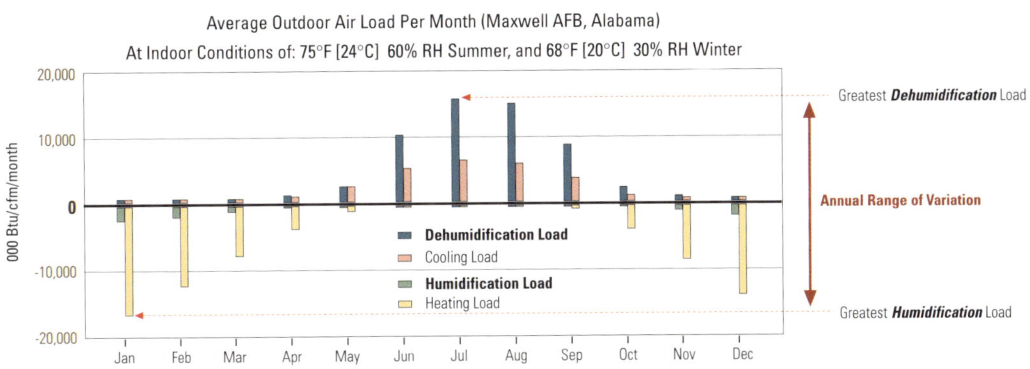

The next section shows simple diagrams for integrating humidity control into major classes of commercial HVAC systems. Most commercial buildings are reasonably complex, and contain a mixture of these systems, but the humidity control designer cannot go far wrong if he or she concentrates on providing control of humidity in the ventilation air before it enters the rest of the system.

Systems based on central station air handlers

Central systems are used in all sorts of commercial buildings, but more frequently in larger structures, where the more common and less costly rooftop or room unit systems are impractical. Typical examples include tall buildings, institutional buildings and similar high-grade, large structures. Figure 10.7 shows a central system with dedicated humidity control equipment which works by pretreating the ventilation air.

Central systems are custom-designed using separate, individually-sized and specified components, and are built-up on-site. With such systems the HVAC designer has great flexibility to optimize, fitting the system to the building precisely.

On the other hand, specifying individual fans, filters, coils, chillers, boilers, heaters, sensors, controllers, valves, strainers, flow-setting devices, refrigerant piping, etc. is quite time-consuming and costly compared to designing a simpler system using packaged rooftop or room units. So central systems are usually used where the cost of the overall HVAC system is quite large. In big, expensive systems, customizing and optimizing each component saves much more money in equipment and installation than it costs in design time.

Another advantage of central systems is reduced maintenance. Heating and cooling is supplied from a few components, and hot water or chilled water is pumped around the building. Large chillers and boilers are centralized, so there are fewer maintenance points compared to systems which use rooftop or room units. On the other hand, maintaining briefly-documented but highly complex central systems demands a high degree of technical skill. Such systems need less maintenance, but owners often have difficulty finding and keeping the skilled staff to perform or supervise that maintenance.

Some tips and traps for integrating humidity control into central systems include:

Consider packaged ventilation pretreatment units, but check their moisture removal capacity at peak dew point design conditions

In recent years, manufacturers have developed packaged systems that pre-treat ventilation air and provide humidity control. These offer the advantage of separating temperature from humidity control, placing responsibility for humidity control in one vendor's hands, and eliminating the need to deal with humidity control issues within the more complex central system.

On the other hand, many of these units are designed to merely reduce the load from incoming air, not to control humidity in the entire building. When considering such packaged systems, it is wise to review the entire building's moisture load in each season with prospective package manufacturers. Then the designer can be certain whether the packaged ventilation equipment will have the capacity to remove internal loads in addition to tempering outdoor air.

If the proposed ventilation package does not have that capacity, the designer will need to include additional humidity control equipment and controls in the central system.

Beware of rising chilled water temperatures in summer

Central systems often rely on chilled water coils for dehumidifying air in the summertime. These are economical to operate and are compact additions to outdoor air handlers. This strategy can be effective when a dedicated chiller is set aside for drying outdoor air. Set the chiller to deliver water cold enough to dehumidify incoming air below the dew point desired in the space. Keeping the chilled water temperature low is not always easy. For example, if the coils

Fig. 10.9
Check Pretreatment Unit Capacity
Pretreatment systems are sometimes designed to really control humidity rather than just to moderate it. If the unit can dehumidify below the control condition, there may be no need for additional dehumidification equipment in the rest of the system.

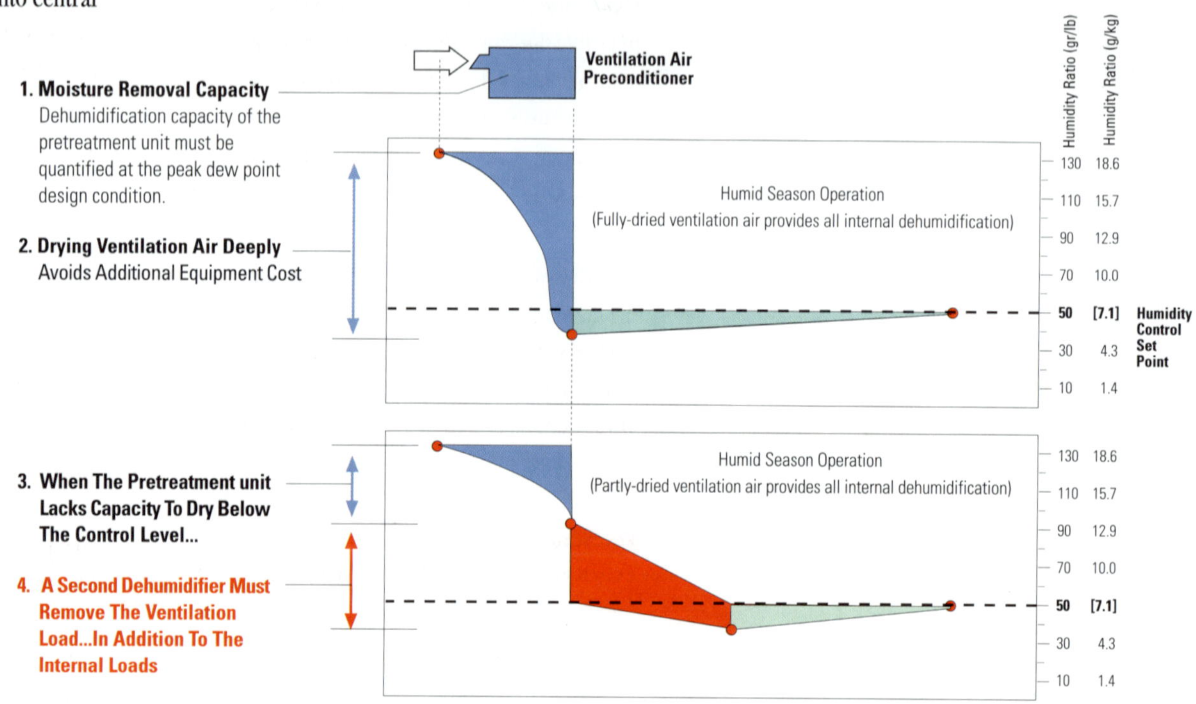

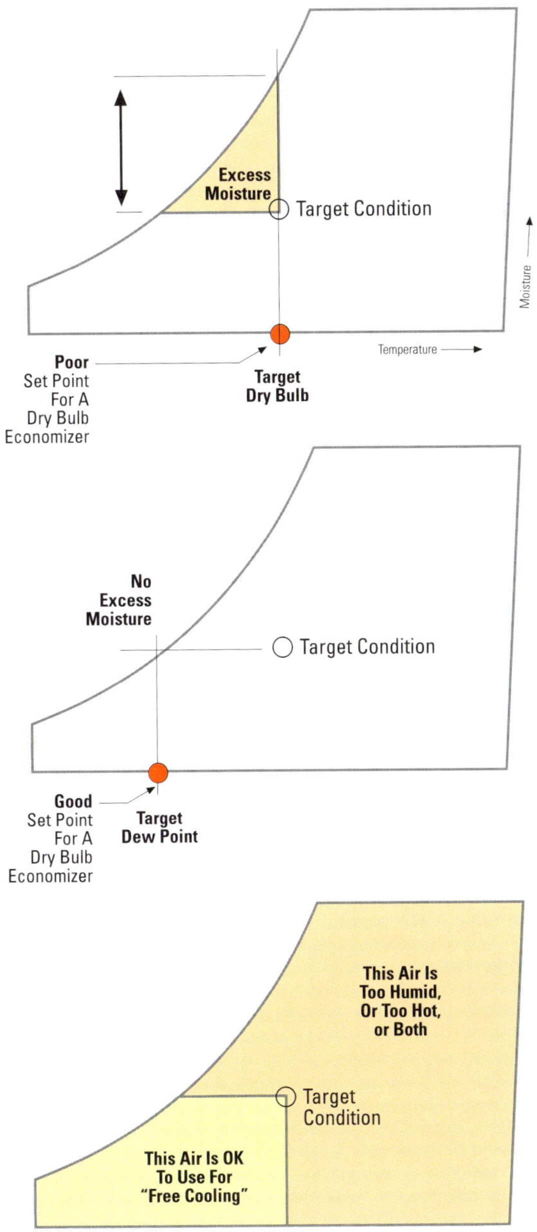

Fig. 10.10
Economizers Must Be Set by The Dew Point
Air-side economizers are often controlled by dry bulb temperature or enthalpy. These will prevent humidity control unless the economizer decision also considers the dew point of the outdoor air.

are served by the main chiller rather than a smaller, dedicated unit, heat loads in other parts of the building may cause the chilled water temperature to rise before it reaches the outdoor air handlers.

Dedicating a chiller to outdoor air avoids this problem, provided the supply water temperature is set low enough, and provided that the piping from the chiller to the coils is in parallel. Chilled water should go directly from the low-temperature chiller to each coil. Piping the coils in series would allow the chilled water to gain heat from the first few coils so it may not be cool enough to dehumidify by the time it gets to the last ones in line.

Be especially cautious with air-side economizers

Large buildings sometimes use an "air-side economizer" as a means of providing low-cost cooling when the air outdoors is cold. These units automatically circulate outdoor air instead of return air when the air outdoors is below the temperature in the building, and when the thermostat calls for cooling. Using cold outdoor air is sometimes called "free cooling."

However, although an economizer may reduce the cooling bill, it severely interferes with humidity control when the decision to bring in outdoor air is based only on dry bulb temperatures. Cool air nearly always contains excess moisture during summer, spring and fall. So while the building might be cooled by the outdoor air, the building will also load up with moisture.

The problem can be avoided by investing in a more complex control system. An economizer may still be effective if the decision to use outside air is based on *both* temperature and dew point. In that arrangement, the controls first measure and compare the desired dew point inside the building versus the dew point outdoors. If the outdoor dew point is lower, then the controls compare temperatures. When the outdoor temperature is also low enough, the system may use outdoor air for cooling.

Economizers also present problems when the building is humidified in winter. Once again, the decision to use outdoor air for cooling must consider the outdoor dew point vs. the indoor set point. If the air is too dry outdoors, using that air for cooling will impose a large, unnecessary load on the humidifiers, which may not be sized to provide that much moisture. Also, even when the humidifiers are sized with ample capacity, the incoming air must be warm enough to actually accept all the moisture needed to reach indoor set point. Air cannot carry more moisture than its saturation value, and that value is very small when air is cold. Finally, the designer should note that in buildings that need winter humidification, the savings from economizer cooling are often more than offset by the *cost* of humidifying that "free" cold air.

While the calculation and sensor capacity for such complex decision-making is quite common in central systems, it is important for the HVAC designer to be sure that the required indoor dew point is not ignored in the "free-cooling" decision.

Beware of coils that cannot drain all the condensate

In large custom systems that never have enough mechanical room space allotted, the designer is tempted to minimize costs by specifying the smallest possible coils, configured in a way that fits the available duct space. Designing under those pressures, the designer often forgets that dehumidification coils need to be *deep* in the direction of air flow, have *wide fin spacing* and *be short* rather than tall.

Deep coils allow enough contact time to cool the air low enough for dehumidification. Wide fin spacing makes it possible to clean such deep coils, and allows rivers of condensate to drain from the coil. Otherwise, the coil could become "water-locked", which means air flow is partly blocked off, increasing fan operating costs and leading to carryover of water droplets into the air. Short coils collect less condensate in lower sections, so water is less likely to blow off the coil and into duct work. A good rule of thumb is to keep the coil height less than roughly 3.3 ft tall [1.0 m] Where taller coils cannot be avoided, a low-cost solution is to split the total cooling surface into two coils, stacked on top of each other, with a second drain pan in between them to catch condensate from the upper coil.

Also, in custom systems there may be no packaged equipment manufacturer to accept responsibility for condensate removal. The system designer must design a condensate drain pan and drain line assembly that can really drain all the condensate produced by every cooling coil. The drain pan must be long enough to catch droplets that might blow off the coil and it must sloped all the way to the drain connection in two axes so that no standing water can remain in the pan. The diameter of the drain line must be large enough to accommodate the maximum flow of water off the coil. Often plumbing codes will specify a minimum drain diameter. In the absence of other guidance, a minimum diameter of 1.25" [32 mm] will allow the use of a "plumbers' snake" to clean out the drain when it becomes clogged. Also, the stem height of the drain line trap must be greater than the pressure exerted by the fan, as shown in figure 10.11.

Beware of bringing outdoor air through fan-coil units that cannot dehumidify

In large central systems, individual rooms are often served by individual fan-coil cooling units controlled by a room thermostat. These are seldom designed with enough cooling surface to cool air deeply enough to dehumidify. When untreated outdoor air is brought into the building through such units, its moisture load makes humidity rise inside the room. This problem has been common in hotels and dormitories, where such equipment is often installed.

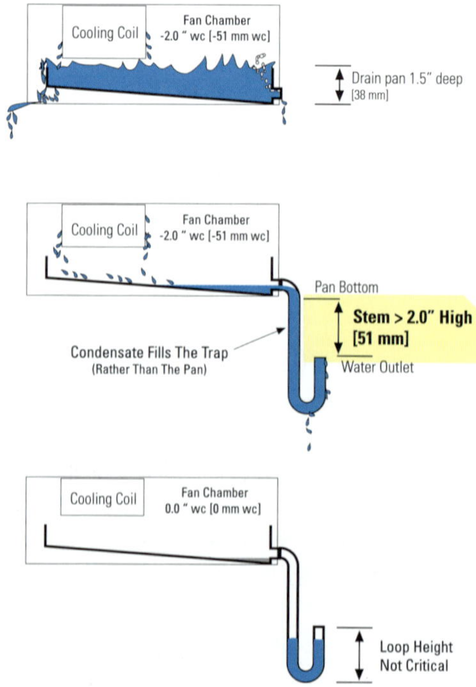

Without a Trap
The fan lifts up the condensate, which spills over if the fan suction exceeds the depth of the pan. Standing water allows microbial growth. Water spilling out of the pan can leak out of the unit and drip into the building.

With a Trap
The condensate drains out of the unit because the fan cannot lift the water out of the stem of the trap. The stem is measured from the *bottom of the pan* to the *top of the water outlet* from the trap.

Note... The Stem Height Is Critical
The height of the stem must be greater than the suction of the fan. eg: A fan with -2.0" wc pressure needs a trap with a stem height *over* 2.0" [A fan with suction of 51 mm of water column needs a drain stem higher than 51 mm].

Loop Height Not Critical - But Deeper Is Better
The distance from the water outlet to the bottom of the loop is not critical. But deeper loops do have a benefit. They take longer to dry out, which prevents the fan pulling in untreated air if the coil is not condensing water.

Fig. 10.11 Condensate Drain Traps
All cooling coils must have a trap deep enough to hold water against the suction exerted by the fan. Otherwise, condensate may not drain freely and might leak into duct work instead.

To avoid this problem, and also to allow for winter season humidification, pretreat the ventilation air centrally and provide conditioned air to each room through a separate supply air duct system. This suggestion adds the cost of a separate duct system, but allows humidity control in all seasons, which is not normally practical with central system-fan-coil designs. It also allows excess cooling capacity to be removed from the room units. The net costs may be the same, providing humidity control without increasing total installed cost.

If the fan-coil units are indeed capable of removing moisture from ventilation air, and the designer intends to use them in that

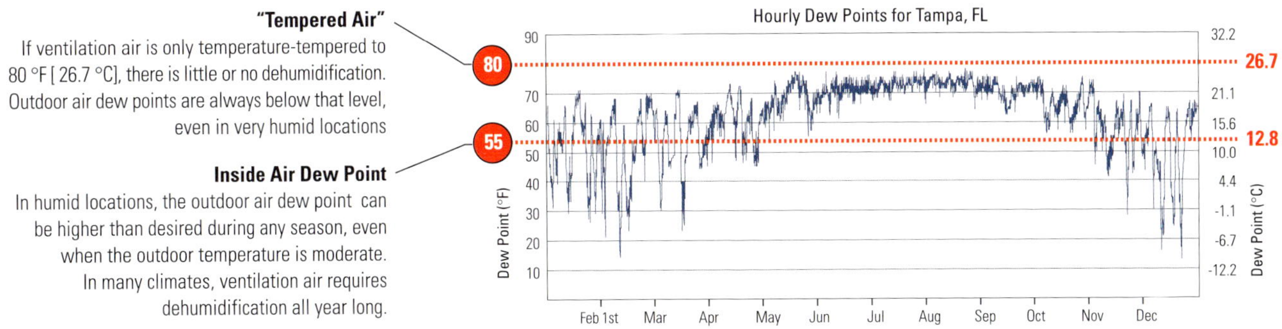

Fig. 10.12
Temperature-Tempering Alone Is Not Enough For Dehumidification
As this graphic shows, ***slightly*** *cooling the outdoor air will not remove its moisture. The incoming air must be deeply dehumidified to avoid loading the internal systems with moisture. (Based on TMY hourly weather observations contained in BinMaker, 1997)*

function, be certain that each unit is equipped with a drain pan and trapped drain designed as described in figure 10.11. Also in that situation, the designer should check the probable chilled water temperature in each unit under peak load conditions. Often, such fan-coil units are piped in series, so the chilled water temperature, although adequate for cooling, rises above the temperature needed to dehumidify the air. Usually, piping units in series means that the ones at the end of the run will not be capable of moisture removal. Then some other method of dehumidification will be needed, such as drying all ventilation air centrally before it gets to each fan-coil unit.

Design the outdoor air system to dehumidify air, not just "temper" its temperature

A common error when designing ventilation air pretreatment subsystems has been to only "temper" the incoming air to a temperature somewhat above the temperature desired in the space. When buildings do not need humidity control, this strategy saves energy and equipment cost, because internal cooling equipment is often oversized, and can remove occasional loads from undersized ventilation coolers. However, when the owner needs humidity control during humid seasons, "temperature-tempering" the ventilation air is not enough. As shown in figure 10.12, cooling incoming air to 80°F [26.7°C] *rarely, if ever, removes any moisture* except in tropical marine climates near the equator. In other parts of the world, the dew point is seldom above that temperature, so just cooling air to that level never actually dries incoming humid air.

To avoid this problem, dehumidify the outdoor air to a dew point below the control level in the space, using either cooling or active desiccant dehumidification equipment as described in Chapter 13.

Consider adding passive desiccant enthalpy recovery to reduce equipment costs

The cost of dehumidifiers and humidifiers can be greatly reduced by using enthalpy recovery heat exchangers to reduce the load from incoming ventilation air. These devices use the energy in the exhaust air to pretreat the incoming ventilation air, which in turn reduces the capacity requirement for humidity control equipment. The overall installed cost can be cut by 20 to 30% whenever the exhaust air can be brought back to the rooftop unit before it leaves the building. The designer can achieve the greatest cost reduction by ensuring the exhaust ducts are located within a few inches [centimeters] of the incoming ventilation ducts. If the ducts are separated, only liquid desiccant systems can be used to move moisture from one air stream to another, increasing the cost and complexity of enthalpy recovery substantially compared to the more widely-used passive desiccant wheels. Such load-reducers are described in more detail in Chapter 15.

Consider using excess boiler heat to remove moisture with an active desiccant system

In some buildings, large boilers sized for winter heating are operated at low firing rates during summertime because of a need for small amounts of steam or hot water, as in hospitals and hotels. In these cases, large amounts of heat are usually available at essentially no cost, because the boiler, sized for winter heat, produces much more heat than needed during warm, humid weather. This excess heat can be used to remove moisture from incoming ventilation air by using an active desiccant system. Desiccants adsorb moisture from the air, but then need heat to remove the moisture from the desiccant. Excess boiler capacity is often used for this purpose, achieving moisture removal at very low cost.

Also, small packaged desiccant systems are frequently used in ventilation systems for commercial buildings. Excess steam or hot water can be used in these units, in addition to the heaters already supplied in the package. This type of equipment is described in Chapter 13.

Place filters upstream, not immediately downstream of dehumidification coils or humidifiers

In packaged equipment, most manufacturers no longer locate filters downstream of coils and humidifiers. Designers of custom central systems should also avoid that practice. When a filter is placed immediately downstream of either a coil or a humidifier, it will become saturated with moisture and grow mold in the dust it collects. Then the mold grows through the filter, releasing spores downstream.

Placing the filter *upstream* of the coil or humidifier avoids this problem, and also protects equipment from accumulations of dirt and dust, reducing the potential for microbial growth.

Systems based on packaged rooftop units

Figure 10.13 shows a typical "rooftop unit" system arrangement. Many self-contained, packaged heating/cooling units are placed on the roof of the building over the zones they serve. This type of system is typical of low-rise commercial buildings such as retail stores, restaurants and office buildings. Each year these buildings account for more than 90% of commercial construction in North America.

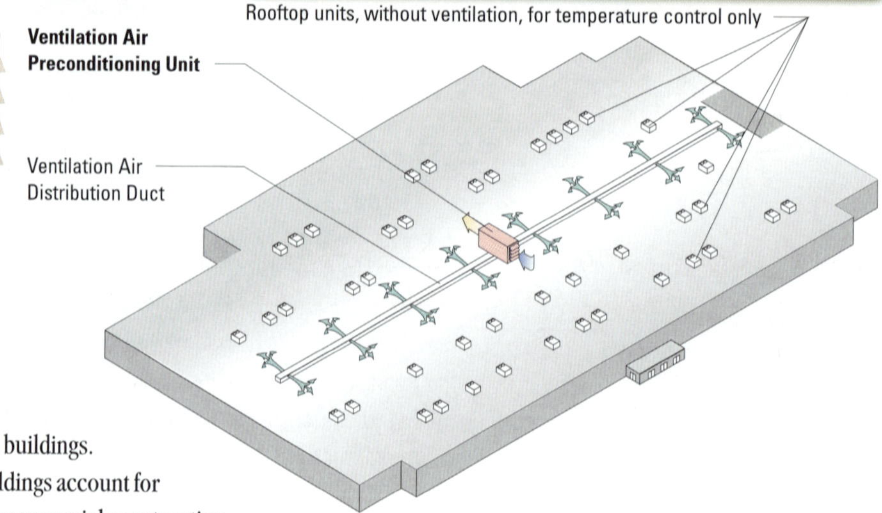

Fig. 10.13 Rooftop Systems
Packaged units are often used on low-rise commercial buildings. To achieve true control of humidity in a defined range (as opposed to simply moderating seasonal extremes) conventional rooftop units need assistance from dedicated dehumidifiers and humidifiers—which are usually provided in outdoor air pretreatment units.

Rooftop units are easy to apply, inexpensive, durable, fault-tolerant and energy-efficient when controlling temperature. They are also available on short lead-times from stock in a wide variety of standard configurations from a wide variety of manufacturers. Consequently they are very popular, and as a result they are easily maintained through a large infrastructure of service technicians and parts distributors. Unfortunately, standard commercial rooftop units are not configured for humidity control. Humidity fluctuates widely in buildings equipped with such units when large amounts of ventilation air must be brought into the building, as in the case of retail stores and restaurants. But it is easy to add humidity control to such buildings.

First, reduce the ventilation air handled by each rooftop unit, or eliminate it entirely. Then provide ventilation air pretreatment through

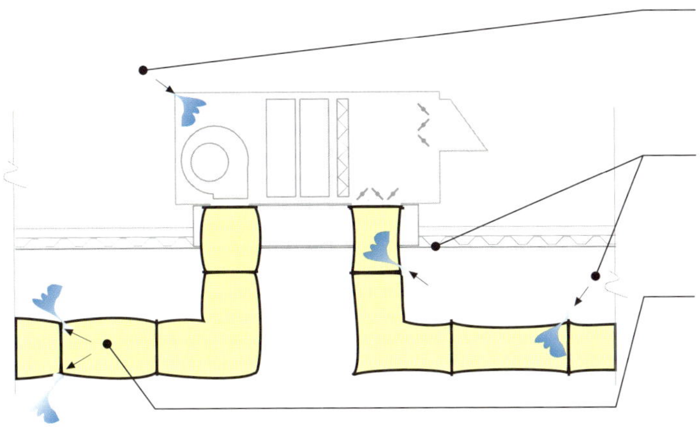

1. Eliminate Air Handler Leaks
Most commercial-grade air handlers leak a great deal of air. Seal open seams with tape and mastic, and make sure all access panels are gasketed, with all screws in place.

2. Seal Return Ducts
If return duct work leaks, it creates low pressure in building cavities, which sucks untreated air into the wall from the outdoors. Sealing return duct work helps avoid condensation from untreated air in these cavities.

3. Seal Supply Ducts
When supply duct work leaks in summer, cold air leaks into building cavities, cooling surfaces, which condense moisture. In winter, warm supply air carries moisture into cold building cavities, where it condenses once again. Sealing supply air ducts prevents these problems.

Fig. 10.14 Seal Units & Duct Work
In most commercial buildings, little attention is paid to air leaks. In humidity-controlled buildings, air leaks must be sealed to avoid control problems and to avoid actual damage to the building itself. (Harriman et al, 2000)

packaged units designed specifically for that purpose. The treated ventilation air is either injected into each rooftop unit, or distributed to the occupied spaces directly through a separate duct system.

Some of the tips and traps for adding humidity control are prompted by the nature of typical rooftop units. Others are cautions prompted by the common construction practices for buildings that usually use this class of equipment.

Check capacity of ventilation pretreatment units

If the humidity control is accomplished by specialized ventilation pretreatment units, it follows that their dehumidification capacity must be checked against moisture loads from both ventilation air and moisture loads from the space. This issue arises because in many cases, ventilation air pretreatment units are only intended to moderate the load from the outdoor air, not to achieve complete removal of all humidity related loads.

The designer often has a choice of different equipment options for treating outdoor air. It is important to be aware of how much moisture each option adds or subtracts to the building under peak load conditions. If the ventilation pretreatment unit cannot remove all the loads, then additional equipment must be added at other points in the system if humidity must be kept within a defined range.

Consider adding passive desiccant wheels to reduce equipment costs

The cost of rooftop units can often be reduced by using enthalpy recovery heat exchangers (passive desiccant wheels) to reduce the load from incoming ventilation air. These devices use the energy in the exhaust air to pretreat the incoming ventilation air, which in turn reduces the capacity requirement in the rooftop unit. The overall installed cost can be cut by 20 to 30% provided that the exhaust air can be brought back to the rooftop unit before it leaves the building. In many applications these devices reduce the load enough to allow conventional rooftop units to bring humidity within acceptable ranges year-round without the need to add dedicated dehumidifiers and humidifiers.

However, one caution is needed with respect to *passive* desiccant wheels. They only remove moisture from ventilation air when the exhaust air is dry, and they only humidify incoming air when the exhaust air is humid. So when loads rise above the capacity of other equipment in the system, humidity fluctuates. Passive desiccants can only moderate the load; they cannot control humidity by themselves. When control is needed within a defined range, dedicated dehumidifiers and humidifiers will also be required. The combination of passive desiccant wheels with dedicated dehumidifiers and humidifiers is especially powerful. Combined systems can keep humidity within very close tolerances at much less cost than if no enthalpy exchanger were present to recover exhaust energy.

Be especially cautious of economizers

Manufacturers usually offer a "dry bulb economizer" as a standard option on rooftop units. These units automatically switch to 100% outdoor air instead of return air when the air outdoors is below the temperature in the building, when the thermostat calls for cooling. Using cold outdoor air is sometimes called "free cooling".

However, although an economizer may reduce the cooling bill, it severely interferes with humidity control when the decision to bring in outdoor air is based only on dry bulb temperatures. Cool air usually contains excess moisture in summer, spring and fall. So while the building might be cooled by the outdoor air, the building will also load up with moisture.

The solution is to eliminate "free cooling" from buildings which require humidity control, or to invest in a more complex control system. An economizer may still be effective if the decision to use outside air is based on *both* temperature and dew point. In that arrangement, the controls first measure and compare the desired dew point inside the building versus the dew point outdoors. If the outdoor dew point is lower, then the controls compare temperatures. When the outdoor temperature is also below the desired condition indoors, the units may use outdoor air for cooling. (See Fig. 10.10)

Economizers also present problems when the building is humidified in winter. Once again, the decision to use outdoor air for cooling must consider the outdoor dew point vs. the indoor set point. If the air is too dry outdoors, using that air for cooling will impose a large, unnecessary load on the humidifiers, which may not be sized to provide that much moisture. Also, even when the humidifiers are sized with ample capacity, the incoming air must be warm enough to actually accept all the moisture needed to reach indoor set point. Air cannot carry more moisture than its saturation value, and that value is very small when air is cold.

While the calculation and sensor capacity for such complex decision-making is quite common in central systems, buildings with rooftop units are not always equipped with such controls. When the system lacks the ability to compare dew points indoors and outdoors, the designer is generally wise to avoid the use of air-side economizers when humidity control is important to the owner.

Be sure that drain pans actually drain all water

Rooftop cooling units are usually equipped with condensate drain pans under the cooling coil. But that pan is not always designed to prevent standing water. Also, the water will not actually drain out of that pan unless the drain line has a trap. So these are two items of concern in all buildings, but which require particular attention when the owner is concerned with humidity control.

The drain pan must be sloped all the way to the drain in two directions. If there is a "lip" between the drain line and the bottom of the pan, water will collect and grow mold, mildew and bacteria. Sloping the pan all the way to the edge of the drain pipe prevents these problems. Also, if there is no trap, or if the stem of the trap is too short, the fan will keep condensate in the pan instead of the trap, and water can be pulled into the duct work and into the conditioned space. The stem height is the distance from the bottom of the pan to the top of the outlet leg of the trap. That height must be greater than the pressure exerted by the unit's supply fan, as shown in Fig. 10.11.

So the designer has two concerns regarding condensate in conventional rooftop units. First, the units must be manufactured with sloped drain pans. Secondly, all rooftop units must be installed with traps on the condensate drain lines, and the stems of these traps must be high enough to hold water in the trap in spite of the pressure exerted by the fan.

Don't add humidifiers without a double-wall enclosure

Many rooftop units have fiber-faced insulation inside the casing instead of a second sheet of metal to cover that insulation. It is not wise to add humidifiers to such units. Moisture produced by a humidifier can saturate fibrous insulation and allow mold and bacteria to grow. Be sure to specify that the rooftop units should have a double-wall enclosure if humidifiers will need to be added by the contractor.

Quantify & design for unit leakage

In normal commercial construction, air leakage in rooftop units is rarely a concern, because unit sizing is usually "generous enough"

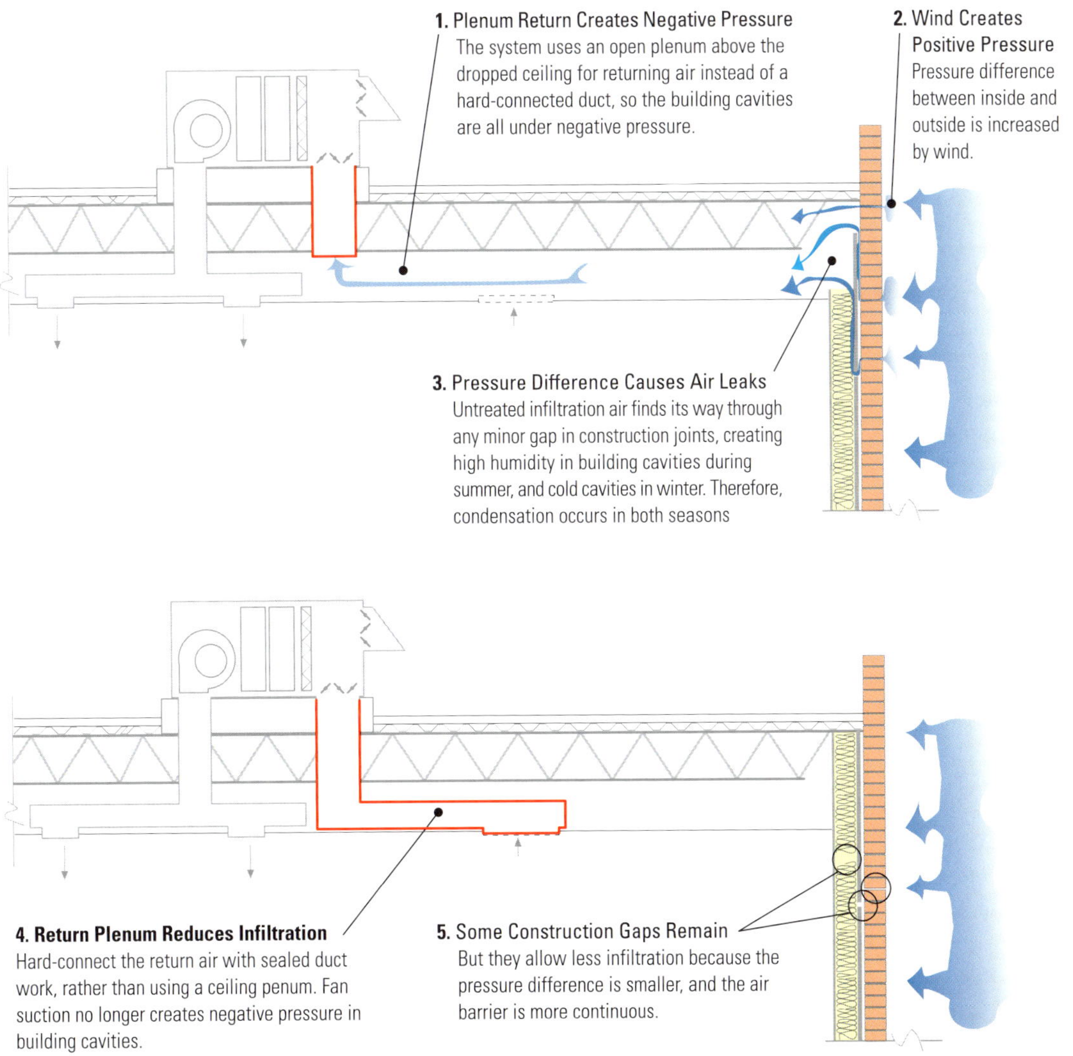

Fig. 10.15 Avoid Plenum Returns

Sealed return ducts are very important in humidity-controlled buildings. With unsealed plenum returns, suction pulls untreated air through the wall, creating control problems and causing condensation. (Cummings et al, 1996)

to cover the reduction in cooling and heating capacity that comes with leakage. But for buildings that need humidity control, the rooftop air leakage rate becomes an important variable, especially when many units are installed . Large retail super stores and movie theater complexes may use 20 to 40 rooftop units on a single building.

Rooftop units are optimized for fault-tolerance, low-cost and high thermal efficiency. They are not made to be airtight. Sometimes they leak a great deal of air, especially after they have been serviced many times and panels loosen and fasteners go missing. Other times they leak less air, as when they are new or when the manufacturer uses tighter-than-normal construction to satisfy the needs of a particular application or a repeat customer. But depending on pressure differences and unit construction, such units can still leak between 2% and 10% of the supply air into or out of the unit casing. When sizing the humidifier and dehumidifier, this leakage must be included in the load calculation.

A prudent designer should ask the manufacturer of the rooftop equipment to define the leakage air flow into and out of the unit casing, and require documentation from the tests that determined that leakage rate.

Seal all duct work

Without mastic over the joints, the duct installation is less expensive, which is often more important to the builder than the cost of the energy wasted through air leakage. But when humidity control is important, the situation changes.

If cold supply air leaks out and chills surfaces above ceilings, humid air will condense moisture, leading to mold growth. If warm, humid supply air leaks into cold building cavities in winter, moisture will condense and cause similar problems.

In most low-cost, low-rise commercial buildings, duct work is not sealed, and therefore it leaks quite a bit of air—At least 48 cfm per 100 ft^2 of duct surface at a pressure difference of 1 in.w.g. [2.6 l/s . m^2 at 249 Pa]. (SMACNA, 1989) In fact, recent field measurements of installed duct work in 70 low-rise commercial buildings suggest the actual leakage rate for unsealed seams can be 10 times that amount. (Cummings et al, 1996)

So when humidity is a concern, the duct work should be sealed to at least SMACNA Seal Class B. SMACNA is the Sheet Metal and Air

Conditioning Contractor's Association. Seal class B requires that all longitudinal and transverse seams be sealed with durable material.

Don't use open plenum returns

In low-cost HVAC systems, sometimes the designer uses the open space above the dropped ceiling to return air to the rooftop units. This saves the cost and time associated with installing a hard duct between the return grills and the unit. While this is never good practice, it is especially important to avoid such "open" return plenums in humidity controlled buildings. The problem is suction created by the rooftop units as they try to pull air from the space up through the open plenum.

That suction does pull air from the conditioned space, but it also pulls air from the outdoors though any cracks in the exterior building envelope. That infiltration greatly increases the heat and moisture loads on the equipment. More important, the suction pulls humid air into a cool building in summertime, so moisture condenses inside the building envelope. (Harriman et al, 2000) Also, during winter, cold infiltrating air chills internal surfaces, and moisture from *inside* the building migrates to cold building cavities and condenses.

Using open plenum returns has even threatened human life in buildings with swimming pools. Roofs have collapsed from snow loads when highly humid air corrodes structural elements hidden behind dropped ceilings in cold climates.

There is no way to avoid the significant problems caused by open plenum returns, so these should not be used in buildings where humidity must be controlled.

Seal rooftop units to the roof curb

A surprising number of rooftop units are simply placed on top of the supporting roof curb without a full air and water seal between unit and curb. Air and water can leak into the building through that gap. If the gap does not leak a great deal of actual rain water, the construction error may not be noticed for years, because sizing of cooling and heating units is usually generous enough to hide the added load. But with humidity-controlled buildings, such gross leakage creates major problems.

Units on buildings with fast-track construction schedules and on speculative commercial buildings often have these gaps. The supporting roof curbs are usually sized for the maximum-size rooftop unit that could conceivably be needed in that part of the building. But the actual rooftop unit may be purchased at a later stage by the tenant, or by the developer when the potential occupancy becomes better defined, or when cash flow improves on the project. Mismatched curbs and rooftop units are therefore quite common. They need not be a problem. The designer and contractor must simply ensure that the gap is weather-tight and airtight.

The designer needs to specify that the rooftop units be sealed-to and not just placed-on-top-of the support curbs. The HVAC contractor must ensure that the seal is actually accomplished.

Seal duct work to the unit

Air leaks through duct seams in proportion to the pressure difference. The higher the pressure difference, the more air will leak through cracks and holes. In a rooftop system, pressure differences are largest inside the unit itself—between the fan compartment and the rest of the unit. But after that, the pressure differences are greatest where the duct work connects to the rooftop unit. So it is important to specify that both supply and return air ductwork be sealed to the unit.

Seal return grills to the return duct work

The joint between the return grill and the return duct work is frequently a slip-fit connection held by a few sheet metal screws. Sometimes, the duct does not quite fit the collar on the grill, and there is a gap between duct and grill. These gaps—which seem minor—often cause problems. The gap occurs behind the finished wall or ceiling, which means the fan suction pulls air out of those building cavities into the return duct. Two problems result: less air is circulated through the occupied space, which reduces system capacity. Also, suction at the exterior wall pulls untreated air into the building, leading to condensation on cool surfaces in building cavities. So it is important to specify that return duct work be sealed to the return grills in humidity-controlled buildings.

Specify an air balance report and include that contractor in early design decisions

In most low-rise buildings, oversized equipment ensures that if the system is not installed carefully... the building will still be heated and cooled comfortably. The system may leak air and use more energy than anticipated, but the client will be satisfied. However, when the building must be humidity controlled, the everyday errors and loose tolerances in rooftop systems will cause problems. If air does not flow in the quantities expected to the locations defined by the designer, humidity will fluctuate beyond the range needed to satisfy the purpose of the project.

To avoid these problems, specify an air balance report that measures and records the air flow at all supply and return grills, at the outdoor air inlet and the air flow through dehumidifiers and humidifiers. That report is the starting point for validating equipment performance and for troubleshooting the system as a whole. Without those measurements, there is no assurance that humidity control will be achieved as the owner and designer intend.

Fig. 10.16 Room Units

Room cooling/heating units are not designed to control humidity. But pretreating outdoor air with a separate system allows the low-cost buildings that use room units to gain humidity control simply and easily.

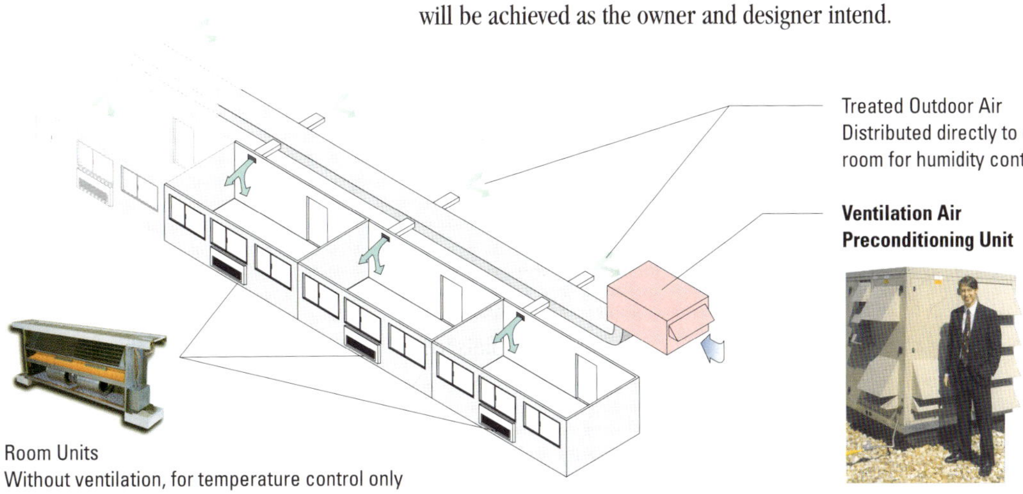

Best practices in this regard are those used in large building design and construction. The air balance contractor is determined in advance of design documents, so experts from that firm can participate in detail design. With early participation, the designer can avoid overlooking the key dampers and flow-measuring devices needed to balance the system. Also, the designer can help the air balance process by making sure there are long, straight duct runs available before and after air flow measuring points. Without the essential measurement and control devices in-place, the air balance report is merely a catalog of design and construction errors—a very discouraging and unhelpful document.

Systems that use room units

Figure 10.16 shows an HVAC system built with independent units that control temperature in each room. Each unit contains a heater, a cooling unit, filters and a supply fan. The cooling unit might be a conventional air conditioner, but heat pumps are also common, rejecting or gaining heat from water piped to a geothermal loop or to a cooling tower and small boiler. Room units are very common in hotels, motels, eldercare facilities, condominiums and dormitories. Larger room units are also quite common in school classrooms.

These systems are low-cost and provide independent temperature control for each room as well as interchangeability, which reduces downtime and simplifies maintenance. No unit operates unless there is a load in that particular room, so the overall system can

be quite energy-efficient. Also, service can be accomplished quickly by replacing the entire unit, avoiding any need to interrupt teaching in a school, and avoiding unusable rooms in a hotel or dormitory.

Although room units have many advantages, humidity control is not one of them. Cooling units are controlled by thermostats. When the cooling load is satisfied, the compressor shuts off. Condensed moisture re-evaporates from the coil surface and returns to the room. In winter, there is no provision for humidification.

Humidity control suffers especially when these units pull ventilation air directly from outdoors. Ventilation is often a continuous requirement. So during hours of moderate temperature but high humidity, the room loads up with moisture. That moisture load is not removed by the unit since there is no demand for cooling.

Adding humidity control to buildings with room units is very easy. The ventilation air is simply pretreated with a central system. Dry or humid air is delivered to each room through duct work dedicated to that purpose. Some tips and traps for adding humidity control to room units include:

Consider room units with passive desiccant wheels if they are configured with dehumidification controls

Passive desiccant wheels, also known as enthalpy heat recovery wheels, can use the energy from exhaust air to pretreat incoming ventilation air. They reduce the load on the room unit by transferring heat and moisture from one air stream to the other. For humidity moderation, room units equipped with such wheels are a great improvement over conventional units. Room units with passive desiccant wheels are often used in classrooms, where the required ventilation air flow for each room is larger than for hotels or dormitories.

However, the cooling unit must be configured with a dehumidification mode of operation if the passive desiccant wheels are to remove moisture. If exhaust air is humid, the wheel cannot rid itself of the moisture it collects from the incoming air. The poor humidity control of wheel-equipped cooling units that do not have a dehumidification mode is shown in figure 10.17.

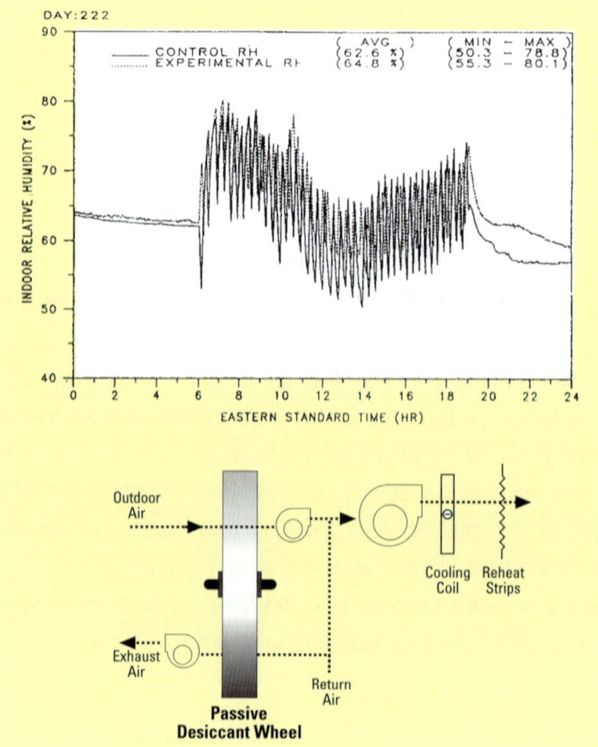

Fig. 10.17
Passive Desiccant Wheels Alone Are Not Sufficient
Passive desiccant wheels are often available as options on large room units. They reduce the moisture load for many of the hours of the year. But to ***control*** *humidity, the cooling units must be controlled by humidity as well as by temperature. Otherwise, as shown by this example, the cooling compressor shuts off when the thermostat is satisfied. Then the exhaust air becomes too humid to dry the incoming air. Indoor humidity goes out of control. (Shirey, 1998)*

When the unit is equipped with a dehumidification mode, the cooling compressor operates even when the thermostat does not call for cooling. The coil condenses moisture, so the exhaust air is dry and the passive desiccant wheel can remove moisture from incoming air. The designer should confer with the proposed manufacturer to be certain that the unit can dehumidify even when the thermostat does not call for cooling. Also, whenever moisture is to be removed by a cooling coil, the manufacturer must be sure to include a sloped drain pan and the contractor must install a well-designed condensate trap as described in the section on central systems.

Although room units with passive desiccant wheels and a dehumidification mode do not control humidity within a narrow range, they can eliminate the cost of a separate ventilation system when simply moderating humidity will satisfy the owner's purpose.

Consider installing thermostats without numbers

When humidity is held below 50% rh in the summertime, people tend to be more comfortable at higher temperatures. Similarly, when humidity is held above 30% rh in winter, people can be comfortable at lower temperatures. (See chapter 4 - Human Comfort) However, thermal sensation is not the only influence on human behavior. People who are accustomed to only cooling and heating tend to set the thermostat to values that proved comfortable in the past.

When humidity is controlled, the designer can help the occupants focus on actual sensation by installing thermostats that have set point marks but no numbers. People are more likely to leave the setting higher in the summer and lower in the winter when they do not have reference numbers that conflict with past experience and therefore override actual sensation. Removing numbers from thermostats helps the occupant avoid overcooling and overheating the room, which saves energy and results in greater comfort.

Reduce oversizing of cooling components

The cost of a central ventilation and humidity control system can be offset by reducing the size of the cooling units to reflect the fact that they no longer have to remove moisture. (Spears & Judge, 1997)

For the sake of simplicity and for easy replacement, room units are often sized the same, even though loads vary between rooms with different external exposures and different occupancy. Extra capacity and large "safety factors" are quite common. For better humidity control, the designer should *avoid any oversizing*, especially when the ventilation system removes the latent heat (the moisture load) from the cooling component. In buildings with many room units, the entire cost of the ventilation system can be offset and the net installation cost reduced by eliminating such waste. Also, cooling compressors last longer, because they do not cycle on and off as frequently as when they are oversized.

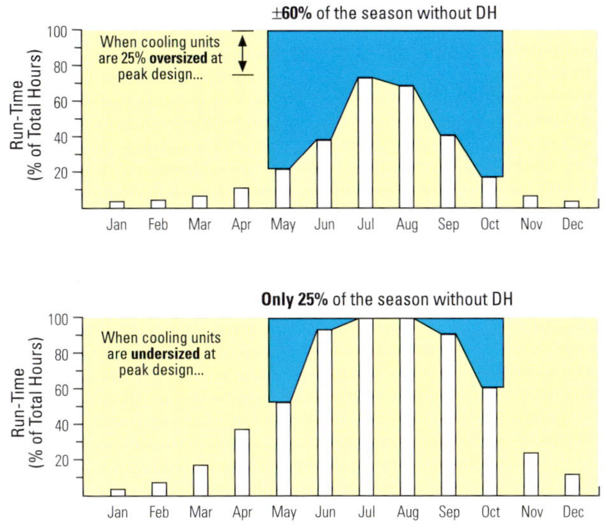

Fig. 10.18 Reduce Oversizing

Oversized cooling units do not operate long enough to remove all moisture loads. Eliminating excess "safety factors" in unit sizing improves humidity control while reducing installed costs.

Seal up the outdoor air paths inside the room units

When ventilation is provided through a central system that does not physically connect to the room units, it becomes important to ensure that untreated air does not leak into the room units by accident. Most units are not constructed to eliminate outdoor air leakage paths, because in many buildings ventilation air is brought into the building through the unit casing.

In the past, problems have occurred in buildings because the HVAC designer logically assumed that if there is no external wall penetration (as in the case of water-source heat pumps), there is no need to be concerned about outdoor air leaking into the building through the room units. Unfortunately, if not sealed internally, the units can create suction in the wall cavity. The suction pulls in untreated air through cracks and holes in the external building envelope at points far removed from the leaking room unit itself. Chapter 16 discusses this common problem in more detail.

The solution is quite simple: when ventilation air is provided through other means, specify that all outdoor air paths inside the unit be covered with sheet metal and sealed with the same material used to seal ductwork.

Seal the return air inlet to the inside surface of the finished wall

Many room units are fitted snugly into the wall cavity, but not necessarily sealed to the inside face of the finished wall. When that happens, the unit can pull air from the wall cavity instead of from the room. Suction in the wall creates infiltration of untreated outdoor air into the building, which has created millions of dollars of damage to buildings. (Harriman & Lstiburek, 2000) Sealing the wall unit, and especially the return air inlet to the inside wall eliminates this potential problem. The specification should be written accordingly.

Summary

For most commercial buildings, the key to low-cost and effective humidity control is to dry and humidify the ventilation air before that air can become a load on the other HVAC systems within the building. When using a dedicated ventilation unit for this purpose, be sure the equipment can remove or add enough moisture to compensate for internal loads as well as the loads from the ventilation air.

If for some reason that lowest-cost strategy is not possible, remember to install dedicated dehumidifiers and humidifiers inside the other HVAC systems to remove the entire load when humidity control is needed to achieve the project purpose. When only moderation rather than control is necessary, consider adding passive desiccant wheels alone. These may reduce the load from the outdoor air to a level that can be removed by the cooling system, eliminating the need for dedicated dehumidifiers and dehumidifiers.

References

ASHRAE, 1997, 2001. Chapter 26, Design Weather Data. *ASHRAE Handbook—Fundamentals*

BinMaker®, 1997. Weather summaries based on TMY hourly observations (Typical Meteorological Year Files, produced by the National Renewable Energy Laboratory for the US Department of Energy). InterEnergySoftware.com, DesPlaines, IL.

Cummings, James B, Withers, C.B, Moyer, N, Fairey, P., McKendry, B, *Uncontrolled air flow in non-residential buildings.* April 15th, 1996, Final Report of FSEC project number FSEC-CR-878-96. Florida Solar Energy Center, 1679 Clearlake Rd, Cocoa, FL 32922.

Harriman, L.G.III, Lstiburek, J, Kittler, R, 2000. "Improving humidity control for commercial buildings." ASHRAE Journal, Oct.ober 2000.

Persily & Grot, 1986. Measured air leakage in eight large office buildings. *Measured air leakage of buildings*, American Society of Mechanical Engineers, STP 904., Philadelphia, PA USA.

Persily, Andrew K. 1999. "Myths about building envelopes." ASHRAE Journal, March, 1999, pp: 39-47.

SMACNA, 1989. *HVAC Duct Systems Inspection Guide* (15D, 1989) Sheet Metal & Air Conditioning Contractors National Association, 8224 Old Courthouse Rd., Tyson's Corner, Vienna, VA. 22182 (703) 790-9890 www.smacna.org

Shirey, D.B., III. 1998. "Evaluation of and enthalpy recovery ventilator in a portable classroom." Proceedings of IAQ & Energy 98, New Orleans. ASHRAE.

Spears, J.W. and J. Judge. 1997. "Gas-fired desiccant system for retail supercenter." *ASHRAE Journal*, 39(10): p.65-69.

US Department of Defense, 2001. *Engineering Weather Data.* USAF Manual 32-1163. Available as CD-ROM from the US National Climatic Data Center, Asheville, NC. Product Number C00515-CDR-A0001 http://nndc.noaa.gov

Donated Photos

Fig. 10.7 © Carrier Corporation

Fig. 10.16 © Enviro-Tec Inc.

Chapter 11
Estimating Dehumidification Loads

By Lew Harriman & Reinhold Kittler

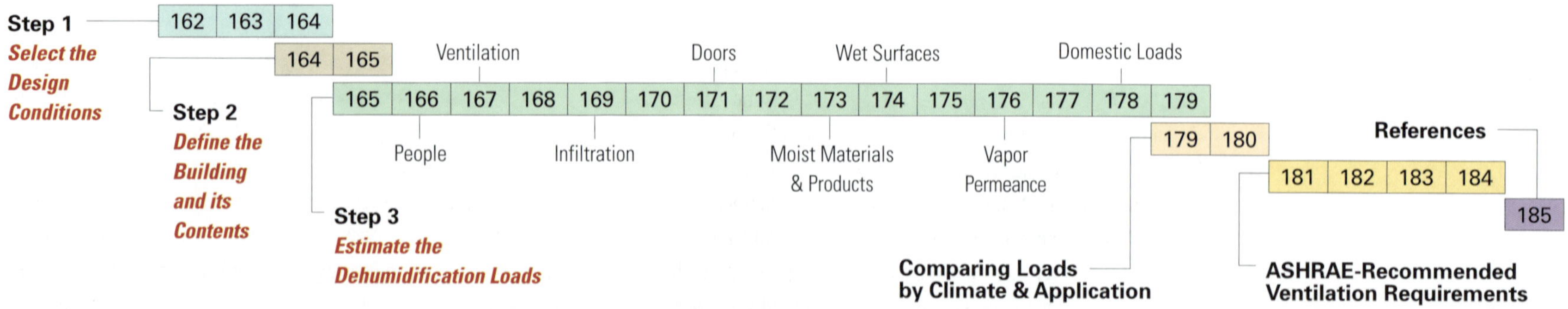

Fig. 11.1 Design Day

For dehumidification, the design day is not hot, but rather warm and raining. On those days, any incoming air is highly humid, and visitors bring moisture into the building on their clothing.

Dehumidification Moisture Loads

In designing dehumidification systems, the first and most important task is to quantify the moisture load. However, moisture load calculations are subject to multiple realities. In spite of the detailed calculations and well-defined tables and graphs provided in this chapter, two equally qualified individuals may arrive at quite different loads for the same building.

Differences arise from differing assumptions. The equation used to calculate leakage through the wall may be the same, but can the designer *really* be certain the contractor will seal the electrical plug box penetrations, and will the air barrier *really* be taped as specified? Will the motel bathroom exhaust be set by dampers as specified, or will the dampers be eliminated as a cost-saving measure and the airflow *greatly exceed* expectations as happened on the last project? The designer must make many assumptions—best estimates—before the load estimate and system design can proceed.

In the final analysis, the moisture load calculation sheet represents an agreement between equipment supplier, system designer, owner and contractors. It defines the assumptions that form the foundation of the system design.

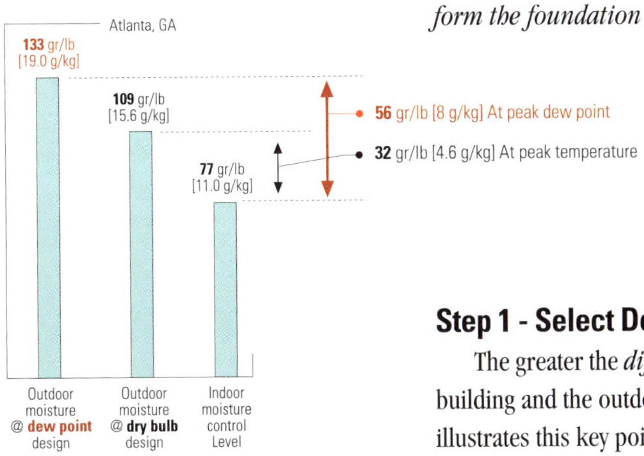

Fig. 11.2 Moisture Difference

The load depends primarily on the moisture difference between indoors and outdoors.

This chapter will discuss the basic elements of dehumidification moisture loads and how they can be quantified. It will also suggest how the designer can assess the relative importance of each load element in different situations.

Step 1 - Select Design Conditions

The greater the *difference* in moisture between the inside of the building and the outdoors, the greater the load will be. Figure 11.1 illustrates this key point. Assume the building is maintained at 75°F and 60% rh (77 gr/lb) and assume the surrounding weather in summertime is 82°F, 133 gr/lb. With those assumptions, every pound of air that infiltrates through the exterior wall brings 56 grains of moisture into the building to be removed by the dehumidification equipment. In contrast, if the designer had assumed the outdoor condition was 93°F dry bulb with a mean coincident wet bulb temperature of 75°F, each pound of infiltrating air would only bring in 32 grains of moisture instead of 56 grains—a 43% lower load at the same infiltration rate.

[In SI units, assume the building is maintained at 24°C and 60% rh (11.0 g/kg) and assume the surrounding weather in summertime is 27.8°C, 19.0 g/kg. With those assumptions, every kilogram of air that infiltrates through the exterior wall brings 8.0 grams of moisture into the building to be removed by the dehumidification equipment. In contrast, if the designer had assumed the outdoor condition was 34°C dry bulb with a mean coincident wet bulb temperature of 24°C, each kilogram of infiltrating air would only bring in 4.6 grams of moisture instead of 8.0 grams —a 43% reduction in load for the same infiltration rate.]

Given this huge difference, the obvious first step in calculating loads is to *define the absolute moisture difference* between the indoor and outdoor environments.

The indoor moisture level depends on the owner's purpose for the project. (See Chapter 10) The control level that satisfies the owner's purpose must be converted to an absolute value before load calculations can proceed. For example, "60% relative humidity" cannot be used for calculating moisture loads because the absolute moisture content at 60% rh varies with temperature. After the owner *also specifies* that the maximum dry bulb temperature is 75°F [24°C], the 60% rh humidity specification becomes meaningful. The relative humidity can be converted to a specific humidity value by consulting a psychrometric chart. In that example, 75°F and 60% rh gives a specific humidity of 77 gr/lb [24°C, 60% rh = 11.6 g/kg]

When the owner's purpose is served by a constant humidity with fluctuating temperature, the designer must pick the temperature that—at 60% rh—will represent the greatest difference in specific moisture between the indoor and outdoor conditions. This will be the lowest temperature within the allowable range. With the *driest* indoor condition established, the designer next defines the *wettest* outdoor condition.

Beginning in 1997, the *ASHRAE Handbook—Fundamentals* provides extremes of dew point in the climatic design chapter. Those values are reprinted in this book in Chapter 32.

ASHRAE research has defined three levels of extreme dew point conditions. The designer and owner must agree on which of these to use according to *how many hours the building can risk being above the humidity set point.* In the most extreme case, they choose the 0.4 % design dew point. Then the team can expect that the moisture outdoors is not likely to exceed the selected value for more than 35 hours in a typical year. (8760 hours in a year x 0.004 = 35 hours) In cases of less extreme need, the designer and owner might agree on the 1% or the 2.5% dew point. In a typical year, outdoor moisture is likely to exceed those values for either 88 or 219 hours.

Selecting the design moisture level has been the subject of much confusion in the past. Before 1997, ASHRAE did not have research to define extreme dew points. The only available data was for extreme dry bulb (with average wet bulb) and extreme wet bulb (without the coincident dry bulb temperature). Neither of those data define the extreme moisture level. The wet bulb without dry bulb is no value of moisture at all—it only defines the temperature at which evaporation occurs. And the average moisture level during hours of extreme temperature is usually 20 to 45% less than the moisture level during the extreme dew point hours.

The magnitude of this common misunderstanding is shown in figure 11.3 for Atlanta, GA. The true peak value for moisture is 135 gr/lb [19.2 g/kg]. But the average moisture level when the *temperature* is peaking is only 100 gr/lb [14.2 g/kg]. If the dehumidification system were designed for loads based on the lower number, the owner could expect that the system would be considerably short of capacity. The building would be above the humidity set point for hundreds, if not thousands, of hours rather than only 35 hours per year.

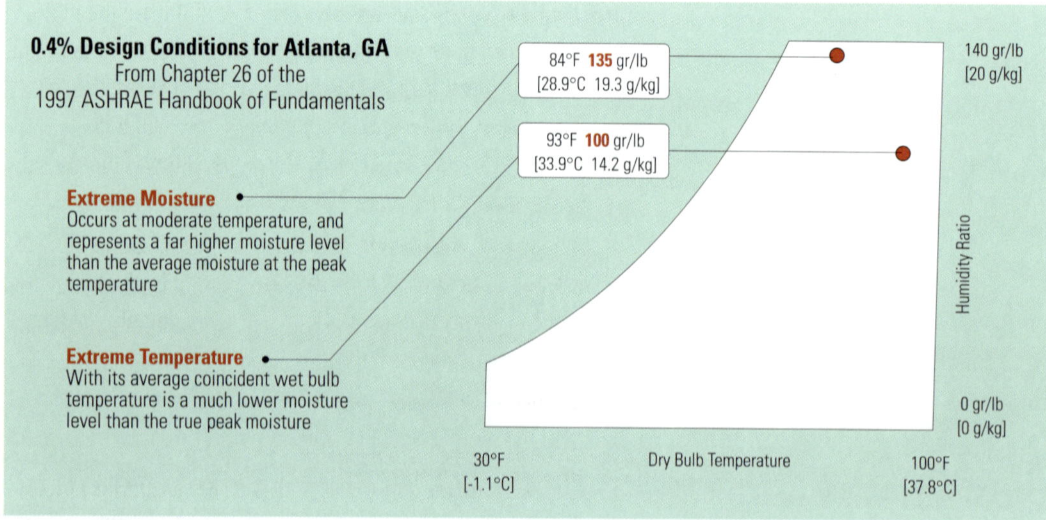

Fig. 11.3
Moisture At Peak Dew Point vs. Moisture At Peak Temperature.
The maximum outdoor moisture occurs during moderate weather—not during the hottest days of the year.

When the building team has agreed on the design conditions inside and outside, the HVAC designer needs some information from the owner and from the architectural designer concerning the building and its use.

Step 2 - Define The Building and Occupants

Moisture loads are larger when the building is larger, and when it is occupied by more people. So the HVAC designer needs to define several key aspects of the building and its use before he can estimate moisture loads. This will require further conversations with the owner and with the architectural designer. For several questions, precise answers will not be possible, and circumstances will change over time. But estimates from the owner and architectural designer are likely to be closer to reality than guesses from the HVAC designer.

Occupants and visitors

Moisture enters the building when people enter the building, when they breath out humid air, and when ventilation air is brought in to dilute the contaminants generated by those people. The designer will need to know roughly how many people will occupy the building at peak load, how often they are likely to enter and leave the building during that peak hour, and what sorts of activity they will be performing. The owner can help by describing groups that usually occupy the building, e.g. 20 teachers, six administrative staff, 375 students, five cafeteria workers, etc. Similarly for a retail store: four cashiers, 10 sales clerks, an average of 200 customers in a peak hour, etc.

Each group is likely to be in particular places, and to be doing similar things. The door activity can be estimated by thinking about the probable activity of these groups, and the total number of people will determine the minimum amount of ventilation air that must be brought into the building.

Exterior wall dimensions & tightness

The HVAC designer must gather details of the exterior envelope to calculate heat loads, so a full set of plans may be readily available for new construction as a matter of course. For old buildings, some dimensions will be needed from as-built plans or a site inspection.

In either case, the dimension and construction of the exterior walls is very important. Air leakage through those walls is likely to be the second largest source of moisture load. During moisture load calculations, the designer will have to decide whether he believes the walls are "tight", "average" or "loose" with respect to air leakage. Also, the designer must decide how much exterior wall surface is likely to be exposed to the prevailing wind on a design day. Those decisions are easier to make after the designer has seen an existing building, or after he has discussed the issues with the architectural designer of a new building.

Exterior doors and vestibules

When wind blows against open doors and vestibules, it carries a considerable load of humid air into the building. That load depends on the dimensions of the doors and the vestibules, and on the time required to close the doors after they have been opened. The architectural designer may have information about "open time" for doors in high-traffic applications like retail stores. This information is very helpful to the HVAC designer, who must estimate the open time in order to calculate the infiltration load.

Exhaust or "ventilation" fans

In new buildings, the HVAC designer will be responsible for all ventilation fans, and will quantify the amount of make up air needed to offset the exhaust. But in existing buildings, he must take care to locate all exhaust fans and to quantify the air they remove from the building. An equal or greater amount of ventilation air must be added to the system and accounted for in the moisture load calculation. Owners are seldom aware of the number of exhaust fans that pull air from the building. A site inspection is wise for any designer who wants to avoid unpleasant surprises after a new dehumidification system is installed.

After the HVAC designer understands the building's occupants and its exterior components, he can proceed to estimating the moisture loads from each load element.

Step 3 - Estimate The Dehumidification Loads

There are eight principal sources of moisture in most commercial buildings:

- People (Their number determines the ventilation load)
- Ventilation
- Infiltration
- Doors
- Products
- Wet Surfaces
- Vapor Permeation
- Domestic Loads

The list above is arranged in order from *most* to *least* concern in typical commercial buildings. But the magnitude of each load element will vary according to the function and construction of the building. In swimming pools, wet surfaces generate much more moisture

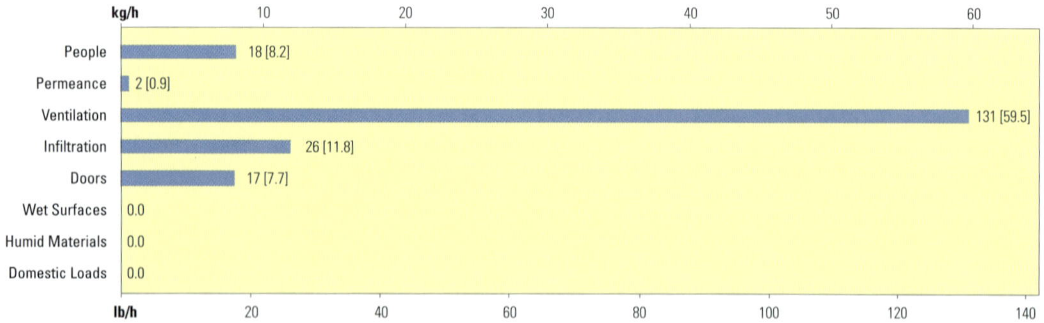

Fig. 11.4
Retail Store Moisture Loads
In retail stores, the designer focuses more attention on estimating the ventilation load than on permeation, because ventilation impacts the design more forcefully.

than people's respiration and perspiration. But in restaurant dining rooms, any wet surfaces create minor loads compared to moisture from the occupants and their clothes.

Figure 11.4 shows the moisture loads for a medium-size retail store. The designer's attention to each element should be proportional to the load it represents. Larger loads demand greater attention and more careful consideration of design assumptions.

People

People release water vapor as they breathe, and they evaporate moisture as they perspire. Also, their clothing absorbs moisture outdoors from both humid air and rain. That moisture becomes a load inside the building as the clothing dries.

Respiration and perspiration

Each time someone breathes out, they release a lung full of air at a temperature of 98.6°F and 283 grains per pound—nearly saturated air at normal body temperature. [37°C, 40.4 g/kg]

The number of breaths per hour depends on the person's physical activity. More active people will breathe more frequently and more deeply. Figure 11.5 provides typical moisture release rates for respiration and perspiration based on different levels of physical activity.

Moisture desorbed from clothing

In retail stores, restaurants, theaters and other high-traffic, high-density buildings, the moisture collected outdoors by clothing can

Activity	Typical of	lb/h	kg/h
Seated, at rest	Theater patron	0.10	0.05
Seated, very light work	Hotel or restaurant patron	0.15	0.07
Seated, moderately active	Offices, retail cashier	0.19	0.09
Standing, light work, walking	Offices, retail patron	0.19	0.09
Walking, standing	Offices, retail floor clerk	0.24	0.11
Seated, light work	Electronic assemblers	0.45	0.20
Moderate dancing	Dancing, nursing care	0.52	0.24
Walking briskly with loads	Restaurant servers	0.60	0.27
Light exercise	Bowling, slow treadmill	0.83	0.38
Heavy work with lifting	Factory, health club machines	0.92	0.42
Athletics	Basketball, heavy excercise	1.04	0.47

Fig. 11.5
Respiration And Perspiration Moisture Loads
As activity increases, moisture loads rise as people breath more frequently and perspire to release metabolic heat.

Respiration & Perspiration Load Per Hour (lb or kg) **Equation 11.1**

n_1 = Group @ activity level 1 (Number of people)
rp_1 = Moisture release per person (Grains or grams per hr)
n_2 = Group @ activity level 2 (Number of people)
rp_2 = Moisture release per person (Grains or grams per hr)

$$W_{rp} = (n_1 \times rp_1) + (n_2 \times rp_2) + \ldots$$

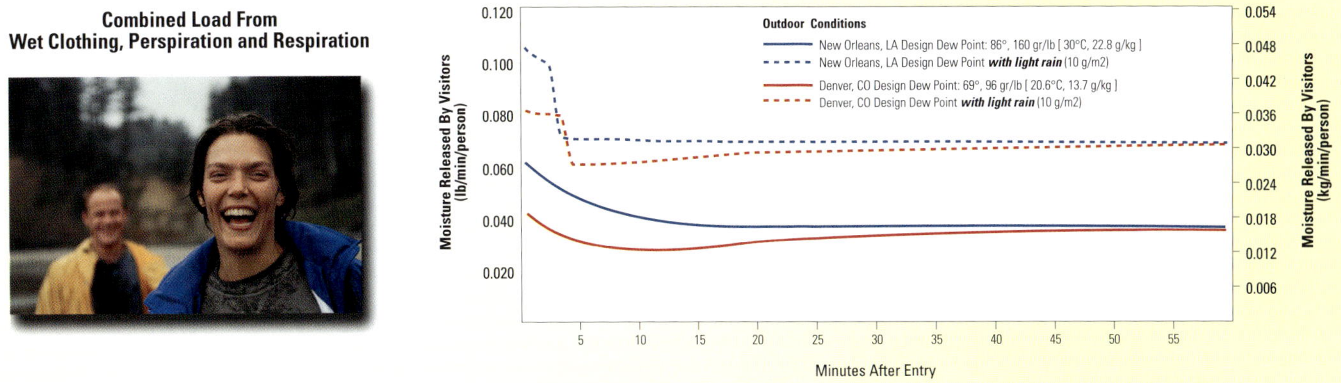

Fig. 11.6
Total Load From People
In addition to moisture from respiration and perspiration, visitors bring moisture into the building in their clothing after they walk in humid or wet conditions outdoors. (Jones et al, 1994.)

be a significant load. That moisture is released by evaporation when the person walks into a humidity-controlled building. The load depends on the amount of adsorbed water, the moisture level indoors and the duration of the visit. The lower the indoor humidity ratio, the more moisture will be released, and most of the clothing moisture load is released during the first 10 minutes.

Figure 11.6 shows the load released by adults dressed in cotton sweat clothes as they enter a humidity-controlled building after walking outdoors in design moisture conditions for Denver, Colorado (a dry climate) and New Orleans, LA (a highly humid climate).

The designer can decide which moisture release rate to assume by asking the owner to estimate the average length of a typical visit to the building, and by taking the average of the initial and final release rates for a visit of that duration. For example, if an average "take-out" patron in a quick-service restaurant in Tampa, FL usually stays in the building for 5 minutes, the initial release rate is appropriate to use in the calculation. But if a patron in a family restaurant stays for 45 minutes, the designer could use the average of the initial and the 45-minute release rates in the equation. If several different "visit lengths" are typical—as in a restaurant with both take-out and sit-down meals—then the designer might choose to solve the equation

Clothing Load Per Hour (lb or kg) **Equation 11.2**

- n: Visitors (Number of people per peak hour)
- t: Average length of visit (Minutes)
- R_{r1}: Initial moisture release rate (lbs/h or kg/h)
- R_{r2}: End-of-visit moisture release rate (lb/h or kg/h)

$$W_c = n \times \frac{t}{60} \times \frac{(R_{r1} - R_{r2})}{2}$$

separately for each category of visitors and combine the results to obtain the total clothing moisture load. (Jones et al, 1994)

Ventilation

Ventilation is usually the largest element in the moisture load in commercial buildings. Outdoor air is brought into the building to:

- Provide fresh air for people to meet building code requirements
- Make up for air exhausted from the building
- Ensure a slight excess of internal air pressure to reduce infiltration of humid air into building cavities

Ventilation for people

Outdoor air dilutes internally-generated pollutants that would otherwise compromise indoor air quality. To help decide how much air is appropriate for that task, ASHRAE provides guidance through Standard 62.1 (Ventilation for acceptable indoor air quality in commercial buildings). But local building codes define the minimum amount of outdoor air required by law. When these two documents disagree, the local code governs the design.

On the other hand, the designer should keep in mind that codes and ASHRAE recommendations are minimums. Given the complexity of buildings in the real world, the current science behind those recommended values is weak. An owner always has the choice of using more than the minimum ventilation air. As with structural steel decisions, some owners prefer a conservative design with more ventilation, staying away from the edge of legal minimum requirements.

The current ASHRAE-recommended ventilation rates are displayed in figure 11.21, located at the end of this chapter. Note that those recommendations vary according to building type and they vary between functions within that building. For example, offices currently have a suggested minimum ventilation rate of 20 cfm [9.4 l/s] per person of outdoor air to dilute the concentration of infectious particles, body odors and objectionable gases. In contrast, retail stores are ventilated at a much higher rate of 0.3 cfm/ft^2 [0.01 l/s/m^2], primarily to dilute the volatile organic vapors emitted by merchandise and its packaging.

But these recommendations are subject to change. When local codes leave the designer in doubt, a current copy of ASHRAE Standard 62.1 can be obtained from the society's headquarters.

Air to make up for exhaust

When air is exhausted from toilets, kitchens and pools, an equal amount of *conditioned* air must be brought into the building to replace that exhaust air. Otherwise, *unconditioned* replacement air is dragged through the exterior envelope where its moisture will condense and damage the building. Insufficient conditioned make up air is a frequent contributor to moisture damage and humidity control problems.

Calculating the amount of make up air required for the building requires knowledge of the exhaust fans. The flow through each exhaust fan must be measured and controlled to the designer's specifications—which requires flow-setting devices in the ductwork.

Assuming that the exhaust air can be controlled, the designer simply adds the total of all exhaust fans to obtain the minimum outdoor air needed to offset that loss. In most cases, the ventilation requirement for people is much larger than the amount required for make up air. Ventilating for people often meets the make-up air requirement as well.

However, the designer must examine each area separately, because providing make-up air in one area may not compensate for a high exhaust rate in another, even though the quantities match. For example, a restaurant dining area needs a high ventilation rate for people, which could also make-up for air exhausted from the kitchen. But if the kitchen is separated from the dining area by a single door, all the air would whistle through the doorway at high velocity and make it impossible to carry food between the two spaces. So each area must be considered separately to see if air must be brought to the space directly, or if ventilation air can be transferred to the space from a different area within the building.

Air to pressurize the building

Keeping the building under a slight positive air pressure is very useful for humidity control during the humid seasons. (Note: During cold seasons, a neutral pressure is essential to minimize the amount of humid indoor air drifting into cold building cavities.) The amount of air needed to maintain a positive pressure varies according to the leakage area in the exterior wall. In practical terms, the total crack area in a building—especially one which has not yet been built—can only be estimated. Most designers simply add 10% more air than

Fig. 11.7
Ventilation Dilutes Contaminants
Flushing the space with outdoor air prevents the buildup of infectious or otherwise objectionable particles and vapors.

Fig. 11.8
Exhaust Fans Create a Need For Dehumidified Make-up Air
Otherwise, exhaust fans pull unconditioned air through the building envelope, leading to moisture damage and poor humidity control. Moisture must be removed from makeup air.

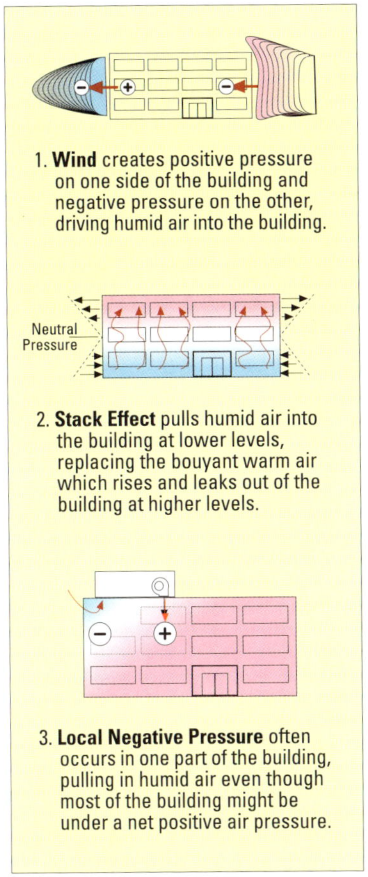

Fig. 11.9
Air Infiltration Into Buildings
Air infiltrates into the building in response to pressure differences caused by wind and by the stack effect. Some infiltration is inevitable, and must be allowed for in load calculations.

Ventilation Air Load Per Hour (lb or kg) — **Equation 11.3**

Density of Standard Air
I-P $d = 0.075$ lb/ft^3
SI $d = 0.072$ kg/l/s

$$W_V = Q \times (60 \times d) \times (M_1 - M_2)$$

- Q — Ventilation air flow (cfm or l/s)
- d — Air density (lb/ft^3/min or kg/l/s)
- M_1 — Outdoor humidity ratio (lb/lb or kg/kg)
- M_2 — Indoor humidity ratio (lb/lb or kg/kg)

is exhausted from the building to the make up air in the hope that the extra air will keep the majority of the building under a slight positive internal air pressure most of the time. Building pressure management is quite complex, but adding some excess air to the ventilation air stream appears to be a good practice. In the absence of better information, assuming 10% additional ventilation air will suffice for load estimates.

Net load from ventilation air

When each of the three air flows above are quantified, the designer calculates the load from ventilation air with equation 11.3.

Infiltration

All buildings leak. (Persily, 1999) Air entering through those leaks carries moisture which must be removed by the dehumidifier.

The amount of air leaking into the buildings depends on the pressure difference between indoors and outdoors. More air leaks in when wind blows against a wall, as shown in figure 11.9. Low pressure on the leeward side of a building pulls air out of the wall which is replaced by air leaking in on the windward side. Leakage varies greatly from one moment to the next as wind pressure changes. It also varies greatly between one part of the building wall and another, creating inward leakage in one part, and outward leakage nearby.

	Probable Leak Rate	
Wall Construction	cfm • ft^2	l/s • m^2
Tight	0.1	0.5
Average	0.3	1.5
Loose	0.6	3.0

Leakage rates are based on field measurements for North American commercial buildings at a pressure difference of 0.3" [75Pa]

Fig. 11.10
Typical Leakage Rates For Commercial Buildings
(ASHRAE Handbook—Fundamentals, 1997)

Because of these complex phenomena, infiltration is notoriously difficult to predict with any accuracy. Although many attempts have been made to correlate the Architect's design and the building code requirements with measured leakage, leakage results are random.

For example, Canadian building codes are considerably more strict regarding building tightness than requirements in most other countries, and contractors are quite experienced with these strict requirements. But one study of high-rise apartment buildings in the prairie region showed that while codes call for leakage of less than 0.02 cfm · ft^2 [0.10 l/s · m^2], measured results showed actual leakage rates between 0.39 and 0.60 cfm · ft^2 [2.06 and 3.15 l/s · m^2] In other words, 200 to 315% more than the expected infiltration. Similarly, measured leakage for 70 commercial buildings in Florida showed conclusively that North American commercial buildings leak a truly astonishing amount of air—in some cases over 10 air changes per hour. (Cummings et al, 1996)

Figure 11.10 shows typical leakage rates for exterior walls of commercial buildings. These values are extracted from Chapter 25 of the 1997 *ASHRAE Handbook—Fundamentals*. Typical low-rise, low-cost commercial buildings have either average or leaky walls. Institutional and high-rise buildings may also have leaky walls, but are more likely to have either average or tight wall construction.

Most technical professionals are understandably uncomfortable using vague terms like "tight" and "leaky" and "typical low-rise buildings" to select values to enter into load calculations. Experts in moisture load calculations are equally uncomfortable. But the only alternative is to perform a whole-building leak test after the structure is complete—an expensive and complex practice, and one which is not possible in the design stage.

Infiltration Load Per Hour (lb or kg) **Equation 11.4**

Density of Standard Air
I-P d = 0.075 lb/ft³
SI d = 0.072 kg/l/s

Wall surface exposed to wind (ft^2 or m^2)
Infiltration rate (cfm/ft^2 or $l/s/m^2$)
Air density ($lb/ft^3/min$ or kg/l/s)
Outdoor humidity ratio (lb/lb or kg/kg)
Indoor humidity ratio (lb/lb or kg/kg)

$$W_i = A \times I_r \times (60 \times d) \times (M_1 - M_2)$$

In the past, this discomfort has led even highly-skilled professionals to simply *ignore* the entire issue of air leakage, frequently with the rationale that "if ventilation air flow is greater than the exhaust air flow then any leakage will be from inside to outdoors, eliminating the leakage moisture load." Unfortunately, not a single field investigation validates that optimistic assumption. In fact, quite the reverse. (Cummings, 1996; Persily, 1999; Treschel, 1986) All buildings pull in *some* outdoor air at some times at some points—even when the *overall average* internal air pressure is positive.

Unless the designer has reason to believe the building will not leak—making it unique among buildings investigated throughout the world to date—engineering judgement must be applied and a decision made as to leakage rates. A leakage rate for the walls exposed to the prevailing winds can be entered into equation 11.4.

As a practical matter, few commercial buildings are controlled to humidity levels below 40% rh, and even fewer are defined well enough at the design stage to merit a time-consuming component-level load estimate. However, when the designer needs to estimate infiltration more precisely, and when great detail is available concerning the construction of the humidity-controlled space, Chapter 25 of the *ASHRAE Handbook—Fundamentals* (1997) contains leakage rates for individual components such as different types of doors, windows, electrical outlets, fireplaces and surface-mounted lights. These values are most useful for calculating loads in very well-defined small spaces that must be maintained at very low humidity.

Return Duct Pressure	Probable Infiltration cfm • ft²	l/s • m²
-0.1 " [-25Pa]	11	5.2
-0.25 " [- 62Pa]	20	9.4
-0.5 " [-125Pa]	31	14.6
-1.0" [-249 Pa]	48	22.6

Leakage rates are based on estimates contained in the SMACNA Duct Inspection Guide

Fig. 11.11 Duct Leakage
New buildings should have sealed duct work, but in older buildings, the designer may need to account for moisture loads when unsealed duct work passes through parts of the building that are not dehumidified.

Rain-soaked masonry walls

Investigations into moisture problems in humid climates have noted that porous, rain-soaked bricks or concrete blocks add moisture to any infiltration air as it traverses the wall. In most cases, the architectural designer will be aware of the need to use water and air barriers on the indoor side of such materials to seal the wall against moisture and humid air leaks.

But if such barriers are not in place for any reason, the humidity control designer can estimate the enormous increase in the moisture load by first estimating the surface temperature of the exterior wall after a rainstorm. Assume that any air entering through that wall will be saturated at the exterior surface temperature, and then recalculate the moisture load from infiltration using equation 11.4. The resulting load will be fearsomely large, which should prompt the architectural designer and contractor to make certain that a continuous air barrier will be in place to limit humid air infiltration.

Leaking return air ducts

In most buildings, return air ductwork passes through conditioned spaces. But where that is not the case, any humid air leakage into the duct adds a moisture load to the system. Good practice suggests that the HVAC designer should specify that all return and supply duct work in dehumidified buildings shall be sealed on all longitudinal and transverse joints, and sealed to the inlets of all air handlers.

Wind Speed Conversion		
mph	**fpm**	**m/s**
1	88	0.4
2	176	0.9
3	264	1.3
4	352	1.8
5	440	2.2
6	528	2.7
7	616	3.1
8	704	3.6
9	792	4.0
10	880	4.4
11	968	4.9
12	1056	5.3
13	1144	5.8
14	1232	6.3
15	1320	6.6
16	1408	7.1
17	1496	7.6
18	1584	8.0
19	1672	8.5
20	1760	8.9

Fig. 11.12
Wind Speed Conversion Table

On the other hand, not all systems are designed and built with good practices. Also, older buildings sometimes need humidity control added long after initial construction. When for any reason ducts have not been sealed, and when the return duct work passes through humid air, figure 11.11 provides estimates of air infiltration rates at different suction pressures. (SMACNA, 1989) These allow the designer to estimate the moisture load from return air duct leaks, using equation 11.4.

Doors

As exterior doors open and close, they allow humid air to enter the building. The amount of air depends on the open area of the door, how long the door remains open, and the air pressure difference between indoors and outdoors.

Also, for tall doors the temperature difference across the wall leads to pressure differences caused by the different buoyancy of the air masses on each side of the door. This is the most important contributor to leakage through doors in cold storage buildings, where doors are tall, and where temperature differences are large during humid months.

Also, exterior doors in tall buildings are subject to large air pressure differences because of the stack effect. Warm air rises inside the building, pulling outdoor air inwards at the base of the building, and forcing indoor air outwards through the exterior wall at the top of the building as shown in figure 11.9.

For low-rise commercial buildings and for typical personnel doors, neither the stack effect nor the temperature difference nor the height of the door is as important as the *exterior wind pressure* exerted when the door is open. When the wind blows against the exterior wall, it forces humid air into the building every time the doors on that wall open to admit people or product. The faster the wind, the more humid air enters during each opening.

The extreme wind speeds for 1453 North American and International locations is presented in Chapter 32. However, extreme winds

Door Load Per Hour (lb or kg) **Equation 11.5**

Density of Standard Air
I-P $d = 0.075\ lb/ft^3$
SI $d = 0.072\ kg/l/s$

$$W_d = n \times A \times V \times d \times \frac{t}{120} \times (M_1 - M_2)$$

- n — Door openings (Number/hr)
- A — Open door area (ft^2 or l)
- V — Wind velocity (fpm or m/s)
- d — Air density ($lb/ft^3/min$ or kg/l/s)
- t — Open time (seconds to close)
- M_1 — Outdoor humidity ratio (lb/lb or kg/kg)
- M_2 — Indoor humidity ratio (lb/lb or kg/kg)

usually occur when the weather is changing rapidly or during storms. Those may not be the periods when a commercial building has the most door openings. Some designers prefer to use the average annual wind velocity for the calculation instead of one of the three extremes. The average annual wind velocity can be obtained from the local airport weather station, or from the U.S. National Climatic Data Center in Asheville, NC. (http://www.ncdc.noaa.gov/) That organization records hourly weather data and summaries for over 10,000 sites worldwide. Annual average wind speeds are often between 25 and 50% of the 5% extreme velocity shown in Chapter 32.

After the designer decides which wind velocity to use and after the owner estimates the door traffic, the designer can estimate the door moisture load with equation 11.5.

Vestibules or air locks

A vestibule or "air lock" is a small chamber between two doors—a common feature of exterior entryways into commercial buildings. Since the entering person does not open both doors at the same moment, wind cannot carry moist air into the building freely—the air halts in the air lock, and only a small amount drifts into the building when the interior door is opened. For doors which must be opened manually, the designer may assume that some fraction of the air in

Vestibule Load Per Hour (lb or kg) — **Equation 11.6**

Density of Standard Air
I-P d = 0.075 lb/ft³
SI d = 0.072 kg/l/s

$$W_{vb} = n \times v \times f \times d \times \frac{(M_1 - M_2)}{2}$$

- n — Openings (Number/hr)
- v — Vestibule volume (ft³ or l)
- f — Percent of vestibule volume that enters the building (Decimal fraction)
- d — Air density (lb/ft³/min or kg/l/s)
- M_1 — Indoor humidity ratio (lb/lb or kg/kg)
- M_2 — Outdoor humidity ratio (lb/lb or kg/kg)

the vestibule is moved into the building with each door opening, and that the moisture content of that air is the average of the indoor and outdoor humidity ratios. Equation 11.6 allows an estimate of the moisture load from vestibules with manual doors.

Sometimes, vestibule doors open automatically, as in supermarkets and any building in which customer convenience is more important than limiting loads. With automatic operation, both doors may be open at the same time, but the net open time to the weather is considerably less than if there were only a single door between the weather and the indoors. For automatic doors on vestibules, the designer can estimate a shorter open time, using the equation for doors.

Door curtains using air or plastic strips

In truck loading docks, "air curtains" are sometimes used to limit the amount of air drifting into the building. An air curtain forces a high-velocity jet of indoor air across the opening. The effectiveness of that air as a barrier depends on its momentum—the product of its mass times its velocity. More air at a higher speed will be a more effective barrier to wind blowing against the opening. For moisture load estimates through air curtains, the designer should consult the manufacturer of the device. These firms can sometimes provide estimates of outdoor air leakage when wind is blowing against the side of the opening at the design velocity. But usually, air curtains are not intended to resist large wind pressures. They function best when the wind is at levels well below design extremes. Leakage at design or even average outdoor wind velocities is likely to be quite large. Buildings which need humidity control seldom use air curtains, because their infiltration load is so large that it greatly increases the size and cost of dehumidification components.

Door curtains made of overlapping plastic strips are often used in warehouses to limit air movement through doorways otherwise open to unconditioned spaces. Estimating the continuous humid air infiltration through these curtains has not yet been the subject of research. The designer is left to make an educated estimate.

If the end of the strips hang above the floor, the designer can estimate the open area, and assume that the humid air moves through that space continuously. If the ends of the strips rest on the floor and bend, leaving openings between strips near the bottom of the curtain, the designer can estimate the amount of open area as a percentage of the door space, and calculate infiltration through that area based on some assumed velocity.

If the doorway is on an exterior wall, the inflowing air velocity will depend primarily on the outdoor wind velocity. If the doorway separates two interior spaces, the velocity of inflowing air depends on the small difference in air pressure across the wall—which is difficult to predict. When both sides are at the same pressure, common practice in dehumidification design in the past has been to assume a velocity of at least 50 fpm [0.25 m/s] through the open area.

Door load reduction through positive internal air pressure

Logic suggests that if the building has a positive average internal air pressure, air should move *out* of the building—not inwards—when the door opens. That may indeed be the case when the air outdoors is still. But usually, wind presses against the exterior wall with a force far greater than any slight counteracting pressure inside the building. The more probable benefit of positive internal pressure is to reduce outdoor air infiltration into building cavities, and also to reduce the total *annual* moisture load which in turn reduces the cost of operation. However, for peak load calculations, most design-

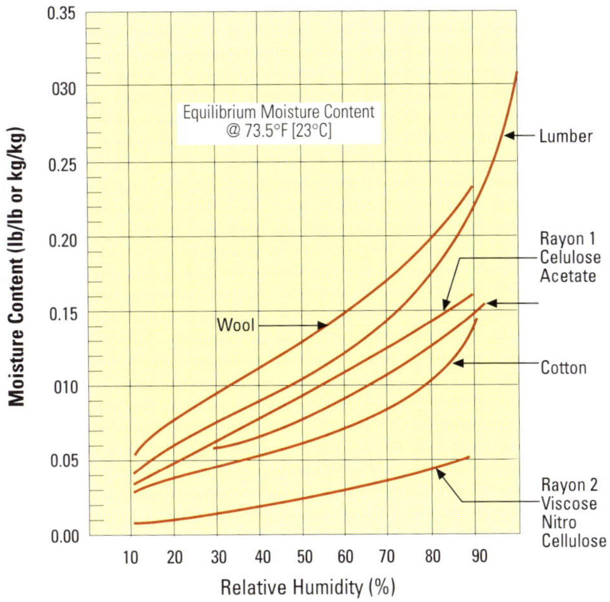

Fig. 11.13 Moisture Equilibrium

Products lose moisture as they move from humid environments to dry ones.

ers do not assume that positive air pressure will measurably reduce the load from door openings. They assume that *some* humid outdoor air will be entering the building every time a door opens.

Moist materials & products

Moisture from wet building materials and from humid packaging is often a load in commercial buildings. These loads are periodic, rather than continuous. So the designer often uses a time-weighted average for these loads, based on their magnitude and frequency.

New product loads

Moisture loads from humid packaging materials are significant in retail stores and warehouses, where a great deal of cardboard packaging may flow through the building in a short amount of time. The outer packaging absorbs moisture during transit to the building, and releases it when the product enters a humidity-controlled storage space. The amount of water given off depends on the sorption

Humid Product Load Per Hour (lb or kg) — **Equation 11.7**

$$W_{pr} = \frac{F_r}{24} \times (MC_1 - MC_2)$$

- F_r — Product flow rate through the building (lb/day or kg/day)
- MC_1 — Moisture content of the product when at equilibrium with high humidity (lb/lb or kg/kg)
- MC_2 — Moisture content of the product when at equilibrium with low humidity (lb/lb or kg/kg)

characteristics of the material, and on the difference between the humid transit environment and the dry storage environment. Figure 11.13 shows the moisture equilibrium curves for a variety of materials. (Harriman, 1990) If the designer knows the nominal weight of the product flowing through the building and the time it resides there, he can estimate the desorption load.

The time required to release the absorbed moisture depends largely on how fast air moves across the moist surface. Stacked cartons will dry quickly on the outside—perhaps in a few hours. But the cardboard between the packages may take days or weeks to dry out. Filing cabinets full of moist papers can take weeks or months to reach equilibrium with the dry air outside the cabinet. When the potential load is small, the designer can apply engineering judgment to estimate the time needed to dry specific materials, and in some cases the owner or the suppliers of the products may be able to assist in these estimates.

When the potential load is really large, the designer can test his assumptions by using an environmental simulation test chamber. These are accessible at reasonable cost through packaging engineering firms and through organizations that provide military-grade stress-testing or industrial material testing services.

Moisture loads from walls, furnishings, papers and books

The building and all its contents adsorb moisture when humidity rises, and release it when humidity falls. Some call this effect the "moisture capacitance" of the building. If the relative humidity indoors stays constant, there is little or no desorption, and therefore no moisture load from this source.

On the other hand, some buildings use cool outdoor air in evenings to precool the building for the following day, making it ready to absorb the high heat loads of the afternoon and reducing the net annual cost of cooling. This technique is often called "free cooling", or "economizer cooling." Unfortunately for humidity control, cool outdoor air usually carries excess moisture into the building. The relative humidity rises, so the building and its contents adsorb moisture which is released later—such as when humidity comes under control later in the day.

In most cases, the load from this source is not large, because without moist outdoor air, indoor humidity does not swing widely over 24 hours. But when the building contains a great deal of hygroscopic material compared to its internal volume—as is the case in libraries—the desorption load can be quite large when "free cooling" is used in humid climates. When a designer needs to estimate such a load, reference material is available in the form of the final report from ASHRAE research project 455 (Bailey et al, 1994). That report is available from ASHRAE headquarters.

Wet surfaces

As water evaporates from recently-cleaned carpets and floors or from pools and spas, it becomes a load on the dehumidification system. The evaporation rate determines the moisture load, and in all cases, evaporation load is greater when the:

- Air flows quickly across the wet surface
- Water is hot (has a high surface water vapor pressure)
- Air has a low water vapor pressure (is dry)
- Surface area for evaporation is extensive

Swimming pools and spas

The moisture evaporating from swimming pools has been investigated frequently. The rate varies widely according to the type of pool, its use, and whether the area includes recreational equipment that increase the surface area available for evaporation.

The lowest load is an unoccupied pool in a private home, used a few minutes a day by a single occupant swimming slowly from end to end. Public indoor wavepools have the highest load, loaded to their capacity with frolicking teenagers, splashing each other and sliding down water slide tubes that induce evaporation air currents. For the same pool surface and other non-activity variables, a quiet private pool evaporates about 33% of the rate of evaporation in the busy waterslide/wave pool. The activity within the space strongly influences the evaporation rate, because it determines how much surface is wet beyond the pool surface alone.

Recognizing this key variable, ASHRAE has added an "activity factor" to the classic wet surface evaporation equation developed by Willis Carrier in 1919. Like the Carrier equation, these activity factors are empirical, based on the experience of Engineers, owners and dehumidifier manufacturers. Factors based on a consensus of these groups are shown in Figure 11.15. (*ASHRAE Handbook—Fundamentals*, 1997) Given those values, equation 11.8 can be used to estimate moisture loads from different types of pools.

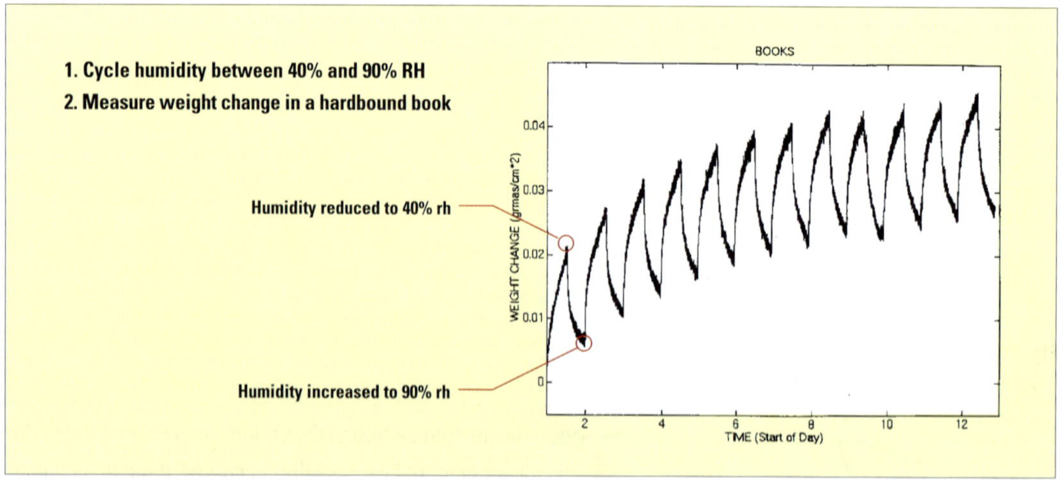

Fig.11.14
Desorption from Building Contents
As humidity changes, the building and its furnishings collect and release moisture. In most cases, this load is small. But when the building contains books, the amount of moisture adsorbed and desorbed in each cycle can be quite large. This graph shows the weight change per unit of exposed surface area of a hard-bound book, when humidity cycles between 40% and 90% rh over the course of each hour.
(Bailey et al, 1994)

Swimming Pool Activity Factors

Type of Pool	Factor
Residential pool	0.5
Condominium	0.65
Therapy	0.65
Hotel	0.8
Public or school	1
Whirlpool, spas	1
Wavepools, water slides	1.5

Fig. 11.15
Swimming Pool Activity Factors

Fig. 11.16
Vapor Pressures at Saturation

Vapor Pressure at Saturation			
°F	in.hg.	°C	kPa
100	1.93	38	6.63
95	1.66	35	5.63
90	1.42	32	4.76
85	1.21	29	4.01
80	1.03	26	3.36
75	0.86	23	2.81
70	0.74	20	2.34
65	0.62	17	1.94
60	0.52	14	1.60
55	0.44	11	1.31
50	0.36	8	1.07
45	0.30	5	0.87
40	0.25	2	0.71
35	0.20		

Swimming Pool Load Per Hour (lb or kg) **Equation 11.8**

$$W_{sp} = 0.1A \times (p_w - p_a) \times F_a$$

- A — Pool surface (ft² or m²)
- p_w — Saturation vapor pressure at water temperature (in.hg. or kPa)
- p_a — Saturation vapor pressure at air temperature (in.hg. or kPa)
- F_a — Activity factor

The designer might also consider the fact that moisture loads vary widely throughout 24 hours, depending on the actual use of the pool. The owner may not be especially concerned about maintaining humidity control in the pool area at peak load conditions. It may be sufficient to calculate the average load over 24 hours, and size the dehumidification equipment for that capacity, reasoning that over time, the equipment will "catch up" with the load. At peak load conditions, humidity will be above the set point, but the owner's purpose may not require continuous control.

Sizing the dehumidifiers for the average rather than the peak load provides substantial savings in both the construction budget and in annual operating costs. Assuming the owner agrees that continuous control is not needed, the designer can calculate the average load by estimating the number of full, partial and zero-use hours, making a calculation for each, summing the loads for all hours, then dividing the total by 24 to obtain the average hourly load. Nearly all manufacturers of swimming pool dehumidifiers provide software that helps a designer estimate these loads.

Still ponds & wet-mopped floors

Research has shown that evaporation rates from still ponds and wet floors are lower than from swimming pools and spas. Unlike swimming pools, floors and shallow ornamental ponds are seldom served with constantly flowing air, nor is the surface agitated. Also, the temperature at the water surface is likely to be near the wet bulb temperature of the air. So when the air velocity across the wet surface is below 25 fpm [below 0.13 m/s], the designer may choose to use an activity factor of 0.25 for better estimating the evaporation rate from such placid surfaces. (Pauken et al, 1995)

Carpet cleaning

Commercial and institutional carpets are often heavily soiled, requiring cleaning by hot water extraction rather than by light shampoo or "dry" absorption techniques. Shampoo, foam and dry absorption leave very little moisture behind on the carpet, and what little remains should evaporate within 1 to 3 hours. In contrast, the hot water technique sprays water onto the carpet at a rate of either 8.3 or 12.5 lbs per minute [3.8 or 5.7 kg/min]. After application and extraction, the water left behind takes between 6 and 24 hours to evaporate.

In the cleaning process, hot water is sprayed onto the carpet and immediately pulled back off the surface by an "extractor head" attached to a vacuum blower. The blower may be located in a portable unit near the operator, or connected by long hoses to a unit located further away in a truck. The amount of water remaining on the carpet after "extraction", depends on the spray rate of hot water and on the amount of water recovered by the extractor head. Those variables depend on both the skill of the operator and the force of the vacuum at the extraction head. Skilled operators that follow standard guidelines will make a second, "dry pass" after each spray application to recover more water, leaving less to evaporate and become a moisture load. Also, truck-mounted vacuum blowers usually recover more water on each pass because they have greater suction than portable units.

Evaporation slows as the remaining moisture approaches the wet bulb temperature of the air, and as less water is available to evaporate. The values in figure 11.17 are based on empirical observations and estimates by four carpet cleaning experts concerning spray rates, extraction rates and typical drying times. Experts all remark on the difficulty of obtaining accurate data on water extraction rates, which

Equipment	Spray Rate gpm	Spray Rate l/s	Extraction Rate	Average Evaporation Rates lb/h • ft² [kg/h • m²]: Small Rooms w. Furnishings	Average Evaporation Rates lb/h • ft² [kg/h • m²]: Large, Open Spaces
Truck-mounted	1.5	0.095	50%	0.05 [0.25]	0.020 [0.10]
			85%	0.01 [0.05]	0.005 [0.02]
Portable	1.0	0.063	50%	0.06 [0.30]	0.020 [0.10]
			85%	0.02 [0.10]	0.006 [0.03]

Fig. 11.17 Hot Water Carpet Cleaning

A considerable amount of water remains on the carpet after "hot water extraction." Estimates in this table assume the remaining moisture evaporates in 8 hours.

vary from 50 to 85% of the water under real world circumstances. They advise that the highest moisture load per hour occurs when the:

- Operator rushes to completion, and fails to make a second, "dry pass" after each "combined" application/extraction pass.
- Equipment has low suction capacity, as in portable units, or when truck units have long hoses or clogged filters.
- Air above the carpet is dry, so that evaporation takes place quickly, in spite of the usual lack of air movement at the surface.

The designer should note the informal character of the estimates in figure 11.17 and adjust his assumptions accordingly.

Wet Carpet Load Per Hour (lb or kg) — **Equation 11.9**

$$W_{cl} = A \times E_r$$

- A — Wet carpet surface (ft² or m²)
- E_r — Evaporation rate (lb/ft²/h or kg/m²/h)

Vapor Permeance Load Per Hour (lb or kg) — **Equation 11.10**

$$W_{vp} = \frac{A \times \text{Perm} \times (p_h - p_d)}{\text{Divide by: 7000 to obtain lb/h; 1000 to obtain kg/h}}$$

- A — Material surface separating dry from humid air (ft² or m²)
- Perm — Perm rating for material (gr/(ft²/in.hg.) or ng/(s/m²/Pa)
- p_h — Vapor pressure on the humid side (in.hg. or kPa)
- p_d — Vapor pressure on the dry side (in.hg. or kPa)

Vapor permeance

Water molecules migrate slowly through solid materials by diffusion, propelled by the difference between water vapor pressures on each side of the material. Humid air has a high water vapor pressure. Dry air has a lower vapor pressure. Water vapor moves slowly through material in response to that pressure difference.

The molecules move slightly faster when the vapor pressure difference is high, and when the material is more porous. That's why fresh bread dries out more quickly when bagged in paper rather than plastic. Moisture passes more freely through porous paper than through dense plastic.

The unit that quantifies water vapor movement through a material with a defined thickness is called the "perm" (the *permeance*). In I-P units, one perm is the number of grains per hour that pass through one square foot of material when the vapor pressure difference is one inch of mercury. (1 perm = 1 gr/hr · ft² · in.hg.)

[In SI units, the permeance is the number of nanograms per second that pass through one square meter when the vapor pressure difference is one pascal.]

The "*permeability*", describes how quickly water vapor moves through a given thickness of a material. In the I-P system, the permeability is the number of grains per hour, per square foot, per inch mercury, *per inch of material thickness*. Permeability can be used

Material	Thickness in.	mm	Permeance gr/h • ft² • in.hg.	ng/s • m² • Pa
Concrete	8.000	203	0.40	23
Brick masonry	4.000	100	0.80	46
Concrete block, hollow	8.000	203	2.40	137
Gypsum wallboard	0.375	10	50.00	2860
Hardboard (Tempered)	0.125	3	5.00	286
Wood, sugar pine	1.000	25	0.4 to 5.4	23 to 309
Plywood (Douglas fir, exterior glue)	0.250	6	0.70	40
Plywood (Douglas fir, interior glue)	0.250	6	1.90	109
Sheet, glass fiber with acrylic resin	0.056	1	0.12	6.9
Sheet, glass fiber with polyester resin	0.480	12	0.05	2.9
Aluminum foil (Thick)	0.0010	0.03	0	0
Aluminum foil (Thin)	0.0004	0.01	0.05	2.9
Polyethelene (6 mil)	0.0060	0.15	0.06	3.4
Polyethelene (8 mil)	0.0080	0.20	0.04	2.3
Polyethelene (10 mil)	0.0100	0.25	0.03	1.7
Polyvinylchloride (PVC)	0.0040	0.10	0.8 to 01.4	46 to 80
Polyester	0.0076	0.19	0.08	4.6
Cellulose Acetate	0.0100	0.25	4.6	263
Vapor retarder paint	0.0031	0.08	0.45	26
Primer-sealer	0.0012	0.03	6.28	360
Vinyl-acrylic primer	0.0016	0.04	8.62	493
Semi-gloss, vinyl-acrylic enamel	0.0240	0.61	6.61	378
Exterior acrylic house and trim	0.0017	0.04	5.47	313
Flat paint (2 coats) on insulation board	-	-	4.00	229
Two coats of enamel on smooth plaster	-	-	0.5 to 1.5	29 to 86
Primers + sealer, on insulation board	-	-	0.9 to 2.1	51 to 120
Primers + flat oil paint on plaster	-	-	1.6 to 3.0	91 to 172
	Typical Density			
	lb/ft²	**kg/m²**		
Built-up roofing	-	-	0	0
Duplex sheet, asphalt, aluminum foil	8.6	0.42	0.176	10
Saturated and coated rool roofing	65.0	3.18	0.24	14
Kraft paper, asphalt-laminated, reinf.	6.8	0.33	1.8	103
Insulation backup paper, asphalt-coated	6.2	0.3	0.6 to 4.2	34 to 240
Asphalt-saturated, coated v. retarder ppr.	8.6	0.42	0.06	3.4
Asphalt-saturated, not coated ppr.	4.4	0.21	20.2	1155
15-lb asphalt felt	14.0	0.68	5.6	320
15-lb tar felt	14.0	0.68	18.2	1040
Single kraft paper, double layer	3.2	0.16	42	2400

Fig. 11.18.1 Vapor *Permeance* of Building Materials and Films

Material	Permeability Perm/in.	ng/s • m • Pa
Air (still)	120	174
Cellular glass	0	0
Mineral wool (unprotected)	116	245
Expanded polyurethane (Board stock)	0.4 to 1.6	0.58 to 2.3
Expanded polystyrene (extruded board)	1.2	1.7
Expanded polystyrene (bead board)	2.5 to 5.8	2.9 to 8.4
Unicellular synthetic flexible rubber foam	0.02 to 0.15	0.029

Fig. 11.18.2 Vapor *Permeability* of Insulation

for calculating moisture movement through materials which vary in thickness, such as air gaps and insulation. On the other hand, most materials sold as vapor retarders are sold in a specific thickness. Such products carry a defined net *permeance* rating, based on their thickness.

The slowly diffusing moisture only becomes a load on the dehumidifier after it enters the inside face of the exterior wall, having traversed the layer with the lowest permeance (highest resistance). Therefore, to size the dehumidification system the designer only needs to calculate the transmission rate through the least-permeable layer—the material with the lowest perm rating.

Permeance or permeability ratings are given for a variety of building materials in figures 11.18.1, and 2. When the permeance and the total surface area are known, they can be used with the vapor pressure difference in equation 11.10 to estimate the moisture load that moves through the building's exterior wall by diffusion.

For dehumidification in commercial buildings, the load from permeance is nearly always so small that it's hardly worth bothering to calculate. Air infiltration is usually one or two orders of magnitude larger than the permeation load. On the other hand, careful calculation of permeance is important in *humidified* buildings such as museums during winter time. In highly humid buildings in cold climates, moisture will diffuse into building cavities from inside the building and condense. Indoor wall permeance is also important in single family homes, where the exterior wall surface area is large in relation to the building's internal volume. In most commercial buildings, however, the exterior surface is much smaller in relation to the enclosed volume. The humid season diffusion moisture load is therefore relatively small, especially compared to the infiltration load carried by air entering through cracks in the exterior wall.

Although the HVAC designer will pay attention to all aspects of load calculations, he can quickly see that for dehumidification, the permeance load is not as important as most other elements, because its contribution to the total load is so small. More benefit will come from good estimates of air infiltration than from determining whether the interior paint *really* has a perm rating of 0.3 or whether it has a more porous 0.5. rating. Either rating is low enough to slow moisture diffusion to nearly inconsequential levels, and neither rating is likely to affect either system design or equipment size.

Domestic loads

In residential buildings like hotels, apartments, condominiums and dormitories, domestic operations can contribute to the moisture load. Normally, however, the duration of these activities is seldom long. Over a 24-hour period the average hourly moisture load from showers, for example, is very small compared to loads from other sources. (Christian et al, 1993) Figure 11.19 shows the typical moisture loads from selected domestic activities.

Load Source	Typical Moisture Load Per Event lb	Typical Moisture Load Per Event kg	Hourly Load lb	Hourly Load kg	Comments & Assumptions
Bath (15 minutes)	0.12	0.06	0.03	0.01	One bath/hr
Shower (15 min)	1.40	0.66	5.60	2.55	Four showers/hr
Breakfast preparation (4 ppl)	0.35 to 0.58	0.16 to 0.27	0.61	0.28	Preparation lasts 20 minutes
Lunch preparation	0.53 to 0.68	0.25 to 0.32	0.80	0.36	Preparation lasts 20 minutes
Dinner preparation	1.22 to 1.58	0.55 to 0.72	0.92	0.42	Preparation lasts 40 minutes
Breakfast dishes (4 ppl, by hand)	0.21	0.10	0.04	0.02	Requires 10 minutes
Lunch dishes	0.16	0.07	0.03	0.01	Requires 10 minutes
Dinner dishes	0.68	0.31	0.23	0.10	Requires 20 minutes
Simmering pot (6 in. diam., per 10 min)	0.13	0.06	0.04	0.02	Simmers for 20 minutes
Boiling pot (6 in. diam., per 10 min)	0.57	0.26	0.19	0.09	Boils for 20 minutes
Floor mopping (dries over one hour)	0.03	0.014	0.03	0.013	Load per hour, per ft2 or m2
Unvented clothes dryer (per load)	4.7 to 6.2	2.13 to 2.90	3.60	1.64	Average load, dry in 40 minutes
Typical total daily load (4 ppl)	**3.5 to 10.9**	**1.6 to 5.0**	**0.3**	**0.14**	**Includes respiration & perspiration**
Green firewood (128 ft^3 over 6 months)	800	363.6	0.18	0.08	Assumes hardwood. Softwood is ±50% less
Crawl space (1000 ft^2, uncovered, per day)	95	43.2	0.004	0.002	Load per hour, per ft2 or m2
Crawl space (covered)	9.5	4.3	0.0004	0.0002	Load per hour, per ft2 or m2
1 ton A/C unit without drain (per day)	6	2.7	0.25	0.11	Moisture re-evaporated from pan
Construction moisture (1st year, per day)	21	9.5	0.88	0.40	Full concrete basement and wood framing.
Damp basement (1000 ft^2, per day)	21	9.5	0.001	0.0004	Load per hour, per ft2 or m2
10 Medium-size ferns (per day)	5.7	2.6	0.24	0.11	Typical of medium-size house plants
One 6ft tall beech tree (per day)	106	48.2	4.42	2.01	Typical of ornamental indoor trees

Fig. 11.19 Domestic Moisture Loads

As with other periodic loads, the designer might consider averaging such intermittent loads over the time the owner might allow humidity to be above the design set point , as opposed to designing the dehumidification equipment to maintain conditions throughout all peak load events. For example, does a hotel room *really* have to be kept below 50% rh while the guest is taking a morning shower? Or can the humidity be allowed to rise for a second hour so a smaller dehumidifier could slowly "catch up" with that one-time load? If the owner agrees to an hour above set point, the moisture load per hour is cut in half and the dehumidification capacity can be smaller and less costly to operate. An owner's preferences may be different in a health club, which could have heavy shower loads for long periods.

Comparing Loads By Climate and Application

Figure 11.20 shows some interesting patterns with respect to dehumidification loads. The peak-hour load is driven by the application more than by the climate. And what drives the dehumidification load higher is the ventilation requirement.

That's why a medium-size hospital located in "dry" Phoenix, Arizona has a peak dehumidification load more than 10 times larger than an office building located in "humid" Atlanta, Georgia. A hospital needs a great deal more ventilation air than an office building, and ventilation air carries an enormous amount of excess moisture.

Also note that despite the common perception, the peak outdoor air moisture in "dry" Phoenix is really about the same as in "humid"

Fig. 11.20
Comparing Peak Hourly Loads

Comparing Peak Hour Moisture Loads By Application & Climate

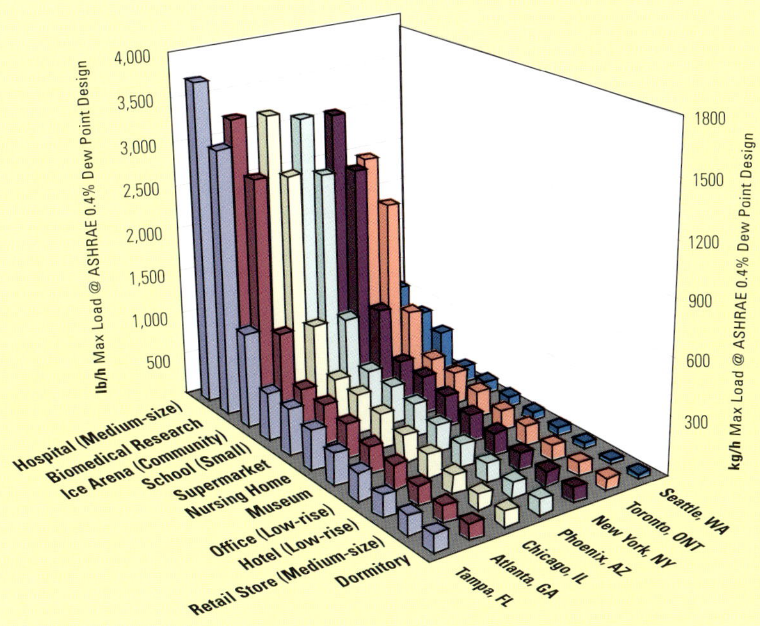

Peak Hour Load (lb/h)

0.4% Design Moisture:	144 gr/lb Tampa	133 gr/lb Atlanta, GA	132 gr/lb Chicago	130 gr/lb Phoenix	129 gr/lb New York	116 gr/lb Toronto	78 gr/lb Seattle
Hospital (Medium-size)	3,710	3,220	3,180	3,080	3,040	2,450	750
Biomedical Research Bldg	3,060	2,640	2,600	2,530	2,490	2,000	560
Ice Arena (Community)	1,090	980	970	950	945	820	460
School (Small)	550	490	480	470	465	390	180
Supermarket	530	465	460	450	440	370	155
Nursing Home	460	400	395	390	380	315	125
Museum	365	320	315	310	303	250	95
Office (Low-rise)	320	280	275	270	265	220	85
Hotel (Low-rise)	254	224	220	215	212	175	75
Retail Store (Medium-size)	220	190	188	185	180	145	50
Dormitory	210	180	180	175	172	145	60

Peak Hour Load (kg/h)

0.4% Design Moisture:	20.6 g/kg Tampa	19.0 g/kg Atlanta, GA	18.9 g/kg Chicago	18.6 g/kg Phoenix	18.4 g/kg New York	16.6 g/kg Toronto	11.1 g/kg Seattle
Hospital (Medium-size)	1,686	1,464	1,445	1,400	1,382	1,114	341
Biomedical Research Bldg	1,391	1,200	1,182	1,150	1,132	909	255
Ice Arena (Community)	495	445	441	432	430	373	209
School (Small)	250	223	218	214	211	177	82
Supermarket	241	211	209	205	200	168	70
Nursing Home	209	182	180	177	173	143	57
Museum	166	145	143	141	138	114	43
Office (Low-rise)	145	127	125	123	120	100	39
Hotel (Low-rise)	115	102	100	98	96	80	34
Retail Store (Medium-size)	100	86	85	84	82	66	23
Dormitory	95	82	82	80	78	66	27

Atlanta—133 gr/lb in Atlanta versus 130 gr/lb in Phoenix . [19.0 g/kg in Atlanta vs. 18.6 g/kg in Phoenix]. Similarly, one is likely to think of Seattle, Washington as being a "humid" climate. Based on rainfall, that is certainly an accurate perception. But based on peak design conditions, "dry" Phoenix has a dehumidification load from ventilation air that is *far* larger than the load in the rainy but cool climate of Seattle.

These facts suggest that the prudent HVAC designer should not assume that "dry" climates will need less dehumidification equipment than "humid" climates. The required amount of cooling and heating capacity will be very different throughout North America, but the required amount of *dehumidification* capacity can be surprisingly similar for many thermally-different climates.

On the other hand, on an *annual* basis the dehumidification loads are indeed much higher in humid climates. In Atlanta, the ventilation load index shows that every cfm of ventilation air carries a dehumidification load of 6.2 ton-hours per year, where in Phoenix, that same cfm carries a dehumidification load of only 1.3 ton-hours per year. (Harriman et al, 1997) In other words, the *annual* load is nearly five times larger in Atlanta. Even though the equipment size will be nearly the same in both locations, the dehumidifier in Atlanta will need to operate for many more hours than will the dehumidifier in Phoenix. This relationship implies that for dehumidifiers, it will be especially useful to install energy-efficient capacity modulation, so that the full capacity needed during peak load periods is not operated wastefully during the rest of the year when loads are much smaller.

Fig. 11.21.1
Minimum Recommended Ventilation Rates For Commercial/Institutional Occupancies
(ASHRAE Standard 62.1 1999)

Application	Estimated Max. Occupancy People/1000 ft2 or 100m2	Outdoor Air Requirements cfm/Person	l/s • Person	cfm/ft²	l/s • m²	Comments
Dry Cleaners & Laundries						
Commercial laundry	10	25	13			Dry cleaning processes may require more air.
Commercial dry cleaner	30	30	15			
Storage & pick up	30	35	18			
Coin-operated laundries	20	15	8			
Coin-operated dry cleaner	20	15	8			
Food & Beverage Services						
Dining Rooms	70	20	10			
Cafeteria/Quick Service	100	20	10			
Bars, cocktail lounges	100	30	15			Supplementary smoke removal may be required.
Kitchens (Cooking)	20	15	8			
Garages, Repair, Service Stations						
Enclosed parking garage				1.5	7.5	
Auto repair rooms				1.5	7.5	
Hotels, Motels, Resorts				cfm/room	L/s • room	Independent of room size
Bedrooms				30	15	
Living rooms				30	15	Installed capacity for intermittent use
Baths				35	18	
Lobbies	30	15	8			
Conference rooms	50	20	10			
Assembly rooms	120	15	8			
Dormitory sleeping areas	20	15	8			
Gambling casinos	120	30	15			
Offices						
Office space	7	20	10			Some office equipment may require local exhaust
Reception areas	60	15	8			
Telecom centers/Data entry areas	60	20	10			
Conference rooms	50	20	10			
Public Spaces				cfm/ft²	l/s • m²	
Corridors & utilities				0.05	0.25	
Public restrooms (cfm or l/s per WC or urinal)		50	25			Normally supplied by transfer air from adjacent spaces
Locker rooms or dressing rooms				0.5	2.5	Local exhaust—with no recirculation—is recommended
Smoking lounge	70	60	30			Supplementary smoke removal may be required.
Elevators				1.0	5.0	Normally supplied by transfer air

Fig. 11.21.2
Minimum Recommended Ventilation Rates For Commercial/Institutional Occupancies
(ASHRAE Standard 62.1 1999)

Part 2 of 4

Application	Estimated Max. Occupancy People/1000 ft2 or 100m2	Outdoor Air Requirements cfm/Person	l/s • Person	cfm/ft²	l/s • m²	Comments
Retail Stores, Sales & Showroom Floors						
Basement and street	30			0.3	1.5	
Upper floors	20			0.2	1.00	
Storage rooms	15			0.15	0.75	
Dressing rooms				0.20	1.00	
Malls and arcades	20			0.20	1.00	
Shipping and receiving	10			0.15	0.75	
Warehouses	5			0.05	0.25	
Smoking lounge	70	60	30			Normally supplied by transfer air. Local exhaust—with no recirculation—is recommended.
Specialty Shops						
Barber	25	15	8			
Beauty	25	25	13			
Reducing salons	20	15	8			
Florists	8	15	8			Plant growth requirements may dictate ventilation rate
Clothiers, furniture				0.30	1.50	
Hardware, drugs, fabric	8	15	8			
Supermarkets	8	15	8			
Pet shops				1.00	5.00	
Sports And Amusement						
Spectatator areas	150	15	8			When internal combustion engines are operated
Game rooms	70	25	13			for maintenance of playing surfaces
Ice arenas (playing areas)				0.50	2.50	increased ventilation rates may be required.
Swimming pools (pool plus deck area)				0.50	2.50	
Playing floors (gymnasiums)	30	20	10			
Ballrooms & discos	100	25	13			
Bowling alleys (seating areas)	70	25	13			
Theaters						
Ticket booths	60	20	10			
Lobbies	100	25	13			
Auditorium	150	18	8			When special effects are used, such as dry ice vapors, mists,
Stages, studios	70	15	8			etc, additional ventilation will be needed to clear the air.
Transportation						Ventilation *within* vehicles may require special considerations
Waiting rooms	100	15	8			
Platforms	100	15	8			
Vehicles	150	15	8			

Fig. 11.21.3
Minimum Recommended Ventilation Rates For Commercial/Institutional Occupancies
(ASHRAE Standard 62.1 1999)

Application	Estimated Max. Occupancy People/1000 ft2 or 100m2	Outdoor Air Requirements cfm/Person	l/s • Person	cfm/ft²	l/s • m²	Comments
						Part 3 of 4
Workrooms						
Photo studios	10	15	8			
Darkrooms	10			0.50	2.50	
Duplicating or printing				0.50	2.50	Installed equipment must include positive exhaust and supply control of undesirable contaminants (toxic or otherwise).
Pharmacy	20	15	8			
Bank vaults	5	15	8			
Meat processing	10	15	8			Normally supplied by transfer air. Spaces maintained at temperatures below 50°F [below 10°C] are not covered by these requirements unless the occupancy is continuous. When occupancy is intermittent, infiltration will normally exceed the ventilation requirement.
Education						
Classroom	50	15	8			
Laboratories	30	20	10			Labs: special contaminant control systems may be required for processes or functions including lab animal occupancy.
Training shop	30	20	10			
Music rooms	50	15	8			
Libraries	20	15	8			
Locker rooms				0.50	2.50	
Corridors				0.10	0.50	
Auditoriums	150	15	8			
Smoking lounges	70	60	30			Normally supplied by transfer air. Local mechanical exhaust —with no recirculation—is recommended.
Hospitals, Nursing & Covalescent Homes Note 1						Special requirements, building codes or pressure relationships may determine minimum ventilation rates and filter efficiency. Also, procedures that generate contaminants may require higher ventilation rates.
Patient rooms	10	25	13			
Medical procedure	20	15	8			
Operating rooms	20	20	15			
Recovery and ICU	20	15	8			
Autopsy rooms				0.50	2.50	Air shall not be recirculated to other spaces.
Physical therapy	20	15	8			
Correctional Facilities						
Cells	20	20	10			
Dining halls	100	15	8			
Guard stations	40	15	8			

Note 1: Always consult local building codes and the newest edition of the American Institute of Architects' "Guidelines for construction of hospital and healthcare facilities" for current guidance regarding ventilation rates and air circulation requirements for these types of occupancies.

Fig. 11.21.4
Minimum Recommended Ventilation Rates For Commercial/Institutional Occupancies
(ASHRAE Standard 62.1 1999)

Application	Outdoor Air Requirements	Comments
Residential facilities		
Living areas	0.35 air changes per hour, but not less than 15 cfm/person [7.5 l/s per person]	For calculating air changes per hour, include the entire volume of all spaces within the conditioned space. Ventilation is normally satisfied by infiltration and natural ventilation. Dwellings with tight enclosures may require supplemental ventilation supply for fuel-burning appliances including fireplaces and mechanically-exhausted applicances. Occupant loading shall be based on the number of bedrooms as follows: first bedroom, two persons; each additional bedroom, one person. Where higher occupancies are known, they shall be used.
Kitchens	100 cfm [50 l/s] intermittent, or 25 cfm [12 l/s] continuous, or openable windows	Installed mechanical system capacity.
Baths, toilets	50 cfm [50 l/s] intermittent, or 20 cfm [10 l/s] continuous, or openable windows	Installed mechanical system capacity.
Garages (separate for each unit)	100 cfm [50 l/s] per car	Normally satisfied by infiltration or natural ventilation.
Garages (common for several units)	1.5 cfm/ft^2 [7.5 l/s • m^2]	

Part 4 of 4

References

ASHRAE Handbook—Applications. 1999. Load Estimation. Chapter 4. Places of Assembly, (Natatoriums), p.7-8.

ASHRAE Handbook—Fundamentals. 1997. Water Vapor Transmission Data for Building Components. Chapter 24, Table 9, p.16-17.

ASHRAE Handbook—Fundamentals. 1997. Commercial Building Envelope Air Leakage Chapter 25, Ventilation and Infiltration, p.16-20 and p. 25-29.

Bailey, D.W., Bauer, F.C., Slama, C., Barringer, C., Flack, J.R., *Investigation of Dynamic Latent Heat Storage Effects of Building Construction and Furnishings.* Final report of ASHRAE 455-RP. June, 1994.

Christian, Jeffrey E. "A Search for Moisture Sources". 1993. Proceedings of the Bugs, Mold & Rot II Conference pp.71-81. National Institute of Building Sciences. Washington, DC (202) 289-7800

Cummings, J.B, Withers, C.R, Moyer, N, Fairey, P, McKendry, B. *Uncontrolled Air Flow In Non-Residential Buildings.* Final Report, FSEC-CR-878-96. April, 1996. Florida Solar Energy Center, 1679 Clear Lake Road, Cocoa, FL. 32922.

Harriman, L.G. III, Ed. *The Dehumidification Handbook*, Second Edition. 1990. Munters Corporation. 79 Monroe St., Amesbury, MA 01913.

Harriman, L.G. III, Plager, D., Kosar, D. "Dehumidification and cooling loads from ventilation air." 1997. ASHRAE Journal, November, 1997, p. 37-45.

Jones, B., Sipes, J., Quinn, H., Mcollough, E., *The Transient Nature of Thermal Loads Generated by People*, Final Report of ASHRAE 619-RP, ASHRAE Transactions, 1994, V.100, Pt.2

Pauken, M.T., Jeter, S., M., Farley, B., Abdel-Khalik, S.A., "An experimental investigation of water evaporation into low-velocity air currents," ASHRAE Transactions, 1995, V. 101, Pt.1

Persily, A., "Myths About Building Envelopes". *ASHRAE Journal.* March, 1999 pp.39-47.

SMACNA. *HVAC Duct Systems Inspection Guide* (15D, 1989 The Sheet Metal Manufacturers and Air Conditioning Contractors National Association 8224 Old Courthouse Rd., Tyson's Corner, Vienna, VA. 22182 (703) 790-9890 www.smacna.org.

Treschel, H., Lagus, P., Editors. *Measured Air leakage of Buildings.* 1986. ASTM STP 904 American Society of Mechanical Engineers, 1916 Race St, Philadelphia, PA 19103 www.astm.org. ISBN 0-8031-0469-3)

Chapter 12

Estimating Humidification Loads

By Bernard Morton and Chris Morton, with Lew Harriman

Fig. 12.1 Peak Load Conditions

In most buildings, the peak humidification load comes when the air outdoors is cold and dry—the winter design conditions.

Sizing Humidifiers

Over many years, manufacturers of humidifiers have used simple methods for estimating the load. Although these methods are very quick, they require judgement on the part of the system designer. People new to the field may not always have the experience that allows for confident judgements.

Humidification experts advise newcomers that most problems with humidification systems come from equipment with too much capacity, combined with controls that cannot properly modulate that large capacity during normal operation. Controls will be addressed in other chapters, but the caution concerning capacity applies here. When in doubt about the load, avoid the all-too-common practice of oversizing equipment. With humidifiers, bigger is not better.

Most commercial applications are comfort-driven. If the humidifiers are undersized, the only consequence is likely to be humidity somewhat below the targeted level for slightly more hours than intended. In contrast, most owners do not gracefully accept the consequences of oversized equipment: wide humidity variations, corrosion, dripping from ducts, mold growth and musty odors.

Narrow-range control & humidity above 35% rh

When owners need precise control of humidification—as in the case of microelectronic fabrication and biopharmaceutical production, or when humidity must not stray outside a narrow range year-round—the system designer is wise to seek assistance and guidance from suppliers of industrial humidifiers and humidification systems. Also, when humidity must be held above 35% in cold climates, the architectural designer and general contractor must be alerted to ensure that the building's exterior wall is designed and constructed to avoid condensation in winter. Most commercial and institutional buildings are not built to this standard because they do not require high humidity levels. The HVAC designer cannot assume that an existing building, or even a new building, has walls that will prevent condensation when indoor humidity is above 35% rh in a cold climate. Collaboration with the architectural designer is essential when humidity must be held above that level.

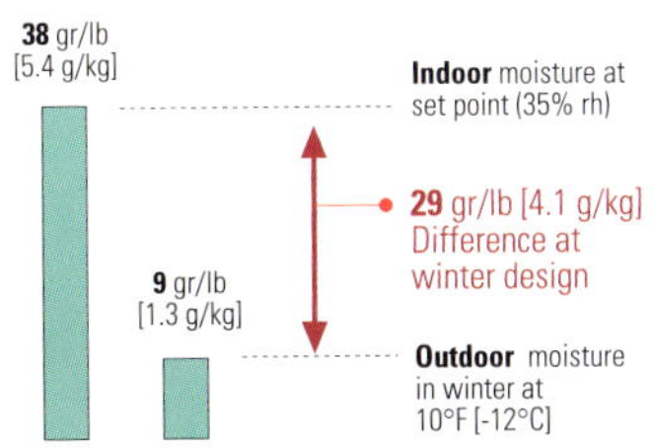

Fig.12.2 Moisture Difference

The humidification load depends on the moisture difference between indoors and outdoors. The load is greatest when the moisture difference is large, and when large amounts of dry, outdoor air enter the building through ventilation or infiltration.

Museums are a common exception, because they are frequently humidified to 40 or 50% rh even in cold climates. When the owner requires those levels, it calls for the utmost care by the architectural designer. The HVAC designer can assist the process by helping the Architect calculate the dew points for each layer inside the exterior envelope to be sure that the wall design will not allow internal condensation. (Consult "Condensation In Exterior Walls" in the reference section of this chapter)

Information in this book is most useful when humidity must be moderated during dry months, raising it to levels between 20 and 35% rh at normal room temperatures and allowing some variation below set point. When the owner's purpose demands a more strict control of dry season humidity, or when humidity must be held above 35% rh, the designer should seek supplemental advice from the suppliers of humidification systems, and should make certain that the Architect and builder fully understand the need for a condensation-free exterior envelope design.

Humidification Loads

As with dehumidification, humidification loads are dominated by outdoor air. Dry outdoor air enters a building through two paths: ventilation and infiltration. Past industry practice suggests that the best strategy is to calculate the loads from both sources, and take the largest of those two loads to size the system's humidification capacity, rather than sizing equipment based on the sum of those loads.

Ventilation

Ventilation air enters the building from outdoors to dilute the concentration of contaminants inside the building. The amount of ventilation depends on how much dilution is required, which in turn depends on how many people occupy the space and how much pollution is released by the building contents. In most cases the local building code will be quite specific, defining the amount of ventilation air required for a given building. In the absence of guidance from local building codes, the designer may find suggestions in either the International Mechanical Code, or in the recommendations of ASHRAE Standard 62, which are reprinted as figure 11.20 in the previous chapter. Note, however, that ventilation recommendations have changed often over the 100-year history of the society. The designer may wish to contact ASHRAE headquarters to ensure that the values of figure 11.20, published in 1999, remain current.

After the designer knows how much ventilation air the building will use, the next step is to define the moisture content of the outdoor air compared to the moisture level indoors. In cold climates, designers often assume that the air outdoors is completely dry. This is a reasonable assumption. In moderate climates, the designer can assume that the air is completely saturated at the winter design dry bulb temperature. Values for the moisture content (humidity ratio) of air at saturation are shown in figure 12.4. In addition to those decisions, the owner and HVAC designer must agree on the target humidity level indoors. Then the humidification load for ventilation air can be calculated with equation 12.1.

Infiltration

Dry outdoor air leaks into a building through cracks and joints and through open doors because of pressure differences caused by outdoor wind and the indoor stack effect. Measurements taken from commercial buildings show that they are surprisingly leaky. This fact stands in sharp contrast to the general perception that new buildings have been built tighter than in earlier years. Such is not the case. Even modern commercial buildings leak

Indoor Humidity Ratio

(gr/lb)

% R.H.	68°	70°	72°	74°
50	51	55	59	63
45	46	49	53	56
40	41	44	47	50
35	36	38	41	44
30	30	33	35	37
25	25	27	29	31
20	20	22	23	25

(g/kg)

% R.H.	20°	21°	22°	23°
50	7.3	7.8	8.4	9.0
45	6.6	7.0	7.5	8.1
40	5.8	6.2	6.7	7.2
35	5.1	5.5	5.8	6.2
30	4.4	4.7	5.0	5.3
25	3.6	3.9	4.2	4.5
20	2.9	3.1	3.3	3.6

- **Under these indoor conditions, the** Architect must take *special* care to build exterior walls that will not condense indoor moisture. (shaded cells: 40–50% R.H.)

Fig. 12.3 Indoor Moisture (W_1)

Outdoor Humidity Ratio @ 100% r.h.

°F	gr/lb	°C	g/kg
-25	1	-30.0	0.2
-20	2	-27.5	0.3
-15	2	-25.0	0.4
-10	3	-22.5	0.5
-5	4	-20.0	0.6
0	6	-17.5	0.9
5	7	-15.0	1.0
10	9	-12.5	1.4
15	12	-10.0	1.6
20	15	-7.5	2.2
25	19	-5.0	2.5
30	24	-2.5	3.4
35	30	0.0	3.8
40	37	2.5	4.5
		5.0	5.4

Fig. 12.4 Outoor Moisture (W_2)

Humidification Load From Ventilation (I-P Units) — Equation 12.1 a

$$\text{lbs/h} = \frac{\text{cfm} \times (W_1 - W_2) \times 4.5}{7000}$$

- **Humidification Load** (Pounds of water per hour)
- Air flow rate (Cubic feet per minute)
- Indoor air moisture content (Grains per lb)
- Outdoor air moisture content (Grains per lb)
- Pounds per hour of air when delivered at one cubic foot per minute
- Grains in one pound

Humidification Load From Ventilation (SI Units) — Equation 12.1 b

$$\text{kg/h} = \frac{\text{L/s} \times (W_1 - W_2) \times 4.32}{1000}$$

- **Humidification Load** (Kilograms of water per hour)
- Air flow rate (Liters per second)
- Indoor air moisture content (Grams per kilogram)
- Outdoor air moisture content (Grams per kilogram)
- Kilograms per hour of air when delivered at one cubic liter per second
- Grams in one kilogram

Wall Construction	Air Changes Per Hour
Tight	0.2 to 1.0
Average	1.0 to 1.5
Loose	2.0

Fig. 12.5 Infiltration Rate
(Humidification Handbook, 1999)

Humidification Load From Infiltration (I-P Units) Equation 12.2a

$$\text{lbs/h} = \frac{\text{ft}^3 \times \text{ac/h} \times (W_1 - W_2) \times 4.5}{60} \div 7000$$

- **Humidification Load** (Pounds of water per hour)
- Interior volume of the building (Cubic feet)
- Estimated infiltration rate (Air changes per hour)
- Indoor air moisture content (Grains per lb)
- Outdoor air moisture content (Grains per lb)
- Pounds per hour of air when delivered at one cubic foot per minute
- Grains in one pound
- Minutes per hour

Humidification Load From Infiltration (SI Units) Equation 12.2 b

$$\text{kg/h} = \frac{\text{m}^3 \times \text{ac/h} \times (W_1 - W_2) \times 4.32}{60} \div 1000$$

- **Humidification Load** (Pounds of water per hour)
- Interior volume of the building (Cubic meters)
- Estimated infiltration rate (Air changes per hour)
- Indoor air moisture content (Grams per kilogram)
- Outdoor air moisture content (Grams per kilogram)
- Kilograms per hour of air when delivered at one cubic liter per second
- Liters in one cubic meter
- Minutes per hour

a large amount of air (See "Measured Infiltration" in the reference section of this chapter).

For humidification load purposes, the amount of air leaking into the building is estimated in terms of "air changes per hour". Figure 12.5 shows typical air change rates in commercial buildings from wall infiltration and door activity. This range of values has been used with good results by the humidification industry for over 40 years. As the reader can see, the numbers are not precise. Any building can be much more leaky or much less leaky than the values shown in figure 12.5. For example, a quick service restaurant has a great deal of exterior wall compared to its internal volume, plus a lot of door activity. The infiltration air change rate for such a building will be very high compared to a retail supercenter, which has much less exterior wall surface compared to its interior volume.

In most cases, the range of estimates shown in figure 12.5 have served the industry well. But when the designer believes the leakage must be estimated more closely, references described at the end of this chapter explain the complex procedures used for that purpose.

When the designer knows the total interior volume of the building and has estimated the air change rate, the humidification load from infiltration can be estimated using equation 12.2.

Loads vs. Equipment Size

After completing the simple load calculations of equations 12.1 and 12.2, the designer chooses the largest of those two loads, and selects humidification equipment accordingly. Also, he must consider whether to use a single large unit, or to use several smaller units.

Using a single unit saves equipment and installation cost. Using multiple units usually allows closer control with a simpler control system, but at a higher installed cost. Where loads are very large, and where the building is also large and internally complex, multiple units are a good choice. The main outdoor air intake can be humidified to a mid-range moisture level. Small supplemental units can humidify each individual zone, allowing multiple control levels.

For smaller buildings or large buildings with wide open spaces a single humidifier may be a good choice. But the designer must make sure the unit's capacity is modulated smoothly when outdoor air moisture content changes. Otherwise, large single units can create wide variations in humidity—usually an undesirable outcome from the client's perspective.

Air-Side Economizers

A "dry bulb controlled, air-side economizer" is an arrangement of dampers, actuators and controls that allow an HVAC system to combine more outdoor air with less return air to help cool the building whenever the outdoor temperature is below the indoor temperature. (See figure 12.6) These devices can save cooling energy cost for many hours each year, in return for a small investment in dampers and controls.

There are three types of dry bulb controlled, air-side economizers in common use. For simplicity of communication, ASHRAE has assigned numerals to each type:

ASHRAE Cycle I

During warm-up in the morning, the outdoor air damper is closed. As the room temperature approaches the thermostat setting, the outdoor air damper is opened fully. The heating source is modulated to maintain the temperature in the building

This cycle is used only where large quantities of outdoor air are needed to replace a continuous exhaust air stream. The largest load will occur at the maximum supply air flow and at the winter design outdoor temperature. The load can be calculated using equation 12.1, by assuming the outdoor air is saturated at the design temperature.

ASHRAE Cycle II

This is the most common arrangement. During warm-up, the outdoor air damper stays closed. As the room approaches the thermostat setting, the outdoor air damper opens to admit a predetermined amount of outdoor air. The amount is based on the minimum ventilation air required by the local building code or on the recommendations in ASHRAE Standard 62. The heat source is modulated to maintain the thermostat setting.

If the room temperature rises above the thermostat setting, the heat source is turned off, and the outdoor air damper opens further, admitting more cool air as long as the outdoor air temperature is below the thermostat setting. The damper can be opened wide enough to admit 100% of the design supply air flow.

This cycle is the most economical because it admits only the amount of outdoor air needed to offset internal heat loads associated with periodic occupancy.

During heating periods, the largest humidification load will occur when the outdoor air damper is admitting the minimum amount of ventilation air required by code. This will occur at peak occupancy of the building—seldom the minimum outdoor air setting at morning start-up. The designer can assume the outdoor air is saturated at the winter design condition, and calculate the load using equation 12.1.

ASHRAE Cycle III

Often, supply air temperatures are fixed and the volume of air is varied to maintain the thermostat setting. This arrangement presents the most complex challenge in estimating the humidification load.

To maintain the constant supply air temperature, the system uses less outdoor air when the outdoor temperature is very low. But consider what happens when the outdoor air temperature rises. Two trends are in conflict, making it difficult to estimate the load :

1. As the outdoor air temperature rises, its moisture content also tends to rise, *reducing* the humidification load.
2. As the outdoor temperature rises, the *amount* of outdoor air rises compared to the return air. Using more outdoor air *increases* the humidification load.

Figure 12.7 shows that these conflicting trends make the humidification load highest when the mixed air temperature set point (the supply air temperature) is lowest. Humidification load *rises* as the desired supply air temperature *falls*. Also, of course, the load goes up as the desired indoor humidity level rises.

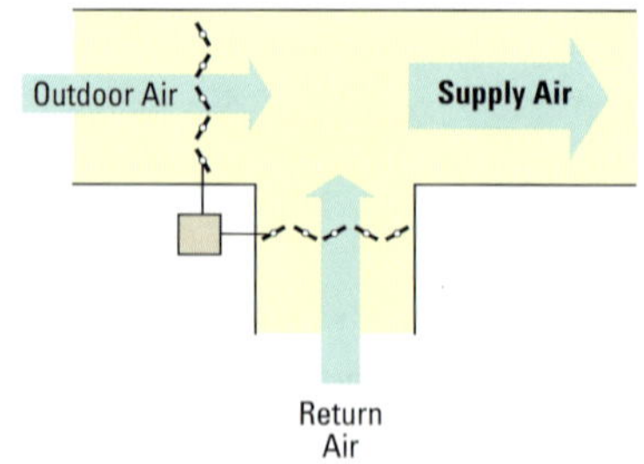

Fig. 12.6 Economizer Cycle

When the temperature outdoors is below the temperature indoors, outdoor air can be used to cool the building. When temperature alone governs economizer operation, HVAC designers face a challenge to estimate the maximum humidification load because the mixture of humid return air and dry outdoor air changes constantly.

Fig. 12.7
Humidification Loads for Systems With ASHRAE Cycle III Economizers

ASHRAE Cycle III Economizer - Maximum Humidification Load

lb/h • 1000 cfm of Supply Air

R.H. Set Point at **68°F** Indoor Temperature

Supply Air (°F)	20%	25%	30%	35%	40%	45%	50%
65	0.5	0.7	0.9	1.2	1.6	2.1	3.2
60	1.4	1.8	2.4	3.3	3.6	5.7	8.4
55	2.2	3.0	4.0	5.4	7.0	9.3	12.2
50	3.1	4.1	5.5	7.4	9.6	12.9	16.1

R.H. Set Point at **70°F** Indoor Temperature

Supply Air (°F)	20%	25%	30%	35%	40%	45%	50%
65	0.9	1.2	1.7	2.1	2.9	3.7	5.8
60	1.9	2.4	3.4	4.3	5.8	7.4	10.9
55	2.8	3.7	5.1	6.4	8.7	11.1	14.8
50	3.7	4.9	6.8	8.6	11.6	14.8	18.6

R.H. Set Point at **72°F** Indoor Temperature

Supply Air (°F)	20%	25%	30%	35%	40%	45%	50%
65	1.3	1.8	2.5	3.2	4.3	5.6	8.4
60	2.3	3.1	4.2	5.5	7.4	9.6	13.5
55	3.3	4.4	6.0	7.9	10.4	13.5	17.4
50	4.2	5.7	7.7	10.2	13.5	17.4	21.2

R.H. Set Point at **74°F** Indoor Temperature

Supply Air (°F)	20%	25%	30%	35%	40%	45%	50%
65	1.8	2.5	3.3	4.3	5.8	7.4	10.9
60	2.7	3.9	5.1	6.6	9.0	11.6	16.1
55	3.7	5.3	6.9	9.0	12.2	15.4	19.9
50	4.7	6.7	8.8	11.3	15.4	19.3	23.8

- Under these indoor conditions, the Architect must take *special* care to build exterior walls that will not condense indoor moisture.

kg/h • 500 l/s of Supply Air

R.H. Set Point at **20°C** Indoor Temperature

Supply Air (°F)	20%	25%	30%	35%	40%	45%	50%
65	0.3	0.3	0.5	0.6	0.8	1.1	1.6
60	0.7	0.9	1.2	1.7	1.8	2.9	4.2
55	1.1	1.5	2.0	2.7	3.5	4.7	6.2
50	1.6	2.1	2.8	3.8	4.9	6.5	8.2

R.H. Set Point at **21°C** Indoor Temperature

Supply Air (°F)	20%	25%	30%	35%	40%	45%	50%
65	0.5	0.6	0.9	1.1	1.5	1.9	2.9
60	1.0	1.2	1.7	2.2	2.9	3.8	5.6
55	1.4	1.9	2.6	3.3	4.4	5.6	7.5
50	1.9	2.5	3.4	4.4	5.9	7.5	9.5

R.H. Set Point at **22°C** Indoor Temperature

Supply Air (°F)	20%	25%	30%	35%	40%	45%	50%
65	0.7	0.9	1.3	1.6	2.2	2.9	4.2
60	1.2	1.6	2.1	2.8	3.7	4.9	6.9
55	1.7	2.2	3.0	4.0	5.3	6.9	8.8
50	2.1	2.9	3.9	5.2	6.9	8.8	10.8

R.H. Set Point at **23°C** Indoor Temperature

Supply Air (°F)	20%	25%	30%	35%	40%	45%	50%
65	0.9	1.3	1.7	2.2	2.9	3.8	5.6
60	1.4	2.0	2.6	3.4	4.6	5.9	8.2
55	1.9	2.7	3.5	4.6	6.2	7.8	10.1
50	2.4	3.4	4.5	5.8	7.8	9.8	12.1

- Under these indoor conditions, the Architect must take *special* care to build exterior walls that will not condense indoor moisture.

Humidification Load For Cycle III Economizer (I-P Units) **Equation 12.3 a**

$$\textbf{lbs/h} = \frac{\text{cfm} \times \text{Load}}{1000}$$

- **lbs/h** — **Humidification Load** (Pounds of water per hour)
- **cfm** — Total Supply air flow rate (Cubic feet per minute)
- **Load** — Maximum Cycle III Load (lb/h/1000 cfm)

Humidification Load For Cycle III Economizer (SI Units) **Equation 12.3 b**

$$\textbf{kg/h} = \frac{\text{l/s} \times \text{Load}}{500}$$

- **kg/h** — **Humidification Load** (Kilograms of water per hour)
- **l/s** — Total Supply air flow rate (Liters per second)
- **Load** — Maximum Cycle III Load (kg/h/500 l/s)

For most purposes, the values shown in figure 12.7 will be sufficient to size the humidification equipment for systems with an air-side economizer. Insert the values from the table into equation 12.3 to estimate the peak humidification load for an ASHRAE Cycle III economizer.

The values in figure 12.7 assume a constant relative humidity for outdoor air. If more precise estimates are required, consult the reference section for guidance on constructing a table with different outdoor air assumptions.

Cautions concerning air side economizers

Economizers are installed because they save money in cooling. But they may actually increase net operating costs when the building must be humidity controlled. Outdoor air is seldom at the desired indoor moisture content, even though its dry bulb temperature might be ideal from a cooling perspective. During humid months, outdoor air carries excess moisture into the building, which must be removed by dehumidification equipment. In dry months, the incoming outdoor air must be humidified. Both operations consume energy, often at a cost which exceeds the cooling savings of the economizer.

To say that humidity cannot be controlled if the building has dry bulb economizers is probably too extreme. But it is fair to say these devices have been responsible for many humidity control problems, and that the cost of humidity control with an economizer may exceed any savings they accomplish in cooling. The designer should consider the hourly cost of operation for humidity control against the hourly savings expected from the economizer before he decides that an economizer will really provide any net benefit to the owner of a humidity-controlled building.

References

Measured Infiltration

Air infiltration through building envelopes is difficult to measure. Ventilation air flow must be shut off, and measurements must extend over time, gathering long-term averages rather than simply recording what may be transient extremes of a few moment. One excellent compilation of field studies is *Measured Air leakage of Buildings,* edited by Heinz Treschel and Peter Lagus. This 438-page book was published by the American Society of Mechanical Engineers in 1986 (ASTM STP 904, 1916 Race St, Philadelphia, PA 19103 www.astm.org. ISBN 0-8031-0469-3) It contains chapters with measurements from different types of buildings, including aircraft hangers, office buildings and many varieties of residential buildings. It also includes chapters which discuss measurement techniques.

Another useful reference is *Air Change Rate and Air Tightness in Buildings*, ASTM STP 1067, published in 1989 (ISBN 0-8031-1451-6). It deals with similar issues, but is more concentrated on residential construction, including multi-story army barracks, manufactured housing, high-rise apartments and single-family dwellings. It also addresses the complex problem of interzonal air movement between rooms and floors inside such "cut-up" buildings—which is the key to understanding ventilation effectiveness. Simply putting outdoor air into a building at a single point does not mean the fresh air will automatically find its way through the HVAC system to all spaces.

To find highly practical advice based on field experience, the reader might wish to consult *Indoor Air Quality and HVAC Systems*, written by David W. Bearg. (Lewis Publishers, CRC Press, 2000 Corporate Blvd, Boca Raton, FL 33431. 1993 ISBN 0-87371-574-8) Through a wide variety of case histories of field investigations, Mr. Bearg explains the many challenges as well as successful techniques for quantifying air leakage into and out of buildings.

The references above were written quite some time ago. But the leakage characteristics of commercial buildings built in North America have not changed substantially since those references were published. This assertion is supported by an article that appeared in the March, 1999 issue of the *ASHRAE Journal*, titled "Myths About Building Envelopes" by Andrew K. Persily. Using measured results of field tests, the article demonstrated that air leakage rates have no correlation with the date of the building's construction. Buildings constructed in the 1980's and 1990's are not "tighter" than those built in the 1950's, 60's and 70's.

Condensation in Exterior Walls

To find techniques for predicting condensation in exterior walls, the designer can consult Chapter 22 of the 1997 edition of the *ASHRAE Handbook—Fundamentals*. (Thermal and moisture control in insulated assemblies) That chapter shows how to calculate the temperature of each surface in the wall, which can then be compared to the dew point temperature at the same points. If the dew point exceeds the surface temperature, condensation is certain.

Computer programs are also available to reduce the time needed for these dew point analyses.

One of these is "MOIST", written by Douglas Burch and J. Chi at the US National Institute of Standards and Technology. For more comprehensive analyses of wall moisture behavior including wetting and drying rates and profiles, a program called "WUFI" is available, written by Hartwig Kunzel of the Fraunhofer Institute of Building Science in Germany and Achilles Kargiozis of the Oak Ridge National Laboratory in the US. It allows the designer to predict not only internal condensation, but also drying behavior of walls that have experienced condensation or wetting from wind-driven rain, based on a complete annual simulation of hourly weather at the site. In most cases, the key issue is not whether the wall will condense moisture for a few hours at extreme design conditions, but rather how much moisture will condense, causing what moisture content in materials, over what amount of time and with what degree of automatic drying through normal operation of the building and normal climate variations. This program provides answer to such complex questions in both quantitative detail and compelling graphic display.

Both of these computer programs are available on the CD-ROM that is included with the ASTM publication titled *Manual 40 - Moisture Analysis & Condensation Control in Building Envelopes* (ISBN 0-8031-2089-3) The book with its CD-ROM is available from the American Society of Testing & Materials, 100 Barr Harbor Drive, P.O. Box C700, West Conshohocken, PA, 19428-2959, USA, www.astm.org

More Precise Load Estimates for Cycle III Economizers

To calculate the values contained in figure 12.7, one must first assume a value for the moisture content of the outdoor air. The values shown in that table assume that the outdoor air moisture content is always 50% rh at any dry bulb temperature. This assumption is usually precise enough for systems with air-side economizers, because these are not usually used on critical applications.

But where the designer feels the need to optimize the humidification load calculations using different assumptions about outdoor air moisture content, he can be guided by the following procedure, outlined in more detail in *The Humidification Handbook* , published by Dri-Steem Humidifier Co. in Eden Prairie, MN (612) 949-2415. Through repeated calculations, this procedure allows creation of the values in figure 12.4.

1. Calculate the amount of outdoor air used by the economizer for each outdoor dry bulb temperature.
2. Calculate the moisture deficit (humidification load) at each outdoor air dry bulb temperature, using the designer's preferred assumptions about the average moisture content of the outdoor air at those temperatures.
3. Choose the highest load based on the series of values calculated in step 2.

Chapter 13

Dehumidifiers

By Lew Harriman & Reinhold Kittler

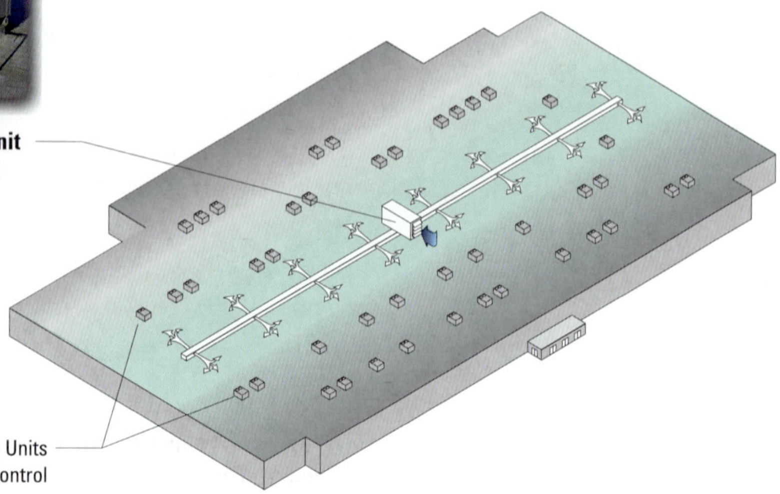

Fig. 13.1 Dry Out The Ventilation Air

Ventilation air carries the greatest part of the moisture load in nearly all commercial buildings. Using a dehumidifier to keep that incoming air dry allows excellent humidity control indoors, without any need to change the rest of the HVAC system.

Dehumidification

Dehumidification is simple. Put enough dry air into the building to absorb any excess humidity.

Given that simplicity, the general public is sometimes surprised and annoyed when air conditioning systems have difficulty controlling humidity during wet weather. But an HVAC designer can easily understand the confusion. Commercial buildings have cooling equipment that has a modest dehumidification effect during some hours, reducing the worst extremes of humidity. However, such equipment is dedicated to cooling and controlled by temperature. It will only *moderate* the humidity—not control it.

When the owner needs to control humidity within a defined range, the building needs equipment dedicated specifically to removing excess moisture—a dehumidifier and its associated controls. This chapter describes dedicated-purpose dehumidifiers used in commercial buildings. It is not concerned with the intermittent dehumidification effects of a building's cooling system.

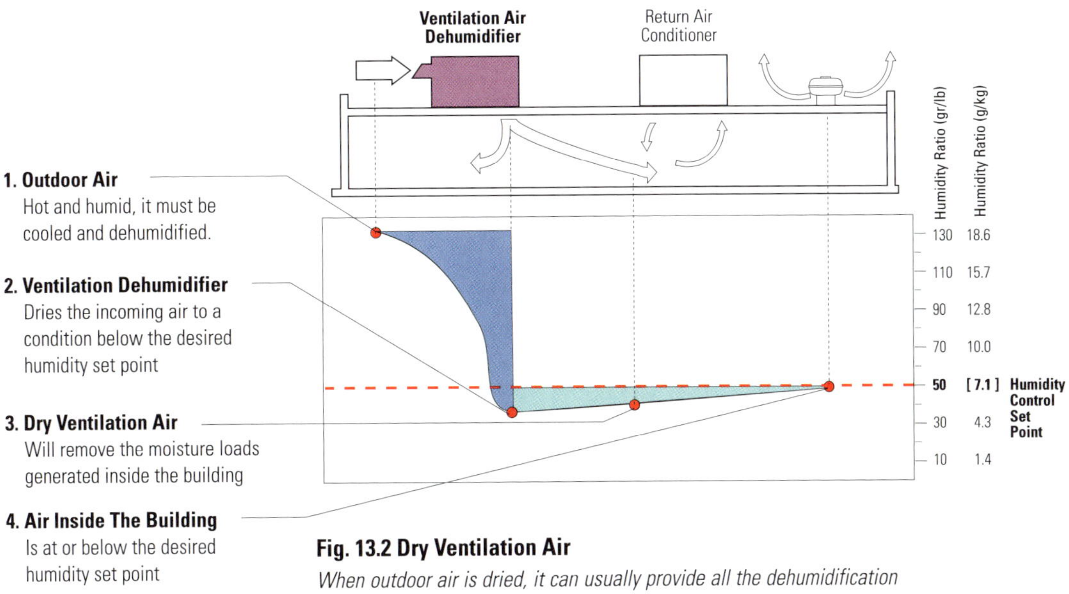

Fig. 13.2 Dry Ventilation Air

When outdoor air is dried, it can usually provide all the dehumidification capacity the building needs.

Designing Dehumidification Systems

There are some suggestions that apply to all dehumidification systems in commercial buildings.

1. Dry the ventilation air first

The bulk of the moisture load in most commercial buildings is carried by the incoming ventilation air. Since that air stream has a higher concentration of water vapor than the return air, drying the ventilation air may be the only dehumidification equipment the building needs to have. The ventilation air is also a smaller air stream than the return air. Locating the dehumidifier on the incoming air means the equipment will be relatively small and control will be excellent.

In a few applications, drying only the ventilation air may not remove the internal moisture load. This is likely to be true if:

- The dehumidifier does not dry the ventilation air deeply.
- The amount of ventilation air is small compared to the total supply air flow (under 10%).
- The humidity control level is very low—as in the case of a supermarket or ice rink.
- The internal moisture load is very high compared to the ventilation load—as in the case of a swimming pool.

To gain the lowest costs in *those* applications, the designer should first size the dehumidifier for the blended loads of the return and outdoor air. Then subtract the outdoor air moisture to see if removing that portion of the load reduces the size of the primary dehumidifier enough to pay for a secondary dehumidifier to treat the ventilation air alone.

However, applications such as these are unusual. In all other commercial buildings, simply drying the ventilation air will provide excellent humidity control at minimal cost.

2. Dry part of the supply air, not all

Dehumidifiers are optimized to remove moisture, so they dry air more deeply than the cooling system. In most cases, it is not neces-

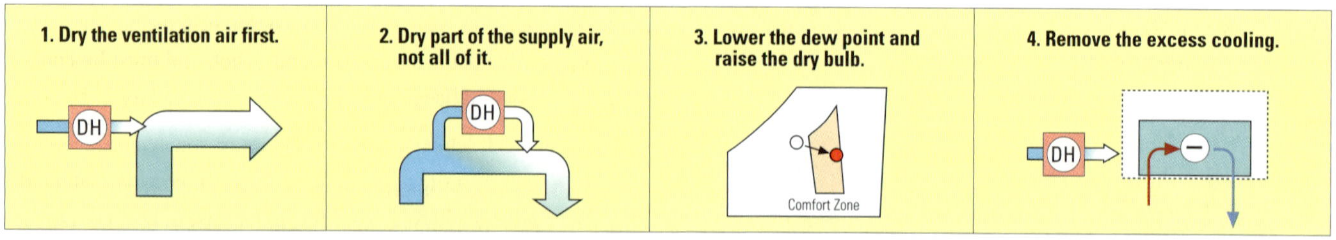

Fig. 13.3 System Design Suggestions For Improving Humidity Control At The Lowest Cost

sary to dry the entire air stream with the dehumidifier. Drying a *portion* of that air *deeply*, then blending the dry air back into the supply air is the usual technique. It provides dehumidification at a lower cost than drying all the air less deeply.

3. Lower the building's dew point and raise its temperature

When the building contains a dedicated dehumidifier, the designer can save cooling costs without reducing comfort. Keep the dew point low, but allow the dry bulb temperature to rise.

When relative humidity is held below 45%, occupants have the same perceived comfort level at 78°F as they would at a temperature of 74°F and 50%. In fact, at 78°, 45% rh more people are likely to be comfortable than if the temperature were 72°F with 60% rh.

[When the humidity is held below 45%, occupants have the same perceived comfort level at 25.5°C as they would have at a temperature of 23.3°C and 50%. More people are likely to be comfortable than if the temperature were 22.2°C with 60% rh.]

If the dry bulb temperature is allowed to rise to that level, the owner saves money in two ways:

- Less cooling equipment is needed because the temperature difference across the exterior wall is reduced. Installing less equipment saves construction cost.
- Since the temperature in the space is higher, there are fewer hours in the year when cooling of any kind is needed, so operating costs are reduced. One field study suggests that savings may be greater than 13% of annual energy expense in a central U.S. climate (Judge et al, 1996).

4. Downsize cooling equipment

When the building is equipped with a dehumidifier, there may be little or no need for the cooling system to remove latent heat (moisture). Usually, cooling systems are sized under the assumption that they will need to remove moisture. Subtracting all or part of that latent cooling load from the cooling units allows them to be downsized. They will operate more effectively and also cost less to install, reducing the construction budget.

5. Define the dehumidifiers' moisture removal capacity (MRC)

The capacity of commercial dehumidifiers is defined by the number of pounds of water they remove per hour [kg/h]. Their capacity changes with the inlet conditions. The designer must not make the mistake of thinking that a "100-lb/hr unit" will remove that much moisture when the air starts out dry instead of humid. As figure 13.4 shows, the unit's capacity depends on the control level, and on how dry the air leaves the dehumidifier. If air leaves the unit at 35 gr/lb, it removes 19 lb/h from an office, but only 6 lb/h from an ice rink. [If air leaves the unit at 5 gr/kg, it removes 8.6 kg/h from an office, but only 2.7 kg/h from an ice rink.]

To check a given unit's capacity, define the temperature and humidity ratio of the inlet air. Then consult the unit's technical specifications or ask the manufacturer to confirm how much moisture the unit will remove at that *specific entering condition*. Also remember to calculate the entering condition based on the weather during peak moisture load hours rather than during peri-

Fig. 13.4 Moisture Removal Capacity

The unit's capacity depends on the humidity set point, and on how dry the air leaves the dehumidifier. The same unit has different capacities, depending on how dry the space must be maintained.

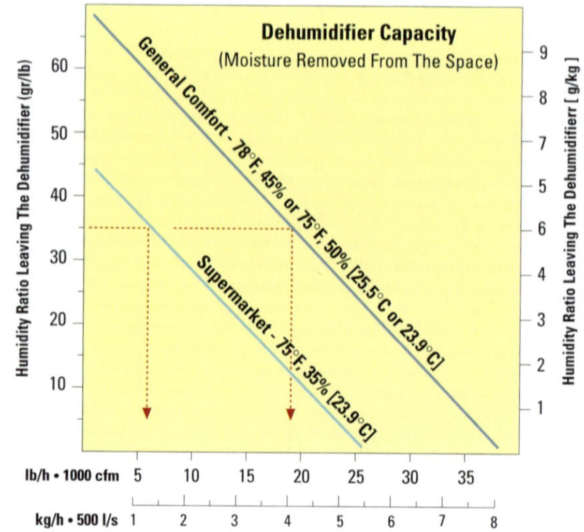

ods of peak temperature. Outdoor moisture is highest when the temperature is moderate rather than when it is extreme.

Mechanical vs. Desiccant Dehumidifiers

Dry air can be produced in two ways. Air can be cooled to force some of its moisture to condense, or water vapor can be attracted out of the air stream by a desiccant.

Both processes are shown in figure 13.5. Mechanical dehumidifiers use the cooling principle. They chill air below its dew point, and then reheat that air with recycled heat recovered from the initial cooling process. The equipment uses electrical power.

Desiccant dehumidifiers attract moisture from the air through a difference in vapor pressures. As air is dehumidified in a desiccant unit, its latent heat (moisture) is converted to sensible heat (temperature). Air leaving a desiccant unit is hot, and must be cooled through a heat exchanger. The process is powered largely by heat.

Fig. 13.5 Types of Dehumidifiers

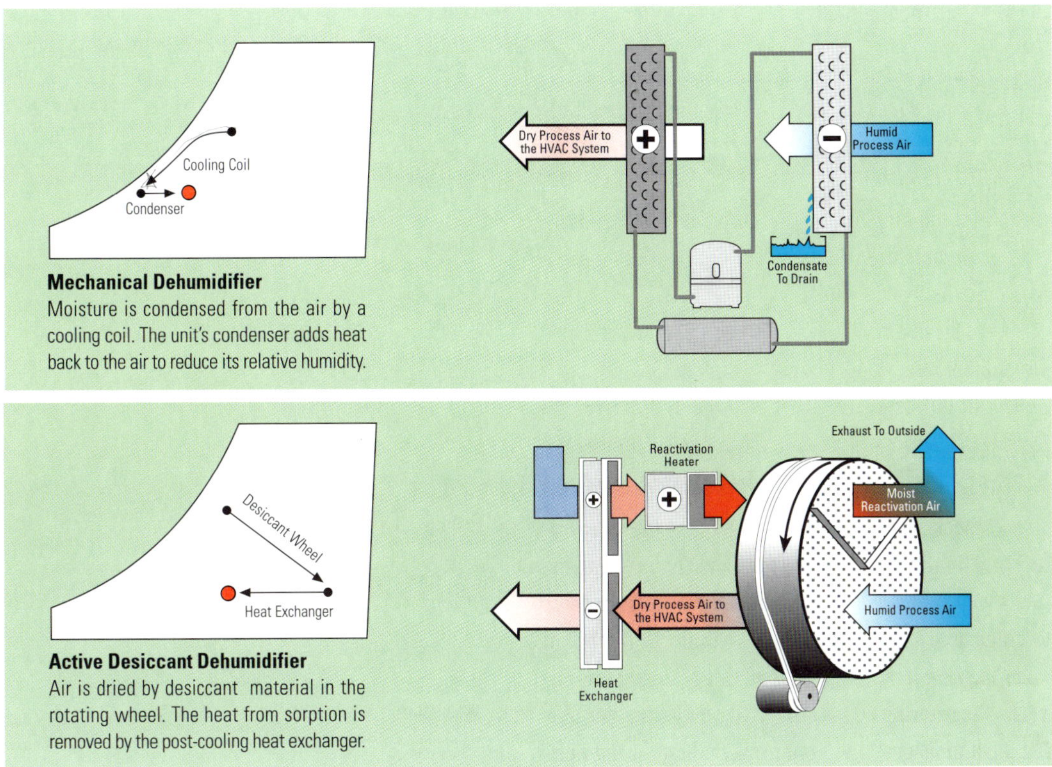

Mechanical Dehumidifier
Moisture is condensed from the air by a cooling coil. The unit's condenser adds heat back to the air to reduce its relative humidity.

Active Desiccant Dehumidifier
Air is dried by desiccant material in the rotating wheel. The heat from sorption is removed by the post-cooling heat exchanger.

Mechanical dehumidifiers are the more widely-used technology, especially in residential buildings. Units for commercial buildings have much higher capacity than units used in residences, and are often customized to some extent for each project. Mechanical dehumidifiers are very energy-efficient compared to other methods of removing moisture from air. They have strong advantages at higher humidity control levels or where cooling is needed in addition to dehumidification. Also, the equipment is similar to cooling equipment and is therefore familiar to most field service technicians.

Desiccant dehumidifiers have been used more widely in industrial applications than in commercial buildings. However, their use has expanded into commercial buildings where there is a benefit to lower-than-usual humidity, such as supermarkets, ice rinks and refrigerated loading docks. They have also been used in ventilation-intensive buildings like schools, theaters and restaurants. Desiccants use much more energy than mechanical dehumidifiers. On the other hand, most of the energy requirement is for heat. They can use waste heat, from distributed power generation and from natural gas, all of which are very inexpensive during humid seasons. They gain advantages as the required dew point goes down to levels where mechanical dehumidification begins to freeze its condensed moisture, or for project locations that have both low-cost heat and high-cost power.

In recent years, owners have benefitted from vigorous competition between mechanical and desiccant dehumidifiers in commercial buildings. Both technologies have been successfully applied to all the buildings types described in this book.

Mechanical Dehumidifiers

Mechanical dehumidifiers use the cooling principle to remove moisture from air. The process is illustrated on the psychrometric charts shown in figure 13.7.

Fig. 13.6 Mechanical Dehumidifier
Such units are available in many sizes as standard products, Also, they are often customized to meet the specific needs of the project at somewhat higher cost.

When air is cooled down, the relative humidity increases (Point A to B). As the air cools below its saturation point at point B (100% rh), the air can no longer hold all of its original moisture. Any cooling below the saturation point (the dew point) causes the air's moisture to condense on the fins and tubes of the dehumidification coil.

As moisture condenses from air, it releases the heat that was originally needed to evaporate the moisture. That heat, called the latent heat of the air, is removed by a refrigeration system. The system also removes some sensible heat from the air. The total heat (the sum of the latent plus sensible heat) is transferred to a refrigerant gas inside the cooling coil.

After the refrigerant gas absorbs the heat, it is compressed and sent to a second coil. The second coil is placed in the cold, dry air stream leaving the dehumidification coil. Inside the second coil, called the condenser, the hot, high-pressure refrigerant gas condenses back to a liquid, releasing heat which transfers back to the air stream. The dehumidified air has the same enthalpy as the entering air, but its moisture content is less, and its dry bulb temperature is higher.

This design is called a "closed loop" system because all the heat stays within the air stream. The path taken by the air is shown in figure 13.7 (Point A - B - C - D) In that configuration, the final air temperature is above the original temperature, because the compressor added some heat to the refrigerant during compression. The amount of extra heat depends on the efficiency of the refrigeration cycle. A less efficient cycle adds more heat.

When excess heat is not desirable in the dry air stream, as is often the case in commercial buildings, a second condenser coil can be added to the refrigeration system and mounted outside the building. Hot refrigerant is sent to condense and release its heat outside the building through the outdoor condenser coil. Then the supply air can leave the dehumidifier at point C rather than point D on figure 13.7. In that configuration the dry air can also be used for cooling the building.

With two condenser coils, the refrigeration system can split the flow of hot refrigerant gas, modulating the heat released in each condenser. This is sometimes called the "zero reheat option" or "isothermal dehumidification," because air enters and leaves the dehumidifier at nearly the same temperature.

Performance of mechanical dehumidifiers

The humidity ratio of air leaving a mechanical dehumidifier depends on the apparatus dew point of the coil, which is principally a function of the refrigerant pressure inside the coil. By adjusting and then controlling refrigerant pressure, a manufacturer can optimize a dehumidifier for a given application.

The HVAC designer defines the temperature, humidity ratio and flow rate of air entering the unit, along with the amount of water that the dehumidifier must remove (expressed in lbs or kilograms per

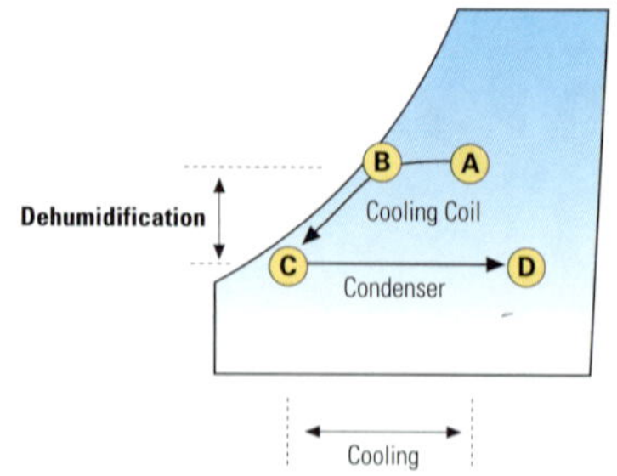

Fig. 13.7
Psychrometrics of Mechanical Dehumidifiers

hour). Based on that information and on the designer's preference for condensing the heat from dehumidification, the manufacturer will configure the dehumidifier to the project requirement.

Performance standards

General-purpose commercial-grade mechanical dehumidifiers are currently without performance testing standards, partly because these units are often custom-built to fit the application. However, dehumidifiers used in swimming pools are rated for moisture removal according to a standard published by the Air Conditioning & Refrigeration Institute. That testing standard (ARI-910) can also be used to validate the *moisture removal capacity* (MRC) of any mechanical dehumidifier in terms of lbs or kilograms of water removed per hour at the air conditions typical of swimming pools, which are 82°F, 60% rh [27.8°C, 60% rh]. Design conditions for outdoor air in most parts of North America are usually higher in both temperature and moisture content, but the test results for the swimming pool condition can provide some basis of comparison between competing units.

Testing to ARI standard 910 also establishes the unit's moisture removal *efficiency* (MRE). This is calculated by dividing the moisture removal capacity (MRC) by the power input (kW).

The industry does not rate mechanical dehumidifiers by either their coefficient of performance (COP) or by their energy efficiency ratio (EER). To do so would create confusion, since cooling units are tested by these criteria, and they do a very different job. Cooling units are optimized and tested for removing sensible heat, not moisture. The familiar COP and EER ratings reflect sensible heat removal. On that basis, mechanical dehumidifiers would produce very low numbers that would confuse rather than enlighten the public. In fact, higher EER test results for cooling equipment are generally achieved by sacrificing dehumidification performance.

To estimate the dehumidification potential of a cooling system, ask the manufacturer to define its sensible heat ratio (SHR) when operating at the inlet air conditions of the system in question. The cooling process removes two forms of heat—latent heat and sensible heat, which added together indicate the total heat removed. The SHR expresses the amount of sensible heat removed by the equipment, divided by the total heat it removed:

SHR = (Sensible heat removed ÷ Total heat removed)

A high sensible heat ratio indicates the unit performs well with respect to *cooling*. A low sensible heat ratio reflects good performance with respect to ***dehumidification***. For example, an SHR of 1 indicates the unit removed only sensible heat, with no moisture removal at all. At the opposite extreme, an SHR of 0.1 indicates that only 10% of the heat removed is sensible heat, which means that 90% of the unit's capacity was used for dehumidification. Most commercial cooling units have sensible heat ratios above 0.8. In contrast, a mechanical dehumidifier typically has a sensible heat ratio between 0.4 and 0.7—far greater moisture removal than a typical cooling unit.

Components of mechanical dehumidifiers

A packaged mechanical dehumidifier contains many components that are familiar to those who work with cooling units, but the hardware is optimized for removing moisture rather than cooling the air:

Air filters

When the dehumidifier is drying recirculated air, medium-efficiency filters are adequate. When treating ventilation air from outdoors, two filters are placed in series to remove the greater particulate load. Filtration is especially important in front of dehumidifying coils. The coil has an extensive wetted surface area, and the coils are "deep" in the direction of air flow. So they are excellent air washers, collecting any particulate in the air stream. Over time, particulate sludge would flow downwards and clog the coil, reducing its effectiveness. Consequently, good filtration is essential.

Filters must be changed regularly so that air flow is not reduced, because reduced air flow leads to performance problems.

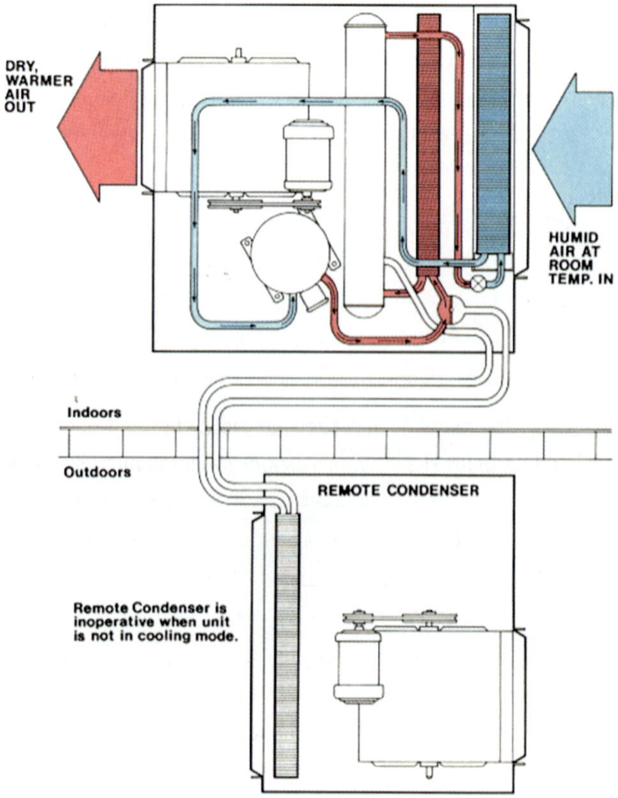

Fig. 13.8 Mechanical Dehumidifier Components

Dehumidifying coil

The evaporator coil is designed very differently from the coil in a typical air conditioner. In a dehumidifying coil, air must be cooled to the lowest possible temperature without allowing moisture to freeze. So the coil is deep in the direction of air flow (6 to 8 rows of evaporator tubes) and operates at low face velocities. It also has fins spaced wide-enough apart to allow water to drain down to the bottom of the coil without blocking air flow. To allow cleaning of this rather deep coil, most manufacturers provide access to both the entering and leaving faces of the coil.

Most dehumidifying coils are short (under 3 ft. [1 metre]). If the coil were tall, a great deal of condensate would collect near the bottom of the fins and clog the air flow passages. Then water would blow off the coil into the air. When a unit must be tall instead of wide, two short coils are stacked on top of each other with a drain pan in between. This catches the condensate from the upper coil before it drains to the lower one.

Finally, the dehumidifying coil seldom dries all the air flowing through the unit. Part of the air is bypassed around the coil and simply serves to remove heat from the condenser coil (the reheat coil) downstream of the dehumidifying coil.

Reheat coil

The reheat coil follows the dehumidifying coil. It is much thinner in the direction of air flow, and its fins are spaced more closely together. The coil uses the heat rejected from the refrigeration process to reheat the dry air to somewhat above its original temperature.

Blower with motor

The dehumidifying coil is deep, and is followed by the reheat coil. So the air flow resistance through the device is higher than through an air conditioning unit. The blower and motor are sized to offset this somewhat larger resistance.

Refrigerant expansion valve

Thermostatic expansion valves with external equalizers are used to meter the refrigerant liquid into the dehumidifying coil. The arrangement ensures that the gas pressure inside the coil stays constant in spite of changing loads on the air side.

The capillary tubes that distribute refrigerant in residential mechanical dehumidifiers do not work well in commercial units, because the loads are much higher and more variable. Commercial units take out much more water per unit of air flow because they operate at colder temperatures. In this range, capillary tubes are not effective at preventing freeze-up, so the more robust combination of expansion valve and distributor is used instead.

Refrigerant compressor

The compressors in dehumidifiers are the same as those used in other cooling systems. That makes service and replacement a simple matter if needed at some point during the life of the equipment.

Optional secondary refrigerant condenser

To avoid reheating the air after dehumidification, some units are equipped with secondary condensers. These reject heat to outdoor air or to a water circuit.

Water-cooled condensers are highly recommended. They have high heat removal capacity in a small space. Often the heated water can be used productively somewhere else in the building, and they are usually mounted in or near the unit. When mounted nearby, little or no refrigerant need be added to fill long pipe runs.

Air-cooled condensers are much less desirable. They reject heat to the outdoors, so are usually mounted remotely. This complicates the refrigerant piping, adds to the weight and cost of the refrigerant charge and increases the likelihood of refrigerant leaks over time.

Optional dry coolers

Dehumidifiers equipped with these devices are similar to water cooled systems, except that the fluid contains an antifreeze and is recirculated through a remote dry cooler. The dry cooler looks similar to an air cooled condenser. Dry cooler systems have a slightly lower efficiency during design operating conditions since the condensing pressure is slightly higher. This is due to the effect of the addition of the secondary heat exchange fluid.

Control panel

Dehumidifiers have more modes of operation than typical air conditioners. These include:

- Dehumidification with reheat
- Dehumidification without reheat
- Dehumidification with air cooling

The internal flows of air and refrigerant are different in each mode. Consequently, most dehumidifiers are equipped with programmable microprocessor-based controls rather than simpler but less-flexible electromechanical relays.

Internal controls

Dehumidifiers are designed to operate automatically, independent of other control systems in the building. They respond to a humidistat, and factory-installed internal controls ensure proper operation. These controls ensure smooth internal operation and also operate the blower continuously, which avoids many problems.

Cycling the blower would allow air stagnation and pockets of high humidity in the space, as well as wide swings in humidity as the blower turns on and off. Dehumidifiers are also quite sensitive to air flow fluctuations. Too much flow prevents the air from being cooled and dried as deeply as expected. Too little air and the coil surface gets so cold it may freeze up the condensate.

Constant blower operation helps avoid these problems, and also allows the humidistat to be located in the return duct near the dehumidifier instead of in the space. In the return duct, air flows across the sensor at high velocity. This improves the response of low-cost sensors. Such sensors respond slowly, and are less reliable when mounted in the space where air velocity is much lower than in the duct work.

In commercial cooling equipment, continuous blower operation has a limitation: when the compressor cycles off, the constantly-flowing supply air re-evaporates any moisture remaining on the coil, which raises humidity in the controlled space. However, manufacturers of mechanical dehumidifiers minimize the effect of this limitation by using hydrophobic coatings and "short" coil designs that allow condensate to drain off the coil surfaces much more quickly than in equipment optimized for cooling. On balance, continuous blower operation has more advantages than limitations.

Many units are also equipped with air flow sensors. These shut down the refrigeration system to protect its components if there is no air flow.

Field-installed controls

Most manufacturers also offer additional controls for mounting and wiring in the field. Remote-mounted display and control panels are popular. They repeat the functions of the unit's main control panel and display the current status in the office of the system operator.

More sophisticated systems are equipped with a modem, which allows the manufacturer or third-party service company to monitor and adjust unit operations from any remote location. This feature is especially useful for ventilation air dehumidifiers, which are capable of staging their capacity and can be reset to maintain precise dew point control as seasons change.

When the dehumidifier must cool the air as well as drying it, a remote-mounted air cooled condenser may be provided. Additional controls are provided and wired in the field to start the condenser fan when the system calls for air cooling.

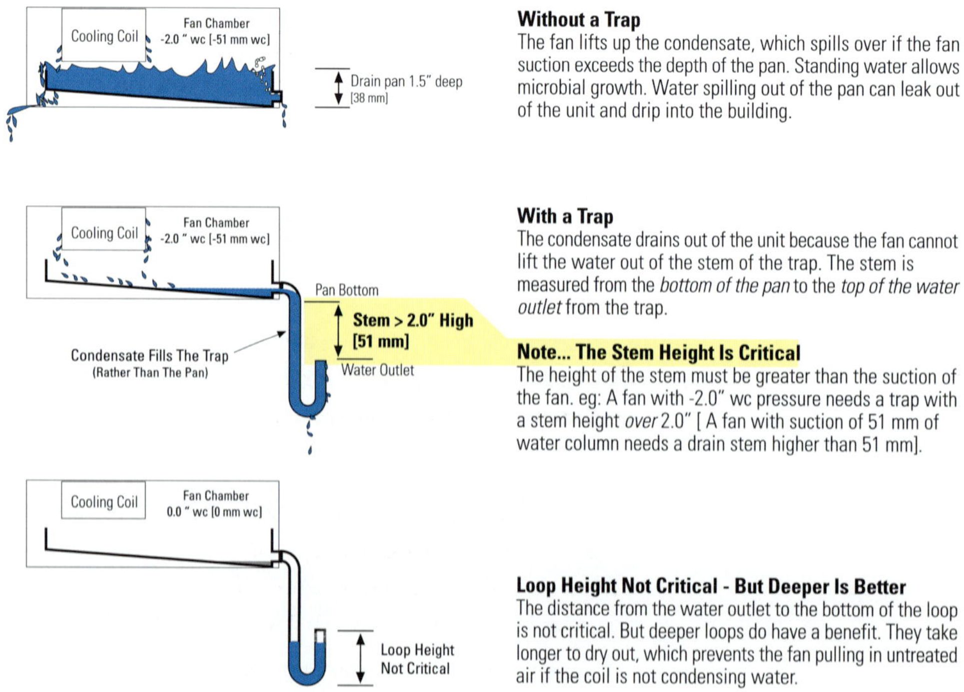

Fig. 13.9 Condensate Trap Is Critical

Capacity modulation

The cost and sophistication of capacity modulation is proportional to the precision requirements. When examining the precision requirements, we must first distinguish between make-up air and recirculation air dehumidification. The basic difference is that make-up air dehumidifiers must cope with varying entering conditions that may change dramatically by the hour and provide a constant or near constant leaving condition. In contrast, dehumidifiers in a recirculation application normally operate at a *near constant inlet condition* and provide supply air at a variable temperature and humidity ratio, depending on the internal load.

Make-up air dehumidifiers

These dehumidifiers are usually arranged as stand-alone units. Single compressor units have limited means of capacity control. As with other refrigeration systems, hot gas bypass is most commonly used to prevent coil freeze-ups at low load condition. However, 2-speed compressors are available for higher energy efficiency. Larger units can use multiple compressors or compressor unloading for capacity control. Since evaporator coil loading depends on the total heat difference between entering and leaving air, the preferred staging controller uses an enthalpy sensor. Less expensive equipment uses dry bulb temperature sensors at the air inlet.

Space dehumidifiers

These dehumidifiers are either stand alone or are connected to the return air of a ducted cooling system. Space dehumidifiers require less sophisticated controls. The flywheel effect of the conditioned space often allows the use of on-off controls, sensing space humidity.

As a general rule, a refrigeration system is most efficient when it is fully loaded and when it cycles as little as possible. In other words, it is important to resist the temptation to oversize the equipment. For control and for durability, bigger is not better—it is worse.

System design suggestions & field experience

Equipment manufacturers provide guidance for integrating dehumidifiers into HVAC systems. In most buildings, adding a dehumidifier is quite simple. Size the unit (or units) for the moisture load, and let them respond to humidistats independently from the operation of the cooling system. Experts have some suggestions for the first-time designer, including:

1. Access

Mechanical dehumidifiers are precision refrigeration systems. Regular maintenance is essential for trouble-free operation and peak performance. There is no maintenance unless the designer provides access *to* the unit, and provides floor space *around* the unit.

Even the most reliable equipment must anticipate the eventual replacement of coils or blowers. Service personnel need to get to the equipment safely while carrying a tool box and possibly a refrigerant cylinder or replacement compressor. So access to a suspended unit via a ladder and manhole-size hatch is not adequate. There needs to be clear floor space around the unit as specified by the manufacturer.

As a minimum, there must be unrestricted access equivalent to the unit's width from at least one side . The unrestricted access to the electrical control panel is governed by local codes, but should never be less than 3.5 ft. [at least one metre].

Condensate drain line trap

The drain trap is critical, but is often overlooked by the designer and by the installers. The weight of the column of water in the trap counteracts the fan suction, allowing water to drain out of the unit. Without the trap, condensate stays inside the unit, leaking out into the ductwork and out of the unit's seams. So the unit *must* have a trap on its drain line, and the stem height of that trap must be greater than the suction of the fan, as shown in figure 13.9.

Duct layout

Place the dehumidifier between the longest possible lengths of straight duct entering and leaving the unit. Air must flow smoothly into and out of the dehumidifier. Otherwise the unit will not perform as expected.

If there are sharp bends near the inlet, the air flow through the dehumidifying coil and bypass is turbulent and unstable, which leads to too much or too little air going through the coil, reducing performance. Turbulence at the inlet can also lead to coil freeze-up or moisture carryover into the supply air. On the leaving side of the unit, sharp bends prevent the air from regaining static pressure, so less dry air reaches the space. Also, with such flow restrictions, it may not be possible to obtain the specified amount of supply air, despite blower speed adjustments. To avoid these problems, locate the unit between the longest and straightest possible runs of ductwork.

3. Outdoor condenser location

When an outdoor condenser is added to the refrigeration circuit, locate that device as close as possible to, and slightly above, the dehumidifier.

When the dehumidifier is operating in the reheat mode it does not use the outdoor condenser. Then the outdoor circuit fills up with liquid refrigerant. The dehumidifier must still have enough refrigerant in the active circuit to function properly. A long distance between dehumidifier and remote condenser means the system must be charged with a great deal of refrigerant. Shorter distances save money and make the system more reliable.

Locating the condenser *above* the dehumidifier allows the liquid refrigerant to simply return by gravity rather than requiring more expensive field piping.

Finally, the manufacturer must be made fully aware of the distance between the condenser and the dehumidifier so the contractor can be advised of the correct refrigerant charge and the correct piping diameters for the system.

Commissioning mechanical dehumidification systems

Commissioning the system can be accomplished by a factory technician or any fully-trained refrigeration technician. Before the technician is called to commission the system, the installer should ensure that:

- The air moving system has not operated during the construction. Construction dust such as that produced by sanding drywall seams solidifies inside coils and blowers. There is no practical means of removing the solidified mass; therefore manufacturers warranties are void under those circumstances.
- The system air flows have been balanced (measured and set to the correct values). The balancing report must be available to the commissioning technician and must show that the flows conform with unit specifications.
- The balancing report shows that the temperature and humidity ratio of air entering the unit is within the operating range specified by the manufacturer. The report must also define blower amperage and voltage on all phases and these must be checked against the name plates.
- If a water cooled condenser is used, water flow is established as specified.

When these items are complete, the dehumidifier can be started up by a qualified refrigeration technician. Start up procedure will vary between models with different features. But in general, the procedure is likely to include the following steps:

1. If a crankcase heater has been installed, it must be energized for several hours before start-up or as specified by the unit manufacturer. This rids the crankcase of refrigerant that would interfere with lubrication.
2. If an outdoor condenser has been added in the field, bring the refrigerant charge in the system to the amount specified by the manufacturer.
3. Compressor operation should be verified using standard refrigeration technician procedures. This includes measuring voltage, amperage, suction line pressure and superheat, discharge gas pressure and temperature and oil level (if equipped with a sight glass). All must be within manufacturer's specifications.
4. With the refrigeration system in operation for at least 30 minutes, the temperature of air leaving the evaporator coil should be near 50°F [10°C] or as specified by the manufacturer.
5. The performance of a mechanical dehumidifier can be measured fairly accurately by measuring the condensate over a period of time. Moisture Removal Capacity (MRC) is measured in lb/h or kg/h.

The unit's air discharge temperature will vary according to its operating mode. In the dehumidification/reheat mode, the temperature is likely to rise by about 20°F [11°C] through the unit. In the zero-reheat mode, the air leaving temperature should be very close to the air entering temperature. During air cooling, the leaving temperature should be close to the temperature leaving the evaporator. (Some heat is picked up from the blower.) A slight temperature rise is actually desirable because saturated air supports mold and mildew growth that creates indoor air quality problems.

To initiate the warranty, some equipment manufacturers request the return of a completed start-up check list. This document becomes the basis of later communication about the system. Start-up technicians should avoid the all-too-human practice of recording *assumed* rather than *measured* values on the start-up sheet. Guesses during start-up do not serve the interests of the owner.

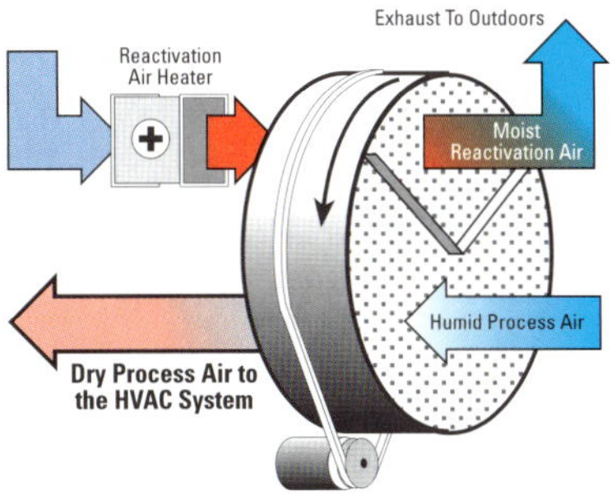

Fig. 13.10 Desiccant Unit Diagram

Fig. 13.11 Desiccant Dehumidifier

Active Desiccant Dehumidifiers

Air can also be dried by flowing it through desiccant materials, which attract water out of the air as a vapor rather than as condensed liquid. Humid air has a high water vapor pressure. Dry desiccant has a low vapor pressure. Propelled by this pressure difference, water molecules move out of humid air to the desiccant.

In a dehumidifier, the desiccant is impregnated or formed in-place in a honeycomb matrix. The desiccant-loaded honeycomb is machined into the shape of a wheel, which is then mounted between two air streams separated from each other by air seals as shown in figure 13.10. The desiccant wheel rotates slowly (6 to 20 rph) between these two air streams.

The incoming humid air stream is called the process air. It flows through the desiccant wheel and gives up its moisture to the desiccant. The dried process air then joins the supply air and is distributed throughout the building.

As the desiccant wheel rotates within the process air it collects moisture until it becomes almost saturated. To restore its capacity, the desiccant continues its rotation and enters the second air stream, called the reactivation air. The reactivation air has been preheated. Hot air heats the saturated desiccant, drying it out. Moisture released by the desiccant is absorbed into the reactivation air, which is sent out of the building.

Desiccant units with *heated* reactivation air are called "active" desiccant dehumidifiers. This distinguishes them from the "passive" desiccant wheels which use the building's *dry exhaust* air instead of heated air for reactivation. Passive desiccant wheels remove much less moisture, but also use much less energy than active desiccant dehumidifiers. (Passive wheels are described in more detail in Chapter 15 - Load Reduction Equipment).

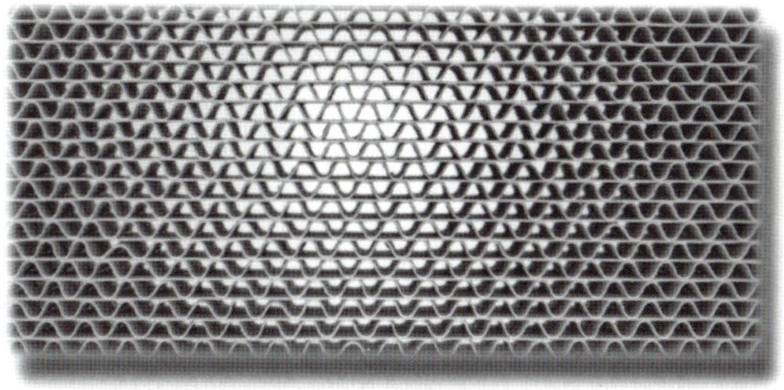

Fig. 13.12 Honeycomb Desiccant Matrix

Usually, an active desiccant dehumidifier dries only a small portion of the supply air. Active desiccants dry air very deeply. A small amount of *very* dry air removes moisture from the building quite effectively, while minimizing the size and cost of the dehumidifier.

Performance of desiccant dehumidifiers

The amount of moisture removed by a desiccant dehumidifier depends on many factors such as the depth of the wheel, its rotation speed and the specific sorption characteristics of the desiccant. But in commercial practice, these variables are fixed by the manufacturer to provide the designer with a simpler device.

To predict performance, the HVAC designer only needs to define the temperature, humidity ratio and volume of the entering process air, the size of the unit and the temperature of the reactivation air. These variables interact to produce the process outlet moisture levels shown in figure 13.13. From that diagram, the reader can see that the moisture content of process air leaving the unit depends on the inlet conditions:

- The *drier* the air entering, the drier it will leave the unit.
- The *cooler* the air entering, the drier it will leave.
- The *hotter* the *reactivation air*, the drier the process air will become.
- The *slower* the *process air velocity*, the drier the air will become.

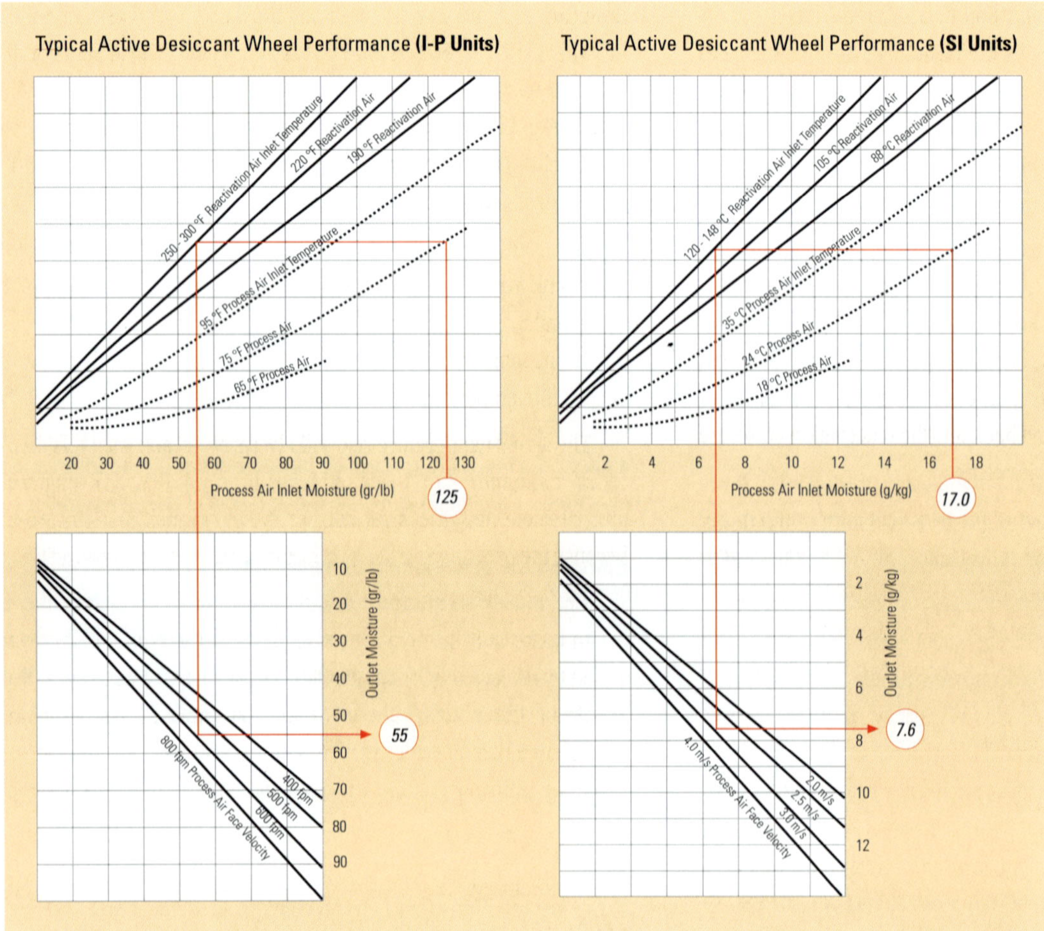

Fig. 13.13 Typical Performance of a Desiccant Dehumidifier

Fig. 13.14 Typical Moisture Removal and Temperature Rise

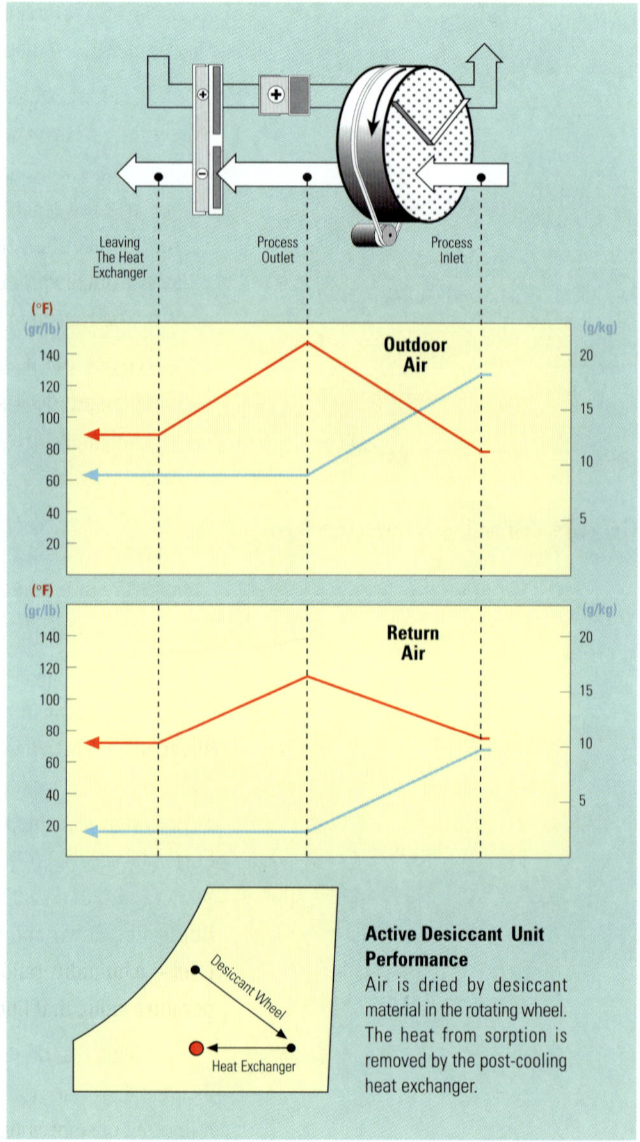

Process air temperature rise

As water vapor is pulled out of the process air, the temperature of the air rises. This phenomenon may be more easily understood by considering the reverse process: evaporative cooling.

As a child runs through a sprinkler in the summertime, his body cools by evaporation. It takes heat to evaporate moisture into the air, and that heat comes from the child's body. So his body cools down as some of its heat is used to evaporate the thin film of water on his skin. Dehumidification produces the reverse effect.

As moisture is *removed* from the air, it *releases* the heat that was used to evaporate it originally. The amount of heat depends on the amount of water removed from the air. More dehumidification releases more heat.

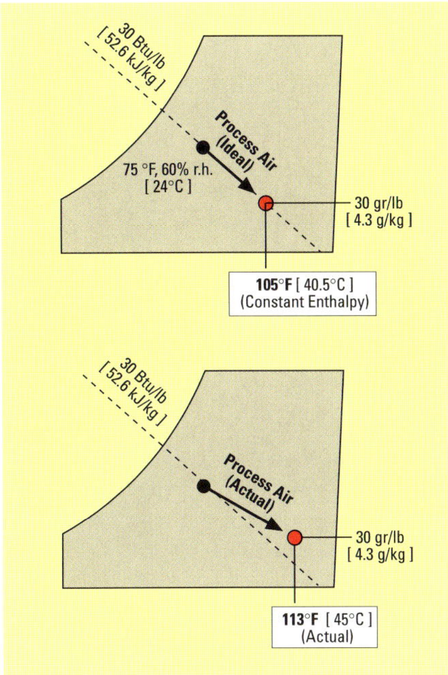

Fig. 13.15 Estimating Temperature Rise

In addition to dehumidification heat, a bit more heat can be carried into the process air by the desiccant wheel as it rotates out of the hot reactivation sector. That "carryover" heat usually accounts for less than 10% of the total temperature rise. The exact amount depends on the wheel rotation speed, the mass of the wheel and the temperature of the reactivation air.

In many equipment designs, the carryover heat is eliminated entirely by a purge sector located between the reactivation and process sectors. In a purge sector, cool air sweeps heat out of the wheel after it rotates out of reactivation and before the wheel rotates into the process air. Purge air carries the wheel's heat away from the process air and back into the reactivation air where it can be reused to dry the desiccant.

Figure 13.14 shows an example of typical moisture removal and temperature rise through a desiccant unit. The exact values will be specific to each manufacturer's hardware. But regardless of the cleverness of the equipment design, the temperature rise must be *at least* as much as the heat released by the dehumidification process. With a psychrometric chart, the designer can quickly estimate the minimum leaving air temperature as long as he knows the inlet temperature and moisture along with the outlet moisture.

A simple temperature-estimating procedure is shown in figure 13.15. Assume that there is no energy change as the process air passes through the desiccant wheel. In other words, the process air follows a line of constant enthalpy, converting all moisture removed by the desiccant to sensible heat. Assume also that the process air enters the wheel at 75°F, 60% rh and leaves at 30 gr/lb. Those inlet conditions represent an enthalpy of 30 Btu/lb. On the chart, follow the 30 Btu/lb line down and to the right, stopping at the outlet moisture level—a humidity ratio of 30 gr/lb. The corresponding dry bulb temperature along that line of constant enthalpy is 105°F.

[Assume that there is no energy change as the process air passes through the desiccant wheel. In other words, the process air follows a line of constant enthalpy, converting all moisture removed by the desiccant to sensible heat. Assume also that the process air enters the wheel at 24°C, 60% rh and leaves at 4.3 g/kg. Those inlet conditions represent an enthalpy of 52.6 kJ/kg. On the chart follow the 52.6 kJ/kg. line down and to the right, stopping at the outlet moisture level—a humidity ratio of 4.3 g/kg. The corresponding dry bulb temperature along that line of constant enthalpy is 40.5°C.]

So for this set of circumstances, we can assume that the process air leaving the desiccant will have risen from 75°F to *at least* 105°F at the outlet [from 24° to 40.5°C]. If there were less moisture removal, the temperature rise would be less. If the air were dried more deeply, the temperature rise would be greater.

The actual temperature rise will always be greater than this theoretical minimum. For closer estimates of the process air temperature rise, the designer must consult the equipment manufacturers because the exact performance depends on the operating conditions as well as the hardware configuration.

In some industrial applications and in supermarkets, the heat added by desiccant dehumidification can be a useful addition to the supply air. But in most comfort-driven commercial buildings, extra heat is not a benefit, so it must be removed by a post-cooling device in the desiccant subsystem or located somewhere else in the system.

Post-cooling

Desiccant equipment manufacturers use many methods to remove excess heat from the dry air, including:

- Air-to-air heat exchanger
- Indirect evaporative cooling
- Vapor-compression cooling

These alternatives are listed in increasing order of cost and cooling power. Often, the desiccant equipment includes only a small amount of post-cooling. The HVAC system designer uses cooling capacity located elsewhere in the system to remove any heat remaining from desiccant dehumidification.

This strategy makes use of the fact that moisture loads are at their peak during the early morning and evening, when the sensible heat loads on the building are relatively low. The surplus cooling capacity available at those times removes heat produced by dehumidification, eliminating the need to add more equipment and cost to the desiccant subsystem.

Figure 13.16 shows an example of the simplest form of postcooling. An air-to-air heat exchanger is mounted downstream of the desiccant wheel to cool the hot, dry process air. On the other side of the heat exchanger, relatively cool building exhaust air carries the excess heat out of the building. Often the exhaust air is evaporatively cooled *before* it enters the heat exchanger to provide a greater cooling effect for the process air.

Exhaust air provides postcooling at the lowest operating cost. But in many buildings the exhaust outlet cannot be located near the supply air inlet. Then outdoor air is used on the other side of the heat exchanger. In that case, indirect evaporative cooling becomes even more helpful, since the outdoor air is not as cool as exhaust air from the building.

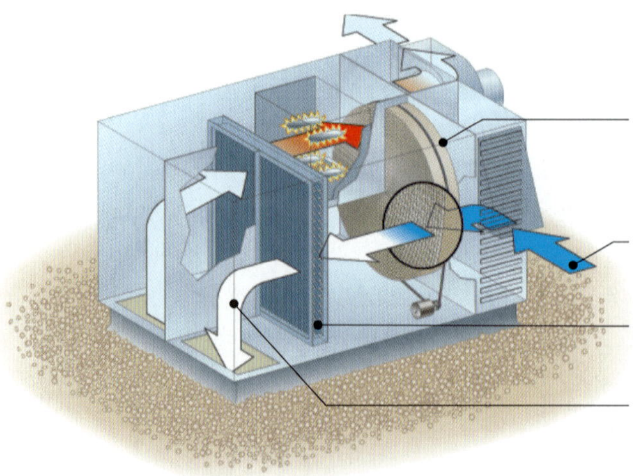

Fig. 13.16 Desiccant Dehumidifier for Ventilation Air

The unit shown in figure 13.16 is typical of desiccant dehumidifiers that dehumidify ventilation air for low-rise commercial buildings where packaged rooftop units are the primary HVAC equipment. When a building has a cooling tower, the desiccant system can use a tower water cooling coil for post-cooling, replacing the heat exchanger shown in figure 13.16.

Controlling desiccant dehumidifiers

As in conventional systems, controls for desiccant systems are a mixture of components provided by the manufacturer and the control subcontractor. We will divide the discussion into two sections—controls needed for internal system operation, and external controls needed to instruct the equipment to provide a particular supply air temperature and humidity.

Internal controls

Inside the system, the equipment manufacturer generally provides controls to ensure that:

- The reactivation heater does not turn on if there is no reactivation air flow.
- The reactivation fan continues to operate for a short period after the reactivation burner has turned off to allow the burner to cool down.
- The reactivation burner is shut down if the air temperature entering or leaving the desiccant wheel is excessive.
- An alarm or fault indicator is tripped if the desiccant wheel is not rotating.

Also, manufacturers often provide controls inside the system to modulate reactivation energy in response to changes in moisture load. Manufacturers generally suggest that it would be unwise and uneconomical for the design engineer to perform these internal control functions with an external building automation system. On the other hand,

Custom Industrial System

Custom Commercial System

Ventilation Air Dehumidifier

Portable Desiccant Units

Fig. 13.17 Types of Desiccant Dehumidifiers

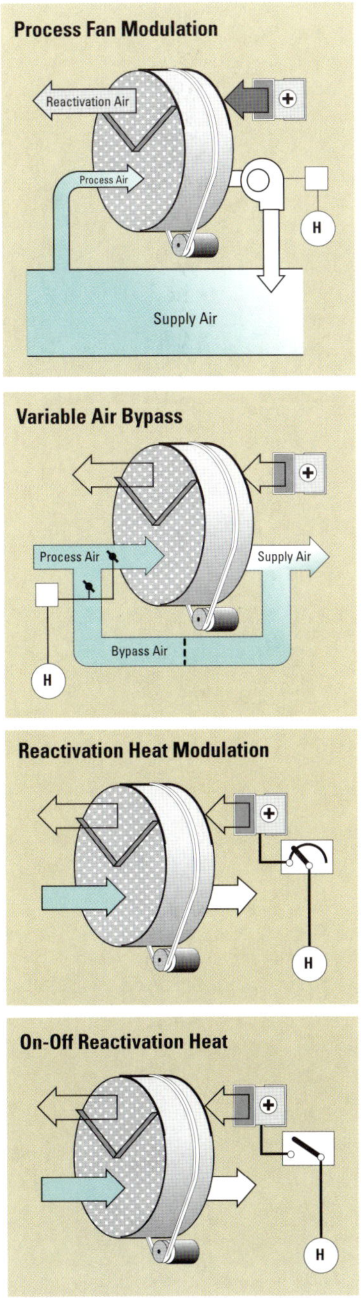

Fig. 13.18 Capacity Control for Desiccant Dehumidifiers

it is common for the desiccant equipment's controls to communicate alarm or status conditions to an external monitoring system.

Dehumidification capacity controls

Controlling dehumidification capacity is often a shared responsibility between equipment manufacturer and the control subcontractor. Figure 13.18 shows four methods for controlling desiccant dehumidification capacity:

- Process fan modulation: *Closest tolerances, lowest operating cost*
- Variable air bypass: *Close tolerances, needs periodic maintenance*
- Reactivation heat modulation: *Moderate tolerances, low equipment cost*
- On-off reactivation heat control: *Loose tolerances, lowest equipment cost*

Each of these methods is effective, depending on the degree of precision needed for the humidity control level in the building, and depending in one case on the characteristics of the desiccant. Specifically, units that use lithium chloride do not use either reactivation heat modulation or on-off reactivation heat control. That powerful salt would absorb too much moisture when the heater is off or operating with low heat. Then the lithium chloride would go into solution and drip out of the matrix. To avoid that problem, methods one and two are used to control capacity of lithium chloride units, or the unit is shut off entirely rather than allowing humid process air to flow through the wheel after the humidistat is satisfied.

Process fan modulation

Varying the speed of the process air fan may be the most precise and energy-efficient way to control capacity. In that arrangement, a variable-speed fan is sized to the largest air flow expected through the process side of the dehumidifier. Then the unit and process fan are mounted next to the supply air duct. On a rise in humidity in the space, the process fan begins to pull air from the supply air duct, drying the air and injecting it back into the supply air downstream of the dehumidifier. As humidity in the space falls below the set point, the process fan slows down, reducing the amount of dry air sent to the space.

This method provides smooth and accurate control without the continuous resistance to air flow that comes with the bypass dampers described below. A bypass damper arrangement must have equal pressure loss through both the desiccant unit and the bypass. A desiccant unit may create a pressure loss between 1.0" and 2.5" w.c. [between 249 and 622 Pa]. With a variable speed process fan, this resistance is encountered only by the process air, not the entire supply air. Also, the resistance only occurs during the hours when the system calls for dehumidification rather than all 8760 hours per year. The annual cost savings can be considerable, and the building occupants enjoy very steady humidity control.

Variable process air bypass

Another precise method of controlling humidity requires a bypass air duct and variable-position dampers for the face of the desiccant wheel and for the bypass duct.

When humidity in the building rises above the set point, the bypass damper closes and the face damper opens, forcing more air to flow through the desiccant wheel. When the humidity control level is satisfied, the bypass damper opens to allow more air to flow past the desiccant wheel so the building is not over-dried. The key to smooth modulation is to make certain the pressure drop in the bypass air is equal to the pressure drop through the desiccant unit itself. This means the bypass must have a sliding orifice plate or fixed-position damper to adjust the resistance in the bypass after the system has been installed.

This method is often used for industrial applications. Bypass control is used less often in commercial buildings, partly because it costs more to push air through the system and partly because the dampers and linkages require some degree of maintenance attention. Also, commercial buildings do not usually need precision control. When control within ± 5% or 7% rh is sufficient to meet the needs of the building, less expensive methods of control are usually adequate.

Reactivation energy modulation

In this scheme, a humidistat varies the amount of heat produced by the reactivation heaters. When the building rises above set point, the controller adds heat to the reactivation air, which provides more drying power for the desiccant. When the humidity in the building falls below set point, the reactivation heat is reduced, so the desiccant does not remove as much moisture.

This method has the advantage of low cost and mechanical simplicity. On the other hand, the control system must have lag-time built in so that the heater does not over-respond to a rise in humidity. For example, if the humidistat in the building calls for more dehumidification, but does not immediately see a fall in humidity, it may continue to call for more dehumidification long after the heater input has been increased. Then the dehumidifier may "overshoot" the desired control level because it is still being fed with extra reactivation heat.

These problems can be avoided when the manufacturer sets a high rotation speed, so the increased dehumidification effect of the higher temperature is sensed more quickly in the controlled space. Also, a step-function can be built into the control, so that the humidistat calls for a small increment of extra heat, then waits for a few wheel rotations before calling for additional heat.

Winter operations add one additional consideration. When supply air flows through the wheel without reactivation, moisture will build up in the desiccant wheel. Excess moisture supports fungal growth within the core material. In springtime, as the humidistat calls for dehumidification, the warm, humid wheel can give off odors reminiscent of dirty laundry until the heat kills microbial growth accumulated through the winter.

But this problem can be avoided by periodically rotating and reactivating the desiccant wheel through any season when humidity is too low for the humidistat to call for dehumidification. If this wheel rotation and reactivation is performed once every eight hours for ten minutes, the problem is unlikely to occur in the spring.

On-off reactivation control

This method is a variation on the modulating reactivation heat scheme. Instead of modulating the reactivation heat according to the moisture load, the heaters are simply turned on or off according to control signals from a humidistat mounted in the building.

The on-off method is even less expensive than the modulated reactivation method, and gives good results in many situations. For example, when the desiccant system is mounted on the incoming ventilation air, the large mass of return air buffers the effect of changes in outside humidity.

However, the same cautions as used in the reactivation modulation method apply: the wheel should be rotated and reactivated several times a day even in periods of low humidity to avoid microbial growth during inactive seasons.

System design suggestions & field experience

Desiccants can dry the air very deeply, and they use heat for reactivation. These two characteristics allow the HVAC designer to produce a better and more economical desiccant system when he follows the suggestions described below and in figure 13.19.

1. Use low-cost surplus heat or natural gas for reactivation

The principal operating cost for a desiccant dehumidifier is the heat used in reactivation. The HVAC designer should use the lowest-cost heat sources available in the building. In most cases, surplus heat from sources like engine cooling jackets, refrigeration condensers and steam condensate will not be hot enough to accomplish *full* desiccant reactivation, or may not be available during all the hours the units need heat. But when available, such low-cost energy can be used to *preheat* the air used in reactivation, which reduces

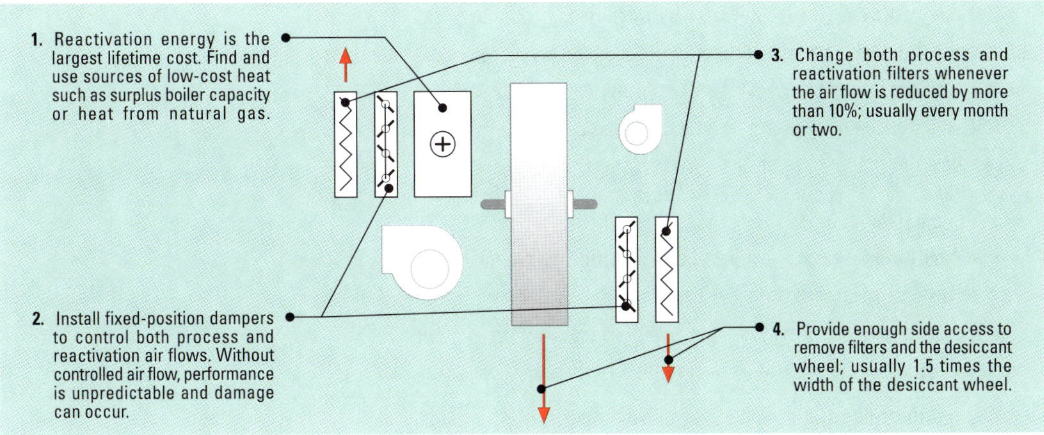

Fig. 13.19 Design Suggestions for Desiccant Systems

the amount of more expensive heat that brings the air up to full reactivation temperature. Most commercial desiccant units are reactivated with air at temperatures between 180° and 220°F [between 82 and 105°C]. Low dew point applications may require reactivation air as hot as 250° to 275°F [121° to 135°C].

In some buildings a steam boiler must operate during the summer season. Without the heating demand of cold weather, the boiler may be venting excess heat that could be used to reactivate a desiccant unit. The reactivation energy cost from that source is essentially zero. This source has frequently been used in hospitals where local codes or operational practices require summer, spring and fall boiler operation under low heating-load conditions.

In North America, natural gas is often the least expensive reactivation energy source after using all available surplus heat. Natural gas consumption is lowest during summer months. The large supply and low demand keeps summer gas prices low. The cost advantage produces significant savings over using electrical heat, even when a gas line must be run to the building or up to its roof.

2. Control and monitor air flow through both process and reactivation

Surprisingly, not all desiccant dehumidifiers are sold with the necessary adjustable air dampers. The reasons for this omission are known only to manufacturers. All logic suggests, and field experience demonstrates, that controlling air flow is essential to proper operation of desiccant equipment. Failure to control air flow is the single most common reason for performance difficulties in the field.

Consequently, the HVAC designer must take care to specify that air flow control dampers shall be provided for the process and reactivation air streams of the desiccant unit, and that the start-up technician must set the air flows to the specified values *after installation*.

If air flows are not set to specifications, the unit will not perform as expected. Too much *process* air overloads the wheel with moisture, so the leaving air dew point will not be as dry as expected. Too much *reactivation* air reduces the air temperature after the heater, so the wheel will not be completely dried. With water in the wheel, the process air will not be dried as deeply as the designer expects.

In the equipment submittal, the manufacturer defines the pressure drop through the wheel at the specified air flow. This pressure drop is usually a reliable indication of the air flow rate. So setting the air flow properly is very simple, provided that there is some means of *measuring* that pressure drop, and some means of *adjusting* the flow so the pressure drop across the wheel matches the manufacturer's specifications.

3. Change filters regularly

Many owners incorrectly assume that filter maintenance is to protect the wheel, and therefore a loaded filter would not be all bad, since it will catch more particles. Also, the instruction to change filters regularly comes from manufacturers so often that it becomes easy to overlook—like a mother's admonition to eat all your vegetables and to always brush your teeth after every meal.

In fact, the wheel does not need much protection. A desiccant wheel is very tolerant of particulate. Also, it is "blown out" many times every hour as it rotates between the counterflowing process and reactivation air streams. The reason for changing filters is to make sure there is *enough air* flowing through the unit.

As a filter slowly clogs with particulate, air flow is reduced below specifications. On the process side, that means the unit will not be delivering the amount of dry air expected, so humidity in the space will rise. On the reactivation side, less air means the heaters will be operating at higher temperatures, which lead to high-temperature faults or to heaters burning out. Both conditions shut down the unit.

In summary, filters should be changed about every month, to make sure the air flow is not reduced below the design values.

4. Provide side access for wheel and filter replacement

While the maintenance staff must change the filters, the system designer must make it possible. Desiccant units need side or top access to filters. They also need side access for wheel replacement.

Like a compressor, most desiccant wheels are warranted for five years and expected to last between 10 and 15 years. But at some point in the life of the equipment, maintenance technicians may need to replace a wheel. Therefore, clear side access is essential when the designer lays out the system.

Commissioning desiccant systems

Heat and moisture on the dry air side of desiccant equipment is balanced equally by the heat and moisture on the reactivation side. To confirm that a desiccant system is operating as designed, the commissioning technician must measure air flow, temperature and moisture on each side to calculate a mass balance.

It follows that if the dehumidification unit does not include the means, the system should be designed to let the technician take the readings essential to commissioning and troubleshooting. He or she must measure the flow rates, temperatures and humidity ratios of air streams as they enter and leave the wheel. Some cautions follow for taking these readings at different points in a desiccant system:

1. Air flows

Air flow instruments measure actual cubic feet per minute or actual liters per second, which must be converted to standard cfm or litres per second in order to calculate mass flow. Since the temperatures in a desiccant system are often well above or below standard temperature, these corrections are essential.

2. Air temperatures

Air streams in desiccant systems can range between 0°F and 250°F [Between -18 and +121°C]. Also, there are wide temperature variations and stratification as the air leaves the desiccant wheel. Air temperature readings must be averaged across the duct for accurate calculations. Readings taken after a fan tend to be more uniform, but corrections must be made for the heat added by the fan.

3. Process air moisture leaving the wheel

Air leaving the desiccant wheel is quite warm as well as being very dry. The technician's instrument will be reading very low humidities—usually below 20% and often below 10% rh Most low-cost instruments have very poor accuracy when measuring below 15% rh, and all but the most costly instruments have an error of ±3% rh at that level. Consequently, technicians will have to use very accurate instruments at this measurement point.

Manufacturers of desiccant equipment recommend the use of manual dew cups, optical dew point hygrometers or low-dew-point aluminium oxide sensors for moisture measurements at this location. ASHRAE Standard 41.6-1982 provides a description of these instruments and describes procedures for their proper use.

When circumstances do not allow the use of dew point instruments, other methods may be necessary. But technicians should be aware of some essential cautions with respect to wet bulb readings at low dew points. Low humidity readings are difficult to take with wet bulb thermometers, because the wet wick dries out very quickly, often before the true wet bulb reading is reached. So wicks must be monitored (somehow) for wetness. Also, when the wet bulb temperature is below the freezing point of water, readings take much longer, which can again allow the wick to dry out. Most experienced technicians avoid the use of wet bulb readings in the air leaving a dry

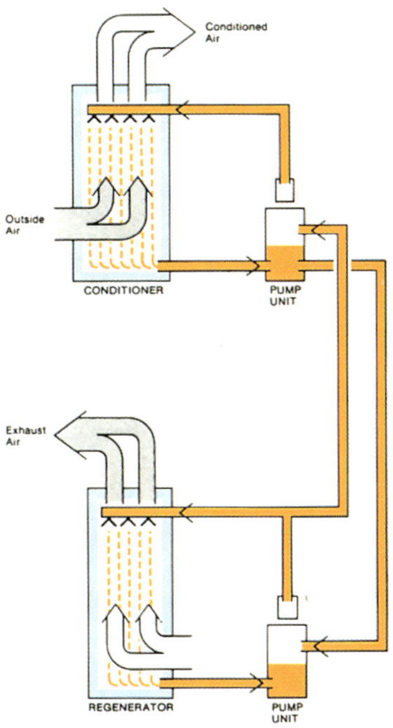

Fig. 13.20 Liquid Desiccant Diagram

desiccant bed, partly for these reasons, and partly because of the difficulty and time required to obtain average readings across the whole wheel.

Like the air temperature, the moisture level leaving the wheel will vary considerably, so if taken close to the wheel, readings must be averaged to obtain a true value for the whole air mass.

4. Reactivation/regeneration air moisture leaving the wheel

Air leaving the reactivation side of the wheel is quite warm and close to saturation. If the sensor is at ambient temperature, moisture may condense on its surface, distorting the reading. It is good practice to warm the sensor, perhaps by taking the moisture reading in the warm dry air of the process-leaving air stream before reading moisture in reactivation. If a wet bulb instrument is used, the water for the wet bulb must be at or above the dry bulb temperature of the air, or the instrument will read lower than the true wet bulb temperature of the air.

Liquid desiccant systems

Some desiccants can be presented to an air stream as a liquid instead of a solid. While this type of equipment has been less common in commercial buildings, it provides some unique advantages over other means of dehumidifying air.

Fig. 13.21 Liquid Desiccant Unit

In a liquid system, a solution of lithium chloride coats a cooling coil or an extended surface similar to cooling tower media. This part of the system is called the conditioner. Process air flows up through the conditioner and gives up its moisture to the liquid desiccant. The dry air is sent to the building. The diluted liquid desiccant drips down into a sump.

The dilute liquid is sent to a "regenerator" to have a portion of its moisture removed. In the regenerator, the dilute liquid is first heated, then flowed across an air contact surface. A small outdoor air stream also flows through the regenerator. The hot desiccant releases its moisture to the regeneration air, which is then vented back to the outdoors. The reconcentrated liquid desiccant flows back to the conditioner, where it can absorb more moisture from the process air.

When the temperature and concentration of the solution are controlled internally, process air leaves the unit at the precise temperature and moisture needed in the supply air, without the post-cooling needed by a desiccant unit that uses a honeycomb wheel. Also, water can be *added* to the liquid before it enters the conditioner, so that the device behaves as a humidifier for the process air instead of drying it.

Liquid systems have been used in larger commercial and institutional buildings where the cost of the equipment and its installation is offset by energy savings from using low-temperature heat for regeneration and low-cost cooling tower water for cooling. As the attention of the HVAC design community focuses on dehumidification more frequently in the future, additional hardware configurations for liquid systems are likely to appear in commercial buildings. An example of such newer designs is shown in figure 13.21. The photo shows a nominal 2000 cfm [1000 l/s] liquid desiccant unit that dries and also cools the ventilation air for a commercial building.

References

Chapter 21 Air Cooling & Dehumidifying Coils. *ASHRAE Handbook—Systems & Equipment*. 2000.

Chapter 22 Desiccant Dehumidification & Pressure Drying Equipment *ASHRAE Handbook—Systems & Equipment*. 2000.

Standard 910-99 Indoor Pool Dehumidifiers. Air Conditioning & Refrigeration Institute, 4301 North Fairfax Drive, Arlington, VA 22203 (703) 524-8800 www.ari.org

Standard 910-98 Desiccant Dehumidifier Components. Air Conditioning & Refrigeration Institute, 4301 North Fairfax Drive, Arlington, VA 22203 (703) 524-8800 www.ari.org

Standard 1060-97 Rating Air-To-Air Energy Recovery Ventilation Equipment. Air Conditioning & Refrigeration Institute, 4301 North Fairfax Drive, Arlington, VA 22203 (703) 524-8800 www.ari.org

Harriman, Lewis G. III. *The Dehumidification Handbook*, 2nd Edition. 1990. Munters Corporation. 79 Monroe St, Amesbury, MA 01913 (978) 241-1100 www.muntersamerica.com

Harriman, Lewis G. III. *Desiccant System Application Guide*. 1997. American Gas Cooling Center, Inc. 400 N. Capitol Street, NW. Washington, DC 20001 (202) 824-7141 www.agcc.org

Judge, James and Spears, John W., "Gas-Fired Desiccant System for Retail Super Center." ASHRAE Journal, October 1997. p.65-69.

Donated Photos and Diagrams

Fig. 13.1 - © Munters USA

Fig. 13.6 - © Dectron Internationale

Fig. 13.8 - © DessertAire

Fig. 13.20 - © Kathabar Inc.

Fig. 13.21 - © DryKor USA

Chapter 14
Humidifiers

By Bernard and Chris Morton, with Lew Harriman

Fig. 14.1 Design Conditions
The peak load for humidifiers usually occurs on a cold, windy day in winter. Outdoor air is very dry. Ventilation and wind pressure forces that dry air into the building, where it becomes a load on the humidification equipment.

Humidification

Humidification is simple. Put enough humid air into the building to equal the moisture that is lost when dry outdoor air flows through the building. The HVAC designer accomplishes humidification by:

1. Creating enough humid air in the most economical way, considering both initial cost and operating cost.
2. Controlling the humidity and temperature of that air so it flows through the ductwork and the building *without condensing.*

For most commercial buildings, the humidification requirement is usually prompted by concerns for general comfort, and for avoiding health problems rather than curing them. Most humans are quite adaptable. For those purposes, precision or tight tolerances are seldom necessary, and relative humidity levels in winter do not usually have to rise above 35%.

Hospitals and museums are two exceptions. They are more challenging because winter relative humidity may need to be above 45%. Later chapters will address the unique concerns of such high-humidity buildings in more detail. This chapter deals with the simpler concerns of more common buildings like schools, offices and retail buildings, where humidification enhances comfort and general health rather than acting in a therapeutic role or as protection for valuable artifacts.

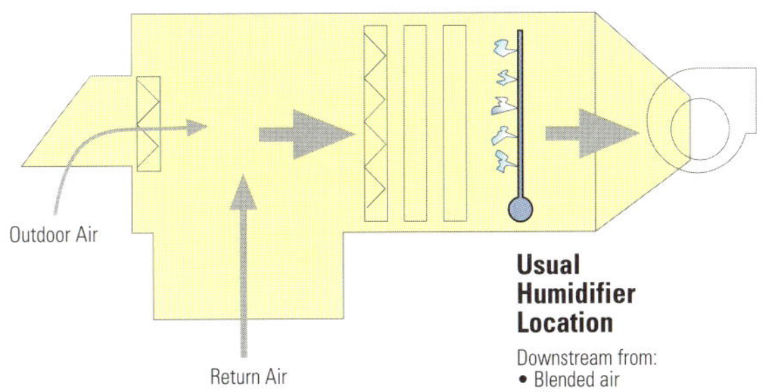

Fig. 14.2 Humidifier Location

In most buildings the humidifier is located downstream of the blend between outdoor and return air. In this position the unit provides maximum capacity with less risk of condensation and less need for preheating the air.

When designing either dehumidification or humidification systems, the designer must focus attention on the ventilation air. That air is excessively humid during the summer and excessively dry in winter. Dehumidifiers are usually placed to dry the ventilation air before it blends with the return air. In contrast, most humidifiers are located ***after*** *the outdoor air has blended with the return air. In that position, the humidifier can add larger amounts of moisture to the supply air with less risk of condensation and subsequent microbial growth.*

In some cases humidifiers are also located at the ventilation air inlet. For example, when very large amounts of outdoor air are used or when indoor humidity must be very high, ventilation air will probably need a separate unit. The ventilation humidifier raises the moisture level of the outdoor air, reducing the load on internal humidifiers and allowing greater precision and higher humidity levels inside the building. But for the more typical case of a building that only needs winter humidity raised to 35% rh, the load can usually be met by a humidifier located after the blend point between ventilation air and return air.

There are two basic types of humidifiers and each type produces a different air temperature. Using **adiabatic** humidifiers, the air cools down. In **isothermal** humidifiers the air temperature stays nearly constant.

Adiabatic vs. Isothermal Humidifiers

Humidification occurs only after the water has absorbed enough heat to evaporate. About 1,000 Btu's are needed to evaporate each pound of water. [2326 kJ/kg] This is true regardless of whether that water is in a reservoir, on the surface of a bank of media, or in the form of water droplets floating in the air. This essential heat is called the latent heat of vaporization. The industry divides humidification equipment into two categories, depending on which source provides the heat to evaporate the water. Adiabatic humidifiers use heat *from the air* being humidified. Isothermal humidifiers use heat that is added to *the water* being used for humidification.

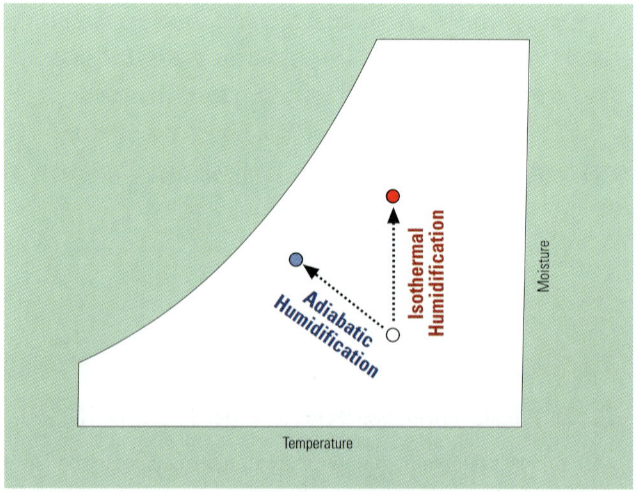

Fig. 14.3 Adiabatic vs. Isothermal Humidification

In adiabatic units, the air cools down. In isothermal units, the temperature remains nearly constant.

In adiabatic humidifiers, the air cools down because its own sensible heat is used to evaporate the water it is absorbing. On the psychrometric chart, this process follows the wet bulb line as shown in figure 14.3. As the air absorbs moisture and moves up the wet bulb line, the rh and the air temperature are both changing, but the total heat content (the enthalpy) remains unchanged. The air temperature goes down and the rh goes up as sensible heat transfers from air to water, as that water evaporates.

The word adiabatic describes processes like this one, which proceed by themselves without needing energy from outside the mixture of air and water vapor. This natural phenomenon is sometimes used to save energy costs by augmenting mechanical refrigeration in warm, arid climates. In many parts of the world, buildings are cooled and humidified simultaneously using this principle. It is also used in cold climates when humidification is needed in buildings that have surplus internal heat from equipment or electronic gear.

In contrast to adiabatic units, isothermal humidifiers use heat added to the water being evaporated. Typically, water is boiled and the resulting steam is added to the air. The steam contains the heat needed for evaporation.

The word isothermal describes a process occurring at constant temperature. In fact, however, an "isothermal humidifier" usually warms the air by several degrees. This is due to the temperature of the steam. To be completely accurate, the process would be described as *nearly* isothermal, because the air temperature will vary by a few degrees depending on steam temperature. But for simplicity of communication, the industry shortens the description of this type of equipment to a single word—isothermal.

In small and medium-size buildings, isothermal humidification systems are more widely used than the adiabatic type. In those size ranges, isothermal units offer lower installed cost, fewer and simpler controls and smaller physical size.

In large installations, agricultural applications and for lower humidity levels, adiabatic humidification units are often used. They are simple and inexpensive. The units can also use sensible heat from the return air stream, which may reduce operating costs if excess heat is being produced inside the building. Adiabatic units are also available in very small sizes for personal use. Each technology has its advantages and limitations, depending on the needs of the building.

Isothermal Humidification Systems

Isothermal humidifiers are the most common units used in commercial buildings, and are made in a wide variety of hardware configurations. They have two principal components: the unit that cre-

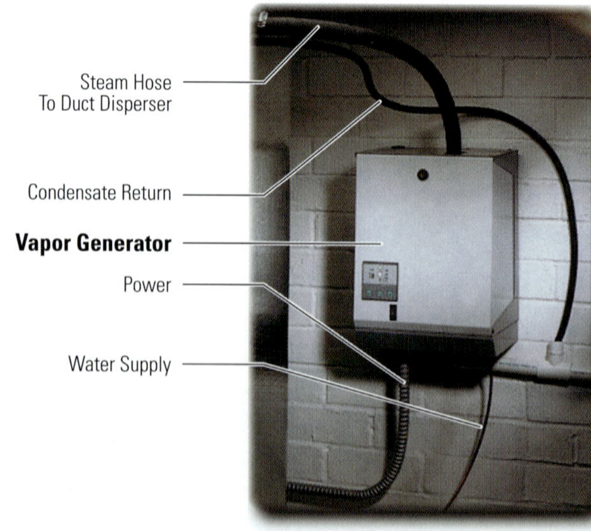

Fig. 14.4 Isothermal Humidifier

This heated vapor generator and other isothermal-type humidifiers are used widely in small and medium-sized commercial buildings . Such low-pressure vapor generators are made with several types of heaters, and provide high capacity for their physical size.

ates the vapor, and the device that distributes the vapor into the air stream. Units that create the vapor include:

1. Steam Boilers, which supply pressurized steam directly to steam dispersers in the ducts.
2. Steam Heated, Secondary Steam Boilers, which also provide pressurized—but cleaner—steam.
3. Heated Vapor Generators, which supply non-pressurized vapor to ductwork or open spaces. In these, water will be vaporized by one of five popular heat sources:
   - Replaceable plastic evaporator
   - Electric resistance heater
   - Electric infrared lamp
   - Steam heat exchanger
   - Hot water heat exchanger

Steam boilers

This is perhaps the oldest method of providing commercial and institutional humidification. The building's main boiler produces steam, which is mixed into the air by steam-jacketed dispersers or steam dispersion panels. The method allows good absorption of the steam by the air and accurate control of relative humidity in the building, even when humidity levels are quite high. Application considerations for steam boilers include:

1. No mineral dust is produced.
2. Very little maintenance is needed and when required is confined to one location (the boiler).
3. Accurate and high rh humidity control is achievable.
4. Duct disperser noise may be objectionable.
5. Humidification may be required at times when the boiler may not be needed for heating (spring and fall).
6. Boiler water treatment chemicals may be unhealthy.

In spite of the benefits of this method, a building's main boiler should seldom be used for humidification because of the health risks associated with steam containing boiler water treatment chemicals. Equipment described below avoids this problem while retaining boiler steam's great advantages of low energy cost and high capacity.

Steam-heated, secondary steam boilers

In this system, humidification steam is generated in a simple pressure vessel called a secondary steam boiler. Instead of a fuel burner, the secondary boiler is supplied with steam from the building's main higher-pressure boiler. Steam transfers heat to the secondary boiler water and creates low-pressure steam (under 15 psig or 100 kPa).

Demineralized makeup water is used in the secondary boiler. It offers the dual advantage of providing high-purity steam and eliminating mineral buildup on the heating surfaces of the tube bundle.

The low-pressure steam is distributed throughout the building via stainless steel piping to steam-jacketed steam dispersers or dispersion panels located in the duct air stream. To protect against the corrosive effects of demineralized water, all components in contact with the water and steam (secondary boiler, valves, piping, steam traps, and steam dispersers) are typically constructed of series 300 alloy stainless steel.

Fig. 14.5 Secondary Steam Boiler

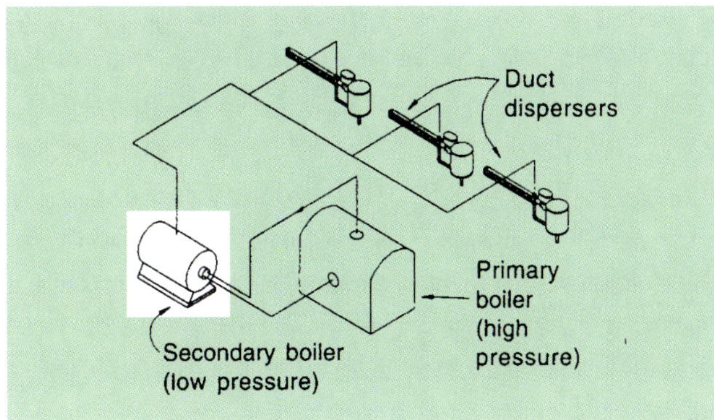

Because of the relatively high equipment cost, this system is used most widely in large, sophisticated humidification systems such as those in pharmaceutical manufacturing clean rooms, museums and hospitals, which need large amounts of high purity steam and close control of relative humidity. Application considerations for secondary steam boiler systems include:

1. Accurate humidity control and high relative humidity are readily achievable.
2. Equipment has long service life and low maintenance.
3. Steam is free of contaminants.
4. Duct disperser noise may require extra care.
5. Installed costs are quite high.

Heated-vapor generators

Figure 14.4 shows a heated vapor generator—the most common type of humidifier used in small to medium-size commercial and institutional buildings. Vapor generators are made in many different sizes and with many heat sources. The equipment consists of a small, non-pressurized boiler/water reservoir containing a heating device. That heating device can be an electric resistance heater, electrodes for passing current through the water, a steam heated tube bundle, a hot water heated tube bundle, or a gas burner with firing tubes.

These units produce vapor at very low pressure—just high enough to overcome the static pressure of air in the duct, plus the pressure drop through the tube conveying the vapor, plus the resistance of the vapor disperser. This combined resistance is normally less than 1/2 psi—12 inches water column [Less than 3 kPa].

To keep that resistance to a minimum, the generators are installed at or very near the point of use, and rigid thin-wall metal tubing or steel pipe (preferably insulated) is used to carry the vapor to the duct mounted disperser. In small capacity, low-budget applications, steam hose is sometimes used instead of metal tubing as the steam conductor as shown in figure 14. 4.

These vapor generators are fully automatic. An internal control system monitors the water level, replenishing it as required and providing low-water protection for the heater element. Vapor output is controlled by a simple on/off humidistat or a more sophisticated programmable microprocessor that modulates the output to provide very tight control.

Systems without water pretreatment usually bleed-off some water through slight overfilling. This continually dilutes the concentration of dissolved solids in the reservoir to minimize fouling caused by precipitation of minerals. Each time the makeup water valve opens, a slight overfill occurs and immediately bleeds off to a drain. To accommodate water with higher concentrations of dissolved minerals, some models incorporate a timer and a drain valve which periodically drains the reservoir and refills it with fresh water.

When the hardness (the mineral content) of the makeup water exceeds 10 grains per gallon, pretreatment with an ion exchange resin is usually recommended by manufacturers. This "softening" significantly reduces mineral precipitation in the generator, usually allowing several seasons of operation without cleaning.

When the owner needs a system with no cleaning requirement or vapor with zero contaminants, demineralized make-up water is used. With such nonconductive water, the flow of make-up water is controlled by a float-operated make-up valve and a low water cutoff. With untreated water, the flow of make-up water is controlled by an electronic probe which uses the conductivity of the water to sense a fall of the level in the reservoir.

Special versions of these units are available for non-duct applications. A fan is mounted on the vapor generator itself to disperse the water vapor directly into the surrounding space.

Application considerations for heated vapor generators include:

1. Versatile, small physical size, easy to locate.
2. Relatively simple to maintain, especially when softened, deionized or reverse-osmosis make up water is used.

3. Close control of humidity is achievable.
4. Equipment cost is low.
5. Operation is quiet.
6. Units are usually located next to or near the supply air duct. The interior finish detail and the exact component location *must* allow easy access for service.

Vapor generators use many different forms of heaters. The most common types are described below.

Electric hot element vapor generators

This version is perhaps the most widely used simply because there are so many applications for which it answers the need. These vapor generators are most common in small to medium size buildings, because their small physical size, relatively simple maintenance, and relatively low installed cost provide the best solution in spite of the comparatively high cost of electrical energy.

These units are available with a capacity of up to 300 pounds of water vapor output per hour [135 kg/h]. When the building needs more capacity, multiple units are ganged together and operated by a single controller. Small versions of these units are used extensively in computer-room applications, where they are installed within packaged computer room air conditioners.

Fig. 14.6
Electric Hot Element Vapor Generator

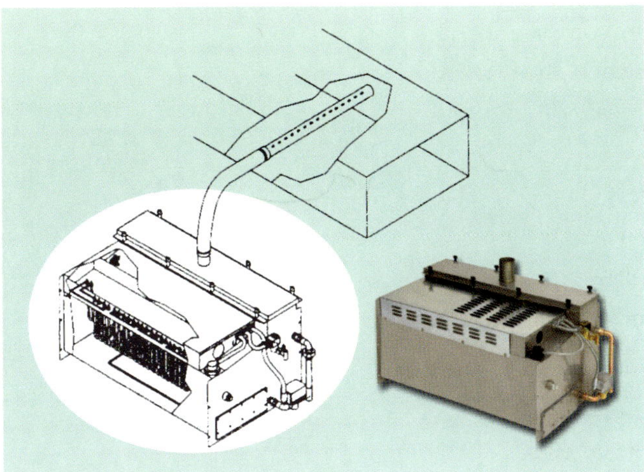

Another version is designed for installation in finished spaces. It has a vapor generator mounted in a metal fan cabinet with a circulating fan. It draws air from the room, humidifies it and discharges it back into the conditioned space.

While these electric units are widely used, owners of buildings with large humidification loads or high electrical power costs will prefer vapor generators that use less expensive heat sources.

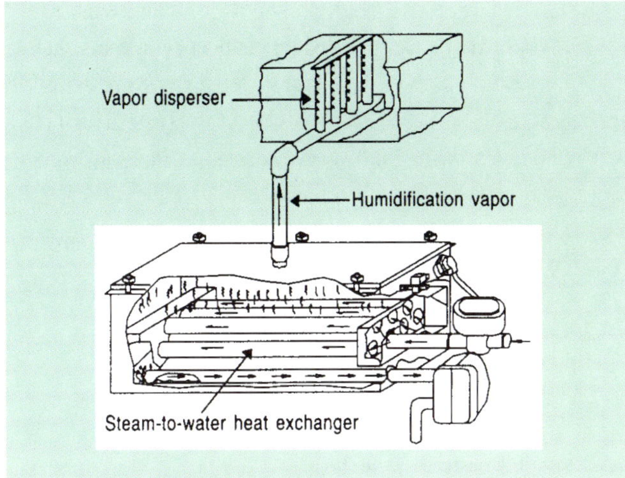

Fig. 14.7 Steam-heated Vapor Generator

Steam-heated vapor generators

This version is popular in buildings that have boiler steam but do not wish to use it for humidifying because it contains water treatment chemicals.

Steam-heated vapor generators are usually considerably less expensive than the secondary boiler system described earlier. The vapor generator operates at much lower pressures and its steam is sent only a short distance to a low-pressure vapor distributor. In contrast, a secondary boiler system has extensive piping and more expensive, higher-pressure steam distributors in many locations.

Steam-heated vapor generators are available in capacities up to 1600 pounds of vapor per hour [725 kg/h]. A construction cost issue is that steam must be piped from the building's main boiler to the vapor generator. However, they have a large operating cost advantage over electric vapor generators. High-pressure steam heat is usually produced with low-cost natural gas or oil at far less cost than electrical heat, saving a great deal of money over the life of the equipment.

Hot water heated vapor generators

This version is used where high temperature hot water is available (pressurized water, at temperatures above 240°F/115°C). This heat source is not common in schools, retail stores or offices, but is sometimes available on some college campuses and municipal district heating systems. Like steam-heated vapor generators, low energy cost is a significant advantage with this method. The units are available in capacities over 500 lbs. per hour [225kg/h].

Gas burner-heated vapor generators

This system offers significant operating cost savings over other energy sources because of its comparatively high overall efficiency and its very low-cost fuel. The savings are especially advantageous in larger systems.

In this vapor generator, one or more firing tubes provide the heat for evaporation. Each tube contains an infrared gas burner. The burners use forced-draft air, and burn either natural or liquefied petroleum (LP) gas.

More firing tubes provide more capacity. This type of unit is available in size ranges from 75 to over 300 pounds of water vapor output per hour [from 35 to over 135 kg/h vapor output].

In return for the great operational economy of the units, the designer must locate them so the gas burner exhaust is vented outside the building—sometimes a challenge in taller or more complex buildings.

Disposable electronic vapor generators

This equipment covers a broad capacity range and is widely used in commercial and institutional type buildings. It contains one or more low-cost plastic, bottle-shaped containers called generators, mounted inside a metal or plastic cabinet. Each generator contains water heating electrodes and maintains its water level by means of a solenoid-operated fill valve.

Electric current flows through the water between the electrodes, causing the water to boil. The resulting vapor is conducted through

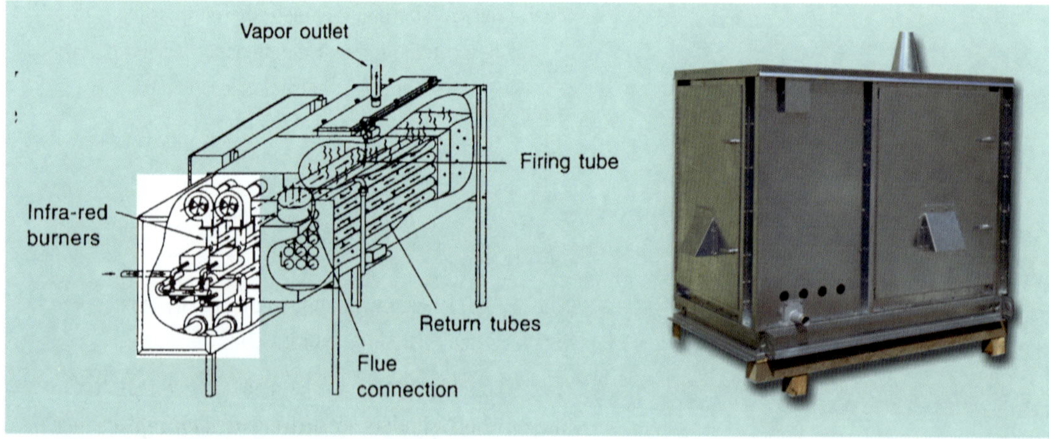

Fig. 14.8 Gas-fired Vapor Generator For Ventilation Air

a steam hose from the generator to the supply air stream. Ductless versions are also manufactured. These are provided with a fan unit for dispersing the vapor directly into the space.

To extend the service life of these generators, an automatic drain system dilutes the concentration of dissolved minerals. Eventually, mineral residue collects on the electrodes and in the containers, rendering them unusable. At that point they are simply discarded and replaced with new generators.

These units must be supplied with untreated water which conducts electricity. Demineralized water, being nonconductive, cannot be used with this system.

Also, care must be taken when using softened water in this equipment. Sodium is a by-product of the softening process. The concentration of sodium—and therefore the conductivity of the water—increases as water evaporates. Eventually the electric current may arc inside the generators, burning out the electrodes.

These units may be the most widely used type in commercial buildings, and they have a few application considerations that do not apply to other types of vapor generators:

Fig. 14.9 Disposable Electronic Vapor Generator

1. Small physical size and no need for venting means they are easily located, although maintenance access is critical since generators must be replaced regularly.
2. They use electrical power—the most expensive source of heat energy.
3. Ongoing cost of replacing generators keeps operating costs high.
4. Time lag between power input and vapor production can reduce the precision of humidity control.
5. Vapor output is reduced if conductivity of the water is too low, or when cylinders are nearing need for replacement due to mineral build up.

Infrared lamps with evaporator pans

This equipment consists of a pan with the water level maintained by an automatic fill valve. Infrared lamps radiate heat downward towards the water, causing it to evaporate. The vapor is then carried away by the air stream flowing above the pan.

These units are most often installed in specialized computer room air conditioners. They work best for small spaces with low humidification loads (little or no outdoor air ventilation) and where they are cleaned regularly.

If they are not kept clean, operating efficiency falls off quickly. Mineral scum forms on the surface of water. The scum absorbs much of the radiant energy from the lamps, converting it to heat at the scum surface. Since heat stays at the surface, most of it is carried away by the airstream rather than evaporating the water as intended. Also, the now-wasted heat increases the cooling load on the air conditioner, reducing operating efficiency even more. Consequently, there are several application considerations for this type of vapor generator that do not apply to other types, including:

1. Small physical size—generally an advantage inside packaged equipment.
2. Low output capacity—best used when the only need is for fairly low relative humidity levels in small, tightly enclosed spaces.
3. Regular cleaning is essential to effective operation.
4. Output falls off rapidly and operating costs increase when precipitated minerals begin to form surface scum.
5. Poor humidity control when loads increase with any ventilation or air infiltration.

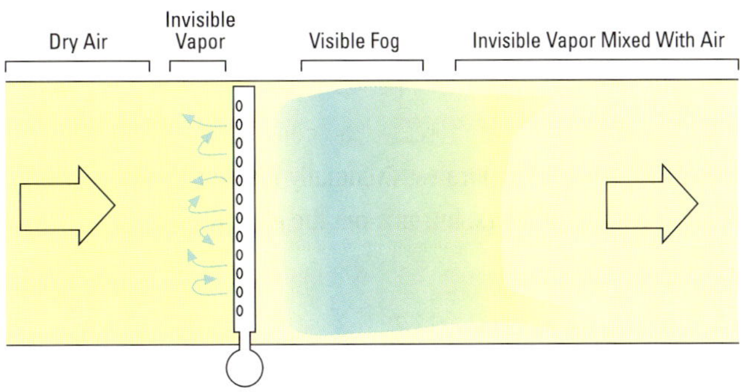

Fig. 14.10 Stages Of Steam Absorption

Introducing heated vapor to an air stream

While it is extremely important that the vapor-producing device be properly designed and applied, effective *delivery* of the vapor to the ducted air stream is equally important. When the vapor is not absorbed, condensation and moisture "fallout" happen. Wet ducts encourage fungal growth and corrosion. These are significant problems which cannot be allowed to occur. They can be avoided by matching the dispersion system to the application and by understanding *how* steam is absorbed by an air stream.

Figure 14.10 shows a top view of water vapor being added to an air stream by a steam dispersion tube with multiple steam ports, injecting steam against the airflow.

As steam emerges from the ports and moves downstream along with the air, vapor is not visible for about the first inch (25mm) of it's travel. Then the steam becomes a visible fog, expanding and becoming less dense as it travels downstream. Then the steam becomes invisible once again after the vapor is absorbed into the air. Two changes of state have occurred.

As the steam emerges from the tube, it is 100% saturated vapor. It immediately begins mixing with the colder air. The mixture cools and the air becomes saturated. Some of the vapor condenses in the air stream, forming the visible fog. During this first change of state—condensation—heat is released. That heat immediately raises the air temperature by as much as 30°F [17°C].

As the fog moves further downstream, another change of state begins taking place. Plumes of mist fan out and become intermingled with pockets of air that are not saturated. Heat flows back into the mist from these pockets, evaporating the mist so it disappears.

After absorption is complete, the air has cooled back nearly to its starting point. But of course the final air temperature will always be one or two degrees higher than its initial value, because the temperature of the steam was higher than the temperature of the air before they mixed. Several observations are now in order:

1) The more uniform the initial mixing of steam with the air ,the more rapid will be the two changes of state. This *shortens* the visible vapor travel—the absorption distance.
2) The lower the initial air temperature, the higher will be the relative humidity of the air during mixing. This will *lengthen* the distance needed for complete absorption.
3) Visible mist will collect on any objects it contacts in the duct such as fans, filters, straightening vanes and dampers. This is called fallout, and it must be avoided to limit the hazards of fungal growth and the problems of water dripping from duct work. The designer must take care to locate any objects well downstream of the steam absorption zone, and must *estimate the length of that zone based on the lowest expected air temperature* upstream of the humidifier.

Manufacturers of steam dispersers have developed tables to estimate absorption distances for various combinations of air temperature and humidity upstream and downstream of the disperser. The HVAC designer can use these tables to lay out a system that will inject the necessary water vapor without wetting the duct.

Duct steam dispersers

Dispersers usually consist of one or more metal tubes that span the width of a duct. The tubes have small ports or a slot along their length from which the steam is discharged into the passing air stream. The metal tube must not drip condensed steam into the duct. Also, the disperser should allow complete absorption of the steam before it collides with duct surfaces downstream.

Two general types of steam dispersers are in common use: steam-jacketed dispersers and vapor dispersers. They differ in design because of the pressure of the steam they use. For "spit-free" operation, *steam-jacketed dispersers* must use steam at a pressure of 2 psig or more. [14 kPa or more] *Vapor dispersers* are used when steam pressure is below that level.

Fig. 14.11 Steam-Jacketed Disperser

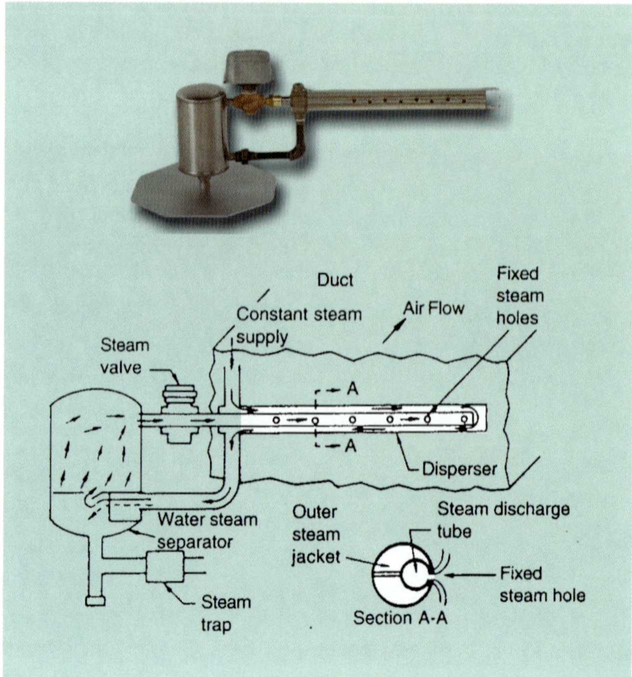

Steam-jacketed dispersers

Figure 14.11 shows this device, which consists of a water/steam separator, a steam jacketed duct tube, a control valve and steam trap.

Steam holes or slots in the duct tube are sized by the manufacturer for the maximum steam flow for which that tube diameter and length can be used. The steam disperser tube usually spans the width of the duct. Surrounding that disperser tube is a second, "jacketing" tube. The jacket tube contains steam at line pressure—several degrees hotter than the atmospheric pressure steam being discharged from the inner tube. The heated jacket isolates the distributor tube from the air, preventing discharge of condensate into the ductwork.

These units typically include a water-steam separator made of stainless steel or cast iron. This removes condensate that might otherwise enter the disperser from the steam main. Following the separator, the dispersers are often equipped with a normally-closed modulating valve to control capacity and a filter/noise suppressor.

Application considerations for jacketed dispersers include:

1) The steam-heated jacket tube constantly heats the air stream. Factory-insulated jackets are available.
2) The disperser should be installed well upstream of filters, fans, or dampers to avoid condensation and wetting.
3) Humidity control is usually quite consistent, even when air must be raised to high relative humidities.
4) To limit noise in the controlled space, locate the distributor well upstream of supply air registers.
5) For minimum noise and optimal modulation of capacity, the steam pressure at the disperser should be between 2 and 12 psig [between 14 and 83 kPa].
6) Condensate formed in the separator and disperser jacket is under line pressure (high pressure) and can normally be returned to the boiler.
7) Because of their high mass, cast iron separators generate large amounts of condensate during warm-up from a cold start. They should be equipped with a temperature switch to prevent spitting during warm-up.
8) A single tube is likely to provide adequate absorption as long as the final rh of the supply air is below 60% or where the tube is installed at the beginning of a long distance of straight duct.

Fig. 14.12
Single-Tube Vapor Disperser

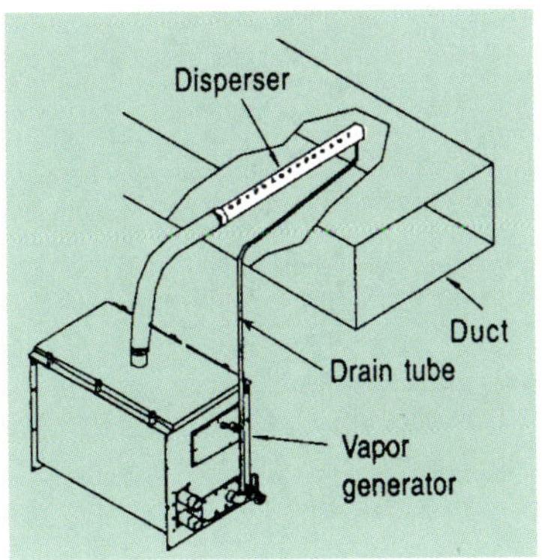

Multiple tube steam-jacketed dispersers

In this variation, several jacketed tubes are located one above the other, usually in large, tall air ducts. They attain absorption in a shorter distance of air travel compared to that provided by a single-tube installation by dispersing the steam into more planes of the air flow. These units cost more than single-tube dispersers, but provide larger capacity and faster absorption, shortening the absorption zone and reducing the risk of condensation downstream.

Vapor dispersers

This equipment is used to feed an air stream with very low pressure vapor, such as that produced by the vapor generators described earlier. They typically operate at less than 2 psig [less than 14 kPa].

The disperser consists of a metal tube that usually spans the width of the duct. Vapor is conveyed to the duct disperser from the vapor generator through a hose or rigid metal tubing.

The disperser has vapor emission ports that are punched or drilled along one side. Higher quality dispersers have high-temperature plastic tubelets as vapor emission ports. Tubelets are pressed into the disperser tube and extend into the center of that tube. They prevent the condensate that forms on the wall of the disperser tube from being spit into the air along with the vapor.

Condensate forms in the duct disperser and must be removed. The unit needs a tube to allow the liquid condensate to be drained away from the duct.

Upfed multiple-tube dispersion panel

Cooler or faster-moving air, or buildings that need large amounts of humidification are greater challenges for the vapor disperser. The answer is to use more disperser tubes in an array across the duct.

In the upfed design, a horizontal header at the bottom of the duct supplies vapor to a row of vertical disperser tubes that reach from the bottom to the top of the duct. The spacing between tubes depends on the length of the absorption zone. When the available absorption distance is short, the tubes must be closer together.

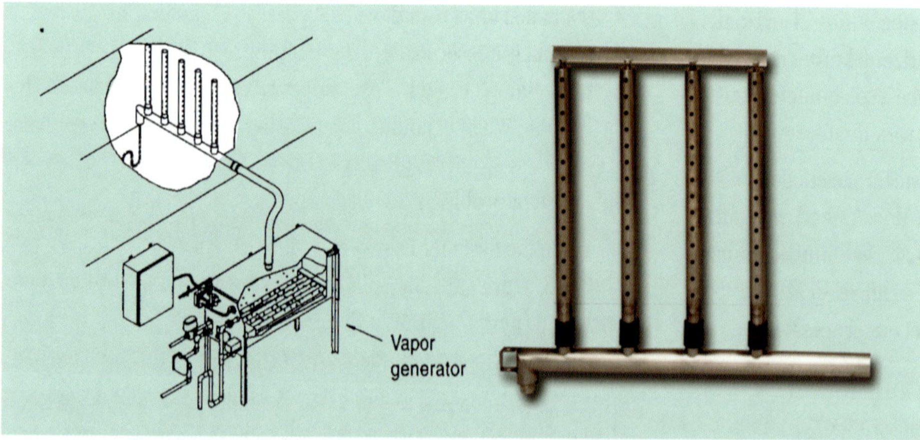

Fig. 14.13
Upfed, Multiple-Tube Dispenser Panel

Openings in the tubes are arranged in opposing vertical lines. They discharge water vapor towards each other, across the air stream. In the better performing designs, each steam discharge opening consists of a nonmetallic tubelet inserted through the tube wall and extending to the center region of the vertical tube. Condensate adheres to the inside tube wall and flows down to the header for removal. The tubelets prevent that falling condensate from being discharged into the air along with the steam.

Because the vapor flows up and the condensate flows down, the capacity of this design is limited. For example, if the upward velocity of the vapor were high (high humidification rate), it would prevent the downward flow of condensate. The tube would clog with water, which would spit out of the tube along with the vapor. The dual-header dispersion panel described below is used to achieve higher humidification capacity.

Dual-header dispersion panel

This is a more sophisticated version of the upfed panel described above. It is used where the absorption distance requirements cannot be satisfied with the simpler upfed design.

This design includes two headers, one for supply and one for return, instead of a single header that would have to serve both functions. With separate paths, the vapor and condensate move in the same direction (downward), which allows higher steam velocities and therefore much higher capacities. This design achieves thorough mixing and therefore rapid steam absorption.

Additionally, the dual-header arrangement allows the vapor disperser tubes to be mounted horizontally. That orientation would not work well with a single header design because condensate could not

Fig. 14.14
Dual-Header Disperser Panel

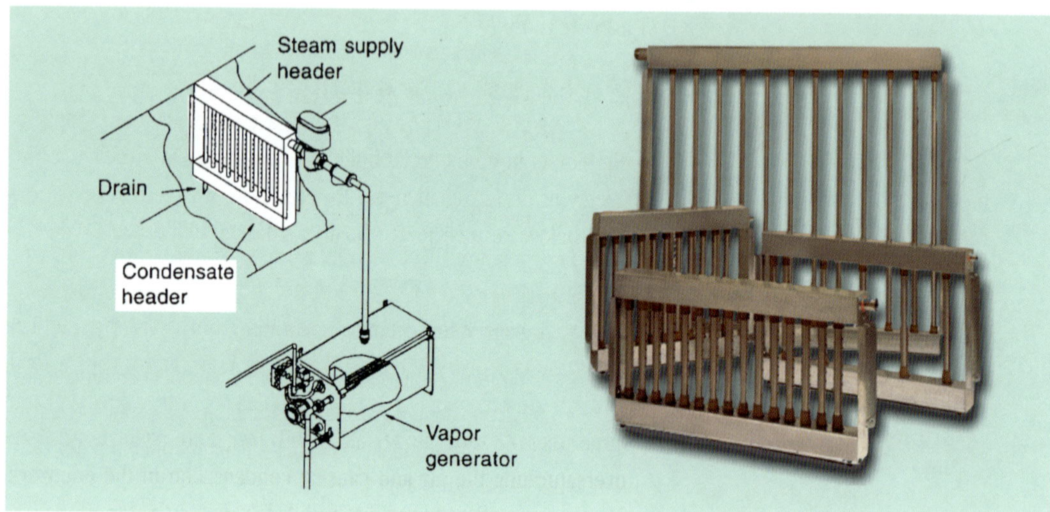

drain back to the single header against the steam flow. Horizontal tubes are useful in wide, flat ducts. Fewer horizontal tubes cover the cross section as effectively as many vertical tubes, so equipment costs are reduced with horizontal tubes equipped with dual headers.

This equipment is often used instead of steam jacketed dispersers for boiler-steam applications because it doesn't add unwanted jacket heat to the air stream. Steam jacketed disperser tubes remain hot at all times, and when multiple tubes are required to accomplish rapid steam absorption, the heat gain can become excessive.

Vapor dispersers for open areas

Some spaces do not have ducts into which water vapor can be supplied for distribution by the air handling system. Examples could include buildings served by hydronic systems or large, open spaces like open-plan offices, churches, wood-floor gymnasiums or retail stores. For these applications, a vapor generator suspended in the space can be a cost-effective solution.

Eye-pleasing units are available in which the vapor generator is mounted within an attractively-finished metal cabinet. Inside the cabinet a blower draws in room air, passes it over an internal vapor disperser where it is humidified, and discharges the humid air back into the room.

Design suggestions for steam humidification

Steam and hot vapor humidifiers are the most commonly-used equipment in commercial buildings, which means they face a wide variety of design challenges. Experts have provided some suggestions below for designers who do not design such systems every day.

Unit size vs. load

Experts agree that the most common problem with humidification systems is oversized units. When the unit is too big for the load, or for its air stream, it is nearly impossible to control. The unit pushes large slugs of vapor into the air, rapidly overshooting the target level, oversaturating the air and causing condensation in the ductwork. Then it turns off and humidity rapidly falls below the target level.

Oversized units usually reflect a designer's natural insecurity: "If a small unit is questionable, I'll specify the next-larger size to be sure of having enough capacity." But ***oversizing is the problem***—not the solution. When in doubt, undersize the unit rather than oversizing it. Few if any problems are caused by undersized units. Very severe and constant problems are typical of oversized units.

When the designer is unsure of the load but still believes there is a need for extra capacity, split the work between one large and one small unit, perhaps using a division of 30%/70% of the maximum load. The smaller unit handles the work for nearly all the operating hours and will be able to modulate its capacity smoothly as seasons change. The second, larger unit will only be needed at peak load conditions which occur for fewer than 35 hours in a typical year.

Zone control with primary/secondary units

When buildings are large and complex, or when they need high relative humidity, the designer may consider a two-stage system where the main supply air duct is equipped with a unit to humidify *all* the air to some baseline humidity level. Smaller, second-stage humidifiers are added downstream in the supply ducts leading to each zone. The main unit is controlled by humidity in the supply duct. The second-stage units are controlled by humidistats in each zone.

The two stage system costs more, because more units are needed, along with more controls and wiring. In return for that investment, the system provides excellent control, and makes it much easier to maintain a high humidity level without causing condensation.

Also, when the load reduces, the main humidifier can be shut off entirely. The secondary units will maintain the zone control levels. This approach provides much more consistent control than if a single, large unit were trying to modulate down to produce only 10 or 20% of its capacity. Such a unit is likely to overshoot and undershoot the desired control level. The heater control or steam valve will be operating near the bottom of its throttling range. Any small demand for humidity produces a larger amount of moisture than the need, as the large valve tries to crack open by only a tiny amount.

In contrast, the primary/secondary strategy costs more initially, but saves operating costs and delivers a more consistent control level all year round.

Steam pressure vs. cost of the disperser

Any steam disperser will emit more vapor at higher pressures. So using a high steam pressure allows the use of a smaller, less expensive disperser. But there are two reasons the designer should avoid economizing in this way—noise and control.

As steam pressure increases, so does the velocity of the vapor as it leaves the disperser. Higher velocity steam jets create more noise. If the disperser is located well away from the supply air diffusers, noise may not be an issue. But if the dispersers are close to the occupied spaces, extra noise is likely to annoy occupants.

Also, higher pressure is more difficult to modulate smoothly at low-load conditions. Remember that by definition, lower-than-maximum load conditions occur for 99.6% of the building's life.

Using lower steam pressure and larger dispersers will improve control consistency for 99.6% of the operating hours, and will reduce noise. To most building owners, the slight increased cost seems a modest investment to ensure these useful benefits.

Noise

If a large amount of vapor must enter the air stream from a single, short tube with few openings, the steam velocity will be very high, and therefore it will generate a great deal of noise. To reduce the noise, use more dispersion tubes, such as the dispersion panels shown in figures 14.13 and 14.14.

Condensate return from jacketed steam dispersers

Any steam flow creates some liquid condensate in either the steam line or the disperser, or both. Condensate is usually returned by gravity to the boiler. But in some cases, the boiler is above the level of the disperser, so the liquid cannot return by gravity.

For vertical distances less than 5 ft. [1.5 metres], the steam pressure itself can be used to lift the condensate back to the boiler, avoiding the need for a condensate pump. Every pound of steam pressure will lift the condensate by about 6 inches. [150 mm of lift for every 7 kPa of steam pressure].

The method requires a steam trap that cycles between full-open and full-closed, such as the inverted bucket or disk-type traps. Modulating traps like the float or thermostatic types do not provide the sudden surge of steam pressure that lifts the condensate. Also, the diameter of the condensate return line must be small enough that steam acts like a piston to push up the liquid, rather than just bubbling through a wide column of water. Usually the line should have an inside diameter less than 0.5 in. [under 13mm]. Finally, the return line needs a check valve following the steam trap, to prevent the condensate from flooding back into the trap under the pressure of the column of liquid in the vertical return line.

Disperser location In ductwork and air handlers

The disperser or disperser panel must be located so that the vapor is absorbed before it can impinge on objects in the air stream. In locating the disperser, the designer must first know the absorption distance that will be necessary for the situation. This information is available from the manufacturer, provided the designer can define the temperature and moisture content of the entering air, its duct velocity and the maximum allowable relative humidity of the air as it leaves the humidifier.

Shorter absorption distances—which allow more flexibility in locating the disperser—are possible when:

- A dispersion panel with many tubes is used in place of a single disperser tube.
- The air velocity is slower.
- The entering air temperature is higher.
- The air approaching the disperser is well-mixed with a uniform temperature across the duct and an even velocity profile, avoiding extremes of temperature or velocity on one side of the duct or the other.

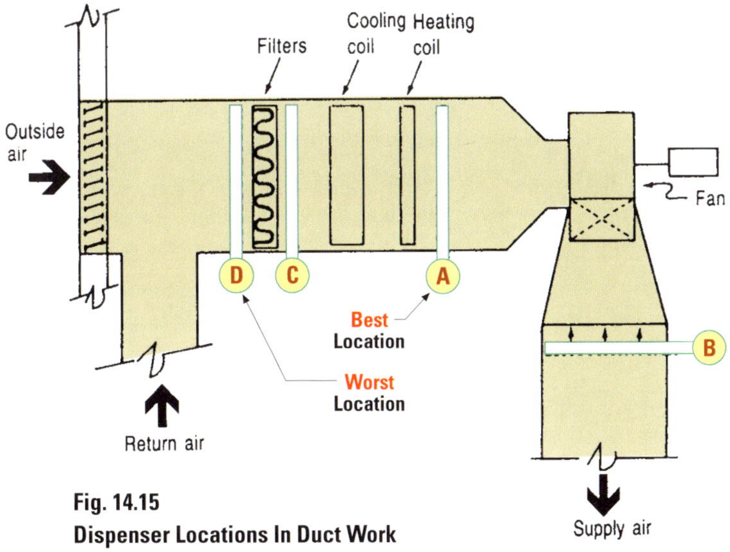

Fig. 14.15
Dispenser Locations In Duct Work

Figure 14.15 shows a typical arrangement of components in a system, and is marked with four possible locations for the vapor disperser. **Location A is usually the best choice.** All the others have more limitations. At location A:

- The air is likely to be warm.
- Access to the inside of the duct work for cleaning and service is likely to be easier for the designer to ensure.
- Connection points for steam, power and condensate drainage are likely to be nearby.
- The approaching air has traversed the coils and filters, and is likely to be flowing straight, and with minimal variation in velocity across the face of the duct. The air absorbs moisture evenly, with less risk of pockets of overly humid air which could condense downstream.
- There is usually enough height and width to accommodate the disperser—including disperser panels, which allow more rapid absorption therefore a shorter absorption zone.
- The distance to the next component—the fan—is likely to be long enough for full absorption, minimizing the risk of condensation.

Location **B** can be used when the duct work immediately downstream is long and straight. Absorption from location B is not as fast as from location A because:

- The air velocity is very high, and the velocity profile across the duct is very uneven with slow velocity at the duct wall and very high velocity in the center. Vapor will not mix uniformly with the air until well downstream.
- The cross sectional area of the duct at this point is likely to be small, so there may not be enough space to locate a dispersion panel. Using a single tube disperser limits the amount of moisture that can be injected into the air.

Location **C** is less preferable, although its dimension allows a full range of large dispersion panels to be used and distance from the fan allows plenty of time for absorption. At location C:

- Quite a bit of moisture could be lost when the cooling coil operates. Even when it is not cooling, its surface provides an obstruction that will condense moisture from the air, reducing humidification capacity.
- Access for inspection and cleaning is likely to be limited.

Avoid Location D. The air is likely to be too cold and not well-mixed, which makes condensation probable for most of the hours of its operation. Also, the filters will become saturated. Since filters trap dirt, mold and mildew and bacteria will grow in them, creating a potential health risk if they are not changed frequently.

Figure 14.16 shows an elbow in the duct work, along with two possible locations for the disperser. Location A is the better choice, because there is less risk of condensation on the turning vanes. At the same time, however, more air will flow along the outside of the curve. Vapor will be absorbed more quickly if most of the steam is discharged into the air that flows along the outside of the elbow.

Fig. 14.16
Disperser Locations Near An Elbow

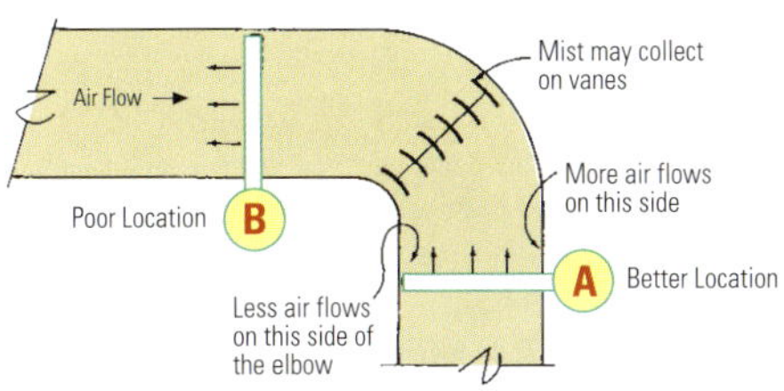

Adiabatic Humidification Systems

Adiabatic humidifiers are frequently used where the supply air must be cooled as well as humidified, or where there may be excess sensible heat in the return air that can be used for evaporation. In those situations their operating costs are lower than isothermal units. The more common types of adiabatic humidifiers include:

1. Wetted media
2. Pressurized water atomizer
3. Compressed air fogger
4. Electric oscillation atomizer (ultrasonic)

Wetted media

In these very simple devices, dry air passes through a bank of wetted cells, absorbing moisture. The fibrous fill or cell media is made of glass fibers, metal, or natural fibers. Water cascades down over the cells to keep them wet. Any water that reaches the bottom of the pad without evaporating is returned to the top of the pad by a pump. Application considerations for these units include:

a) Operating costs are very low when surplus internal building heat is available.

b) No dust is produced. Any dissolved minerals stay in the media or are flushed down the drain.

c) Units are short (with respect to the air flow direction). Evaporation takes place in the media. No additional distance is needed after the device. In most other types of humidifiers, vapor or water droplets must travel some distance though the duct work before the water is completely absorbed by the air.

d) Possible algae and microbial growth can be minimized by ozone treatment of the water or chlorination when chlorine is permitted in the air stream.

e) Close control is achievable, but needs a high-integrity control system that compensates for the time lag between the signal for more moisture and the relatively slow increase in evaporation from the media, and the response of controls that compensate for the change in air temperature.

f) When local water has a high dissolved solids content, extra water flows through the unit and out through the drain to flush mineral deposits off the media. Therefore, water costs and availability may need to be examined.

g) In cold climates, freeze-up protection may be needed to avoid damage to water supply piping.

Atomizers and foggers

The distinction between atomizers and foggers comes from the size of the water particles they produce. Foggers produce smaller droplets and more of them. Smaller droplets are desirable because they evaporate quickly, providing high capacity in a short distance with less risk of fallout. In contrast, atomizers produce larger droplets which evaporate more slowly, using more travel distance in the duct. But atomizers are usually simpler and less expensive to purchase than foggers.

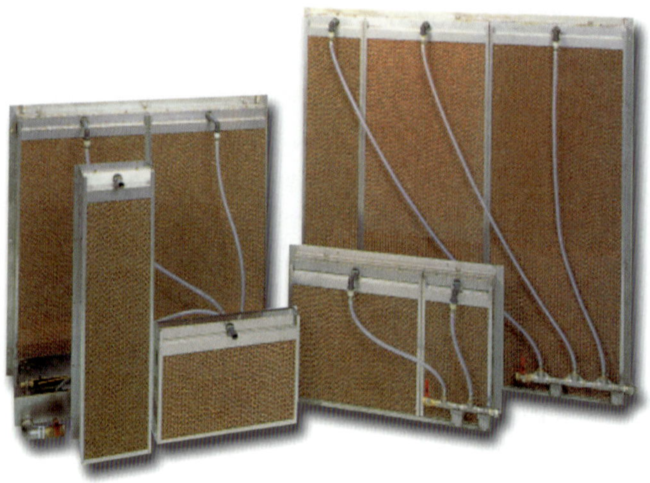

Fig. 14.17
Wetted Media, Cooler/Humidifiers

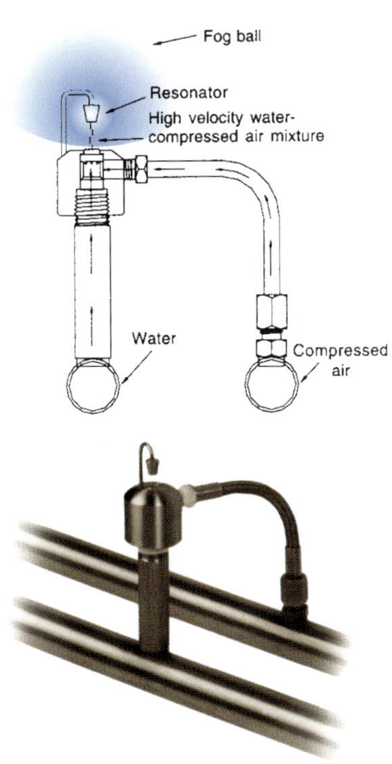

Fig. 14.18 Compressed Air Fogger

Fig. 14.19 Ultrasonic Fogger

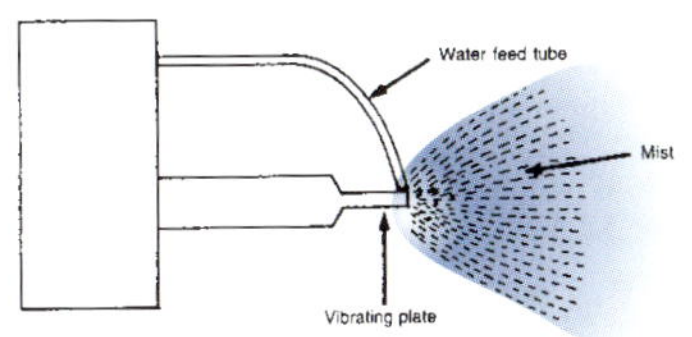

Atomizers include mechanical pressure atomizing nozzles, rotating disks, and rotating cones. These devices are limited in their ability to produce a high percentage of fine particles. Consequently, more "fallout" of water occurs in the duct work and longer distances are needed downstream of the unit for complete absorption. When only short absorption distances are available, the air must be preheated to speed evaporation.

Pressurized water atomizers

This method utilizes spray nozzles arranged in a bank inside a duct or air handler. The water is supplied by a booster pump. Some systems use pressures up to several hundred psi. [Pressures up to 20 kPa] Higher pressures produce smaller sized mist particles. Application considerations for pressure atomizers include:

a) Any dissolved minerals will be concentrated as the droplets evaporate, and will precipitate as dust. To avoid this, remove dissolved solids from water before it is sent to the atomizer.

b) When discharging mist, avoid placing objects nearby because they will get wet.

c) Avoid using atomizers in cold air streams. The rate of evaporation depends on droplet size, air temperature and relative humidity. The lower the air temperature and the higher the rh, the lower the rate of evaporation and the greater the risk of moisture fallout.

d) In duct applications, install mist eliminators and low-point drains to remove water accumulations from the duct. These tend to foul, so they should be located for convenient service access and the staff should be prepared for regular cleaning.

e) Avoid fungal growth in ducts. Atomizers should not be installed in duct sections with open glass fiber insulation or porous sound attentuators. Those materials hold the dust which feeds the mold that grows on moist surfaces.

Compressed air foggers

This type of humidifier is shown in figure 14.18. It uses compressed air to create much smaller water particles than atomizers. In the better-performing fogger heads, a high-velocity stream of compressed air passes at right angles to a small stream of water. The water is sheared into very fine droplets resembling fog. The resulting mixture of compressed air and mist is then driven against a conical block called a resonator, forming a "fog ball" in the duct air stream, where it is carried downstream and evaporates.

The small size of the mist particles and the mixing action with the duct air stream improves the evaporation rate from the fogger, making it usually superior to other types of atomizers.

Typically, the fogger heads are mounted in the air handler and arranged in a grid pattern *downstream* of the preheat coil and *upstream* of the cooling coil. Also, any dampers, straightening vanes, sensors and filters must all be located well downstream. Ductwork must be externally insulated if it passes through cold building cavities, and drain pans are recommended inside the duct to remove any water that falls out of the air.

A compressed air supply and a water supply are required, along with motorized valves and a system to control the fogging rate. Periodic maintenance will include regular cleaning and adjustment.

This equipment is used most frequently in large air handlers serving industrial applications such as semiconductor manufacturing, textile manufacturing, and printing. Some other application considerations include:

a) A high volume of compressed air will be needed.

b) Some foggers may produce a high-pitched whistle.

c) Installations require ample space and access for maintenance personnel to adjust and maintain the system.

d) Non-demineralized water may foul the orifices.

e) When using non-demineralized water, a filter may be required downstream of the foggers to trap mineral dust.

f) The entering air temperature and final supply air relative humidity affect evaporation performance. Low air temperatures and high relative humidities limit evaporation.

Electronic oscillation (ultrasonic) foggers

In these units, a piezoelectric transformer converts electrical current to mechanical vibration, causing water to oscillate at high frequency and breaking it into a fine mist that evaporates easily.

The output capacity of this equipment is rather low compared to its cost, so it is most frequently used in special-need applications where high capacity is not the primary criterion for evaluation. Special needs might include a requirement for a low noise level, compact size or low energy consumption (when compared to compressed air foggers). Application considerations for ultrasonic humidifiers include:

a) Oscillators produce finer mist particles than rotating disk or cone atomizing methods, providing more rapid evaporation.

b) Demineralized water must be used.

c) In a duct installation, an effective moisture eliminator may be necessary to prevent mist from being carried too far downstream, especially if entering air temperatures can drift below the usual average.

d) The small room-type, free-standing units and others with water reservoirs must be disinfected regularly.

Introducing cold mist into an air stream

An atomizing or fogging apparatus produces water mist particles ranging in size from 10 to 50 microns. Such equipment is most commonly used in open spaces with plenty of room available for full absorption. Placing this equipment in an air handling system requires some care to avoid oversaturation and excess water in the duct work.

Humidification system designers sometimes underestimate the degree of difficulty in converting atomizer-produced fog to water vapor in an air handling system. Although steam and cold fog are very similar in appearance, cold fog requires much more travel distance for absorption unless the air is very hot. It is easy to overlook this issue and fail to provide enough heat for evaporation. Without this heat, water falls out of the air and drips through the duct work.

At lower humidity levels, the issue is not quite so complex. But when using a cold fog system to maintain humidity above 35% rh, managing the air temperature and relative humidity in the fog zone becomes very important. In a fogger system the evaporation rate is controlled by :

1. The vapor pressure differential between the fog particles and the surrounding air. Dry air speeds evaporation.
2. The surface area of the water exposed to the air. Fine fog particles expose a far greater water surface, and will evaporate far more readily, than the same amount of water in the form of raindrops.
3. The thoroughness of intermixing between fog particles and air molecules. Better mixing speeds evaporation.

As air flows into the fogger zone and becomes permeated with fog, evaporation and cooling of that air begins. As it proceeds through the fogger zone, the air temperature drops and its relative humidity rises. At the same time, the vapor pressure difference between the air and water is declining, which slows down the evaporation.

If it were possible to create an ideal fogger system, the entering air temperature and the rate of fog discharge would be exactly balanced. The last fog particle would evaporate as the air left the fogging zone, and both the air temperature and relative humidity would leave the duct work at the levels needed to control temperature in the space.

Unfortunately, it isn't quite that simple. First, the system design must provide for an air temperature entering the fogger zone that is high enough to supply the heat required to evaporate the necessary amount of fog, and provide for enough heat left over to maintain a sufficiently high vapor pressure difference between air and fog. Without that extra heat, the evaporation rate would fall off to almost nothing. The last few fog particles would have to float in the air for a long time before they are absorbed, which would create an excessively long absorption zone.

Second, the design must provide for more fog than will be evaporated. Extra fog is needed to compensate for a falling rate of evaporation. Evaporation slows down because the total water surface area declines as droplets get smaller during evaporation. Without extra heat and fog, the time required and the length of the fogging zone would be excessive.

These issues present fewer problems when the building only needs to be humidified to 35% rh. In that situation, the final relative humidity in the supply duct can still remain quite low, and the available heat in air entering the fog zone is quite high compared to the energy needed for evaporation. But as the need for relative humidity rises above 35% rh, the designer must usually provide for a surplus of heat and extra fog in the entering air. Also, the air in the absorption zone will need to be very well-mixed to fully absorb the moisture within a short distance.

After humidification, any excess heat must be removed in order to deliver the air to the space at the desired supply air temperature. Also, excess fog must be removed to prevent wetting the duct system downstream of the fogger zone.

Removing the excess heat usually requires a cooling coil downstream of the fogger zone. However, as that cooling coil removes heat, it may also remove moisture. In some cases the temperature of air leaving the preheat coil and the air temperature leaving the cooling coil must be interrelated by the control system so the relative humidity and temperature of the supply air will stabilize at the appropriate levels.

The excess fog is removed by a mist eliminator located at an appropriate distance downstream of the fogger heads. The fogging zone is the distance between the fogging heads and mist eliminators. These eliminators, being continuously wet, can present an air-contamination problem unless they are regularly cleaned of dust and other nutrients for fungus and bacteria.

To avoid the risk of compromising indoor air quality, wetting of downstream ducts should not be allowed to occur. The leading fogger system manufacturers have created duct test systems and offer published design data. The data allows the engineer to design systems that will function without wetting the duct system.

References

Morton, Bernard W. *The Humidification Handbook, Second Edition*. 1998. Dri-Steem Humidifier Company, Eden Prairie, MN. (612) 949-2415

Chapter 19. Evaporative Air Cooling. *ASHRAE Handbook—Systems and Equipment.* 1996.

Chapter 20. Humidifiers. *ASHRAE Handbook—Systems and Equipment.* 1996.

Donated Photos & Diagrams

Figure 14.4 - © Carel USA

Figures 14.5 through 14.8 - © Dri-Steem Humidifiers Inc.

Figure 14.9 - © Carel USA

Figures 14.11 through 14.16 - © Dri-Steem Humidifiers Inc.

Figure 14.17 - © Munters USA

Figures 14.18 and 14.19 - © Dri-Steem Humidifiers Inc.

Chapter 15
Load Reduction Equipment

By Lew Harriman

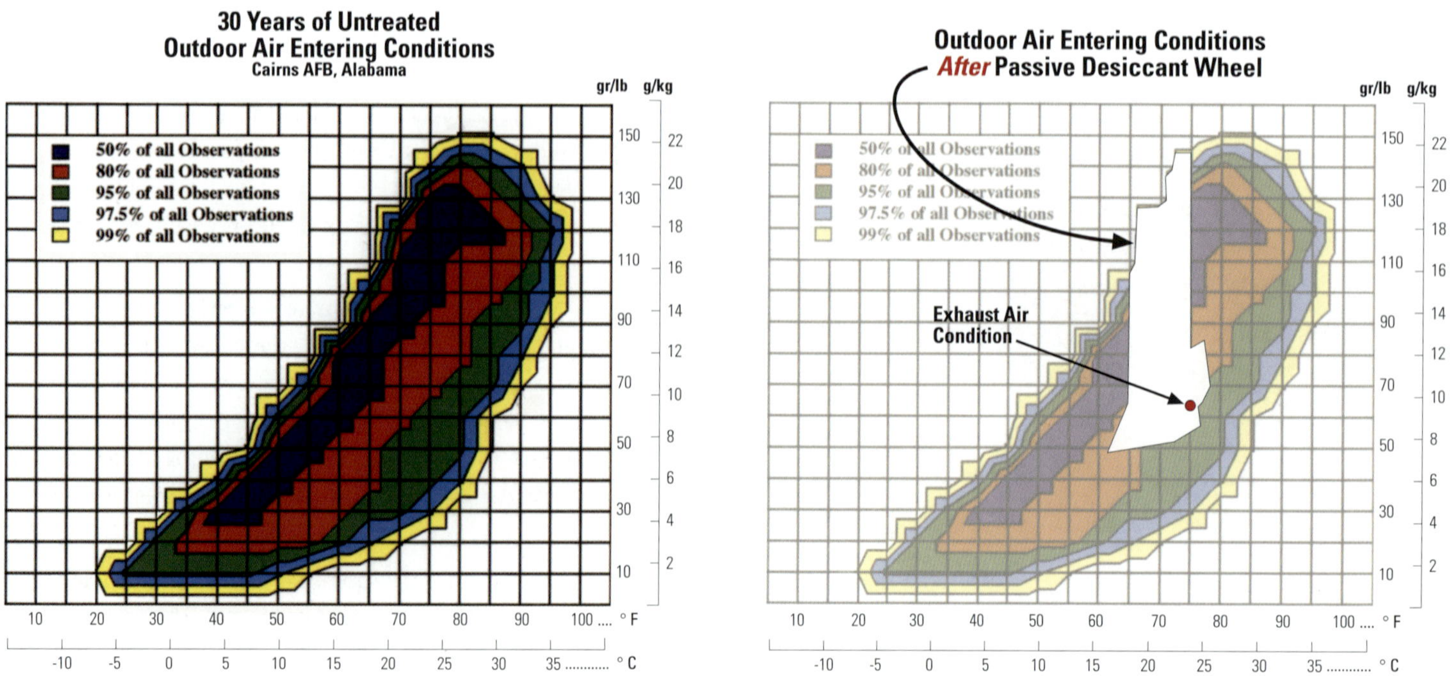

Fig. 15.1
Ventilation Loads Before & After a Passive Desiccant Wheel
Moisture loads from ventilation air are not only high, but they vary widely throughout the year. Adding a passive desiccant wheel to the system reduces both the magnitude of the loads, and also its variation. This load reduction equipment saves money in both construction and operating costs.

Fig. 15.2 Passive Desiccant Wheels

Load Reduction Equipment

The moisture load determines the size and therefore the cost of buying and operating the building's dehumidifiers and humidifiers. Consequently, the HVAC designer will be interested in any equipment which reduces the size of the load.

The equipment discussed in this chapter does just that. These are air-to-air heat exchangers. First, they can use the energy contained in the building's exhaust air to precondition incoming ventilation air, reducing the amount of remaining work for the humidity control equipment. Secondly, they can reheat or post-cool the air leaving a dehumidifier, which reduces or eliminates the need for cooling or heating equipment in that location.

Air-to-air heat exchangers accomplish an enormous amount of work at very low operating cost. For example, a passive desiccant wheel working with 10,000 cfm [4,719 l/s] in a humid climate can recover about 133 kW of cooling and dehumidification effect in return for an energy input of about 1.7 kW. That means its coefficient of performance (COP) at peak design is roughly 75. This compares very favorably with the COP of air conditioners, which usually ranges between 4 and 5.

Fig. 15.3
Passive Desiccant Wheel Operation

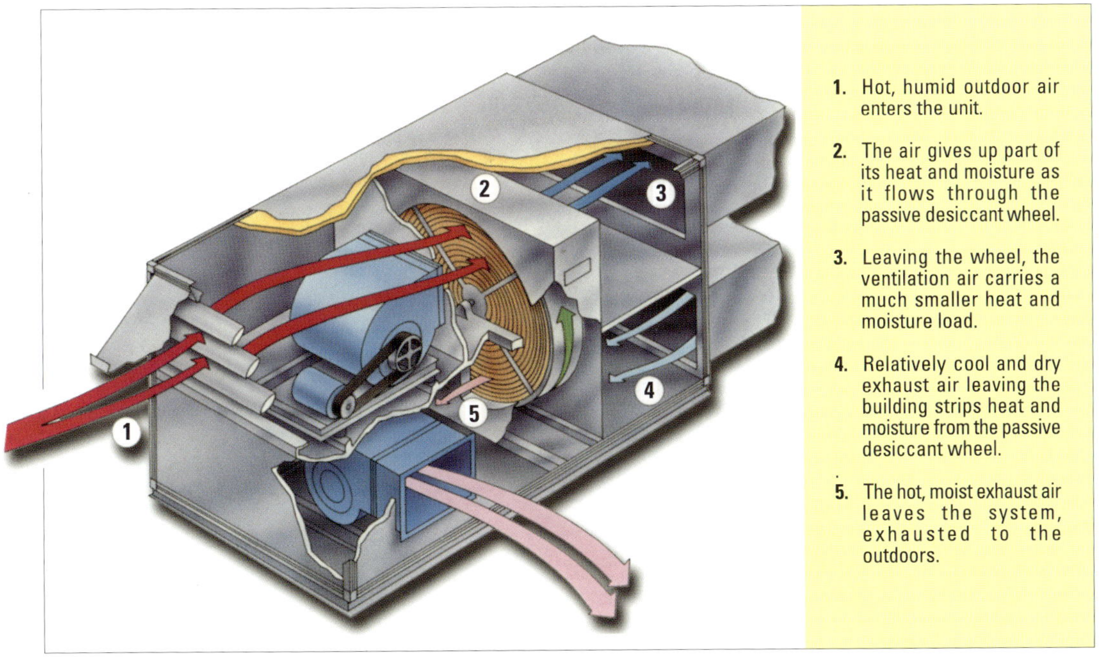

Air-to-air heat exchangers usually reduce the construction cost as well. They reduce the loads so far that the cost reduction in other components pays for the heat exchanger and its installation, with extra savings besides.

In summary, air-to-air heat exchangers are enormously useful in any HVAC system that must control humidity. However, the designer must keep in mind that air-to-air heat exchangers do not actually control humidity. They only reduce the load by recycling the dehumidification and humidification effects of other equipment in the system. Heat exchangers are not a substitute for dehumidifiers and humidifiers—but they are very useful, and they reduce the system's overall costs.

Passive Desiccant Wheels

As described in earlier chapters, the incoming ventilation air represents the largest moisture load. In the winter, the air is too dry. In the summer, spring and fall, the air is usually too humid. A passive desiccant wheel uses the energy contained in the exhaust air stream to reduce the load created by the ventilation air, and works during *both* winter and summer weather.

The wheel consists of honeycomb-form heat transfer media made of plastic, metal or glass fibers. The honeycomb is coated with a desiccant material, so the media can absorb and release both heat and moisture. Air can flow freely down through the length of the honeycomb flutes and out the other end, encountering very little resistance. The honeycomb wheel is mounted between the exhaust and ventilation air streams, and it rotates continuously at about 20 rpm.

First, consider the wheel's operation during summertime. The incoming ventilation air is hot and humid. The exhaust air leaving the building is relatively cool and dry. As ventilation air flows through the rotating honeycomb, it gives up part of its heat and moisture to the wheel and its desiccant coating. Then the wheel rotates into the exhaust air stream, which strips both heat and moisture from the media. Therefore the ventilation air enters the rest of the system cooler and dryer than it began, which greatly reduces the dehumidification and cooling loads.

Next, consider winter operation. Incoming ventilation air is cold and dry compared to the building's exhaust air. As the exhaust air leaves the building, it gives up part of its heat and moisture to the wheel and its desiccant coating. This heat and moisture is stripped off the wheel by the incoming ventilation air as it flows through the honeycomb. Ventilation air enters the rest of the system warmer and more humid than it began, which greatly reduces the heating and humidification loads.

Passive desiccant wheels can transfer between 10 and 90% of the heat and moisture difference between two air streams, depending on the wheel's rotation speed and other factors.

These devices are called "passive" desiccant wheels to distinguish them from "active" desiccant dehumidifier wheels described in Chapter 13 (Dehumidifiers). Active wheels use *heated* air to strip water vapor from the desiccant, and they dehumidify the air very deeply. Passive desiccant wheels use *dry air* instead. While passive wheels dry the air less deeply, they do not use any supplemental heat to accomplish that drying. Passive desiccant technology is quite remarkable for its economy and simplicity.

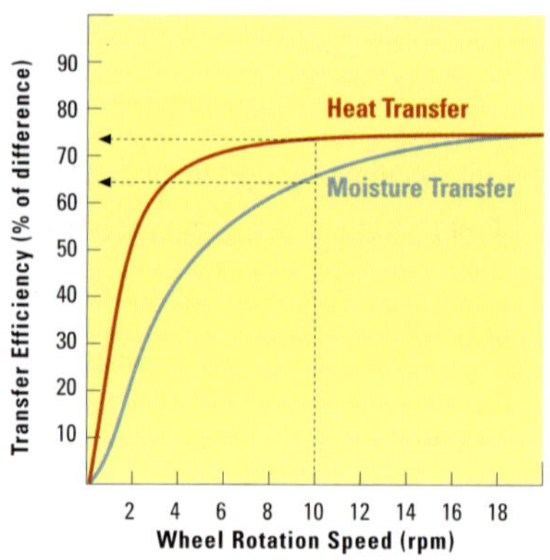

Fig. 15.4
Transfer Efficiency vs. Wheel Speed
More of the moisture difference is transferred at higher rotation speeds

Moisture transfer performance

The amount of moisture transferred between exhaust and ventilation air depends on how much desiccant material is rotating between the air steams (the depth of the wheel and its rotation speed), and how fast the air moves through the wheel. More heat and moisture are transferred when the:

- Wheel is deep in the direction of air flow
- Rotation speed is high
- Air flows through the wheel slowly

Wheel depth

Commercial units use wheels that vary in depth between one inch and eight inches [between 25 and 400 mm]. Deeper wheels transfer more moisture and allow for the very robust construction needed when the diameter of the wheel is very large (when it must handle a great deal of air). Thinner wheels cost less, and can partly compensate for their smaller amount of desiccant surface by rotating at a higher speed. In general, mechanical rather than thermodynamic considerations govern wheel depth. Larger-diameter wheels tend to be deeper, and smaller-diameter wheels tend to be thinner.

Rotation speed

At about 20 rpm, wheels coated with molecular sieve, aluminum oxide or silica gel reach their near-maximum moisture transfer performance. Above that speed, little benefit is gained. The more powerful lithium chloride desiccant reaches its maximum at about 10 rpm, but that deliquescent salt is less popular because it is not as fault-tolerant as solid desiccants.

Below 20 rpm, moisture transfer falls off rapidly, a shown in figure 15.4. In contrast, heat transfer stays near its maximum until rotation speed falls below about 10 rpm.

Fig. 15.5
Transfer Efficiency vs. Face Velocity
More of the moisture difference is transferred at lower face velocities.

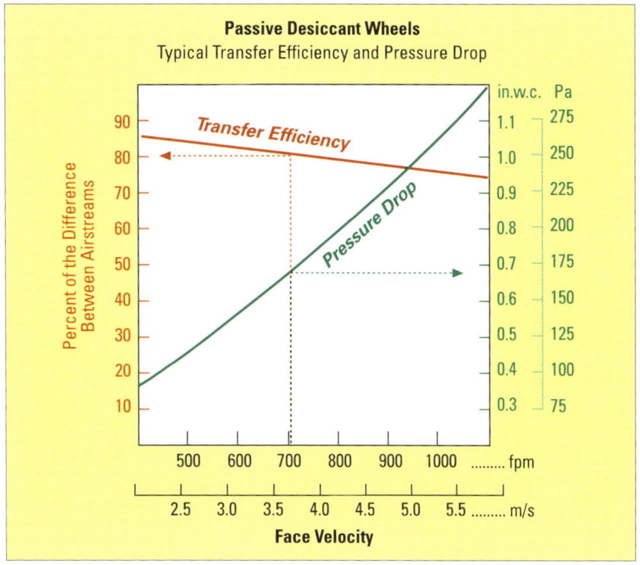

Air velocity

Commercial face velocities are usually between 400 and 1000 fpm. [between 2.0 and 5.0 m/s] At lower air velocity and higher rotation speed, a passive desiccant wheel can transfer about 90 % of the moisture difference between the two air streams. When the air velocity is in the upper range, the moisture transfer potential drops to about 70% of the moisture difference between the two air streams.

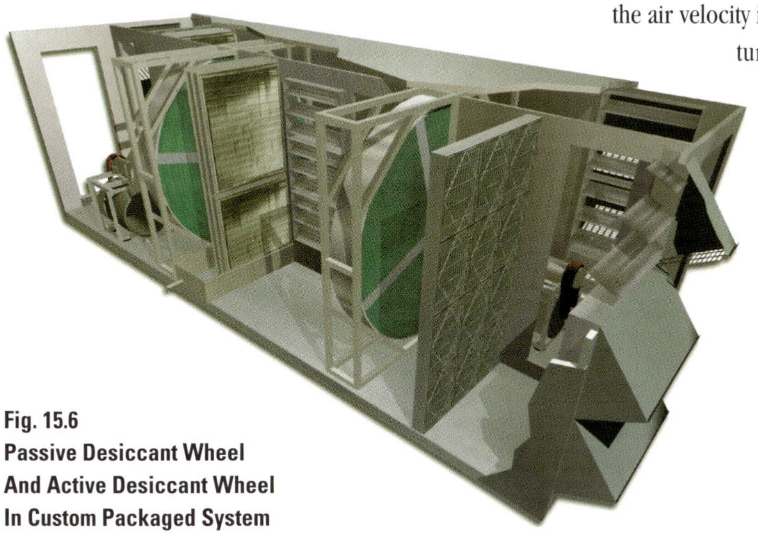

Fig. 15.6
Passive Desiccant Wheel And Active Desiccant Wheel In Custom Packaged System

Hardware variations

Passive desiccant wheels have been in use since the early 1960's, with many evolutions in mechanical design and desiccant sorption characteristics over that time. The technology is now available in many forms.

Stand-alone wheels

Larger structures such as hospitals and tall buildings often have mechanical systems that are built-up on-site from individual components. Passive desiccant wheels are used in these systems as free-standing components, complete with wheel, seals, support casing, drive motor and controls for mounting in the duct work. Single units are readily available in sizes from 3,000 cfm to 80,000 cfm. [1400 to 38,000 l/s] When even larger air flows are needed, or where system height should be minimized, several wheels are set together in an array. A built-up system allows the designer great flexibility, but also costs more than other systems and demands excellent mechanics and excellent technical supervision at the job site. These systems are usually used where the extra cost of engineering and construction is more than offset by the operating cost efficiencies made possible by optimizing components for that single building.

Custom-designed packaged systems

Rather than building-up a system on-site, the designer may choose to have a custom packaged unit designed and built in a factory, trucked to the site and lifted into place as a single unit or as modules. This alternative saves costs over site-built systems, but still provides the designer with considerable flexibility. Stand-alone passive desiccant wheels are purchased and installed by the package manufacturer according to the HVAC designer's specifications.

As options inside standard packaged air conditioners

Over the last 20 years, the cost of passive desiccant wheels has come down, and their mechanical fault-tolerance has reached the high levels needed for poorly-maintained commercial buildings.

Passive desiccant wheels are now available as options within low-cost, packaged rooftop air conditioning units. This equipment is used for low-rise commercial buildings, where packaged units have always been popular and cost-effective. When a passive desiccant wheel is added to these units, they can provide the building with a much larger proportion of outdoor air without increasing the size of other components. This saves both construction cost and operating cost.

Packaged units with passive desiccant wheels can also moderate the moisture loads during many of the hours of the humid season. However, for full control during all hours of the humid season, some form of dedicated dehumidifier will also be needed. There are many hours that a passive desiccant wheel should not rotate, for reasons described below under the section on wheel speed control.

Typical design features

Air is prevented from leaking between the two air streams by two key design features: a purge sector and effective air seals. Also, at some load conditions the wheel must be slowed down to avoid over-recovering heat from the exhaust air stream. These features are important to realizing the benefits and minimizing the limitations of passive desiccant wheel technology.

Purge sector

When the honeycomb passages move from the exhaust to ventilation air, they are filled with a small amount of exhaust air. A purge sector uses some of the incoming fresh air to flush the exhaust air out of the wheel—before that contaminated air can carry over and mix into the rest of the incoming fresh air.

The purge sector is a small, pie-shaped duct mounted on the indoor face of the wheel between the exhaust and the ventilation air sectors. As the wheel rotates out of exhaust and into the ventilation air, the first ventilation air that fills the wheel is forced over to the exhaust side through the purge sector. The purge air leaves the building along with the rest of the exhaust air. That way, the wheel media does not carry contaminated air into the fresh ventilation air.

Seals. leakage and fan position

Air can leak from one air stream to the other through the wheel casing, or across the air seal that divides the air streams from each other. So the design and maintenance of the seals is important, and so is the placement of the fans.

The air seals are part of the equipment design, and so is the fan placement when wheels are integrated into packaged equipment. The HVAC designer must simply be aware of the potential for air leakage, and should confirm the amount of air leakage by questioning the manufacturer. Standard tests allow the designer to have confidence in the information provided by the manufacturers (ARI-1060-97).

When the designer is building up a system on-site including wheels as individual components, he must be aware that fan placement is an important consideration. Figure 15.8 shows the recommended alternatives. Placement "A" (draw-through, draw-through) is preferred, because the pressure differences are minimized, which reduces supply air losses to about 6% of the air volume.

Position "B" (blow-though, draw-through) is used when the exhaust air is especially contaminated, so no leakage can be allowed from exhaust to the ventilation side. In that position, air leaks from ventilation towards exhaust, so the supply fan must be sized to include the leakage air, usually between 10 and 15% of the air volume.

Wheel speed control

In some circumstances, rotating the wheel could *increase* costs rather than reducing them. So wheel speed control is often provided with the unit. Figure 15.1 shows the effect on the supply air when the wheel must be slowed to avoid over-recovering heat from the exhaust air. The wheel must be stopped for part of the year.

For example, consider the situation in spring, fall and during summer evenings when air indoors is 75°F and air outdoors is 70°F. To remove internal heat, the supply air condition leaving the system should be no higher that 65°F. [Air indoors at 24°C. Air outdoors at 24°C. Supply air should be no higher than 18.3°C]

Under those circumstances, the wheel should *not* rotate. The outdoor temperature is higher than the desired supply air temperature. If the wheel were transferring heat from exhaust, the supply air temperature would rise even further, increasing rather than reducing the building's cooling load. This and many similar situations call for the wheel speed to be varied in response to changes in the differences between the ventilation and exhaust air streams.

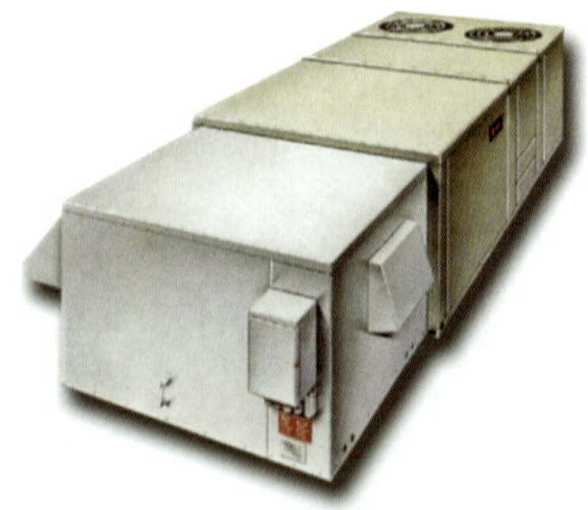

Fig. 15.7
Passive Desiccant Wheels in Modular Addition to Rooftop Packaged System

Fig. 15.8 Preferred Fan Positions

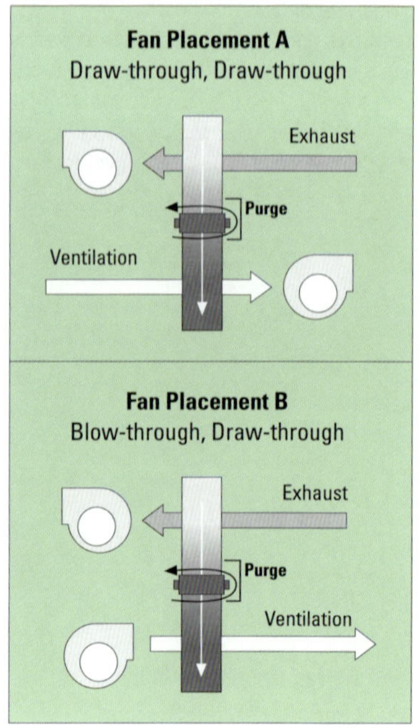

Moisture Load Without Cooling Load
Outdoor Air - TMY-2 Hourly Data
Typical Year - West Palm Beach, FL.

Dry Bulb °F		65	67	69	71	73	(75)	
Dry Bulb °C		18.3	19.4	20.6	21.7	22.8	23.9	Hours
g/kg	gr/lb	0	0	0	0	0	30	30
18.2	128	0	0	0	0	7	93	100
17.5	123	0	0	0	0	47	136	183
16.8	118	0	0	0	0	97	121	218
16.1	113	0	0	0	11	86	113	210
15.4	108	0	0	10	32	124	52	218
14.6	103	0	0	20	67	138	68	293
13.9	98	1	14	62	101	116	85	379
13.2	93	8	33	62	50	65	45	263
12.5	88	34	70	64	56	67	79	370
11.8	83	38	51	39	55	88	71	341
11.1	78	39	43	22	48	65	37	254
10.4	73	33	41	21	52	64	32	243
9.6	(68)	16	27	42	47	57	25	214
							Annual Hours:	3317

Fig. 15.9
Hours Without Load Reduction
When the outdoor temperature is between the indoor temperature and the desired supply air temperature, the wheel stops. During those hours, humid air enters the building untreated. That's why the system still needs a dedicated dehumidifier to remove incoming moisture and to control humidity.

For best control, the system must measure moisture differences as well as temperature differences to decide what wheel speed is appropriate. For example, consider an exhaust air stream at 75°F, 65 gr/lb and an outdoor air stream at 68°F, 100 gr/lb with a desired supply air temperature of 65°F [Exhaust air at 24°C, 9.3 g/kg, outdoor air at 20°C, 14.2 g/kg, with a desired supply air condition of 18.3°C] In that situation, it could be useful to keep the wheel rotating, using dry exhaust air to remove moisture from the incoming ventilation air. But if the wheel were turning, it would bring excess heat back into the building, increasing the cooling load. So the wheel must not operate.

The optimal control system recognizes both temperature and moisture differences, and slows or stops the wheel if either the moisture load or cooling load would be increased by doing so.

Suggestions for system design

Passive desiccant wheels can improve the economics of *any* HVAC system as long as the exhaust air can be brought to the entry point for the building's ventilation air. The designer using this equipment for the first time may wish to consider these suggestions:

Quantify air leakage and losses inside the unit

Passive desiccant wheels leak some air. Not much, but enough to be significant in some situations—such as when pressure differences are large. Equipment manufacturers have worked with the Air Conditioning & Refrigeration Institute to develop a series of tests that quantify air leakage as well as heat and moisture transfer performance. The designer should ask the manufacturer for the results of tests performed in accordance with ARI Standard 1060-97 (Rating air to air energy recovery ventilation equipment). With that information, the designer will know what net air flow to expect, and can design accordingly.

Change filters frequently

Changing filters frequently ensures the system actually delivers the ventilation air required by the design.

This issue is especially important when designing with low-cost, packaged units. That class of equipment uses forward-curved fan wheels for maximum economy, and some provide very little external static pressure (seldom more than 0.5" w.c. or 125 Pa). When filters are dirty, these fans fail to provide enough pressure to overcome the filter's resistance, so ventilation air flow falls off rapidly—to as little as 30% of the design value. When that happens, the ventilation air fan pulls exhaust air through the wheel casing from the *other* side of the wheel. Exhaust air is recirculated through the building instead of fresh air.

For example, one field study measured the concentration of CO_2 in a classroom served by a packaged air conditioning unit equipped with a passive desiccant wheel. (Bayer, Crow & Fischer, 1999) When the classroom was occupied, the CO_2 level averaged 2500 ppm. But after clean filters were installed, the level dropped within 10 minutes to 850 ppm—comfortably below the usual1000 ppm maximum.

Filters should be changed when their air flow resistance increases from a typical clean value of 0.09" w.c to a clogged value of 0.5" w.c. [from 22 to 125 Pa]

Also, the designer should select fans which will deliver the design air flow even when dirty filters create a resistance of 0.5" w.c. [125 Pa] It may be necessary to specify the next-larger size equipment if the unit cannot provide the design air volume when filters are dirty.

Filters also protect the wheel from clogging with large debris such as insects and feathers, but in fact, most wheels are quite tolerant of fine particles. The air passages are blown out 20 times every minute by counterflowing air streams. Any accumulation on the entering face is blown off when the wheel rotates through the second air stream that flows through the wheel in the opposite direction.

Expect load reduction, not humidity control

The most common misimpression among first-time users is that passive desiccant wheels control humidity. They don't. They just make it easier for other equipment to do so. Your system will still need equipment to dehumidify the air, and to humidify it. The passive wheel simply recycles the dehumidification or humidification work performed by other equipment.

Save money by removing excess capacity

In packaged equipment, the capacity of components has usually been reduced to take advantage of the load reduction accomplished by the passive wheel. But in built-up or custom factory-built systems, first-time designers sometimes forget to—or prefer not to—downsize the capacity of heating and cooling components.

In some cases, this reflects the designer's natural preference to "play it safe," and retain spare capacity in case of problems with the desiccant wheel. But this practice *causes* problems rather than avoiding them. Passive desiccant wheels can reduce the outdoor air load by 30 to 50%. When equipment is *not* downsized, that huge extra capacity becomes very difficult to control. Both humidity and temperature swing widely as the oversized equipment starts and stops. This rapid cycling also damages mechanical components, causing unreliable operation and premature failure.

If for some reason the designer has less confidence in the desiccant wheel than the confidence he has in other components, he can remove only part of the now-excess cooling and heating capacity. But if the designer truly believes the system is mission-critical and that redundant capacity is appropriate, a better plan is to install two wheels instead of one, or split the load between two smaller wheels to provide half capacity in case of some mishap.

Specify calibrated humidity sensors for speed control

Wheel rotation speed is usually controlled to make sure the system does not over-recover heat or bring in humid air without dehumidifying it. However, the sensors that send signals to the control system are not usually calibrated as a matter of course by either the manufacturer or installer.

When humidity sensors are not calibrated, the control system can make poor decisions, or may operate the wheel so it actually *increases* rather than reduces ventilation air loads.

To gain any benefit from the equipment, humidity sensors *must* be calibrated. The designer should specify that sensors must be calibrated on-site after installation, with controls adjusted accordingly. Also specify that the calibration points along with each sensor's deviations are documented in the system's commissioning report. Finally, regular calibration must be performed over the life of the system to minimize operational costs.

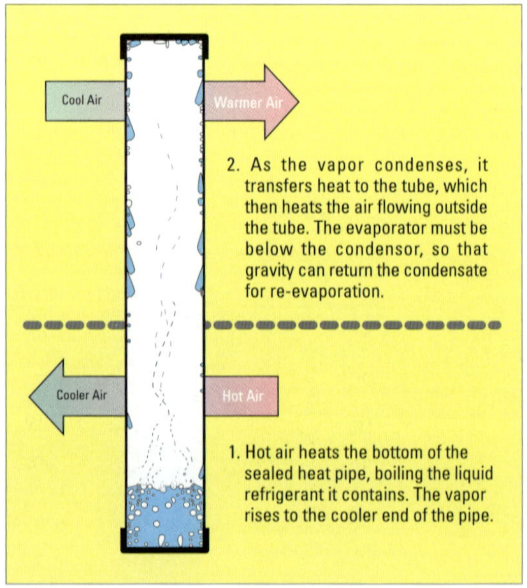

Fig. 15.10 Heat Pipe Operating Principle

Heat Pipes

Heat pipes can be used to improve the operation of either desiccant or mechanical dehumidifiers. They can postcool the hot dry air leaving an active desiccant wheel. Also, they can precool and reheat air passing through the cooling coil of a mechanical dehumidifier.

Heat pipes are sealed tubes containing some liquid and a gas at low pressure. Because the internal pressure is below ambient, the liquid inside the tube can boil at a low temperature. The lower end of the pipe is placed in the hot air stream. The upper end is placed in a colder air stream.

The liquid inside the bottom of the tube boils as it absorbs heat. This cools the air around the bottom of the tube. The boiling vapor drifts upwards inside the tube and condenses, releasing its heat to air surrounding the top of the tube.

Heat pipes assisting desiccant dehumidifiers

In commercial heat pipes, many tubes are arranged in an array with fins, making the device look like a cooling coil with a divider plate halfway up the tubes. The lower half is mounted in the hot air, and the upper half extends into the cooler air. The divider plate keeps

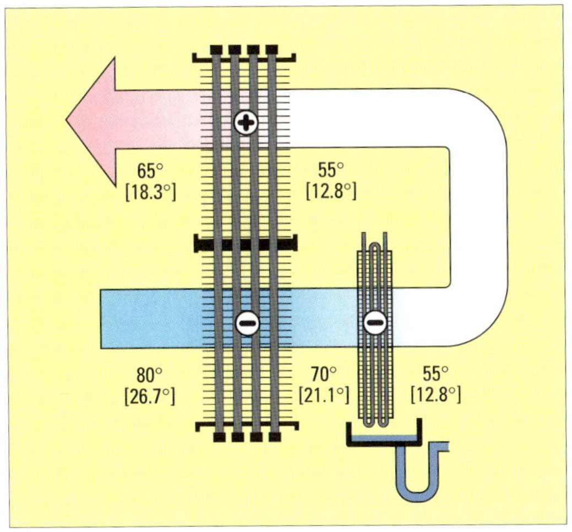

Fig. 15.11
Heat Pipe Assisting a *Mechanical Dehumidifier* by Pre-Cooling and Post-Heating Dry Air

the air streams separated so they do not leak into each other. This is the form of heat pipe used to postcool air leaving an active desiccant wheel. Usually, the heat pipe array is mounted nearly horizontally, so the air streams can be side-by-side. The upper half of the pipe is only slightly above the level of the lower half, but the arrangement still allows condensate inside the pipe to return to the lower half by gravity.

Heat pipes assisting mechanical dehumidifiers

A clever design variation allows the air to pass through a mechanical dehumidifier in a straight path. The heat pipes are bent into a u-shape and mounted nearly horizontally, wrapping around a cooling coil. The lower part of the pipe is in front of the coil. The upper part of the pipe is behind the coil, and arranged so it is slightly above the level of the lower part of the tube. Once again, the tubes are arranged in an array with fins. The assembly looks like two connected coils "wrapped around" the cooling coil.

As air enters the lower part of the heat pipe, it cools down. Then the air cools much further as it passes through the cooling coil. After it leaves the cooling coil, air passes through the upper end of the heat pipe, reabsorbing the heat it lost originally.

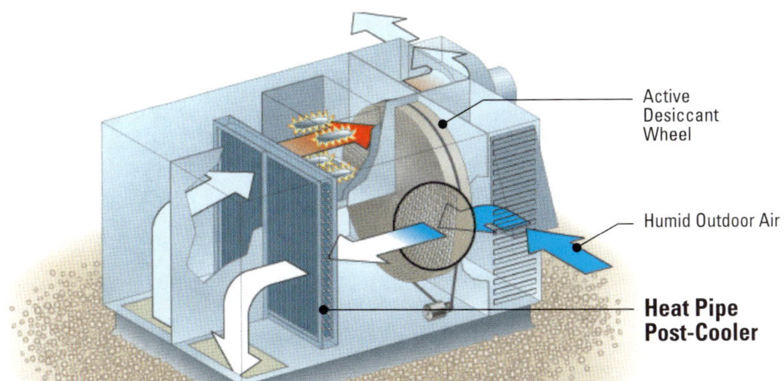

Fig. 15.12
Heat Pipe Assisting an *Active Desiccant Dehumidifier* by Post-Cooling Dry Air

This wrap-around design improves the dehumidification performance of the cooling coil, since it cools air more deeply than without the heat pipe. The device also avoids the cost of reheat energy. The leaving temperature is close to the entering temperature, with no need for supplemental heat.

Heat pipe performance

In commercial units, heat pipes are usually designed to transfer between 45 and 70% of the temperature difference between two air streams. Figure 15.11 shows typical performance for a wrap-around heat pipe assisting a mechanical dehumidifier. Like a cooling coil, the transfer effectiveness is influenced by the number of rows of pipes in the array, the fin spacing and the velocity of air as it passes through both sides of the array. More heat is transferred when:

- There are more rows of pipes in the array
- The fin spacing is closer together
- Air velocity through the array is slower

Controlling heat pipe capacity

Unlike the wheel-type heat exchangers, there are no moving parts in a heat pipe. Heat transfer happens automatically, any time there is a temperature difference between the two air stream. This is an advantage in terms of maintenance, but a challenge in terms of control.

Capacity can be controlled by bypassing some air around one end of the heat pipe array. If less air flows through the heat pipe, less heat is transferred.

The bypass is arranged with a variable-position damper across the bypass duct. When less heat should be transferred, the bypass opens slightly, and less air flows through the heat pipe. For smooth control, the pressure drop through the bypass must be equal to the pressure drop through the heat pipe. Otherwise, when the bypass damper opens, most of the air will follow the path of lesser resistance through the bypass. This would force the mixed temperature to swing widely within a few seconds, making it very difficult to maintain stable temperatures downstream of the heat pipe.

In heat pipes for side-by-side or upper-lower duct arrangements, the manufacturer often supplies a bypass within the casing of the heat pipe array. The designer simply needs to allow for more width around one side of the duct work.

For wrap-around heat pipes, the capacity control must be accomplished by the HVAC designer. Lay out a bypass duct with dampers around the entire heat-pipe-and cooling-coil assembly, taking care once again to make sure the bypass is equipped with a fixed-position damper or an orifice plate. With a damper, the bypass resistance can be adjusted to be the same as the duct section containing the heat pipe and cooling coil.

Hardware variations

Heat pipes can be purchased in a wide variety of configurations. The relatively specialized manufacturers have thrived by being extremely flexible and oriented towards custom fabrication. This provides the designer with the advantage of optimizing the device for a given application, in return for a somewhat higher cost than what higher-volume, standard products would allow.

Custom-fabricated heat pipes

For built-up and factory-built custom systems, the designer simply specifies the volume, temperature and moisture levels for the two air streams. The manufacturer will then provide performance data for heat pipe arrays with different numbers of rows and fin spacings. The designer chooses the unit based on how cold or hot the air stream must be made.

Wrap-around heat pipes can be supplied for both new and retrofit installations. The manufacturer will fit the device to the face area of the cooling coil in the system. The designer simply needs to allow enough space on both entering and leaving sides of the coil to install the heat pipe, plus space *between* the devices for cleaning.

Standard components inside dehumidifiers

Heat pipes are popular options inside dehumidifiers as shown in figures 15.11 and 15.12. Desiccant units are often equipped with heat pipes for post-cooling. Mechanical dehumidifiers may come equipped with wrap-around heat pipes to increase the dehumidification capacity of a given compressor size.

Suggestions for system design

Heat pipes that are preinstalled in dehumidifiers have been arranged by the manufacturer for minimal cost and maximum simplicity from the HVAC designer's perspective. But when wrap-around heat pipes are installed in built-up systems, the HVAC designer using these devices for the first time may wish to consider these suggestions from manufacturers and experienced designers.

Clearance and access for cleaning

Dehumidification coils are deep in the direction of air flow so they can cool the air way below its entering dew point. When a wrap-around heat pipe is added to that cooling coil, the entire assembly may be as deep as 12 to 16 inches. [300 to 400 mm] It is very difficult to clean into the center of such a deep coil from either end. Over time, the lower half of the coil will eventually clog as dirty condensate flows down from the upper part of the coil.

To avoid this problem, install access doors for cleaning both in front of and behind the heat pipe. Also, make sure the manufacturer designs the wrap-around heat pipe with *long* connecting tubes, so you can provide doors for access to clean both the front and the back of the cooling coil that is surrounded by the heat pipe.

Size fans for extra pressure drop

The pressure drop through dehumidification coils is fairly high, because they are deep, and because their surface is covered with condensate. When a wrap-around heat pipe is added to such coils, the pressure drop is even higher. HVAC designers expecting the characteristic low pressure drop of sensible cooling coils do not always allow enough fan capacity to pull air through this deep assembly.

Be sure to consult the manufacturers of both the dehumidification coil and the heat pipe to be sure of the total pressure drop when the devices are producing maximum condensation from the air stream. Size the fans accordingly.

Seasonal bypass

In most cases, ventilation is continuous but the need for dehumidification is not. During cold and dry seasons, the dehumidification coil and heat pipe create pressure loss that is not balanced by any benefit. This loss can be quite expensive in terms of fan power, especially for large systems.

In more polar locations such as Canada or Argentina where the dehumidification season is short, economics can be improved by installing a full air bypass around the coil assembly. In this case, the bypass is not for control, so its air flow resistance does not need to equal the resistance of the coil. Indeed, that's the whole point; there should be very little resistance in the bypass, so less fan energy is needed to operate the system during non-humid hours of the year.

Change filters frequently

In a heat pipe system, filters keep the coil from clogging. Without filters, dirt ends up on the fin surfaces, where it flows down with the condensate and clogs the coil. From that perspective, a loaded filter is a more efficient filter, so many operators make the mistake of assuming that longer periods between filter changes save money and do a better job of protecting the coil.

While that may be true, it misses the real point of having a dehumidification system, which is to produce a defined amount of dry air. When filters are dirty, they present more resistance, so the air flow is reduced. With less dry air, the humidity level in the building rises above set point during humid seasons. Many dehumidification systems that "are not performing" simply have dirty filters that do not allow enough air to flow through the unit.

Typical 30% efficiency pleated filters have a resistance of about 0.09 in. w.c. when they are clean. Dirty filters have a resistance of roughly 0.5 in. w.c. [20 and 125 Pa] Filters should be changed *before* their resistance gets as high as 0.5 in. w.c. [125 Pa] This is likely to occur after one or two months of use. Clean filters will save fan energy that helps pay for the cost of the filter, and will help ensure that the dehumidifier delivers the expected amount of dry air.

Fig. 15.13
Plate-Type Heat Exchanger in a Packaged System

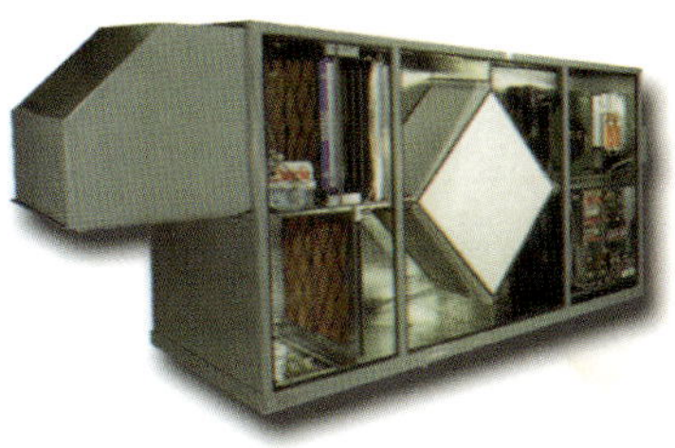

Plate-Type Heat Exchangers

Plate-type heat exchangers provide the same functions as heat pipes, but they usually cost less. On the other hand, they do not transfer as much heat as heat pipes, and they do not allow as much flexibility to the designer of the duct system.

In a plate exchanger, the hot and cold air streams are separated by thin plates. Air flows through the device in an "x" or a "z" pattern. The cold air absorbs heat from the hot air on the other side of the plates. These are perhaps the simplest and least expensive heat exchangers used in humidity control systems. Like heat pipes, these are static devices without moving parts apart from optional bypass dampers.

While these units could be used to post-cool air from a desiccant wheel, they are seldom used for that purpose. This may be because plate exchangers tend to be larger than the heat pipes, and the heat pipe usually transfers more heat as well.

Plate exchangers are used more frequently to precool and reheat air in a mechanical dehumidifier. While not as efficient as a heat pipe, the plate units are usually less expensive, and can be made more easily by more manufacturers.

Heat transfer performance

Higher heat transfer efficiency is gained by adding more plates (more heat transfer surface), or by slowing the air down as it flows through the unit (lower face velocities).

When too many plates are added, manufacturing cost goes up, and the air friction through the unit becomes prohibitively expensive to overcome with fan power. Likewise, when the air is slowed down,

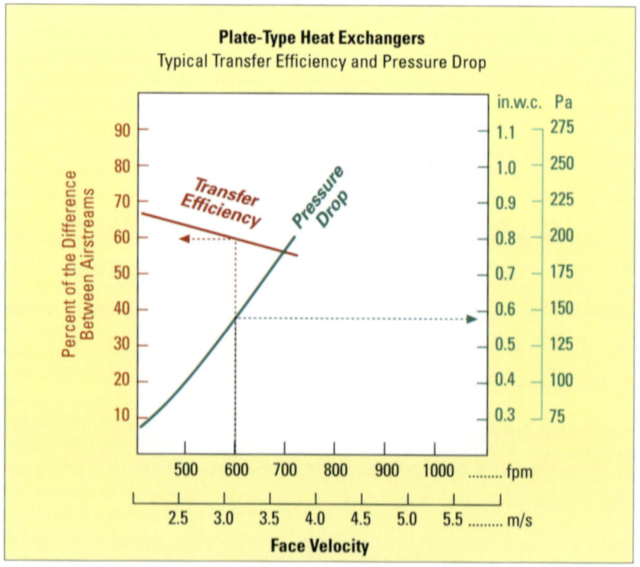

Fig. 15.14 Plate-Type Heat Exchanger Performance

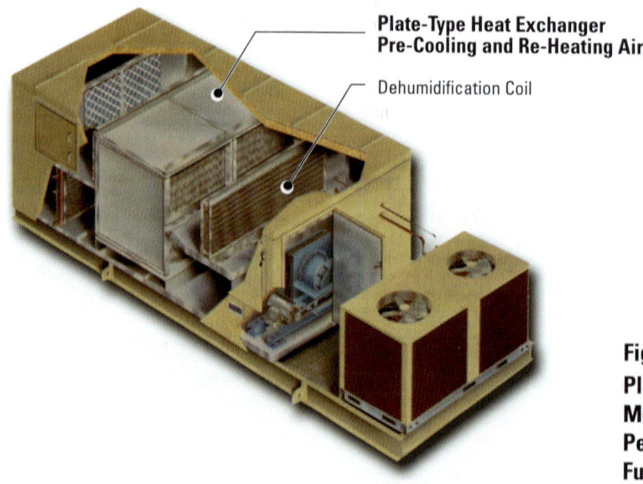

Fig. 15.15
Plate-Type Heat Exchanger Inside a Mechanical Dehumidifier, Performing Pre-Cool and Re-Heat Functions

more heat transfer surface is being used per unit of air flow, which means the unit becomes very large for a given flow rate.

For these reasons, most configurations of plate-type exchangers will transfer between 60 and 65% of the temperature difference between two air streams. These rates are usually achieved by flowing the air through the unit at face velocities between 300 and 600 fpm. [3 and 1.5 m/s]

Suggestions for system design

Usually, plate type heat exchangers are installed inside mechanical dehumidifiers, so the HVAC designer does not need to be especially concerned with the details of integrating the device into a system. Where stand-alone units are used, as in larger built-up systems, the designer can be guided by the suggestions provided in the section describing heat pipes.

References

ARI Standard 1060-97 Rating Air-to-Air Energy Recovery Ventilation Equipment. 1997. Air Conditioning & Refrigeration Institute, 4301 North Fairfax Drive, Arlington, VA 22203 (703) 524-8800 www.ari.org

Bayer, C, Crow, S. & Fischer, J. *Causes of Indoor Air Quality Problems In Schools*. 1999. U.S. Department of Energy, Oak Ridge national Laboratory report ORNL/M-6633. National Technical Information Service, Springfield, VA. (800) 553-6847, www.ntis.gov.ordering.

Chapter 44. Air-to-Air Energy Recovery. *ASHRAE Handbook—Systems & Equipment*. 2000.

Donated Photos & Diagrams

Figure 15.1 is based on the "Long-Term Psychrometric Summary" of hourly weather observations for Cairns Air Force Base, Alabama, published in the US Department of Defense Handbook 32-1163, Engineering Weather Data (CD-ROM).

Figures 15.2 and 15.3 - © Semco Manufacturing

Figure 15.4 - © Flakt Inc.

Figures 15.6 and 15.7 - © Semco Manufacturing

Figure 15.13 - © Venmar Ltd.

Figure 15.15 - © DesChamps Laboratories

Special Thanks

The author is deeply grateful to more-experienced professionals for information contained in this chapter. Among these are Kirk Mescher, who first suggested the idea that resulted in figure 15.1; Ken Bowlen and Jack McGuinness, early pioneers who first introduced me to the technology and application imperatives of passive desiccant wheels; and John Fischer, who, through thousands of installations, has been responsible for many imporant advances in the state of the art over the last 25 years.

Chapter 16
Building Pressure Management

By Lew Harriman

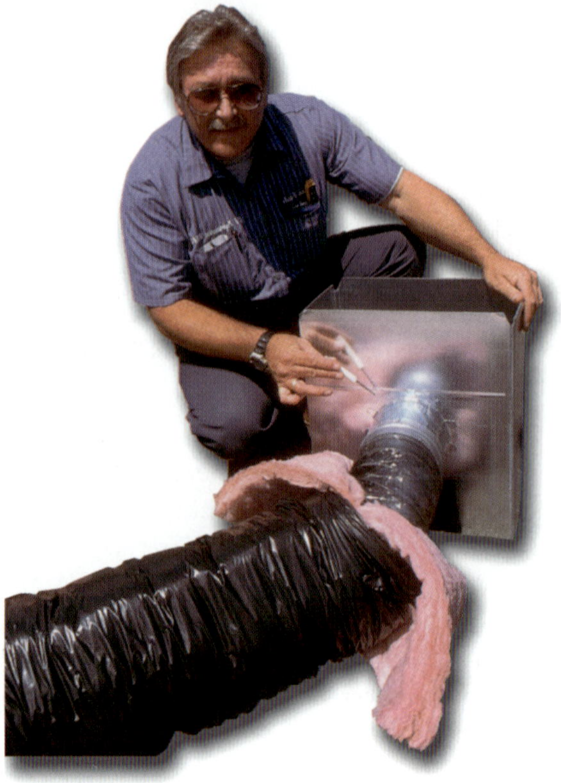

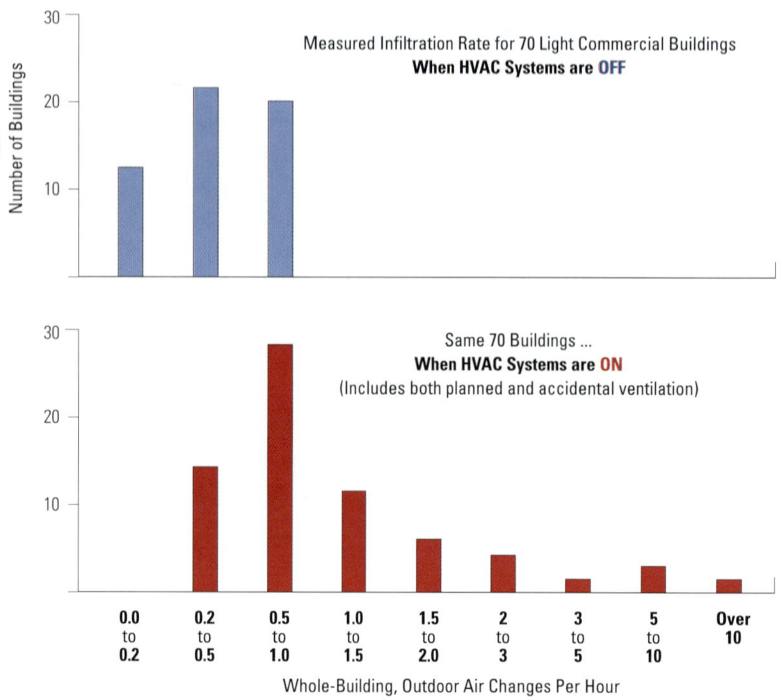

Fig. 16.1 Unsealed Connections
The duct connection at left appears reasonably tight, but it's not. Unsealed seams, especially in return duct connections, are responsible for the whole-building leakage shown in the graphs. (Cummings et al, 1996) Such leakage leads to damage from condensation in building cavities in both hot and cold climates.

Fig. 16.2 External Wall Joints

Anywhere one building component meets another, the joint can leak air. Joints leak whenever the HVAC system or the wind creates even a small air pressure difference.

Commercial Buildings Leak Water and Air

Any small difference in pressure across an exterior wall forces air into building cavities. In those hidden places, moisture condenses on cool surfaces, creating problems for human health, reducing the indoor air quality, damaging the interior finish and reducing the structural integrity of building materials.

Since these problems are often caused by condensed moisture, building owners frequently respond by instructing the HVAC designer to "improve humidity control" in the building. While working to avoid or resolve such problems, the owner, architectural designer, HVAC designer and contractors should keep in mind that:

1. The most severe moisture-related problems in buildings are caused by water leaks through joints in the exterior wall. But the problem is not usually the leak—it's the fact that water becomes trapped inside the wall. The design and construction of exterior walls must not allow outdoor water to become trapped inside building cavities. This is the responsibility of the architectural designer and the general contractor. Inadequate attention to air pressures across the exterior wall can accelerate problems caused by water leaks, but even flawless air pressure management cannot eliminate them.
2. The HVAC portion of the problem is almost always caused by leaking duct connections and leaking fan plenums inside equipment—not by a surplus or deficit of humidity in the occupied space.
3. Regardless of how well the building wall is designed and built, *it will leak air*—usually far more than quality-minded construction professionals would expect. The HVAC designer must keep this in mind when calculating loads, when specifying duct work, when designing the connections between the system's duct work and its equipment, and when specifying the connections between the duct system and the interior walls or ceilings.

To avoid building-related moisture problems caused by the HVAC system, the HVAC designer can specify that all duct connections be sealed, all fan chambers inside HVAC equipment also be sealed and any connection between the indoor ceiling or walls and ductwork or equipment be either sealed or gasketed. The HVAC contractor can help avoid moisture-related problems by ensuring that these specifications are really followed in the field.

How Much Air Leakage?

Air moves from one place to another because there is a difference in pressure between those two places. So when trying to control air movement, look for locations where the air pressure changes over a short distance—that's where air movement will occur. In a building, there are always pressure differences between indoors and outdoors, and between the duct system and the space around the system.

Exterior envelope

Air is pushed into the exterior wall by wind, which raises the air pressure outside the building compared to the pressure indoors. Air can also be pulled into the building by suction created by the HVAC system inside the building.

A few examples from a field investigation of 70 buildings in the southern United States will help the designer visualize the magnitude of these forces in typical low-rise commercial buildings. (Cummings et al. 1996)

In this field investigation, each building was pressurized internally with a blower, so that the average pressure difference across the exterior wall was 0.2" w.c. [50 Pa]. Once that pressure difference was achieved, the air flow through the blower into the building was measured with a calibrated venturi nozzle. The air leakage from 70 low-rise buildings is shown in figure 16.1. Some buildings leaked more than 50 complete air changes per hour. Others leaked less than 5 air changes per hour. The average leakage for all buildings was about 20 air changes.

A pressure difference of 0.2" w.c. [50 Pa] is created whenever the wind blows against the outside wall at a velocity of 25 m.p.h. [11.2 m/s]. As we will discuss shortly, outdoor wind pressure is never uniform around the whole building. But even allowing for lower average pressure differences, the test results show that typical light commercial construction leaks a great deal of air. That's why the HVAC designer and contractor must take care not to create internal suction near the building wall.

Duct connections & equipment casings

The same field investigation measured the outdoor air quantities that entered the buildings *with* and *without* the HVAC system in operation. This measurement was accomplished with tracer gas released inside the building. A faster decay in the tracer gas concentration means more outdoor air is flowing into the building.

At rest, the 70 buildings leaked at an average rate of 0.4 air changes per hour. But turn on the HVAC systems, and they pull in outdoor air at an average rate of 0.9 ac/h. This is far in excess of the intended amount of ventilation air. The excess comes in the form of accidental leakage caused by suction produced when fans pull air into ducts through joints that are not sealed. That suction acts on the building wall, pulling outdoor air into the building.

These field measurements stand in sharp contrast to leakage values shown by standard industry references. For example, the SMACNA Duct System Inspection Guide (SMACNA, 1989), suggests that unsealed duct work should be expected to leak about 1.2 cfm per 100 ft^2 of duct surface. Given the amount of ductwork in typical commercial buildings, the total system leakage would then be about 5 cfm per 1,000 ft^2 of gross floor space. However, field measurements on 70 completed air handling systems show they actually leak at a rate of 341 cfm per 1,000 ft^2 of gross floor space. In other words, typical unsealed air handling systems leak at a rate *68 times larger than what the industry expects* based on leakage tests in the laboratory.

[Unsealed duct work might be expected to leak about 0.063 l/s per square metre of duct surface. Given the amount of duct work in typical commercial buildings, the total system leakage would then be about 0.027 l/s • m^2 of gross floor space. However, field measurements on 70 completed air handling systems show they actually leak at a rate of 1.82 l/s • m^2 of gross floor space. In other words, typical unsealed air handling systems leak at a rate *68 times larger than what the industry expects* based on leakage tests in the laboratory.]

Some of the unexpected leakage comes from air handlers that are not well-sealed, and additional leakage comes from gaps between the ductwork and the walls and ceilings. If the return grills are not gasketed to the wall or ceiling surface, air will be pulled *not* from the space, but instead from the area *behind* the wall or ceiling. Also, while long duct runs may be well-sealed, the installing contractor may neglect to seal the straight runs to the transitions that draw air from the occupied space. Once again, that leaking joint pulls air from the building cavities.

Another classic leakage point is the casing surrounding the fan in a cooling unit mounted inside a building wall—as in packaged terminal air conditioners (PTACs) and wall-mounted fan/coil units. If those casings leak, their fans pull air from the wall cavity, creating suction that pulls outdoor air into the exterior wall. The same problem can occur even when the casing itself is tight, if the installing contractor has not sealed the return air inlet of the unit to the inside wall surface. Then air is pulled into the unit from the wall cavity, just as if the casing itself were poorly sealed.

Finally, exhaust ducts are very seldom sealed. Many HVAC designers and contractors reason that any leakage will be *into* rather than out of the duct, so contaminants would not escape into the building from the exhaust. Unfortunately, pulling excess air into the exhaust duct creates suction in building cavities, which eventually pulls untreated air into the building from outdoors.

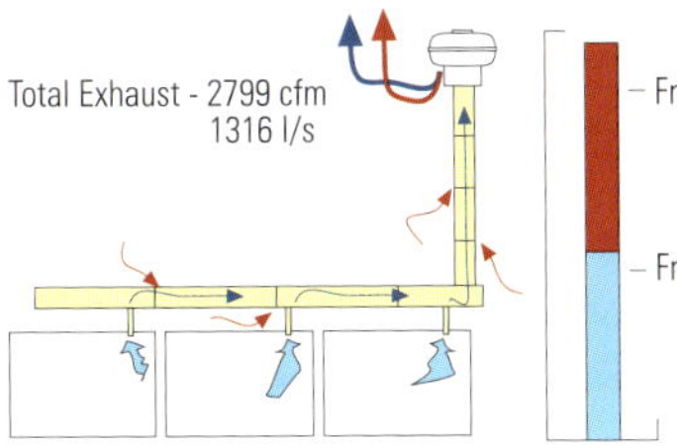

Fig. 16.3 Unsealed Exhaust Duct

When the exhaust duct leaks, it pulls air from building cavities, and eventually from outdoors. In this case, leaks allowed the fans to pull more air from cavities than from the bathrooms. (Cummings, 1996).

An example of this sequence of events and its consequences is shown in figure 16.3. In a small hotel, unsealed exhaust ducts pulled air through the outside wall continuously. The air carried moisture, which condensed behind vinyl wall covering in the cool hotel rooms. The mold created the musty odors so familiar in humid climates, and the building's interior walls had to be pulled out and replaced to eliminate health risk from that mold.

In the Cummings study described earlier, the field investigators measured the air flows entering and leaving one unsealed central toilet exhaust system in a business hotel. A single fan on the roof drew air from exhaust grills in 40 guest rooms. The total of air flows entering all 40 exhaust grills was 1324 cfm. The air flow *leaving the fan on the roof* was 2799 cfm. [622 vs. 1316 l/s] In other words, the leaking exhaust ducts pulled more air from building cavities than from bathrooms. (fig. 16.3)

Air pulled from building cavities often comes from the outdoors, flowing easily through open steel stud walls. Since the exhaust operates continuously, the infiltrating humid air carries a tremendous amount of water vapor. In central Florida, the annual moisture load from one scfm of untreated infiltrating air is 203 lbs per year. (Harriman et al, 1997) So the leaking exhaust duct system in this case

Fig. 16.4 Consequences of Exhaust Duct Suction

Humid outdoor air flows through the building's walls, drawn by suction created by the leaking exhaust duct. Moisture condenses inside cool walls, and mildew grows.

Each hotel with this problem spends millions to correct it. The much less expensive solution would have been to seal all exhaust duct connections at time of construction. (Lstiburek, 1999)

brings a total of 299,425 pounds of excess water vapor through the hotel walls over a year's time. [An annual flow of 43 kg in every litre per second, totalling 135,639 kg/yr for this case]. That water vapor flowing through a building certainly helps explain the mold and mildew problems of hotels in humid climates, as seen in the hotel shown in figure 16.4.

Sealing all the duct connections in the building, and sealing all connections between the ducts and the interior finish is the least costly way to avoid suction and infiltration of humid air into cool or cold building cavities. The HVAC designer who wants to limit the potential for moisture-related building damage should write the duct specification and the test and balance requirements to ensure that the system is *tightly sealed at all joints*, and that the total system *leakage must be quantified* before the system goes into operation.

Indoor & outdoor air pressures are not uniform

In elementary physics, we were taught that air pressure is equal throughout all parts of an enclosed volume like a bottle or storage tank. From this fact, we often make the assumption that the same is true for a building, which may seem like a single enclosed pressure vessel. Unfortunately, that assumption is not correct.

Buildings consist of a complex assembly of connected chambers that leak at different rates. The pressure differences—therefore the air leakage volume and direction—are quite different along each part of the building's walls and roof. Contributing to further complexity, the outdoor air pressure varies greatly along the outside edges of the building, as shown in figure 16.5 (Bearg, 1993).

On the windward side of the building, the average outdoor air pressure is high compared to pressure inside the building, and it varies with wind speed, so the outdoor pressure is highest at the top of the building, which is furthest away from the ground friction that slows the wind. But on the downwind side of the building, the average outdoor air pressure is low compared to the average air pressure inside the building. These pressure differences force outdoor air though the building.

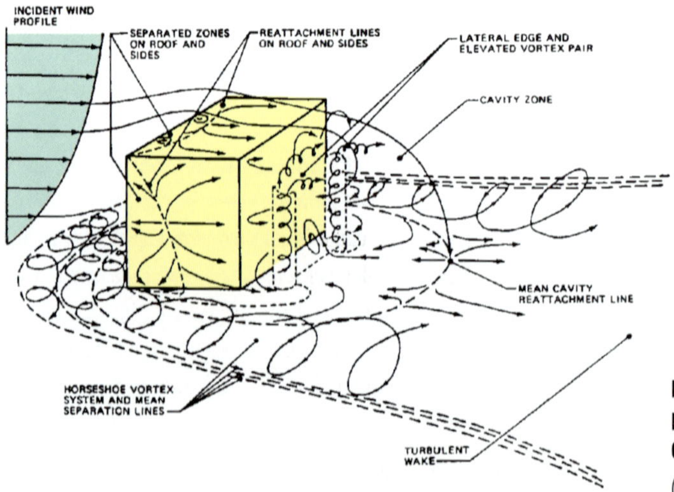

Fig. 16.5
Pressure Varies Widely Around The Outside Wall Of All Buildings.
(Bearg, 1993)

When the weather is windy, air is *pushed into* the building through cracks on the windward side, and *pulled out* through cracks on the downwind side. But as the diagram shows, the local pressure differences can vary widely within that basic relationship between overall average pressures. Local suction at a single point on the downwind side could pull air in through cracks, even though the *average* downwind air pressure is below the *average* indoor air pressure.

From these facts, one can easily see that maintaining a positive air pressure inside the center of a building does not guarantee that the building pressure will be positive at *all* points on the exterior wall. That's another reason to seal the duct connections and unit casings to avoid accidental suction at the perimeter of the building.

Leakage Mechanisms In Tall vs. Low-rise Buildings

Buildings pull in unwanted infiltration air through three principal mechanisms:

- Outdoor wind pressure
- Indoor stack effect
- Leaks at duct connections.

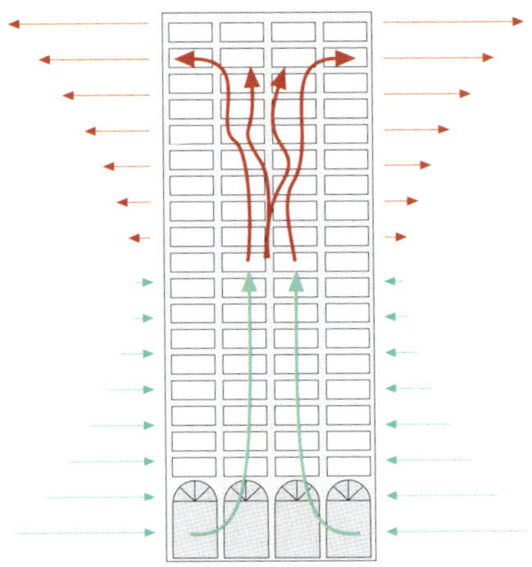

Fig. 16.6 The Stack Effect

Warm air rising inside the building leaks out at the top, which pulls air in at the bottom. The construction tarp in the photo shows the effect—air leaking out of the building pushes the tarp outwards at the top, and air being pulled into the building pulls the tarp inward at the bottom of the building.

Most of the leakage in tall buildings comes from the wind and from the stack effect. In low-rise buildings most of the leakage comes from duct connections.

Tall building leakage is dominated by the stack effect

The stack effect is what pulls smoke up a chimney and out of a house. Warm air rises because it is less dense, and therefore lighter than cold air. When a sealed vertical tube like a stack or a tall building isolates a mass of warm air from the surrounding cooler air, the warm air stays together, unmixed and very light. It therefore rises rapidly, flowing out the top of the stack and pulling in more air at the bottom of the stack.

The volume of air is proportional to the temperature difference between the air at the top and bottom of the stack, and also proportional to the height of that stack. Large temperature differences and taller stacks pull in more air at the bottom.

In buildings, the stack effect makes air flow *outward* through crack at the top of the building, and *inward* through cracks at the bottom of the building. All buildings do this, but the taller the building and the colder the outdoor temperature, the more air leaks through the walls because of the stack effect. Figure 16.6 illustrates this effect with both a diagram and a photo.

Air tightness is most important at the top floors. Air leaks in the upper stories generate the most flow, because as air leaks out of the top of the building, it generates great suction, which pull air inwards at the bottom levels. In addition, external wind pressure is highest at the top of a building, because that's where the wind blows hardest. Friction from contact with the ground slows the wind at lower levels. So the way to minimize air leakage in tall buildings is to seal the exterior wall very tightly, with special attention given to the inside surfaces of the upper floors.

In addition to tight sealing of the exterior envelope, the architectural designer and general contractor can reduce stack effect leakage by increasing the resistance to air flowing upward between floors. Three paths allow vertical air flow:

- Elevator shafts
- Fire Stairs
- Floor penetrations for pipes and conduits

In most cases the elevators are given a great deal of attention from both the Architect and the elevator supplier, which minimizes leakage through this path. Less attention is given to ensuring tight gaskets around the doors that lead to the fire stairs. This leakage path can be greatly reduced by good door and sill gaskets. In older buildings, the gaskets tend to work loose, which leads to increasing leakage through the stack effect. Replacing the gaskets or improving them is usually less costly than improving the air tightness of the exterior envelope.

Floor penetrations are another source of vertical air movement. To keep fires from spreading upwards to other floors, codes demand a fire seal around all pipe and conduit penetrations through floors in high-rise buildings. But in many cases, the seals are designed to expand with heat, and may not be air tight before a fire. Adding air sealant to the penetration detail can help reduce vertical air movement, and therefore reduce air infiltration through exterior walls.

The more common problem with floor penetrations is that the fire seal is entirely missing. It is often difficult to schedule the workers who place the seals to immediately follow the workers who run the pipes and conduits through the penetrations. Sometimes the seals are not set in place before the penetrations are concealed behind interior walls. Any unsealed penetrations are, of course, a serious fire risk. And they also allow vertical air movement through the building and therefore they increase the air leakage through the exterior walls through the stack effect.

Leakage from duct connections in tall buildings can also create pressure differences that generate air infiltration. But in general, as the size of the building goes up, the duct pressures also increase. With high-pressure ductwork, the HVAC designer and Contractor are less likely to allow duct-related leakage through inattention. Leaking duct joints in high rise buildings are quickly obvious and are likely to be fixed before they create long-term problems. Duct connection leaks are a more frequent issue in low-rise construction, where their negative consequences are not immediately apparent.

In summary, to minimize condensation problems in tall buildings it is not enough to control the indoor humidity and to place enough insulation and avoid thermal bridges inside the exterior walls. The architectural designer and the general contractor can minimize the stack effect by:

1. Building a tight exterior envelope, especially at the inner surface of the upper stories.
2. Sealing pipe and conduit penetrations through floor slabs.
3. Ensuring that all doors entering onto fire stairs are tightly gasketed, including at the threshold below each door.

The HVAC designer can help minimize infiltration by keeping the elevator shafts and fire stairs under a net positive air pressure with respect to the outdoors. Detailed guidance for pressure management in tall buildings is beyond the scope of this volume, but a modest amount of additional assistance can be found in the text and references of Chapter 25 of the 2001 *ASHRAE Handbook—Fundamentals* (Ventilation and infiltration).

Low-rise leakage is dominated by duct & equipment joints

The stack effect also creates leakage in low-rise buildings, but the magnitude of the effect depends on the height of the building, so the effect does not generate as much air flow. Low-rise buildings are likely to have much more infiltration from leaking duct connections.

Information presented earlier in this chapter described air infiltration caused by leaking duct connections. Because of that potential leakage, the author strongly recommends that all duct connections and connections to equipment be sealed tightly. The reader should also consider the problems created by using the open space above a ceiling as a return plenum rather than using return air ductwork.

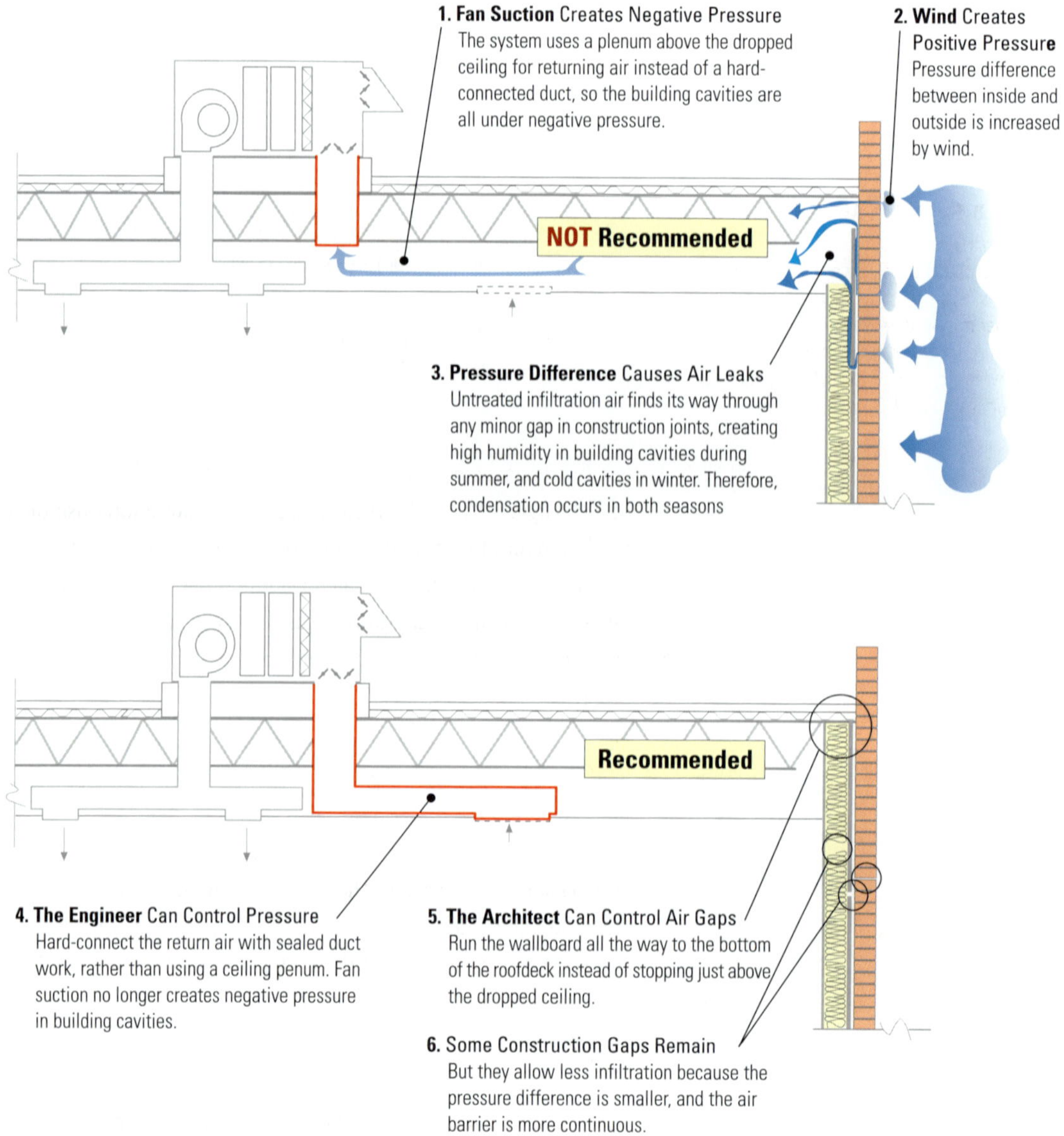

Fig. 16.7 Avoid Using Building Cavities As Return Air Duct Work.

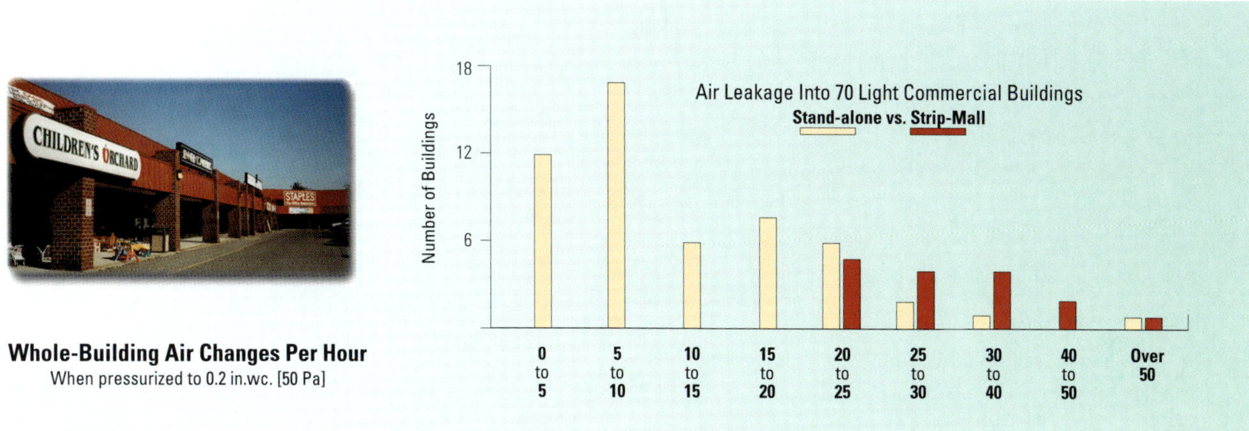

Fig. 16.8 Strip Mall Air Leakage
Stand-alone buildings are usually tighter than attached buildings. This may be because the owner's emphasis on costs tempts the designer to eliminate return air duct work in favor of ceiling plenum returns. These usually leak, allowing AC units to pull in outdoor air through the walls. (Cummings, 1996)

In low-cost commercial buildings such as those found in "strip malls" (figure 16.8), the HVAC designer may not always use duct work to return the air to the roof-mounted packaged air conditioning units. Instead, sometimes an open grill is placed in the dropped ceiling grid, and the air slowly finds its way from the conditioned space to the inlet of the cooling unit through the ceiling plenum.

At a first look, this design detail saves construction money and should not greatly increase energy expense. Insulation is above the roof deck, protecting the open return plenum from excess heat and cold. On closer inspection, however, one realizes that the design creates suction throughout the plenum space, and therefore creates suction at the exterior wall. The only way to avoid that problem is to treat the entire plenum as a sealed duct, by making sure interior finish extends up to the roof decking as shown in figure 16.7. The cost of return air duct work is usually less than the cost of sealing all that interior finish to the roof decking.

The field investigation described earlier (Cummings et al, 1996) showed that buildings in strip malls have much greater air leakage than stand-alone buildings. On-site research showed that leaks between buildings in strip malls are very common. Suction created by a plenum return in one building pulls air from the next building. Figure 16.8 shows the impact of these common imperfections.

Assuming the owner has expressed a concern and identified a budget for better control of humidity, the HVAC designer would be prudent to invest part of that budget in well-sealed supply and return air duct work. This will provide more predictable and favorable results compared to *hoping* there will be no holes in fire walls and *hoping* that all interior finish will be tightly sealed to the roof deck.

Cold vs. Hot Weather

In hot, humid weather, internal pressure should be ***positive***. In cold weather, internal pressure should be ***neutral***. Negative internal pressure is never useful. The outdoor air temperature and humidity should govern whether the average pressure inside the building is kept positive or neutral. Here's why.

Cold climates

In cold weather, any infiltrating air chills the surfaces inside the walls. Then, cold surfaces produce condensation from the *indoor* air, which is likely to be relatively humid. So negative internal pressure—suction—is not good, because it creates cold surfaces inside the walls. On the other hand, positive internal pressure can be even worse. When that pressure forces humid indoor air into the building cavities, moisture condenses in large amounts, supporting mold and mildew and leading to structural damage.

So the designer's goal in cold weather should be to keep the internal pressure of the building neutral with respect to the outdoors. That's not easy. In fact, it's probably not possible to keep pressure neutral along *all* parts of the walls on *all* sides of the building, for the reasons described earlier in this chapter. Some cold air will leak into the building wall, and some humid air will leak outwards into the wall cavities. But the designer's goal should be to minimize those flows by keeping the average internal pressure neutral.

On the other hand, guidance in cold climates can vary by location. For example, in some Scandinavian countries, regulations require that buildings be kept under negative air pressure rather than neutral. That guidance is based on the tight construction also mandated in those countries. Tight construction avoids the steady inward leakage of large amounts of cold air that would occur in the more loosely-built buildings of the United States. The designer must always consult local building codes in case they carry instructions that relate to building air pressure.

The controls can call for neutral air pressure any time the outdoor air temperature falls below some preset value. There is no common agreement among building scientists on what outdoor air temperature calls for a neutral building pressure. Most experts would agree that pressure should be neutral when the outdoor temperature falls below freezing. But others argue that pressure should be neutral long before the outdoor temperature falls that low. Here are some considerations that have been mentioned by people who argue that building pressure should become neutral at outdoor temperatures considerably above freezing.

Big problems can occur in highly-humidified buildings

If the indoor dew point is higher than usual in the winter, then a great deal of water will condense if indoor air is forced into a cold wall. So highly-humidified buildings such as museums should probably maintain a neutral pressure all year long.

Thermal lag extends the risk period for condensation

In theory, condensation will not occur in the building wall when the outdoor air temperature is *above* the indoor air dew point. If all materials were at steady-state equilibrium with air in that situation, no wall surface temperature should be cold enough to condense moisture from the indoor air. But the mass of the building stays cold for much longer than the outdoor air. Consider a warm morning after a cold and windy night. The building has been well chilled all night long, so the internal cavities stay cold even though the outdoor air temperature in mid-morning might be quite warm.

Therefore, some believe it would be wise to set the neutral pressure switchover temperature fairly high—perhaps at the building's thermal balance point temperature, which is likely to be well above the indoor dew point in winter.

Hot and humid climates

When the dew point outdoors is high, the building pressure in the outside wall must be positive. The goal is to have *dry air leaking out* of the building rather than having humid air sucked in.

So in tropical and sub-tropical regions where the outdoor dew point is nearly always very high, and where no cold weather chills the outside wall, the building should remain under positive air pressure with dry air all year long.

Mixed climates

The question of internal pressure is more complex in mixed climates that have extreme conditions during both summer and winter seasons—as seen in locations midway between the earth's equator and its poles. Then the designer will need to decide *at what point* to have the building switch from neutral to positive pressure. As in cold climates, the building research community has not yet provided the HVAC designer with clear, undisputed advice for the switchover point.

The authors suggest that any time the outdoor air dew point is above the desired indoor dew point, infiltration generates an extra

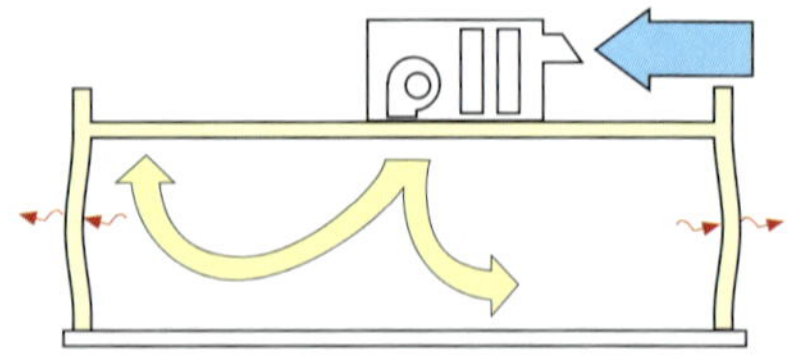

Humid Weather = Slight Positive Pressure
Ensures that dry indoor air rather than humid outdoor air will fill the building cavities, keeping them safe from condensation, mold and mildew.

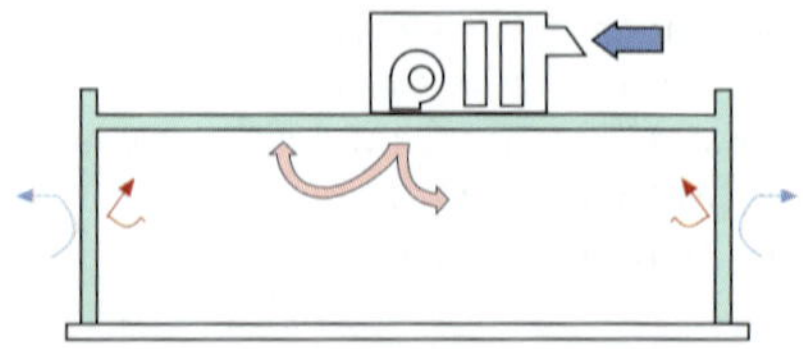

Cold Weather = Neutral Pressure
Try to keep BOTH cold outdoor air and humid indoor air OUT of the building's wall cavities.

Fig. 16.9
Suggestions For Building Pressure

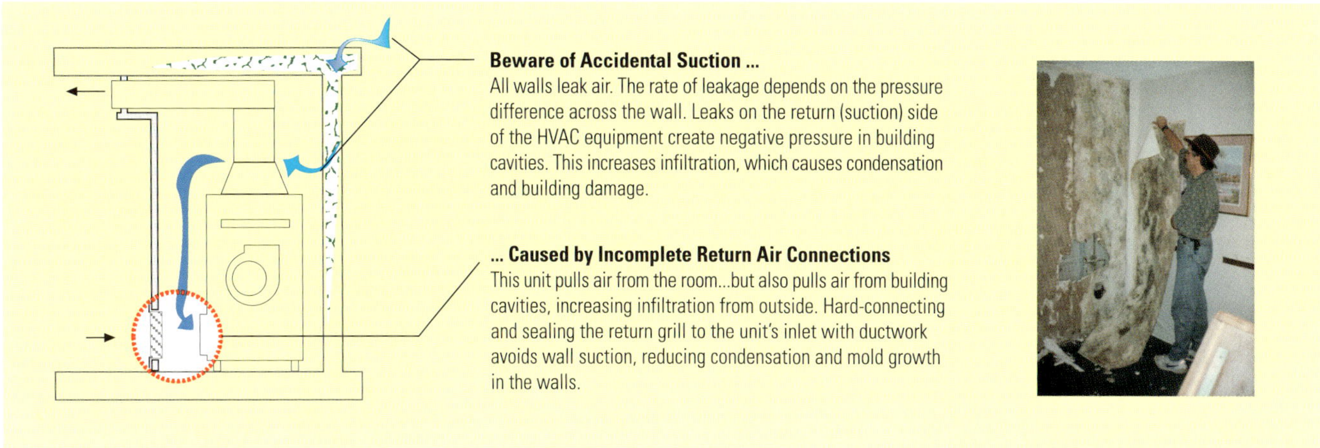

Fig. 16.10 Wall-mounted HVAC Units

If the return air inlet is not sealed to the interior finish, the unit can pull air from building cavities instead of just from the room. The fan suction will pull in humid outdoor air, supporting the fungus seen in the photo. (Lstiburek, 1999)

Fig. 16.11. Smoke Puffer

These disposable units help show the location of air leaks, and show the direction of air movement through doorways. Each bottle produces smoke on-demand for about 24 hours after its internal reaction is started.

and undesirable dehumidification load. So for buildings equipped with dew point control, the system could be set to switch from neutral to positive internal pressure *when the outdoor dew point rises above the indoor dew point limit*. Also at this time, the control system should stop using outdoor air for "free cooling." So the control system might already be set to signal a change in fan operation or damper positions when the outdoor dew point rises above the indoor moisture control level.

For buildings with temperature-controlled outdoor air economizers instead of dew point control, the internal pressure switching point could simply be the outdoor dry bulb temperature that equals the maximum indoor dew point temperature. In other words, the building switches from neutral to positive indoor pressure whenever the outdoor dew point *could possibly* rise above the indoor dew point limit. This is more conservative and will result in more hours of positive pressure than if the actual dew points indoor and outdoors were compared. Those extra hours generate more fan cost—but they also reduce the risk of damage. The extra fan cost could be avoided by installing a dew point control rather than using temperature control to modulate the outdoor air volume.

In the final analysis, the switchover point simply determines the number of *hours of risk* of condensation per year, and the probable *volume of condensation* when it occurs. If the system waits until the outdoor dew point climbs very high before switching to positive pressure, there will be more hours and higher volumes of potential internal condensation. If the switchover point is based on a lower outdoor dew point, there will be fewer hours of risk each year, and a lower total volume of condensate if a problem should occur. So the designer could save fan cost by setting the switchover point higher than the maximum internal dew point. But with that less conservative approach, the owner should ask the HVAC designer to be completely certain there will be no cold surfaces within the building structure. (See figures 16.4 and 16.10 , which show consequences of failing to realize that overcooled rooms produce condensation behind walls.)

Locating & Measuring Air Leaks

Designers and contractors can limit air leaks by ensuring tightly-sealed duct connections and exterior wall joints. But not all construction teams have been careful. Many building owners have structures that—for whatever reason—leak air. Here are a few suggestions for locating and quantifying air leakage.

Exterior walls

To locate leaks on the exterior walls, focus on the seams between different building components. Those are usually where the largest air gaps occur. Leaks can be located from inside or outside the building by using hand held "smoke" puffers such as those shown in figure 16.11.

When working inside the building, first make sure the building is under a negative air pressure. This can be accomplished by shutting off the make-up air while allowing the exhaust fans to continue operating. Then move the smoke puffer slowly along the edges of windows and the edges of any wall penetrations, such as those made to accommodate wall-mounted air conditioning units, and those around electrical outlets. An air leak will blow the smoke away from the wall rapidly. This test is simple and can be performed after working hours without any great disruption to building occupants. It locates the inside end of the worst air leaks. But to locate the outside end of the leak, two people will be needed.

Working outside the building, first put the building under a positive air pressure by increasing the amount of make-up air and reducing the exhaust air flow. Then, working in pairs an observer is stationed outside the building while the partner slowly moves a smoke puffer around window frames and other wall penetrations from inside the building. The outside partner notes any appearance of smoke leaving the building, which indicates a crack on the exterior.

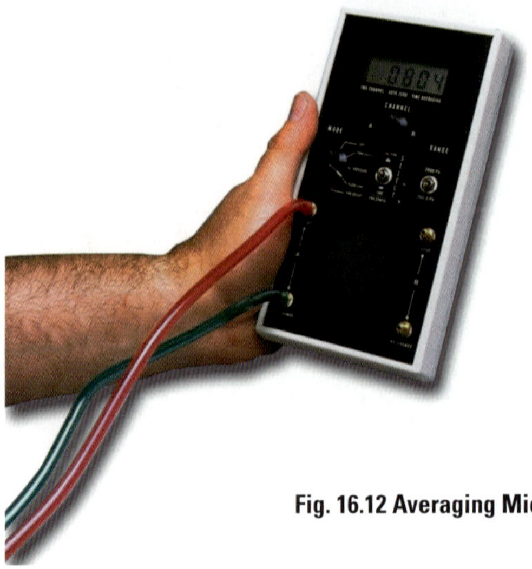

Fig. 16.12 Averaging Micromanometer

Quantifying the leakage through a single wall section is not usually done, because it demands an airtight seal that isolates that wall section. Usually, the leakage of the whole building is quantified as explained below.

When the HVAC systems are being tested and balanced, it is often useful to quantify the local pressure difference between the room and the building cavity, and between the cavity and the outdoor ambient. The goal is to have no pressure difference in cold climates, and a very slight positive internal air pressure in humid climates. These pressure differences are very small, and therefore they can be reversed easily through normal air movement and atmospheric pressure changes. So a digital micromanometer with a time-averaging feature is the most useful tool for this job. (figure 16.12)

The manometer must be very sensitive, because the pressures are so small that the process has been compared to "measuring the force of a butterfly's cough." Relevant pressure differences are likely to between 0.002 and 0.016 in.w.c. [between 2 and 5 Pa]

From the manometer, one tube is placed outside the building, and the other placed inside the wall cavity. The micromanometer must average many readings during each second or minute to obtain the average pressure difference over that short time span. The air handling system can be adjusted using a few such readings, and local pressure excursions that reflect local leakage can be identified through many such readings taken near external wall penetrations. (Odom et al, 1993, 95, 96, 97)

The same device can used for quantifying the pressure difference between the wall cavities and the conditioned space. Odom and co-workers suggest the use of a gasketed metal pie plate, taped to the inside surface of the wall so that it covers an electrical outlet. One of the manometer ports is open to the room. The other is connected through a tube to the inside of the pie plate covering the elec-

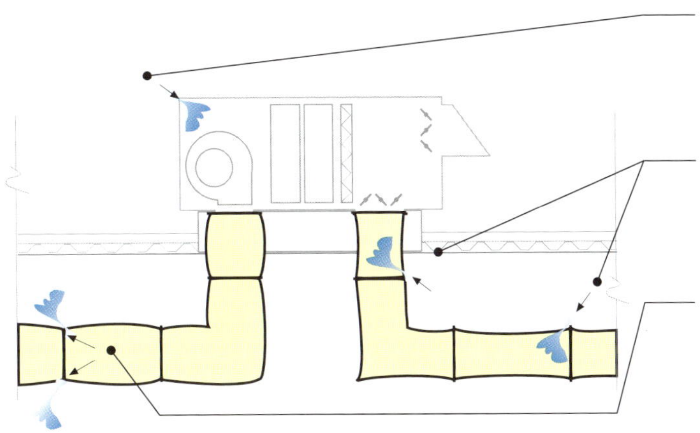

Fig. 16.13 Air Handler Connections
The potential for leakage is greatest where the pressure difference is large. That's why air handler doors and duct connections demand more attention and tighter seals than other parts of the system.

1. Eliminate Air Handler Leaks
Most commercial-grade air handlers leak a great deal of air. Seal open seams with tape and mastic, and make sure all access panels are gasketed, with all screws in place.

2. Seal Return Ducts
If return ductwork leaks, it creates low pressure in building cavities, which sucks untreated air into the wall from the outdoors. Sealing return duct work helps avoid condensation from untreated air in these cavities.

3. Seal Supply Ducts
When supply ductwork leaks in summer, cold air leaks into building cavities, cooling surfaces, which condense moisture. In winter, warm supply air carries moisture into cold building cavities, where it condenses once again. Sealing supply air ducts prevents these problems.

trical outlet. That second manometer port will sense the air pressure in the wall at the point immediately behind the electrical outlet.

Duct connections

The largest amount of leakage is likely to occur where the pressure difference is the greatest. That will be where the supply and return ducts meet the air handling unit, and at the doors and seams of the air handling unit itself. Using a hand held smoke generator, a technician can move slowly over the unit's seams and door jambs and over the supply and return duct connections to locate points where air leakage is especially large. The smoke will be pulled rapidly into, or pushed rapidly away from, a leaking seam.

Where the supply ducts can be accessed visually, leaking duct connections can be located by using a theatrical "fogger" to flood the system with a fine mist of visible particles. Where that fog escapes the ductwork on its way to the supply registers, the location of any leak is relatively easy to see.

However, in most installations, the supply and return ducts will be covered by external insulation, so that leaks in long duct runs will not be visible. On the other hand, long duct runs are often well sealed, or at least do not leak a great deal of air. The evidence one way or the other should be visually clear by looking at the insulation jacket when the system is in operation. If there is a great deal of air leakage under the insulation, the return air insulation will be sucked tighter onto the duct, and the supply air insulation will be puffed-up and pushed away from the duct. Leaking sections may be identified this way, assuming the insulation jacket is sealed itself.

With well-covered duct systems, the first places to inspect are all the connections between the ductwork and the surface of the interior finish. Then inspect the seams where the straight runs meet the transitions to and from the space. These seams are often leaky, and sealing them may eliminate the need to strip away insulation from difficult-to-reach sections of duct further into the system.

In most large buildings with complex duct systems, however, there are very few practical ways to locate leaks after the obvious leaks have been sealed at the air handlers and at the entry and exit points from the occupied spaces. Where hidden duct leakage remains large and spaces are not accessible, the owner may wish to simply seal the entire system internally. Work done at the US Department of Energy's Lawrence Berkeley Laboratory has shown that fogging technology can be used to seal up duct leaks from the inside, using a fine mist of suspended adhesive polymer particles. (Modera, 1996) As the adhesive mist escapes through any leaking connection, it quickly clogs that seam, sealing up the air leaks.

Fig. 16.14 Duct Leakage Test

To help the owner decide whether it makes any economic sense to invest money in sealing leaks in the duct system, technical services are available—often through test and balance contractors—to quantify the amount of air leakage in each part of the duct system. As shown in a residential situation by figure 16.14, a calibrated fan is attached to the duct system and all other inlets and outlets are sealed with temporary adhesive tape. The fan speed is increased until the pressure in the duct work is raised to about 0.1 in.w.c. [about 25 Pa]. The amount of air passing through the fan to maintain that pressure (and therefore the amount of air leaking out of the ducts) is measured by recording the pressure difference across a smooth orifice that forms the fan casing. (Nelson, 1998 and Sherman, 1997)

Whole buildings

A useful first step in tracking down and sealing up leaks is to determine whether the building as a whole is under positive or negative air pressure. When the doors are cracked open, does air leak inwards (negative internal pressure) or does air leak outwards (positive internal air pressure). This test is very fast and inexpensive. It can be done with a strip of lightweight facial or toilet tissue paper, held near the door jamb when the outside door is cracked open slightly. The technician could also use a theatrical smoke generator or a smaller, easily available source of visible particles such as a lit cigarette or incense stick. The direction of the air leakage should be most obvious in the early morning, when the outdoor air is relatively still and not yet windy.

Another crude but simple procedure for densely-occupied buildings to measure the average CO_2 concentration in the return ducts when the building is occupied, and then measure that same concentration at short intervals as the building is being vacated at the end of the day. The rate of decay in average CO_2 concentration is a rough indicator of the overall ventilation rate. (Measuring CO_2 concentration can also be used for more precise and detailed investigations, as described below.)

Accurately quantifying whole-building leakage is an expensive and complex task. The larger the building, the more difficult it will be to locate and quantify the air leakage. For smaller buildings, the whole-building blower door test gives the quickest and most economical results. For taller, larger or more complex buildings, tracer gas tests are usually the most practical alternative.

Fig. 16.15 Blower Door Test

Blower doors

As shown in figure 16.15, a blower door includes a fan mounted in a panel that fits tightly into a doorway. After fitting the panel and sealing it with gaskets, the technician starts the fan, which pushes air into the building. (Nelson, 1997). The fan speed is slowly raised until the pressure difference between indoors and outdoors reaches a defined limit. Standard test pressure differences are either 0.016 or 0.2 in.wc. [4 or 50 Pa]. When the test pressure is reached, the technician reads the air flow rate being produced by the fan from the blower door's instrument panel. The blower's air flow rate is the amount of air that is leaking through the building envelope at the pressure difference selected for the test. The procedure provides a useful indication of the magnitude of the leakage at a defined pressure difference.

The total leakage number does not tell the building owner exactly *where* the leakage is occurring. But when a blower door puts the building under a positive pressure, technicians can find many leaks by working in pairs and using indicating smoke. One technician working inside flows smoke around the window frames and other exterior wall penetrations. The other technician, working outside, can sometimes see the smoke leaking out through cracks in the exterior walls, which locates the larger, more obvious leaks.

The basic blower door technique has also been used to quantify leakage in tall buildings (Bahnfleth et al, 1999). Each floor is isolated by sealing up all openings other than the door to the fire stairs. That doorway is fitted with a blower, and the floor is put under positive pressure. While the procedure has been used successfully, it requires complex technique and great care to separate the air leakage through the exterior wall from the air leakage to different floors within the building. Bahnfleth concluded that tracer gas tests are usually more practical and less disruptive for quantifying and locating leaks in tall or complex buildings.

Tracer gas

Gas concentration inside a building is slowly diluted by outdoor air flowing into that building. In a tracer gas test, a known volume of an inert gas, usually sulfur hexafluoride, is released in the space and the time is noted. At periodic intervals, the technician pumps a sample of the indoor air into a vapor-tight plastic pillow and seals it securely. Later, in the laboratory, the sulfur hexafluoride concentration of each sample is measured and recorded. Since the volume of the building and the initial volume of gas is also known, the air infiltration rate can be calculated from the decline in the tracer gas concentration over the known time interval.

This technique can be used to show air exchange rates in many isolated parts of the building over the same test interval. So compared to a blower door, the tracer gas test makes it easy to quantify leakage in different parts of a building. Also, the test can be run with or without the HVAC systems in operation. Comparing the results of two tests, the technician can calculate how much outdoor air infiltrates into a 'passive" building, and how much outdoor air enters when the system is in operation.

On the other hand, the test takes time, and requires lab equipment. To address those shortcomings, equipment has been developed to monitor tracer gas concentrations in real time from many parts of a building (Bearg, 2000). The equipment consists of a gas analyzer equipped with a suction pump and tube/valve manifold. Each tube runs to a different part of the building, and the suction pump draws a small sample of air through all tubes continuously. Every few seconds, the valve manifold switches, flowing a different tube's air through the gas analyzer. That way, different tracer gas concentrations from all over the building can be recorded in a matter of a few minutes. With this arrangement, the measurements are taken so frequently that transient phenomena such as door openings and short meetings can be seen in the gas concentration records. Some variations of the gas analysis equipment are also equipped with CO_2 gas sensors and dew point sensors. That additional data provides useful information about the movement of air within the building throughout normal daily operation.

An example of a report from a multipoint gas analyzer is shown in figure 16.17. In this figure, the issue under investigation was ventilation effectiveness. The CO_2 concentration was recorded in different parts of the building over a typical 24-hour operating day.

The graph shows how room 107 is very well ventilated, since its gas concentration is nearly the same as outdoors. In contrast, rooms 178 and 112 are considerably under-ventilated compared to their occupancy—a fact that is not apparent when measuring the average CO_2 concentration in the return air, which seems quite adequately low. This same technology is used to find and quantify outdoor air leaks in different spaces when the air system is turned off and the building is unoccupied.

Fig. 16.16
Multi-point Tracer Gas Analyzer and Recorder

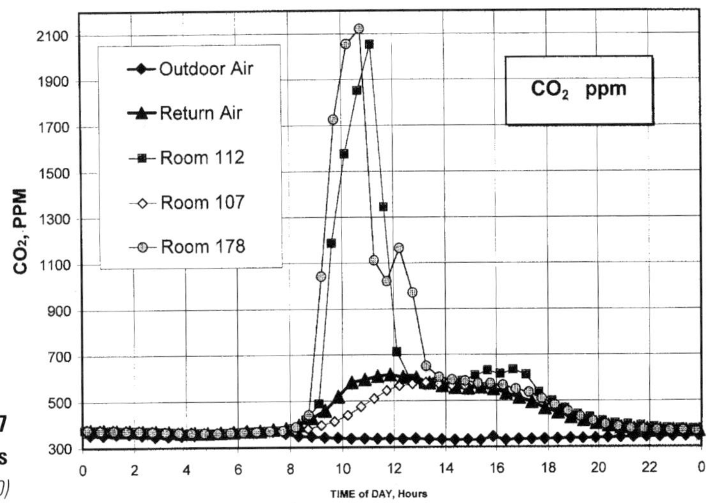

Fig. 16.17
Multi-point Tracer Gas Analysis
(Bearg, 2000)

Summary

Air leakage into the building envelope from both inside and from outdoors is one of the main reasons that moisture-related building problems occur. Although the HVAC designer cannot control water leakage, he or she can certainly help keep the building cavities free from condensation.

Duct connections must be sealed not only within the system but also where the duct work meets the air handlers and where it meets the interior finish. The designer and the contractors must pay special attention to sealing fan cabinets of wall-mounted AC units to the interior finish, and to sealing exhaust ductwork so that no suction is created inside building cavities that connect to the exterior wall. Finally, in cold weather the building pressure should be *neutral* with respect to the outdoors. But when outdoor weather is hot and humid, the building must be under a slight *positive* pressure.

With these minimal precautions, the HVAC designer can prevent a large proportion of the humidity-related problems uncovered by building scientists and forensic Engineers over the last few decades.

References

Air Infiltration & Ventilation Centre. ***AIRBASE CD-ROM***. 1999. Computer database of techniques and field research results regarding measuring and locating air infiltration into buildings. AIVC, University of Warwick Science Park, Unit 3A, Sovereign Court. Sir William Lyons Rd., Coventry CV4 7EZ, United Kingdom. www.aivc.org

American Hotel & Motel Association. *Mold & Mildew in Hotels and Motels* (The Survey) 1991. AH&MA, Washington, DC.

Bahnfleth, William P., Yuill, G.K., Lee, B., "Protocol for field testing of tall buildings to determine envelope air leakage rate." 1999. ASHRAE Transactions, V.105, pt 2.

Bearg, David W, *Indoor air quality and HVAC Systems* 1993, ISBN 0-87371-574-8, Life Energy Associates, 20 Darton St. Concord, MA 01742

Bearg, David W., "Monitoring for ventilation and air tightness." ASHRAE Transactions, V. 106, Pt.1, 2000.

Clarkin, Michael, and Brennan, Terry M., "Stack-driven moisture problems in a multi-family residential building." 1998. ASHRAE Transactions, V.104, Pt.2.

Cummings, James B, Withers, C.B, Moyer, N, Fairey, P., McKendry, B, *Uncontrolled air flow in non-residential buildings.* April 15th, 1996, Final Report of FSEC project number FSEC-CR-878-96. Florida Solar Energy Center, 1679 Clearlake Rd, Cocoa, FL 32922.

DeFrees, J.A., and Amberger, R.F., "Natural infiltration analysis of a residential high-rise building." Proceedings of the 1987 ASHRAE IAQ Conference, Arlington, VA, May 18-20. ISBN 0-910110-51-4. ASHRAE.

Handigord, Gustav. Chapter 18 New High Rise Commercial and Residential Construction, *Manual for Moisture Control in Buildings.* 1994, Am. Society for Testing & Materials, Philadelphia, PA.

Kudder, Robert J, Lies, K.M, Hoigard, K.R, "Construction Details Affecting Wall Condensation." 1986. Proceedings of the Symposium on Air Infiltration, Ventilation and Moisture Transfer. NIBS.

Lstiburek, Joseph, "Humidity Control in the Humid South." *1993.* Proceedings of the 2nd Conference on Bugs, Mold & Rot. National Institute of Building Sciences, (NIBS) Washington, DC.

Lstiburek, Joseph. "The Pressure Response of Buildings." 1999. Proceedings of the 7th Conference on Thermal Envelopes, pp:799-817. ASHRAE

Modera, Mark. *Repairing your duct work*. 1996. Aeroseal, Inc. 75 Fairview Ave. Oakland, CA 94610 (510) 601-8575 http://www.aeroseal.com/repair.htm

Nelson, Gary, Nevitt, R, Tooley, J., Moyer, N., "Measured Duct Leakage, Mechanical System Induced Pressures and Infiltration in Eight Randomly Selected New Minnesota Houses." 1993. Proceedings of the 1993 Energy Efficient Buildings Association Conference. www.energyconservatory.com/articles

Nelson, Gary. *Air leakage testing of buildings.* 1997. The Energy Conservatory, 2801 21st Ave. South, Suite 160 Minneapolis, MN 55407 www.energyconservatory.com/airtight.html

Nelson, Gary. *Duct leakage testing.* 1998. The Energy Conservatory, 2801 21st Ave. South, Suite 160 Minneapolis, MN 55407 (612) 827-1117 http://www.energyconservatory.com/duct.html

Odom, J, David, III, and DuBose, George, and Fairey, P.W., "Why HVAC commissioning procedures do not work in humid climates." 1993. ASHRAE Journal, December, 1993. pp. 25-36.

Odom, J, David, III, and DuBose, George, *Summary Report, Indoor air quality evaluation of the Palm Beach County judicial complex*. 1995. CH2M/Hill Publications, 225 East Robinson St, Suite 405. Orlando, FL 32801 (407) 423-0030.

Odom, J, David, III, and DuBose, George, *Preventing indoor air quality problems in hot, humid climates: Problem avoidance guidelines.* 1996. CH2M/Hill Publications, 225 East Robinson St, Suite 405. Orlando, FL 32801 (407) 423-0030.

Persily, Andrew K., "Myths about building envelopes." 1999. ASHRAE Journal, March, 1999, pp: 39-47. Using measured results of field tests, the article demonstrated that air leakage rates have no correlation with the date of the building's construction. Buildings constructed in the 1980's and 1990's are not "tighter" than those built in the 1950's, 60's and 70's.

Sherman, Max H. *Air Change Rate and Air Tightness in Buildings, ASTM STP 1067*, 1989 (ISBN 0-8031-1451-6).

Sherman, Max. *The use of blower door data.* 1997. Lawrence Berkeley Laboratory, Energy Performance of Buildings Group, Berkely, CA. PDF document published at: http://epb1.lbl.gov/blowerdoor/

Treschel, Heinz , Lagus, Peter. *Measured Air leakage of Buildings.* 1986. American Society of Mechanical Engineers (ASTM STP 904, 1916 Race St, Philadelphia, PA 19103 www.astm.org. ISBN 0-8031-0469-3) This book contains chapters with measurements from different types of buildings, including aircraft hangers, office buildings and residential buildings.

Tsongas, George and Nelson, Gary. "A Field test for correlation of air leakage and high moisture content sites in tightly-built walls." 1991. ASHRAE Transaction, V. 97, pt. 1

Donated Photos & Diagrams

Figure 16.1 - © Michael Easley and the HAC&R NEWS, BNP Publishing.

Figure 16.4 photos - © Dr. Joseph Lstiburek

Figure 16.5 - © David Bearg

Figure 16.6 photo - © Christopher Makepeace and the Alberta Infrastructure Technical Resources and Standards Branch.

Figure 16.10 photo - © Dr. Joseph Lstiburek

Figures 16.12, 14 and 16 - © The Energy Conservatory

Figures 16.16 and 17 - © David Bearg and AirExpert Systems.

Special Thanks

Managing building air pressures is one of the more controversial and under-defined elements of building science. This Author approaches the subject with humility, and is deeply grateful for the information and assistance supplied by more experienced field investigators and design practitioners. If there are errors they are mine. If the chapter is useful, it is because of their extensive efforts and excellent publications. Thanks especially to Joseph Lstiburek of Building Science Corp, Andrew Persily of the National Institute of Standards and Technology, James Cummings of the Florida Solar Energy Center, Gary Nelson of The Energy Conservatory, David Bearg of Life-Energy Associates and both J. David Odom III and George DuBose of CH2M/Hill.

Chapter 17

Humidity Sensors & Controls

By Lew Harriman

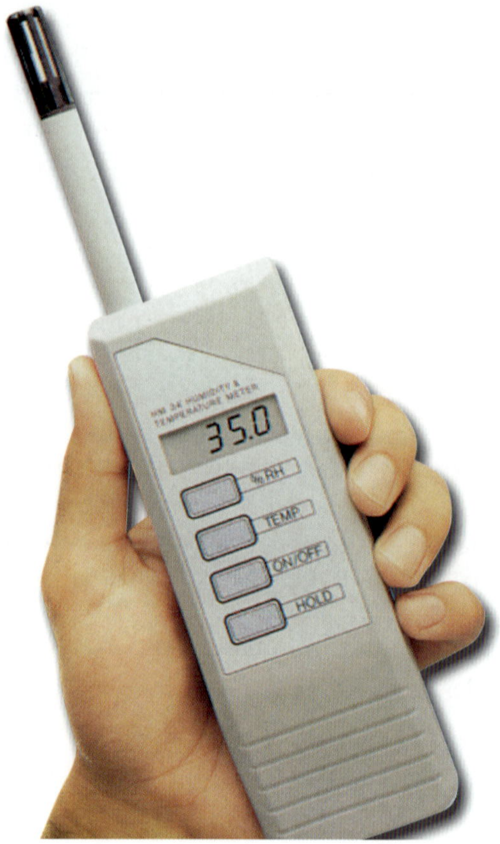

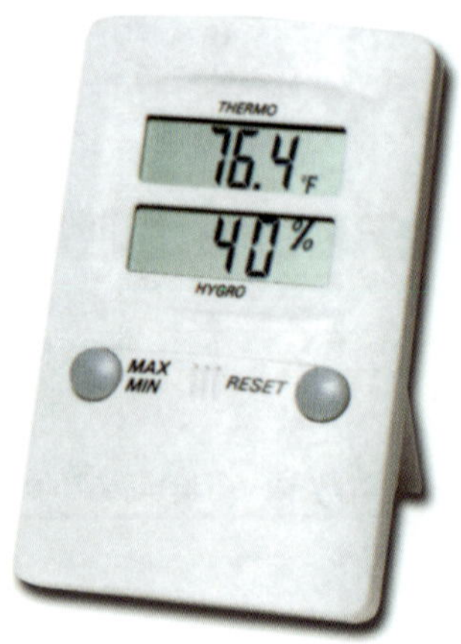

Fig. 17.1 Humidity Sensors Often Disagree

The low-cost sensor at right is digital, but is not as accurate as the higher-cost device at left, which can be calibrated. Humidity is difficult to measure with precision. Regular calibration is essential to successful humidity control.

Measuring Humidity

In commercial buildings, measuring temperature is simple and inexpensive. In contrast, measuring humidity is more challenging, and costs a bit more.

The principal challenge is to make sure the humidity sensor's signal really represents conditions in the controlled space. Two important factors influence the usefulness of any sensor signal: its calibration and its location. Humidity control problems are usually caused by failing to adjust the controls to the sensor's calibration after installation, or because the sensor is mounted in a poor location—one which does not represent the moisture level in the space.

Successful control requires more than just a sensor

Low-cost sensors are seldom calibrated by the manufacturer. Their calibration must be quantified after installation, and the controls adjusted accordingly during commissioning. Also, the system will tend to keep humidity most constant at the sensor location. For example, if the sensor is located at the discharge of the supply air duct, the humidity at that point will be fairly constant. But when the loads in the space vary, as in a building with intermittent occupancy, a constant supply air moisture level will not create a stable humidity in the space. A sensor location inside the controlled space would provide more consistent results.

The cost of sensors depends on their tolerance and their response time. Tighter tolerances and faster response times cost more. A sensor accurate within ±2% rh often costs 10 times more than one which measures with a tolerance of ±5% rh. Further tightening of tolerances increases costs yet again. Measuring humidity within ±1% rh will probably cost at least 50% more than measuring within ±2% rh. Similarly, a sensor with a 30-second response time is usually more expensive than one which needs 5 minutes to respond to the same change. The designer should expect low-cost sensors to have wide tolerances and slow response times. More costly sensors can be expected to have tighter tolerances and faster response times.

Additionally, the range of variation in moisture and temperature at the sensor location can strongly influence the cost and accuracy of the measurement, and it will suggest the best type of sensor for the job. For example, a sensor located at the intake of the building's ventilation air will see daily or hourly fluctuations through a very wide range of temperatures and relative humidities—from 10% rh to 100% rh and between temperatures of -40°F to 110°F [-40°C to 43.3°C]. That sensor will have condensation on its surface frequently, along with contaminants that will shift its calibration and lead to unreliable signals. The outdoor air intake location calls for a rugged sensor. Even the expensive, high-quality sensors should not be expected to be more accurate than ±3% rh between calibrations in this demanding location. In contrast, if a sensor is located in a controlled space, the range of variation will be quite narrow—a nearly constant temperature with humidity staying between 30 and 65% rh. In that forgiving location, a low-cost, slow-responding, ±5% rh sensor can be made much more accurate by adjusting the controller to the sensor's calibration. In stable conditions and within a narrow range of humidities, a low-cost sensor may achieve a ±2% rh tolerance at a fraction of the cost of a more sophisticated sensor.

Finally, the tolerance of the sensor is only a small component of the tolerance of the entire system. The tolerances of each component are combined to give the tolerance of the system as a whole. So from a practical perspective, investing in high-cost sensors alone will not achieve tight tolerances in the space. Other components in-

Fig. 17.2 Location & Calibration

No matter how costly or inexpensive the sensor, its performance depends on its location and its calibration.

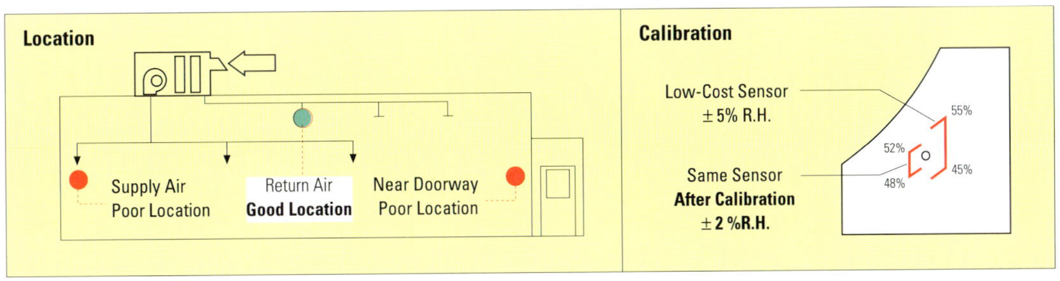

clude the lag-time inherent in the sensor's response, the tolerance of the electronics, the tolerance of the controls within the dehumidifiers and humidifiers that respond to signals from the control sensor, the tolerances of the cooling and heating components and their control sensors, and most importantly the time lag of all these components acting as a system while temperature and moisture loads change.

In summary, *sensor* tolerances are much tighter than *system* tolerances, and the system will tend to control humidity most closely *at the sensor location*. The designer should remain aware of these facts so he can guide discussions with the owner and the architectural designer. Such awareness helps keep the client's expectations about humidity control tolerances in line with his or her budget.

Which Variable To Control—Dew Point or RH?

The designer must consult with the owner to decide whether to control the building to a constant rh or to a constant dew point. Defining the owner's purpose is the foundation of that important decision (See Chapter 2). In commercial buildings, owners have usually been more concerned about relative humidity because they are most familiar with that term, and because most low-cost sensors report and display that variable. But in most cases, the dew point is more important to achieving the owner's purpose.

For example, assuring human comfort and health is the most common reason for controlling humidity in schools, office buildings, retail stores, restaurants and residential buildings. Most owners will measure the success of the project by how steady the relative humidity is maintained. But in fact, human comfort is a function of the difference in water vapor pressure between the skin surface and the air. Vapor pressure is determined by the air's dew point, not the relative humidity. So when comfort is the criterion, the designer may wish to control the dew point even if the relative humidity is also displayed and reported for the convenience of showing the occupants a familiar variable. This usually saves money, and provides a better result.

The problem is that relative humidity is easy and cheap to measure, but difficult to make uniform throughout a building. This is because the temperature will vary considerably throughout that building, but the absolute amount of moisture in the air tends to equalize by vapor diffusion.

For example, in a large, open retail store with high ceilings, it may be quite common to have a 10°F [5°C] temperature difference between the floor and ceiling during hot weather. But the dew point is not likely to vary by more than 1°F [0.5°C]. This means the relative humidity will be higher at the cold floor level than at the hot ceiling level. Eliminating this difference in relative humidity requires mixing the air more uniformly, circulating more air at higher velocity to reduce the temperature difference. But since the dew point will not change, comfort is not going to be improved even after that costly increase in air flow. Also, with more uniform indoor temperatures, cooling costs will rise. In this example, the higher temperature at the ceiling had been limiting the heat transfer from the hot roof deck. When the air temperature at the ceiling level is cooler, more heat moves through the roof into the building, increasing cooling costs.

So in most commercial buildings where human comfort is the principal concern, controlling the dew point has more advantages than controlling the relative humidity.

On the other hand, most low-cost sensors measure relative humidity. To control the dew point, the rh signal must be combined with a temperature signal in the control system. In recent years, even the lowest-cost building automation systems are able to handle this task, allowing the designer to measure and control dew point for the very modest additional cost of adjusting the software rather than the much higher cost of sensors that measure the dew point directly. Since the building automation software should be adjusted to reflect the actual calibration of the relative humidity sensor, obtaining a dew point signal for control adds nearly no cost.

Although less familiar until recently in buildings like offices and schools, this technique has been used extensively in supermarkets.

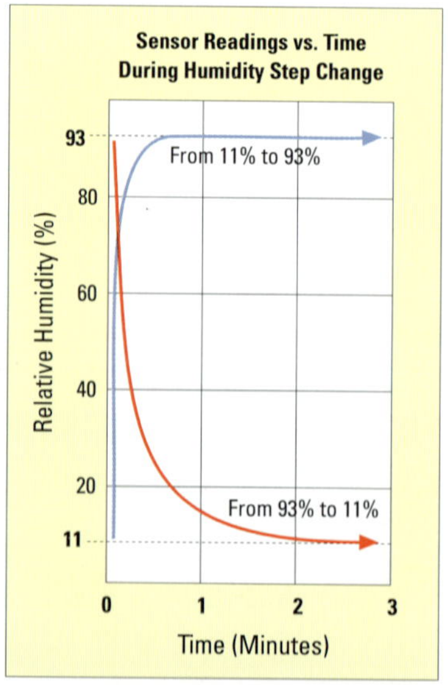

Fig. 17.3 Response

In general, faster response costs more money. In all cases, a step change from high to low humidity takes longer to register—in this sensor, about 1.5 minutes longer than a step change from dry to wet.

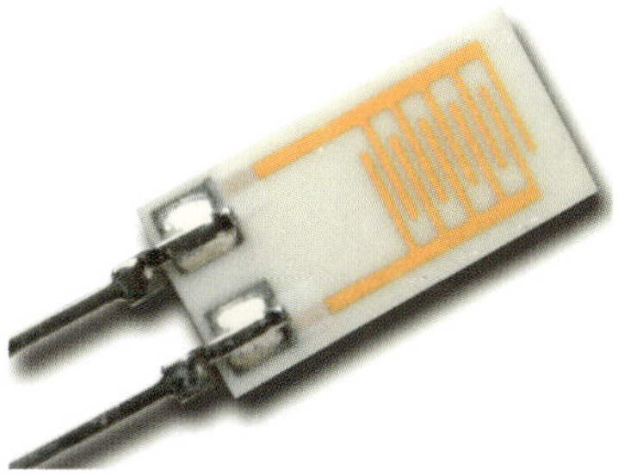

Fig. 17.4 Resistance RH Sensor

These sensors are durable and tolerant of periodic condensation, making them especially useful for outdoor air intake locations as well as for more forgiving indoor environments.

In those buildings the dew point is the principal concern because it governs the amount of condensation on display cases and on frozen product. Supermarket control systems usually measure relative humidity, and then combine that value with the dry bulb temperature at the same location. Software converts these signals to the dew point value, which is used to control the display case door heaters and dehumidifiers.

Understanding how humidity sensors operate helps the designer fit the instrument to each application with minimal cost and maximum reliability.

Relative Humidity Sensors

When the relative humidity is high, all materials absorb moisture. They release that moisture when the surrounding relative humidity declines. Also, water molecules are excellent electrical conductors. When materials absorb water vapor, they conduct electrical current more readily. The changing electrical properties of hygroscopic materials are the basis of most relative humidity sensors used in commercial buildings.

Bulk resistance sensors

In resistance sensors, a small electrical current flows from metal electrodes through a hygroscopic polymer as shown in figure 17.4. The resistance to that flow is measured by comparing the voltage drop through the polymer to the voltage drop through a precision resistor. The difference in voltages is directly proportional to the resistance through the polymer, which in turn reflects the amount of water vapor absorbed by the material. The voltage transmitted across the polymer can be correlated to the relative humidity.

The response shown in figure 17.5 shows a large and nearly linear correlation between relative humidity and voltage drop, which indicates that the sensor will provide an excellent representation of the relative humidity between 20 and 90%. However, below 20% rh and above 90%, the voltage change is smaller as well as less linear, which means the sensor will have more difficulty representing humidity accurately. In commercial buildings, these extremes are seldom important. The advantages of bulk resistance sensors are very strong compared to other types of sensors in similar price ranges:

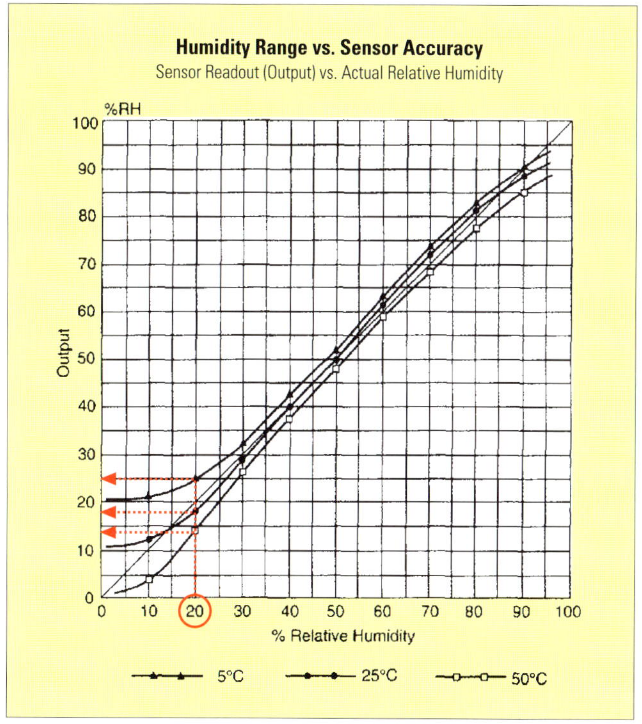

Fig. 17.5 Typical RH Sensor Readout vs. True Humidity

Accuracy of typical rh sensors is less consistent below 20% rh because at those levels the output signal varies more widely with temperature. (Wiederhold, 1997)

- Accuracy is even better than required in most commercial buildings (usually ±2% rh) and they are inexpensive compared to more accurate types.
- Very little hysteresis. Nearly identical values for the same relative humidity, regardless of whether the sensor has been absorbing or desorbing water vapor.
- Tolerant of periodic condensation without a major drift in calibration over years of use.

Given these advantages, bulk resistance sensors are frequently used for measurements at the outdoor air intake. In that location, a wide variety of contaminants combine with frequent condensation to produce unreliable readings in less-durable sensors. But although bulk-resistance sensors perform well in that and many other situations, they still have limitations, which include:

- Response time is slower than capacitance-type sensors. Absorbing and releasing water vapor from the polymer takes minutes, not seconds. Although response is often less than two minutes, response slows when the relative humidity change is small, or when there is liquid water on the polymer surface, or when the air flow across the sensor is very slow and humidity is falling slowly. In those situations, response time can be as long as 15 minutes.
- More expensive than lower-accuracy capacitance-type sensors. Capacitance-type sensors described in the next section are usually more costly, but they can also be quite inexpensive if tolerances are wide. Bulk resistance sensors may be more accurate and sometimes more expensive than necessary for locations where humidity always stays between 30 and 60% and where neither corrosion nor condensation are present, or where tolerances can be wider than ± 3% rh
- Calibration still drifts over time in harsh environments. When outdoor air contaminants combine with frequent condensation, resistance characteristics will eventually change as the polymer absorbs conductive ions into its bulk that are not released by drying. The sensor must be recalibrated periodically when mounted in the highly corrosive air typical of dense urban areas, oceanfront structures, indoor swimming pools or near major highways. (This limitation is only relative to the designer's expectations of accuracy. All other sensors are much less tolerant of such harsh environments.)

Capacitance relative humidity sensors

Many manufacturers use sensors containing variable capacitors to measure the relative humidity in the air. Such humidity-sensitive capacitors are used in very precise and costly sensors, but are also used in low-grade sensors that cost 1/50th of the cost of more accurate devices. Since the technology covers the full range of costs and tolerances, the capacitance measurement method—by itself—does not imply any particular level of accuracy or cost.

A capacitor stores an electrical charge. The amount of that charge depends on several factors, one of which is the dielectric constant of a thin film that separates flat foil electrodes in the capacitor. A dry film has a high dielectric constant, so the capacitor can store a larger charge. A film that has adsorbed water molecules is more conductive—it has a lower dielectric constant. That prevents the capacitor from storing a large charge because part of the charge leaks away through the dielectric film to the grounded electrode.

As shown in figure 17.7 capacitance humidity sensors consist of a supporting substrate, a lower electrode, a thin dielectric film and a porous upper electrode. At high relative humidities, water vapor passes through the porous upper electrode and is adsorbed into the dielectric film, reducing it's dielectric constant. An electronic circuit continually measures the impedance of that capacitor (how large a charge it can store, given a constant voltage and current difference between the electrodes). Low impedance means the film has absorbed more water vapor, which happens when the surrounding humidity is high. Another circuit converts the impedance value to either a voltage or a milliampere output. The greater the output value, the higher the relative humidity surrounding the sensor.

In more costly capacitance sensors, the upper electrode is made very uniformly, and with a large number of very small pores. Also, the dielectric film has a more exact thickness and uniform chemical composition. Consequently, the expensive capacitors respond to a humidity change very quickly. Also, any given sensor is likely to have nearly identical characteristics to other sensors of the same design.

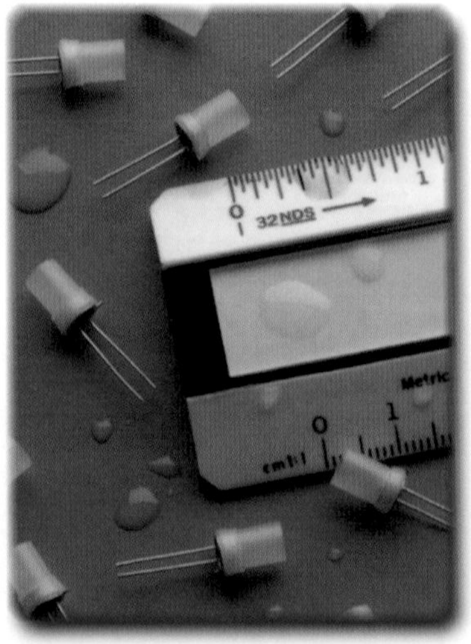

Fig. 17.6 Capacitance RH Sensors

Instruments that use the capacitance principle are used in both high- and low-end devices. In high-cost instruments, capacitance sensors have a reputation for very fast response and solid performance between 85 and 5% rh For instruments that cost 10 times less, expect a much slower response (minutes instead of seconds) and a more narrow range for humidity measurements, as well as a looser tolerance within the measurement range.

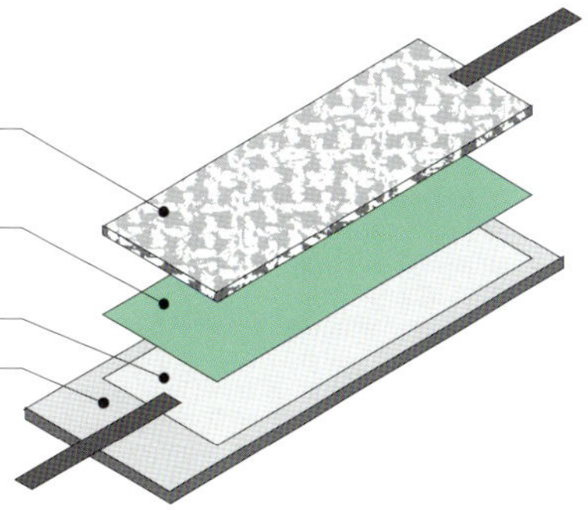

Fig. 17.7
Capacitance Sensor Construction
The quality of the measurement depends on the uniformity of the components. The speed of response depends on the pososity of the upper electrode.

In contrast, inexpensive capacitance sensors have less uniform upper electrodes with fewer pores, and the composition and thickness of the dielectric may not be consistent between sensors. Batches of these lower-cost sensors are not likely to give a consistent value for humidity, and will respond very slowly to a humidity change. For example, the author performed an informal comparison of low-cost humidity sensors (such as that shown in figure 17.1) from the inventory of a local electronics store. Relative humidity values reported by 10 units arranged on the retail counter top ranged from 16 to 29% rh, when a newly-calibrated, higher-accuracy device indicated the relative humidity around the sensors was 21%. The less-consistent instruments cost about $30 each. The higher accuracy, calibrated device costs about $500. Both types use the capacitance principle for measuring humidity.

Advantages of capacitance sensors are usually associated with the *higher-cost* instruments:

- Broad range of hardware choices to fit any budget and any tolerance range.
- Sensors respond very quickly to a humidity change—on the order of a few seconds.
- Sensors are very repeatable over time and quite rugged, as long as they do not encounter condensation.
- Sensors can be used at levels down to 5% rh.

The major limitations of the technology are associated with the *lower-cost* devices:

- Digital output implies sensor accuracy that is optimistic at best, and grossly misleading at worst.
- Sensor calibration can drift substantially over time as the less durable upper electrode weathers with condensation and contamination. Some sensors, when exposed to relative humidity above 85%, simply never recover their original characteristics.
- Sensor probes from the same manufacturer which may appear physically identical provide greatly different measurements in the same air stream.
- Response time is very slow (several minutes), especially after condensation on the upper electrode or where air is not flowing across the sensor at high velocities.

Dew Point Sensors - Chilled-Mirror Principle

Chilled-mirror hygrometers use the condensation principle to measure the moisture content of air. The technology has a reputation for high accuracy but also high costs.

In these devices, a sample of air flows through a small sensor chamber equipped with a light source, a photocell and a chilled mirror. Light reflects off the mirror towards the photocell. When condensation forms on the mirror, the light is scattered rather than reflected directly. The photocell records that reduction in the light level, which indicates the presence of condensation. At the moment of condensation, a temperature sensor records the surface temperature of the mirror. That surface temperature is the dew point of the air flowing across the mirror.

The surface of the mirror is chilled continuously by an array of semiconductors known as Peltier junctions, which form a thermoelectric cooler that can be controlled by varying its electrical current. A control circuit modulates the current passing through the

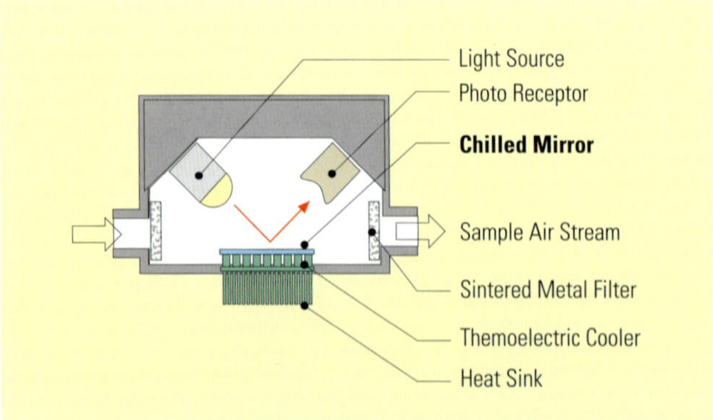

Fig. 17.8 Chilled Mirror Dew Point Sensor

These costly, but very accurate devices are often used to calibrate lower-cost and less-accurate sensors that are more common in commercial buildings.

semiconductors, keeping the temperature of the mirror constant at the dew point of the air. Chilled-mirror sensors are not as widely used in commercial buildings as relative humidity sensors, mostly because of their cost. But they are highly accurate, and are often used to calibrate lower-cost devices. Advantages include:

- Measures the dew point directly. Controls can be set based on the instrument's output signal without the need to calculate the dew point based on temperature and rh.
- Accurate. Typical dew point tolerance is ± 0.4°F [0.2°C].
- Wide range. Even with a single stage of mirror cooling, the close tolerance measurement can easily be maintained between temperatures of 0 and 100°F with coincident dew points between -20 and 80°F [18 to 27°C dry bulb with dew points between -28 and 27°C].

Limitations of chilled-mirror sensors include:

- Cost. The lowest-cost versions are about twice as expensive as close-tolerance rh sensors, and the broader-range chilled-mirror devices cost about 5 times more than the lowest-cost versions.
- Contamination. The mirror surface must be kept clean and free of hygroscopic dust that creates condensation at a temperature *higher* than the true dew point temperature of the air. The air sample must be filtered, and the filter must be replaced regularly when particulate loading is especially heavy.

Aluminum oxide dew point sensor

Aluminum oxide sensors use the capacitance principal to measure the moisture content of air. Aluminum oxide is extremely hygroscopic. It pulls water molecules into an air gap between two capacitor plates. The air acts as the dielectric layer in the capacitor. Because of the sorption characteristics of the oxide, the capacitance of such sensors varies nearly linearly with the dew point rather than with the relative humidity, as in other capacitance-based sensors.

This technology is used most frequently in dry industrial process air or in pressurized gas or liquid streams, where dew points below -40° must be measured reliably, and where the temperature may be well above ambient levels. It is seldom used in commercial buildings, primarily because of high cost, and the fact that few hardware variations are available compared to relative humidity sensors. But like the chilled-mirror devices, aluminum oxide instruments are used to check calibration of other, less accurate sensors. Advantages of aluminum oxide sensors include:

- Measures the dew point directly. Controls can be set based on the instrument's output signal without the need to calculate the dew point based on temperature and rh
- Measures very low dew points very accurately (±0.5°F or ±0.3°C), even when air temperature is over 100°F [38°C].

Their limitations are mostly commercial:

- Cost. Aluminum oxide instruments cost more than double the cost of a high-accuracy relative humidity sensor.

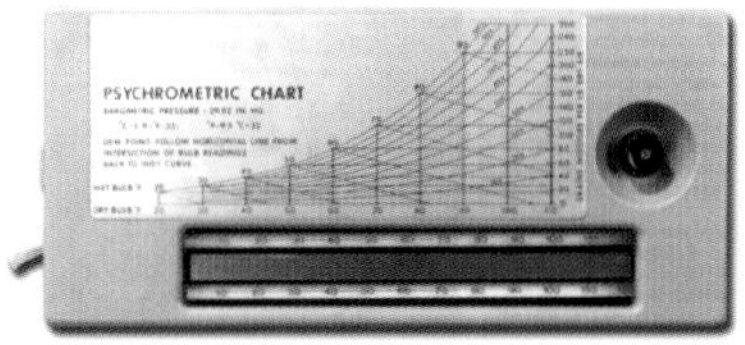

Fig. 17.9 Aspirated Psychrometer

A psychrometer uses the fundamental principle of wet bulb and dry bulb measurements, along with a psychrometric chart to determine the moisture content of air.

With care, the aspirated-type hygrometers such as this one can be used to cross-check readings from low-cost, low-accuracy humidity instruments such as mechanical hygrothermographs used to informally record humidity in museums. However, higher-accuracy devices should be used for checking calibration when humidity levels must be known within ±2% rh.

- Hardware configurations are limited. Most aluminum oxide sensors are designed for rugged industrial environments or pressurized gases and liquids. There are fewer designs optimized for mounting in commercial duct work and in rooms.
- Sensitivity to condensation. The oxide layer slowly changes composition over time, but more quickly when the sensor surface is soaked by condensation. Two problems occur. The sensor sorption characteristics will change, requiring recalibration. Also the oxide grips liquid water so firmly that it may take hours for the sensor to dry out after condensation, sending false signals in the meantime. The device is most reliable and responsive where relative humidity will not exceed 80%. Locate them in rooms or in return air ducts, not in outdoor air streams and not downstream of either cooling coils or humidifiers.

Psychrometers

A psychrometer measures humidity by taking both a wet bulb and a dry bulb temperature reading. With those two values known, the other properties of the air—including its moisture content—can be determined by computation or by reading a psychrometric chart. In commercial buildings, psychrometers are seldom if ever used for control, but they are still occasionally used to check the calibration of humidistats or relative humidity sensors.

Sling psychrometers are a very poor choice for that purpose, because the wet bulb measurement is so difficult to make accurately, in spite of its apparent —and therefore misleading—simplicity. The error is very large unless the operator is quite experienced.

Slow air velocity, inadequate water coverage of the wick, radiational heating of the wet bulb and contamination of the wet wick are compounded by the difficulty of being sure the wet bulb reading is at its minimum while the thermometer is swinging. These problems mostly come from incomplete evaporation from the wet bulb wick, so the measurement error is always ***above***, rather than below the true wet bulb reading. In other words, poor measurements from sling psychrometers will always ***overestimate*** the true moisture content. (The only exception occurs when cold water rather than ambient-temperature water wets the wick. In that case it is possible to underestimate the true humidity level with any psychrometer.)

Aspirated psychrometers with clean wet wicks using distilled water are more reliable than sling-type units. Shown in figure 17.9, an aspirated psychrometer combines low cost with the comfortably fundamental measurement principle of wet and dry bulb readings. For typical humidity ranges of commercial buildings (30 to 60% rh at 68 to 75°F [20 to 24°C]) aspirated psychrometers provide a low-cost way to check readings from low-accuracy sensors.

In an aspirated psychrometer the wet and dry bulb thermometers are mounted inside a plastic case which contains a battery-powered fan. The fan draws air across both dry and wet thermometers at a constant, high velocity to provide uniform evaporation. The case prevents radiation from influencing the temperature of the thermometer bulbs. The wick must be changed regularly ***with gloved hands*** to prevent skin oils and air stream particulate from affecting evaporation, and only ambient-temperature distilled water can be used to wet the wick. Further, the wick must remain completely wetted until the wet bulb temperature has stopped dropping. As long as all those precautions are followed, aspirated psychrometers can be useful to cross-check readings from low-accuracy sensors. The advantages of aspirated psychrometers include:

- Low cost. These devices can be purchased for about half the cost of a similarly accurate electronic device.
- Reasonable accuracy in indoor environments. A tolerance of +5% of the wet bulb reading can be achieved in careful operation in middle and upper-range humidity levels.
- Portable. The instrument can be brought to a room sensor location easily.

The limitations of wet bulb readings must also remain clear:

- Requires a psychrometric chart. To obtain humidity values, the operator must carefully plot the point and read values on an accurate psychrometric chart. Plotting and reading introduce two major sources of error. Poor results from aspirated psychrometers usually come from incautious plotting and reading of the psychrometric chart after the wet bulb and dry bulb readings are obtained.
- Difficult to use in ducts. The device must draw air only from the duct and not from the air outside that duct. It's difficult to avoid air mixing when opening an access door, and difficult to read the results inside a dark duct.
- Difficult to use in low-relative-humidity air. Wet bulb temperature readings below the freezing point of water are difficult to obtain because it takes a long time to cool the wick low enough to freeze the water, and a long time to stabilize the temperature after an ice layer has been formed. These precautions are seldom taken outside of a carefully-controlled laboratory test rig. That means psychrometers are seldom useful in low-humidity air streams where sub-freezing wet bulb temperatures are common.
- Subject to error in reading the thermometers. For accurate results, the operator cannot neglect to define what fraction of a degree the thermometer is sensing. Reading fractions of a degree from small thermometers requires care, good light and good eyesight.
- Subject to errors of contamination. In the day-to-day reality of building operations, the wet bulb wick is not always kept clean of particulate, and is often wetted with mineral-laden water or handled by bare skin which adds oils. All of these raise the wet bulb reading, increasing the measurement error so the operator overestimates the true humidity.

In summary, there are a wide variety of humidity sensors available. But in the relatively forgiving, mid-range humidity control levels typical of commercial buildings, the type of sensor is much less important than its location, and its regular calibration.

Wiring & Calibration

Humidity sensors typically transmit a variable, low-power current signal to a controller or building automation system that modulates the capacity of the humidifier or dehumidifier. The variable signal, usually 4 to 20 milliamperes, represents 0 to 100% rh It is carried by a "two-wire loop" between the sensor and the controller. The first wire carries 24 volt power from a power supply in the controller to the sensor, and the second wire returns the milliampere signal from the sensor back to that controller. This loop of two wires is therefore carrying both the power and the signal between the sensor and controller. This makes the on-site wiring much simpler and less costly than if power and signal were carried by separate pairs of wires. The arrangement also allows signals to be carried for long distances without degradation. If necessary, the sensor can be separated from the controller by several thousand feet [over 1,000 meters].

With that typical configuration in mind, the designer must consider how the sensor will be calibrated. This thought process is often overlooked at the specification stage, which results in the contractor purchasing sensors that cannot, as a practical matter, be calibrated at all. Lack of calibration guarantees that the system will never respond as the designer intends and as the owner requires.

Calibration

Calibration should occur at least annually. But frequent calibration will be needed if the sensor encounters condensation regularly, or if it is likely to encounter temperature swings towards the limits of its specified range, or if the air carries a more-than-usual amount of contamination. Such conditions are typical for outdoor air sensors and for sensors mounted downstream of heating or cooling coils. In

Fig. 17.10 Calibration

This technician is examining the sensor to see if its signal can be read out and the calibration adjusted locally... or if he must make the long trek back to the building automation system to enter a correction factor in software. The designer's sensor selection will allow easy calibration, or make it a practical impossibility.

Fig. 17.11 Portable Calibrator

These devices allow a technician to read out the sensor's signal while also viewing a more accurate signal from the calibrator. The sensor in this photo allows adjustment at the hardware, so its signal can be shifted to match the signal from the calibrator.

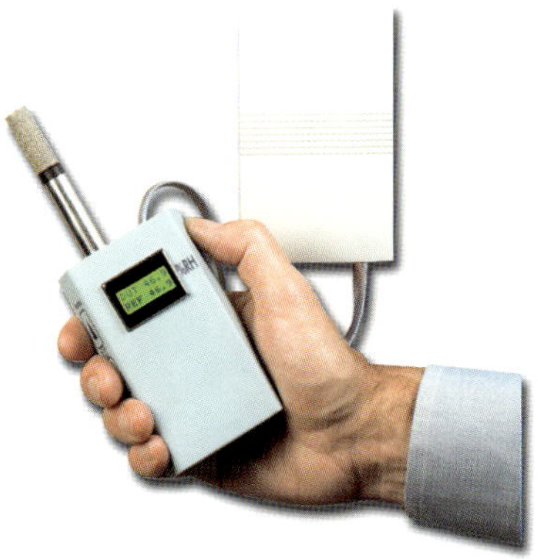

these locations, sensors need more frequent calibration. But in most buildings, sensors mounted in the room or in the return air ducts will do an adequately-reliable job with only annual calibration.

Calibration requires three elements:

1. Knowing the *true* value of the local humidity while the sensor is operating.
2. Knowing what humidity value the sensor is *transmitting* to the controller.
3. Some means of *adjusting* the sensor output or the controller software to reflect the correct local value of humidity.

The first element, knowing the "true" local humidity, can be accomplished by bringing a portable humidity instrument of high accuracy and recent calibration to the sensor location, and allowing a few minutes for the high-accuracy device to come to equilibrium with the air being measured by the sensor undergoing calibration.

The second element requires disconnecting the sensor from the controller, providing that sensor with 24 volts and reading out the value of the milliampere signal that would normally be carried back to the controller. This can be accomplished by a portable, battery-based power supply with digital readout. Several manufacturers offer such devices, which are usually combined with the high-accuracy humidity sensor used as the measurement standard.

But the third element demands the designer's attention at the specification stage. How will the sensor output or the controller software be adjusted so the sensor signal matches the true humidity? Certainly, the most convenient method is to specify a sensor with a calibration adjustment screw accessible without opening the sensor casing. Not all sensors have that feature. Without a means of adjusting the signal at the sensor, the calibrating technician must make a note of the difference between actual and reported humidity, and enter that offset value into the controller's software at a later time (assuming the controller has that feature).

In summary, the HVAC designer's sensor specification will allow or prevent the sensors from being calibrated. Ideally, sensors should have calibration adjustment screws that are easily accessible, as in the sensor shown in figure 17.11. In the absence of that feature, the controller or building automation system must at least have the ability to accept offsets to adjust the sensor's calibration.

Locating Humidity Sensors

The system will tend to control the humidity most closely at the location of the sensor. It follows that the sensor location should be chosen based on the owner's reasons for controlling humidity.

For example, if the purpose of the project is to protect wooden panel paintings in a museum, locate the sensor on the wall near the paintings, not in the supply duct or the return duct as has been done too often in the past. Humidity swings in the duct work really don't matter. The purpose of the system is to keep the rh constant near the paintings so they don't crack because of humidity cycling.

Similarly in a winter humidification project for a school, the sensor should be located in the classrooms, or in a return duct from those classrooms rather than in the hallway near the entry doors. The entry way will have much lower moisture levels than the classroom, especially in the morning as high traffic pulls in dry air.

In all cases, the designer must consider the sensor location carefully, and ensure the contractor actually installs the device in the specified location. Otherwise the system is not likely to behave as expected. Manufacturers of humidity sensors have provided some suggestions:

Avoid Exterior Doorways

Sensors located near exterior doorways will measure wide swings in humidity during both summer and winter. Then the control system will force humidifiers and dehumidifiers to maximum capacity in short bursts, which means humidity throughout the building will swing widely. Therefore, keep sensors away from exterior doors.

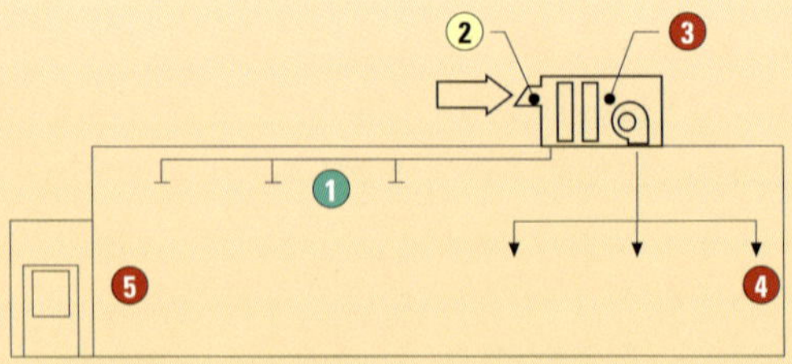

1. **Return Air Inlet - Usually a Good Location**
   In this position, moisture loads from the space are well-averaged by air mixing.

2. **Outdoor Air Inlet - Caution**
   Repeated condensation and wide swings demand a rugged sensor. Regular calibration is required.

3. **Near Cooling or Heating Coils - Avoid**
   Humidity is extreme, therefore difficult to measure. Also, humidity swings widely.

4. **Supply Air Outlet - Avoid**
   Fast response here, but humidity does not reflect the room condition.

5. **Near Doorway - Avoid**
   Humidity swings widely with door openings, System responds with pulsed, uneven control.

Fig. 17.12 Locating Humidity Sensors

Consider two-stage humidity control, with separate sensors

When the humidity must be kept especially high or low, the designer may want to consider a two-stage system. For example, in a museum, one stage could treat the outdoor air so that the incoming ventilation air is brought to a constant dew point year-round, controlled by a duct-mounted dew point sensor located after the blend between the outdoor and return air streams. Then individual zones are "trimmed" with separate humidifiers and dehumidifiers downstream, controlled by sensors located in each zone, or located in the return air duct from each zone.

Ensure constant air flow around the sensor

If the humidity sensor is placed on a wall behind a clothing display rack in a retail store, it will be slow to respond to changes in the rest of the building. Locate the sensor so that air flows constantly, but gently over the sensing element. This will improve the response of the system and therefore improve the uniformity of moisture levels throughout the building.

Locate the sensor near the return duct

Humidity in the supply duct has very little relationship to humidity in the controlled space, except for the few hours each year when there is zero load in that space. So locating the sensor in the supply duct will create humidity swings in the space as loads change. A better location is on the wall near the return duct, where the effect of all loads have been averaged by air mixing.

In applications other than museums, locating the sensor in the return duct itself can be effective, but it is important to allow access for cleaning and calibration, and to make sure the return duct seams are tightly sealed. That way the sensor measures air that represents the condition in the space, rather than a mixture that includes untreated air from outside the controlled space, pulled into the duct work through leaking seams.

In museums, there can be a large difference between the relative humidity near the objects under protection and the location of the return air. Locate the humidistat near the objects, not the return air.

For outdoor air, use a rugged sensor and expect to recalibrate frequently

Outdoor air sensors seldom control the system, but sometimes provide useful data to the building automation system. Sensors in this location are notoriously inaccurate, because they are easily contaminated by pollutants, particulates, condensation and even frost. Locate such sensors downstream of filters, and invest in the most rugged, cleanable sensor the budget will allow. If the humidity signal is important rather than simply informational, be prepared to invest in recalibration on a regular basis—at least annually. In general, do not expect that these signals will always be reporting within the manufacturer's tolerances.

Avoid outdoor air locations near condensing units and bushes

Sometimes, the most aesthetically attractive location for an outdoor air sensor is behind something. But be careful not to place the sensor near bushes or an air conditioning condensing unit. Near the bushes, the humidity will be higher than the true outdoor condition, and near the condensing unit the air temperature will be too high. A good location for outdoor air sensors is at least one story above ground level, under a rain and radiation shield, and located on the northern side of the structure when the building is located north of the equator, or on the southern side when the building is located south of the equator.

For systems controlled by rh, avoid locating the senor on the exterior walls or in sunny exposures

For most buildings, the authors suggest that the designer control the system to achieve a constant dew point in the space. However, this is not always possible. For systems that must be controlled by relative humidity, be sure to keep in mind the influence of the surface temperature of the sensor.

For example, consider an office building with a need for winter humidification. If the rh sensor is located on a sunny inside wall, the surface temperature of the sensor may be quite high because of radiational heat gain. Then the sensor would report a low relative humidity, and the humidifier would add more moisture to the supply air. Excess moisture could then condense on cold surfaces inside exterior walls, even through the relative humidity as reported by the warm sensor is still quite low.

Similarly, consider the same building in a light wind with the humidistat located on an exterior wall near a leaking window frame. Cold air leaks into the sensor, chilling it. Then the sensor reports an excessively high rh, even though the moisture content of the air in the occupied space is still too low for comfort.

In short, when the system is controlling to a constant relative humidity, the sensor location must be chosen so it senses a *representative air temperature* as well as its average relative humidity.

Locate the sensor far from heat and moisture sources

A surprising number of otherwise fine systems fail to operate properly because the humidity sensor is located above a moisture source such as the office coffee pots, a decorative fountain or near the door to a locker room with a shower. Also, equipment like copiers, computers, lab instruments or hot plates can create small "heat islands" which distort a moisture reading that is based on the signal from an rh sensor. Locate the humidity sensor well away from sources of heat and moisture.

Remember that humid air rises

Humid air is lighter than dry air. Humid air contains relatively large but light oxygen atoms, which displace some of the smaller, heavier nitrogen atoms that would otherwise occupy the available volume. So when left at rest in the absence of convection currents or forced air distribution, moist air rises to the ceiling.

This is one reason that structural steel corrodes above dropped ceilings above swimming pools, even though the steel surface may not be cold enough to actually condense moisture.

The system designer should keep this fact in mind when locating the humidity sensor. If the ceiling is high and air circulation is limited, as in the case of an open, warehouse-style retail store, a tall office atrium or indoor arena, the absolute moisture level at the ceiling will be higher than at the floor.

Avoid multiple sensors in the same location

Building owners and occupants sometimes install additional humidity indicators to display the humidity in the controlled space. They do not always understand why a second sensor is unlikely to match the value from the primary sensor, especially when both have a confidence-inspiring digital display. The subtleties of sensor tolerances versus calibration and sensor costs are seldom obvious to the non-technical professional.

The designer can avoid needless annoyance for all parties by providing a local display that reads out the humidity value being reported *by the controlling sensor*. This avoids the lower-cost, but guaranteed-to-irritate alternative of the owner or occupant installing an inexpensive wall-mounted humidity instrument, which will nearly always display a different humidity value than the controlling sensor.

Install high-limit sensors downstream from duct-mounted humidifiers

Condensation in ductwork must not be allowed to occur. Water combines with dust to allow the growth of fungus and bacteria, which is never healthy and can even become quite hazardous. When the designer installs a humidifier in the ductwork, the device needs a high-limit humidity sensor downstream, just past the vapor absorption zone. Independent of other controls, this sensor should limit the capacity of the humidifier so that the relative humidity in the duct work does not exceed 70%. If this limit means the air stream cannot provide enough moisture to satisfy the control condition in the space, then additional humidifiers should be added in other air streams, all limited to ensure they do not saturate any air above 70% rh.

Cut-up spaces may demand looser tolerances

Often, the humidity control requirement is not critical, so a single sensor is expected to represent an average condition from several zones. The sensor will control the system more evenly if it can be located in a common return duct from several zones. But frequently there may be no common return between zones, as in the case of hotel rooms or classrooms served by packaged terminal air conditioners. Short of equipping each room with separate humidifiers and dehumidifiers, the only alternative is to provide a constant humidity in the ventilation air system that *supplies* a group of rooms. Humidity will vary according to the load in each room.

Avoid duct locations near cooling and heating coils

Downstream of a cooling coil the air is nearly at 100% relative humidity. Downstream of a heating coil, the relative humidity may be below 10%. In both cases, relative humidity sensors are not likely to provide accurate readings, especially since radiation from the heater's surface will also affect the sensor temperature. Avoid these locations.

Allow time for portable sensors to equilibrate

Portable sensors are often used to provide measurements used in calibrating wall or duct-mounted sensors. But any sensor has inertia. Sensors must be allowed time to reach the same temperature as the air being measured. The classic problem is a costly, highly-accurate sensor transported in a service technician's air-conditioned truck, then carried through hot, humid outdoor air to the rooftop, where it is placed into a return air duct. The sensor surface would be cold enough from the truck ride to condense water on its way to the rooftop, greatly slowing response time because the air must evaporate the condensed moisture on the sensor. Portable sensors need to be insulated from extremes of temperature and humidity so they can respond quickly and accurately when moving between different environments. But if an instrument becomes unavoidably hot or cold, remember that a considerable time must be allowed before a new reading is valid at a different extreme. The delay will be longest when moisture must be desorbed from the sensor, such as when the sensor has condensation on its surface, or when it simply moves out of very humid air into drier air.

Humidity Control & Building Automation

The large majority of commercial buildings built in recent years are equipped with building automation systems that control equipment operation centrally. These systems can reduce the cost of humidity control and tighten tolerances, because they can integrate many signals and make logical decisions at high speed and low cost as occupancy and loads change.

On the other hand, the implied accuracy of a digital display has not always been a positive force for improving humidity control. When the system operator reads a computer screen that indicates the relative humidity in the building is 41.26%, the extra digits generate credibility that is not deserved. Relative humidity signals are *never* more accurate than ±1%, and commercial-grade low-cost sensors are usually no closer than ±5 to 7%. So although the display might indicate 41.26%, the actual rh could be anywhere between 35 and 48%.

The value of precise and speedy internal calculations is limited by sensor input. In the immortal words of the computer programmer: "Garbage in = garbage out." The building automation system cannot make good control decisions unless it has been adjusted for the sensor's calibration after the sensors have been installed and the system is operating. Also, the building automation system will continue to make poor decisions even with calibrated sensors if they are not located according to the advice in the previous section.

In summary, for reliable results, humidity sensors must be carefully located and calibrated frequently. Otherwise the owner's costly investment in building automation is largely wasted.

References

Huang, Peter H., Ph.D., "Humidity measurements and calibration standards." ASHRAE Transactions, 1991. V.97, Pt. 2.

This article will assist the reader who needs to understand the basis of calibration by the National Institute of Standards and Technology.

Lafarie, J-Patrick. "Accuracy of relative humidity instruments." *Sensors* Magazine. June, 1987. Helmers Publishing, 174 Concord St., Peterborough, NH 03456. USA.

A useful magazine article that explains the sources of error in measuring relative humidity. Written by a manufacturer, the article will help the reader appreciate the difficulty of making sensors live up to the industry's specifications and a designer's expectations in real-world situations.

Thomas, Robert M., P.E., "Humidity sensors in heating, ventilating and air conditioning (HVAC) systems." ASHRAE Transactions, 1992. V.98, Pt.2.

Mr. Thomas measured the actual performance of a variety of humidity sensors in carefully-controlled field tests. The results show that manufacturers' tolerances are somewhat optimistic and seldom achieved in the field. The article helps the less-experienced HVAC designer understand the practical limits of humidity measurements in commercial buildings, which helps avoid unrealistic expectations for the precision of any humidity control system.

Wiederhold, Pieter R. *Water Vapor Measurement.* 1997. Marcel Dekker Inc., 270 Madison Avenue. New York, NY 10016 USA. ISBN 0-8247-9319-6.

A very comprehensive 350-page reference book detailing all types of humidity instrumentation in great depth, with an emphasis on precision measurements in laboratory and industrial environments.

Special Thanks

Special thanks are due to Frank Caporale of General Eastern Instrument Corp, for sharing his 25 years of experience as a field technician, design engineer and instrument product manager. Also, other useful points were contributed by Doug Lee of DLA Associates, Chris Morton, formerly of Dri-Steem Humidifier Corp and Steve Brickley of Munters USA.

Donated Photos & Diagrams

Fig. 17.1 (Portable sensor) - © Vaisala Ltd.

Fig 17.1 - (Desk-mounted sensor) - © Davis Instruments

Fig. 17.3 - © General Eastern Instruments

Fig. 17.4 - © General Eastern Instruments

Fig. 17.5 - © Pieter Wiederhold

Fig. 17.6 - © Panametrics Instrument Corp.

Fig, 17.9 - © Davis Instruments

Fig. 17.11 - © General Eastern Instruments

Chapter 18

Schools

By Lew Harriman

Fig. 18.1
Schools Exist To Serve Students....
To Do That, Humidity Control Extends Beyond The Classroom
The designer's focus is rightly on the classrooms. But controlling humidity is also important in libraries, gymnasiums, locker rooms and swimming pools.

Humidity Control In Schools

School buildings contain many activities. The purpose of humidity control varies with the function of each zone and varies according to the desires of the owner, which must be clearly defined before beginning the project (see Chapter 2). There are at least five different zones where humidity control may be important to an owner:

- Classrooms - Human Comfort
- Library - Mold Prevention
- Gymnasium - Floor Protection
- Showers & Locker Rooms - Mold & Corrosion Protection
- Swimming Pool - Condensation & Corrosion Protection

In smaller schools, the designer may not have the luxury of separate systems for each area. But regardless of the building size and budget, it is useful to understand the reasons for humidity control before deciding what sort of system and equipment is appropriate.

Classrooms

The usual purpose of controlling humidity in classrooms is to create a healthy, comfortable environment for teaching and learning.

Typical control levels & tolerances

Humans are surprisingly tolerant of a wide range of humidity. There are no firm standards for classrooms, but research that supports ASHRAE Standard 55 (Human Comfort), suggests that 80% of a typical adult population will not object to the environment if humidity is held above a 30°F dew point during dry months, and below a 60°F dew point during humid months. [Above -1°C and below 15.6°C dew points.] Regarding tolerances, there is no compelling need to constrain humidity within a narrow range, as long as the humidity stays within those fairly loose high and low limits.

Note that human comfort is relative, not absolute. No single temperature and humidity range will ensure that *all* of the rooms' population will be comfortable *all* the time. For example, in an elementary school, the larger, more active teachers are likely to prefer a lower humidity during warm months, so they can release metabolic heat through perspiration. The seated students generate less heat, and also have more surface area compared to their mass. The students can release more heat more easily, and are likely to be more tolerant of high humidity.

In winter, people who wear contact lenses will be more sensitive to extended periods of low humidity. In dry air, the protective liquid film on the surface of the eye dries out more frequently, so that the contact lens has less lubrication and can become an irritant.

No single range will satisfy everybody all of the time, although keeping humidity above a 30°F dew point and below a 60°F dew point will satisfy *nearly* all people *nearly* all the time. [Above -1°C and below 15.6°C dew points.] However, in this age of high expectations for total perfection and complete comfort, the designer would do well to discuss comfort expectations with the owner, to avoid future misunderstandings. Chapter 4 contains detailed information on human comfort and humidity to guide this discussion.

Apart from comfort, health can be affected if the humidity is allowed out of the comfort range for extended periods. High humidity allows the growth of dust mites and fungus, which produce allergens and can even produce toxic byproducts. At the other extreme, low humidity favors the longevity of infectious viruses, and therefore their transmission between occupants—a familiar phenomenon to parents of school children in cold climates.

Problems caused by high and low humidity are discussed in more depth in Chapters 7, 8 and 9 (Fungus, Dust mites and Bacteria & Viruses), but unless humidity is *extreme* for *extended* periods, adverse health effects are not likely. Short excursions outside of the human comfort zone seldom lead to health problems except in extremely sensitive people.

Principal loads in classrooms

Ventilation, infiltration and people generate the principal moisture loads in a classroom. Compared to these very large sources, the designer can be much less concerned about the minor amounts of water vapor gained through vapor permeation or wet surfaces.

Quantifying and controlling the amount of **ventilation air** is especially important. ASHRAE Standard 62 currently recommends a minimum of 15 cfm per person [7.4 l/s/person]. But the "correct" amount of ventilation air has remained one of the most controversial subjects in HVAC engineering since the founding of ASHRAE in 1898. The society's guidelines have varied between 5 and 25 cfm per person over the last 100 years [between 2.4 and 12.3 l/s]. Reflecting this variation and the timing of different revisions, ventilation codes in the U.S. vary considerably between different code jurisdictions.

The local code will define the minimum amount, but the owner may prefer to design above that lowest standard. In any case, the outdoor air carries a great deal of excess moisture during humid months, and a large moisture deficit during dry months. So the designer *must* be certain of the required amount of ventilation air before planning a humidity control system.

To minimize operating costs and improve humidity control, the designer should also find out if local codes allow reduction of ventilation when rooms are not occupied. The year has 8760 hours but classrooms are seldom occupied for even 30% of that time. Local codes will govern the minimum *continuous* ventilation rate when the classroom is empty, which may be much less than the rate required when the classroom is fully occupied.

People generate the second-largest moisture load in classrooms. The load depends on their activity level, and the load is only present during occupied hours. Designers sometimes overlook the moisture load from wet clothing shed by students and teachers entering the building during rainy weather. Chapter 11 contains information to allow the designer to account for this large—but short and infrequent—moisture load.

Infiltration is the next most important moisture load. Commercial buildings leak air. They leak more air than any HVAC designer would normally assume. (See Chapter 11) The amount can be minimized by maintaining a building under a slight positive air pressure when the structure is located in a warm climate. In a cold climate, the internal pressure must be neutral to avoid condensation inside walls. But regardless of air pressure, the designer must assume there will be some humid air leakage through exterior walls during windy and rainy weather, and make allowance for that water vapor in the load calculation.

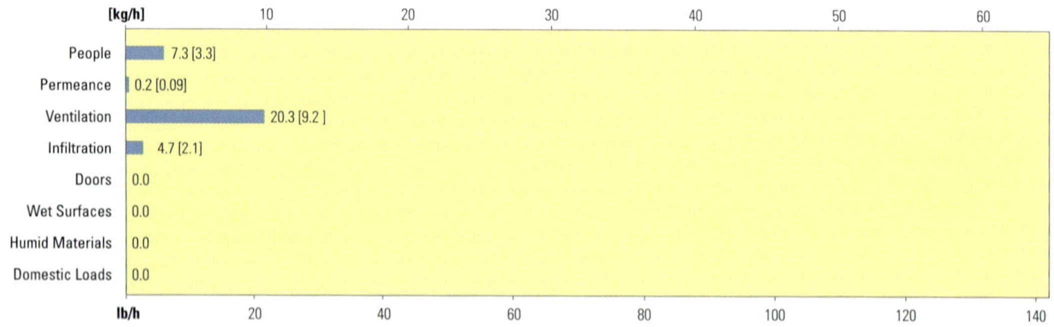

Fig. 18.2
Moisture Loads For A Single Classroom
Note the size of the ventilation and people loads. When the room is unoccupied, reducing the ventilation air flow will nearly eliminate the loads. It becomes very easy to control humidity during vacations and weekends, as long as a dehumidifier is installed in the system. This avoids the odor and mildew problems so common at the end of the summer shutdown.

To summarize, the most important load is contained in the ventilation air. That load component will be nearly continuous, and large. So the system must be designed to remove that load continuously; otherwise the designer cannot hope to keep humidity within the target range. In contrast, the other load components are intermittent. If the system lacks the capacity needed to remove *all* those loads at peak periods, there will be some hours outside of the comfort zone. But that excursion is not likely to last long as long as the system has the capacity to keep up with the continuous ventilation load.

Suggestions & cautions for classrooms

1. Provide continuous humidity control

Many classroom HVAC systems have continuous temperature control, but rely on the intermittent dehumidification effect of the cooling system to moderate humidity in hot and humid weather. In those cases, neither the designer nor the owner should expect humidity to be controlled within the comfort range. Continuous ventilation demands continuous dehumidification, because there are thou-

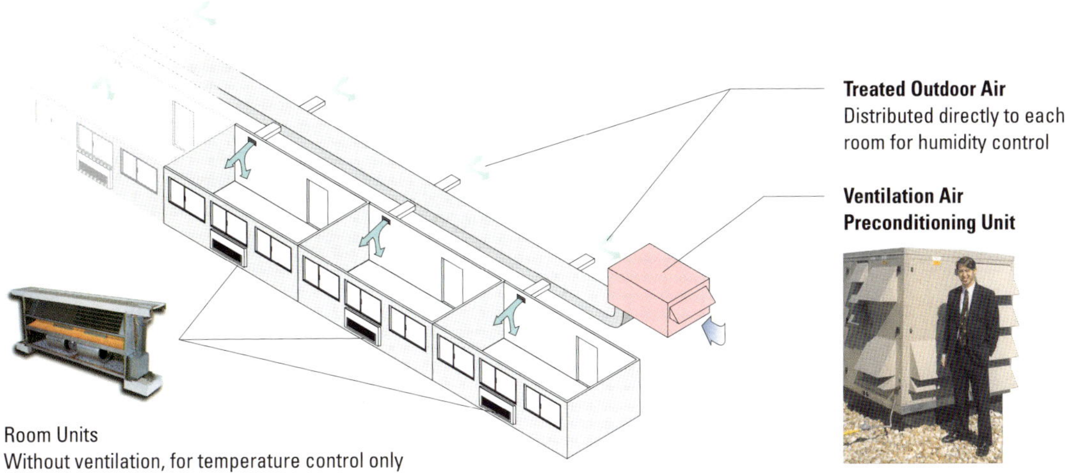

Fig. 18.3
Pre-Treat Ventilation Air
*The ventilation air carries the largest moisture load in humid months, and greatest moisturte deficit in dry months. For best humidity control, install a separate system to bring the moisture content of the ventilation air to the indoor control level **before** the air enters the classrooms.*

sands of occupied hours when the outdoor air carries a moisture load but no sensible heat load.

When the cooling system alone is expected to control humidity, it must have a feature that allows a dehumidification mode in response to a humidistat. With constant ventilation, thermostatically-controlled cooling systems do not reduce humidity during hours of moderate outdoor temperature combined with high humidity. A much better solution is to install a separate system to control humidity.

2. Install a separate, ventilation/humidity control system

Since the largest moisture and humidification loads come from ventilation air, the designer can achieve humidity control by dehumidifying and humidifying the incoming ventilation air to a dew point within the comfort zone. An example is shown in figure 18.3. The system supplies dry or humid air to the classrooms to control humidity, while satisfying the ventilation code requirement.

This strategy avoids the need to install dehumidification and humidification equipment in each classroom, reducing maintenance and construction costs. It also removes the noise of constant ventilation fan operation from the classroom. The room units need only turn when the room's dry bulb temperature needs adjusting.

On the other hand, when classrooms are unoccupied, codes usually allow a reduction in the ventilation air flow rate. If the ventilation unit is the only source of humidity-controlled air, reducing its flow or eliminating it entirely also eliminates the humidity control. To avoid the problem, ensure that the ventilation air/humidity control unit has a *recirculation mode*. Then it can still add or subtract moisture from the classrooms even when its outdoor air flow is reduced, avoiding the problem of odors in the classrooms after summer vacations.

3. Avoid moisture in ductwork

Moisture in ductwork leads to growth of fungus. The problem becomes evident when "musty odors" are present. When allowed to continue, fungal growth can become hazardous. The problem is most common in ductwork downstream of cooling coils that are overloaded with moisture and not really designed to dehumidify air. From such coils, water droplets can escape and be carried into the air stream, which saturates any dirt or dust downstream.

To avoid this problem, ensure that all cooling coils have a drain pan with a p-trap to remove condensate. Also, consider installing a dedicated dehumidifier in the system to remove moisture in a predictable way at a well-controlled location as described above in suggestion two.

4. Design the dehumidification for peak dew point conditions—not peak dry bulb temperatures

Since temperature control is more important than humidity control, the designer sometimes sizes all equipment based on the peak outdoor temperature, neglecting to size the dehumidification equipment at the peak dew point condition.

This common error leads to poor humidity control during the warm, rainy days when the moisture load is at its peak. On those days, the cooling equipment—sized for peak sensible loads—does not run for long periods, so it does not dehumidify the classrooms. The dehumidifier must be sized and controlled so that it will operate during the moderate-temperature, highly-humid hours when moisture loads are at their maximum—the design dew point conditions shown in Chapter 32 of this book.

5. Consider installing load reduction equipment

In many cases, the exhaust air can be captured and brought back to the point where the ventilation air enters the system. When the exhaust leaves the building at the same point as the incoming supply, it provides an excellent opportunity to reduce the load by installing a passive desiccant wheel—one which uses dry air rather than heat to remove moisture from the desiccant. Such passive desiccant wheels are sometimes called enthalpy wheels or total heat wheels.

These devices can greatly reduce the summer and winter loads on the humidity control system. They transfer excess moisture from one air stream to the other, along with heat. In the summer, they act as dehumidifiers. In winter, they humidify the incoming air. Their capacity allows the designer to reduce the overall cost of the system, in spite of adding that extra piece of equipment.

One caution, however, is important to note. Passive desiccant wheels can only dry incoming humid air *if the exhaust air is dry*. These devices do not control humidity—they only reduce the load. So there will still need to be another component dedicated to dehumidification somewhere inside the building. Otherwise, humidity inside the building will slowly rise until the exhaust air is too humid to remove moisture from the incoming air. See Chapter 15 for additional information about load reduction equipment.

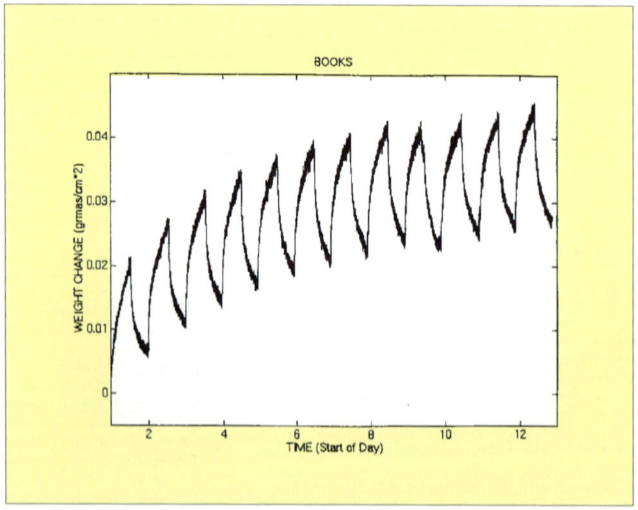

Fig. 18.5 Books Accumulate Moisture

Unlike newspaper and other thin-layer materials, books accumulate moisture and do not give it up easily. In this experiment, books were cycled 40%r h and 90% rh every hour. They steadily gained water even after exposure to periods of 40% rh. To dry books effectively, air must be well below 40% rh. (Bailey et al. 1994)

School Libraries

In a library, humidity is controlled to protect books and magnetic tapes. The large moisture storage capacity of the books will either work in favor of humidity control or against it, depending on the designer's decisions.

Typical control levels & tolerances

The same humidity-related comfort control limits apply to the library. That is to say, humidity should be kept between dew points of 30°F and 60°F to avoid discomfort for 80% of the typical occupant population. [Between dew points of -1° and 15°C]

However, controlling humidity in this range does not prevent books from absorbing excess moisture. The relative humidity governs moisture absorption. Although people tolerate a high relative

Fig. 18.4 Moldy Books

This example came from an elementary school in Houston, Texas. The air cooling system was re-set to a high temperature during summer vacation, but ventilation continued. The incoming humid air was never dehumidified, because the temperature was set too high and there was no dedicated dehumidification component. Moisture accumulated in the building. Thousands of books were either cleaned or replaced after summer vacation.

and absolute moisture level, books and magnetic tapes will not. Books take up moisture quickly—but lose it slowly, if at all.

Figure 18.4 shows the effect of cycling a book between 40 and 90% relative humidity. Note that even 40% relative humidity is not low enough to completely remove all moisture from the book. The paper slowly collects more moisture in spite of dry periods. Eventually, it will collect enough moisture to support mold and mildew growth, which generate the familiar "old book odors" while breaking down the paper fibers. To avoid odors, allergic reactions to fungal emissions and damage to books, keep the humidity below 55% at all times, regardless of the temperature. For fungal protection, drier is better, but humidity below 20% rh will increase the number of electrostatic shocks, add to eye irritation and annoy the occupants.

Regarding tolerances, there is no need to constrain humidity in school libraries within a narrow range. To ensure that books do not have fungal damage, keep the dew point above 30°F [-1°C] and make sure the relative humidity does not rise above 55%.

Principal loads in libraries

Ventilation and infiltration are the principal load sources in a library. Infiltration is usually a smaller load, as not all walls are likely to be exposed to the weather, and the wind does not force air through the exterior walls continuously. As in classrooms, ventilation air is the principal source of moisture and humidification loads.

Ventilation rates for libraries are also similar to those for classrooms, so that discussion need not be repeated in this section, except to note again that ventilation requirements during unoccupied hours may be considerably lower than for occupied hours. This provides the designer with an opportunity to save operating costs by reducing air flow, even if there is no change in the *size* of the equipment needed to meet peak ventilation loads.

Suggestions & cautions for libraries

School libraries in humid climates in the US have become somewhat notorious for mold, mildew and odor problems. The authors speculate that this may be because HVAC designers and owners seldom give any special thought to humidity control in the library beyond the casual concern given to classrooms. But libraries deserve more attention with respect to humidity control, because they contain a huge amount of fungal food in the form of books and papers. Here are some suggestions which can help avoid problems:

1. Dedicated dehumidifier and continuous dehumidification

To save energy, many schools allow the temperature to rise during summer vacations and weekends when the library is not occupied. That strategy causes no problems, provided that the maximum humidity is still limited to 55% rh. Unfortunately, designers do not always install dedicated dehumidification equipment, relying instead on cooling equipment to remove excess moisture. When cooling equipment does not operate (as when the temperature is reset to higher levels), there is no dehumidification. Any solution costs extra money. Either install a dedicated dehumidifier so humidity is controlled below 55% regardless of temperature, or accept the high operating cost of cooling equipment to remove moisture during unoccupied periods. A dedicated dehumidifier is the lower-cost solution.

2. Ventilation humidity control—with recirculation mode

The majority of the moisture and humidification load is created by the incoming ventilation air, so it makes sense to dehumidify and humidify that air stream to achieve humidity control at the lowest cost. On the other hand, when the library is unoccupied, codes usually allow a reduction in the ventilation air flow rate. If the ventilation unit is the only source of humidity-controlled air, reducing its flow or eliminating it also eliminates the humidity control for the library.

A better solution is to ensure that the ventilation air unit has a recirculation mode. Then it can still add or subtract moisture from the library even when its outdoor air flow is reduced.

3. Dryout service companies

When books have absorbed moisture, it is very difficult to dry them out. Humidity levels well below 30% rh are required. After construction or other periods of high humidity such as vacations, the owner may wish to consider drying out the library. Otherwise, the HVAC system—never built for that task—will have great difficulty controlling moisture. Over the past 20 years in North America and Europe, the insurance industry has built up an infrastructure of water damage service firms. These companies are expert in drying out buildings after floods, fires and disasters, using portable high-powered dehumidifiers. Drying out a library and its books is a simple, quick way to prevent mold and mildew growth if the material has become loaded with humidity after construction or after unoccupied periods without humidity control.

Gymnasiums With Maple Floors

Most gymnasiums are not humidity-controlled, since ideal comfort is not expected by either the spectators or participants during sporting events. But if the facility has a *wooden* sports floor, the designer and owner should consider installing humidity control equipment to avoid costly damage.

Typical control levels and tolerances

High-quality sports floors for basketball are made of wood—more specifically, maple flooring strips. Because basketball is a common sport in elementary and secondary schools, maple floors are quite common in school gymnasiums.

In the United States and Canada, the Maple Flooring Manufacturers Association (MFMA) specifies that the humidity at the floor level "shall be maintained between 35 and 50% relative humidity at all times"—or the installer's warrantee will be void. When the owner wants to ensure the warrantee stays in force, the designer should contact the MFMA for current guidance.

HVAC designers will be pleased to note that MFMA guidelines allow humidity excursions for the relatively short periods of public events. An 8-hour humidity excursion will not damage the floor. But if humidity is allowed to rise above 50% over the whole summer, or fall below 35% during the winter, the floor will be damaged by excessive expansion or contraction.

Expansion is the more common cause of a flooring problem, because it leads to irreversible compression of the wood fibers. The flooring strips compress during humid months, but no problem is apparent. The damage shows up during dry months. The strips contract, opening cracks that will never fully re-expand because their fibers have been crushed during the initial expansion. After the floor has "opened up," no amount of humidification will solve the problem. The only way to avoid such costly damage is to ensure the floor has a humidity-controlled environment continuously—including during it' installation. The MFMA specification is clear: maintain the relative humidity between 35 and 50% *at all times*, regardless of temperature, except for short periods during public events.

Principal loads in gymnasiums

Ventilation, infiltration and people create the largest moisture loads in gymnasiums, although door openings towards the weather or towards shower rooms can also be significant for short periods.

Ventilation overwhelms all other loads when operating at full capacity during public events. At that point, the cost of controlling humidity becomes so high that it is not practical. The designer should design instead for the peak load during normal daily operation of the gym. That will include the load associated with minimum ventilation, plus the moisture load from door openings and people. Infiltration is not likely to be a continuous load, unless the gym is under negative air pressure accidentally, as from exhaust fans.

Suggestions & cautions for gymnasiums

In gyms with wooden floors, failure to control humidity may have more costly consequences than failure to control temperature. On the more positive side, humidity control is fairly simple to accomplish during normal operation and during unoccupied periods.

Fig. 18.6 Humidity-warped Floors

Excess humidity makes wood expand. Although sports floors are engineered to accommodate expansion, they have limits. When humidity is uncontrolled during hot months, the floor can be destroyed, as in these examples.

Excessive expansion during humid months crushes the fibers, so that gaps open wide between flooring strips during the dry season. After the fibers are crushed during one humid season, no amount of humidification will ever remove the gaps between strips.

1. Install dedicated dehumidifiers and humidifiers

The least expensive and most effective way to meet MFMA specifications is to install dedicated dehumidifiers and humidifiers, separate from the heating and cooling equipment. That way, the humidity control equipment can operate continuously, regardless of what the building staff may need to do with re-setting temperature during unoccupied periods to reduce operating costs. Humidity needs to be constant year-round. There is no similar requirement to control temperature within a constant narrow range.

2. Install the sensor at the floor—not the ceiling

The humidity control sensor must be located near the floor, not high on the wall or near the ceiling. This is a bit challenging. For a sensor, the floor level is a difficult environment—full of dust and subject to kicks and dropped heavy objects. But at the ceiling, the humidity is essentially irrelevant. On an outside wall, falling convection currents during cold months can distort the reading. Near the doors, the sensor would be influenced by air from outside the gymnasium. The best location is in a caged and filtered housing, mounted on an inside wall near the floor, but away from doors.

3. Avoid cold floors

Since the relative humidity at the floor must be controlled, the designer must consider the effect of changing temperature on the relative humidity. If air stagnates and cold air collects near the floor, the relative humidity rises at that level. The dehumidifier may turn on, using energy even though the air could be dry enough if it were mixed with air at a slightly higher temperature from a short distance above the floor. Arrange fans to blow a *gentle* current of *mixed* air across the floor to avoid needless energy consumption.

Showers & Locker Rooms

In showers and locker rooms, the usual purpose of humidity control is to dry out the room and its wet surfaces quickly, to limit corrosion and to limit the growth of bacteria and fungus that generate odors. Low humidity is rarely, if ever, a concern.

In most schools the budget does not allow a dehumidification system for locker rooms. Usually the exhaust air is increased, so that comparatively dry air is "transferred" from other areas of the building to replace the exhaust air, along with "temperature-tempered" outdoor air to make up for the exhaust. This strategy has the advantage of low construction cost, and can be effective during dry weather. But it seldom achieves effective drying during most of the year. Lockers rust, their contents smell and slime accumulates in corners. But occasonally, an owner has the budget and feels the need for more rapid and effective drying of the locker room.

Typical control levels and tolerances

The showers themselves are designed for continuous wetness, so there is seldom any concern in that location. In the locker rooms, the goal is not really to control humidity at a specific level but rather to speed the drying, and to extend that drying to all wet surfaces.

The lower the dew point of the air, the faster it will remove water from wet surfaces. But the exact time needed to dry is seldom critical. Often the budget determines the dew point. The humidity level tolerance of the application is quite wide, so that cost and other commercial considerations usually govern the dew point of air provided by the dehumidifier. As a general rule, select a dehumidifier to supply the lowest dew point the budget will allow, but do not expect rapid drying from air delivered at dew points above 50°F [10°C].

Principal loads in locker rooms

The amount of incoming air and its dew point governs the moisture load, and that air quantity is determined by the exhaust air flow. Codes determine the minimum exhaust flow. Since toilets are often combined with the shower/locker area, exhaust flows can be large.

The makeup air is often "temperature-tempered" outdoor air, along with some conditioned air transferred from inside the building near the locker rooms. While there are certainly moisture loads from people, wet surfaces and humid air from the showers, the makeup air load is far larger than all other loads combined. So that element needs the most attention from the designer. The code should be examined carefully to make sure the minimum amount of exhaust air is used, especially during unoccupied hours.

Suggestions & cautions for locker rooms

1. Locate an exhaust register above the passageway to the showers.

Most of the exhaust will be pulled from above the toilets. But the designer can speed-up the drying considerably if hot, saturated air from the shower is exhausted before it can drift into the locker and toilet areas. A long, narrow exhaust register at the ceiling will serve this purpose, extending across the passageway to the shower.

2. Locate an exhaust trunk above the lockers

Odors from lockers are difficult to avoid, but if dry air flows across the face of the lockers, internal drying will be faster, which will reduce the magnitude of the odors. The designer might consider locating exhaust registers above the lockers to both remove odors directly, and to encourage dry air flowing up the face of the lockers.

3. Size the dehumidifier for moderate outlet dew point. Then recirculate to produce lower dew points when room is unoccupied.

When the incoming ventilation air flow is at its peak, it carries a tremendous amount of moisture into the room. It would be very costly to size a dehumidifier to produce a low dew point with such a high air flow. A more cost-effective design might be to dry only a portion of the incoming air, reducing the dew point in the entire ventilation air stream only as low as the upper limit of the human comfort range. Then when the room is unoccupied, the exhaust and makeup air flows can be reduced. During unoccupied hours, recirculate air from the room back through the dehumidifier. The unit will slowly reduce the dew point, drying out the room. When recirculating, control the dehumidifier by a humidistat set at 40% rh. The continuing flow of dry air through the room will help minimize odors from damp clothing drying slowly inside the lockers.

4. Consider supplemental circulation fans

In locker rooms, the architectural designer is seldom generous in allowing space for supply and exhaust duct work. This will often force the HVAC designer into very poor duct arrangements that simply cannot provide enough airflow for adequate drying.

When well-designed air distribution is a practical impossibility, the designer might consider installing supplemental air circulation fans. Mount these inside the room, aimed to encourage dry air to find its way into wet corners. The fans could activate during night hours when the room is unoccupied, so that their high-velocity air currents do not chill wet young athletes. Night drying has another advantage. It will increase the moisture load on the dehumidifier at a time of day when the cost of electrical power is low, saving operating costs in areas where time-of-day utility pricing is in effect.

Swimming Pools

Swimming pool areas are discussed in more detail in Chapter 27, which is specifically dedicated to that application. It is usually wise to separate the HVAC system for the pool from the other systems in the school. Pool activity generates a very large moisture load as well as a substantial amount of chlorine-related IAQ contaminants. Ideally, these should be isolated from the rest of the building.

References

Bailey, D.W. et al. *Investigation of dynamic latent storage effects of building construction and furnishings* (ASHRAE Research Project 455-RP). 1994. Project report with Appendices.

Humidity Control Guidelines for Maple Flooring. 1998. Maple Flooring Manufacturer's Association. 60 Revere Dr. Suite 500 Northbrook, IL 60062 USA. (847) 480-9138

Special Thanks

Humidity problems in schools have been thoroughly investigated on behalf of the U.S. Department of Energy's Oak Ridge National Laboratory by Dr. Charlene Bayer of the Georgia Tech Research Institute, Professor Sidney Crow of Georgia State University and Mr. John Fischer. The author is deeply grateful to these institutions and these helpful people for their assistance in understanding both the nature of the problems and their implications for HVAC design. Also, Jim Rice and Chuck Hammock of Andrews, Hammock & Powell in Macon, Georgia provided an especially useful dose of reality, based on their experience of designing humidity control systems that must fit into tight school construction budgets. Finally, Gene Ostrom of Rochester, Minnesota, an expert in drying buildings after floods, fires and disasters, generously donated the photos shown in figure 18.6.

Chapter 19

Office Buildings

By Lew Harriman

Fig. 19.1 Human Comfort & Machinery
Office productivity and profits depend on comfortable people and on smooth-functioning machinery. Humidity control assists both requirements

Humidity Control in Office Buildings

Given the economics of the marketplace, HVAC systems for low-rise commercial construction are likely to consist of high-efficiency, packaged rooftop heating/cooling equipment. These units are simple to apply, easy to maintain, inexpensive and very fault-tolerant—perfect for serving thrifty owners.

On the other hand, conventional packaged rooftop units are not optimized for humidity control. They achieve their high efficiency rating as a result of their sensible cooling and heating performance—not their ability to control humidity.

This chapter provides suggestions for controlling humidity where packaged rooftop units are the preferred system. Adding humidity control to a low-rise office building will not reduce construction cost, nor will it save operating costs unless the owner is willing to raise the thermostat set point. For the designer's work to be productive, the owner must recognize these facts, and be willing to invest more money than usual in the HVAC system when he or she needs to control humidity.

Fig. 19.2
Optimal Humidity Ranges For Comfort & Equipment Operation

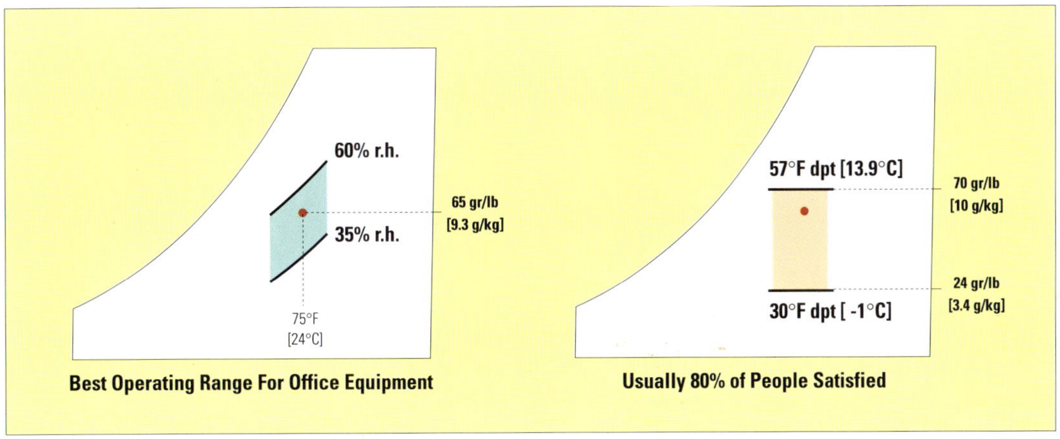

Typical purposes of humidity control

The owner's usual purposes for investing money in humidity control include:

- Assuring the comfort and productivity of workers
- Allowing office machines like printers and computers to operate smoothly, without paper jams or costly damage from electrostatic discharges.

The human comfort range is wider than the optimal humidity range for equipment. But both sets of goals can be satisfied by keeping moisture levels between dew points of 30° and 57°F [-1° and 14°C].

Control levels & tolerances

Humans are surprisingly tolerant, able to be comfortable over a wide range of humidity levels. Research that supports ASHRAE Standard 55 (Human Comfort), suggests that 80% of a typical adult population will not object to the environment if humidity is held above a 30°F dew point during dry months, and below a 57°F dew point during humid months. [Above -1°C and below 14°C dew points.] Regarding tolerances, there is no compelling need to constrain humidity within a narrow range, as long as the humidity stays within those high and low limits.

Note that human comfort is relative, not absolute. No single temperature and humidity range will ensure that *all* of one buildings' population will be comfortable *all* the time. For example, consider a training room in an interior zone of an office building, occupied by both women and men, with both dressed formally. Men are likely to be wearing suits while women may be dressed in skirts with blouses—much less skin coverage. Although the men may be comfortable at a low temperature, the women are losing much more heat, and therefore are uncomfortably chilly at the same room temperature.

Similarly, different people have varying degrees of tolerance for low humidity. For example, people who wear contact lenses will be more sensitive during winter. In dry air, the protective liquid film on the surface of the eye dries out more rapidly, so their contact lenses have less lubrication and can irritate their eyes.

Consequently, no single range will satisfy everybody all of the time. In this age of high expectations for total perfection and complete comfort, the designer would do well to review this point with the owner, to avoid future misunderstandings. Chapter 4 contains detailed information on human comfort and humidity that can help guide this discussion.

Health

Apart from comfort, health can be affected if the humidity is high enough to allow the growth of mold and mildew, which produce allergens and can even produce toxic by-products. At the other extreme, low humidity favors the longevity of some infectious viruses, and therefore favors their transmission between occupants—a familiar phenomenon to parents of school children in cold climates.

Problems caused by high and low humidity are discussed in more depth in Chapters 7 and 9 (Fungus, and Bacteria & Viruses), but unless humidity is *extreme* for *extended* periods in the office area, adverse health effects are not likely. In office buildings, the more common mold and mildew problem occurs in ductwork, where both dirt and high humidity are present for long periods, allowing mold to grow and reproduce. Short excursions outside of the human comfort zone *in the occupied space* do not lead to health problems except in extremely sensitive people.

Electrostatic discharge

Printers, computers and other electronic equipment operate with fewest problems when the humidity is held at dew points between 40° and 60°F [4.5° and 15.6°C]. At lower moisture levels, electrostatic shocks become a problem. When the air dries out, the frequency of shocks rises considerably, not because of the dry air itself, but rather its drying effect on fabric surfaces.

Normal walking on commercial carpets produces about three times as many shocks at 30% rh compared to the number of shocks at 35% rh The phenomenon is governed by the relative humidity of the air because for nearly all materials, the sorption of water vapor is a function of relative humidity. As the relative humidity rises, more water molecules are adsorbed onto fabric surfaces, which makes them more conductive. But with dry surfaces created by low humidity air, the charges created by normal movement charges are not easily dissipated. They build up to high levels and are released all at once, creating uncomfortable shocks for people and damaging shocks for electronic equipment.

Even though the electrostatic problem depends on relative humidity, the designer can still design in terms of dew point. At a typical winter building temperature of 70°F, a 40°F dew point is roughly 35% rh— the level at which the frequency of electrostatic discharges falls off sharply. [At 21°C, a 4.5°C dew point is roughly 35% rh]

Condensation inside walls

Expressing the control level in terms of dew point is useful, because it helps the designer remain aware of the potential for condensation inside building cavities during cold months. According to manufacturers of humidifiers, more problems come from excess internal humidity than from lower-than-optimal humidity. Condensation inside wall cavities comes from putting too much water vapor into a building with conventional walls that are not designed to avoid condensation.

If the building needs winter humidification above a 40°F dew point [4.5°C], the architectural designer must be alerted to pay close attention to the likelihood of condensation. The HVAC designer should perform a dew point analysis of the proposed wall sections to assist the architectural designer in avoiding a condensation problem. Conventional walls will be damaged by high internal humidity in cold climates. Also, the HVAC designer must take care to keep the building pressure neutral rather than either positive or negative with respect to the outdoors. Cold air leaking *in*, or humid air leaking *out* will cause severe damage in heavily-humidified buildings.

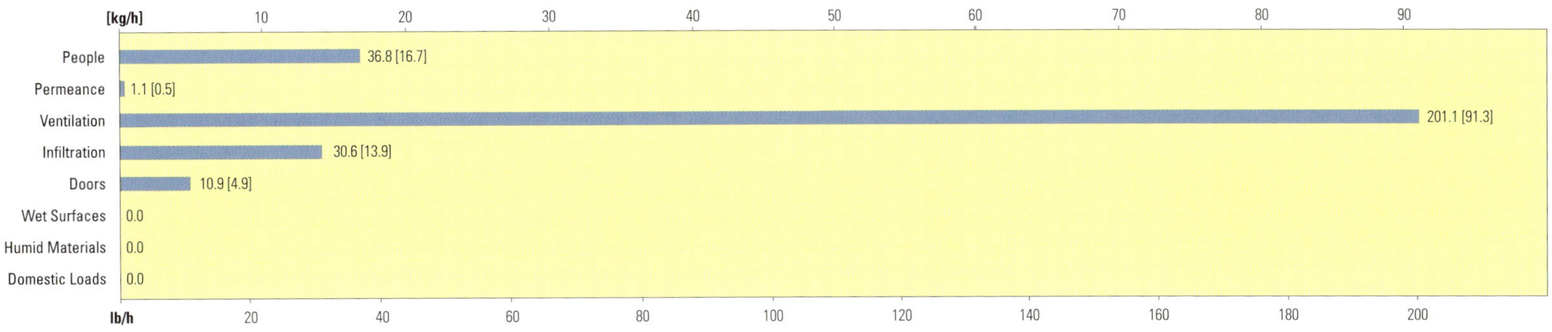

Fig. 19.3 Office Moisture Loads
Note how the ventilation air dominates the moisture load. Usually, the most cost-effective strategy is to dry that incoming air ***before*** *it can mix with the return air.*

Paper and printing

High-speed printers are also likely to have problems when the humidity is low. The rapid separation of paper from the rubber traction roller generates a large electrostatic charge, causing the paper to cling to the roller, and perhaps to jam the mechanism. But the more common office printing problems occur at high humidity.

Paper adsorbs moisture from high humidity air, and the resultant curling and buckling at the paper edge interferes with smooth feeding of humid sheets through the printer.

Also, high humidity interferes with copiers and laser printers, because these depend on maintaining separation between areas of high and low electrostatic charge on their photosensitive drums. If the charges leak from one part of the drum to another, the sharp printing detail cannot be held, and any large areas covered by toner are likely to have thin spots that look like "blotches". In those areas the toner coverage is uneven because part of the electrostatic charge leaked away in the high humidity, reducing the amount of toner attracted to the surface of the drum.

Finally, in high humidity, ink-jet printing is likely to appear fuzzy. When paper adsorbs water vapor, the ink spreads out on the humid paper, obscuring detail and blending adjacent areas of different color. This problem also occurs *after* printing when ink-jet images are stored in humid filing areas. A few months later, text has become blurred and the graphic images have "bled" at the edges. Adsorbed water vapor acts as a vehicle to spread the ink deeper into the paper, and across its surface.

Suggested set points

In summary, if eliminating electrostatic discharge is an especially important goal, keep the moisture level in dry months up to a 40°F dew point. But in winter, human comfort needs can be met and many problems with electrostatic charges can be eliminated by the lower level of a 30°F dew point [4.5°C dew point to minimize electrostatic discharge, where a dew point of -1°C is usually sufficient for comfort]. During humid months, keep the dew point below 57°F [13.9°C] for comfort, which will also minimize problems with laser printers, copiers and ink-jet printers.

Principal Loads

In an office, the people visiting and working inside the building are responsible for most of its continuous moisture load. Not only do people exhale and perspire moisture, but they enter and leave through exterior doors. Most importantly—they require outdoor air for ventilation. Figure 19.3 shows the impact of these three sources of humid-season moisture load for a three-story office building with roughly 45,000 ft2 [4,180 m2] of floor space, occupied by 250 people and located in Atlanta, GA.

Ventilation

Note how the ventilation load is more than twice as large as all other sources combined. Clearly, the designer must pay close attention to the ventilation load. This example assumes an outdoor air code requirement for 20 cfm per person [10 l/s/person]. If the client prefers to use more outdoor air to dilute indoor contaminants or if the code requirement is less, the size of the required dehumidifier will rise or fall significantly.

Given the importance of the ventilation moisture load, the designer might consider some means of reducing that load while still meeting the letter and intention of the code. Outdoor air is used to dilute the concentration of gaseous contaminants and infectious particles. When the building is unoccupied or only partly occupied, the code may allow the ventilation air quantity to be reduced. The controls to reduce the outdoor air flow would add cost, and the installed dehumidification capacity would *not* change—but the annual energy cost savings could be very large.

In some locations, ventilation codes allow the outdoor air flow to be reduced if the concentration of carbon dioxide in the building falls below 1,000 parts per million. In other locations, gas-phase filtration combined with high-efficiency particulate filtration can be used to clean the recirculating air so that less outdoor air is required. This strategy can be especially useful where the outdoor air is not especially clean—such as in dense urban areas or near congested highways. When ventilation air can be reduced without reducing the quality of the indoor air, humidity control equipment will be less expensive to operate, and control levels will be more stable.

On the other hand, overemphasis on saving operating costs was partly responsible for the notorious indoor air quality problems in the U.S. during the 1980's and 1990's. Ventilation air is used for the important purposes of maintaining comfort and ensuring health. Minimizing outdoor air is useful—but eliminating it or reducing it too much can be against code and hazardous.

Infiltration

Next, consider the infiltration load. This example assumes that the building will be much tighter than "average" office buildings, so that when wind blows humid air against the longest exterior wall, that wall will not leak more than 0.1 cfm/ft^2 [0.5 l/s·m^2]. In most locations, this leakage rate is a conservative assumption. The actual peak load is likely to be lower in most climates. Here's why.

By definition, the peak design moisture level is not likely to be exceeded for more than 35 hours in a typical year, and when that does occur, the wind is not necessarily blowing at full strength against the longest wall. So to include this highly variable and intermittent load in the capacity of the dehumidifier will ensure the unit has some unused capacity in many climates. To need full dehumidification capacity, the building would have to be fully-occupied during a hot rainstorm with a stiff breeze blowing against the longest wall. On the other hand, in areas like Central Florida, such conditions may occur for an hour or two during every day of the rainy season!

So in hot, rainy climates, the designer may choose to assume the exterior walls will have "average" or "loose" construction to avoid being short of dehumidification capacity. According to the *ASHRAE handbook—Fundamentals* (1997), the average construction of commercial buildings probably leaks at 0.3 cfm·ft.^{2}, and loose construction leaks at about 0.6 cfm·ft^2 [1.5 and 3.0 l/s·m^2, respectively]. In other words, assuming average or loose wall construction multiplies the infiltration estimate by three or six times, respectively. Chapter 10 contains more details concerning infiltration assumptions.

People

Turning now to the load from people's respiration and perspiration, this example assumes that most people will be seated, doing light work, and that only a few will be walking or lifting, and that nobody will be entering the building with rain-soaked clothing. So the load estimate from people could be much larger if the office were a municipal building with a regular flow of citizens needing permits of some kind.

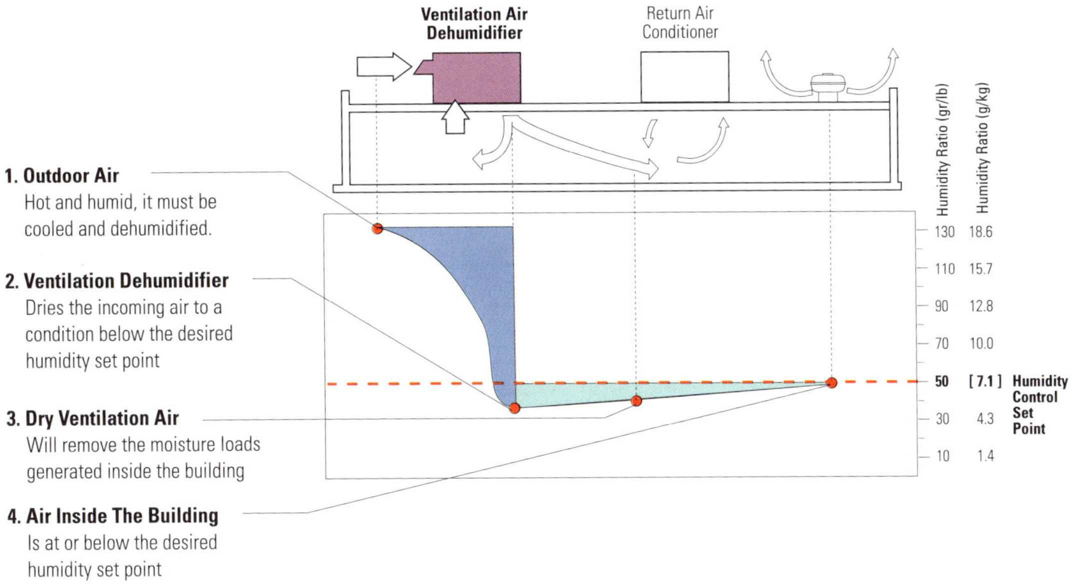

Fig. 19.4

Using Dry Ventilation Air To Dehumidify The Whole Building

By over-drying the incoming air, dry ventilation air can absorb the internal moisture loads, freeing the cooling units to remove sensible heat more efficiently. But that ventilation dehumidifier needs a recirculation mode so that dehumidification can continue when ventilation air flow is reduced during unoccupied hours.

Door Activity

The load from the doorways assumes that there is a vestibule at the main entrance to the building—two doors in series, separated by a small distance. That configuration keeps the moisture load to a minimum. If there were no vestibule to keep wind from blowing humid air into the building, the load from door openings would be about 30% larger for the same number of openings per hour.

Vapor permeation

Finally, note the very small size of the load from vapor permeation. Although this load element is well known to engineers and much attention has been focused on it in the past, one can see that its contribution to the total load is nearly negligible. The HVAC designer should focus much more time and attention on the larger load elements such as the ventilation requirement and the owner's estimate of the door traffic per hour. Similarly, the architectural designer would do well to focus on reducing infiltration by installing a continuous air barrier—rather than spending time and money to specify building materials with low vapor permeance ratings. The perm rate on any material without air holes is already so low that there is no benefit to reducing it further in typical office buildings. Sealing up the duct connections to air handlers and sealing all duct seams will reduce the moisture load much more than adding a more vapor-retardant layer to the wall.

Suggestions & Cautions for Offices

After the loads are well-quantified, the designer can proceed to laying out the system. Looking at the distribution of those loads, one can quickly see why controlling humidity in offices is largely a matter of controlling the moisture content of the ventilation air.

1. Consider drying the ventilation air centrally

If the ventilation air flow is dehumidified to slightly *below* the control dew point during the summer, the air will act as a "sponge", absorbing the moisture loads generated inside the building. Without that preconditioning, the ventilation moisture load becomes a major burden that must be removed by equipment that recirculates air inside the building.

In the example building, the ventilation air carries an excess moisture load of about 200 lbs per hour because the outdoor moisture level is 133 gr/lb compared to the indoor level of 65 gr/lb. But if the outdoor air is dehumidified down to a level of 55 gr/lb (a dew point of 50°F), the incoming dry air can *remove* 30 lbs/hr of the internal load, instead of *adding* 200 lbs to that load.

[In the example building, the ventilation air carries an excess moisture load of about 90 kg per hour because the outdoor moisture level is 19 g/kg compared to the indoor level of 9.3 g/kg. But if the outdoor air is dehumidified down to a level of 7.8 g/kg (a dew point of 10°C), the incoming dry air can *remove* 13.6 kg/hr of the internal load, instead of *adding* 90 kg/h to that load.]

Given dry ventilation air, the balance of the load is small enough so that conventional packaged rooftop cooling/heating units can keep humidity below a dew point of 57°F [14°C], which satisfies requirement for comfort and for office equipment operation.

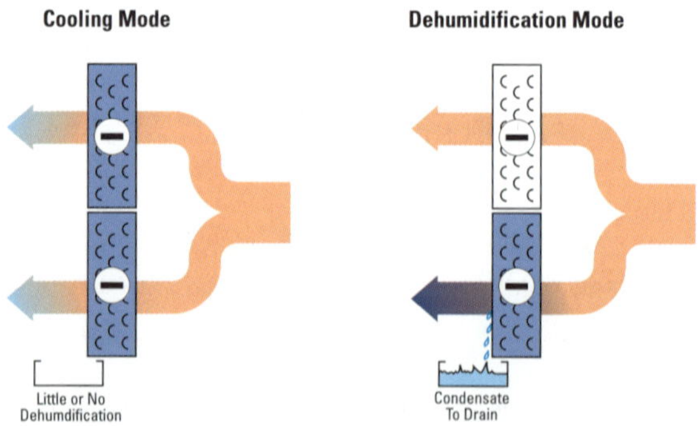

Fig. 19.5 Split-Coil Option
With this option, the lower half of the coil responds to the humidistat as well as to a demand from the thermostat

However, the outdoor air dehumidifier should still be equipped with a recirculation mode. That way, when ventilation air flow is reduced the unit can still act as a dehumidifier for the return air. The other air conditioning units are still relieved of the burden of the internal dehumidification load.

The main difficulty with drying ventilation air centrally is the need to distribute that air evenly throughout the building. The designer has two choices: distribute dry air to each of the other air handling units through rooftop ductwork, or direct to each zone through a second set of duct work. Both choices are expensive. A typical compromise is to serve a *group* of rooftop units from one ventilation dehumidifier.

In the winter, if moisture is added to the ventilation air it will be much easier to keep the dew point up to the desired level inside the building. In cold climates, however, the outdoor air moisture deficit is so great, and the temperature is so low, that additional humidification will still be needed inside the building, in addition to humidifying the outdoor air. When the humidification load is that high, it often costs less to humidify the blend between the outdoor and return air, rather than humidifying each air stream separately as suggested for dehumidification.

2. Consider the split-coil cooling unit option

To gain control of summertime humidity, a dedicated dehumidifier is necessary. But the dehumidifier can be smaller when the cooling units can dehumidify the air even when it does not need cooling. Also, if the owner only needs *moderation* rather than *control* of humidity, such specially-configured cooling units may satisfy the need without the additional dedicated dehumidifier.

To accomplish dehumidification without excess cooling, some manufacturers offer the option of splitting the cooling coil into upper and lower segments served by separate compressors. This arrangement is shown in figure 19.5. The lower segment operates on a command from either the thermostat or the humidistat. The upper segment of the coil only responds to the thermostat. When there is no demand for cooling, but humidity is too high, the lower segment of the coil operates, and chills part of the supply air to dehumidify it. The overcooled dry air then blends with the warmer air that came through the upper part of the coil, and the resulting mixture is dry, but no longer overchilled.

3. Consider installing a load reducer

Since the majority of the excess moisture comes from ventilation air, a passive desiccant wheel can be very helpful in reducing the annual load. A passive desiccant wheel uses the relatively dry exhaust air from the building to reduce the moisture level of the incoming outdoor air. The wheel rotates between both air streams, transferring excess heat and moisture from one air stream to the other, without transferring the contaminants of the exhaust air.

A passive desiccant wheel reduces the peak moisture loads and also reduces the size of the humidification load during winter. With reduced loads, conventional rooftop packaged equipment can often moderate humidity extremes during both seasons without support from dedicated dehumidification and humidification equipment. To achieve actual *control* of humidity, dedicated equipment will still be needed.

Fig. 19.6 Load Reduction Option
Passive desiccant wheels (enthalpy wheels) are available as a standard option on many brands of packaged rooftop cooling units. These devices reduce the moisture load during many of the unit's operating hours, allowing the dehumidification and humidification units to be smaller and less expensive to install and operate.

More information on load reduction equipment is contained in Chapter 15. To take advantage of this equipment, the designer must lay out the ductwork so the incoming ventilation air enters the building at the same point as the leaving exhaust air. The wheel must be placed between those two air streams.

4. Economizers must compare indoor and outdoor dew points

Air-side economizers are a common feature of low-rise office HVAC systems. When the outdoor air temperature is cooler than the desired supply air temperature, these devices open the outdoor air damper wide while closing the return air damper. That way the system uses more "free" cooling from outdoor air, reducing the cost of running the cooling system.

However, when the building needs summertime dehumidification, the designer must arrange the economizer controls to avoid bringing excessively humid air into the building, even when its cool temperature would be useful. Otherwise, the economizer may indeed "save cooling," but that will often come at the expense of losing control of humidity.

With modern building management systems, the economizer control can be arranged to compare not only temperatures, but also the dew point differences between indoors and outdoors. Set the economizer to open when the outdoor temperature is low—but override that signal if the outdoor dew point is above the desired supply air dew point. This reduces the number of hours of "free cooling," but will preserve humidity control—which costs more than cooling.

5. Seal the ductwork to avoid condensation damage to the building.

If cold supply air leaks out of ductwork into unconditioned spaces in the summertime, moisture may condense on cold metal surfaces chilled by the leaking air. This is often the reason for dripping water above dropped ceilings.

Leaking return air ducts and exhaust air ducts can also create condensation problems. The leaking duct creates a slight suction in the spaces it passes through, which often connect directly to the cavities near the exterior walls. When those cavities are under suction, outdoor air is pulled into the building, leading to condensation in the building structure wherever humid air drifts over a cold surface.

The solution to both problems is to seal the ductwork with mastic. Then cold supply air won't blow out into unconditioned spaces, and return air won't create suction that pulls untreated air through the exterior walls.

6. Size and locate duct-mounted humidifiers carefully to avoid condensation

Low-rise office buildings are often built on speculation with severe budget constraints, so their floor-to-floor heights are usually minimized. This reduces the amount of space available for ductwork. The HVAC designer may be forced into using higher-than-recommended duct velocities, because the loads demand a given amount of air, but the ducts must remain very small to fit above the ceilings. Most designers are familiar with the noise created by high velocities, but they may be less familiar with the implications for sizing and locating humidifiers.

Duct-mounted, heated vapor generators are a popular type of humidifier for office buildings. They are usually mounted in the supply air ducts downstream of the air handler, so that the cold outdoor air and the warm return air are well-mixed. Warmer air absorbs water vapor more quickly than cool air. But when the duct velocity is very high, the vapor travels a longer distance through the duct before it is fully absorbed into the air, increasing the risk of internal condensation. Two suggestions may help the designer when he or she must humidify fast-moving air.

First, be sure to allow long, clear duct runs downstream of humidifiers when duct velocities exceed normal limits. Otherwise, vapor can condense on duct obstructions like turning vanes and sensors or even supply air diffusers. Admittedly, this suggestions is usually difficult to follow. When space is limited, long, straight runs of supply duct are not common. So the second suggestion may be more widely applicable.

Consider splitting the load between multiple humidifiers in different air streams. Adding equipment costs more, but smaller amounts of vapor are more quickly absorbed by fast-moving air streams, so condensation is less likely.

6. Avoid oversizing the cooling equipment

Building pathologists constantly identify oversized cooling equipment as one of the leading causes of moisture-related damage and fungal problems in commercial buildings.

Here's the typical failure mechanism: compressors in oversized cooling units operate only a short time to satisfy the thermostat. The compressors shut off quickly, but the supply fan stays on to provide ventilation air and general circulation. The coil surface warms up, halting dehumidification, but ventilation moisture remains. Further, the moisture originally condensed on the coil re-evaporates into the supply air. The thermostat is satisfied, but moisture from ventilation air continues to build up—first to uncomfortable levels and then, over time, to levels that create damage and allow fungal growth.

Most designers will not heed advice to avoid oversizing. It seems counterintuitive and appears imprudent to select the "next smaller" unit rather than the "next-larger" one. But smaller is better when selecting cooling equipment from the perspective of humidity control. A smaller unit will take longer to remove the heat load, which means it will be removing moisture for longer periods.

When the designer considers the question of unit size vs. load, he would do well to remember that most load estimates are inherently conservative. They assume that all loads will peak at the same moment, and that they will all occur when both the solar load and the convective heat loads from outdoors are also peaking. By definition the peak load occurs for less than 1% of the building's life. Oversizing the cooling equipment pretty much guarantees that the occupants will indeed be comfortable for those 1% hours—but only at the cost of being ***uncomfortably cold and clammy for much of the other 99% of the building's life***. Slightly undersizing the cooling equipment provides more hours of comfort.

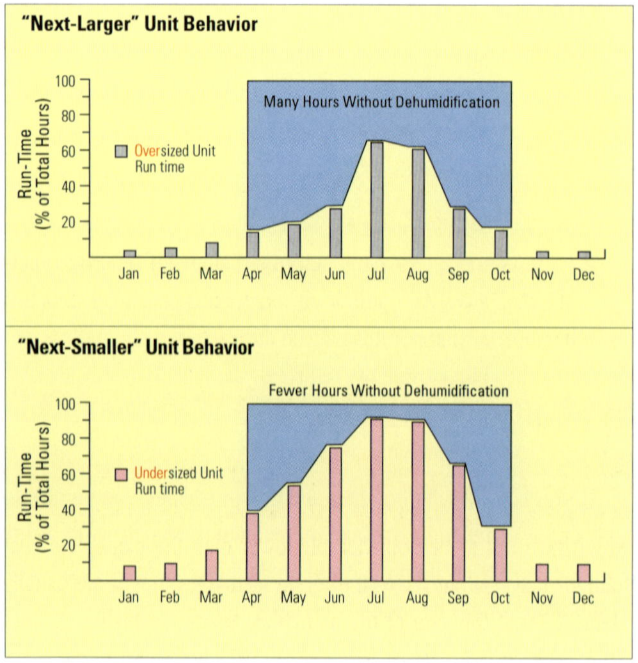

Fig. 19.7 Avoid Oversizing Cooling Equipment

Oversized cooling units run so infrequently that they are not effective dehumidifiers during most of the occupied hours. Reducing "safety factors" and installing the "nearest-smaller" unit for the load, instead of the "nearest-larger" unit will lead to longer run-times, and therefore more dehumidification capacity.

References

Human comfort, ASHRAE Standard 55-2000

Ventilation for acceptable indoor air quality, ASHRAE Std 62-1999

Parker, Danny S, Fairey, Phillip W, III, McIlvane, Janet E.R, "Energy-efficient office building design for Florida's humid climate." ASHRAE Journal, April, 1997, pp: 49-57

Rosenfeld, Shlomo I, "Worker productivity; hidden HVAC cost." Heating, Piping and Air Conditioning Magazine. pp: 46-48. September, 1991. Penton Publishing, Chicago, IL.

References to help estimate heat loads from office equipment

Designers sometimes oversize cooling equipment in offices because heat loads from office equipment can be difficult to quantify. Several papers in recent years have actually measured plug loads, which turn out to be considerably lower than the nameplate heat load for that equipment. With diversity, actual heat loads are usually about 25% of the sum of the nameplate loads. According to field measurements, net actual loads range from less than 0.5 W/ft^2 in typical low-density offices to slightly over 1.1 W/ft^2 for extremely dense installations of computer equipment with many display screens and active laser printers. These articles quantify plug heat loads from many types of computers and office equipment in great detail, while still remaining simple to understand and engaging to read.

Wilkins, Christopher K; Kosonen, Risto; Laine, Thomas. "An analysis of office equipment load factors." ASHRAE Journal, September, 1991. PP. 38-44

Komor, Paul, Ph.D. "Space cooling demands from office plug loads." ASHRAE Journal, December 1997. pp: 41-44.

Wilkins, Christopher, Hosni, M.H, "Heat gain from office equipment." ASHRAE Journal, June 2000. pp: 33-44

Donated Photo

Figure 19.5 - © AAON Inc.

Chapter 20

Retail Buildings

By Lew Harriman with Reinhold Kittler

Fig. 20.1 Comfort = Retail Sales
With dedicated humidity control, the store need not be overcooled or overheated. This provides a more comfortable shopping environment, which increases sales and profits.

Retail Buildings

The longer you stay in a store, the more you will buy. So the store's temperature, humidity and air freshness should be inviting rather than merely tolerable. By controlling humidity, the HVAC designer can create an environment that encourages each customer to linger longer, which increases the merchant's ability to serve a customer's needs through more sales.

At the same time, retail stores need heavy ventilation. They contain large volumes of new products and packaging. New materials release vapors that must be diluted with outdoor air to limit odors and to avoid health problems. But outdoor air carries a very large moisture load.

*The designer can either install a dedicated dehumidifier, or use a passive desiccant wheel to reduce the ventilation load. Simply reducing the ventilation load works best when the owner only needs **moderation** of moisture peaks and valleys, as in general retail stores. Installing a dedicated dehumidifier is the solution when the owner needs to actually **control** humidity within specified limits, as in a supermarket.*

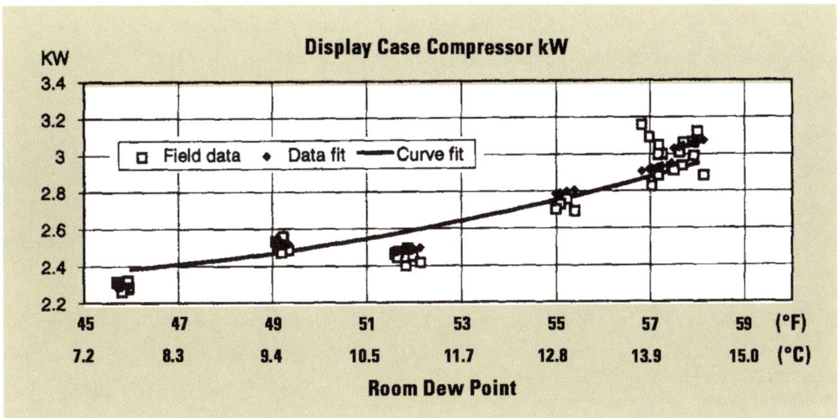

Fig. 20.2 Compressor Power vs. Store Dew Point

When less moisture condenses on display case cooling coils, the load and the power requirement go down. The lower the dew point in the store, the greater the savings in refrigeration. This is one of several reasons why supermarkets are dried more deeply than other commercial buildings (Faramarzi, May 1997)

Supermarkets

In a supermarket, the product display cases cool and dehumidify the store by accident. Display cases spill cold air into the store aisles. Also, their cooling coils operate at very low temperatures, which produces condensation and dehumidification. Neither function is energy-efficient when performed by display cases. The HVAC system can cool and dehumidify the store's air at much lower cost.

Consequently, supermarket HVAC systems are designed to make effective use of any waste cooling that leaks from the cases, and designed to dehumidify the air, so that excess moisture does not freeze on cooling coils or on the products.

Purpose of humidity control in supermarkets

In a typical North American supermarket, annual profits are between 1 and 2% of sales. Energy costs represent about the same percentage of sales. So when all other costs are constant, every dollar of energy savings represents a much larger contribution to profits than for other commercial buildings.

For example, if supermarket energy costs can be reduced by $20,000 per year, that accomplishment has roughly the same profit impact as having sold an additional $1,000,000 worth of groceries. That's one reason why supermarket owners are more concerned about excess humidity than other commercial building owners might be. High humidity creates operational costs for the refrigeration systems and for the anti-condensation heaters on the display cases.

Also, sales are lost when humidity is too high. On closed-door display cases, fog on the glass interferes with product visibility. And on frozen products, excess frost makes the customer avoid the product and choose a different item, resulting in both unsalable product and costs for disposing of those unsalable items.

Fig. 20.3
Dedicated Dehumidification
supermarket HVAC systems are usually custom-designed, packaged rooftop equipment that contain dedicated dehumidification components. This example unit contains a gas-reactivated active desiccant wheel. Others often contain mechanical dehumidification equipment, which competes with the desiccant alternative. (Banks, 1982)

In addition to creating high operating costs, humidity indirectly affects the safety of the food. In the United States, the US Food and Drug Administration now requires that the surface temperature of any refrigerated food product never rise above 41°F [5°C] (USDA Food Code, 1999). This standard is very difficult to maintain when the display case coils must heat-up frequently to rid themselves of large amounts of frost. Frequent defrost wastes energy because it requires both heating and then re-cooling the air in the case. But more importantly, frequent defrost increases the risk of growing hazardous bacteria on or in the food products. During long defrost cycles, the temperature of the product can rise to a level that allows more prolific bacterial reproduction. Reducing humidity shortens the defrost cycles and reduces their number.

In a supermarket, high humidity is a problem. Low humidity is not. In theory, customers could become uncomfortable at very low dew points. But a supermarket is not usually dried down to uncomfortable levels (below a 30°F or -1°C dew point).

In summary, supermarkets are usually dehumidified to achieve three purposes:

1. Reduce the amount of energy wasted by display cases as they defrost and as they operate their electric anti-condensation heaters. (Banks, 1982)
2. Reduce frost on product, and speed evaporation of condensation that obscures the view through glass display case doors. (Adams, 1992)
3. Avoid frequent or lengthy defrost cycles that might allow the product temperature to drift above 41°F [5°C] into a more risky temperature range. (Faramarzi, May 1998 and USDA Food Codes of 1993 and 1999)

Typical control levels in supermarkets

The lower the humidity, the greater the benefits for the product and for the refrigeration systems. On the other hand, the HVAC operating costs increase when the store is dried deeply. The optimal dehumidification control level usually depends on whether the owner prefers to use desiccant or cooling-based equipment.

Using desiccant dehumidifiers, the store is typically controlled at 35% to 40% rh at 75°F—a dew point between 45 and 50°F—which represents a humidity ratio between 45 and 52 gr/lb. [Humidity ratios between 6.4 and 7.4 g/kg]

With cooling-based dehumidifiers, owners usually prefer a less difficult humidity level of 45 to 55% rh at 72°F—a dew point between 50 and 55°F—which represents a humidity ratio of 53 to 65 gr/lb. [Humidity ratios between 7.4 and 9.2 g/kg.]

The different control levels reflect the economics of each technology. Desiccants can achieve the lower dew point without adding

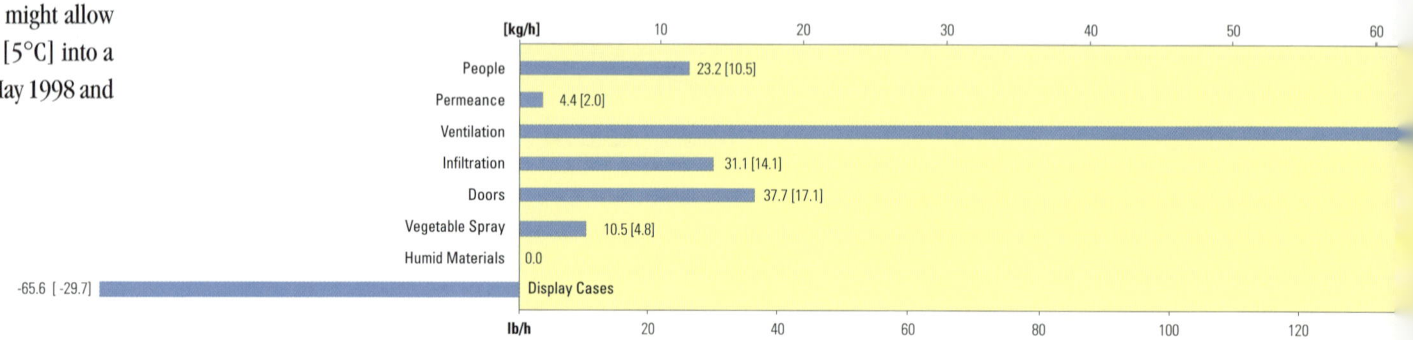

Fig. 20.4
Supermarket Moisture Loads
Note the display case refrigeration coils remove moisture even when the dew point is kept low.

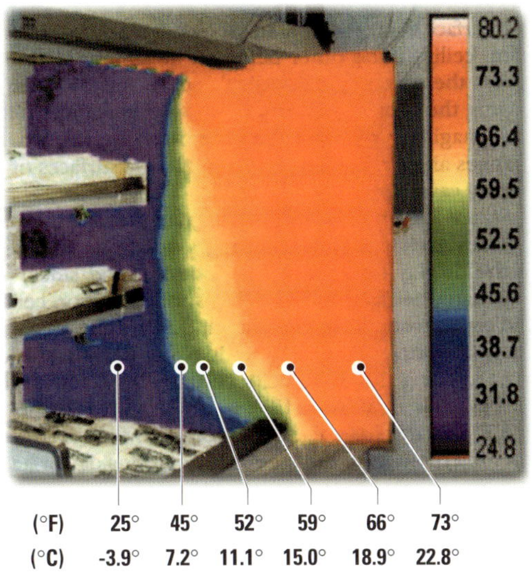

Fig. 20.5 Cases Help Cool The Store

Cold air spilling out of the cases creates a "case credit," a reduction in the store's cooling load. The amount of load reduction depends on the number and type of cases in the store. (Faramarzi, 1999)

construction cost. On the other hand, cooling-type units can achieve lower HVAC operating costs by working at higher control levels.

Owners who favor cooling-based dehumidification generally believe that a lower humidity level does not provide enough savings in refrigeration to offset the higher HVAC operating cost. In contrast, those who prefer desiccants generally believe that since desiccant equipment does not cost more to install, there are both economic and comfort benefits to operating at lower dew points and higher dry bulb temperatures.

There are persuasive technical arguments and field tests in favor of both technologies. So some owners use both desiccants and cooling based units at different locations, making their decision based on factors like the annual moisture load at each site (Harriman, 1997). Others may decide based on site-specific utility costs for natural gas vs. electrical power.

But in the end, most owners decide to use either desiccant or cooling dehumidifiers because they have had more reliable results with one or the other technology. The humidity control level follows naturally from the capabilities of the preferred equipment.

Principal moisture loads in supermarkets

Moisture and heat loads in supermarkets are a bit unusual. Figure 20.4 shows a typical load estimate for a supermarket with a store area of about 45,000 ft^2 [4,180 m^2].

Note the effect of the refrigerated display cases, These act as dehumidifiers, since they collect frost on their evaporator coils. The net moisture load is reduced by this dehumidification.

Next, note the amount of moisture brought into the store by the ventilation air. This large load reflects the outdoor air flow recommendations of ASHRAE Standard 62-1999. A lower standard of ventilation may be permissible in some code jurisdictions. But regardless of the exact amount, it is clear that the ventilation air carries the largest moisture load. That's why a supermarket HVAC system almost always removes that load by dehumidifying the outdoor air before it can enter the controlled space.

Door openings generate another large component of the moisture load. To minimize the door load, the building is usually kept under a slight positive air pressure. About 10% more dry ventilation air can be brought into the building compared to the amount of air exhausted from the cooking areas and from the toilets.

This "10% extra" design practice is used less often in supermarkets than in other commercial buildings. The supply air quantity in

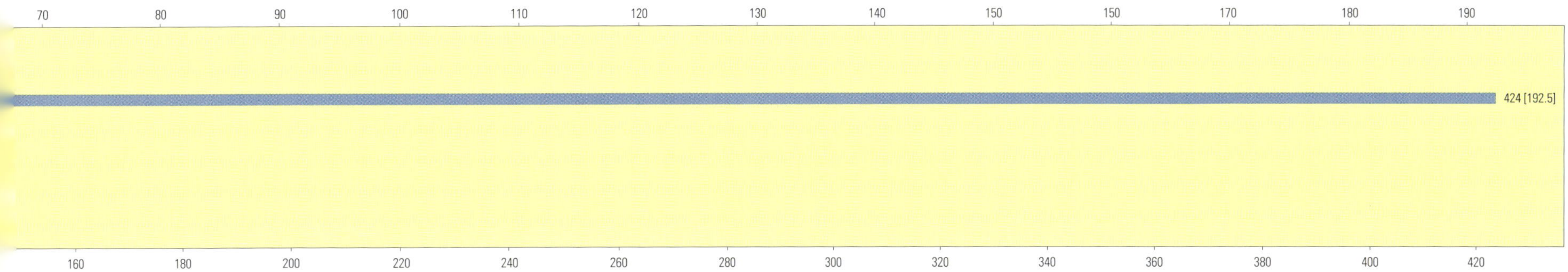

supermarkets is so high that 10% of that amount is usually considered excessive for "extra" pressurization air. Using an alternative approach, enough dry ventilation air is brought inside to provide a net outward air velocity of 50 fpm through the open area of the entry doors. [An outward air velocity of 0.25 m/s]

With either of these design practices, some owners believe there will be no significant net moisture load from door openings. Other owners prefer to leave a reduced door moisture load component in the calculation for a more conservative approach, which results in larger dehumidification equipment.

The same debate continues with respect to the load component from air leakage. Some owners and designers believe there is no net infiltration load if the building is maintained under an average net positive air pressure. But there are excellent field measurements that support the opposite point of view. (Cummings et al, 1996) All commercial buildings leak air, for the reasons described in Chapter 16 (Building Pressure Management). So some allowance must be made for humid air leaking into the building through cracks if the low humidity control level in a supermarket is to be maintained reliably. Chapter 11 contains more discussion of these issues.

Suggestions & cautions for supermarkets

In a supermarket, there are constant and complex interactions between the HVAC systems and the refrigeration systems in the display cases. The designer who may be more experienced with other buildings will be most successful in supermarkets by keeping such interactions in mind.

Use the case credit—Avoid expensive oversizing and discomfort

Most commercial buildings do not contain business equipment that removes heat. But that's exactly what happens in supermarkets; display cases spill cold air into the aisles, as shown in figure 20.5.

The amount of cooling contributed by the cases depends on their design. Closed-door cases spill less air, while open vertical cases spill more air. But all cases spill enough air so that on average, between 40 and 65% of the total store's refrigeration evaporator capacity becomes a "case credit." In other words, the cases remove a great deal of cooling load from the HVAC system—between 30 and 50 tons of case credit in typical large supermarkets. [Between 105 and 175 kW of cooling case credit in large stores] It would be a costly error to duplicate that cooling capacity in the HVAC system.

Figure 20.5 shows estimated case credits for an actual building, built in the upper Midwest of the United States during 1999. The four sources did not agree on the amount of cooling spilled by the cases. Each display case manufacturer suggested different credits. Next, the ASHRAE Handbook contained data for "generic" cases, so those values did not agree with any specific manufacturer. Then the dehumidifier manufacturer had a different estimate which was based on hourly monitoring of similar stores. But as seen in figure 20.5, all sources other than the "New Consultant" agreed that quite a bit of cooling would be accomplished by the cases.

The Consultant's HVAC design was not chosen by the owner. That design would have cost a great deal more to install, operate and maintain compared to systems that realistically reflect the cooling effect of the cases. Also, with less oversizing, the HVAC system does a better job of avoiding overcooling the customers—a constant problem in supermarkets.

Locate return air ducts below the cases

Since cases spill cold air into the aisle, the HVAC system often pulls cold air away from the customers by placing return air ducts under the cases. That way, the cold air falling out of the cases is extracted before it can chill the air in the aisle. In most stores, about

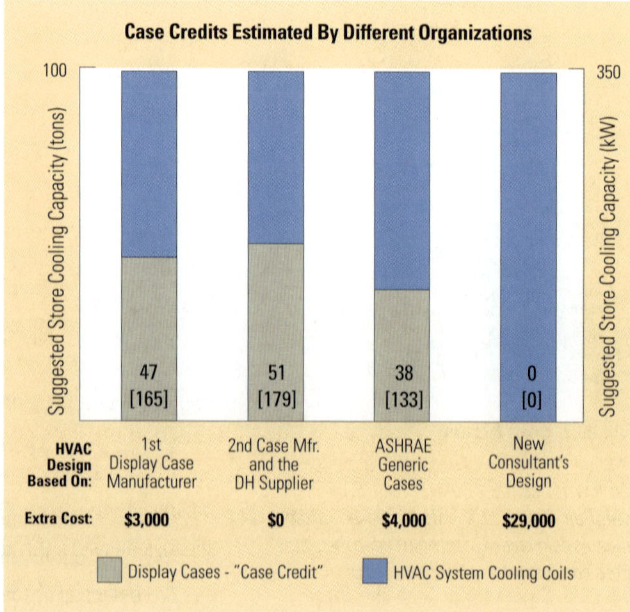

Fig. 20.6
Store Case Credits Should Not Be Ignored
When designing the HVAC system, it is important to ***subtract the case credit from the cooling load.*** *Otherwise, as seen in this example, the system will have excess cooling capacity, which costs more than necessary, and makes it difficult to keep the store's temperature up to comfortable levels.*

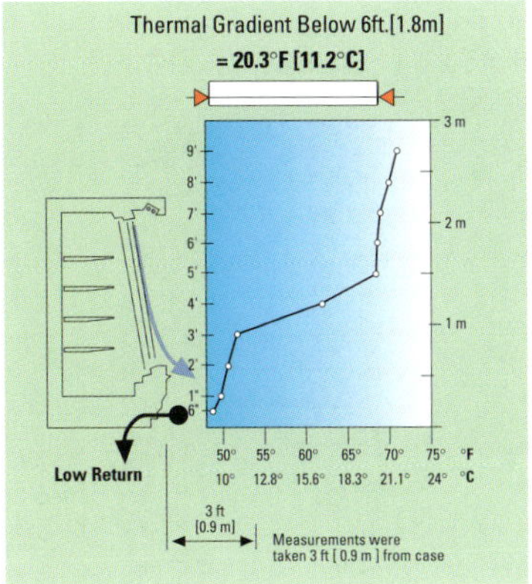

Fig. 20.7 Cold Aisles

Cold air from cases keeps the temperature too low for comfort below waist level. Dry, warm air from dehumidifiers can be mixed into this cold layer, which helps keep the aisle more comfortable. (Harriman, 1996)

Fig. 20.8 Below-Case Supply Air

By mixing warm dry air into the aisle from below the display cases, the air temperature in the aisle becomes not only warmer, but more uniform in the occupied zone. (Harriman, 1996)

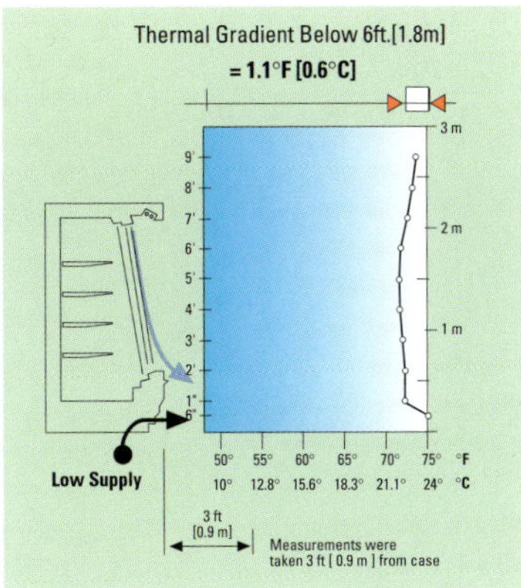

30 to 60% of the total return air volume is pulled from under the cases. The rest of the air is pulled from other parts of the store, to ensure that air does not become stagnant or stale in those other locations. Figure 20.6 illustrates the cold aisle problem. It shows measurements taken from a frozen food aisle in a supermarket located in Minneapolis, Minnesota on a day in July, 1995. (Harriman, 1996) The figure shows that even when using below-case returns, the aisle temperature gets very cool below knee level. Without the below-case returns, that lower part of the aisle would have been even colder and less comfortable.

Control on dew point or humidity ratio, not relative humidity

Controlling the dew point provides the most consistent benefits because the owner's purposes for humidity control all involve minimizing condensation. On the other hand, most commercial humidity sensors measure the air's relative humidity.

To control an absolute amount of moisture in the air, the building's energy management system can be used to calculate the dew point from temperature and relative humidity signals. An alternate approach is to install a dew point sensor and measure that variable directly. This is a superior solution from a technical perspective, but as of the publication date of this book, dew point and humidity ratio sensors are more costly than relative humidity sensors.

Install the humidity transmitter in a thermally stable location

In many commercial buildings, the humidity transmitter is located in the return air duct, so it can report the average relative humidity from different parts of the building. But that's not a good practice in a supermarket. The return air is likely to be rather cold, because air returns from the cold aisles below the display cases. In theory, measuring rh at low temperatures should not present a problem. The energy management system should integrate the rh signal with the temperature signal and provide an accurate dew point value for control. But at cold temperatures and high relative humidities, any small temperature change can quickly skew the sensor's reading, which makes the control system "hunt" for the right humidity level. When that happens, humidity levels fluctuate widely as equipment turns on and off rapidly in response to a fast-changing dew point calculation from the controller.

So it is best to locate the humidity transmitter in a location with a more stable temperature and moisture level—***certainly not***:

- In the return duct
- Near the entry doors
- Near the supply air diffuser

Most owners place the humidity sensor on a column near the display cases, at a height just out of reach of a curious customer, about a third of the way towards the back of the store from the entry doors—and mounted so that the supply air does not blow directly onto the sensor.

Reduce the defrost cycles

In the early 1990's, some owners who installed dehumidification were disappointed to find they did not achieve the expected operational savings. A closer look at these buildings showed that the designers and owners sometimes forgot to shorten or reduce the number of defrost cycles, and/or neglected to reset the anti-sweat heaters to reflect the fact that the store was being kept dry. Both functions must be reset to ensure savings, which requires some thought and attention on the part of the designer of the refrigeration equipment. (Tassou & Datta, 1999)

As of the publication date of this book, most defrost controls remain rather primitive with respect to gaining benefits at lower humidity levels. Defrost is often initiated by a time clock on a set schedule, regardless of the actual amount of frost on the coil. These controls often terminate the cycle based on the air temperature in the case, regardless of when the frost actually melts off the coil. With such overly simplistic controls, it is critical to monitor a store, investigate the actual need for defrost frequency at different dew points, and then adjust the number and duration of the cycles accordingly.

Some owners have controls that automatically reset the defrost schedule to longer intervals when the store stays below a certain dew point for a certain length of time. Others have put their experience into manual re-programming of the cycle—such as setting it for two rather than three defrost cycles in 24 hours. One hopes that better defrost controls will emerge in the future.

Regardless of the method, the owner or refrigeration designer must adjust the frequency and duration of the defrost cycles to reflect the level of dryness in the store—otherwise costs will stay constant in spite of dehumidification.

Reduce anti-sweat heater on-time

Display cases are equipped with thin heaters to warm their surfaces so that moisture does not condense and drip. On closed-door cases, these heaters evaporate condensation from the glass doors. On "coffin" cases, the heaters are mounted around the upper edge of the case so condensation does not wet the customer's clothes as she leans into the case to retrieve the product. In a typical store, these heaters may consume about 30 kW of electrical power every hour. Like the defrost cycles, the heater on-time must be reduced to reflect the level of dryness in the store. Once the dew point falls below a certain level, there will be no need for heaters at all. During periods of mid-range dew points, the heaters may only need to operate half the time or less.

For anti-sweat heaters, good controls are available. A sensor measures the dew point in the store, and adjusts the heater on-time automatically. But whether automatically or manually, the anti-sweat heater on-time must be reduced if the owner is to realize savings from keeping the store dry.

Close off outdoor air on deli and bakery cooling units

Small rooftop heating/cooling units are often installed above areas with high sensible heat loads, such as the deli area or the bakery. These units, however, do not have the dehumidification capacity to remove moisture from incoming ventilation air. The ventilation air for these areas, as well as for the rest of the store, is dried and supplied by the large, main air handlers that contain the dehumidification units.

Therefore the designer should specify that any outdoor air inlets on the deli and bakery units must be sealed up—so they do not accidentally pull humid outdoor air into the building.

Size the dehumidifier for the moisture load—not the supply air volume

The design process for most HVAC system begins with a sensible heat load calculation—which usually determines the amount of air in the system. The cooling coils are sized to handle that amount of air. But the dehumidifiers should not be sized to handle the entire supply air flow.

This is because both mechanical and desiccant dehumidifiers remove much more moisture than conventional cooling coils. Since they produce much drier air, they do not need to supply large amounts of that very dry air to remove the moisture load.

The best practice for sizing supermarket dehumidification is to:

1. Determine how dry the dehumidifier can make the air, based on the mixed air temperature and moisture content entering the unit.
2. Then calculate how much of that air is needed to remove the moisture load inside the store.
3. Size the dehumidifier for that smaller amount of air—not the entire amount of the supply air.

This procedure produces smaller, more efficient and much less expensive dehumidifiers than if the designer simply tried to push the entire supply air quantity through the dehumidifiers. Most of the supply air can bypass the dehumidifier, reducing both installed costs and operating costs.

Take advantage of dry air to downsize HVAC

Owners with extensive experience with dehumidification have usually reduced the amount of air they circulate through a store.

This reduces the cost of ductwork, and saves the cost of excess fan power for the entire life of the building.

Before dehumidification, supermarket HVAC designers found that to moderate the humidity, large amounts of air needed to be circulated through conventional cooling units. Since these did not remove much moisture, more air was needed to remove the moisture load. A very high circulation rate of 1 cfm/ft² [0.044 l/s • m²] became standard practice. The high air flow also helped to mix the air in the cold aisles with warmer supply air, so that customers were less uncomfortable.

But with dedicated dehumidifiers, owners found that less air was needed to remove the internal moisture load. They also developed better ways of distributing the air, so that temperatures could still be kept up to comfortable levels without excess air circulation.

Dehumidifier manufacturers usually suggest a circulation rate of 0.5 cfm/ft²—half of the older practice. Some owners disagree with that much smaller amount, and prefer to circulate between 0.5 and 0.7 cfm/ft² total supply air. But all owners with experience with dehumidifiers are likely to agree that the old practice of 1 cfm/ft² would be wasteful of installation and operating costs when air in the store is kept dry by dedicated dehumidifiers.

[Dehumidifier manufacturers usually suggest a circulation rate of 0.022 l/s • m²—half of the older practice. Some owners disagree with that much smaller amount, and prefer to circulate between 0.022 and 0.031 l/s • m² total supply air. But all owners with experience with dehumidifiers are likely to agree that the old practice of 0.044 l/s • m² would be wasteful of installation and operating costs when air in the store is kept dry by dedicated dehumidifiers.]

Supply warm, dry air into cold aisles

In both mechanical and desiccant-type dehumidifiers, dry air leaves the unit warmer than it entered. In desiccant-type units, this temperature rise is much larger than in cooling-based, mechanical dehumidifiers. But this heat—so often a disadvantage in conventional commercial buildings—is very useful in supermarkets.

As discussed earlier in these suggestions, frozen food aisles in supermarkets are always a comfort challenge because the air from the cases overcools the aisles. The designer can arrange the air distribution so that warm, dry air goes from the dehumidifier *directly* to the frozen food aisles. This has two beneficial effects: the dry air reduces the frost on the product and on the case refrigeration coils, and the warmth of the air helps keep the aisle at a more comfortable temperature.

When the designer chooses to use this strategy, it will be important to remember that hot air wants to rise rather than fall into the aisle. The air velocity from ceiling-mounted supply diffuser will need to be fairly high if the warm air is to mix well with the cold air near the floor.

Another air distribution strategy uses below-case supply air ducts. With that design, the hot dry air mixes directly with the air near the bottom of the aisle, and comfort is achieved without high-speed jets of hot air falling from above, as shown in figure 20.8.

References for supermarkets

Adams, Paul Jr., 1992 "Merchandising vs. energy consumption in the supermarket." *Heating, Piping and Air Conditioning* Magazine, April, 1992. Penton Publishing Co., Chicago, IL.

Banks, Nancy MacDonald, 1982. "Utilization of condenser heat for desiccant dehumidifiers in supermarket applications." ASHRAE Transactions, V.88, Pt.1.

Calton, Dean, 1985. "Application of a desiccant cooling system to supermarkets." ASHRAE Transactions, V.91, Pt.1

Calton, Dean, 1992. "Distributed air conditioning in supermarkets." *ASHRAE Journal*, May 1992, pp.40-44.

Cox, R.L, Haberl, J.S, and Claridge, D.E, 1993. "A study of energy use in grocery stores." ASHRAE Transactions, V. 99, Pt.1.

Faramarzi, Ramin, and Kemp, Kimberly. Aug. 1999. "Comparing older and newer refrigerated display cases." *ASHRAE Journal*, August, 1999, pp. 45-49.

Faramarzi, Ramin. 1997. "Learning more about display cases; Space relative humidity can have a more severe impact on performance than dry bulb temperature." *Engineered Systems* Magazine, May 1997, pp.40-43. Business News Publishing, Troy, MI.

Faramarzi, Ramin and Woodworth, Michele. 1998. "The case of the cold case; Detecting the effects of the FDA's 1993 food code on the thermal performance of a refrigerated display case." *Engineered Systems* Magazine, May 1998, pp.72-77. Business News Publishing, Troy, MI.

Faramarzi, Ramin, Nov. 1999. "Efficient display case refrigeration." *ASHRAE Journal*, November, 1999, pp. 46-54.

FDA Food Code 1999, Sections 3 and 5. U. S. Food and Drug Administration, Washington, DC. http://www.fda.gov/

Grzanka, Laura E., 1996. "Big savings in aisle one. Controlling humidity in supermarkets using waste heat." *Engineered Systems* Magazine. May 1996, pp. 70-73. Business News Publishing, Troy, MI.

Harriman, Lewis G, III. 1994. "Field experience with desiccant systems." *Engineered Systems* Magazine. January, 1994. pp.63-68. Business News Publishing, Troy, MI.

Harriman, Lewis G, III. 1996. "Under-case supply air for desiccant target zones." Proceedings of the 1996 FMI Energy Conference. Food Marketing Institute, Washington, DC. www.fmi.org.

Kosar, Douglas R. 1990. "Development of a gas-fired desiccant air conditioner." Proceedings of the 1989 Gas Research Conference. pp. 888-901 Government Institutes, Rockville, MD.

Khattar, Mukesh and Henderson, Hugh I., Jr., 1999. "Impact of HVAC control improvements on supermarket humidity levels." ASHRAE Transactions, 1999. Vol. 105, Pt. 1.

Litwak, David. 2000. "Food Safety enters the 21st century." *Store Equipment & Design* Magazine. June 2000. pp. 33-40.

Manley, Diane L, (Griffiths), Bowlen, Kennard and Cohen, Barry., 1985. "Evaluation of gas-fired desiccant-based space conditioning for supermarkets." ASHRAE Transactions, Vol. 91, Pt.1

Reid, David S., 1999, "Food preservation." *ASHRAE Journal*, November, 1999, pp. 40-45.

Tassou, Savas A., and Datta, Debra. 1999. "Influence of supermarket environmental parameters on the frosting and defrosting of vertical multideck display cabinets." ASHRAE Transactions, Vol. 105, Part 1.

Special Thanks

The Authors are indebted to the creative and tenacious early pioneers of dedicated dehumidification for supermarkets. In particular, Nancy Banks and David Parkman, two of the originators of the concept from the manufacturing side generously shared their 20 years of field experience, as did many owners, including Ralph Williams of WalMart, Mike Malone of HEB, Ken Welter of Stop & Shop and Charles Beach of KMart. Also, the industry itself is indebted to Ramin Faramarzi, who has spent several years carefully measuring and publishing the true effects of different dew points on the performance of supermarket refrigeration systems.

Donated Photos & Diagrams

Fig. 20.2 - Ramin Faramarzi

Fig. 20.3 - © Munters Dehumidification Division

Fig. 20.5 - Ramin Faramarzi

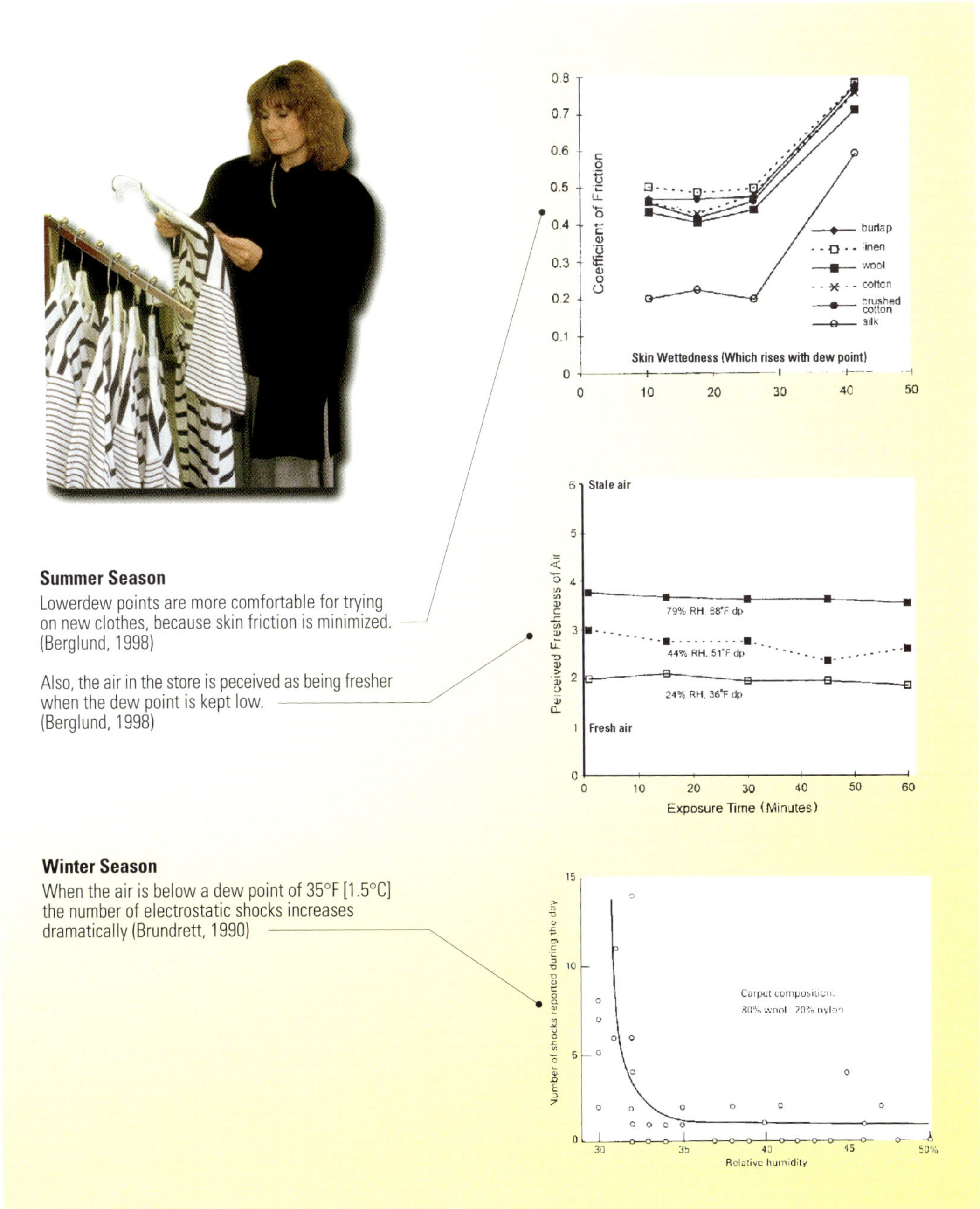

General Retail

Retail stores are built quickly and inexpensively. Unlike supermarkets, the mechanical equipment in a general retail store does not have a central role in the success or failure of the commercial enterprise. Consequently, the Senior Management of a retail enterprise at the chain or at the store level is not likely to be well-informed about equipment alternatives and the subtleties of equipment behavior. They are also unlikely to perceive any significant benefit to increasing the construction budget to provide more efficient or more durable systems. The cost of the mechanical system's operation and its effectiveness is much less important to these managers than making sure the store is staffed economically, and ensuring that the merchandise is well-stocked, competitively priced and well-presented.

On the other hand, retail managers become concerned with humidity control if customers and staff are uncomfortable. The HVAC designer can help minimize such discomfort by moderating the extremes of humidity that are typical of retail stores, which are highly-ventilated with outdoor air.

Purpose of humidity control in general retail

At an annual U.S conference for retail store planning, engineering and construction held in 1998, Al Powers, the Manager of Mechanical Systems for a major US clothing retailer explained the underlying principles of his HVAC designs: "Ladies and gentlemen, our company is in the business of convincing you to take *your* clothes off and put *ours* on... so our store environment must help achieve that goal."

All retailers have similar goals. They want customers to be comfortable, so they will linger longer and purchase more goods. All retailers also share the problem of staff turnover. In North America, retail staff are seldom life-long employees. They come and go frequently, creating high costs for training. So employee comfort can be

Fig. 20.9 Comfort in Retail Stores

nearly as important as customer comfort, because the retailer seeks to minimize staff turnover costs.

Temperature and humidity certainly influence the comfort of customers and the staff. But except for unusual circumstances in extreme climates, the retail store owner is seldom willing to spend extra money to achieve a higher level of comfort through tight control of humidity. Simple moderation of humidity extremes is generally sufficient to achieve the owner's purpose of providing a comfortable environment for shopping.

Typical control level and tolerance in general retail

Human comfort is an elusive goal, and one which changes according to each individual's physical characteristics, activity level and expectations.

In a retail environment, the store visit is not likely to last long, which means the customers' comfort expectations are also influenced by the immediate thermal alternatives: the environment outside the store and inside the car or other means of transportation to the store. The store is likely to be much more comfortable than the thermal environment outside the building. So for short visits, the HVAC designer's work is likely to create a favorable impression regardless of the indoor humidity level. The indoor air temperature in the summertime is more influential. It should not be so cold that it "shocks" a lightly-dressed customer entering the building. Some retail store HVAC managers in North America find that it is best to keep the indoor temperature no more than 15°F [8°C] below the outdoor temperature to avoid thermal shocks for most customers. But opinion varies on this issue, and consensus is difficult to achieve among equally-experienced retailers.

For longer customer visits and for the staff, humidity influences comfort more than it does during short visits. Ideally, humidity should stay within the human comfort range of dew points between 30° and 57°F [-1° and 14°C]. Above that range the body requires more time to release metabolic heat through evaporation, which creates the sensation of discomfort in the meantime. Also above that level, fabrics absorb enough moisture so that—in combination with a high level of skin-wettedness—customers will find that clothing will pull and catch, making it less comfortable to try on new clothing. Below that target humidity range, eye irritation increases and electrostatic build-up makes fabrics cling to the customer uncomfortably as clothes are pulled on and off the body.

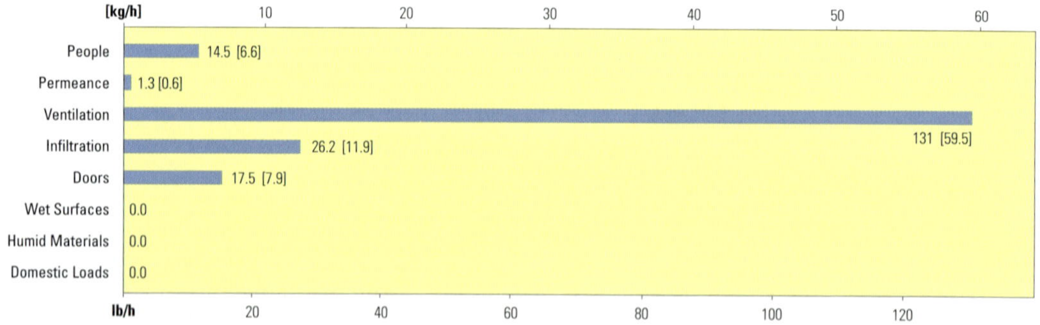

Fig. 20.10
General Retail Moisture Load
As in most commercial buildings, the ventilation air contributes the largest moisture load. To moderate humidity, reduce that load. To control humidity, dry the ventilation air to eliminate the largest portion of the load entirely.

Humidity moderation vs. humidity control

Next, the client must decide the tolerance—how many occupied hours the humidity can be allowed to remain above or below the comfort range. If the client defines a tolerance, the HVAC designer knows that the project will involve actual *control* of humidity, which implies adding a dehumidifier and humidifier to the mechanical system.

If the client does not define the number of hours that humidity can escape the comfort zone, it usually means that client has no need for control, and therefore only humidity *moderation* is required. On the other hand, no defined tolerance could reflect simple inattention or unfamiliarity on the client's part. So the prudent HVAC designer asks the client to express the need in terms of the number of hours the humidity level can be beyond the comfort zone. If the answer is more than 220 hours (more than 2.5% of the hours in a year can be uncomfortable), then simply moderating the extremes with passive desiccant wheels may be sufficient, avoiding the cost of dedicated dehumidifiers and humidifiers. Otherwise, the client will need to accept the cost of additional equipment to maintain control rather than only moderating the humidity extremes.

Principal loads for general retail stores

Figure 20.10 shows a typical moisture load summary for a medium-sized, free-standing consumer electronics store with a retail area of about 10,000 ft^2 [930 m^2], located in Atlanta, Georgia. The load estimate assumes the store meets the ventilation recommendations of ASHRAE Standard 62-1999. The graph shows that moisture contained in ventilation air dominates the total load.

Ventilation

For retail stores, ASHRAE Standard 62 recommends an outdoor air ventilation rate of 0.3 cfm/ft^2. Many experts believe this high ventilation rate is needed to dilute the concentration of the volatile organic vapors released from the retail merchandise itself, or from its packaging. Others believe the emission rates are not so high as to require such high ventilation rates. When building code officials can be convinced to agree, these owners prefer to ventilate at lower rates based on the average occupancy over a four-hour period, or at rates that vary based on the concentration of CO_2 that builds up in the space. Concentrations above 1000 ppm are often taken as an indication of poor indoor air quality.

But even when the ventilation air quantity falls below ASHRAE minimums, the ventilation moisture load is still likely to be the largest component of the total load. That's why—when the owner needs true control—a retail store HVAC system must be designed to remove that entire ventilation moisture load rather than just "tempering" the temperature of that incoming air. Where *control* is the goal, the designer usually places the dehumidifier so that the incoming ventilation air is dried before it enters the building.

When simple *moderation* is the only requirement, the designer can just reduce the size of the ventilation load by using a passive desiccant wheel to remove part of the ventilation air's moisture before the air enters the building. Since the outdoor air load is reduced, the building's cooling equipment can usually remove enough moisture to keep the store humidity from becoming extreme, without the need for a dedicated dehumidifier and humidifier.

Infiltration

Humid air leaking into the building is the next largest moisture load component. This load is driven by two forces: external wind pressure and internal suction caused by the HVAC system itself. Many designers make the common mistake of assuming the infiltration load only occurs when the wind is blowing hard outside. Since that seldom occurs for long periods, the load from infiltration is easy to ignore at the design stage. But measurements from commercial buildings show that infiltration occurs whenever the HVAC system is operating (Cummings et al, 1996). Leaks in return air duct work, and leaks in air handlers allow air to enter the duct work from cavities in the building envelope. That air is replaced by untreated outdoor air flowing through cracks around joints on the exterior walls. Typical infiltration rates in small retail buildings rise from 0.2 - 0.5 air changes per hour when the systems are off, up to 2.0 to 10.0 air changes per hour when HVAC systems are operating.

In other words, there will be some leakage into the building at *all* times, even when the wind is not blowing hard outdoors. The designer would be unwise to assume the infiltration moisture load will be zero. Experts might dispute the exact amount, but *some* dehumidification capacity will be needed to remove moisture from infiltrating humid air.

Doors

Humid air enters the building when doors are opened to admit customers. This load estimate assumes there is an air-lock vestibule at the main doors, and that only 20% of the mixed air in that air lock enters the store when the doors open. The estimate also assumes that the air pressure is neutral across the door opening.

If the building were under an average *negative* air pressure with respect to the outdoors, or if the building had many floors instead of all being at the same level, the humid air load coming through the doors might be considerably larger. A designer can minimize this load component by taking the steps assumed in this case: use an air lock vestibule, and keep the building under an average slight positive air pressure by bringing a slight excess of dry air into the building.

In cold climates pressure management during the winter is more complex. It is important to make sure that humid *indoor* air is not forced into cold building cavities where it would condense moisture and cause problems. In very cold weather, the building should be under neutral or slightly negative air pressure. But for warm and humid weather, the prudent design uses a slight excess of dry air to pressurize the building.

People

People release moisture from their wet clothing, and they exhale humid air with every breath. The load estimate in figure 20.10 assumes that 200 people enter the store every hour, and it assumes they are entering from cool, rainy outdoor weather so their wet clothing releases moisture. If the customer traffic is higher or lower, the load would be adjusted accordingly. There is little the designer can do to minimize this load. After all, the whole point of a store is to attract *more* customers, not fewer! In most cases, the retail store owner will have a reasonably accurate estimate of the average and peak number of customers per hour, because the retail economics for a new store cannot be calculated without an estimate of customer traffic.

Permeance

As in most commercial buildings, the moisture load from vapor permeation is so small as to be negligible. The designer should focus on better estimates of the door and people loads, rather than spending a great deal of time estimating the permeation load.

Suggestions & cautions for general retail

When a retail client asks for better humidity control, the designer should spend some time discussing the issue with the client. In many cases, the client does not really want *control*, nor does he intend to pay extra for that feature—he simply wants to avoid drips from cold supply diffusers in the summertime, or wants to keep the relative humidity below 70% most of the time, or wants to reduce the frequency of electrostatic shocks during the winter. In other words, what is wanted is simply moderation—not control. With that in mind, the authors have provided two groups of cautions and suggestions, the first for moderation, and a second set for when true control is required.

To simply moderate humidity:

1. Avoid oversizing the cooling equipment

Building pathologists constantly identify oversized cooling equipment as one of the leading causes of moisture-related damage and fungal problems in commercial buildings.

Here's the typical failure mechanism. Compressors in oversized cooling units do not have to operate for long periods to satisfy the thermostat. So the compressors shut off after a few minutes, but the supply fan stays on to provide ventilation air and general circulation. The coil surface warms up, halting dehumidification, in spite of a constant ventilation moisture load. Further, the moisture originally condensed on the coil surface re-evaporates into the supply air.

When units are greatly oversized, the compressors may only operate for a few minutes out of every hour. The thermostat is satisfied, but moisture from people and from ventilation air continues to build up—first to uncomfortable levels and then, over time, to levels that create damage and allow fungal growth.

Most designers will not heed this advice to undersize equipment. It seems counter-intuitive and appears imprudent to select the "next smaller" unit rather than the "next-larger" one for the estimated load.

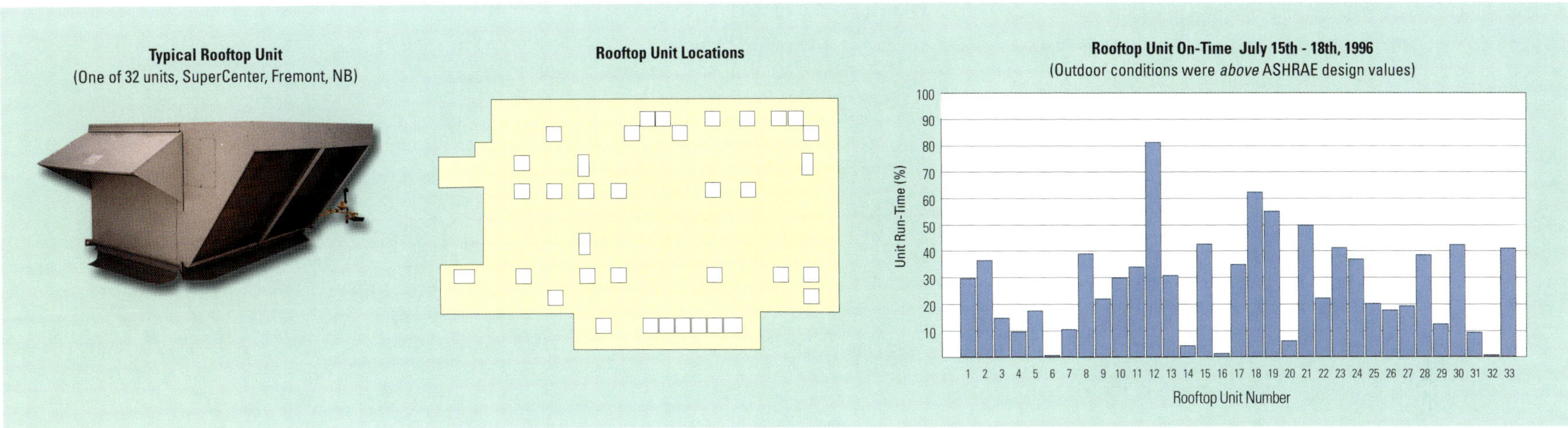

Fig. 20.11

Avoid Oversizing Cooling Equipment

The photo and graph show a retail SuperCenter in Fremont, Nebraska. It has too many cooling units on the roof. They cool the building so quickly that they turn off before effective dehumidification can begin, even on a day when temperature is well above the ASHRAE design temperature. Run time even at peak load is very low, as shown in the graph.

Undersizing the cooling units would provide more run-time, increasing the number of hours of dehumidification—one way to somewhat reduce the humidity extremes of such a system. (GRI, 1997)

But smaller is better when selecting conventional equipment from the perspective of humidity control. A smaller unit will take longer to remove the heat load, which means it will be removing moisture for longer periods.

When the designer considers the question of unit size vs. load, he would do well to remember that most load estimates are inherently conservative. They assume that all loads will peak at the same moment, and that they will all occur when both the solar load and the convective heat loads from outdoors are also peaking. By definition the peak load occurs for less than 1% of the building's life. Oversizing the cooling equipment pretty much guarantees that the occupants will indeed be comfortable for those 1% hours—but only at the cost of being *uncomfortably cold and clammy for much of the other 99% of the building's life*. Slightly undersizing the cooling equipment will result in more hours of comfort.

2. Consider installing a load reducer

Since the majority of the excess moisture comes from ventilation air, a passive desiccant wheel can be very helpful in reducing the annual load. A passive desiccant wheel uses the relatively dry exhaust air from the building to reduce the moisture level of the incoming outdoor air. The wheel rotates between both air streams, transferring excess heat and moisture from one air stream to the other, without transferring the contaminants of the exhaust air.

A passive desiccant wheel reduces the peak moisture loads and also reduces the size of the humidification load during winter. With reduced loads, conventional rooftop packaged equipment can often moderate humidity extremes during both seasons without support from dedicated dehumidification and humidification equipment. More information on load reduction equipment is contained in Chapter 15.

Fig. 20.12

A Passive Desiccant Reduces The Load

Passive desiccant wheels (enthalpy wheels) are available as a standard option on many brands of packaged rooftop cooling units. These devices reduce the moisture load from ventilation air for many operating hours.

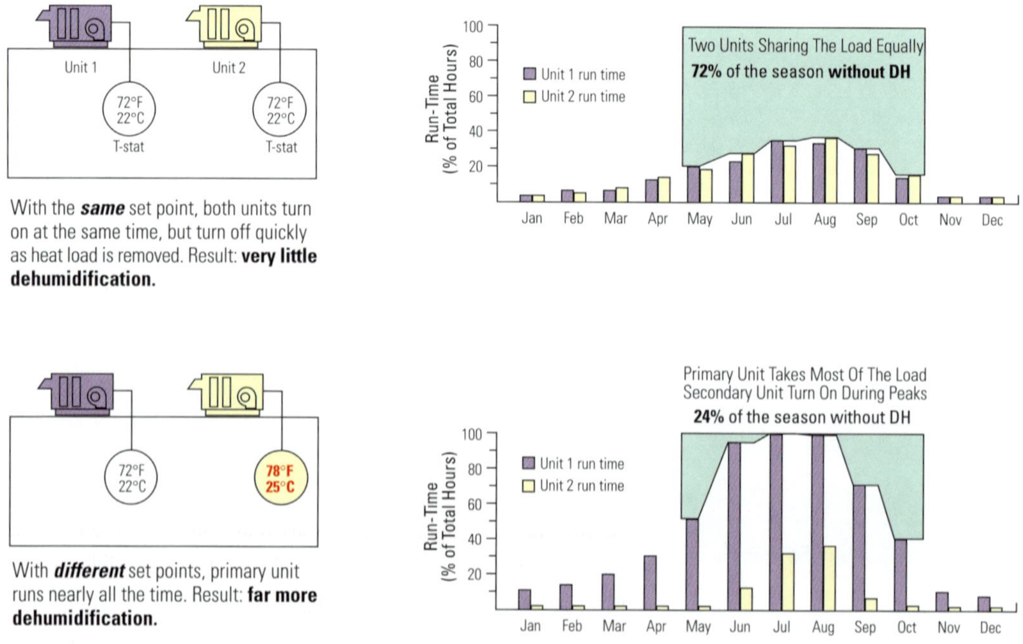

Fig. 20.13
Multiple Thermostat Set Points Helps Avoid Long Periods Without Dehumidification

To take advantage of this equipment, the designer must lay out the ductwork so the incoming ventilation air enters the building at the same point as the leaving exhaust air. The wheel must be placed between those two air streams. Also, unlike active desiccant wheels that are reactivated by heat, passive desiccant wheels need dry exhaust air. That means they will not remove moisture unless the exhaust air is dry. So the cooling units inside the space must operate in response to a humidistat as well as in response to a thermostat.

3. Consider the split-coil cooling unit option

To gain control of summertime humidity, a dedicated dehumidifier is necessary. But the dehumidifier can be smaller when the cooling units can dehumidify the air even when it does not need cooling. Also, if the owner only needs *moderation* of humidity, such specially-configured cooling units may satisfy the need without the additional dedicated dehumidifier.

To accomplish dehumidification without excess cooling, some manufacturers offer the option of splitting the cooling coil into upper and lower segments served by separate compressors. The lower segment operates on a command from either the thermostat or the humidistat. The upper segment of the coil only responds to the thermostat. When there is no demand for cooling but humidity is too high, the lower segment of the coil operates, and chills part of the supply air to dehumidify it. The overcooled dry air then blends with the warmer air that came through the upper part of the coil, and the resulting mixture is dry, but no longer overchilled.

4. Consider multiple thermostat set points

Retail stores are usually wide open spaces, served by many rooftop heating/cooling units that control separate zones which overlap a bit. In that situation, the designer can arrange the ductwork so that the zones overlap more heavily, and downsize the units so they share part of the load from the overlapping zones. Then one or two of the units can have thermostat set points below the other units.

The units with the lower temperature set points will operate longer, so that some dehumidification is being accomplished at nearly all times. The idea is diagrammed in figure 20.13. The second group of units only turn on when the temperature in the zone rises above the higher set point. These units provide the "peaking capacity" needed to keep the temperature within acceptable limits. The first units provide the bulk of the dehumidification.

5. Economizers must compare indoor and outdoor dew points

Air-side economizers are a common feature of retail HVAC systems. When the outdoor air temperature is cooler than the desired supply air temperature, these devices open the outdoor air damper wide while closing the return air damper. That way the system uses more "free" cooling from outdoor air, reducing the cost of running the cooling system.

However, when the building needs summertime dehumidification, the designer must arrange the economizer controls to avoid bringing excessively humid air into the building, even when its cool temperature would be useful. Otherwise, the economizer may indeed "save cooling," but that will often come at the expense of losing control of humidity.

With modern building management systems, the economizer control can be arranged to compare not only temperatures, but also dew point differences between indoors and outdoors. Set the economizer to open when the outdoor temperature is low—but override that signal if the outdoor dew point is above the desired supply air dew point. This will reduce the number of hours of "free cooling," but will preserve humidity control—which costs more than cooling.

6. Seal the ductwork to avoid condensation damage to the building.

Leaking return air ducts and exhaust air ducts create condensation problems. The leaking duct creates a slight suction in the building cavities it passes through, which connect to the cavities near the exterior walls. When those cavities are under suction, outdoor air is pulled into the building, leading to condensation in the building structure wherever humid air drifts over a cold surface.

Also, if cold supply air leaks out of duct work into unconditioned spaces in the summertime, moisture may condense on cold metal surfaces chilled by the leaking air. This is often the reason for dripping water above dropped ceilings.

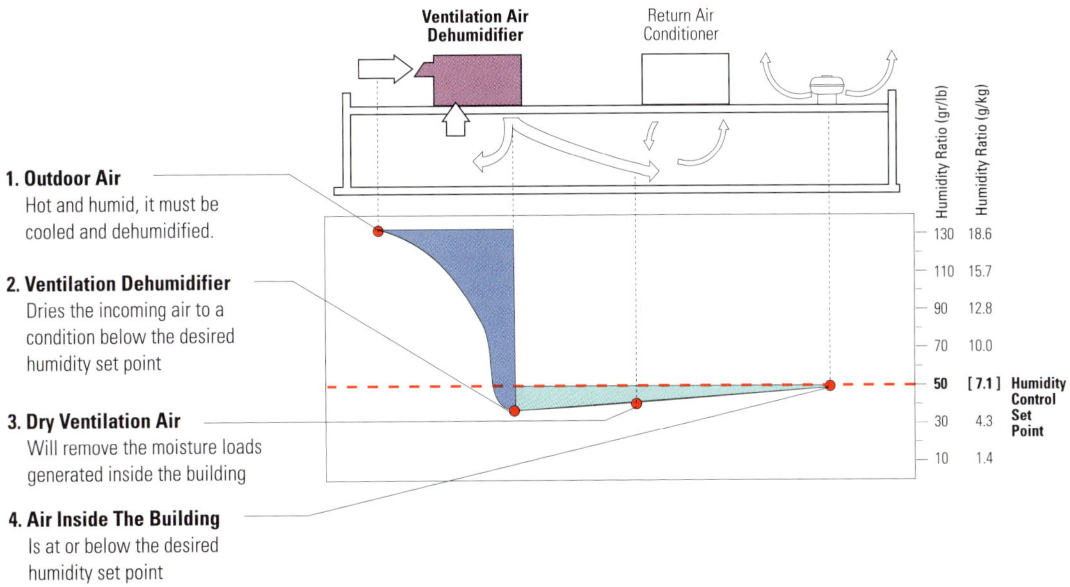

The solution to both problems is to seal all the duct work and all its air handler connections with mastic. Then cold supply air won't blow out into unconditioned spaces, and return air won't create suction that pulls untreated air through the exterior walls.

7. Specify air balance to avoid suction and excess outdoor air

Low-rise retail HVAC systems are not always balanced by the installing contractor or by an air balance company. Where humidity is not important, this saves cost. But where the owner cares about moderating the humidity, air balance and testing is a very worthwhile investment. The single most common problem for systems which only moderate humidity is air leakage and excess outdoor air. When outdoor air is set carefully at the design value, and when all leaks in the supply and return duct work are eliminated, humidity is easily moderated by cooling equipment without the need for more costly dedicated dehumidifiers.

To control humidity within a defined range:

1. Dry the ventilation air centrally

If the ventilation air flow is dehumidified to slightly *below* the control dew point during the summer, the air will act as a "sponge," absorbing the moisture loads generated inside the building. Without that preconditioning, the ventilation moisture load becomes a major burden that must be removed by equipment that recirculates air inside the building.

In the example building, 3,000 cfm of ventilation air carries an excess moisture load of about 140 lbs per hour because the outdoor moisture level is 133 gr/lb compared to the indoor level of 65 gr/lb. But if the outdoor air is dehumidified down to a level of 55 gr/lb (a dew point of 50°F), the incoming dry air can *remove* 19 lbs/hr of the internal load, instead of *adding* 140 lbs to that load.

Fig. 20.14 Dry The Ventilation Air Deeply

To ***control*** *rather than just moderate humidity, dry the ventilation air deeply. That way the incoming air can absorb excess moisture inside the building, rather than adding excess moisture to the indoor air.*

[In the example building, the ventilation air carries an excess moisture load of about 64 kg per hour because the outdoor moisture level is 19 g/kg compared to the indoor level of 9.3 g/kg. But if the outdoor air is dehumidified down to a level of 7.8 g/kg (a dew point of 10°C), the incoming dry air can *remove* 8.6 kg/hr of the internal load, instead of *adding* 64 kg/h to that load.]

Given dry ventilation air, the balance of the load is small enough so that conventional packaged rooftop cooling/heating units can keep humidity below a dew point of 57°F [14°C], which satisfies requirement for comfort and for office equipment operation.

In that design, the outdoor air dehumidifier should also be equipped with a recirculation mode. With that feature, when ventilation air flow is reduced the unit can still act as a dehumidifier for the return air. The other air conditioning units are still relieved of the burden of the internal dehumidification load.

The main difficulty with drying ventilation air centrally is the need to distribute that air evenly throughout the building. The designer has two choices: distribute dry air to each of the other air handling units through rooftop duct work, or direct to each zone through a second set of duct work. Both choices are expensive.

A typical compromise is to serve a *group* of rooftop units from one ventilation dehumidifier, and another group can be served by a second such unit, and so on. In the winter time, if moisture is added to the ventilation air during dry months it will be much easier to keep the dew point up to the desired level inside the building. In cold climates, however, the outdoor air moisture deficit is so great, and the temperature is so low, that additional humidification will still be needed inside the building, in addition to humidifying the outdoor air. When the humidification load is that high, it often costs less to humidify the blend between the outdoor and return air, rather than humidifying each air stream separately.

2. Size and locate duct-mounted humidifiers carefully to avoid condensation

Strip mall retail buildings are often built on speculation with severe budget constraints. So their ceiling-to-roof-deck heights are usually minimized, which reduces the amount of space available for duct work. The HVAC designer may be forced into using higher-than-recommended duct velocities, because the loads demand a given amount of air, but the ducts must remain very small to fit above the ceilings. Most designers are familiar with the noise created by high velocities, but they may be less familiar with the implications for sizing and locating humidifiers.

Duct-mounted, heated vapor generators are a popular type of humidifier for office buildings. They are usually mounted in the supply air ducts downstream of the air handler, so that the cold outdoor air and the warm return air are well-mixed. Warmer air absorbs water vapor more quickly than cool air. But when the duct velocity is very high, the vapor travels a longer distance through the duct before it is fully absorbed into the air. Two suggestions may help the designer when he or she must humidify fast-moving air.

First, be sure to allow long, clear duct runs downstream of humidifiers when duct velocities exceed normal limits. Otherwise, vapor can condense on duct obstructions like turning vanes and sensors or even supply air diffusers. Admittedly, this suggestions is usually difficult to follow. When space is limited, long, straight runs of supply duct are not common. So the second suggestion may be more widely applicable.

Consider splitting the load between multiple humidifiers in different air streams. Adding equipment costs more, but smaller amounts of vapor are more quickly absorbed by fast-moving air streams, so condensation is less likely.

References for general retail

Berglund, Larry G. 1998. "Comfort & Humidity." ASHRAE Journal, August, 1998, p. 35-41

Brundrett, Geoffrey W. 1990. *Criteria for Moisture Control.* Butterworths Publishing, London.

Chasar, David A, and Shirey, Don B., III, 1998. "Impact of ASHRAE Standard 62-1989 on Florida Retail Stores." Proceeding of the 1998 ASHRAE IAQ Conference, ISBN 1-883413-68-0

Cummings, J.B, Withers, C.R, Moyer, N, Fairey, P, McKendry, B. *Uncontrolled Air Flow In Non-Residential Buildings.* Final Report, FSEC-CR-878-96. April, 1996. Florida Solar Energy Center. 1679 Clear Lake Road. Cocoa, FL. 32922.

Gas Research Institute 1997. *ASHRAE Standard 62 compliance in a retail store using desiccant systems.* Report No. GRI-97/0035 Gas Technology Institute, Des Plains, IL. USA

Special Thanks

Much of the information about the behavior of oversized cooling units and dedicated dehumidification systems in retail environments was gathered from extended field research conducted by James Judge, John Spears and Stephe Yborra on behalf of WalMart Stores, the US Department of Energy and the Gas Technology Institute. Also, the Authors are grateful to Al Powers and Margaret Manning of TJ Maxx stores (TJX Corp.), who provided their perspective on retail store humidity control, based on hourly logs from their stores located throughout the US.

Donated Photo

Fig. 20.12 - © AAON Inc.

Chapter 21

Hotels

By Lew Harriman, Don Gatley, Joseph Lstiburek & J. David Odom III

Fig. 21.1 Preventing Mold & Mildew In Hotels

Humidity control is important—but avoiding building suction is even more important in preventing these problems.

(Note that this photo was taken years before the industry realized that all mold remediation workers need protective masks that prevent inhalation of the fungal spores released when wall covering is stripped off a moldy surface.)

Hotels

The principal concern of hotel owners has been to avoid problems caused by excess humidity. So although there can also be concerns in cold climates, this chapter is focused primarily on preventing mold and mildew in hot, humid or mixed climates.

Near the end of the long North American construction boom of the 1980's, the Executive Engineers' Committee of the American Hotel & Motel Association surveyed their members regarding the growing problem of mold and mildew in new buildings. Their investigation required cooperation over several years between different hotel companies, as well as all members of their construction teams and equipment suppliers—a courageous, unusual and very effective way to track down the complex causes of a serious problem. The results of those AHMA investigations were published as a special report of the Executive Engineers committee in late 1990 (AHMA, 1990). The research showed that the two most common moisture problems in humid weather are:

- Moisture and mold trapped behind an inappropriately located vapor retarder such as interior vinyl wall covering.
- High indoor relative humidity and mold caused by infiltration of humid outdoor air, and/or ventilation with makeup air that has not been dehumidified.

21.2 The Mildew Square

Virginia Peart of the University of Florida suggests that the "mildew square" helps us remember that fungus needs four elements for growth. The HVAC designer can help control the amount of water that reaches the material. That's the focus of humidity control in hotels. Chapter 7 contains more detailed information about mold and mildew.

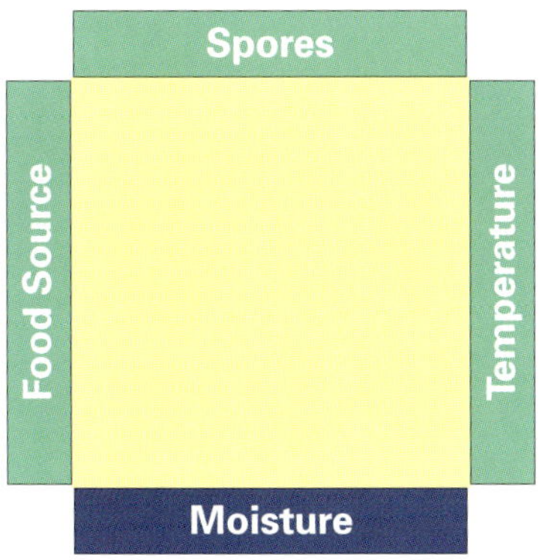

A comprehensive solution to these problems requires attention from the architectural, interior and HVAC designers plus all of the installing contractors. No single team member can solve the problems alone. The architectural and interior designers must provide forgiving walls that dry to the interior. These must not have a vapor retarder material on the cool indoor surface of walls exposed to the high humidity of infiltrating outdoor air. Also, the exterior wall must not leak water. Preventing such water leakage is a joint responsibility of the architectural designer and the general contractor.

Next, the HVAC designer and contractor must help keep the building dry by continuously supplying dry makeup air, in amounts that exceed the total exhaust air volume. They must also make certain that the HVAC system does not create suction inside the building or inside the walls. Avoiding negative internal air pressure is just as important as adequately drying all the ventilation air.

So in hot and humid climates, four HVAC elements are essential for avoiding mold and mildew:

1. *Keep the entire building under a slight positive air pressure using dry makeup air, so that humid air is not pulled into the building through cracks in the building envelope.*
2. *Seal the ductwork and inlet boots of the toilet exhaust system, and seal all return air duct work, so that suction is not created that would pull humid outdoor air into wall cavities.*
3. *Seal up any vertical, wall or closet-mounted AC units so that air cannot be pulled into the unit from the wall cavities, which would lead to suction at the outside wall.*
4. *Gasket the "sleeves" of horizontal, through-wall AC units to the exterior and interior wall finishes. Also seal the units to their sleeves so that humid outdoor air is not pulled into wall cavities by the unit's fans.*

Purpose Of Humidity Control

Travellers seldom choose a specific hotel because of its pleasant indoor air. The building's location, the room cost, its furnishings and other building amenities are much more influential in the guest's decision than whether the room and the sheets seemed cold and damp during a previous stay. Also, our expectations regarding hotel comfort are rather low, perhaps because experience tells us that in general, hotel environments in hot climates are seldom comfortable in all respects.

But guests *do* object to staying in a room that smells unpleasant, or one which has stains on the walls or furnishings. On some subconscious or conscious level most people realize that something is wrong with such a room. Some of us may insist on being moved into a different, "cleaner" one—which means the hotel loses the original room's rent for that night. So the typical purpose of controlling humidity in a hotel is more to avoid mold and mildew growth than to improve comfort.

This is a very important distinction. Comfortable temperature and moisture is always useful. But if the HVAC designer focuses on the temperature and moisture of air in the rooms—as in most other buildings—the project is likely to fail. The real purposes of humidity control in hotels are:

1. Keeping moisture out of structural and decorative components of the building.
2. Keeping fabrics and carpets dry.

To accomplish these goals, the HVAC designer must expand his concern, and focus on managing air pressures so that humid air does not infiltrate into building cavities. Also, ensure that air inside the occupied spaces is dry enough so that fabrics do not absorb enough moisture to grow mold.

Fig. 21.3
Leaky Exhaust Ducts Cause Mildew
Avoiding suction in building cavities is just as important as maintaining low humidity in the rooms. (Harriman & Lstiburek, 2000)

Control Levels

The most important control level is neither temperature nor moisture—it is the pressure of air in the building cavities, compared to the pressure of air outside the building. The building cavities must be under an *average positive pressure with respect to the outdoors*. And the air inside those cavities must be dry enough to prevent condensation on the cold surfaces of pipes and supply air ducts, and on the cool inside face of the wall board.

Very little positive pressure is needed in the building cavities. An average positive pressure of 0.004 in.wc. [1 Pascal] is sufficient to prevent air infiltration from becoming a problem. (Odom, DuBose & Nelson, 1993) That's nearly impossible to measure. Also, such a small pressure difference reverses its direction many times in a few seconds as outdoor breezes raise and lower the pressure acting on the outdoor side of the wall. But the goal is not really to keep any defined pressure difference—it's to *prevent a continuous flow* of humid air into the building. The designer is simply ensuring that the building cavities are *not* continuously sucking in humid air from outdoors.

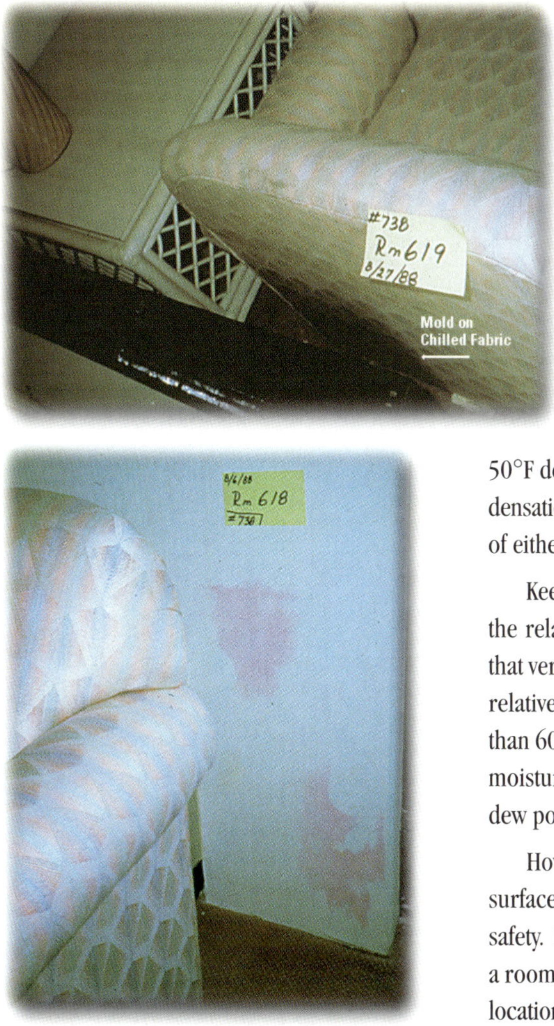

Fig. 21.4
Cold Surfaces = High rh + Mold
The average rh of the air in the room may be acceptably low for comfort. But when cold supply air chills the surfaces of walls and fabrics, the local rh becomes very high. Chilled surfaces will absorb enough moisture to support mold and mildew.

There is some debate concerning the appropriate humidity level for preventing materials from absorbing moisture that supports mold and mildew. The science shows that in this respect, drier air is always better. But drier air costs more. The question is further complicated by the fact that even if one part of the wall is safe from condensation, another part of the wall may remain cold enough to condense moisture.

Allowing for all the probable circumstances, keeping the indoor air below a 55°F dew point will prevent most problems. Keeping it below a 50°F dew point provides an extra level of assurance that condensation will not occur. [Keep the indoor air below dew points of either 13° or 10°C.]

Keeping the dew point below 55°F [13°C] at least keeps the relative humidity of an overcooled room low enough so that very little moisture is adsorbed by fabric and carpets. The relative humidity of air at 70°F and a dew point of 55°F is less than 60%—a comfortable margin of safety to avoid excessive moisture adsorption from the air. [21°C dry bulb with 13°C dew point is still less than 60%rh].

However, to be certain of preventing *condensation* on cool surfaces, the lower dew point provides a greater margin of safety. For example, consider the cold walls near the outlet of a room AC unit. Figure 21.4 shows what can happen in such a location. Cold supply air from the cooling unit is bounced backwards towards the wall by furniture or draperies. The wall surface cools rapidly to nearly as low as the unit's supply air temperature. The surface can easily fall below 55°F [13°C] for extended periods, which leads to the problems shown by figure 21.4. To prevent such condensation at or behind the wall surface, the air in the building cavities should probably be kept below a 50°F dew point. [10°C dew point]

Even at that low dew point, mold and mildew on or behind walls is not ruled out entirely. The relative humidity of air in actual contact with the wall surface will be over 90%—so the wall paper or gypsum board facing paper could still adsorb enough water from humid air in building cavities to support fungal growth. But as long as gross liquid condensation does not occur regularly, short periods of occasional vapor sorption and desorption is not likely to lead to major problems.

Avoid vinyl wall covering on exterior walls

One key point is useful to mention for the benefit of architectural and interior designers. Building envelopes that trap moisture are likely to have mold problems, no matter how cleverly the HVAC system is designed. If stray moisture can move freely out of the wall towards the dry air in the rooms, then a dehumidification system can eventually remove that water. But *if the inside surface of the walls is nearly impermeable* to vapor flow—as when covered by most forms of vinyl—then stray moisture from rain drips, lawn sprinklers or from concrete blocks soaked during construction will stay inside the building cavities. Mold will grow behind the wall covering no matter how dry the air is made inside the occupied spaces.

Field experiments with full-scale, operating hotels filled with very dry air over more than a year have proven that positive pressure with dry air is very useful in avoiding problems (Banks et al, 1993, 1994). But those observations also show that even very deeply-dried air in the occupied spaces *cannot* completely overcome problems caused by ongoing water leakage, or the negative effect of sucking humid outdoor air into the building through exterior walls that are covered on the inside with vinyl.

Impermeable wall coverings act as an accelerator for minor problems, trapping moisture and rapidly creating major problems of mold and mildew. Unfortunately, this scientific reality is not obvious to hotel owners, who find the material useful in resisting scuffs and bumps from luggage, as well as being easy to clean compared to a painted surface. Also, no owner expects his new building to leak, and few designers have the courage to point out that no exterior wall is totally air-tight or completely water-tight.

Owners and interior designers who incline towards impermeable wall coverings in spite of this caution should be informed that vinyl is very unforgiving of normal construction flaws. Using the material is somewhat analogous to the excitement of sky diving—it's very impressive as long as the parachute opens properly. Similarly, vinyl wall coverings demand that the building be built—and perform—flawlessly. Flawless performance of all other building components is not a realistic expectation. It is far better to design a fault-tolerant interior finish than to pay for removal of wallboard through costly mold remediation while the hotel is trying to operate.

An intelligent compromise is to finish the inside of the exterior walls with vapor-*permeable* paint. Porous paint allows minor amounts of moisture to diffuse into the room air instead of being trapped in the wall. The remaining interior walls can be covered in vinyl with minimal risk of mold, *provided that*: interior walls are sealed off from the exterior envelope by a combination air/vapor barrier, exhaust and return ducts are tightly sealed, and the building is kept under an average positive pressure with dry air.

Moisture Loads

Figure 21.4 shows an estimate of the moisture loads typical of a small hotel of 90 rooms on three floors built near Atlanta, GA in the Southeastern US. As shown in figure 21.5, the ventilation load is over twice as large as all other loads combined.

Ventilation

Most building codes in North America require air to be exhausted from toilet rooms. Consequently, there is a large and constant moisture load from makeup air. The designer must arrange the system to include enough dehumidification capacity to remove the incoming moisture load. That ventilation air should:

1. Be dehumidified at least as low as the control dew point level indoors—continuously.
2. Exceed the volume of the exhaust air by at least 10% to help keep the building under a positive air pressure with dry air.

Buildings with mold and mildew problems have usually failed to follow these two suggestions. In many locations, codes allow make up air to be sent to the hotel corridors, and then pulled into the guest rooms by the exhaust fans in the toilets. In the past, some HVAC designers reasoned that since corridors are not occupied, the makeup air could be simply "temperature-tempered" instead of dehumidified. Following that temperature-driven logic, the incoming makeup air might only be cooled to 77°F [25°C]. But humidity-driven logic exposes the problem with that practice. Figure 21.7 shows why.

If outdoor air is never cooled below 77°F [25°C], that means it will *never* be dehumidified, at least in Northern parts of Europe, Asia and America and in southern parts of Africa and Australia where there are no outdoor design dew points above that level. Air entering the building at 77°F will *always* be carrying its full moisture load, as

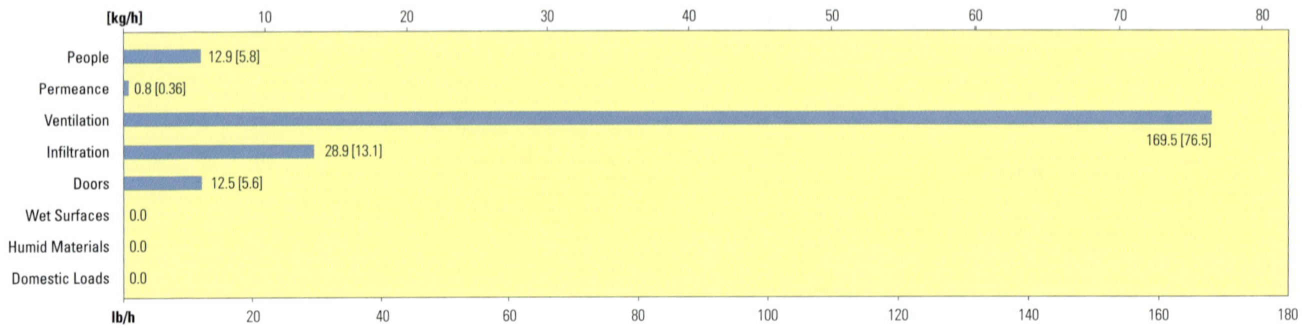

Fig. 21.5 Typical Hotel Moisture Loads

The ventilation air contains the largest moisture load. That's why it's best to dehumidify ventilation air before it enters the building, as shown in figure 21.6.

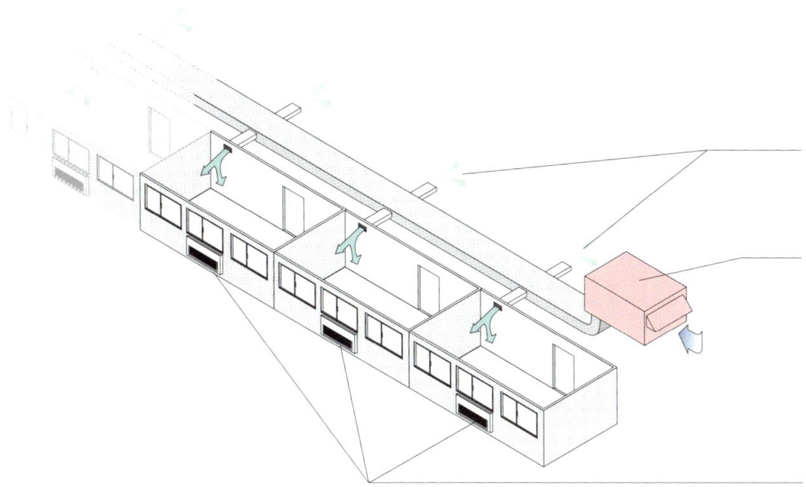

21.6 Ventilation Air Dehumidifier

shown in figure 21.7. To avoid bringing that huge excess of moisture into the building, makeup air must be dehumidified down to at least the building's control dew point—or preferably below. In hotels, the control dew point is likely to be either 50 or 55°F [10 or 13°C] as discussed earlier. So makeup air must be dried *at least* to that level before it enters the building. Drying the ventilation air *below* the indoor dew point will provide dehumidification capacity to absorb some of the internal moisture loads as well.

Infiltration

The amount of infiltration estimated here assumes that only one wall is subject to wind pressure, and that the wall will be "tightly" constructed. The infiltration represents a significant load, but it will not be continuous, as in the case of the ventilation load. Infiltration occurs when the pressure difference is large: in other words, when a strong wind is blowing. So the load estimate here probably represents a worst case of relatively short duration. (Except when the exhaust and return duct work is not sealed up tightly. Then the infiltration load would indeed be continuous, and will be much larger than shown here—a very harmful turn of events.)

Consider figure 21.8 which shows the results of field research that measured air flowing out of an unsealed hotel toilet exhaust system in Florida. (Cummings et al, 1996) The research team measured the air flowing into the system from each bathroom, and compared the total with the flow measured at the fan itself. The measurement showed that more than half of the air leaving the building did not come from the bathrooms—it leaked into the exhaust duct through joints in the sheet metal. In other words, air was pulled *from the building cavities* through which the ducts passed. That means that much of the replacement air probably came from the outdoors through the exterior wall, pulled into the building by the suction created by that leaky exhaust duct.

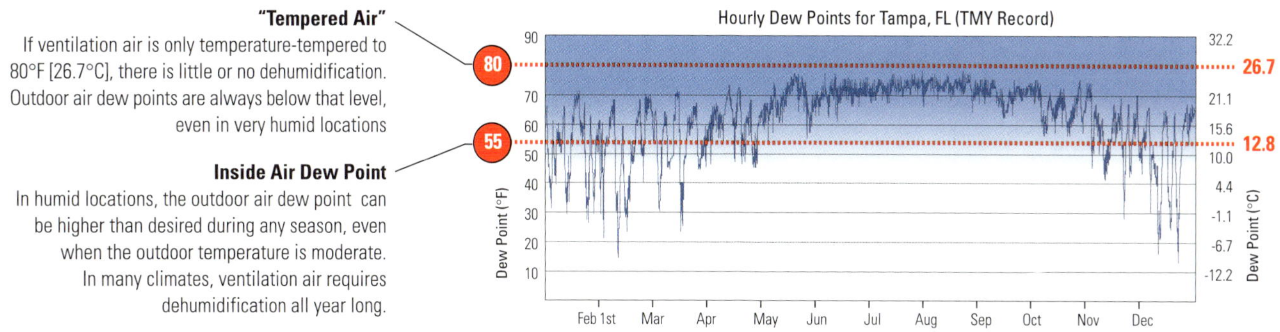

Fig. 21.7 "Tempering" Caution

If only the temperature of outdoor air is controlled, that air may be carrying excess moisture into the building.

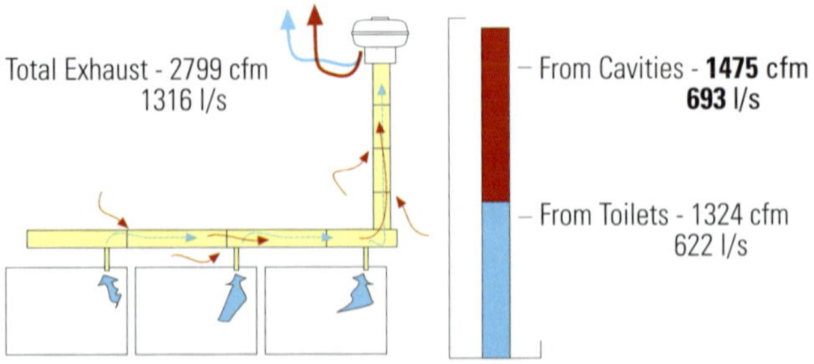

Fig. 21.8 Unsealed Exhaust Ducts Pull Air From Cavities

When the exhaust duct leaks, it pulls air from building cavities, and eventually from outdoors. In this hotel, leaks allowed the fans to pull more air from cavities than from the bathrooms. (Cummings, 1996).

The load estimate shown here assumes that the designer and HVAC contractor have sealed up the exhaust and return duct work, so that such an enormous extra load does *not* occur.

People & Doors

The loads from people and door activity are about the same. Both are fairly small and neither is continuous. So if the load estimate from these sources is erroneous, it will not make much of a difference in the success of the project. After all, the goal is not comfort as much as preventing fungal damage. Short excursions above the design dew point can happen at periods of peak occupancy or during special events. But those will not be long enough to create a fungal problem.

Permeation

As in other commercial buildings, the permeation load is so small as to be negligible. There is little to be gained by trying to reduce that load through special paints or vapor retarders. The designer's time and budget are much better spent by making sure the exhaust and return ducts do not leak.

Domestic loads - showers

In the early stages of the AH&MA's investigation of mold and mildew in hotels, much attention was given to the issue of moisture loads from showers. Because many mold and mildew problems occur in bathrooms, Engineers assumed that moisture from showers was contributing to the growth.

However, the more important contributors to mold were nutritive soap films left on the shower walls and curtains, combined with a lack of dehumidified makeup air that would normally have removed residual moisture. When bath surfaces were clean, and kept dry by air flowing through the bathroom, mold was not a problem. The moisture load from the bathroom is significant only while a shower is occurring. Some studies estimate the total evaporated load from a 15-minute shower is 1.5 lbs—or in other words an hourly load of 0.37 lbs. [0.68 kg in 15 minutes, or 0.2 kg over an hour.]

Since the shower load only occurs once or twice in 24 hours, many designers discount it entirely, believing that at worst, shower evaporation will cause a rise in room humidity for less than an hour. After that, dry makeup air removes the load, carrying the humid air out the exhaust stack. The authors are inclined to agree with that logic, which is why there is no moisture load from domestic activity shown in figure 21.5.

Suggestions & Cautions

The most important suggestions for hotel humidity control have been stated in several different ways in this chapter: dry the ventilation air, seal the exhaust and return duct work and keep the overall building and all its cavities under a net positive air pressure using dry ventilation air. Beyond those obvious steps, experts have additional suggestions as described below.

Encourage the Architect to design a forgiving exterior wall

The building needs an exterior wall that drains liquid water to the outside and can release any trapped water and water vapor to the inside during humid weather. The architectural designer should not allow an inappropriately located vapor retarder. In a humid climate the use of vinyl wall coating as the interior finish on an exterior wall is clearly an inappropriately placed vapor retarder.

Also, the exterior of the wall must not absorb or store liquid water. This element requires careful design and detailing and the maintenance of the wall over its life. It may require recoating the wall every three years with a clear hydrophobic coating to encourage the runoff of water as opposed to the absorption of water.

Also, the best exterior wall designs keep the rain off the exterior surfaces with projections and overhangs, and they "drain the rain on the plane." That is to say, the wall may occasionally allow some liquid to penetrate through construction cracks in the outermost layer, but there is always a water-tight drainage plane located *inside* from the outermost layer. The drainage layer is equipped with low-point drains that carry any liquid water back outside the wall.

Finally, if vinyl wall covering is used on interior walls that connect to exterior walls or attic, place a combination air/vapor barrier across the connecting joints to block air flow. The realities of field construction suggest that such barriers are unlikely to be air-tight, but a good barrier will help reduce the risk posed by vinyl.

A dew point analysis of exterior walls is not useful for buildings that pull-in outdoor air

Many Architects and Engineers are familiar with the technique of dew point analysis. In this procedure, the designer calculates the probable temperature of each layer in the exterior wall, together with its vapor diffusion permeability. If any layer *traps* moisture, its surface must remain above the dew point temperature of the air it traps. Otherwise, moisture will condense onto that cool surface, and eventually soak the inside of the wall. Such analyses are especially useful in highly-humidified buildings such as swimming pool enclosures and museums in cold climates. Dew point analysis helps avoid a fatally-flawed wall design in cold climates.

However, dew point analysis is focused on vapor diffusion. It does not take into account *air leakage* that results from cracks in the exterior wall combined with internal suction caused by leaky ducts and excessive exhaust. Humid air leakage is *far* more likely to cause mold problems than vapor diffusion. The traditional dew point analysis takes the designer's attention away from more serious issues, and can provide a false sense of security in hot and humid climates. Better to focus attention on avoiding suction by sealing up all duct work and by providing enough dehumidified ventilation air to keep the building under positive pressure.

Deliver dry ventilation air directly to rooms

Ducting makeup air to corridors, hoping that toilet exhaust fans will pull dry air through the guest rooms, does not work well.

Toilet exhaust fans pull a suction on the entire room. So when the exterior wall is leaky, the exhaust fan pulls air into the room through the electrical outlets as well as from the corridor. That's especially true when the corridor is supplied with air at one end instead of from the middle. At the end of a long corridor, there may be very little positive air pressure to help push dry air into the guest room. So the suction force of the exhaust fan pulling air *through the exterior walls* into the room can be significant. The resulting problems are seen in figure 21.3.

Also, air flows along the path of least resistance. When the bathroom is near the doorway to the corridor—as is typical—ventilation air flows into the entryway area and out through the bathroom without ever having traversed the bedroom portion of the guest room. In other words, the corridor, entryway and bathroom are effectively ventilated, but the actual guest area is not. A better solution is conceptually simple, but costs more.

Duct the dry ventilation air to each room directly, distributing the fresh air into the sleeping area as shown in figure 21.6. With that design, the actual occupied area has an excellent flow of clean, fresh air at all times. And the room itself is less likely to be under suction, which reduces the risk of humid outdoor air being pulled into the building cavities.

This solution costs more than the corridor air distribution method. The building will need an extra set of duct work to carry ventilation air into each room. On the other hand, the design provides effective ventilation, where the corridor air distribution method does not. Also, ducting ventilation air directly to each room improves fire safety. Each room contains a self-contained circuit of supply and exhaust air independent from the other rooms. This reduces the chance that fire will spread into the guest rooms, following a path of air flowing into the room from a pressurized corridor.

Swimming pools need negative air pressure

Unlike other parts of the building, the swimming pool room should be under a slight *negative* air pressure with respect to the rest of the building.

Swimming pools release a very large amount of moisture, which must be kept out of the rest of the building. In fact, the odor of chlorine is a low-cost way for the staff to determine if the pool area has positive or negative air pressure with respect to other spaces. If one can smell chlorine in the rest of the building, the swimming pool area is *not* under enough negative air pressure, so the air balance of the pool's HVAC system should be adjusted. Any perceptible odors indicate that excess water vapor is being pulled from the pool area into other parts of the building—never a good idea in any climate.

Balance kitchen exhaust systems

Kitchen HVAC systems are notorious for poor air balance, often leading to negative air pressure throughout the entire building rather than just the kitchen. The exhaust air flow leaving the kitchen through the grease hoods is very large.

Make sure the air balance contractor sets the kitchen exhaust fan speeds carefully, so they are not simply pulling all the air they are capable of pulling, rather than the correct amount. This will be even more important as time moves on and makeup air filters clog, which reduces the volume of makeup air.

Change filters frequently on ventilation air pretreatment units

Even well-designed ventilation and exhaust air systems will have difficulty keeping the building under positive air pressure when the outdoor air filters are not changed frequently.

The air flow resistance of filters rises as they fill up with particulate. As the resistance increases, the air flow reduces. In some cases, simply installing clean filters will restore the building's positive air pressure, as well as reducing the cost of fan operation.

It is difficult to say for certain how often outdoor air filters should be changed, because the particulate load varies by location and by season. But most service firms find that outdoor air filters need to be changed at least every month. In areas with heavy dust loading, like near highways, in dense cities or near construction, ventilation air filters need replacement even more frequently.

The operating staff that can afford a more technical approach can measure the initial air flow resistance of a filter with a manometer. When the resistance of the filter is greater than the maximum specified by the HVAC designer, a clean filter should be installed. This will usually be when the resistance is double its initial value.

Special attention for EIFS buildings

EIFS buildings are named for the "externally-insulated fiberglass system" that covers the outside walls. With this system, rigid insulating board is applied to the outside of the building over steel studs, and then covered by smooth vinyl paste, reinforced by a fiberglass mesh. The system produces a nearly impervious layer on the outside wall, which keeps rain water out. The system has become very popular in North America because it is quick to install, economical, aesthetically pleasing and provides an excellent thermal barrier between the building and the outdoor environment.

However, some of these buildings have become notorious for mold and mildew problems. If the wall has any cracks at all, so that even a small amount of water *can* enter the building, the impervious character of that wall becomes a liability instead of an asset. With the EIFS on the outer face, any moisture will certainly not dry towards the outdoors. Then if the wall cannot "dry to the inside" because vinyl wall covering is used on the inside surfaces, trapped moisture feeds mold growth.

The architectural designer and general contractor who choose EIFS cladding must be especially careful to install the product as the manufacturer directs, paying close attention to joints between the EIFS system and other building components like windows, doors, balconies and the roofing. The seams above and below windows and air conditioning units are often "out of sight and out of mind" for the installing technician. But any rain or sprinkler water will enter such cracks as it flows down the wall.

The HVAC team members can help avoid problems by ensuring that the building is under positive air pressure, and also by alerting the other team members to any difficulties in maintaining positive indoor air pressure—which could mean there are gaps in the EIFS cladding. Also, the seams around the outside of though-wall AC units have been classic water leakage sites in EIFS walls. The HVAC designer who uses such units in EIFS buildings needs to consult with the unit manufacturer to be absolutely certain how to detail the connection between the unit and the exterior wall to avoid water leakage. Pay special attention to the joints under the units, and to joints hidden under the lips of decorative air louvers, where installers are not likely to be as careful as in more visible areas.

Special attention for concrete-block buildings

The AH&MA survey of mold and mildew in hotels showed that problems often occurred in buildings with concrete block and/or brick as the exterior wall. The authors make two observations about such buildings.

First, during construction, open concrete blocks are soaked with rain on a regular basis. In central Florida, for example, rain occurs daily. If the building is closed in before that water can dry out, the soaked blocks serve as a reservoir for moisture, slowly releasing large amounts of water into building cavities as the building heats and cools. If the HVAC system does not have effective dehumidification capacity, the excess water released by the drying materials can cause mold and mildew problems after construction is complete.

Secondly, if the exterior masonry retains construction moisture or rain water and the building is under negative air pressure, the infiltrating air coming through the soaked porous masonry will be humidified *well above* the moisture level of the ambient air. In that situation, the infiltration air may carry twice or three times its normal moisture load, and carry that moisture directly into the building cavities where it can do the most damage.

Once again, the HVAC team is not directly responsible for problems created by excess water in the exterior masonry. On the other hand, the owner can look to the HVAC team for help. Keep the building under positive air pressure, and make sure the ventilation air system reduces the problem by delivering dry air to the building rather than air which is above the control dew point. That way the HVAC system reduces rather than contributes to mold and mildew problems created by wet materials.

Consider drying moist structures after close-up

The realities of construction dictate that parts of the building are likely to be wet as the building is closed-in and made weather-tight. In most situations, a small amount of excess moisture is not a problem. It is released slowly, or more quickly when the HVAC system is turned on. On the other hand, sometimes parts of the building are *really* soaked. Or there is some reason the HVAC system cannot or should not be turned on to dry out the moist material. In those circumstances, the owner and general contractor can take advantage of the services of building drying companies.

Since the late 1980's, a large group of contractors has developed in North America to serve the disaster water damage market. These firms dry out buildings after floods, fires and disasters in order to reduce the cost of restoring the buildings to their original functions. They use large, portable industrial dehumidifiers, fans and heaters to accelerate drying under difficult temporary circumstances.

HVAC professionals seldom know about these firms, since they neither sell nor install equipment. But their services can be enormously cost-effective by speeding construction of water-soaked buildings, which helps the general contractor avoid post-construction mold and mildew problems. Such companies can be found easily by consulting the local telephone directory under the "water damage" or "restoration" categories, or by a search using similar key words on the internet.

References

American Hotel & Motel Association. *Mold & Mildew in Hotels and Motels* (The Survey) 1991. AH&MA, Washington, DC.

Banks, N.J, 1992. "Field test of a desiccant-based HVAC system for hotels." AN-92-18-4. ASHRAE Transactions, V. 98, Pt. 1.

Banks, N.J, 1994. *Development of desiccant-based air conditioning for hotels and motels.* Gas Research Institute, Marriott, Disney. GRI report #5090-243-1963. Gas Technology Institute, Des Plains, IL.

Busweiler, Ulrich, 1993. "Air conditioning a hotel with a combination of radiant cooling, displacement ventilation and desiccant cooling." DE-93-2-3, ASHRAE Transactions V. 99, Pt. 2.

Cummings, James B, Withers, C.B, Moyer, N, Fairey, P., McKendry, B, *Uncontrolled air flow in non-residential buildings.* April 15th, 1996, Final Report of FSEC project number FSEC-CR-878-96. Florida Solar Energy Center, 1679 Clearlake Rd, Cocoa, FL 32922.

Gatley, D.P, 1992. "Designing for comfortable cooling season humidity in hotels." ASHRAE Transactions, V.98, Pt. 1.

Harriman, L.G.III, Lstiburek, J, Kittler, R, 2000. "Improving humidity control for commercial buildings." *ASHRAE Journal*, October 2000.

Lstiburek, Joseph. "The Pressure Response of Buildings." 1999. Proceedings of the 7th Conference on Thermal Envelopes, p. 799-817. ASHRAE

Odom, J, David, III, and DuBose, George, and Fairey, P.W., 1993. "Why HVAC commissioning procedures do not work in humid climates." *ASHRAE Journal*, December, 1993. pp. 25-36.

Odom, J, David, III, and DuBose, George, *Preventing indoor air quality problems in hot, humid climates: Problem avoidance guidelines.* 1996. CH2M/Hill Publications, 225 East Robinson St, Suite 405. Orlando, FL 32801 (407) 423-0030.

Peart, M.V, and Gatley, D.P, 1992. "Introduction to hotel mold and moisture problems in warm, humid climates." Proceedings of the 5th conference on the thermal performance of building envelopes, December 1992, p. 531-537. ASHRAE.

Shakun, Wallace. 1992. "The causes and control of mold and mildew in hot and humid climates." AN-92-18-2. ASHRAE Transactions, V. 98, Pt. 1

Wall Street Journal. 1999. "The dish on hotel air." July 18th, 1999, pp:W1 and W14. Dow Jones, New York, NY.

Special Thanks

The Authors are grateful to all the owners, building scientists, engineers and research institutions that began the difficult field work which defined the mechanisms of mold and mildew growth in hotels. In particular, much of the early work was accomplished by the American Hotel & Motel Association's Executive Engineers Committee, supported by the Gas Technology Institute. In turn, the AH&MA was first focused on the problem by the concern and financial support provided by Mr. Jones of MMI Hotels of Jackson, Mississippi.

Photos

Fig. 21.1 - © Joe Lstiburek & Neil Moyer

Fig. 21.2 - Virginia Peart

Fig. 21.3 -© Joe Lstiburek

Fig. 21.4 - © James Kimbrough

Chapter 22

Restaurants

By Lew Harriman

Fig. 22.1

Lower Dew Point + Higher Temperature = Better Comfort

When humidity is kept under control, the dry bulb temperature can be raised. This improves comfort for lightly-dressed patrons, without making the hard-working staff too hot to be productive and pleasant.

Restaurants

Restaurants use lots of air. To begin with, an enormous amount of air flows through the kitchen. Next, if it has a dining area, a restaurant needs a great deal of ventilation air for people. Finally, the sensible heat loads in glass-walled quick-service restaurants are very high, so the supply air flows in dining areas are large as well.

All that air is difficult to manage in a small space. Any change in the damper positions or in the operation of fans or in the filter loading can have big consequences in the relatively small cubic volume of a restaurant. Unless air flows stay carefully balanced, many familiar problems occur:

- Customers have difficulty opening the doors, which also tend to slam shut.
- The pilot lights on gas burners blow out, and flames from burners can reach out of equipment and into the building in dangerous ways.
- Regardless of the thermostat setting, the dining area stays cold and clammy—or stuffy, smoky and hot.
- Cold condensation drips from supply diffusers onto customers or tables, or makes the floors slippery.

Dripping condensation is likely to prompt a restaurant owner to ask for better humidity control. So this chapter will focus on ways to avoid excess humidity in hot and humid weather. The most useful and lowest-cost improvements in humidity control can be accomplished by careful attention to setting air flows rather than just focusing on the humidity control equipment and its control level.

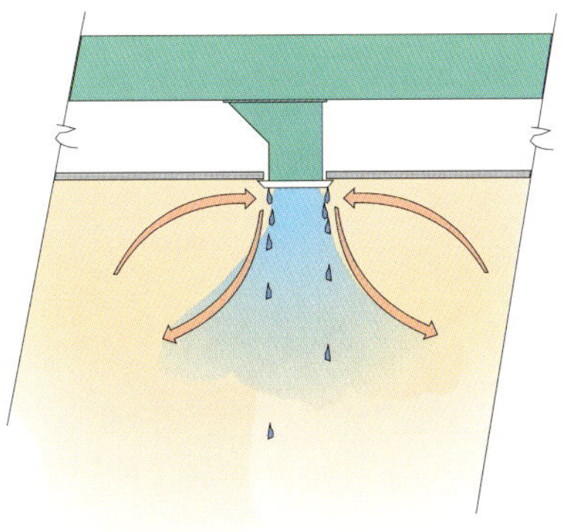

Fig. 22.2 Dripping Diffusers
Often, the main purpose of the project is to prevent drips from diffusers and ductwork onto floors and customers.

Purpose Of Humidity Control

In the US, the owners most anxious to improve humidity control have been those that operate quick-service restaurants in hot and humid areas of the country. Equipment suppliers that serve these owners report that the most common complaint is condensation dripping onto diners from ductwork and from cold supply air diffusers.

Sealing and insulating the ducts prevents condensation on those surfaces. But supply air diffusers at the ends of those ducts are located over the dining areas, and effective insulation of their cold surfaces is not so practical. To avoid drips entirely, the dew point in the restaurant should be held slightly below the supply air temperature—which will also be below the temperature of the cold metal.

While the restaurant operator is likely to be most concerned with avoiding drips, the solution to that problem can bring other potential benefits—provided that the operations staff allows an adjustment in the thermostat setting. By *lowering* the dew point and *raising* the dry bulb temperature, both patrons and staff will be more comfortable, and energy costs can be reduced. Here's why.

Most of us are familiar with the restaurant comfort problem from the customer's perspective. Female customers tend to be lightly-dressed in the summertime, and are often cold in restaurants because the staff sets the thermostat at the lowest setting the manager will allow. That's because the hard-working staff are very active, and need to release large amounts of metabolic heat to stay comfortable. Colder air helps the staff stay more comfortable—but it makes the customers quite uncomfortable because they are sitting quietly instead of running from table to table.

The store manager responds to complaints from the staff, keeping the temperature as low as possible until enough customers complain of uncomfortable cold drafts. Customer complaints usually come from patrons seated under the cold air diffusers. In those locations, the coldest air combines with the highest air velocity to "quick-chill" the seated diners.

Fortunately, when the store is kept dry, the staff will be able to release heat more efficiently through evaporation. The evaporation rate depends on the difference between the amount of water vapor in air at the skin surface, compared to the amount of water vapor in the room air. The greater this difference, the faster the body can release its heat (assuming other comfort factors stay constant). This fact allows a beneficial chain of events.

At low dew points, the staff can be comfortable at a higher dry bulb temperature. When the dry bulb temperature is higher, there is a smaller temperature difference between indoors and outdoors, which means the cooling load goes down and equipment can be downsized. This leads to savings in both construction cost and operating cost on the cooling side which can help offset the cost of the dehumidification equipment.

As a practical matter, these potential savings are small, and they do not occur if the operations staff refuses to raise the thermostat set point. On the other hand, if the owner feels the need to invest in dehumidification to prevent drips, the system will at least have the potential to improve comfort for patrons by keeping temperatures higher. So if the staff becomes concerned with customer comfort, some savings may accrue on the operating side even if the designer keeps the cooling equipment at its original size in the construction budget.

To summarize, the typical goal of most restaurant dehumidification projects is to prevent drips. Other goals are often secondary.

Control Levels & Tolerances

To avoid condensation, the dew point of any air that contacts the cold surface of the supply air diffuser must be *below* that metal's surface temperature.

The supply air itself is not the problem. That air will never be above 100% relative humidity, so its dew point will always be below its dry bulb temperature. The real problem is humid air in the *rest* of the room that is pulled into contact with the cold metal when the supply air rushes out of the diffuser, as illustrated in figure 22.2. To prevent condensation, the designer should keep the dew point in the occupied space below the dry bulb temperature of the supply air.

The air in the restaurant does not have to be much drier than the supply air temperature. If the two are exactly the same, some minor condensation will occur, but the amount will be so small that it will be difficult to see, let alone large enough to cause drips. Frequent drips only occur when the air is *so far* above the surface temperature that large amounts of condensate collect rapidly. In other words, the target control dew point need only be slightly below the dry bulb of the supply air.

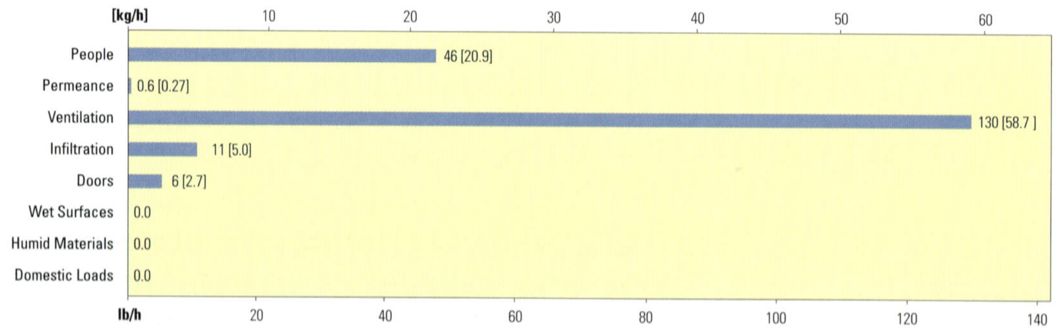

Fig. 22.3 Dining Area Moisture Load

To keep air in the dining room that dry, the supply air must be dry enough to absorb the moisture loads in the building. So the supply air must be delivered to the space at a drier condition than the control dew point.

It also follows that the *greater the moisture load in the space, the drier the supply air must be delivered.* In other words, as the internal moisture load goes up, so does the size and cost of the dehumidification equipment. Consequently it's useful for both designer and operator to focus on keeping the air in balance—so that no extra humid air is pulled into the building. The designer and building operator must clearly identify, quantify and control the amount of air entering and leaving the building. Otherwise the dehumidification project will not succeed.

Moisture Loads

Figure 22.3 shows the typical moisture loads in the dining area of a medium-size casual dining restaurant located in Atlanta, Georgia that seats 198 people. Two of the load components stand out: ventilation and people.

Fig. 22.4 Casual Dining Restaurant

This building is typical of those that benefit from better control of summertime humidity. It is the basis of the load estimates for a dining area shown in figure 22.3.

Ventilation

This load estimate assumes the dining area will meet the recommendations of ASHRAE Standard 62, providing 15 cfm per person of outdoor air for a total of 2,970 cfm [7 l/s per person, for a total of 1,396 l/s of ventilation air]. Two items are worth noting about this quantity of ventilation air.

First, there is a great deal of additional ventilation air supplied to the kitchen area to make up for exhaust from the kitchen hoods and toilet exhausts. The total ventilation air for the building will be about 6,000 cfm—far in excess of the 2,970 cfm supplied to the dining area. [A total of 2,870 l/s compared to only 1,396 to the dining area] In short, the ventilation air is a very large proportion of the total air flow in the building. That means the flows must be carefully balanced and dampers must be set very carefully to avoid overloading the system by excess extraction of exhaust air.

Secondly, it is very important to remove the ventilation load before it gets into the building. Otherwise, the humidity level in the dining area would rise dramatically. Figure 22.5 makes this point graphically. It shows what happens if the kitchen exhaust fans are not properly balanced, and they pull an extra amount of ventilation air into the building through doors and cracks in the exterior wall. Since the system is not designed with excess dehumidification capacity, the humidity level would rise to the level where condensation would form on the cold supply air diffusers. Although the designer assumes that systems will be built and operated according to plan, such out-of-balance conditions are unfortunately common in restaurants, as noted in field investigations. (Cummings et al, 1996)

The consequences of poor air balance are exceptionally important in restaurants, because the flows are so large compared to the internal volume of the building. Small flow variations create big problems in a few minutes whenever the outdoor humidity loads are high. That fact suggests that the designer and operations staff should invest more-than-usual attention and equipment in measuring and controlling exhaust and ventilation air flows when humidity control is important to the owner. Controlling flows is less expensive and more effective than buying larger dehumidification equipment to keep humidity under control when flows go out of balance.

People

The occupant density is fairly high in a restaurant, and all these people release moisture through respiration and perspiration. The staff releases about three times the moisture released by the customers, because the staff is more active. On the other hand, during rainy weather, customers bring extra moisture into the building in their damp clothing. Finally, a small amount of moisture is released from the food being consumed by these customers.

Moisture loads from people are another reason that the incoming ventilation air should be dried to a level below the control condi-

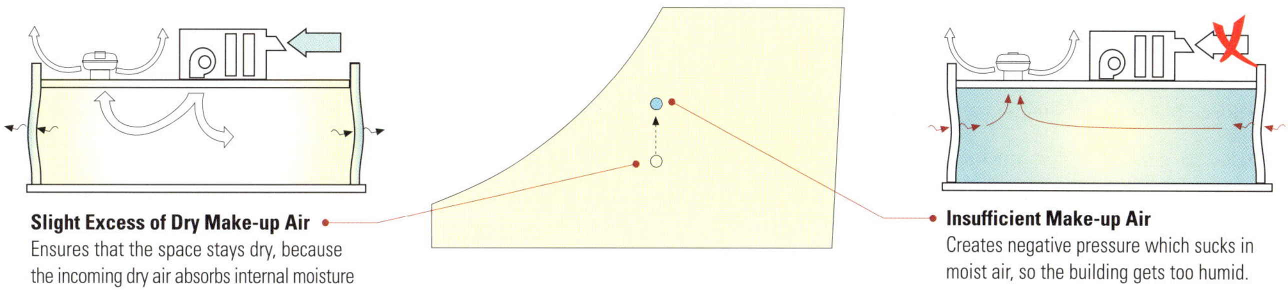

Fig. 22.5
Without Enough Dry Ventilation Air, Humidity Rises

tion in the space. Then the incoming dry air can act as a "sponge" to absorb moisture being released by the people and moisture entering the building through the effect of wind pressure.

Infiltration

In this load estimate, the infiltration is quite low. That's because it assumes that ventilation and exhaust air flows will be balanced so that the dining area is under a slight positive air pressure with respect to the outside of the building. Under those ideal circumstances, the infiltration load is minimal, and humid air only enters the dining area when the wind outside blows against the building with a force greater than about 5 mph [2.2 m/s].

On the other hand, this estimate may be too optimistic. Unless the designer specifies enough dampers to keep the flows in balance, and unless the filters are changed on a regular basis, the infiltration load can be *far larger* than shown here. In the worst case, the ventilation load will become the infiltration load. In other words, the infiltration load increases from 11 to 130 lbs/hr [5 to 59 kg/h] if the ventilation air system is blocked, or if the system does not dehumidify the incoming air that makes up for the air exhausted from the kitchen and bathrooms.

Therefore, some designers prefer to estimate the infiltration air at a higher value to cover small imbalances that might occur over time—such as when ventilation air filters are not changed. (When that happens, the dirty filter blocks the ventilation air inlet, so the exhaust fans suck more humid air into the dining area through walls).

But if the designer assumes the internal load is higher—as in the worst case of unbalanced flows—he will have to invest a great deal more money from the construction budget in a larger dehumidifier. The more cost-effective alternative is to invest in dampers and the commissioning to ensure that flows will not be out of balance.

Doors

The estimate for the moisture load from door opening also assumes the building will be in balance. The estimate also assumes the door does not open directly onto the weather, but has a vestibule which helps limit any humid air blown into the dining area when the customers enter the building. As with the infiltration load, some designers choose to assume a much larger moisture load through doors, because any imbalance in air flows will drag more humid air into the building through this path of low resistance.

Permeance

A very small amount of moisture will enter the building by vapor diffusion through the floors, walls and roof. But as can be seen from this estimate, the amount of moisture that enters through this path is negligible. The designer can gain better control of humidity by ensuring the air flows are balanced rather than spending time searching for a less permeable material for walls or floors.

Suggestions & Cautions

Designers that have experience controlling humidity in restaurants usually suggest, as in this chapter, that the principal focus should

Fig. 22.6 Dry The Ventilation Air

Drying the ventilation air separately allows smaller cooling units, which cost less to buy and operate. The ventilation dehumidifier then provides superior humidity control, but usually adds cost to the construction budget.

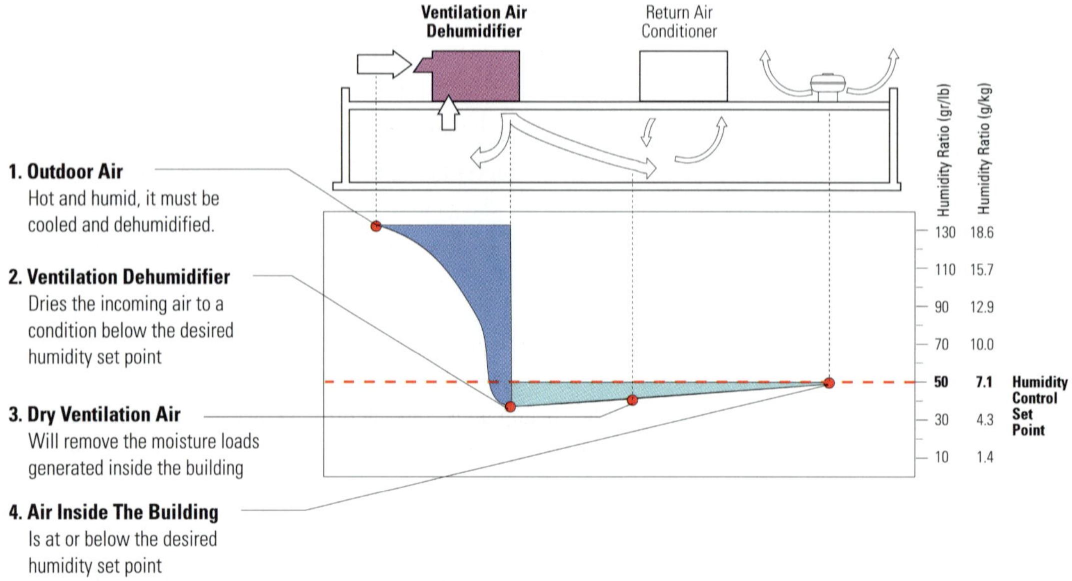

be on making sure the exhaust air does not exceed the dried ventilation air. Both the designer and the operations staff have important roles in keeping humidity under control. The team may wish to consider these suggestions:

Install enough dampers so that air flows can be controlled

Most restaurants use packaged rooftop air conditioning equipment for cooling and heating. These units do not always come equipped with dampers to set all three air flows (outdoor air, return air and supply air), relying instead on the installing contractor to use the flow-setting devices he feels are needed for the application. The designer needs to clearly specify that all air flows in all units must be controllable separately through a damper on each air flow.

In fact, all flows will change slightly in reaction to all pressure differences in all parts of the system. But without individual dampers on each air flow, there will be no hope of balancing the system and the building as is necessary for humidity control.

Use dry ventilation air twice

The kitchen exhaust requirement is immense, and there is no need to exhaust air from the dining area separately. Consider installing a transfer grill in the wall between the dining area and the kitchen. The conditioned air helps keep the kitchen cooler and drier than it would be if all the makeup air for the kitchen hoods was untreated.

With a transfer grill, the dry air can flow to the kitchen without going through the doorway that connects the two spaces. This avoids high air velocity through the doorway, which could be annoying or even hazardous if a door swings shut too quickly.

Don't dehumidify needlessly

Consider installing automatic dampers on the outdoor air inlets of the rooftop units, and interlocking them to close only when the ventilation air dehumidifier starts up in response to a high-humidity signal from the dining area.

In field tests of ventilation dehumidifiers in humid climate restaurants, investigators noted that while dedicated dehumidifiers were needed for the outdoor air during most of the occupied hours, there were still thousands of hours in the year when the rooftop units were perfectly capable of keeping humidity under control by themselves. In less humid climates, the rooftop units can do an acceptable job of humidity control for even more hours. (Yborra & Spears, 2000)

Consider arranging the system so that air can either be brought through the dedicated ventilation air dehumidifier, or brought through the usual outdoor air inlets of the rooftop units (figure 22.7). On a rise in humidity above set point, the outdoor air dampers on the rooftop units will close, and the ventilation dehumidifier will provide the dry air needed to control humidity during periods of high moisture loads. This arrangement saves the cost of operating that ventilation unit, and pulling the outdoor air through it.

Be careful when considering load reducers

In commercial buildings with large amounts of exhaust air, the designer should consider installing load reducers such as enthalpy heat exchangers (passive desiccant wheels).

Fig. 22.7 Don't Dehumidify Needlessly

By allowing for two paths for ventilation air, the designer can save operating costs without disturbing humidity control. (Yborra, 2000)

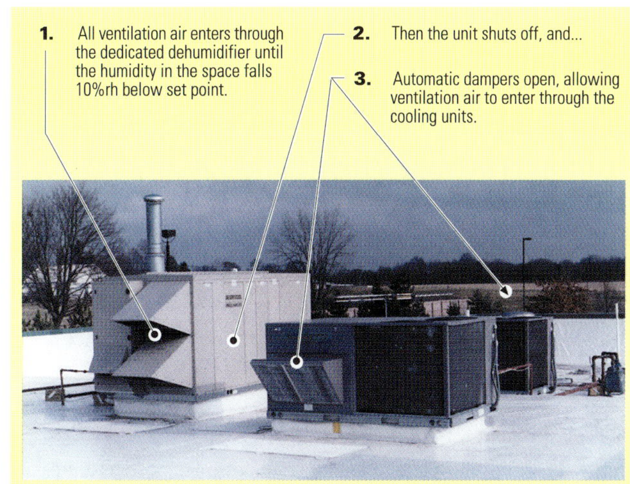

In a kitchen, the exhaust from the hoods is too hot, humid and grease-laden to be of any use in reducing the cooling and dehumidification load. But the designer might consider using clean exhaust from the dining area to reduce the makeup air load.

This strategy has been used successfully. It can help moderate the humidity in the dining area, just as long as the compressors on the cooling units will operate on a high humidity signal as well as on a high temperature signal from the control system.

However, such load reducers will not work all the time, because they are frequently stopped to avoid over-recovering heat. In fact , the peak moisture loads occur when the outdoor temperature is near the desired indoor temperature. During those hours, operating the passive desiccant wheel will overheat the store.

Some designers consider that there is not much point to using load reducers in the dining area, because exhausting the air in that location simply means that more air will have to be brought into the building in the kitchen area and treated at that location. These professionals reason that it is more economical to use the dry air twice: once in the dining area and then in the kitchen—rather than having separate circuits in both parts of the building.

Other equally-experienced designers believe that separate dining and kitchen systems are better for air balance, and that kitchen staff do not expect to have cool and dry air. Under those assumptions, there is no need to use air from the dining area to improve comfort in the kitchen. Then the dining room exhaust can be passed through load reducers to pre-treat the incoming air.

Both groups have persuasive economic arguments. In the end the owner will decide what value to place on cool dry air for the kitchen, and that value will probably make the decision. If better kitchen comfort is useful, then there is no point to using load reducers in the dining area. If kitchen comfort is less important, then separate circuits can save money, and load reducers make sense in the dining area.

Watch for signs of excess exhaust

If the total amount of exhaust air exceeds the amount of dried ventilation air, it will be difficult or impossible to keep humidity down when the outdoor moisture level rises. The operational staff can be alerted to a potential problem *before* humidity rises in the store if they remain aware of signs that the building is "starved" of dry make up air.

For example, when odors from toilets are apparent in the dining area or in the kitchen, it means that air is being pulled out of the bathrooms by the kitchen exhaust fans instead of by the toilet exhaust fans. That means there is not enough make up air in relation to the kitchen exhaust.

Other warning signs include:

- The pilot lights won't stay lit on gas appliances.
- Flames come out of the boiler combustion area.
- It's difficult to open the exterior doors, or they tend to slam shut.
- The kitchen doors are left open constantly to prevent these problems.

Any of these events indicate there is too much exhaust, or too little dry ventilation air—or both. The first remedy is to replace dirty ventilation air filters on the rooftop units with clean ones. If that doesn't work, call the maintenance staff so they can re-adjust the air balance.

Change outdoor air filters monthly

In well-designed systems, the most common reason for high humidity is untreated air leaking into the building instead of being dried by the ventilation dehumidifier. That's usually because the filters at the ventilation air intake are very dirty. When the filters are dirty, they create too much resistance to air flow, so humid air finds other ways to get into the building which do not involve so much resistance. Often the only "repair" needed to bring humidity under control is to replace the filters so that the incoming air goes through though the unit and can be dehumidified.

In most parts of North America, monthly replacement of outdoor air filters will be frequent enough to avoid problems of too much air resistance. In some cases, filters will need more frequent replacement, such as when the air is loaded with large amounts of particulate from nearby grease hoods, construction, highways or trees.

When the owner chooses to save money by replacing filters by need rather than simply on a time schedule, he can determine the need by measuring the pressure drop across the filter. When resistance (the pressure drop) across the dirty filter is double the resistance of a clean filter, the old filter should be replaced.

Don't exhaust any air needlessly

If the kitchen is not operating, there is no need for the exhaust hoods to be throwing expensive dry air out of the building. Turn off the exhaust hoods when the grills and stoves are not in use. This avoids the needless expense of drying all that make-up air.

This suggestion comes from field investigators who survey the actual operation of quick-service restaurants. They point out that employees often get into the bad habit of turning on the hoods upon arrival and only turning them off as they leave the building. The actual cooking may take place for only 70% of those hours. Turning the exhaust hoods off saves money, and is appropriate any time that cooking is not in progress.

References

Cummings, James B, Withers, C.R, Moyer, N.A, McKendry, B.B., 1996. "Field measurement of uncontrolled airflow and depressurization in restaurants." ASHRAE Transactions, V.102, Pt.1.

Rojeski, Peter Jr, Turner, C, Gonzalez, M, Mobley, C., 2000. "I'll take some fries—hold the smoke." Field measurement of ventilation effectiveness in 18 restaurants in North Carolina. *Air Conditioning, Heating & Refrigeration News.* June 19th, 2000. Business News Publishing, Troy, Michigan USA.

Smith, Vernon A, Young, R, Spata, A, Fisher, D., 1999. "Virtual vs. real; Modeling the energy performance of a quick-service restaurant." ASHRAE Transactions, V.105, Pt. 1.

Yborra, Stephen and Spears, J., 2000. *Field tests of ventilation dehumidification strategies in quick-service and casual dining restaurants.* Gas Technology Institute, Des Plaines, IL, USA.

Special Thanks

Comprehensive and ground-breaking research was performed with the support of the Gas Technology Institute by Stephe Yborra and John Spears, on behalf of the Burger King and Applebees restaurant chains. The results of that research were especially useful in writing this chapter. Also, Mike Ireland of Efficient Systems and Tom Willett of Whataburger Restaurants were very generous in sharing their experiences with humidity control in humid climates.

Donated Photos

Chapter 23

Museums, Libraries & Archives

By William Lull and Lew Harriman

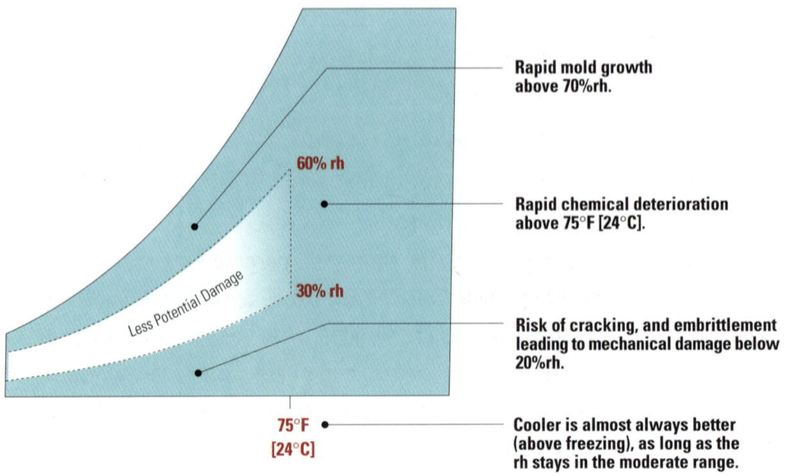

Fig. 23.1 Mostly Harmless

When humidity is kept between 30 and 60% rh, most objects and documents have a much longer life than when humidity goes beyond these limits. But it is very important to limit mechanical stresses by slowing seasonal humidity changes so they occur over days or weeks rather than minutes.

Humidity Control In Museums, Libraries & Archives

The purpose of a museum, library or archive is to provide access to information and artifacts from the past. Nothing lasts forever. But curators, librarians and archivists want to make sure that objects and documents under their care will last as long as technology and public access will allow.

They care deeply about human knowledge as expressed by artifacts and documents. Robert Herskovitz is a conservator at the Minnesota Historical Society. At the 2000 annual meeting of ASHRAE, he spoke to HVAC designers, explaining the passionate attitude of his colleagues: "People don't join our profession because they want to get rich. We gain personal satisfaction from preserving objects and the lessons of history they represent. We tend to be detail-oriented and we *take no prisoners* when we feel that artifacts or artwork are being threatened by an HVAC system!"

The HVAC designer who serves these clients should remember that no matter how thin their budgets may be in relation to their requirements, museum clients are thinking about preservation in terms of centuries and millennia, not years or decades. This attitude stands in sharp contrast to owners of other commercial buildings such as hotels and restaurants, which may be vacated, rebuilt or sold to different owners within a few years. Staff members in museums, libraries and archives have a longer view.

The lifetime of historic artifacts can be measured in thousands of years, as evidenced by objects preserved in Egyptian tombs. So museum professionals know that such longevity is possible. On the other hand, the Egyptian tombs did not have hundreds of people walking through the burial chambers every day during those thousands of years, and of course the low humidity in the tombs preserved some objects while destroying others. Keeping *all* objects safe for centuries in an active museum environment is a more complex challenge than preserving *most* of the objects in a sealed tomb. Especially since the Architect and HVAC designer seldom have access to the budgets and manpower available to the ancient Egyptian Pharaohs.

The HVAC system can speed up or slow down deterioration. To a great extent, deterioration of materials depends on the relative humidity of the air around them, as well as how often and how quickly that humidity cycles up and down. Most artifacts and documents last centuries longer when humidity stays between 30 and 60% rh, and when that humidity swings *gradually* between those extremes over a period of weeks rather than over a period of minutes or hours. On the other hand, a few objects can be damaged beyond repair even within this range.

Different types of material need different levels of humidity. And collections with several types of objects require humidity compromises. Using a generalized control range which is "typical for museums" or "just like the last five projects," the HVAC designer may produce a smooth-functioning system, but one which is best for a different institution than the current client. The designer must rely on the conservators of the collection and their consultants for guidance concerning the most appropriate humidity range for the specific objects protected by each HVAC system.

Typical Purposes Of Humidity Control

In these buildings, there are two long-term purposes of humidity control that apply to all objects and documents:

1. Ensure that materials do not absorb or condense enough moisture to accelerate adverse chemical reactions or allow biological attack.
2. Avoid desiccating the materials to the point where they would lose either mechanical strength or flexibility.

Further, for composite objects such as carvings, furniture and framed paintings, the short-term purpose for control may be even more important: *limit the rate of change in humidity*. This reduces the speed of moisture sorption, which reduces expansion and contraction, which in turn limits the magnitude of mechanical stresses. Such stresses can damage composite objects irreparably in a period of hours.

Control Levels & Tolerances

Preferences on control levels and tolerances vary widely throughout the international community of conservation professionals. Over the last quarter century, there have been vigorous arguments about the optimal humidity control levels in these environments. The humidity specifications greatly affect both the budget and the system alternatives, so the economic consequences of different specifications adds heat to the debate.

Because there is no universal "best" humidity level, some history of conservation and discussion of professional ethics may help the HVAC designer to understand why some clients prefer one humidity specification, while equally accomplished professionals demand a very different level, either of which may seem unreasonable or risky to the other group.

Among others, Garry Thomson, in his landmark book *The Museum Environment*, noted an interesting historical fact. During World War II in Britain, paintings from the National Gallery and the British Museum were stored for safety in Welsh coal mines, deep underground at cool temperatures and where the relative humidity stayed nearly constant between 45 and 55% rh. After the war, conservators noted that the paintings had survived with no detectable damage, in contrast to the considerable damage noticed before the war, after central heating had become popular in galleries during the first half of the 20th century. In the absence of more complete research, this observation prompted many in the museum community to conclude that keeping humidity at precisely 50% all year long in all climate regions (or "flatlining" as the practice came to be known) might be a useful compromise between the competing humidity needs of different materials. A control level of 50%rh and a tolerance of ±5% rh thus became enshrined in museum literature.

As conservation professionals became more aware of humidity effects, and as budgets became available to actually attempt to achieve "flatlining," problems occurred in some buildings. In cold climates, humidifying old buildings to 50% rh during the winter created damage to exterior walls. Condensation in building cavities corroded metal fasteners and created ice, which could expand to loosen stone and masonry facades. Also, poor decisions on equipment and controls often meant that humidifiers had difficulty reducing their output during swing seasons. This allowed condensation to collect in ductwork to feed mold and mildew, producing musty odors and fungus. Many in the museum community began to question the need for heavy humidification in cold climates.

In recent years, the museum community has performed a great deal of research to determine the micromechanical, biological and chemical behavior of materials. The results of some of this research are included in this chapter under the heading of "Humidity Effects on Artifacts and Documents." The research shows there are no easy or simple answers from a technical perspective. Some collections will survive longest in a "flatlined" range of 50% rh ±5% rh. Other collections have a broader tolerance as long as the change occurs over weeks instead of hours to minimize the mechanical stresses created by sorption and desorption. Each collection has its own composite "half-life response." For example, large pieces of highly-finished furniture respond to humidity changes over weeks or months, while thin canvas paintings may respond within minutes.

Also the effect of a given humidity level varies according to the objects. A collection dominated by ancient armor will benefit from humidity below 35% rh. But if the same room must contain a carriage with a steam-bent wood panel, the humidity might need to be kept above 40% to avoid damage, in spite of the potential benefit of lower humidity for the metal armor.

Further, the museum community in particular is heavily influenced by the critical economics of special exhibitions, which in turn may dictate humidity specifications. Here's why.

Museums frequently struggle financially, receiving intermittent large donations at unpredictable intervals. Special exhibitions attract large numbers of patrons without the immense cost of acquiring new objects or works of art. On the other hand, museums are not likely to lend their treasures for special exhibits in other parts of the world if there is any potential for damage. To reduce the risk, the lending institution may demand environmental control that is much tighter than usual for either the lending or receiving institution. To support these requirements, the receiving institution must sometimes invest in temporary humidity control during a special exhibit. They may therefore conclude that any new building should have very tight control, to avoid the future expense and difficulty of a temporary humidity control installation. These financial factors can influence the clients decision on control levels for new projects as much as any material science research.

Further, some interpretations of the Museum Code of Ethics can influence budgets, and therefore the humidity control levels.

Many HVAC designers are puzzled by the fact that even though artifacts are incredibly valuable and in many cases irreplaceable, the building and HVAC systems that protect them are often completely inadequate. An Engineer might reason that if the material is important and in danger, it might be prudent to sell one or two pieces to obtain the funds needed to save thousands of other deteriorating treasures. But that line of reasoning is not popular.

The 2000 Code of Ethics for Museums, published by the American Association of Museums, states that: "... Proceeds of the sale of non-living collections are to be used consistent with the established standards of the museum's discipline, but in no event shall they be used for anything other than the acquisition or direct care of collections." Some believe that this guidance tolerates the use of such funds for staff and supplies, but not for "indirect benefits" provided by new buildings or environmental control systems, even if those are needed for "care of collections."

The issue and the management's position sometimes depend on the type of museum and its size. Some institutions are more sensitive than others. The possibility of appearing less than ethical during an accreditation review can make some clients reluctant to de-accession any object to protect others, even at the risk of continuing deterioration of a valuable collection. Apart from the blow to the institution's prestige, without accreditation an institution would have difficulty obtaining loans of artwork for special exhibitions.

Not all museum professionals subscribe to this unique interpretation of ethical behavior. Smaller institutions with collapsing buildings often feel differently. But where this particular opinion is in force, the HVAC community may sometimes be confronted with the distressing sight of valuable collections rotting or cracking without adequate environmental protection.

Consequently, any given institution may have many compelling but non-technical reasons for establishing the humidity control levels they have specified. Also, the specification is usually the product of a collaborative process such as that suggested by the Association for Preservation Technology (APT Bulletin, Vol. 27, no.3, 1996). HVAC designers should pay careful attention to understanding the client's entire situation before offering any opinion or advice concerning humidity control levels. Usually, the most helpful role is to make the client aware of the cost implications of different specifications.

Fig. 23.2
Sorption Is Fast, Drying Is Slow
In these tests, fabrics were exposed to saturated air, and then to air at 60% rh Note how moisture uptake was very quick compared to drying—hours vs weeks. This is one of many reasons why even brief excursions to high humidity are more damaging than one might expect. (Ballard, 1996, Reference 12)

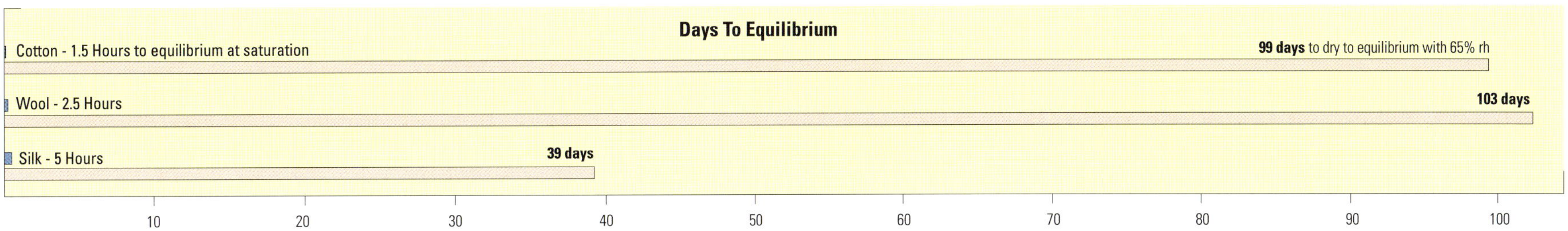

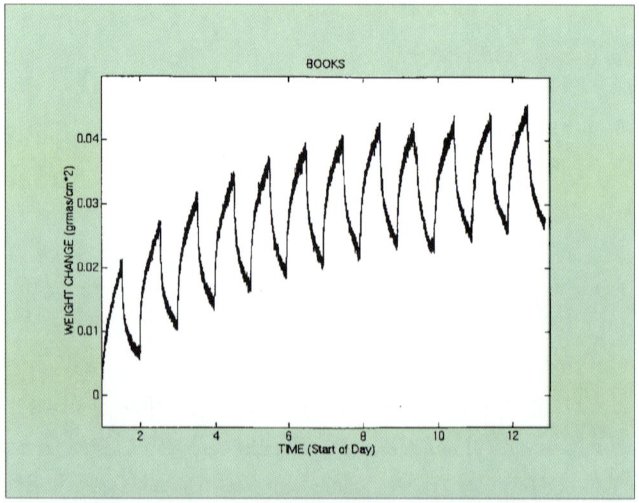

Fig. 23.3 Books Adsorb, But Do Not Easily Release Moisture
In this research, books were exposed alternately to 90% and 40% humidity, changing environments every hour. Note how the moisture content steadily rises rather than returning to dry levels. (Bailey, 1994) Keeping libraries at moderate and low humidities prevents the damage shown in figure 23.4

On the other hand, there is no controversy on the subject of stability. All agree that the more stable the relative humidity, the better. Fast, large changes in humidity generate large mechanical stresses in hygroscopic materials, and each rise-and-fall cycle shortens the life of the object. Also, when a humidity excursion is extreme, some types of objects can become damaged irreparably because of a single excursion beyond the established safe range.

For example, when humidity is near saturation, fabrics can absorb a great deal of moisture within hours, which then may take weeks to dry out even after humidity falls (See figure 23.2). Paper reacts in a similar way, as seen in figures 23.3 and 23.4. This behavior explains why it is important not to let the humidity rise to high levels even for short periods. Some materials can take on enough moisture to sustain adverse chemical reactions and support fungal or insect infestation indefinitely, even after humidity in the air comes back down below 55% rh.

Similar irreparable problems can occur at very low humidity. Once a wooden statue or an ethnological artifact made of bone cracks open because the air reached very low humidity, there will be no way to completely repair the object.

Moisture Loads

The load estimate displayed in figure 23.5 is intended to show the relationship of the load elements to one another, rather than suggest an absolute load estimate for all museums. The assumptions behind this particular estimate include:

- 33,000 ft^2 floor space [3,070 m^2]
- 200 visitors indoors, with 10 staff members
- 25 cfm/person ventilation [12 l/s/person]
- 200 door openings/hr, through a vestibule
- Weather for Atlanta, GA

As the figure shows, the moisture load is dominated by the ventilation air required by the occupancy. The number of visitors drives the load much more than any other factor.

Ventilation

The ventilation load is about 2.5 times larger than all other loads combined. It's a function of the number of people expected to occupy the museum during peak periods. Clearly, if this estimate is in error, the required ventilation is different, which greatly affects the need for dehumidification capacity. It is also clear that after the visitors and the staff leave the building, there is little need for such enormous amounts of ventilation air. If the ventilation air quantity is reduced, the moisture load is smaller. The operating costs for dehumidification will be reduced dramatically without compromising the control level.

So the designer might consider controlling the amount of ventilation air based on the indoor CO_2 concentration if the local codes allow that strategy. That gas is a product of human metabolism. As such, is a good indicator of actual occupancy, which affects many

Fig. 23.4
Eventually, mold and mildew grows
Given time and high moisture content, mold and mildew grows on books. This photo shows the effects of an energy conservation measure at a school library in Houston, TX. During summer vacation, the thermostat was re-set to 85°F [29°C] to save operating costs. But there was no separate dehumidifier, and minimum ventilation continued to bring in humid outdoor air. A dedicated dehumidifier operating independent of temperature could have avoided the $50,000 cleaning and replacement costs.

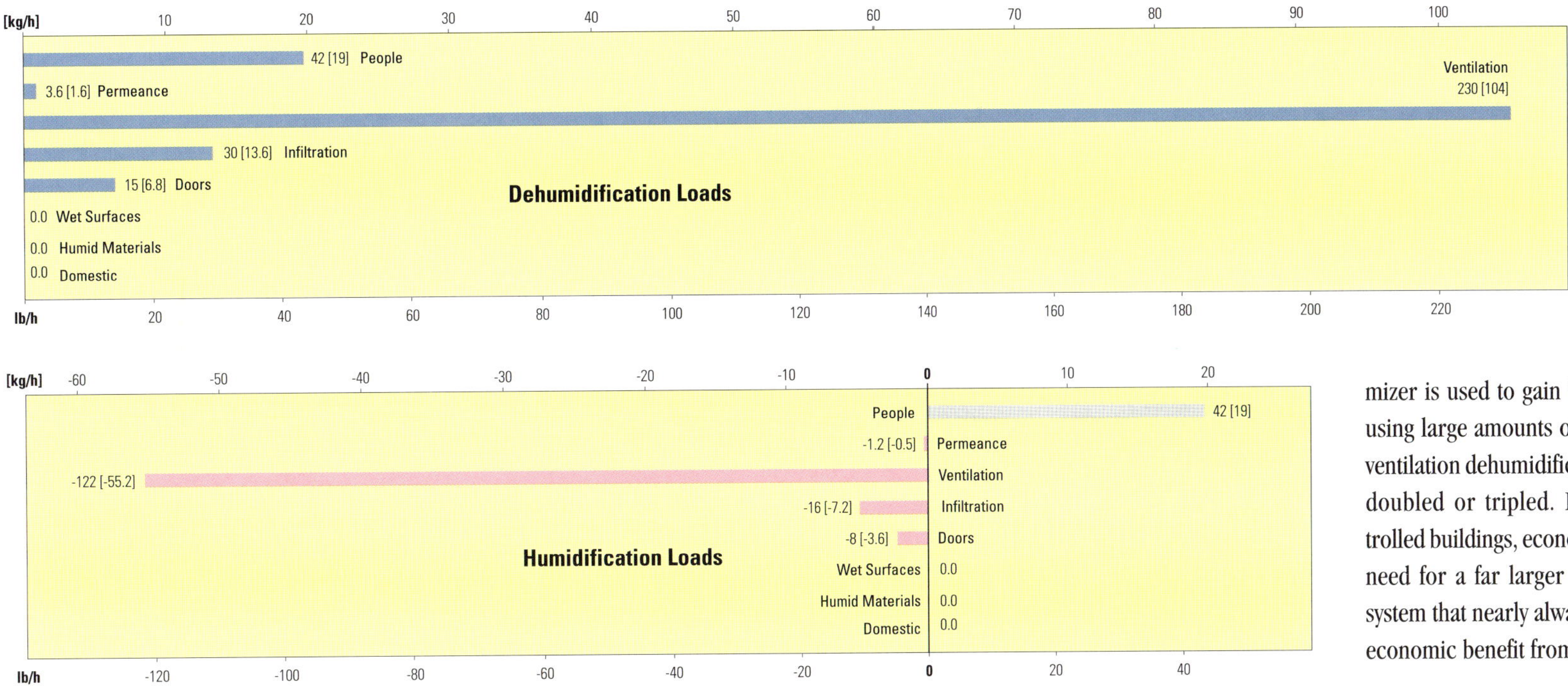

Fig. 23.5 Load Components

Ventilation brings in about 2.5 times more load than all other sources combined, which suggests preconditioning that air separately.

aspects of the indoor air quality. When the concentration of CO_2 rises 600 ppm above the background concentration outdoors (above an indoor level of about 1,000 ppm), it will probably be appropriate to bring more outdoor air into the building to dilute IAQ contaminants. As the indoor concentration falls, less outdoor air is needed to achieve the same dilution.

Most codes will demand a minimum quantity of outdoor air per unit of floor area. But above that minimum, the air volume may be negotiable based on occupancy. In those locations, the designer can greatly reduce operating costs by providing variable-volume ventilation based on CO_2 concentration.

Note that this load estimate is based on ventilation for people, not the much larger ventilation that occurs when cool and humid outdoor air is used for cooling the building. When an air-side economizer is used to gain "free cooling" by using large amounts of outdoor air, the ventilation dehumidification load can be doubled or tripled. In humidity-controlled buildings, economizers create the need for a far larger dehumidification system that nearly always eliminates any economic benefit from "free" cooling.

People and door opening loads

Visitors and staff release moisture into the air through respiration, perspiration and from their damp outer clothing. The designer has no control over the number of people visiting the building. In fact if the museum is successful, the number of people visiting the building will be greater than the estimate—not fewer.

But the architectural designer can help maintain the control level during peak periods by isolating the traffic-related loads with interior walls and doors that are normally closed. For the system which serves the entryway and cloakroom zone, the HVAC designer may wish to provide greater moisture removal and greater humidification capacity compared to systems that serve interior zones. Ideally, the system that serves the entry areas will have the capacity to remove those large loads before they can penetrate to interior zones. That way, the internal humidity will be more stable.

Infiltration

Some air will leak into the building when wind blows against the exterior walls. This estimate assumes that the wall will be leaking at the low rate of "tight construction," as opposed to leakage rates for either "average" or "loose" buildings. The HVAC designer is probably justified in assuming tight construction, as long as the exterior wall is robust enough to avoid condensation from forced humidification in cold weather. But if that is not the case, or if the museum building already exists and had served a different purpose previously—such as an office building or manufacturing plant—then the designer should assume average or loose construction and increase the system's dehumidification and humidification capacity.

Permeance vs. infiltration

As figure 23.5 makes clear, the moisture load from vapor permeation is very small. It "gets lost" in the error of the overall estimate. From a dehumidification point of view, there is no great need to be concerned about permeation.

On the other hand, permeation is very important in cold climates. Moisture can diffuse outward from the humidified exhibit spaces into cold building cavities and condense, causing serious damage. In cold climates, the inside layers of the exterior wall will need a continuous vapor retarder layer. It is even more important to keep the inside face of the exterior wall air-tight, and to keep the building under neutral pressure with respect to the outdoors, so that humid air is not forced outward into cold building cavities.

The permeation load is not important with respect to humidity control equipment, but in humidified museums in cold climates it is extremely important to keep water vapor inside the space rather than allow it to diffuse outward into the exterior walls. Several members of the construction team are responsible for preventing condensation damage in a humidified building.

The architectural designer must design a vapor-tolerant exterior wall for the humidified space, and the general contractor must build it that way. The HVAC designer's principal concern should be with air infiltration rather than vapor permeance. Air pressures must be maintained so that humid air does not enter the wall cavities. That requires maintaining positive internal pressure in the humid season, and neutral pressure during cold weather.

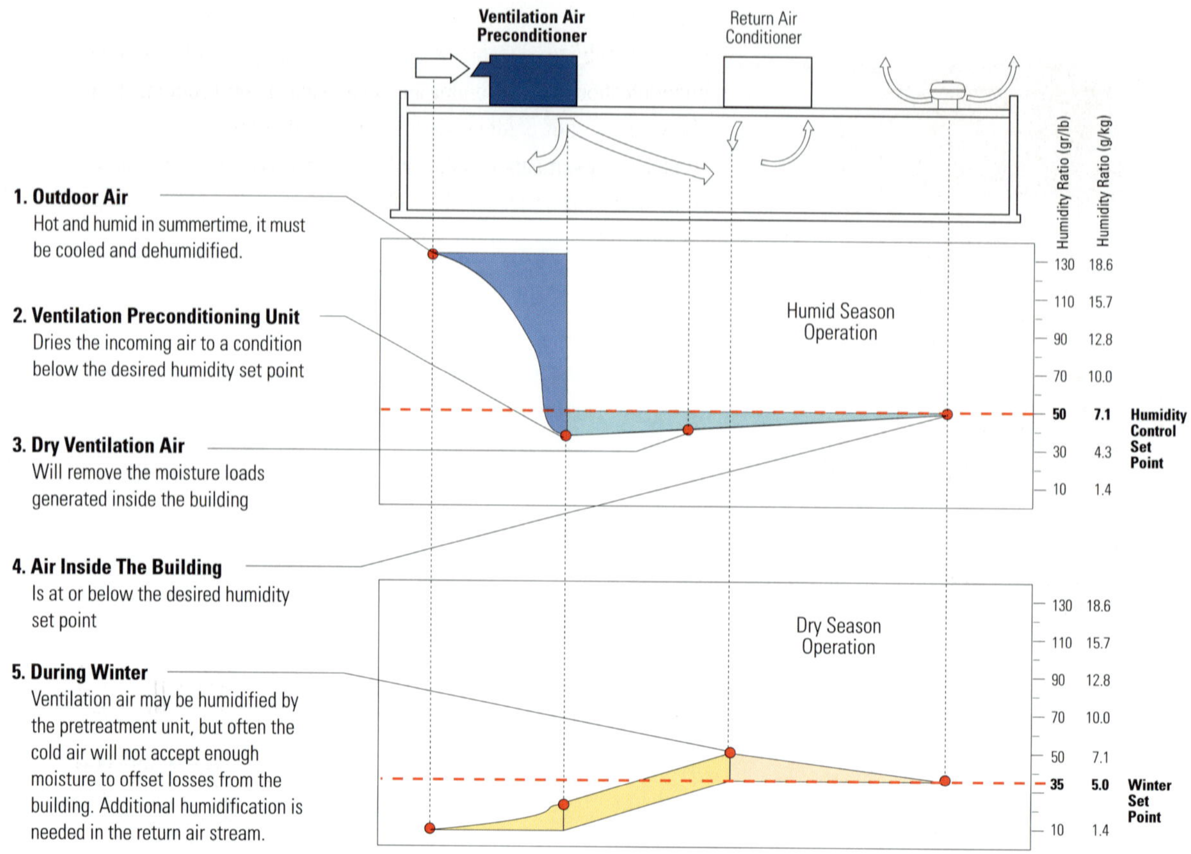

Fig. 23.6 Preconditioning Ventilation Air

By adjusting the moisture content of ventilation air, internal humidity can become very stable, avoiding the fast humidity changes that are so damaging to collections.

Humidification loads

During the cold season, the humidification loads are a mirror image of the dehumidification loads from ventilation, infiltration and doorway activity. In other words, the greatest humidity deficit is created by bringing ventilation air into the building. The next-largest load is the deficit created by infiltration and the doorway activity provides the third significant humidity deficit.

Designers are often tempted to use an air-side economizer (sometimes called "free cooling") during the winter or swing seasons, because the building may need cooling even when the weather is quite cold outdoors. Or perhaps *one* of the zones might need cooling while all the others need heating. But since the building needs stable humidity all year long, an air-side economizer greatly increases the use of the humidifiers during the cold seasons. Humidification and dehumidification of the incoming air usually cancels any economic benefit to using cool outdoor air to avoid running the cooling equipment. Hourly simulation of the building's operation is a useful way to evaluate the advantages or limitations of economizers in buildings where humidity must be held within a specific range.

Humidity Effects On Artifacts & Documents

Materials change their moisture content and undergo physical changes based on the relative humidity of the air around them, not based on the absolute amount of moisture in that air (the humidity ratio).

If the relative humidity is low a material releases moisture to the air; if the relative humidity is high it gains moisture. This moisture exchange continues until an equilibrium is reached between the object and the air. The objects are affected by the relative humidity rather than the absolute humidity of the air since the sorption and desorption of moisture is based on the *relative* moisture content of the object compared to that of the air—not on the absolute moisture content of the air.

Museum, library and archive collections often consist of different types of objects, books, manuscripts, maps, artifacts, photographs and films. Most collection objects are made from organic materials that will last longer when kept at stable relative humidity. However, some objects are more sensitive—will deteriorate faster—under a particular set of environmental conditions.

For example, mold can form on paper in high relative humidity, but the low humidity that suits silver-gelatin microfilm would cause permanent damage to vellum. The mix of different components of a collection, and each component's relative value to the institution, will determine the best environment. The needs of different materials may also suggest segregating the collection, providing different

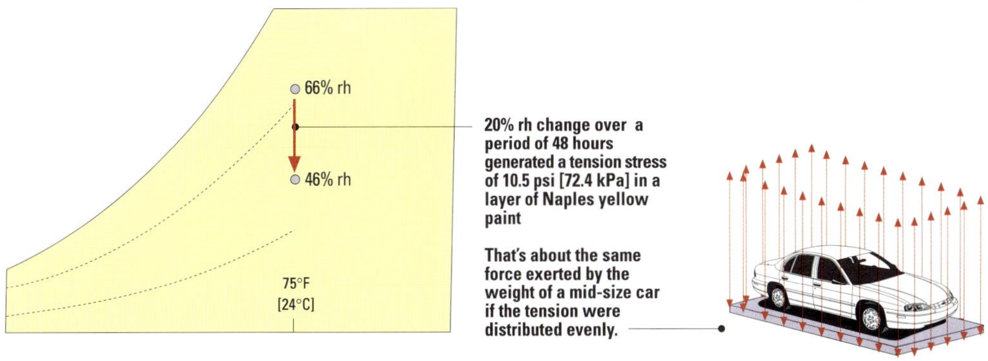

Fig. 23.7

Mechanical Stress From Changes In Relative Humidity

When a painting is switched rapidly from 66% to 46% rh, the tension stress in the paint layers can be about 4.8 psi [33.1kPa]. (Mecklenburg, 1991) That's roughly the force exerted by the weight of a mid-size automobile if distributed evenly. Avoiding fast changes in humidity helps keep paint layers from cracking under tension.

environments for different types of objects with different needs. This is sometimes easier to accomplish in storage areas, where there is less need to place physically different, but culturally-related objects together for the viewing public. As the moisture content of objects changes, they can have dimensional changes that lead to damage.

Low humidity

Low humidity can dry out and desiccate objects, subjecting them to physical damage immediately or when handled. Depending on the materials involved and the previous humidity exposure, low humidity may cause shrinkage and checks in wood, embrittlement of leather, and permanent embrittlement of vellum.

High humidity

High humidity can facilitate oxidization (rust or corrosion) of metals, acidification and general degrading reactions in paper-based materials. High humidity also increases the damaging effects of particulate and gaseous contaminants, and allows fungus and insects to feed directly or indirectly on collections. Periods of high humidity can cause salt migration and crystallization in organic materials, such as skin and bone, and in archaeological and geological specimens.

Stability

In most cases the humidity of a conservation environment should remain as stable as possible. Fluctuations in relative humidity can cause more damage to collections than stable humidities that are moderately high or low. Without control, the relative humidity can fluctuate dramatically over the period of a day, particularly if heating or cooling is turned off or reset at night. Over a year, the relative humidity in a building may range from nearly 80% on rainy days in summer if no dehumidification is provided, to below 10% on very cold days in winter if heating is provided without humidification.

Mechanisms of environmental damage to collections

To understand the damaging environmental factors, and to set environmental goals, it is often helpful to understand the mechanisms that cause environmental damage to collections. The forms of

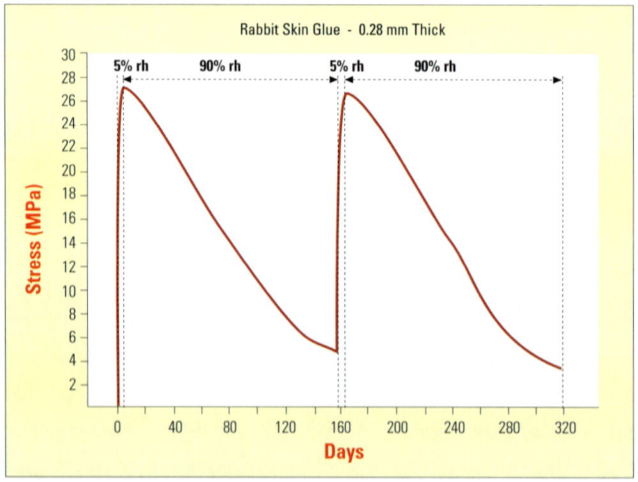

Fig. 23.8 Stress Relaxes With Time

The graph shows the nearly instant increase in stress in a thin layer of rabbit skin glue as relative humidity falls from 66% to 5%. After peak stress, the humidity was raised to 90%, and the stress was relieved slowly over 160 days. Then the sample was desiccated once again to 5% rh. (Mecklenburg, 1991)

Time allows mechanical stresses to relax, which is the principal reason for keeping humidity changes slow. (Note that this graph shows a worst-case; a huge humidity change surrounding a thin homogeneous layer of hygroscopic material. Large, complex objects do not respond this quickly to smaller humidity changes except in their surface layers.)

humidity-related deterioration in typical organic collection materials may be summarized as follows:

1. Mechanical or physical damage

This is usually damage to the object by direct physical contact with a foreign object - the result of mishandling or shipping. It can also result from differential expansion within the object or between layers of construction - the result of temperature, humidity or chemical action on the materials. This is usually seen as splitting, tearing, abrasion, checks, and delamination in composite materials.

Mechanical damage can be best controlled by preventing humidity extremes beyond those to which the objects have previously been exposed. It is also critical to ensure that humidity does not swing rapidly between two different extremes. Figure 23.8 shows the stresses that accumulate with rapid humidity change, which are then released slowly over time as the object adjusts to the new humidity level. In short, more time between changes creates less stress, and therefore less risk of mechanical damage.

2. Chemical damage

Some have observed that chemical damage to materials is governed by their temperature and by the surrounding humidity. Tem-

Fig. 23.9 Paper After Acid Hydrolysis
These paper shreds result from exposure to high humidity, which accelerates acidification that breaks down the bonds between fibers.

perature influences the speed of deterioration. A lower temperature slows that deterioration. The *type* of reaction—the type of damage—is often influenced by how much moisture is available in the material. At lower relative humidity there is less moisture available in the material (the water activity is lower) so reactions are fewer and slower.

Bonds that hold molecules and materials together can suffer chemical breakdown. In considering chemical damage there are three significant subgroups that have to do with environmental factors: kinetic, photochemical and pollutant-induced. Of these, kinetic deterioration is facilitated by both high temperature and high humidity.

Kinetic chemical reactions are those that proceed because of a difference in concentration between components in that reaction. For example, higher moisture concentration in paper accelerates the damage reaction known as acid hydrolysis. Kinetic reaction is also the primary factor in leather rot, dye fading, loss of strength in textiles, and "sticky shed" syndrome in magnetic tape.

Chemical kinetic damage can only be reduced by storing collections at lower temperature and in lower humidity. Recent research and publications have given valuable information for predicting the probable life of objects, based on chemical stability at various combinations of temperature and humidity. Reference 6 can be used for paper-based collections, Reference 7 for acetate film collections, and Reference 8 for color film and photographs. Using these publications, projecting an extended life for a collection can be used to justify improved environmental conditions by avoiding or delaying the cost of reproduction, treatment or loss of collection elements.

3. Biological damage

Biological infestation can damage the collection directly or indirectly. If conditions are sufficiently warm and humid, then mold, mildew, insects and vermin can infest a collection. Biological factors create the most common, rapid and acute damage to objects. Figure 23.4 shows how mold on books grows quickly at higher temperatures and humidities. But given enough time, fungal damage would occur at less extreme conditions.

Biological damage can best be controlled by preventing temperature/humidity conditions that allow biological growth, and by reducing access to the collection by insects and vermin.

4. Humidity affects all damage mechanisms

These three mechanisms may cause different types of damage under the same environmental conditions. Many assume a one-to-one relationship between environmental factors and damage, such as fungal damage being the only result of inappropriate humidity. This can lead to the false assumption that if fungal damage is a minimal concern for the particular collection type, then humidity is also of minimal concern. This is false. Humidity is a factor to some degree in all major forms of deterioration.

Usually, the first priority is to avoid acute risks from biological attack, fire and water damage. After protection from such acute damage, the next most-effective method to prolong the life of a collection is to reduce the rate of chemical deterioration by reducing temperature, humidity, light and contamination.

For some objects, variations in humidity may induce mechanical stresses and damage. As the collections are damaged by these mechanisms they may lose their informational or research value, or may simply be in too poor condition to be used without the risk of further and aggravated damage from handling. Some may be reformatted to recover informational value, while others may be permanently compromised from loss of research or artifactual merit.

In the case of mechanical damage, most objects undergo stresses and damage when they are *first* exposed to an extreme in humidity, or once they return to "normal" conditions after exposure to high humidity. Once "proved" to a given humidity range, repeated exposure to the same humidity extremes or dynamic seldom causes additional damage, provided the change does not come too quickly. Of course, if the object were conserved to repair the damage, the damage may recur within the same humidity range.

With most *existing* systems there is a clear practical tradeoff in temperature and humidity—you can usually decrease one or the other. Reducing *both* temperature and humidity is harder, increases operating costs, and probably involves new equipment.

Paper

Most paper-based collections need a stable relative humidity of around 30 to 40%. The level recommended as recently as ten years ago was as high as 55%, set more by inductive reasoning from the needs of museum collections rather than the specific needs of paper. Recent research and analysis of the chemistry involved has indicated that even lower levels may be better. There are indications that levels as low as 30% may be superior for preservation, as proposed in Reference 6. Low humidity levels are reported to reduce the rate of embrittlement due to acid hydrolysis, to reduce the rate of sorption and damage from gaseous contaminants, and to increase the stability of the paper. On the other hand, somewhat higher humidity helps maintain the mechanical properties of the bindings, which must stay flexible. Once again, the best level is a compromise.

For paper, Chandru J. Shahani of the Preservation and Testing Office of the Library of Congress, suggests that daily variations of humidity between 40 and 60% do not have the effect of the average humidity, 50%. Rather, they have the weakening and acidification effects on paper of the higher, 60% rh level.

Too-low humidity can desiccate materials such that they become less supple and more brittle, leading to handling damage. For example, glues dry out and break down when air is very dry. Also, it is fairly clear that paper itself is too inflexible at humidities below 20%. In very dry air, paper becomes more vulnerable to mechanical stresses, such as those generated when humidity fluctuates quickly.

Other materials in paper-based collections may need slightly higher moisture levels. For example, humidity below 25% may cause permanent moisture loss and embrittlement to vellum and similar animal skin book bindings. Some institutions keep their vellum-bound books above 45% rh, even though the paper would last longer at lower levels. Conversely, humidities above 60% may invite damage from mold and pests, if not dimensional change and mechanical damage from differential movement.

Estimating the life of paper-based materials

Sebera's work in Reference 6 shows the theoretical relationship of temperature and humidity upon the life expectancy of paper through what he calls "Isoperm diagrams."

These numbers represent "relative permanence values." As the temperature and humidity decrease, the permanence value goes up; and as they increase, the permanence goes down.

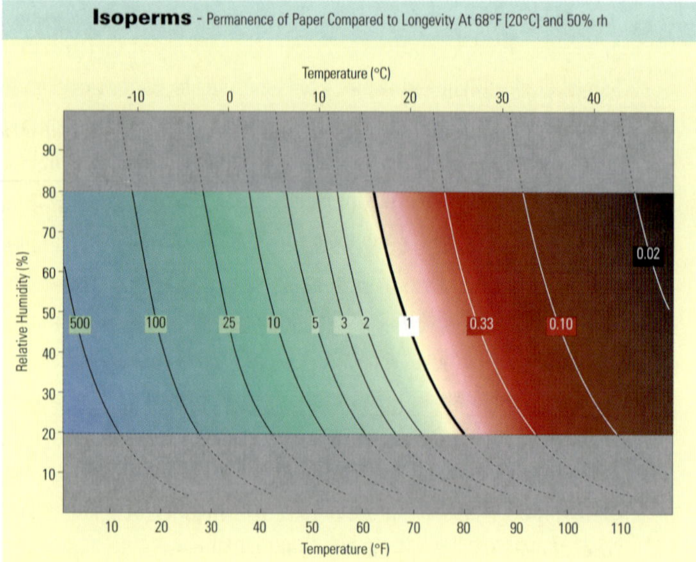

Temp.		Relative Humidity (%)						
°F	**°C**	**20**	**30**	**40**	**50**	**60**	**70**	**80**
41	**5**	26	18	13	11	8.9	7.6	6.7
50	**10**	12	7.9	5.9	4.7	3.9	3.4	2.9
59	**15**	5.5	3.63	2.7	2.1	1.8	1.5	1.3
68	**20**	2.5	1.7	1.3	1	0.83	0.71	0.6
77	**25**	1.2	0.8	0.6	0.48	0.4	0.34	0.3
86	**30**	0.59	0.39	0.29	0.23	0.2	0.17	0.15
95	**35**	0.29	0.2	0.15	0.12	0.1	0.08	0.07

Fig. 23.10
Isoperms— Zones of Relative Chemical Permanence
Sebera's isoperm diagram shows how chemical deterioration of paper slows down at low temperatures and low humidities. (Sebera, Reference 6)

In addition, note that all damage mechanisms are simultaneous. When relative humidity is over 70% or under 20%, biological or mechanical damage will shorten the life of paper much more quickly than the effects of chemical deterioration.

Figure 23.10 shows values from the isoperm equation for paper, based on using Sebera's 20°C [68°F] with 50% rh as perm value of 1.0.

Consider the immense difference between a perm value of 18 at 41°F, 30% rh compared to a perm value of 0.1 at 95°F, 60% [5°C vs. 35°C] In other words, the cooler and drier environment is about 180 times as effective at preserving the collections compared to the warmer and more humid conditions. Isoperm values can be used to estimate the relative effects of different environmental specifications.

For example, assume that the working life of common copier paper is 50 years when held at 68°F [20 °C] and 50% rh. Given that 50 years is the life of paper at a perm value of 1, Sebera's chart can be used to estimate the relative life expectancy for any other value of temperature and humidity. For example, at 77°F [25°C] and 60% rh, the permanence value is 0.40. So the life expectancy for the paper will be 0.4 x 50 years, or 20 years. On the other hand, if the temperature/humidity is reduced to 59°F [15°C] and 30% rh, the permanence value is 3.6. The paper's life expectancy is therefore 3.6 x 50 years, or 180 years.

Although Sebera's isoperm chart displays the probable life of paper above 70% rh and below 20% rh, factors other than chemical deterioration will reduce its life outside those limits. Below 20%, the paper's brittleness, when combined with any mechanical stress will destroy the paper much more rapidly than chemical reactions. Likewise, above 70% rh, fungal degradation will destroy the paper very quickly compared to the usually slow rate of chemical reactions.

Natural history specimens

The moisture content of objects (skins, furs, feathers, wood, paper and artifacts) are determined by the ambient relative humidity. Their moisture content is critically important. Skins have split, teeth have delaminated, and hair has loosened or been lost when humidity changed quickly. Breakage in feathers, baskets, shells, and split skins have all been attributed to humidity near or less than 20% rh.

Also, because of their high salt content, archaeology ceramics need a stable humidity between 35 and 45% rh to avoid excessive moisture absorption. Shale, shell and subfossil bone are layered specimens and are very sensitive to differential movement from humidity extremes. These must be kept at a stable humidity between 40 and 60% rh.

Herbarium collections (Dried plants)

Based on discussions with conservators, low humidity can be expected to cause damage to collections if they are handled when very dry. But low humidity is not a problem for the specimens, although it might dry out any glues or paper on which specimens are mounted. Conservators report that the primary threat to herbarium collections is rehydration of the specimens, which occurs when humidity rises to high levels and stays there for extended periods. As soon as vegetable materials absorb enough moisture, they become vulnerable to fungal growth and can be eaten by insects—which is what happens outdoors as part of nature's recycling process.

Metals

Metal corrodes so slowly as to be difficult to measure when humidity can be kept below 35% rh. Metal objects that can be separated from other materials can last almost indefinitely when stored under such dry conditions. However, even trace amounts of residual salts lower the safe relative humidity, because salts attract moisture out of the air and onto the metal surface even when the air is relatively dry. Above 50% rh, metallic corrosion accelerates rapidly, catalyzed by other gaseous or particulate contaminants on the surface of objects from years of exposure to the air or soil.

Film and photographs

The life of photographic media is particularly affected by temperature, with dramatic improvements if temperatures are reliably kept below 70°F [21°C]. Some vaults are kept as cold as 0°F [-18° C] for an expected film life which may exceed 500 years.

The upper humidity limit for film collections is 50% rh, but ideally the humidity is held at 30%. However, as long as the humidity level does not exceed 50% rh, lowering the temperature below 70°F [21°C] extends film life much longer than would reducing the humidity to 30%. The optimal humidity for a particular collection can vary with the film type and its housing. See References 4, 7 and 8 for guidance. Cold storage vaults for film must also held at low humidities, generally around 30% rh.

Acetate film collections

Recent investigations into film deterioration have shown that all forms of acetate film are at risk of "vinegar syndrome." In this reaction, acetic acid is released. Acetic acid is the primary constituent of vinegar, hence the term "vinegar syndrome."

When acetate films have a clear vinegar odor, they have begun to deteriorate. Acidic deterioration of the acetate film support base can lead to wholesale loss of film rolls and sheets as the acetate shrinks, wrinkles and separates from the emulsion. This deterioration of acetate film is an irreversible, inherent characteristic of acetate films.

The only control mechanisms are reducing temperature and humidity, and, removing any ambient gaseous acids. Unfortunately, the acid gases released during this deterioration will initiate or accelerate the deterioration of other nearby film. This is an autocatalytic reaction. The acid molecules that cause the breakup of the acetate media molecule are not consumed in the reaction. In fact, they release additional acid molecules, which break more acetate molecules.

This chemical deterioration is usually initiated by temperature, and accelerated by humidity. The only practical options are to keep the film under cool conditions until it can be copied to a stable medium, or to keep the film cold and dry enough to serve its intended life. The cost to copy acetate film to polyester has to be weighed against the cost of maintaining temperature and humidity conditions that would extend its life.

The projected life of acetate film collections is quantified by the Acetate Film Guide, Reference 7. Using the Guide, one can predict the remaining life of fresh as well as degrading film. As with isoperms, the relative life of acetate film collections can be estimated for different temperature and humidity conditions. Knowing the storage conditions, one can estimate the number of years left before the next level of damage occurs. Decision-makers can judge the value of short-term cold and dry storage for a longer life, against the eventual need for a permanent solution of copying to a more stable medium.

Color film collections

Preservation of original colors is a much greater challenge than safe storage of monochrome images. Most color media are chromogenic rather than dye-transfer, and are subject to "dark fading." That means they loose color even when stored in the dark. While nothing can reverse this chemical process, like vinegar syndrome it can be slowed by a reduction in temperature.

Most black and white acetate film collections are stored at 60°F to 40°F [16 to 4°C], which slows deterioration to a rate that allows reformatting to polyester. Once on polyester, the black and white image is relatively permanent. On the other hand, color media, even on polyester base, may require storage below 32°F to as low as -4°F, to slow color loss to an acceptable rate. [Store below 0° or -20°C to gain an acceptable storage life.] Given such seldom-found and costly storage requirements, the unfortunate fact is that most color film collections will not survive in their original color balance.

Similar to acetate film, the projected life to manage the preservation of color photographic media is quantified by the Storage Guide for Color Photographic Materials, Reference 8. For further advice on storage of color photographic media also consult Reference 9.

23.11 "Vinegar Syndrome"
Older film may have an acetate base rather than polyester. Moisture catalyses a reaction in the acetate base that creates acetic acid, which has the same smell as vinegar. The reaction is irreversible and inevitable, but can be delayed for decades by storing the film at low humidity and low temperature.

Suggestions & Cautions For Humidity Control

A typical project will consist of many different types of spaces, such as galleries, reading rooms, search rooms, laboratories, conference rooms, stacks, storage rooms, restaurants or cafeterias, auditoriums, rare book vaults, offices, lounges and study rooms, to name a few. Not all of these spaces will hold environmentally sensitive objects. The areas where the objects or collection items are kept will need a special environment for preservation of the "collection."

The HVAC system for a preservation environment must usually establish and maintain relative humidity, maintain temperature and air movement, filter the air, and perform these tasks evenly throughout the space with minimum risk of damage to the collections, at a cost that the institution can sustain.

In most cases special HVAC system indications fall into two general categories: museums and libraries. Museums include galleries and other spaces where environmentally sensitive objects are kept or displayed. Libraries include archives and other spaces where primarily paper-based collections are stored and used, with some additional concern for film and other media.

In most museums and libraries, decision makers have high expectations for performance, along with low or limited budgets. To make the best use of the budgets, our suggestions and cautions begin with a section on overall design.

Project design

1. Consider the functional organization

An effective preservation environment involves more than just the HVAC system. The basic architectural design (such as windows and vapor barriers) and building operation (24-hour operation and availability of tempering sources) must complement these purposes—and not pose impractical challenges to the HVAC design.

To avoid impossibly expensive systems, the HVAC designer should be involved in the very early project planning stages to ensure that the space layout does not present unnecessary problems. For lowest HVAC costs, collections are stored separately from visitors, users, staff and all other functions. Where this is not practical, processes and activities that pose a threat to collections should be physically and mechanically separated from the collections. Separate systems for collection and non-collection areas allow isolation of environments, and will reduce project costs for non-collection areas.

A typical issue, particularly in fine art museums, is whether to treat executive offices as collection spaces. This should be carefully considered, not only for the added capital and operating cost, but the risks to the collection if the offices are not so treated but are nonetheless used for collections display. It is always unfortunate when a priceless object cracks into pieces during a board meeting in the Director's office.

One of the most environmentally disruptive elements is a frequently used entrance, such as the lobby or loading dock. The designer should take care to manage the enormous heat, moisture and particulate loads from these spaces, perhaps using separate systems, so those loads do not penetrate to the primary collection areas.

2. The building envelope must tolerate high winter humidity

Winter humidification should be a high priority for a preservation environment in a cold or temperate climate. But during cold, dry weather, high humidity is difficult to contain. The building must be able to tolerate high humidity without damage. Without special attention, the building envelope, the walls, roof and windows, will have problem condensation in winter at interior humidities as low as 25% rh—either on single-glazed windows, on window frames, or within the exterior wall or roof. When indoor humidity is very high, this condensation can cause cosmetic or even structural damage to building components. To avoid problems, the designer has three alternatives:

1. The indoor winter humidity must be kept below the dew point where problem condensation will occur, or;

2. The space temperature must be reduced so that the low dew point represents a higher relative humidity, or;
3. The building envelope must be modified to tolerate a relatively high indoor dew point—the most costly solution.

If the collection that needs the higher humidity does not dominate a building, but only occupies a manageable minority of the spaces, then separate humidified storage containers, cases, cabinets or rooms are an alternative. These must be carefully sealed and isolated to preserve their internal environment and to protect the building from the damaging moisture.

3. Without partitions, humidity travels

In open areas, humidity equalizes quickly. The owner and Architect must not expect areas of high or low humidity to stay that way unless there are doors or solid partitions to contain the humid air.

4. Protection must extend to all hours, all year

In most cases the systems should operate 24 hours a day. Galleries and reading rooms can have a high occupancy that only occurs at some times. Some gallery occupancies are as high as 10 square feet per person, in contrast to stack or storage areas where the density may be 1000 square feet per person, or less. [90 m^2 per person or less] Not only should the system be designed to handle this load, but it should be able to deal efficiently with the more common part-load condition.

5. Protection demands fast repairs

A critical aspect of HVAC system performance is reliability. Most collections can tolerate several hours of lost conditions without major damage, but many days of system malfunction could be very hazardous. The Engineer should evaluate equipment failure scenarios against collection sensitivities and likely maintenance efforts to see what reasonable precautions can be taken to minimize downtime and damage to the collection. In some cases spare equipment may need to be kept at the project site to allow timely repairs.

6. Exhibit cases should remain independent from the HVAC system

Exhibit cases should be designed for protecting the collection from environmental extremes and excesses. Among the typical approaches that might be used for environmental control are sealed cases and vented cases. Sealed cases rely on isolation from the ambient environment in the exhibition room, and usually require passive or special conditioning systems independent of the regular room air. Since sealed cases are subject to buildup of gaseous contaminants, they should be made of inert materials. Where exhibit cases are inside, built-in, or back up to a properly conditioned space, the cabinets might instead be vented to the conditioned space.

Exhibit cases should not be conditioned by blowing HVAC supply air into the cases, nor by drawing return air through the cases. Supply air will be at varying temperature and humidity conditions as it corrects space conditions and can cause extremes in temperature and humidity if blown into a case. Even if temperature and humidity conditions are sensed inside the case as part of the control system, the typical high ratio of supply air to the case volume has made virtually all such treatments problematic in the past. Problems are usually caused by poor design, or the drastic effects of an HVAC system component failure; acute temperature or humidity conditions from the supply air are undiluted and immediately visited on what are usually the most sensitive objects in a collection. While return air is more stable, most return air paths accumulate particulate contamination, accentuated by the sweeping effect of the supply-return air flow, leading to particulate accumulation in the display case.

This is not to say that active conditioning exhibit cases has not been used successfully—but this has been done by using special equipment designed specifically for this purpose (Note 1). Such systems are not configured like typical HVAC systems, and have various features, such as desiccant beds to stabilize the supply air humidity, that are part of their successful application. Case conditioning is most often used in cold climates where conditioning large galleries has been problematic, and case conditioning is often the only solution.

Moreover, the stability of conditions achieved with such systems has been less than ideal, and certainly less stable than sealed exhibit cases with passive conditioning agents. Their primary value is where many cases need to be conditioned where rigorous sealing and re-conditioning passive agents are impractical.

7. Optimal care of film usually involves special vaults

Projects often require cold storage vaults to extend the life of materials particularly sensitive to thermal deterioration, such as acetate film and color photographic materials. These usually require special equipment and systems, and most such successful systems (Note 2) are provided by experienced turnkey or design-build vendors.

Suggestions and cautions for system functions & components

1. Avoid large temperature differences

In older buildings, there is a temptation to use extra-cold air for summer cooling, because then more cooling can be done with less air, so the ductwork is smaller and easier to hide. But in museums, that strategy is very dangerous for the objects. Big temperature differences create big differences in relative humidity, which create mechanical stress for artifacts and artwork. The best practice uses *larger* than usual air flows and *smaller* than usual temperature differences for better stability. The owner and Architect must be informed that this will mean the HVAC designer will need more than the usual amount of space for ductwork.

2. Circulate air constantly

Libraries and museums rely on air flow to filter the air, assure even temperature and humidity and to suppress mold growth by keeping materials dry—not just for cooling and heating. Air should be constantly circulated at sufficient volume to ensure good circulation throughout the collections space, regardless of space-temperature and humidity needs.

3. Avoid intermittent hot and cold spots

Usually, perimeter radiation and other hydronic, sensible-only heating or cooling components should not be used in the controlled spaces. They cause local extremes in *relative* humidity which puts any nearby collection items at risk of mechanical damage. The relative humidity swings widely as equipment turns on and off, and such cycling generates mechanical stresses for hygroscopic materials.

4. Override cooling and heating if stable humidity is threatened

The system should provide the necessary cooling and heating, but must subordinate temperature control, particularly excessive or capricious heating, to the need for stable relative humidity. Cooler conditions are often selected for superior preservation, and this may require more cooling capacity. Most of the challenges with cooler conditions relate to dehumidification rather than cooling.

5. Insist on week-long graphic records of relative humidity

Building automation systems often display current temperature and humidity at an impressive number of locations, and produce a text file with the values that have been recorded over time. But unless the designer specifies it, not all automation systems produce week-long graphic records of relative humidity.

Graphic records are essential for understanding the recent stresses on the objects, and for helping the operating personnel make the real-time adjustments needed to keep relative humidity constant. The display needs to reflect the long-term history, not just the last few hours, and the data must be recorded frequently enough to reflect short-term fluctuations as well as long-term changes.

Ideally, the terminal display and printed reports should contain a graph of the relative humidity over the most recent seven days, recorded at five-minute intervals, using time and humidity scales that allow the viewer to perceive (at a glance) the following conditions:

- Any relative humidity outside the control limits
- A variation in relative humidity of more than 2% within any single hour

6. Ensure even modulation of humidifiers and dehumidifiers

Dehumidifiers and humidifiers that serve museums have to be big. They must remove a great deal of moisture from ventilation air in the humid season, and add it back during the dry season. But that large capacity must be evenly controlled. Simple on-off control of capacity will cause humidity to swing widely up and down, because the equipment has a large capacity in relation to the size of the space.

To avoid swings, the control system must modulate the capacity of these components smoothly. For example, steam valves that regulate humidifiers must modulate, not simply turn steam on and off. And if the valve body has a convex parabolic plug rather than a linear-taper plug, modulation near the end of the throttling range will be smoother. Similarly, face and bypass dampers are often used to force air through a cooling coil or a desiccant dehumidifier, or allow the air to pass around the device. But if the dampers have only two positions, there is effectively no modulation. Face and bypass damper positions must vary continuously, according to how much dry air is needed in the space.

Further, the system must be equipped with humidity sensors that prevent the humidifiers from over-saturating the air streams they serve. But those sensors must not shut off the humidifier entirely. They should modulate the unit's capacity, otherwise the humidity will drop rapidly in the conditioned space.

Dehumidifiers and humidifiers also work best when modulated in response to a dew point sensor mounted in the ductwork downstream, after the dry and wet air have been well-mixed. That way, the moisture in the supply air is relatively constant. Humidity sensors in the controlled space can *reset* the duct-mounted dew point controller when air with a different dew point is required by the space.

7. Use extra caution designing humidification subsystems

As noted in several other places in this chapter and this book, *extreme care is required* when humidifying buildings in cold climates. Improvements to the building envelope will probably be necessary to make it tolerant of the high indoor dew points in winter.

Humidification should be provided by steam injected into the air, or by water evaporated from pads. Systems must be selected and designed to prevent condensation or standing pools of water, and otherwise should follow good humidification design as described further in Chapters 1 and 20 of the *ASHRAE Handbook—Systems and Equipment*, as well as Chapters 12 and 14 of this book.

Humidification system options start with a drain-through pad, wetted belt or wetted drum humidifier; this type will only provide modest performance unless additional heat is added to the air to facilitate evaporation, and may require monthly maintenance and cleaning.

Most humidification for cold climates uses a steam generator and some form of steam injection in a ducted supply air system. The steam can come from non-pressurized sources, such as electrode canisters or evaporative pans; these types of humidification can have problems with cycling at part-load conditions. Better control comes from the use of a pressurized clean steam source, delivered in a preheated manifold, directly modulated with a steam valve. Whenever steam is used there is the potential for over-humidification. Steam systems should have a high-limit control to positively protect from runaway humidification.

Also, most buildings will require a dedicated steam generator for humidification rather than being able to use the main boiler's steam directly. Steam for heating systems is treated with chemical inhibitors to keep the pipes from corroding. Most such compounds are not healthy for people or for collections (Note 5). Avoid this problem by using secondary boilers or steam-to-steam vapor generators. These can use a large boiler's heating capacity and favorable operating economics without the risk of steam contamination.

Virtually all buildings have separate temperature control zones, but for museums and libraries, humidity control is even more of a concern than temperature control. Zone humidifiers are often necessary, even when the same humidity level is desired in each zone on the same system. This is because loads can vary widely between rooms in the same zone. As when, for example, a single zone includes both outdoor entry areas along with interior, isolated exhibit rooms.

Maintaining widely different conditions in zones off the same air handler can be wasteful of energy and may be impossible. For the lowest-cost solution, all spaces within the same zone should have the same humidity ratio—the same absolute moisture content. Then reheat can be used to trim the relative humidity for each of the separate spaces.

This approach is much easier and less costly than adding individual humidifiers to meet each room's slightly different needs. In fact, it's a useful strategy for the entire building—humidify the whole building to a single, defined dew point, and then trim the relative humidity with reheat in each zone. Although this strategy requires some flexibility in room temperature (usually less than ±3°F or ±1.5°C), the humidifier is much easier to control. It's less likely to overshoot the need or oversaturate the air in the duct , and a single array or a few central units are less expensive to install and much easier to maintain than many distributed small zone humidifiers.

8. Dedicated dehumidification is required

The most common problem in museums and libraries is inadequate or ineffective dehumidification. Modest dehumidification can be achieved with most cooling systems, limited by the apparatus dew point at the cooling coil, and requiring adequate reheat. Most problems can be traced to compromises in cooling medium temperature, or the lack of reheat. Some chilled water systems may not be able to reliably deliver chilled water that is sufficiently cold. The chiller may also reset the chilled water temperature automatically to save energy, or the system may have cooling coils too shallow for effective dehumidification. Finally, some reheat coils require a heating source unavailable in the summer or too expensive to operate.

There are three (Note 5) typical approaches to more cost-effective dehumidification that can be added to the basic heating and cooling systems. In order of *increasing* capacity to reduce the indoor dew point, these alternatives include:

1. Low-temperature chilled water, usually based on a glycol solution. This has the advantages of economical, familiar operation and stable control, along with the ability to serve several air handlers with the same chiller. It has some disadvantages, including glycol management, the need for reheat, piping and valving expense and high costs for very small or remote air flows.
2. Mechanical dehumidifiers that use direct expansion (DX) cooling. Commercial-grade mechanical dehumidifiers far outperform the lightweight residential units familiar to home owners, with greater air flow capacity, better reliability and reduced maintenance. They are used most frequently to reduce the dew point to moderate levels and have very favorable operating economics.
3. Desiccant dehumidifiers. These units are best-known for their ability to maintain very low dew points regardless of the temperature or weather. But for most moderate-level dew points they usually cost more to buy and to operate than mechanical dehumidifiers, and their economy of operation is very sensitive to the cost of the reactivation heat source.

Also, to maintain comfort conditions, all desiccant dehumidifiers need supplemental cooling. Some units provide that cooling internally. In a few cases, desiccants are used without cooling for applications where the temperature can swing up and down as long as relative humidity or dew point stays constant, as in storage areas. And for other projects that require very cool and dry conditions—such as film archives—a desiccant system may be the only practical alternative.

8. Avoid outdoor air economizers

Since the goal is to maintain a close-tolerance indoor environment, large amounts of outdoor air for "air-side economizer" cooling present a problem. Outdoor air is almost always at other-than-desirable temperature and humidity, and can introduce particulate and gaseous pollution. Economizer cooling is often the primary cause of humidity fluctuations in museums and libraries.

Air-side economizers should not be used, except where an analysis of hourly outdoor air dew points shows the air to have a favorable moisture content for a sufficient number of hours to be helpful, and unless the dew point of outdoor air can be reliably measured and the incoming air flow properly modulated. Outdoor air should be minimized, providing only the necessary fresh air for occupants, and enough air to pressurize the collection spaces, which reduces infiltration of pollutants and particulates.

Relief air—indoor air exhausted from the building or the system—is seldom needed, unless high occupancy results in so much required outdoor ventilation air that internal air pressures would become excessive. Since peak occupancies are rarely the norm, outdoor air can be kept to an appropriate minimum by modulating its volume in response to the indoor concentration of carbon dioxide, or other appropriate gases indicative of poor indoor air quality.

Preconditioning outdoor air, usually with a precooling coil, often provides major benefits to humidity control and improves operating economics. Dry ventilation air can reduce or eliminate the need for internal dehumidification, and may reduce the static pressure on the supply fan by keeping the cooling coil for the recirculation air dry. It can also reduce or eliminate the need for costly reheat if the main cooling coil had been used for dehumidification.

9. Consider dehumidifier-only or heater-only systems for storage

In cases where comfort conditions are not required and for objects that are not especially affected by high or low temperature, institutions can save cooling and heating costs by using either a dehumidifier alone, or a heater alone to control humidity. Examples include storage buildings for cars, boats, carriages, aircraft and other objects originally built for outdoor use.

Frequently, storage buildings in cold or moderate climates are only equipped with a desiccant-type dehumidifier, which turns on in response to a humidistat set slightly below the desired maximum relative humidity. In climates that are warm all year long, mechanical-type (cooling-based) dehumidifiers can also be used for this purpose.

Also, in cold climates, humidity-controlled heating (Notes 6, 7, 8) can be used with similar effect. This approach modulates heating within a very broad temperature range to seek stable or moderated humidity conditions; heating is actually used to control relative humidity rather than temperature. A humidistat calls for heating whenever the humidity is above the set point, or whenever the temperature is below a maximum set point. For new unoccupied spaces, heating might be controlled entirely by such a system. For heating in existing unoccupied spaces, such humidity-dependent logic might be added to the control system, allowing the building to operate within a broader temperature range. Humidity can govern heating whenever the temperature is between the minimum and maximum values that prevent damage to the building or its contents—usually between 55° and 75°F [12.8° and 23.9°C].

Suggestions and cautions by type of system

Some types of systems are better than others for achieving control of relative humidity (as opposed to control of dew point or absolute humidity) in large buildings with complex internal spaces. For museums, libraries and archives, constant air flow is important to filter the air, control humidity, and suppress mold growth, not just for heating and cooling the collections space. Minimum recirculation air flows vary from 6 to 8 air changes per hour (Note 9). These needs are usually best met with constant-volume systems. Other cautions include:

1. Avoid distributed piping

Designers frequently overlook the problem with maintenance access and risk to the collection from disruption and leaks from decentralized equipment. Pipes carrying water or steam over and into collection areas always present the possibility of leaks. While some systems can provide full control without running any piping to the zones, other systems require anywhere from two to seven pipes to be run to zone. To reach collection spaces these pipes must usually be run over or in collection areas and are often the pipes that leak. Repairing such leaks and maintaining the distributed components can interrupt effective use of the display spaces. That's one reason why all-air systems are usually specified.

A central air-handling station co-locates filtration, dehumidification, humidification and maintenance outside the protected space, *away from the collection*. The investment in added overhead space, and the cost of the more elaborate duct system provides major benefits in reduced disruption to the collection spaces, lower-cost maintenance, and in dramatically extended service life to the air distribution system.

Unlike most commercial buildings, "tenant turnover" and use changes are infrequent, so re-zoning the HVAC system is rarely needed. In some museums, 60-year old multizone duct systems have been reused. Renovating the old system was quite economical, with the majority of the changes confined to the mechanical rooms. This stands in sharp contrast to changes in common-duct distribution systems (such as terminal reheat, dual- duct and variable-air-volume systems), where upgrading to satisfy preservation criteria often requires a new duct system and terminal equipment. Changes in common-duct systems are costly and disruptive, since they require demolishing the old system, installing the new duct system and re-installing architectural finishes.

2. Constant-volume reheat systems need centralized components

A constant volume reheat system can present problems if components are located near the space. In many institutions, terminal reheat with steam or hot water coils located near or over collection spaces have developed chronic problems from steam and water leaks. Efficient zone-level humidification will often suggest the humidifier be located downstream of the reheat coil. But if the reheat coil is located near or over collection spaces, preventive maintenance on humidifiers further complicates the maintenance problems. Constant-volume reheat systems can be used quite effectively by locating the reheat coils and humidifiers entirely within the mechanical room rather than at the terminal, feeding through what is effectively a "multizone" air distribution system.

3. Multizone systems need zone-specific heaters and humidifiers

A multizone air handler with zone reheat and zone humidification can provide a stable and relatively energy-efficient solution. However, multizone systems *without individual zone reheat and zone humidification* have proved problematic in many institutions, requiring the retrofit of zone equipment for stable humidity control. With the proper layout and equipment, a multizone system can reduce the amount of reheat and be among the most energy efficient.

An objective comparison of systems (Note 10) shows the preference for constant volume and multizone systems in museums and libraries. When other system types are used, the client should be made fully aware of the compromises in performance, cost and serviceability that may come with the specified system.

4. Multizone and dual-duct systems require a dedicated dehumidification device

An important feature for multizone and dual-duct air handlers is a separate dehumidification coil upstream of both the hot and cold decks. This separate cooling coil, distinct from the cooling coil in the cold deck, is used when there is a dehumidification demand. In this way air might be cooled to dew point even if it eventually flows through the hot deck. Without this feature moist return air can be warmed in the hot deck and delivered back to the room without ever being dehumidified. An alternative is to locate a single cooling coil upstream of both decks, where the cold deck simply bypasses the hot deck, although this will add energy expense during operation.

5. Systems and components that are seldom appropriate

While all HVAC systems have their place in commercial buildings, and most can be used in the offices and the restaurant areas of a museum, not all systems are appropriate for maintaining a preservation environment. For example:

Fan-coil units have proven problematic when used in and above collection areas. Fan-coil units increase and decentralize maintenance, causing disruption in the collection areas as well as adding costs. Since they cool the air, they need condensate pans, drain traps and drain lines, which are seldom installed properly. These assemblies have a history of clogging, leaking, failing to drain and developing microbial slime. As all-water systems, they require four pipes with pressurized water to run to each unit, increasing the chance of piping leaks in collection areas. On top of all these problems, fan-coil units rarely if ever can provide the filtration and humidity control features required for collection spaces. So they are seldom used in museums or archives.

VAV systems used in the past in museums and libraries have a reputation for poor humidity control, inadequate air flow, maintenance disruption, leaks in the collection spaces, and inflexibility in meeting different control levels. Many times the VAV system—perfectly appropriate for other building types— is chosen due to space and budget constraints. But in virtually every case, the cost and space required for a "museum-appropriate" VAV system (full filtration, local humidification, local dehumidification, minimum air volume settings, well-planned piping, easy maintenance access, and well-documented operating instructions) gives no advantages over the more conservative, less problematic, and easier to maintain constant volume systems. While the VAV system can be made to save energy compared to the constant-volume system, it is usually doing so at the expense of the client's collection.

If used, a VAV system should look much like a constant volume reheat system, where there are minimum air flows that prevent mold growth, contamination buildup, and uneven conditions within the conditioned space. Terminal equipment should include reheat for each zone, with the terminal equipment located in mechanical rooms or other spaces where access and service will not place a collection at risk. Such a system can be easily converted to a constant-volume reheat system by simply adjusting controls if its VAV performance should prove problematic.

Fan-powered mixing boxes can help assure air circulation to dry materials and suppress mold growth, they do not provide the opportunity to effectively filter the air for particulates and gases. The local fans increase the maintenance requirement over collection spaces, which disrupts visitor activity. Also, if the boxes have reheat there is added risk—from leaks if the reheat is water or steam, or from fire if reheat is electric. They are generally inappropriate for use in either exhibit or storage areas. Office, restaurant or other non-collection spaces have fewer negative issues for this equipment.

Suggestions and cautions regarding energy costs

The costs for operating a preservation environment can be significant. But for the long-term protection of a valuable collection, the annual costs of a good environment are a wise expenditure. In most cases the increase in the usable life of a collection easily justifies the annual energy and operating costs to maintain the required environmental conditions. For the institution with a small or limited operating budget, some compromises might be warranted, or some initial capital investments can be made in systems to offset recurring annual costs.

1. Minimize the volume of the protected space

One of the greatest opportunities for reducing operating costs is to treat as little of the building as possible. Spaces not requiring the more costly preservation environment should be in separate parts of the building on separate air systems so they can be shut off when those parts are unoccupied.

2. Use waste heat for cool-reheat dehumidification

An efficient method of providing reheat for cooling-based dehumidification is to use condenser heat. This limits the use of new energy for reheat and substantially reduces the energy cost for cooling-type dehumidification.

3. Consider a water-side economizer

Although an air-side economizer can cause problems, a water-side economizer can allow efficient winter cooling on buildings equipped with a cooling tower. Cold water from the cooling tower can be used to cool the returning chilled water directly, rather than cooling the water by running the chiller.

4. Select equipment to avoid high costs at low loads

Museums, libraries and archives are seldom occupied at night, so the loads can be very low when the outdoor temperature is moderate and the ventilation is reduced to its minimum. This suggests selecting several chillers and several boilers rather than just one large unit of each type. That way, most of the excess capacity can be turned off entirely rather than running at costly low efficiencies during each day's low-load period. Similarly, a primary + secondary pumping layout can save operating costs, because the chilled water and hot water pumping loads vary throughout the day and vary between different parts of the building.

5. Daylighting always costs extra energy

The use of natural light via skylights is often proposed for museums and libraries. These features sometimes save energy in *other* types of commercial or institutional buildings. Unfortunately, studies have shown that in *museums*, daylighting always creates a net energy penalty (Note 11). The loads created by daylighting for heating, cooling and humidity systems cancel any savings from electrical illumination.

Often, daylighting is popular as an aesthetic feature in spite of its energy penalty. Where used, any daylighting aperture should be kept to a minimum, and avoided over collection areas. Also, for lower risk of rain and snow leaks and for better-managed illumination, indirect lighting from clerestories is preferred over skylights.

5. Does the system work properly with few maintenance personnel?

A common failing in many system designs is little or no consideration of operation and maintenance. Most HVAC systems will work if properly adjusted and maintained, but many institutions do not have the staff and budget to give a large, complex system the attention it needs. In contrast to many other commercial buildings, building operations personnel in museums, libraries and archives tend to be very well-educated and to have valuable experience, creativity and drive. But there may be very few of them.

As much as possible, the designer should match the required maintenance to the institution's probable manpower and staff capabilities. For example, for large projects in some cities, code-required staffing might suggest that smaller, simpler chillers be used at night to avoid the need for an on-site licensed engineer to supervise operation of large centrifugal units. Similarly, small buildings without HVAC maintenance staff might use packaged equipment that does not require the daily attention of large, centralized chillers and boilers.

Notes

1. Sease, Catherine. 1991. "The Development of the Humidity Control Module at Field Museum," Journal of the American Institute for Conservation, Fall 1991, Vol. 30, No. 2, PP. 187-196.
2. Wilhelm, Henry. 1993. *The Permanence and Care of Color Photographs:* Traditional and Digital Color Prints, Color Negatives, Slides, and Motion Pictures, with Carol Brower, 1993, Preservation Publishing Company, Grinnell, IA.
3. Lull, William P. 1990. *Conservation Environment Guidelines for Libraries and Archives*, New York State Library, 1990.
4. Volent, Paulz.1985. "Volatile Amines Used as Corrosion Inhibitors in Museum Humidification Systems," and Norbert S. Baer; submitted to The International Journal of Museum Management and Curatorship, 6 July 1985.
5. Carolyn L. Rose, et. Al. 1995. *Storage of Natural History Collections: A Preventive Conservation Approach*. Society of the Preservation of Natural History Collections, 1995. (Section II chapter by Andrew J. Sebor deals with HVAC.)

6. Lafontaine, Raymond H. 1982. "Humidistatically-Controlled Heating: A New Approach to Relative Humidity Control in Museums Closed for the Winter Season." Journal of the International Institute for Conservation, Canadian Group 7, nos. 1&2 (1982):35-41.

7. Marchon, Paul J. 1987. "Controlling the Environment within a New Storage and Display Facility for the Governor General's Carriage." Journal of the International Institute for Conservation, Canadian Group 12, (1987):37-42.

8. Stanforth, Sarah. 1984. *Environmental Conservation. Manual of Curatorship*. ed. John M.A. Thompson. London: Butterworths, 1984, pp. 192- 202.

9. National Bureau of Standards, 1983 "Air Quality Criteria for Storage of Paper-Based Archival Records" NBSIR 83-2795 (pages 21 and 25) - air volumes should be six air changes per hour for the minimum air circulation rates, regardless of air flow to meet thermal requirements. 1995 *ASHRAE Handbook - HVAC Applications* (Table 1, page 3.2) indicates air circulation to be 8 to 12 air changes per hour.

10. Presented in "Qualitative Engineering," an article by Mr. Carl Bovill in the April 1988 *ASHRAE Journal*. This article compared typical HVAC distribution systems based on various performance factors. The factors applicable to museums differ from the emphasis of the article. However, the final system analysis diagram (fig.10), shows that without the need for many temperature control zones, but with a requirement for high air quality, the choices are limited to constant volume, multizone with bypass, dual-duct and 4-pipe induction.

11. Ayres, J.M., H. Lau, PhD, PE; and J.C. Haiad. 1990. "Energy Impact of Various Inside Air Temperatures and Humidities in a Museum when Located in Five U.S. Cities," ASHRAE transactions, V.96, pt.2, Although the energy simulations in this research showed a net energy savings with skylights, this was only for the smallest skylight. Further, the only reason the smallest skylight showed a net energy savings is that it was assumed, unlike the larger skylights, to cause no increase in fan sizing and fan energy use. Instead, if you proportionally increase the fan size and energy for the smallest skylight, to be on par with the other skylights, it is a net energy loss in all cases. The only real energy savings from the skylight in NYC is (apparently) from avoided reheat energy, and this does not offset its energy costs.

12. Ballard, Mary W. *Hanging out: strength, elongation and relative humidity—Some physical properties of textile fibers.* ICOM, Committee on Conservation, 1996. Ms. Ballard explains some of the basic relationships between humidity and mechanical properties of different types of fibers likely to be found in museum artifacts and art work.

13. Bailey, D.W. et al. *Investigation of dynamic latent storage effects of building construction and furnishings* (ASHRAE Research Project 455-RP). 1994. Final project report with Appendices. This comprehensive research helps quantify the different weight-change response of a variety of materials to brief changes in surrounding humidity. Although the emphasis is on furnishings, several types of books and other paper-based products are also examined.

References For Humidity Effects On Artifacts & Documents

1. *Air quality criteria for storage of paper-based archival records. 1983,* National Bureau of Standards, NBSIR 83-2795. Criteria for paper and system recommendations, developed by consensus of a select panel of consultants during intensive workshops at NBS in 1983.

2. *The Museum Environment*, 1986., by Garry Thomson, second edition, Butterworths, 1986; ISBN 7506-20412. Written from the museum perspective with good guidelines. Little ad-

vice on the details of equipment, applications, on project management or on how to get things done, and some advice on HVAC, particularly "VAV" systems, could lead to trouble. Pages 184 through 186 describe assessing light damage as a function of color temperature, as further detailed in the appendix herein. Butterworth- Heinemann, 225 Wildwood Avenue, Woburn, MA 01801; 800-366-2665, fax: 800- 446-6520.

3. *Preservation of historical records*, 1986., National Academy Press, 1986. The recommendations are based on a modification of NBSIR 83-2795 (Reference 1), and are the basis for environmental standards at the U.S. National Archives.

4. *American National Standard for photography - Storage of processed safety film, ANSI Standard PH1.43-1985*, American National Standards Institute, Inc., 1439 Broadway, New York, NY 10018.

6. *Isoperms—an environmental management tool*, 1994., Donald K. Sebera, Commission on Preservation and Access ; 1400 16th St, NW, Suite 740, Washington, DC, 20036-2217, 202-939-3400; (based on "A Graphical Representation of the Relationship of Environmental Conditions to the Permanence of Hygroscopic Materials and Composites," Proceedings of Conservation in Archives, International Symposium (Ottawa, May 10-12, 1988), Paris: International Council on Archives (1989), p. 51-75).

7. *IPI Storage guide for acetate film*. James M. Reilly, Director, Image Permanence Institute, Rochester Institute of Technology, PO Box 9887, Rochester, NY 14623-0887, 716-475-5199. Based on laboratory research at the IPI, this guide includes a circular scale for estimating the life of acetate film at different temperature and humidity conditions.

8. *Storage guide for photographic materials.* James M. Reilly, Director, Image Permanence Institute, Rochester Institute of Technology, PO Box 9887, Rochester, NY 14623-0887, 716-475-5199. Funded and published by the New York State Program for Conservation of Library Research Materials.

9. *The permanence and care of color photographs*., Henry Wilhelm, Preservation Publishing Company, 1993. Much of the content applies to microform media.

10. "Investigation of dynamic latent storage effects of building construction and furnishings*."* Bailey, D.W. et al. ASHRAE Research Project 455-RP. 1994. Project report with Appendices.

11. *Mechanical behavior of paintings subject to changes in temperature and relative humidity.* 1991. Marion Mecklenburg and Charles Tumosa. Proceedings of the International Conference on Packing and Transportation of Paintings. Sept. 1991. London, England. National Galley of Art, Washington, DC. ISBN 0-89468-163-x.

12. "Hanging out: strength, elongation and relative humidity—some physical properties of textile fibers*."* 1996. Mary Ballard. ICOM Committee for Conservation. International Council for Museums, Maison L'UNESCO, 1 Rue Miollis, Paris (33), France.

13. *Code of Ethics for Museums*. 2000. American Association of Museums. 1575 Eye St., NW. Washington, DC 20005. www.aam-us.org

14. "Temperature and relative humidity effects on the mechanical and chemical stability of collections." 1999. Marion Mecklenburg and Charles Tumosa. ASHRAE Journal, April 1999. pp:77-90

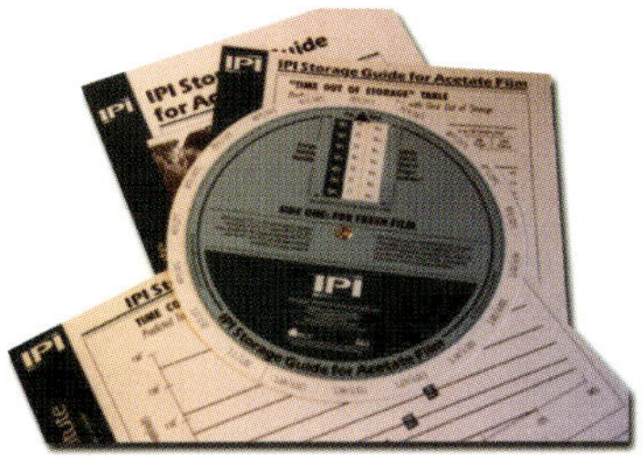

23.12
IPI Storage Guide: Acetate Film
(Reference 7)

Suggested Reading

The following publications may be useful for those wishing to learn more about conservation environments for museums, libraries and archives.

1. "The process of compromise; A team approach to conservation environments." Paul Himmelstein & Barbara Appelbaum. APTI Bulletin, Vol. 23, no. 3, 1996. p. 8-11. Association For Preservation Technology International, Lisle, Illinois. www.apti.org In this brief paper, Himmelstein and Appelbaum describe the logical elements of the collaboration needed to determine environmental conditions for nay given collection. Their suggestions are simple and persuasive, based on years of experience working with teams from institutions of all sizes and collection types.

2. *Magnetic tape storage and handling*, National Media Laboratory, Commission on Preservation and Access, June 1995; 1400 16th St, NW, Suite 740, Washington, DC, 20036-2217, 202-939-3400.

3. *New tools for preservation: Assessing long-term environmental effects on library and archives collections*, Reilly, et. al., Image Permanence Institute, Rochester Institute of Technology, June 1994; Commission on Preservation and Access, 1400 16th St, NW, Suite 740, Washington, DC, 20036-2217, 202-939-3400.

4. *Planning academic and research library buildings*, originally by Keyes D. Metcalf in 1965, second and third edition by Leighton and Weber, American Library Association, 1986, 1999. Although this book is focused on libraries, the discussions of the planning process are helpful on any project. The book is not organized as a reference so it must be read from cover to cover.

5. *Guide to environmental protection of collections*, by Barbara Appelbaum, Sound View Press, 1991.

6. "Relative humidity in museums, galleries and archives; specification and control." 1993. Stefan Michalski, Ph.D. Proceedings of the 2nd Bug, Mold & Rot Conference. National Institute of Building Sciences, 1201 L.St, NW. Washington, DC. 20005. (202) 289-7800. In this brief paper, Michalski explains key aspects of the response of different materials to humidity.

7. "Commissioning a museum and archival storage facility." 1996. Rebecca T. Ellis, PE. ASHRAE Transactions, Vol. 102, Pt.1 The Author, who served as a third-party commissioning agent, describes the process of identifying the causes of system problems, performing verification tests, correcting problems and working to create a team approach to problem-solving following a year of finger-pointing.

8. "Demonstration of efficient humidity-control technology at an art museum." 1993. Don B. Shirey III, ASHRAE Transactions Vol. 99, Pt. 1. In low-cost construction in a humid climate, packaged HVAC equipment can have great difficulty keeping humidity under control. Shirey explains one such building, and what was accomplished with a very limited budget to reduce the problems created by this class of equipment.

Special Thanks

The Authors are grateful to a broad spectrum of experts that have shared the results of their experiences in museum management and conservation technology, especially Stefan Michalski, Marion Mecklenburg and Charles Tumosa, Paul Himmelstein, Barbara Appelbaum and Tim Padfield.

Donated Photos

Fig. 23.7 - © William Lull
Fig. 23.9 - © William Lull
Fig. 23.11 - © William Lull
Fig. 23.12 - © William Lull

Chapter 24
Hospitals

By Lew Harriman

Humidity Control In Hospitals
The HVAC system should "...do no harm."

360 361

361 — **Purpose of Humidity Control**

361 362 363 364 365 — **Control Levels & Tolerances**
- General
- Cold climates
- Duct work and air handling equipment
- Comfort of staff and patients
- Operating rooms

365 366 367 — **Moisture Loads**

367 368 369 370 — **Suggestions & Cautions**

371 — **References**

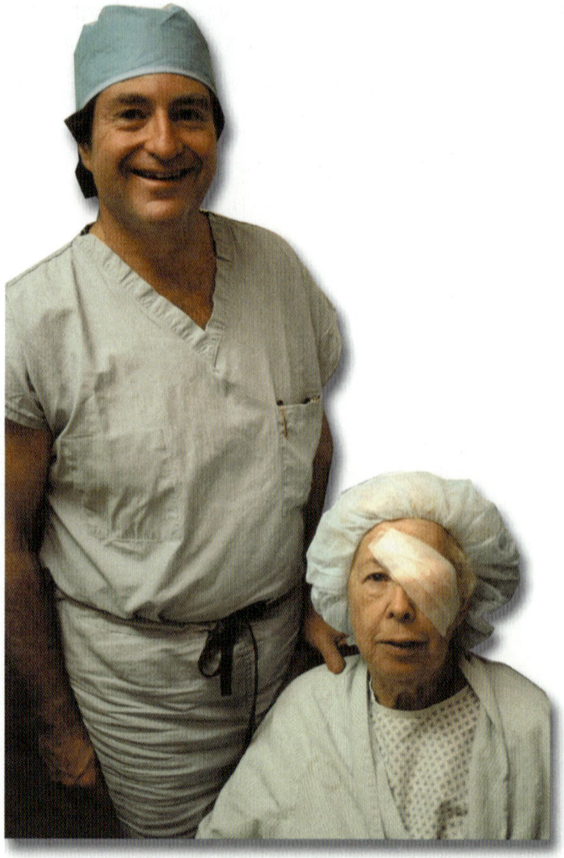

Fig. 24.1 Humidity Control In Hospitals
Although the primary reason for controlling humidity is to minimize health risks, comfort for both active staff and sedentary patients can also be improved by managing the dew point.

Hospitals

Just as the Hippocratic Oath advises Doctors to "First, do no harm...", a hospital HVAC system should not make patients any sicker than they already are. It should also help prevent the transfer of any patient's diseases to the staff. The risk of either problem can be reduced by controlling humidity in the building, and by minimizing moisture in materials inside the HVAC system.

Purpose of Humidity Control

It's no accident that the first 19th century attempts at cooling air with blocks of ice and with mechanical refrigeration were made in hospitals. Keeping air temperatures from being too cold or too hot reduces thermal stress, which can mean life or death to weakened patients. So the design of hospital HVAC systems begins with an entirely appropriate focus on keeping air temperatures within the human comfort range.

Particulate and humidity control are the next concerns. They help prevent health problems that might otherwise originate within the hospital itself—a building that contains a large percentage of people releasing infectious agents.

Most infections enter the body through direct physical contact between people. But some are airborne, and enter through the nose during normal respiration. Airborne infections travel as particles. That's why cleaning the floors and filtering the air are exceptionally important for hospitals.

The human body has some mechanical defenses against airborne infections. The upper respiratory tract produces mucous that traps particles before they can be pulled into the vulnerable lungs. But if air is very dry, the protective mucus dries out and traps fewer particles. That's why humidification is important in hospitals.

Finally, all hazardous microorganisms need water to grow initially, and many of them survive longer when the air is humid. That's why dehumidification is important for hospitals, and equally important inside the HVAC system's ductwork and its components.

Fig. 24.2 Legionella Pneumophila

The bacterium that causes Legionnaire's disease grows in warm water, but when aerosolized it can reach people by travelling through humid air.

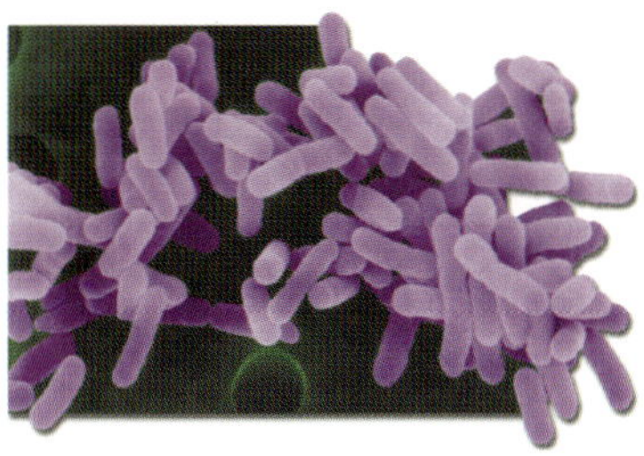

To summarize these functions: Flters reduce the volume of infectious particles that enter the breathing zone. Humidification helps the body resist infectious particles not trapped in filters. Dehumidification reduces the number of microorganisms that exist in the first place.

Control Levels & Tolerances

The optimal humidity level varies according to the function in question. There has been a great deal of debate among scientists as to the ideal humidity for limiting each of the moisture-related problems that can occur in hospitals. Figures 24.3, 24.4 and 24.5 describe some of the effects of different relative humidities on a few phenomena relevant to hospital operations. But in North America the professional consensus is that humidity in most parts of a hospital should be held between a minimum of 30% rh during the winter season and a maximum of 60% rh during the summer season.

Fig. 24.3 Legionella Dies Quickest In Dry Air

One reason to keep air dry in a hospital is to limit the life of this type of bacteria if it should become airborne.

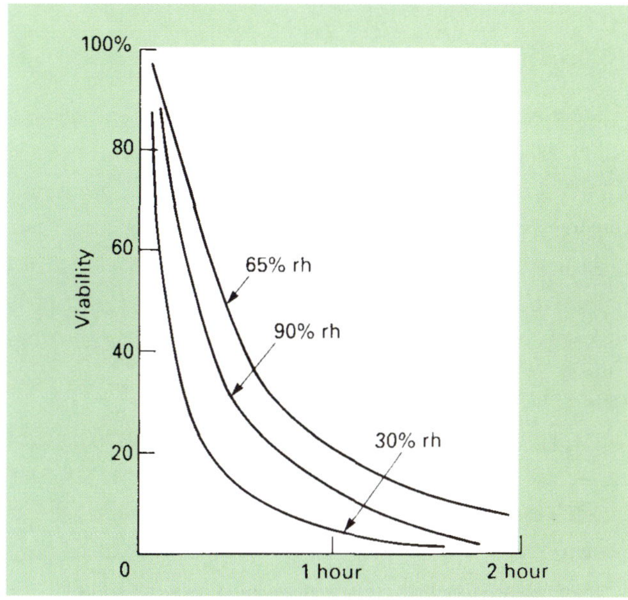

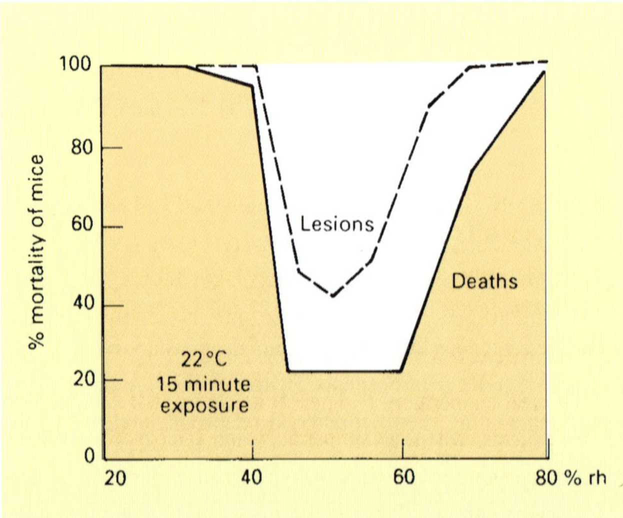

Fig. 24.4 Influenza Is Least Hazardous Between 40 And 60% rh.
This virus survives longest when humidity is either higher or lower than normal. Keeping humidity between 40 and 60% rh reduces its viability, and therefore limits the risk of infection. This study sprayed a broth that included influenza A into cages of mice. Nearly all mice died when the ambient humidity was above 60% or below 40% rh. (Brundrett, 1990)

That consensus is recorded in ASHRAE guidelines (Ninomura, 2001), which are usually adopted by the architectural community (AIA,1997) and by many building codes for hospitals. The humidity requirement may be more strict in "protective environments" such as the nursery or a burn unit. Also, in some cases, owner experience or local codes dictate a different range—often below 55% rh in the summer or a lower range than 30% in very cold weather. Here's why.

Control level in cold climates

In very cold climates the manufacturers of windows recommend setting the maximum indoor humidity according to the insulating qualities of the window, compared to the coldest winter design temperature. (AAMA, 1997) This caution highlights the general problem of condensation inside the building when outdoor temperatures are cold. The entire exterior envelope of a humidified building must be designed (and built) to resist condensation, and to manage condensation gracefully rather than catastrophically if it should occur.

These are largely issues for the architectural designer and the general contractor. But the HVAC designer is an important contributor to the team's efforts to avoid a condensation problem. First, the HVAC designer should provide the architectural designer with a dew point analysis of the exterior walls to determine if condensation is likely to result from outward vapor diffusion (Kunzel, 2000). Second and more important, when the weather is cold, the HVAC designer must make sure the system keeps the building under a neutral or slightly negative air pressure, so that the HVAC system does not force humid indoor air *outwards* into cold building cavities.

The ideal indoor humidity control level in cold climates may be limited by these two issues. If the exterior wall tolerates high humidity, and as long as the HVAC designer has allowed for neutral internal air pressure, then the winter humidity level can be allowed to rise into the 30% to 40% rh range. But if the exterior wall is likely to condense moisture when vapor diffuses outward, or when humid air is forced outward through the wall, then humidity indoors must be held much lower. Humidity levels below 30% rh may have to be authorized by a special exception granted by hospital code officials.

Control level in ductwork & air handling equipment

In the past, ASHRAE Standard 62 (Ventilation for acceptable indoor air quality) contained a caution against relative humidity above 70% in ductwork. That caution became very controversial, because relative humidity immediately downstream of humidifiers and downstream from cooling coils is above that limit. The scientific realities of psychrometrics, together with the economic realities of HVAC budgets, dictate that keeping relative humidity below 70% in *absolutely all* parts of the system is not practical. But here's why it is important to keep *nearly all* surfaces dry nearly always.

Most important and most commonly, dust grows fungus when it gets moist. Chapter 7 of this book explains the process in detail.

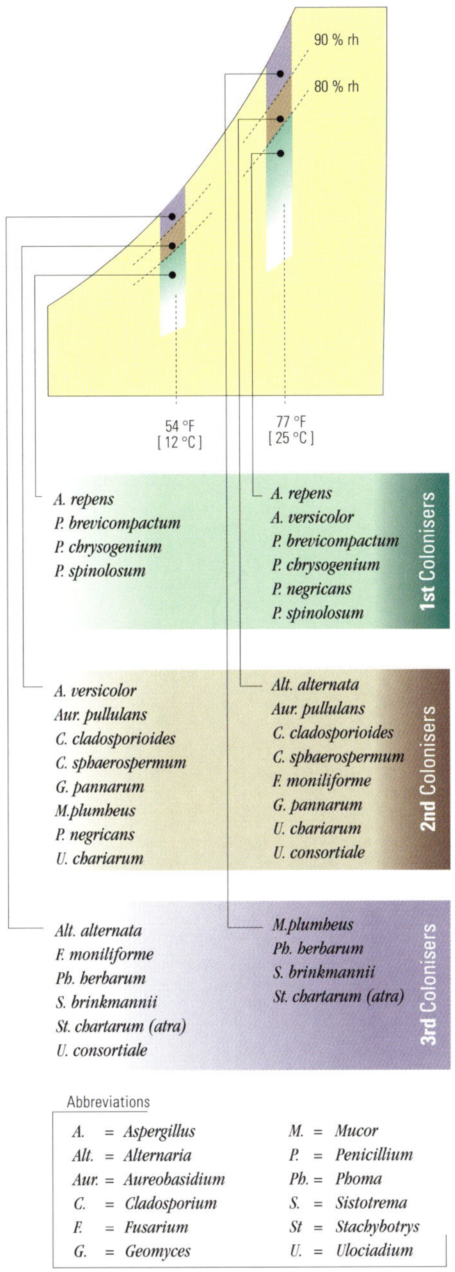

Some members of fungal families like *Aspergillus* and *Stachybotrys* are not good for human health.

Further, standing water supports and amplifies the growth of bacteria. Some of these are hazardous directly, like Legionella. Other bacteria are equally hazardous indirectly, because they contain viruses that infect humans.

To prevent the growth of fungus, duct walls must be kept clean and dry. This is especially important when duct surfaces are porous and therefore trap dust, and especially important for filters, which trap large amounts of dust by design. Cooling and heating coils, dampers, turning vanes and fan blades also collect dust. Smooth metal duct surfaces trap less dust, but protective oils can collect particles, and the oil itself will even support fungal growth. Frequent filter changes prevent dust accumulation, but the designer must keep the air dry.

The definition of "dry enough" to prevent fungal growth depends on the type of fungus, the type of food source and on the temperature. But keeping the relative humidity below 70% will prevent most food sources from absorbing enough moisture to support rapid fungal growth. Ideally, all points inside the ducts and air handling units would have relative humidity below that level. But downstream from cooling coils and humidifiers, that's not possible. The design must prevent the accumulation of dust in that humid zone.

Downstream of those components, specify a smooth, non-porous interior lining. Ideally, *all* interior duct surfaces would be smooth metal. But as a minimum, the length of the smooth lining should extend to the point where air is below 70% rh But good design by itself is not enough. The operations staff must change the filters often enough so that dust does not collect in the wet zone.

Fig. 24.5 Keep Humidity Below 70% In Duct Work

Above 70% rh, many forms of fungi can thrive on moist dust in HVAC ductwork. That's why the designer must take special care with the plenums and ductwork immediately following cooling coils and humidifiers, where humidity is likely to be above that limit.

Control levels to assure comfort for both staff and patients

In winter time, keeping the moisture up to levels between 30% and 40% rh will be sufficient to assure comfort for most occupants. In summertime, 60% rh is likely to be comfortable for most people, but there are reasons to further reduce the humidity.

The ideal temperature is different for active staff compared to passive patients. When air temperature and velocity are constant, comfort depends strongly on each person's metabolic rate. In summer, comfort improves for both groups by keeping the dew point low.

Patients have a very low metabolic rate because they are less active. So they generate less internal heat and therefore become quite chilled even at moderate temperatures. The hospital staff has the opposite problem. They are all very active physically, so they generate excess heat which they release through perspiration (evaporative cooling). So the staff usually prefers a lower temperature than what is comfortable for most patients.

Figure 24.6 shows human comfort modeling that estimates what percentage of a typical population is likely to be thermally *dissatisfied* as a function of their metabolic rate. (Memarzadeh, 2000) Given a supply air temperature of 52°F [11°C] and a metabolic rate of 1.2, only about 13% of the population is likely to be uncomfortable. But at a lower metabolic rate of 0.8 (typical for patients), nearly 96% of the population is likely to feel too cold.

By keeping the dew point low, a moderately high temperature will satisfy a broader range of people and activity levels. Dry air allows more rapid evaporation from the skin surface, and a more rapid transfer of moisture outward through the clothing layers. So at constant temperature, dry air tends to be more comfortable for active people such as the staff. Given a low dew point, the dry bulb temperature can stay a bit higher for patient comfort and to save cooling.

The specific levels of temperature and humidity are likely to be set by code and by patient requirements. But when general comfort becomes the issue, it is useful for the HVAC designer to remember the general effects of humidity levels: *During the summertime, lower dew points will improve comfort. During the wintertime, higher dew points will be more comfortable.*

Operating rooms

When flammable gases were used as anesthetics, humidity in operating rooms was kept above 35% rh to limit the hazard from electrostatic discharges. Modern gases do not present the same explosion hazard. Code-required levels are usually between 30 and 60% rh, primarily to reduce thermal stress on the surgical staff and to prevent dripping condensation.

The surgical patient's thermal well-being is largely a function of temperature, not humidity. But the surgeon has a different problem which can sometimes conflict with the thermal needs of the patient. The surgeon works long hours under high radiant heat loads from bright lights, and is surrounded by electronic equipment that impedes air flow and also radiates heat. In addition, the hazard of blood-borne diseases such as AIDS demands that the surgeon's clothing act as a barrier to liquids. That type of clothing is also relatively impermeable to vapor, which means quite a bit of the surgeon's perspiration can be trapped near the skin. In short, the thermal stress on a surgeon is very high.

To reduce the stress, the surgeon often needs a lower temperature in the operating room, often as low as 68 to 63°F [20 to 17°C]. Those levels can be cold enough to be stressful for the less active members of the surgical staff and for the patient.

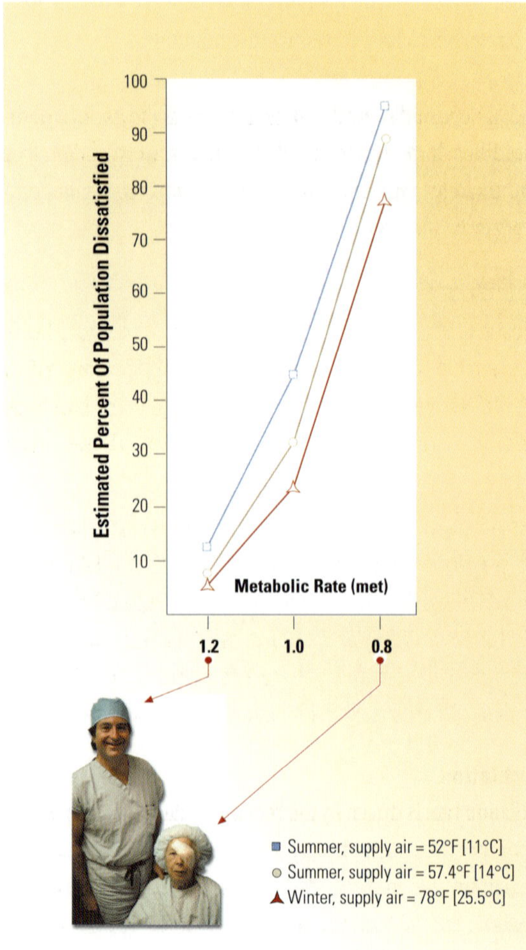

Fig. 24.6
Metabolic Rate Strongly Affects Thermal Comfort
Active people such as the hospital staff can be comfortable at cool temperatures while less-active patients are very dissatisfied. Keeping the dew point low in the summertime can help satisfy both groups using slightly higher temperatures, because active people can release excess heat through evaporation.

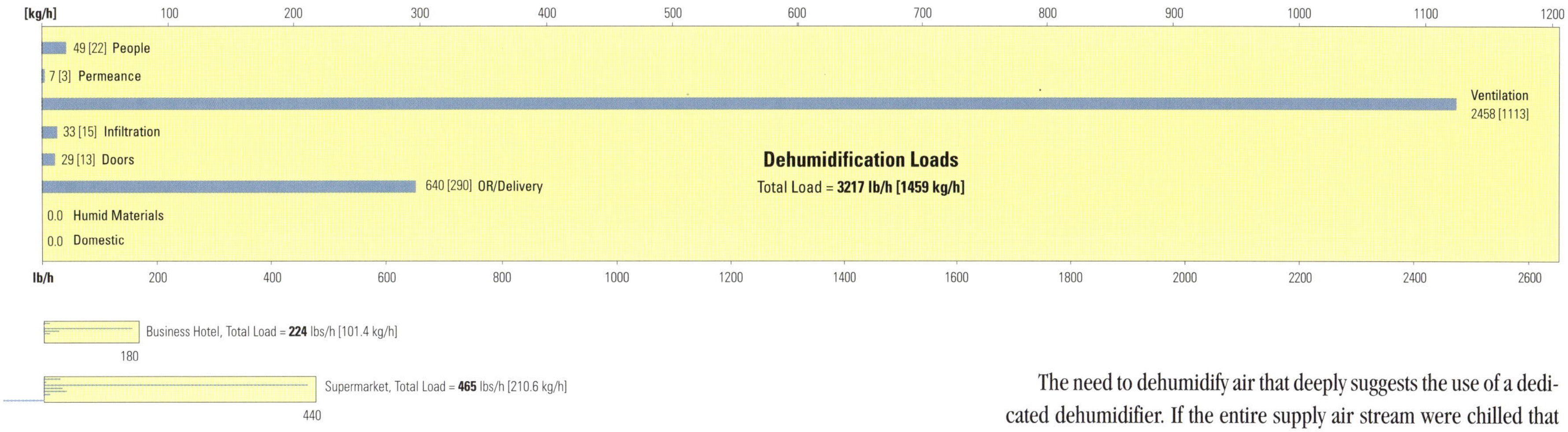

Fig. 24.7 Dehumidification Loads

Moisture loads for hospitals make loads for all other commercial buildings seem very small. But the distribution of the load sources is similar to other buildings—the largest loads come from ventilation air.

An alternate strategy reduces the dew point in the operating room. (Yborra, 2000) The low dew point increases the vapor pressure difference across the surgeon's clothing layers. That difference speeds vapor transfer through clothing and increases the amount of heat that can be released through perspiration. The surgeon can be comfortable at higher temperatures that are less stressful to other occupants of the operating room.

Also, in existing systems, condensation can occur in the room when the system produces very cold temperatures. A cooling system designed to maintain temperature levels between 70 and 74°F [21° to 23°C], will have difficulty keeping the relative humidity below 60% rh at 63°F [17°C]. The very cold and nearly saturated air leaving the supply air diffuser chills surfaces. Moisture released inside the room can raise the humidity to dew points high enough to condense moisture on cold surfaces.

Setting the dew point so the relative humidity is less than 60% will be sufficient to eliminate condensation. But that 60% rh must be achieved at a temperature of *63°F* instead of *72°F* [17°C instead of 22°C]. That means the dew point must be reduced to 48°F instead of only to 57°F [To 9°C rather than only 14°C].

The need to dehumidify air that deeply suggests the use of a dedicated dehumidifier. If the entire supply air stream were chilled that low, the central chiller efficiency would be low, and the annual cost for reheat energy would be very high.

Moisture Loads

Compared to other types of commercial buildings, moisture loads in hospitals are huge. Figure 24.7 shows a hospital as compared to loads in a typical supermarket and school. Loads in those buildings are far smaller than hospital loads, because the hospital uses much more outdoor air.

This estimate assumes a small hospital of 4 stories, 100 beds and 8 rooms for surgery or delivery, located in Atlanta, GA. Like other commercial buildings, the moisture loads are dominated by ventilation required for the building as a whole, and by the ventilation used in the surgical suite. Loads from those two sources are about 26 times larger than all other hospital moisture loads combined.

General ventilation

The ventilation rate is driven by the recommended minimum number of outdoor air changes per hour rather than by the ventilation air needed per person. (Ninomura, 2001) Each area has a different recommendation, but the total volume of outdoor air becomes quite large—about 56,000 cfm in this example [26,426 l/s].

With such an immense amount of outdoor air, the designer would be wise to consider using some sort of load reducer, such as passive desiccant wheels, plate-type heat exchangers or even a simple coil-to-coil loop. Chapter 15 of this book contains further information about load reduction equipment.

But even after a load reducer, the volume of water vapor is still enormous. It makes sense to dehumidify the incoming outdoor air before it can enter the building. It is also wise to consider the operating economics of that equipment very carefully. Cumulative operating costs are likely to equal the system's construction cost within very few years.

In a hospital, there is a significant benefit to drying the ventilation air to a condition *below* the desired indoor dew point. With that approach, there are far fewer hours when the indoor cooling coils will condense moisture. Less liquid is condensed, and condensation occurs for shorter periods. That means the cooling coils and ductwork are less likely to grow large amounts of mold and mildew.

Further, when a cooling-type dehumidifier or low-temperature chilled water is used to dry the incoming air, the cold ventilation air is also useful in meeting the building's internal heat loads. The indoor coils will have less cooling to accomplish, and there will be little need for reheat energy.

Finally, pre-drying the ventilation air means the main chilled water system for the rest of the building can operate at a higher temperature, because there is no need for cooling coils to dehumidify the recirculating air deeply. With a higher chilled water temperature, the chiller will be more efficient and there will be less need for reheat after the cooling coils (assuming that local codes allow reheat). So operating costs are likely to be much lower than if the entire central chilled water loop had to dry air deeply enough to maintain the desired dew point.

Surgical suite ventilation

The comments above apply to the load from outdoor air needed for the surgical suite and obstetrics delivery rooms. Where possible, dry the incoming ventilation air before it blends with any return air, and strongly consider adding a load reducer to the ventilation air system. Such devices can cut annual energy costs by 25 to 50%

Also, although busy for most of the day, the operating rooms are not likely to be needed for all 24 hours. So it may be possible to reduce the outdoor air significantly when the rooms are not being used.

As with the general ventilation air, it makes economic sense to dry the ventilation air to a dew point below the desired indoor humidity level. But with operating rooms, there's another important benefit from that strategy. The temperature and moisture levels in operating rooms must be adjusted to the different needs of each surgeon and each procedure.

When the outdoor air is pre-dried, changing the room temperature can be accomplished in a few minutes. If that air is *not* dried, the cooling and heating coils can take up to an hour or more to adjust the rooms' temperature and moisture levels. The cooling coils will condense more or less moisture than before, and reheat will be required—all of which takes time to return to equilibrium. In many older hospitals, the surgical HVAC system is run all 24 hours simply to shorten the time needed to change the room's temperature and moisture levels before a new procedure can begin.

Predrying the ventilation air allows the system to respond very quickly, raising or lowering temperature and moisture to the levels needed for the next procedure. Reducing the room's downtime between procedures adds revenue, as well as allowing the staff to respond more quickly to medical emergencies.

People and door activity

The load from people is significant, but the designer has very little control over that load source. The largest load from people

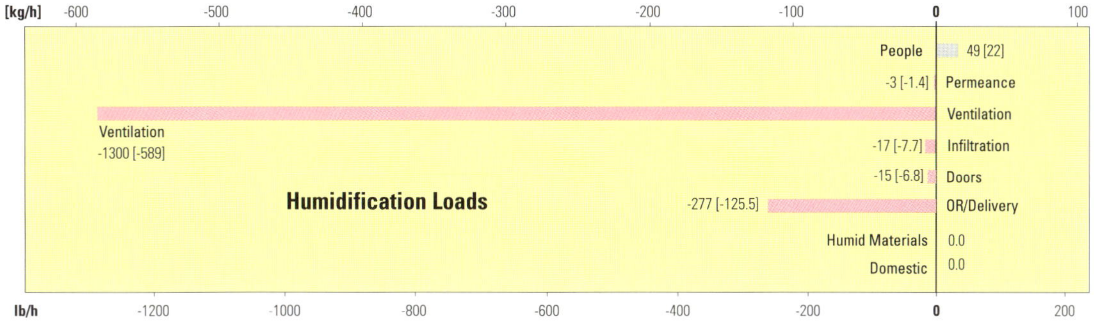

Fig. 24.8 Humidification Loads
Humidification loads are also very large in a hospital. Note that this example assumes a building in Atlanta, GA. In colder climates the humidification loads would be much larger.

Fig. 24.9 Load Reduction Option
Since ventilation loads are enormous, many hospitals use passive desiccant wheels to reduce the dehumidification and humidification loads from ventilation air. Chapter 15 contains more information on this option.

comes as they enter the building wearing rain-soaked clothing. Also, the entrance door or vestibule will allow humid air to enter the building at that location. So when a choice is possible, the designer might consider increasing the moisture removal capacity of the system that serves the most active entryways into the building.

Infiltration

As the wind pushes against the exterior wall, it forces humid air into the building. In this example, we assume that the wind will be acting on the longest exterior wall, and that the construction will be "tight" rather than "average" or "loose."

It's fair to assume that the construction will be "tight." If it were not, the building would have great difficulty keeping humidity high during the cold, dry months. But much of current hospital construction takes place in existing buildings. For most older buildings, the designer might prefer to assume the exterior construction is either average or loose, which increases the moisture load estimate.

Permeance

As in other commercial buildings, the load from vapor permeance is so small that it is difficult to estimate. The designer need not be especially concerned with this load source when sizing the dehumidification equipment.

On the other hand, it is useful for the designer to know the vapor transmission characteristics of the exterior wall. Since the hospital is likely to need humidification, it will be essential to estimate the dew points vs. temperatures at each point through the cross section of the wall, to make sure that condensation does not occur all winter long.

Humidification loads

This estimate assumes the building will be humidified to 30% rh at 72°F [22°C] during cold months. If a higher humidity is needed, the loads will rise accordingly.

It's no surprise that the humidification loads are also dominated by ventilation air. And again, most designers will prefer to add moisture to the incoming air before it enters the building.

In most cases, it will not be possible to humidify the entire building using just the ventilation air. That air will seldom hold enough moisture to make up for all the losses from the rest of the building. So some supplemental humidification will probably be required in places other than the ventilation air pretreatment system.

Suggestions & Cautions

Experienced designers of hospital HVAC systems point out that much healthcare construction involves modifications to existing buildings. Field conditions may not lend themselves to clean, elegant HVAC designs. Perhaps the building's structure and wiring conduits will block the neat, long and straight duct runs necessary for vapor absorption after humidifiers. Or perhaps the need to keep the surgical suite in continuous operation will prevent tearing apart the existing system for six weeks to add a cost-saving load reducer.

So the less experienced designer should carefully investigate any existing physical realities before spending too much time developing design concepts that cannot be implemented in the field. Other suggestions include:

Consider using split coils for dehumidification

When cooling is used to dehumidify the ventilation air, the designer might wish to consider a split coil instead of a face and bypass to modulate dehumidification capacity during cool but humid weather. The split coil takes less space than the face and bypass arrangement.

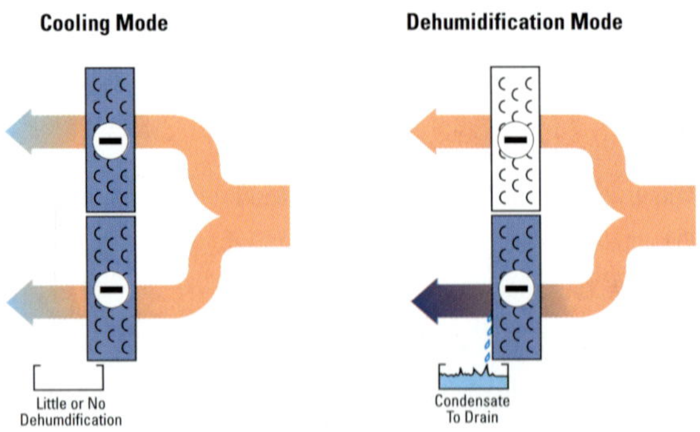

Fig. 24.10 Split Coil Option

Chilled water is circulated through the bottom and top halves of the coil separately. When the indoor dew point is satisfied, chilled water stops flowing through the upper part of the coil. The bottom part keeps cooling and drying the air, and if the dehumidification load continues to decline, the chilled water flow can be reduced.

Don't let chilled water temperature rise too high for dehumidification

Another caution is important when chilled water is being used for dehumidification. In the summertime, extra heat loads can put a strain on the building's cooling capacity. The temptation is to let the chilled water temperature rise slightly to gain cooling capacity. Also, at other times of the year the energy manager may be tempted to let the chilled water temperature rise to maximize energy efficiency. But if the water temperature rises, the cooling coil no longer chills the air *deeply* enough to dry it to the proper dew point.

When the chilled water temperature must rise, it is best to install a separate, independent system to provide reliable dehumidification. That system is best located on the incoming ventilation air. Both mechanical and desiccant dehumidifiers have been used for the purpose, allowing the overall building to stay under humidity control, while at the same time providing energy savings from higher chilled water temperatures.

Avoid tall coils for dehumidification

In a cooling coil, condensate flows down the fins toward the bottom of the coil. When the coil is very tall, enough water can collect near the bottom so that droplets blow off the coil and into the air. For this reason, companies that manufacture mechanical dehumidifiers avoid coils taller than about 3.5 feet high [1 meter high].

Sometimes, space restrictions demand that a cooling coil be taller than that rule-of-thumb limit. In those cases, experienced designers recommend splitting the cooling surface horizontally into two or more stacked coils, each with a separate drain pan and condensate drain. With a shorter distance to travel to the drain, the condensate is less likely to accumulate enough to block air flow or to have droplets blow into the air stream.

Consider using low-cost heat for dehumidification

In a hospital, there is often heat available at low cost during summer months. Sources include natural gas at summer rates, hot water from combined heat and power systems and engine-driven chillers, or steam from boilers that are using only a small part of their capacity during summer months.

Low-cost heat can be used for reheat if a chilled water dehumidification system is preferred, or used to reactivate a desiccant dehumidifier where the designer is drying air to lower dew points. Regardless of the technology, operating costs for dehumidification are always high compared to cooling costs. Using inexpensive heat makes economic sense even if it costs a bit more money in the construction budget to move that heat from one part of the building to another.

Consider smaller, multiple units for a given load

Hospital codes and prudent practice often require backup capacity if critical machinery should fail. Rather than simply duplicating a single large unit, the designer can save money by breaking the load into two or more smaller units, and providing one extra unit of that smaller size as a backup for the two primary units.

This saves money in the construction budget as well as in the operating budget, and also saves space in the mechanical room. It makes the equipment easier to control and to modulate at part-load and low-load conditions—which by definition occurs for 98% of the equipment's operating life. Finally, smoother modulation of capacity means the humidity in the building will be more stable.

Condensate pans must drain freely, completely and continuously

Avoiding standing water is very important to avoiding the risk of bacterial growth. The designer must make sure to specify condensate drain pans under all cooling coils, and specify that the pans be sloped in all directions towards a low-point drain connection. The drain connection itself must not have a "lip" or rim that prevents complete draining, and the condensate drain must have a trap with a stem height tall enough to ensure that water collects only in the trap and never in the drain pan. (See fig. 24.11) The drain line itself must also be a large-enough diameter so that dirt and debris washed off of cooling coils does not clog the drain line. For this reason, drain lines smaller than 1.5 inches in diameter [38mm] are not recommended, even if the volume of condensate is expected to be small.

If either the designer or the contractor neglects these details, standing water is likely in drain pans, which means the air will be in constant contact with a source of bacterial growth. That situation is never appropriate in any building, but is certainly even less appropriate in hospitals.

Avoid placing filters immediately downstream from cooling coils

Air leaving a cooling coil is at a relatively high humidity. When humid air flows through dust trapped in filters, the dust gains enough moisture to support fungal growth in the filter. In fact, fungus can even grow ***through*** the filter and release spores downstream into the final supply air. So it's prudent to locate filters much further downstream. Better yet, locate filters upstream of the coil where the relative humidity is much lower, and where the filter can help keep the coil free of the dirt that supports fungal growth in the coil itself.

Fans "blowing-through" dehumidifying coils have created problems

When a fan is close to a cooling coil that condenses a great deal of water, the force of the fan is likely to blow some of that condensate off the coil and into the airstream. That water can land downstream in dust trapped on duct walls, dampers or turning vanes, which leads to fungus in the air stream as well.

A "draw-through" fan position can help keep the air velocity low and more uniform across the full face of the coil, so that droplets are less likely to be pulled into the air.

This is not to say that a blow-through fan position cannot work. It's only a caution, suggesting that these have caused problems in the past, and that the designer must always be careful to spread the airflow evenly across the face of any dehumidifying coil to avoid droplet carryover.

Fig. 24.11 Condensate Drains

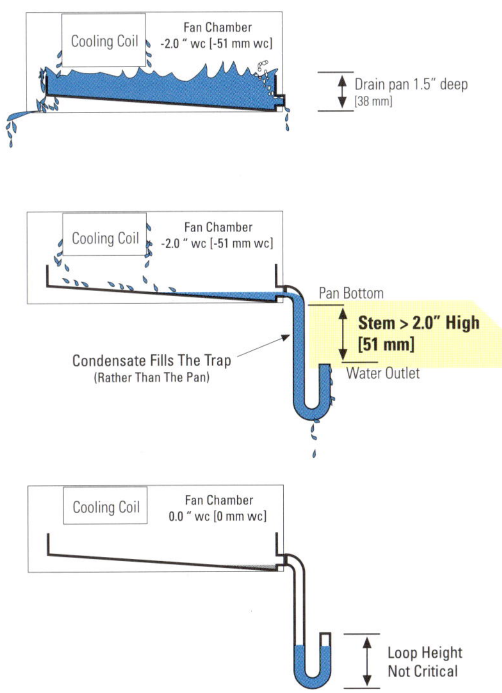

Without a Trap
The fan lifts up the condensate, which spills over if the fan suction exceeds the depth of the pan. Standing water allows microbial growth. Water spilling out of the pan can leak out of the unit and drip into the building.

With a Trap
The condensate drains out of the unit because the fan cannot lift the water out of the stem of the trap. The stem is measured from the *bottom of the pan* to the *top of the water outlet* from the trap.

Note... The Stem Height Is Critical
The height of the stem must be greater than the suction of the fan. eg: A fan with -2.0" wc pressure needs a trap with a stem height *over* 2.0" [A fan with suction of 51 mm of water column needs a drain stem higher than 51 mm].

Loop Height Not Critical - But Deeper Is Better
The distance from the water outlet to the bottom of the loop is not critical. But deeper loops do have a benefit. They take longer to dry out, which prevents the fan pulling in untreated air if the coil is not condensing water.

In-room cooling units are difficult to sanitize

Different patients need different room temperatures. And the heating and cooling loads change on different sides of a building through the day and night. In other commercial buildings, individual room combined cooling/heating units provide a way to accommodate different temperature levels in adjacent rooms. However, these units are difficult to sanitize, especially when they are expected to contribute to dehumidification by condensing moisture.

To minimize the sanitation problem, the designer might consider drying the ventilation air to a dew point just below the dry bulb temperature of the supply air from the room units. That way there will be little condensation to moisten the dust that collects in these units.

No matter how dry the ventilation air is made, however, it will always be prudent to provide a drain pan and condensate drain line for any cooling coil. The ventilation air system might not dry deeply enough to prevent all condensation. And without a pan, the condensate drips onto the floor or into the wall.

One final caution concerns the design of any room units under consideration. Not all equipment is equally well-designed to limit the sanitation problem. Condensate pans in some room units are not designed for easy cleaning access, and not always designed to actually drain water away. Some units have pans intended to simply catch condensate and hold it until it can re-evaporate. These designs virtually guarantee the growth of mold, mildew and bacteria as dust collects in the pan. And since re-evaporation is the intention, they do not actually dehumidify the room. Differently-designed units would be a more appropriate choice for hospitals.

Limit use of air-side economizers by dew point

Operating dehumidifiers and humidifiers is much more expensive than operating a cooling system. That's why it makes economic sense to limit the use of any air-side economizer by the dew point of the outdoor air. Don't use outdoor air to cool the building unless its moisture content is also appropriate when compared to the desired indoor humidity ratio.

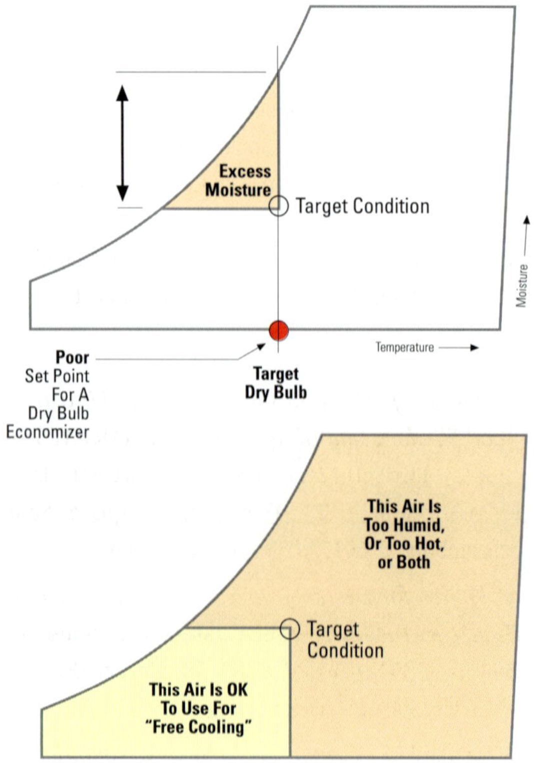

Fig. 24.12
Control An Economizer By Dew Point

In other words, in humid seasons like summer, late spring and early fall, the building automation system should not use excess outdoor air to cool the building unless the outdoor dew point is *below* the indoor dew point. Otherwise the outdoor air *adds* a costly dehumidification load.

In cold seasons like winter, late fall and early spring, the most cost-effective control strategy for an economizer is less clear-cut. In most cases, it probably makes economic sense to minimize the outdoor air any time the outdoor dew point falls below the desired indoor dew point. On the other hand, in some buildings, using an excess of cold and dry outdoor air might avoid turning on the chillers entirely. In that case, it may be economically prudent to spend money to operate humidifiers, even if the outdoor air is very dry and will need a great deal of moisture added as it enters the building.

Here's the key point. In all buildings that are humidity controlled, the economizer control system should make the decision to use excess outdoor air only after comparing the dew points indoors and outdoors. Making a decision based only on temperature or enthalpy differences is likely to generate needless costs for humidification and dehumidification that may exceed savings from "free cooling."

References

AAMA *Voluntary Specifications for Aluminum, Vinyl (PVC) and Wood Windows and Glass Doors (ANSI/AAMA/NWWDA 101/ I.S.2-97)* REVISED 12/99 American Architectural Manufacturers Association, 1827 Walden Office Square, Suite 104, Schaumburg, IL 60173 USA www.aamanet.org

AIA *Guidelines for Design & Construction of Hospitals And Healthcare Facilities*, Revised 2000. American Institute of Architects. 1735 New York Ave., NW, Washington, DC 20006 USA 202-626-7300 www.aiaonline.org

ASHRAE Standard 55 - *Thermal environmental conditions for human occupancy*. ASHRAE, Atlanta, GA.

ASHRAE Handbook—Applications 1999. *Chapter 7 - Health Care Facilities*

Brundrett, Geoffrey W. *Criteria for Moisture Control*. 1990. Butterworth & Co., London, England. ISBN 0-408-02374-4.

Hermans, Richard D, *Health Care Facility Design manual—Room Design*. 2000. ASHRAE Transactions, Vol. 106, Pt. 2

Memarzadeh, Farhad, Manning, Andy. "Thermal Comfort, Uniformity and Ventilation Effectiveness In Patient Rooms: Performance Assessment Using Ventilation Indices." 2000. ASHRAE Transactions, Vol. 106, Pt. 2

Memarzadeh, Farhad and Jiang, Jane; 2200. "Methodology for minimizing risk from airborne organisms in hospital isolation rooms." ASHRAE Transactions, V.106, pt.2.

Ninomura, Paul and Bartley, Judene. 2001. "New ventilation guidelines for health-care facilities." *ASHRAE Journal*, June, 2001 p.29-33.

Künzel, Hartwig M. and Karagiozis, Achilles N. and Holm A., 2000. "WUFI ORNL/IBP; A Hygrothermal Design Tool for Architects and Engineers", *ASTM Manual 40*, Heinz Treschel, Editor. The actual software is developed and published jointly by the Fraunhofer Institite of Building Science, Germany, Dr. Hartwig Kuenzel, 011-49-8024-64345 and by the US Department of Energy, Oak Ridge National Laboratory, Oak Ridge, TN, Dr. Achilles Karagiozis (865) 576-3924.

Streiful, Andrew J, "Health-care indoor air quality: Guidance for infection control." *HPAC Engineering*, October, 2000 p.28-37. Penton Publishing, Chicago, IL.

Yborra, Stephen C, and Spears, John W, *HVAC Solutions For Operating Rooms: Jackson Memorial Hospital*. 2000. GTI-00/0198. Gas Technology Institute, 1700 S. Mount Prospect Road, Des Plaines, IL 60018-1804 USA.

Special Thanks

Several designers, and hospital HVAC researchers generously shared their experiences, which have considerably improved this chapter. Among these are Lynn Werman, Carl Lawson, Andrew Streiful, Anand Seth and Stephe Yborra.

Photos

Fig. 24.2 - © Dennis Kunkel

Fig. 24.9 - © AAON

Chapter 25

Eldercare

By Lew Harriman

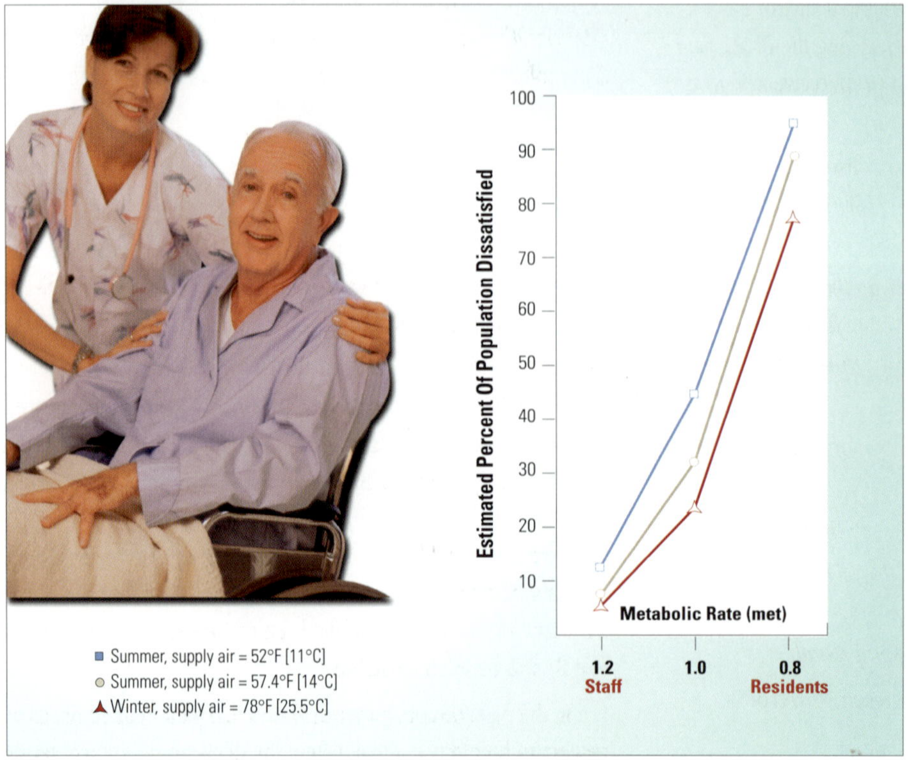

Fig. 25.1 Two Groups, Two Different Needs

As their metabolic rate goes down, people are more likely to be dissatisfied at lower temperatures. Keeping the temperature high and the dew point low can improve comfort for both active staff and passive residents in the same building. (Haghighat, 2000)

Eldercare

During the later stages of life, people often move to buildings designed for the needs of senior citizens, which are somewhat different than the needs of the general population. In most commercial buildings, occupants are visitors who stay for a few hours or a few days. In eldercare buildings, the occupants are residents, and they live most of their lives inside that building.

Older people tend to be less active physically and socially. So the environment inside the building and their personal thermal comfort are large concerns. Residents think about these issues more often, because there is less *other* activity in their lives. Also, comfort is more difficult to achieve, partly because an older circulatory system does not move heat quickly, and also because an older body's automatic thermal sensing and control system does not work well. An older body responds slowly to changes in the thermal environment—and often over-responds or under-responds, leaving the older person uncomfortable when younger people cannot perceive any problem. (Haghighat, 2000)

Further, older people are less able to resist diseases. To a much greater degree than for younger people, the quality of their lives depends on avoiding minor infections.

Given these realities, in eldercare buildings the HVAC designer has a unique opportunity to directly and significantly improve the lives of older people, and to have those improvements endure for 24 hours a day, 365 days per year by:

- *Filtering the air to reduce the number of infectious particles in the breathing zone.*
- *Providing enough ventilation air to keep the building inviting for visitors that improve the social lives of residents.*
- *Keeping the air temperature higher than usual.*
- *Maintaining the relative humidity between 30 and 60% year-round.*

Purpose of Humidity Control

Maintaining the health of residents and staff is the primary reason to control humidity in eldercare facilities, but quantifying that cause and effect relationship is rather difficult. Maintaining comfort for those two groups is secondary, but easier to quantify; and comfort is also important to smooth day-to-day operations. In maintaining both health and comfort, the key to success is to keep the humidity within the moderate range, avoiding extremes.

Also, limiting microbial growth in HVAC components and duct work is a higher-than-usual concern in eldercare facilities, just as it is in hospitals. So the designer needs to remain aware of humidity levels at different vulnerable points within the system, in addition to being focused on controlling humidity within the occupied space.

Control Levels and Tolerances

Comfort issues vary according to the season, and conditions that are comfortable for the staff are not always comfortable for residents.

Summer comfort

In the summer time, the active, hard-working staff prefers a lower air temperature than the less active occupants. The staff has a high metabolic rate, so they must release heat to stay comfortable. The occupants, on the other hand, are producing much less heat, so they need a warmer temperature for equivalent comfort.

Figure 25.1 shows how people's metabolic rate changes their preferences concerning the thermal environment. Given a constant supply air temperature, people who have a lower metabolic rate are more likely to be dissatisfied. For example, when the supply air temperature is 57.4°F [14°C], only about 8% of the active people (with a metabolic rate of 1.2) are likely to be dissatisfied, but at that same temperature, *nearly 90%* of the least active people (metabolic rate of 0.8) will be uncomfortable.

In the past, owners have not always had well-defined needs with respect to humidity control. Often, the decision on control levels is left to the HVAC designer, who usually seeks guidance from local codes and standards, or from ASHRAE suggestions such as those contained in the Health-care Facilities chapter of the *ASHRAE Handbook—Applications*. Those sources allow a rather broad range of acceptable moisture levels from 30 to 60% rh. A summer temperature of 75°F [24°C] with a maximum humidity of 50% rh has been a common compromise between competing comfort requirements. That represents a control dew point of 55°F [13°C]. But the designer might consider a lower dew point within the code-enforced range, which can improve comfort while saving energy.

Since comfort is equally important for the staff and for residents, the HVAC designer can consider keeping the *dew point lower* than usual for the staff, and keeping the *dry bulb temperature higher* than usual for the comfort of the residents. At a lower dew point, the active body releases heat more efficiently through fast evaporation of perspiration. At a higher dry bulb temperature, the older, less active and less responsive body is more comfortable. Given that relationship, the designer could set the control dew point at 50°F [10°C], with a dry bulb temperature set point of 78°F [25.5°C].

Those control levels fall well within the traditional 30 to 60% rh guidelines of many building codes, while providing better comfort for both staff and residents. Also, with a higher dry bulb temperature, the temperature difference across the building envelope is reduced. This reduces the cooling load, which can save both construction and operating costs when the designer can resist the temptation to oversize the cooling equipment.

Winter comfort

Low humidity during the winter season is a significant comfort problem for residents and a labor hour issue for the staff. Unlike younger bodies, the older person's skin lacks a large natural flow of moisturizing oil to keep skin from flaking and reduce uncomfortable friction from clothing. So the lower the humidity, the more frequently moisturizing cream needs to be applied to the skin, which adds work for the staff.

Further, low humidity reduces the effectiveness of the already weaker defenses of an older person against respiratory infections. At low humidity, mucous membranes in the nose dry out, which reduces their ability to trap and neutralize the usual flow of airborne infectious particles from other residents and from the staff.

Finally, very low humidity can work against thermal comfort for residents. At very low dew points, evaporative cooling through perspiration is so quick and effective that residents can be uncomfortable even at relatively high indoor temperatures. Raising the heating temperature increases winter operating costs, because it increases the heating load generated by the temperature difference across the exterior building envelope.

So in the winter months, humidifying the air to at least 30% rh will certainly improve comfort, and may actually reduce operating costs if the owner can keep the temperature a few degrees lower than usual. A dew point of 40°F [4.4°C] provides a minimal 30% rh at 75°F [24°C].

As always with forced humidification, the HVAC designer must be certain of the components of the exterior walls and roof, as well as the placement and thickness of each component. With that information, the designer can complete a dew point analysis, making sure that indoor moisture will not condense on cold surfaces inside the wall as it diffuses outward. (Kunzel & Karagiozis, 2000) The importance of the dew point analysis increases as the indoor dew point goes higher and as the winter design temperature goes lower. So in cold climates such as Southern Argentina, Canada, and Northern Europe, a careful dew point analysis is especially critical to avoid the problem of condensation inside walls at comfortable levels of indoor humidity.

Control level in ductwork & air handling equipment

In the past, ASHRAE Standard 62 (Ventilation for acceptable indoor air quality) contained a caution against relative humidity above 70% in ductwork. That caution became very controversial, because

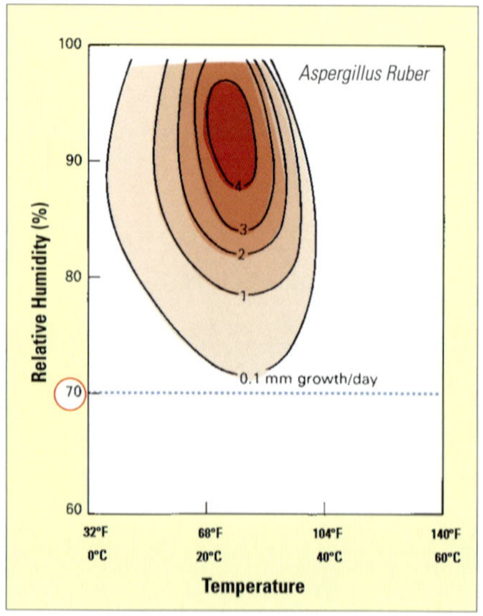

Fig. 25.2 Why Keep RH Below 70% ?
The lower the humidity, the less water is absorbed into the dust that acts as a food source for fungus. Below 70% rh, the growth rate of nearly all types of mold and mildew is very slow, essentially eliminating most health risks from airborne fungus. Chapter 7 contains detailed information on controlling growth of mold and mildew.

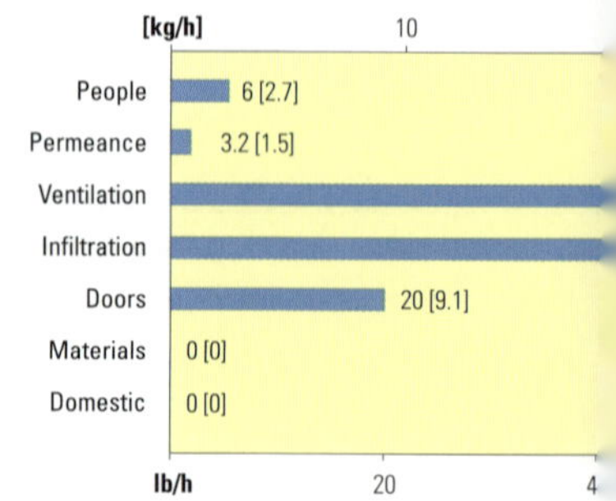

relative humidity immediately downstream of humidifiers and downstream from cooling coils is above that limit. Psychrometrics and the economic realities of HVAC budgets dictate that keeping relative humidity below 70% in ***absolutely all*** parts of the system is not practical. But here's why it is important to keep ***nearly all*** surfaces dry ***nearly always.***

Most important and most commonly, dust grows fungus when it gets moist. Chapter 7 of this book explains the process in detail. Some members of fungal families like aspergillus and stachybotrys are not good for human health.

Further, standing water supports and amplifies the growth of bacteria. Some of these are hazardous directly, like Legionella. Other bacteria are hazardous indirectly, because they contain viruses that infect humans.

To prevent the growth of fungus, duct walls must be kept clean and dry. This is especially important when duct surfaces are porous and therefore trap dust, and especially important for filters, which trap large amounts of dust by design. Cooling and heating coils, dampers, turning vanes and fan blades also collect dust. Smooth metal duct surfaces trap less dust, but protective oils can collect particles that support fungal growth. Frequent filter changes prevent dust accumulation, but the designer must keep the air dry.

The definition of "dry enough" to prevent fungal growth depends on the type of fungus, the type of food source and on the temperature. But keeping the relative humidity below 70% will prevent food sources from absorbing enough moisture to support most fungal growth. Ideally, all points inside the ducts and air handling units would have relative humidity below that level. But downstream from cooling coils and humidifiers, that's not possible. The design must prevent the accumulation of dust in the humid zone.

Downstream of those components, specify a smooth, non-porous interior lining. Ideally, *all* interior duct surfaces would be smooth metal. As a minimum, the length of the smooth lining should extend to the point where air is below 70% rh. But good design by itself is not enough. The operations staff must also change the filters often enough so dust does not collect in the wet zone.

Moisture Loads

The load estimate shown in figures 25.3a/b assumes a 100-person eldercare facility with a mixture of skilled nursing and assisted-living services. The location is Atlanta, GA in the USA. The indoor dew point control levels are a maximum of 50°F during the cooling season and a minimum of 40°F during the heating season. [10 and 4.4°C, respectively]. Ventilation dominates the moisture loads in an eldercare building, just as it does in most other commercial buildings. Unlike other buildings, the ventilation must be continuous, because like a hospital, the building is occupied at all times of the day.

Fig. 25.3a Dehumidification loads

Just like most other commercial buildings, the moisture load in an eldercare building is dominated by the ventilation requirement. That's why load reducers make sense, and why it is generally wise to dehumidify the ventilation air before it can mix with air in the rest of the HVAC system.

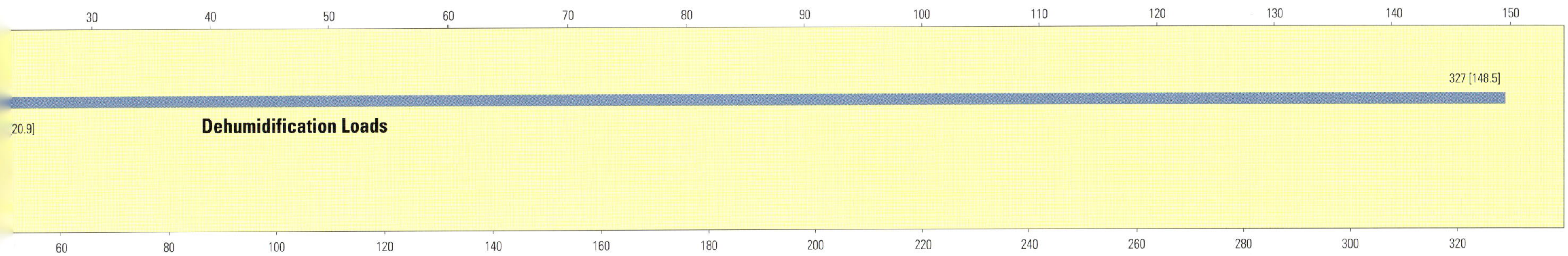

Ventilation

Many local codes in North America follow ASHRAE suggestions, requiring at least 2 air changes per hour of outdoor air for resident rooms and in common rooms. This gives the building a rather high moisture load from humid ventilation air. The load is continuous, unlike in schools or offices, where ventilation can be reduced at the end of the working day when occupants leave the building.

With such a high ventilation load, the designer will immediately appreciate the benefit of a load reducer such as a passive desiccant wheel. Quite a bit of conditioned exhaust air is available from toilets, where the odor dilution is very high. Using a heat exchanger, that exhaust air can be used to pre-treat the incoming ventilation air, greatly reducing the amount of energy and equipment capacity needed to remove the remaining excess water vapor from the air during summer. The same device reduces the humidification load during winter. It is not unusual for these load reducers to cut the ventilation load by more than 50%. Chapter 15 contains more details on how load reducers work, and how they can be applied and controlled.

Further, there are significant benefits to dehumidifying the building by drying the ventilation air *below* the desired dew point for the space. With that strategy, the individual heating and cooling units in each resident's room never have to condense moisture, which greatly reduces risks associated with mold, mildew and bacterial growth. Also, by filtering that incoming air, the rooms are flushed at least twice each hour with air cleaned of nearly all particulate, greatly reducing the risk of airborne infection.

Infiltration

In any building, humid air will be pushed in through cracks between components on the exterior wall. The designer should take this moisture load into account when deciding how dry to make the ventilation air before it enters the building. That air should be made dry enough to absorb the infiltration load, as well as absorbing internal moisture loads from people and from door activity.

This load estimate assumes that the longest wall will be exposed to wind pressure during highly humid times, and that the exterior wall has been built "tight" as opposed to "average" or "loose" construction. The architectural designer needs to make sure the building is indeed tight. Otherwise, the building will not only have greater moisture loads during the summer, but also be subject to condensation in walls during cold seasons.

Door activity

Even allowing for longer open times needed by older people, the load from door activity is much lower than for buildings with high traffic, such as retail stores or hospitals. In eldercare buildings, the moisture load from door activity is not significant in relation to the immense moisture load from ventilation air.

People

The staff releases a considerable moisture load per person because of their active movement and care of the residents. But the load from less-active residents is nearly negligible. The total load from people is still quite small compared to loads from other sources.

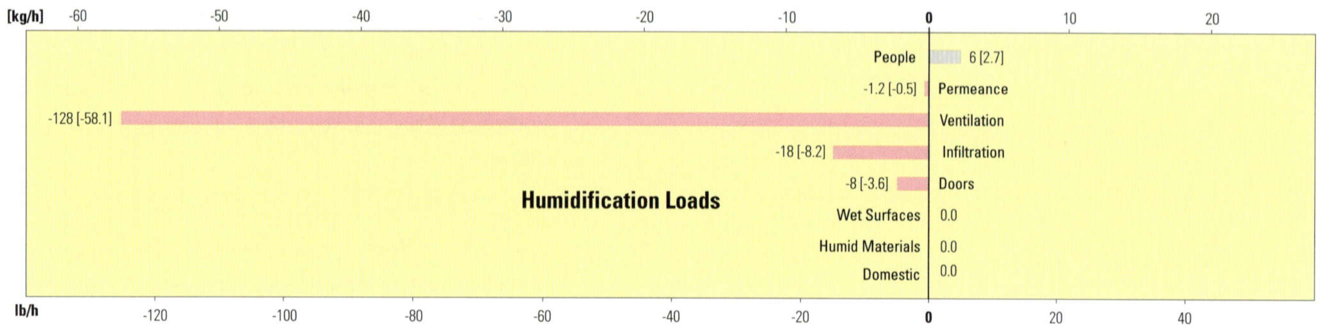

Fig. 25.3b Humidification loads

The humidification load is also dominated by the ventilation requirement. People add moisture internally, but not enough to really matter much. Adding moisture to the ventilation air is the simplest and most cost-effective way to keep the humidity up to comfortable levels during winter months.

Permeation

As in other commercial buildings, the moisture load from vapor permeation through the exterior walls and roof is very small. There is no need to invest extra effort to find impermeable building materials when the ventilation air is properly dehumidified. Likewise, if the ventilation air is *not* dehumidified, an impermeable building will not make any improvement in the control of humidity. Vapor permeation is not a significant component of the total moisture load.

Humidification loads

In winter, the humidification load rises, and its components are similar to the moisture loads during the summertime. The ventilation air is the major component of the load, with infiltration the next largest load element. Pre-humidifying the ventilation air is an excellent way to stabilize the humidity and bring the air up to comfortable levels during dry months.

Note that the humidification load is less than 25% of the size of the dehumidification load. This is true in a location like Atlanta, GA, but is decidedly not the case in colder climates such as Moscow or the province of Alberta in Canada. In those locations the humidification load would be much larger.

Fig. 25.4
Ventilation Air Pretreatment For Humidity Control

1. Outdoor Air
Hot and humid in summertime, it must be cooled and dehumidified.

2. Ventilation Preconditioning Unit
Dries the incoming air to a condition below the desired humidity set point

3. Dry Ventilation Air
Will remove the moisture loads generated inside the building

4. Air Inside The Building
Is at or below the desired humidity set point

5. During Winter
Ventilation air may be humidified by the pretreatment unit, but often the cold air will not accept enough moisture to offset losses from the building. Additional humidification is needed in the return air stream.

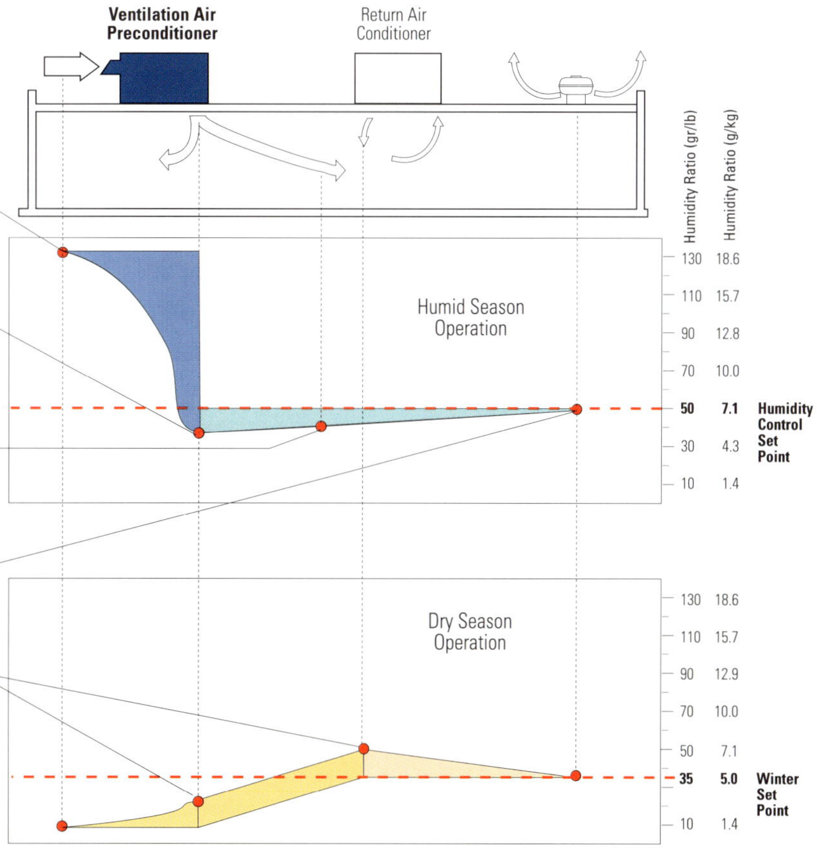

Suggestions & Cautions

Eldercare facilities are similar in many respects to hotels and dormitories. They have many sleeping rooms, and need individual temperature control in those rooms. Also, the buildings have a large volume of exhaust from toilets. As one might expect, some of the cautions and suggestions that apply to hotels and dormitories also apply to eldercare buildings. But since the health of eldercare residents is less robust, and since older people spend more time in their rooms, the designer needs to consider health issues carefully in eldercare buildings.

Don't rely on room units to dehumidify outdoor air

In the United States and many other countries, fire protection codes prohibit using hallways in eldercare buildings for moving air into or out of resident rooms (NFPA 90A). That means that the usual ventilation of 2 air changes per hour of outdoor air must be ducted directly into each room rather than being distributed to the corridors and pulled into the rooms by suction from toilet fans. Also, that ventilation air flow must be continuous.

So the designer has two choices: dehumidify the incoming air centrally and distribute dry air to the rooms, or pull in outdoor air through each room unit. The best choice is to dry the incoming air centrally, and then distribute that dry air through dedicated duct work directly into each room

Room units are not usually capable of dehumidifying air to a low dew point continuously. This is partly because they are usually built for maximum cooling efficiency rather than deep drying, and partly because they are so efficient at cooling that they do not need to cool the air all the time. Cooling operates intermittently, and while cooling is off, the incoming ventilation air continues to flood the room with excess moisture.

Central drying is more effective, and also reduces the health risks associated with bacteria, mold and mildew in drain pans and filters inside the residents' rooms. With central drying, any condensation and most particulate can be removed from the system remotely, so that room units stay dry and relatively clean.

Finally, a central system makes it much easier to humidify incoming ventilation air during winter months. This avoids the need for dozens of small local humidifiers, which rapidly become a maintenance problem.

Oversized cooling units are not the answer for dehumidifying 100% outdoor air

Because conventional packaged rooftop cooling units can partly dehumidify air, the designer might assume that if they are installed with greater cooling capacity, such units could dehumidify an air stream consisting of 100% outdoor air. That's not usually the case.

Any cooling coil needs to be deep in the direction air flow to dry highly humid outdoor air. Also, a cooling-based dehumidifier has special controls to avoid freezing when the air is at a moderate temperature but still too humid to bring into the building without drying. A mechanical or desiccant dehumidifier that has been engineered to dry 100% outdoor air is the best solution for ventilation air.

Carefully seal up any duct or plenum that's under suction

Many of the moisture problems in eldercare facilities can be traced to exhaust ducts, return air ducts and wall-mounted cooling units that create suction in building cavities because they are not sealed up tight. When the building cavities are under suction, they pull humid outdoor air into the building through cracks in the exterior walls. Moisture in that air condenses when it gets inside the building and contacts a cool surface, and mold and mildew problems result, as shown in figure 25.7.

These problems can be avoided by specifying that all ductwork including supply, return and exhaust ducts be sealed with mastic on all transverse and longitudinal seams. Also, specify that the negative pressure plenums of any wall-mounted cooling units be sealed, and their return air connections be sealed to the inside face of the room walls. That way, such units will not accidentally pull air from the walls, creating suction that pulls in humid outdoor air.

Control ventilation dehumidifiers by dew point

Ventilation pre-treatment units are designed to dry air down to a set humidity maximum. The designer must remember to set up the controls that way.

In the past, some cooling units that also supply ventilation air have been controlled by the temperature indoors, rather than by the desired dew point. In other words, if the outdoor air temperature was below the indoor dry bulb set point, the unit would stop cooling the incoming air. That allows humid air to enter the building. This problem has been especially common in hotels, but has also occurred in eldercare buildings.

Regardless of whether desiccant or mechanical dehumidification is used, any ventilation dehumidifier should be controlled based on the desired *indoor dew point*, not the indoor dry bulb temperature. This avoids the problem shown graphically in figure 25.8.

Avoid oversizing room cooling units

When ventilation air is pre-dried centrally, the room cooling units can be downsized because they do not have to remove moisture. Not all designers note this fact, or they prefer not to remove extra capacity from those room units. Excess cooling capacity is a bad idea in

Fig. 25.5 Load Reducers

The ventilation load is very large and continuous, since the building is constantly occupied. Reducing the load greatly reduces costs both in construction and operation. Chapter 15 contains suggestions on how to apply load reduction equipment, such as this passive desiccant wheel.

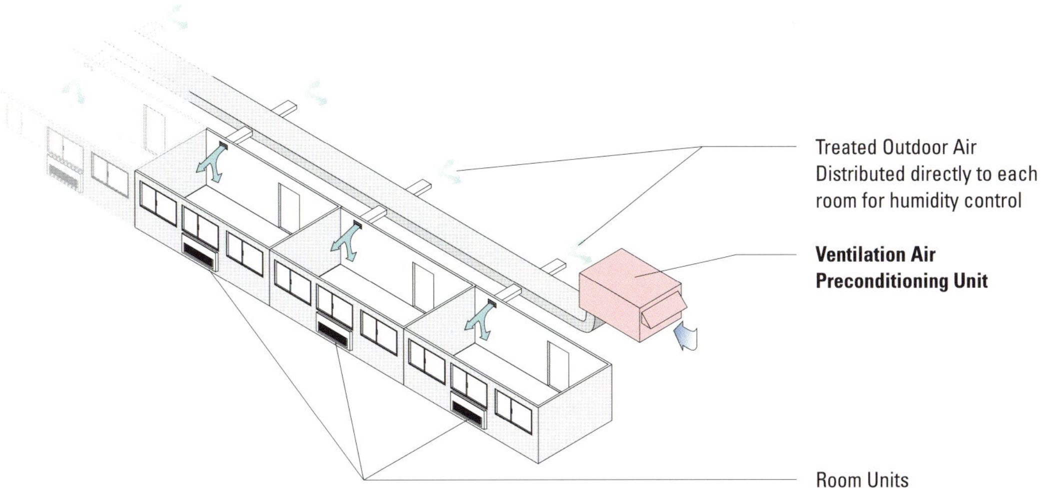

Fig. 25.6 Fire Protection Prohibits Using The Corridor As A Duct
Since ventilation air must be ducted directly to each room, it makes even more sense to dry or humidify that air before it enters the building.

Figure 25.7 HVAC-Driven Suction Can Create Mold In Buildings

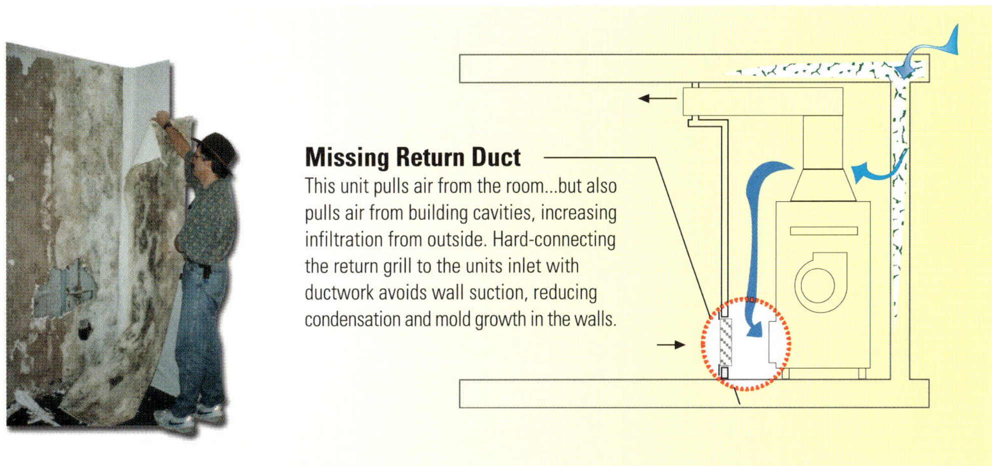

any system, but it is especially poor practice for eldercare buildings where old people must suffer its consequences all their lives.

When cooling units are oversized, they chill the air to very low temperatures very quickly, then they shut off. The room occupants get a strong blast of cold air intermittently, and their slow-responding bodies can't adjust to the new conditions quickly enough to avoid discomfort. In other buildings, designers often leave excess capacity in the system to ensure adequate cooling during periods when weather exceeds design loads. The designer's intent might be noble, but the actual result of "extra" capacity is *continuous torment* for older people—which is generally not what owners of eldercare buildings are paying HVAC designers to achieve.

Avoid high velocity air in resident rooms

The velocity of air across the body has a very great effect on thermal comfort. High velocity removes heat quickly whenever the air temperature is below the body temperature—not what an older person needs in a normally-conditioned building. The designer should keep the air velocities low in spaces occupied by older people. This means, for example, that while it may be an excellent idea to supply ventilation air directly over a bed, it is important to keep the velocity of that air very, very low as it flows across that bed.

References

AIA *Guidelines for Design & Construction of Hospitals And Healthcare Facilities*, Revised 2000. American Institute of Architects. 1735 New York Ave., NW, Washington, DC 20006 USA 202-626-7300 www.aiaonline.org

Brundrett, Geoffrey W. 1990. *Criteria for moisture control.* Butterworths, London & New York.

Chapter 7 - Health Care Facilities. 1999. ASHRAE Handbook—Applications.

Haghighat, Fariborz, Megri, A.C, Donnini, G, and Giorgi, G. 2000. "Responses of disabled, temporarily ill and elderly persons to thermal environments." ASHRAE Transactions, V.106. pt.2.

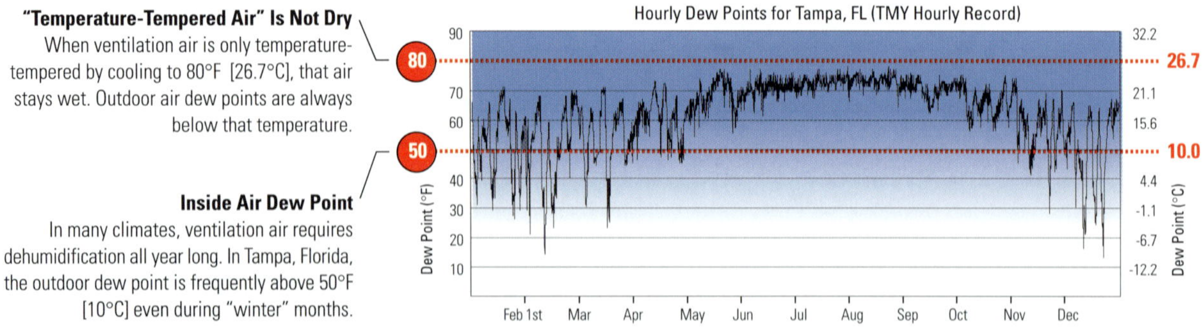

Fig. 25.8
Control Ventilation Pre-Treatment Based On Dew Point, Not Dry Bulb Temperature

Künzel, Hartwig M. and Karagiozis, Achilles N. and Holm A., 2000. "WUFI ORNL/IBP; A Hygrothermal Design Tool for Architects and Engineers", *ASTM Manual 40*, Heinz Treschel, Editor. The actual software is developed and published jointly by the Fraunhofer Institute of Building Science, Germany, Dr. Hartwig Kunzel, 011-49-8024-64345 and by the US Department of Energy, Oak Ridge National Laboratory, Oak Ridge, TN, Dr. Achilles Karagiozis (865) 576-3924.

Larson, Michael L, 2000. "Geothermal system for a skilled nursing facility." *ASHRAE Journal*, June 2000. pp:57-59.

Lotz, William A., 2000 "Expensive moisture problems at several Central Florida and South Texas nursing homes." Proceedings of the 12th Symposium on improving building systems in hot & humid climates. May 2000. San Antonio, TX. Texas A&M University, College Station, TX.

NFPA 90A. *Standard for the installation of air conditioning and ventilation systems, 1999 Edition*. National Fire Protection Association, 1 Batterymarch Park, P.O. Box 9101 Quincy, MA 02269-9101 (617) 770-3000 Fax: (617) 770-0700 www.nfpa.org.

Ninomura, Paul T, and Cohen, M.H, 1999. "IAQ in nursing homes." ASHRAE Journal, February, 1999. pp: 34-38.

Special Thanks

The Author is grateful to several designers and researchers who shared their experiences in dealing with humidity control in nursing homes, especially Stephe Yborra of Energystics, who tirelessly monitored and recorded the everyday performance of alternate technologies over a year in a functioning nursing facility.

Donated Photos & Diagrams

Chapter 26
Dormitories

By Lew Harriman

Fig. 26.1 Densely-Occupied Dormitories

Although comfort is always a concern, the usual reason for investing extra money in humidity control for dormitories is to reduce health risks associated with many people living in a small space.

Dormitories

Dormitories on military bases and those used for student housing are similar in many ways to hotel and eldercare buildings. But dormitories differ in three important respects. Dorm occupants are usually young and in the prime of health. Also, a typical dormitory is occupied more densely, with many residents per unit of floor space. Finally, the comfort expectation in dormitories is lower than for hotels and nursing homes. Living accommodations are not the principal attraction for either a military or an academic institution.

On the other hand, the dense occupancy, high physical and mental stress on occupants and resulting hard wear and tear on facilities suggest other useful characteristics for an HVAC system:

- *High ventilation rates, to help dilute the concentration of both body odors and infectious particles.*
- *Rugged, centralized systems, so less equipment is exposed to accidental or intentional abuse by overstressed occupants.*

The high ventilation rate needed for dormitories generates a significant moisture load during the summer, and in cold climates, ventilation creates a high humidification load. Not all building owners care about comfort implications of high loads—but most are concerned with keeping occupants healthy. As a result, some dormitory owners may wish to keep humidity within a narrower range than is possible with heating and cooling systems alone. This chapter contains suggestions for situations where an owner is prepared to invest extra money in a system that will control humidity, rather than one which simply moderates humidity extremes.

Purpose of Humidity Control

Assuming that the owner is most interested in controlling humidity to avoid health problems, the designer needs to identify which elements of the system and the building design can reduce the risk of humidity-related health problems. That's not easy to do.

In North America, the legal implications of health risks sometimes paralyze rather than enlighten researchers and designers. Following a defensive strategy similar to that employed by ostriches, sometimes designers avoid discussing the health effects of HVAC systems, especially when the conversational group includes members of the legal profession. But some aspects of the health effects of humidity are reasonably clear.

Control Levels and Tolerances

Based on our rather limited understanding, there are at least four opportunities to reduce the risk of health problems through humidity control:

1. Avoid drying out the upper respiratory tract

To defend against airborne infections, the human body traps, filters and neutralizes infectious particles in the mucous membranes of the nose and throat. If these membranes dry out, they provide less protection. Research is not conclusive about the optimal level of humidification for health purposes, and many millions of people survive quite happily in cold climates at humidity levels below 10% rh. But keeping humidity up to a dew point of 35°F [1.7°C] seems to help reduce health problems associated with overdrying the body's defensive membranes.

2. Avoid sorption of moisture into fungal food sources

When fungus (mold and mildew) grows in the HVAC system or in furnishings and personal belongings, it releases spores and volatile organic compounds (VOC's) that can trigger allergic reactions including headaches and "flu-like symptoms." There is also some evidence that heavy exposure to some types of mold can have toxic effects and may partly suppress the immune system, leaving the body more vulnerable to minor infections. (See Chapter 7 for additional information) So keeping the relative humidity below 70% in ductwork and in the occupied space is useful, because below that level, few materials can absorb enough moisture to support fungal growth.

3. Avoid standing water that harbors bacterial growth

When airborne bacteria fall into water, they can thrive and reproduce rapidly. A small bacterial population is quickly amplified into a much larger one. Some of those bacteria are likely to be harmful to humans. To minimize the bacterial population within the air handling system, make sure that all drain pans under cooling coils actually drain water completely and continuously.

4. Avoid sucking humid air into cool building cavities

Mold and mildew growing behind wallboard is unfortunately common in hotels and also occurs in dormitories. The problem is caused by unsealed ductwork and plenums that are under negative air pressure. The suction created by these leaks pulls humid air into cool building cavities through small cracks in the walls. The humid air provides moisture that is adsorbed into cool materials, which then support mold and mildew growth.

This is a humidity control problem with negative health effects—but one which begins in building cavities rather than in the occupied spaces that are usually the designer's focus. Sealing up ***all*** ductwork helps avoid this problem. This means sealing up joints that are seldom given attention by the contractor unless the designers mandates are clear and emphatic in the HVAC specification. Any joint under negative air pressure requires a tight seal. This includes exhaust ducts, negative pressure plenums in air handlers and joints between return air grills and the finished wall surface.

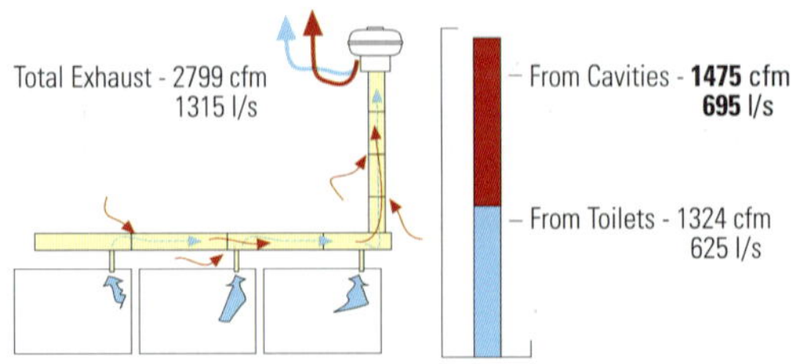

Fig. 26.2 Unsealed Ductwork Creates Moisture Problems

*In this Florida study of a hotel, the exhaust fans pulled more air from building cavities than from the bathrooms. The humid air sucked into the cavities then allows fungus to grow. Avoid this common problem by sealing **all** the ductwork.*

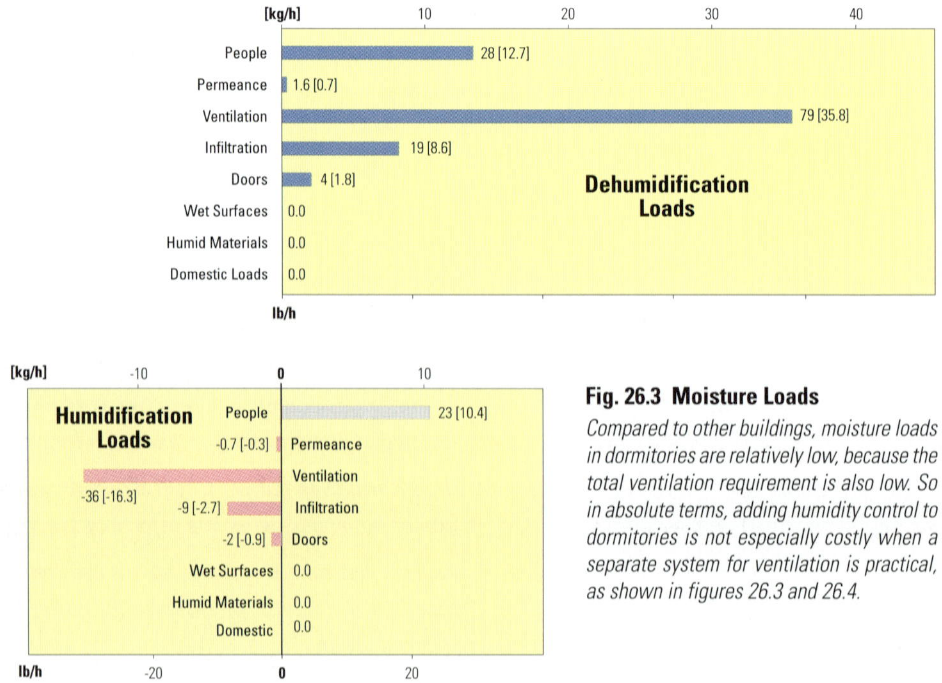

Fig. 26.3 Moisture Loads

Compared to other buildings, moisture loads in dormitories are relatively low, because the total ventilation requirement is also low. So in absolute terms, adding humidity control to dormitories is not especially costly when a separate system for ventilation is practical, as shown in figures 26.3 and 26.4.

5. Suggested dew point limits

Given the health concerns, but also the lack of firm data that suggest specific control levels to eliminate all hazards, the owner might simply consider keeping the dew point in the mid-range of the human comfort zone for summer and above the lower limit of the comfort zone for winter. That would suggest a maximum summer dew point of 57°F [13.9°C] and a minimum winter dew point of 35°F [1.7°C].

Ultimately, of course, the control levels are for the owner to decide. Here is the logic of the suggestions: For summer, if any defined level of humidity control is wanted, there is little difference in equip-

Exhaust air from above the whirlpool for pressure control

Air above the whirlpool is more humid that the rest of the air in the enclosure, because that's where the evaporation rate is highest. Since the enclosure should be held under a slight negative air pressure, the whirlpool is an ideal place for an exhaust fan, which can operate in response to pressure controls. Exhausting that highly humid air also keeps it out of the return duct, reducing the risk of condensation, and also avoiding the rather disgusting build-up of polymerized body oils that is virtually impossible to remove.

Consider the importance of fresh air in the breathing zone at the water's surface

ASHRAE Standard 62 suggests a rather high ventilation rate for pools, primarily to dilute the concentration of unhealthy vapors that sometimes comes from poor balance of pool water treatment chemicals. When water chemicals are out of balance, the concentration of those vapors is highest immediately above the water surface, and unfortunately, that's right where swimmers get their breathing air.

Some have had success using exhaust ducts placed near the floor level to encourage pollutant-laden air to flow away from the pool surface. The exhaust system is switched on during competitions when swimmers can be especially affected by poor air quality.

References

"Thermal and moisture control in insulated assemblies." Chapter 23, *ASHRAE Handbook—Fundamentals. 2001.* ASHRAE, Atlanta, GA.

"Places of assembly." Chapter 4, *ASHRAE Handbook—Applications. 1999.* Natatoriums, pp: 4.5 - 4.7.

Design and dehumidification of swimming pools, 2001. Dectron Internationale, Atlanta, GA.

Kittler, Reinhold 1983. "Indoor natatorium design and energy recycling." ASHRAE Transactions, vol 95, part 1, pp: 521-526.

Künzel, Hartwig M. and Karagiozis, Achilles N. and Holm A., 2000. "WUFI ORNL/IBP; A Hygrothermal Design Tool for Architects and Engineers," *ASTM Manual 40*, Heinz Treschel, Editor. American Society for Testing And Materials, Philadelphia, PA. Note: The actual software is included on a CD-ROM supplied with ASTM manual 40. It was developed and published jointly by the Fraunhofer Institute of Building Science, Germany, Dr. Hartwig Künzel, 011-49-8024-64345 and by the US Department of Energy, Oak Ridge National Laboratory, Oak Ridge, TN, Dr. Achilles Karagiozis (865) 576-3924.

Lotz, William A, 1990. "Considerations for swimming pool building design." *Engineered Systems* magazine. March/April, 1990. Business News Publishing Co. Troy, MI

Lotz, William A, 1995. "Indoor pool design: avoid the potential for disaster." *Heating, Piping and Air Conditioning* magazine, November, 1995. pp: 47-63.

Lotz, William A, 1996. "Are insulation specifications conjured up for magicians?" *Insulation Outlook* magazine, November, 1996. pp: 32-33.

Lotz, William A, 1998. "Insulation failure and moisture problems resulting from inadequate installation of insulation and/or vapor retarders." ASHRAE Transactions, vol.104, pt.2. 1998

Special Thanks

The authors sincerely appreciate the experience shared and the photos donated by William Lotz, who has done so much to improve the industry's understanding of condensation problems and their solutions in swimming pool enclosures.

Donated Photos & Diagrams

Chapter 28
Ice Arenas

By Lew Harriman & Reinhold Kittler

Fig. 28.1 Bruins In The Mist

Before the hockey arena in Boston, Mass. was replaced, scenes like this were quite familiar to local fans and television audiences. Dehumidification eliminates such fog, and prevents the condensation that softens ice and drips onto spectators. Also, keeping the air dry reduces operating costs for ice refrigeration.

Ice Arenas

An ice arena combines the cold storage function of a freezer with the ventilation requirement of an auditorium.

One can imagine the difficulty of designing a refrigerator that would keep its ice cubes frozen while standing outside in the hot sun of South Florida, with its freezer door propped open to allow complete flushing with outdoor air at least once every hour.

In fact, current ventilation requirements for ice arenas are even higher than one air change per hour, and ice arenas are being built frequently in tropical parts of the world. So the visual image of the refrigerator out on the lawn is a fair description of the challenge, and that analogy can help focus the efforts of the design team. *Three aspects of the project require full attention:*

1. *The building must be very well-insulated and sealed, built more like a refrigerator and less like a conventional steel building.*
2. *All ventilation air must be dried deeply before it can enter the building.*
3. *Operating costs can be saved by reducing ventilation air flow whenever the building is unoccupied.*

Purposes of Humidity Control

In an ice arena, owners invest in dehumidification to gain several economic benefits beyond the usual comfort improvements:

Operation becomes practical during humid weather

Ice skating originated in cold climates. The sport season was extended by enclosing the ice rink, and by adding refrigeration for more consistent ice quality for more months of the year. Now winter sports are in demand all year long, and are popular all over the world.

By drying the air above the ice surface, the volume of condensation is reduced, which in turn reduces the refrigeration capacity needed during hot or humid weather. That load reduction makes operation much more practical in economic terms. Given a reasonable building and adequate dehumidification, ice arenas can be located in any part of the world.

Fig 28.2 Modern Arenas

Dehumidifiers reduce the dew point, which minimizes the condensation load on the ice. The radiant heat load is reduced by installing a low-emissivity ceiling and wall coverings.

Spectators can see the action on the ice

Without dehumidification, the usual amount of moisture in the air becomes fog near the cold ice surface. Indoor fog is especially common during the springtime, summer and fall, when the indoor air temperature falls at night while the dew point remains high.

Consider the photographic evidence in figure 28.1. Before Boston Garden was replaced during the 1990's, fog regularly made ice hockey impossible, often during the worldwide attention focused on the playoffs for the National Hockey League championship. Before dehumidification, the common practice during springtime was to have the players skate around the ice for a half hour or so, acting as air mixers to blend warmer air into the cold foggy layer near the ice. Results were not effective, and the audiences and television sponsors were annoyed by the delay.

Also, condensation often obscures the view through the glass safety panels above the dashboards. These panels cool by radiation—they lose heat to the cold ice surface below. Then, when the surrounding air cools to near saturation, dew forms on the cold glass. Dehumidification prevents both fog and condensation easily.

Better skating surface, and more ice time for rent

When moisture condenses on the ice sheet, it liberates heat, which warms and softens the ice surface, making it slower and less durable. Also, without dehumidification, condensation drips from cold roof beams at night. The constant drips create small lumps of ice called "mushrooms" that must be removed for safety reasons before skating can begin in the morning.

An ice surface that is soft and easily-cut can lead to poor performance by athletes during competitions and also means the ice must be resurfaced frequently. That shortens the rentable time for the rink, and interrupts sports for an awkward amount of time in the middle of events. Not only that, but frequent resurfacing with untreated water concentrates minerals that steadily degrade the quality of the ice even when its surface is newly rebuilt (Moore, 1995).

The minerals present in untreated water are most soluble at cold temperatures. But resurfacing water is quite hot, to help melt and flatten the surface. So its minerals are near precipitation rather than fully-dissolved. Further, creating a layer of water at the ice surface lifts minerals from deeper in the ice towards that surface. In other words, the minerals are *most* concentrated at the new ice surface. These minerals lower the freezing point of water, which means extra refrigeration will be needed to complete the freezing of that layer, as well as complete the freezing of any condensation that occurs later from high humidity. Finally, a mineral-laden surface is soft and grainy, which slows the skater, and creates what rink operators call "snow development" at the ice surface during normal use.

Consequently, it is very beneficial to use a dehumidifier to *prevent the condensation that begins this cycle* of frequent interruptions and long-term surface degradation. Keeping the air dry increases rentable hours, and greatly improves the ice quality. These benefits accrue to the owner, but are also appreciated by skaters, especially those involved in high-level competition.

Lower operating costs for ice refrigeration

By keeping condensation on the ice surface to a minimum, less refrigeration capacity is needed to keep the surface at the appropriate temperature. Usually, the best way to take economic advantage of reduced condensation is to raise the refrigerant temperature slightly. Any small rise in refrigerant temperature quickly reduces the power consumption of the compressors.

Less damage to the building

At high indoor dew points, condensation corrodes fasteners and steel roof supports, and feeds steady drips that saturate insulation and corrode exterior cladding. When indoor air is kept dry, it will not condense moisture onto these vulnerable surfaces. At the very least, this means there will be less need for repainting. But the more important benefit is maintaining the structural and thermal integrity of the building so it is less likely to fall down.

Control Levels and Tolerances

Owner preferences for specific temperature and moisture levels depend on which skating activity is planned, and on what the owner believes is economically practical. Usually, the dew point is kept between 35° and 45°F [1.7° and 7.2°C].

Below that range, there are very few benefits to be gained, and operating costs are high. Similarly, when the dew point is above that range, more water vapor condenses on the ice surface, which drives up operating costs for refrigeration, softens the ice surface and increases the amount of condensation on the building structure.

Professional hockey

Because of severe problems with ice quality that have held up or prevented important games in the past, in 1999 the North American NHL (National Hockey League) established minimum standards for

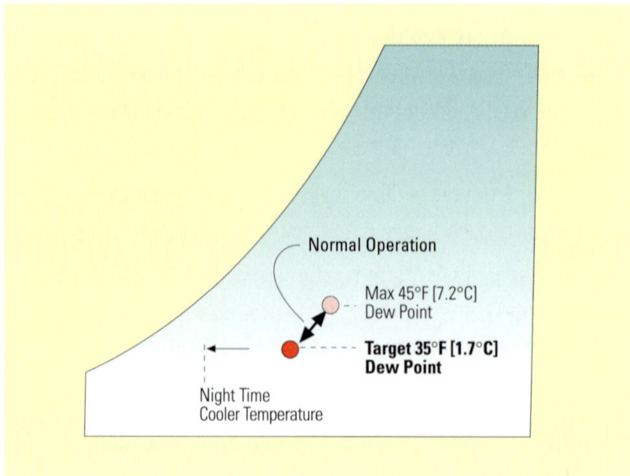

Fig. 28.3 Dew Point Control Levels
Dew points in ice arenas are usually held between 35° and 45°F [1.7° and 7.2°C]. Below that range, there is little benefit. Above that range condensation problems increase substantially, especially at night when air in the arena cools down.

air temperature and humidity that must be achieved at least one hour before game time. These standards (NHL, 1999) state that:

1. "Air temperature must be stabilized at 60°F, and in no event higher than 65°F [Stable at 15.5° and never above 18.3°C]."
2. "Humidity levels must be less than 40% rh."

In other words, to meet the highest North American standard for hockey ice, the dew point should be held below 35°F [1.7°C].

Spectator arenas

In spectator arenas, there is a tremendous ventilation requirement, which makes low dew points very costly to maintain when the arena is occupied. The NHL specification calls for achieving the low dew point one hour before the game starts—not necessarily while it is in progress. So the usual goal during events is simply to prevent fog and heavy condensation. Those goals are much less difficult to achieve, and can usually be accomplished easily by any dehumidification system that has the capacity to meet the NHL pre-game dew point specifcation.

Moisture Loads

The moisture load estimate in figure 28.5 is based on a small recreational ice arena with a spectator capacity of 200 people, similar to the building shown in figure 28.4. Our hypothetical building is located in Atlanta, GA in the southeastern US. Atlanta has a hot and humid climate for most of the year, but has cold months as well.

It's clear from the width of figure 28.5 that the moisture load in an ice rink is much larger than most other commercial buildings. This especially high load comes from high ventilation, along with the need to maintain a very low dew point. Our load estimate is based on keeping the arena down to a dew point of 35°F [1.7°F].

Fig. 28.4 Recreational Arena

The moisture load shown in figure 28.5 assumes such a building. Note how the windows, when necessary at all, are placed on the shady side of the building. Also, any windows should open onto the lobby, not the ice surface.

Ventilation

ASHRAE Standard 62.1 (Ventilation for acceptable indoor air quality) calls for providing 0.5 cfm per square foot of floor space. [0.023 l/s per m^2 of floor space]. Compared to other commercial buildings, this is a great deal of air.

The outdoor air is supposed to dilute the concentration of indoor air pollutants for the health and safety of the occupants. A quick glance at figure 28.5 shows the impact of this amount of outdoor air on the required size of the dehumidifier. With full ventilation, the dehumidification system must remove 982 lbs of water vapor per hour. If ventilation is shut off entirely, the load from all other sources is only 194 lbs/h. Two implications are immediately apparent:

1. *Predrying ventilation air* before it can enter the building is an excellent idea, especially since that can be partially accomplished at very low cost by using load reducers such as passive desiccant wheels (enthalpy recovery wheels).
2. *Reducing the amount of ventilation air* is also an excellent idea, provided the quality of the indoor air remains safe and satisfactory. Many designers use sensors to measure the concentration of CO and CO_2, and then to reduce the outdoor air flow whenever the sensor indicates that indoor air is safe and fresh. ASHRAE Standard 62 allows such an approach, but local building codes will govern what is actually permissible for each project.

Infiltration

Ice arenas are large buildings, with a great deal of vertical surface area exposed to wind and rain. Sadly, they are often poorly constructed with respect to air tightness. Very few meet the standard of being built "tight." This estimate assumes that the building will be built tighter than most ice arenas constructed in the US, so it will be at least of "average" tightness, instead of the more typical "leaky" construction. The graph shows the estimated effect of better or worse air tightness on the moisture load.

In leaky buildings, the most frequent problem is open joints such as those shown in figure 28.6. Open joints that actually show daylight have been observed all too frequently. The moisture load from infiltration in such buildings is very difficult to overcome. Even with dehumidification, there are likely to be problems in such buildings whenever the weather is windy, warm or rainy.

Tight buildings, on the other hand, usually have exterior envelopes constructed of structural panels filled with foam insulation. And the seams between these panels are gasketed and sealed. The many benefits of this type of exterior envelope were highlighted in a energy award for an ice arena built in California (EUN, 1999).

The important point here is that the HVAC designer would be unwise to assume the building will be "tight" unless the architectural designer has specified an exterior wall with both gaskets and sealant on all siding joints, as well as sealant between those high-quality walls and the roof. Air leakage under light wind pressure can quickly overwhelm even the best-designed dehumidification system. So the whole construction team must take care to ensure that air leakage is not excessive.

People

This estimate assumes that about 200 people will be in the building—the sort of attendance one might expect at a hotly-contested hockey game between elementary school students. This load estimate is probably low if all those spectators have to walk through the rain to enter the building, and therefore bring their soggy outer garments into the dehumidified space. But for the more common circumstance of a couple dozen parents watching a childrens' skating class, the load estimate is probably quite conservative.

Experienced designers may prefer to estimate the people load based on an average or time-weighted occupancy, rather than based on the peak load, as shown here. Averaging the occupancy over 24 hours leads to the assumption of a smaller hourly load, and therefore to much smaller dehumidification equipment. That equipment may not have the capacity needed to keep the dew point below the set point during all hours. But given the volume of the space, the humidity is not likely to rise for very long. And if the control level is set low enough in the first place, a short dew point rise during peak occupancy is not going to measurably reduce the desired benefits. Load averaging allows economies in the construction budget that may mean the difference between profitability for the building rather

Fig. 28.6
Wind Blows Humid Air Into Buildings

The team must take care to place gasketing, foam sealants and caulking behind seams like these that will otherwise leak large amounts of humid air under wind pressure.

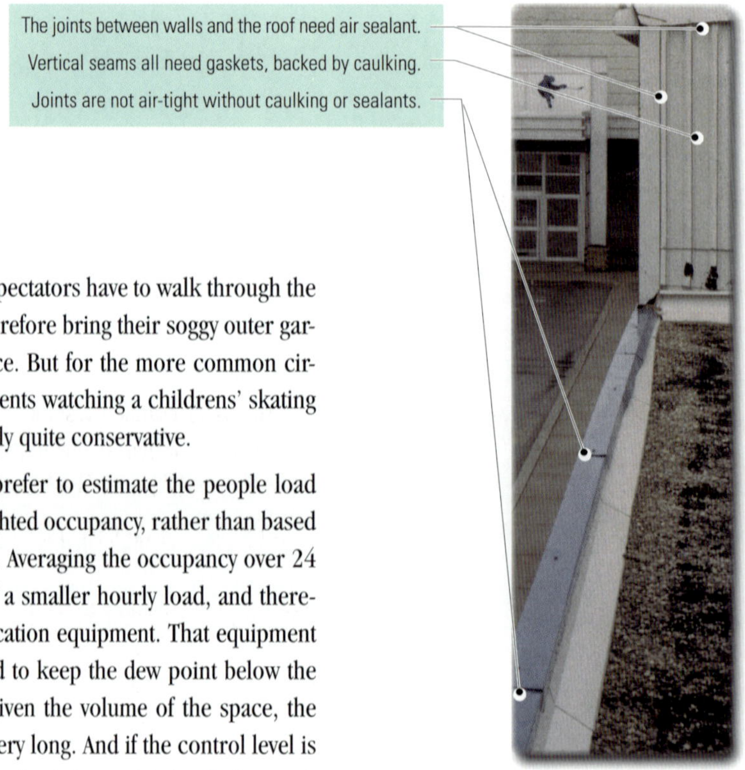

Fig. 28.5 Moisture Load Estimate

The ventilation load is enormous. That's one reason why intermittent, demand-controlled ventilation is popular for ice arenas.

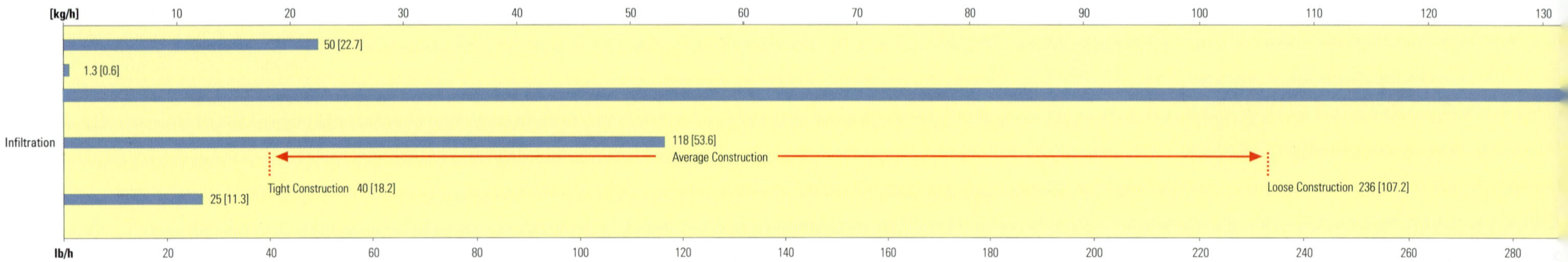

Fig. 28.7 Vesibule & Lobby

Note how the architectural designer has minimized the moisture load from doors by installing both a double-door vestibule and a closed lobby. Humid outdoor air cannot flow easily into the arena itself. (Overall, 1999)

than unsupportable debt. Some professionals with extensive experience in ice rink humidity control (Bradley, 1994) believe that conservatism in the designer's moisture load estimate should not be allowed to bankrupt the project.

Door activity

This load estimate assumes the architectural designer has arranged the floor plan so that normal access to the arena is through a lobby with an air conditioned vestibule, instead of from the outdoors.

Consider two designs: doors opening directly between the spectator area and the outdoors, compared to a closed lobby that also has a vestibule leading to the outdoors. When the wind is blowing, direct-opening doors would allow a massive amount of hot, humid air into the cold space. The closed, conditioned lobby with its vestibule provides three separate barriers to air flow instead of just one. Those barriers dissipate the force of the humid wind before it penetrates into the cold arena. So the moisture load through doors is greatly reduced, to the level shown here.

Ice sheet

Creating a new surface for the ice includes laying down a thin layer of hot water. Some of that hot water evaporates, but not much. In a rink with the usual refrigeration capacity, that resurfacing layer freezes in less than five minutes.

After the water freezes, the ice surface temperature will be below the dew point of air in the space, so the ice sheet will act as a dehumidifier, condensing moisture out of the air rather than adding to the internal moisture load.

The net effect of the ice sheet in a dehumidified space is difficult to determine with any accuracy, so for the purposes of this load estimate, we have assumed there is no net negative or positive effect from the ice sheet. That assumption is very conservative, because the sheet certainly acts as a condensing surface for a much longer period of time than it acts as a water evaporator.

Permeance

As in nearly all other buildings, the vapor permeance is negligible—less than 5% of the total load. The designers should invest time, energy and the construction budget to minimize ventilation and to prevent air leaks, rather than being concerned about vapor permeance, which has no measurable effect on the success of the project.

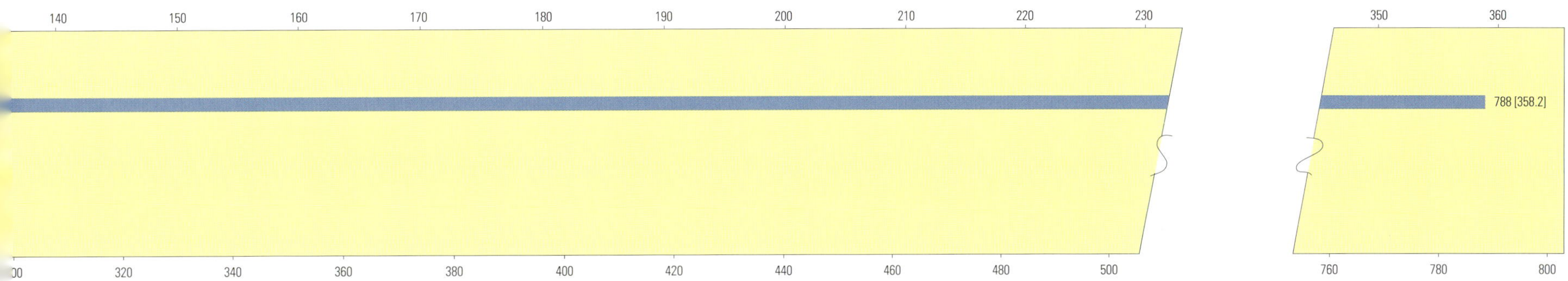

Suggestions & Cautions

We have divided the cautions and suggestions into three subsections, so the HVAC designer can easily highlight and pass the appropriate material along to the owner and the architectural designer.

Architectural design

Build a leak-tight exterior envelope

The building should not leak air. That statement sounds like our parents reminding us to eat vegetables and brush our teeth. But really, when an ice arena leaks air, bad things happen quickly. Energy costs go up, ice quality goes down and condensation corrodes the building and generates mold and mildew, some of which can be toxic.

The best wall system would be built of insulated structural panels, with gaskets and sealant between all joints, similar to the outdoor walk-in refrigerators that serve quick-service restaurants. But that standard of construction is not always affordable for large buildings that contain ice rinks. So the architectural designer must do the best job possible with lower-cost materials such as masonry walls and sheet metal siding.

The key details are at the seams between sheets, joints that connect masonry to steel, and the joints that connect walls to the roof. Outdoor air leaks in easily when those joints are either poorly designed, poorly built, or both.

Metal siding panels should have foam gaskets that run the full length of the seam. After each seam has been bolted together, it *also needs a generous bead of durable caulking* along the entire seam. Further, other joints need air seals. Seams between walls and roof, and between masonry and metal siding need an air seal such as expanding foam or some other means that will ensure that air cannot enter the building through those joints.

Here's the main point: The building should be built like a refrigerator. Metal-clad or masonry-sided buildings are designed to be water-tight, but they are *never* air-tight unless the architectural designer specifies adequate and practical sealing details, and the contractor installs those details with care. Otherwise, the owner suffers and humidity control is very difficult or impossible to achieve.

The vapor retarder goes towards the outside, and must be continuous

Water vapor always enters an ice arena from the outdoors. So the vapor retarder *must be on the outside.*

Some of the most depressing building failures in ice arenas have occurred because the contractor installed the usual plastic-faced fiber glass roll insulation with the non-perforated plastic facing the indoors rather than the outdoors. Given that fundamental error, incoming water vapor flows through the outer wall and insulation. It quickly reaches the plastic layer, which is cold because it's on the inside face in a cold space. Then moisture condenses inside the insulation, soaking it quickly. The soggy insulation is in close contact with the metal siding, which proceeds to corrode and fall off the building when its fasteners rust through.

So the siding is the outer layer. *Then* comes the continuous vapor retarder layer, with all of its seams taped together, and its edges taped to the exterior wall. *After* the vapor retarder comes the insulation, and on top of that, some form of interior finish. (Of course, all interior finish should also be hockey-puck-proof.)

Low-e ceilings are useful

Low-emissivity ceilings are quite useful in ice arenas. They help reduce the amount of heat emitted (radiated) from the warm roof towards the cold ice sheet. Radiating less heat reduces the load on the ice sheet, which in turn reduces the required size of the refrigeration plant as well as its operating cost.

Also, low-emissivity surfaces are good reflectors. So low-e ceilings can act as diffusers of indirect lighting. This provides a more pleasing illumination while reducing the electrical power needed to maintain that lighting level.

Further, when older buildings are retrofitted with low-e ceilings suspended below the rafters, they create an air space between the new ceiling and the bottom of the roof deck. That can be useful in

Fig. 28.8 Low-Emissivity Ceiling
The low-e ceiling is a great help in reducing the radiant heat load from the warm roof.

another way, *provided that the edges are left open*, so that air can move freely in and out of that space. Warm air finds its way to the space above the suspended ceiling so the heat load on the building goes down.

The load reduces because as the temperature of the bottom of the roof deck rises, the temperature *difference* between indoors and outdoors goes down. Less temperature difference reduces the amount of heat transferred into the building. Also, during the nighttime the indoor space cools down, which would usually bring the bottom surface of the roof deck closer to producing condensation. By keeping the air warm near that roof deck, the condensation risk is reduced.

However, all these benefits are replaced by problems if the suspended ceiling is sealed to the walls so that air cannot move freely into and out of the space above the ceiling. If that chamber is sealed, moisture gets trapped inside and condenses. That, combined with the warm temperature leads to quick corrosion and dripping rusty water, which is a structural risk that also generates the "mushrooms" that are a hazard to skaters. So with suspended ceilings, the designer must *avoid sealing the edges to the walls*.

Roof insulation need not be excessive

As long as there is a low-e ceiling installed, there is no great harm in having the indoor surface of the roof deck be warm. In fact, a warm indoor roof deck is less likely to produce condensation and it reduces the rate of heat transfer into the arena. So somewhat surprisingly, there is seldom any benefit to using more than the usual amount of insulation on the roof.

Any extra money being spent to increase roof insulation is probably better-invested by making sure the low-e ceiling is in place, making sure the building is air-tight, and heavily insulating the western wall.

Pay extra attention to the western side of the building

The western wall of the building will receive a very high radiant heat load in the afternoon and evening—the same time that the total sensible heat load is greatest on the building as a whole. The architectural designer can help minimize the costs of this load by paying extra attention to the west wall.

For example, windows in an ice arena add a great deal of equipment cost no matter where they are located, but in the west wall they are an especially poor idea. Radiant heat transmitted through windows causes "hot spots" on the ice, and creates glare that becomes hazardous to skaters and annoying to spectators.

Also, in low-budget arenas the designer might be forced to locate the rink close to one wall of the building. If that is the west wall, radiant heat can make it difficult to keep the ice temperature constant on that side of the rink.

During initial construction, the designer can minimize problems associated with the west wall by:

- Sheltering part of the western side of the building from radiant heat from the sun with landscaping such as trees and earth berms.
- Locating the parking lot on the *other* side of the building, or by planting bushes and trees, or by placing earth berms to keep heat from the hot parking lot from radiating upwards to the building walls.
- Increasing the insulation in the west wall
- Avoiding windows that throw light onto the ice, or at least sheltering them from high solar loads with overhangs and placing them *somewhere other than the western wall*.

Owners that have problems with western sides of existing buildings can reduce the problem by adding plantings outside and by hanging low-emissivity fabric from wires inside, across the western wall.

Exterior access through a closed lobby, with a vestibule

As much as possible, the ice rink space should be isolated from the weather. So it is very useful for the building to have a lobby between the outdoors and the arena space, and to install a double-door vestibule between the lobby and the outdoors. The lobby should also have doors that separate the arena from the warmer lobby area (rather than leaving an open passageway between these spaces).

With this arrangement, the hot and humid wind outdoors cannot easily flow into the cold arena space, which keeps operating costs to a minimum, and helps maintain the quality of the ice surface.

Air distribution

Don't dig holes in the ice

Since ice resurfacing machines often generate carbon monoxide gas which settles inside the rink, there is apparent logic in supplying air downwards to break up the monoxide-laden layer of cold air near the ice. But that's a very bad idea. Dry supply air removes ice by sublimation, carving "dents and trenches" into the surface. Even the gentle flow from ceiling-mounted propeller fans will have a highly-negative effect. Instead flush the monoxide-laden air out of the rink by opening the doors in the dashboards.

Desiccant dehumidification: high supply, low return

For desiccant-based systems, the best air distribution for an ice arena supplies the air *high* and towards the roof peak rather than downwards. The best location for the return is near the ground level.

With this arrangement the warmest air stays high, reducing the cooling load at lower levels. Also, the whole air mass moves generally in a circular pattern from high to low, which provides good ventilation efficiency for spectators while avoiding any high-velocity downward flow that would damage the ice.

Mechanical dehumidification: high supply, directed along the boards

With cooling based systems, the supply air tends to be colder, so aiming it towards the spectators is not usually helpful. Instead, aim the supply air down the length of the boards, away from direct contact with either the spectators or the ice.

Often, two units are installed inside the rink in corners that are diagonally opposed. The supply air is set at a slight angle, so the air in the arena moves slowly in a circular pattern in the horizontal plane for good air mixing. The return air is usually taken into the units at the same high level as the supply. But where it is practical, returning the air from lower in the space is a useful improvement.

Keep warm air off the boards & protective glass

It is not enough to keep warm air off the ice surface. The designer must also take care not to flow warm air against the boards or against the protective glass above those boards. They will conduct enough heat down to the ice to melt part of it, creating a narrow trench of liquid around the edge of the ice sheet. Again, the only useful place for warm air is at the higher levels near the ceiling to limit the potential for condensation.

HVAC components & controls

Avoid controls based on relative humidity.
Dew point controls save money and reduce risk.

In ice arenas, the purpose of controlling humidity is to avoid condensation. Therefore, the system should *control the dew point* of the air, not its relative humidity. In buildings with automated controls, an rh signal can be combined with the air's dry bulb temperature to provide a dew point signal.

Using an rh-based humidistat to control the dehumidification saves a few dollars at the time of installation. But that cost savings evaporates in the first few days of operation, because it results in overdrying the space when the air temperature is cold and underdrying when the air temperature rises. This damages the ice, and certainly damages the owner's finances. Here's how.

Consider an example using the NHL control conditions of 60°F, 40% rh. There is no benefit to drying below that 35°F dew point. But when air in the building cools down at night, a humidistat set at 40% rh will be calling for dehumidification even though the dew point falls way below 35°F. In other words, the humidistat is forcing the system to spend the owner's money every night for no benefit. Similarly, when the temperature rises above 60°F, the humidistat will in-

Fig. 28.9
Control On Dew Point—Not rh
In this building, the relative humidity was below 40%, but the dew point was still high enough to produce plenty of condensation. A dew point signal is the most appropriate for ice arena dehumidification systems. It saves operating costs and improves results.

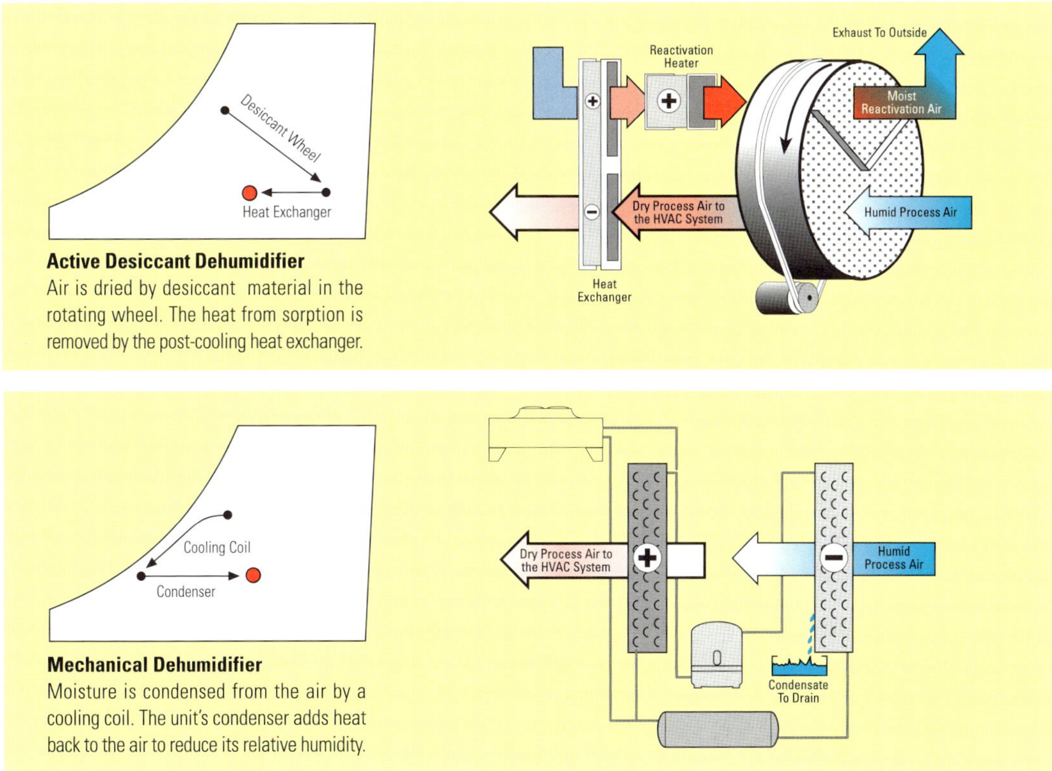

Active Desiccant Dehumidifier
Air is dried by desiccant material in the rotating wheel. The heat from sorption is removed by the post-cooling heat exchanger.

Mechanical Dehumidifier
Moisture is condensed from the air by a cooling coil. The unit's condenser adds heat back to the air to reduce its relative humidity.

Fig. 28.10
Active Desiccant and Mechanical Dehumidifiers
Both technologies have been used widely in ice arenas. In fact they are often used in combination when ventilation loads are high.

struct the system to stop drying. But the indoor dew point keeps rising, increasing the rate of condensation on the ice and increasing the probability of condensation on the structure itself.

If rh-based control is mandated for some other reason, then the relative humidity set point should be *re-set* when the temperature is expected to change, such as each night, and during periods of peak heat load. Resetting an rh control each evening and every morning can minimize its economic damage and limit the risks for the ice.

Avoid using the ice plant refrigeration system for dehumidification

Before the ice rink construction boom in North America during the 1990's, some ice refrigeration systems were equipped with air cooling coils to act as dehumidifiers. The low-temperature ice coolant was circulated through a heat exchanger or a mixing valve to make sure the air temperature in the coil was not below freezing. But using ice plant refrigeration has not been as successful as using dedicated dehumidifiers.

Problems include the fact that the full capacity of the refrigeration system is needed for the ice sheet at the same time as that capacity is needed for dehumidification. Also, using such a system requires extra costs in construction of the expensive refrigeration plant. Finally, it costs much more to remove water vapor with low-temperature refrigeration than it costs to do the same job with a stand-alone dehumidifier. Dehumidifiers cost less to install and to operate than the equivalent ice plant refrigeration capacity.

Active desiccant vs. mechanical dehumidifiers

Both desiccant and mechanical dehumidifiers have been used in ice rinks. Mechanical dehumidifiers have been successful for decades where moisture loads are low. They usually cost less per cfm [per l/s] than desiccant systems. Also, when maintaining air at or above a 45°F [7.2°C] dew point is the owner's preference, they often have operating cost advantages. Further, when the arena is small and loads are low, they can be installed without ductwork, minimizing installation expense.

Desiccant dehumidifiers usually cost more per cfm [per l/s]. But they continue to dry air at sub-freezing temperatures and will always dry it to much lower dew points than mechanical dehumidifiers. Consequently they need less air (fewer units) to accomplish the same moisture removal. So they may be less expensive to install, given the same moisture removal requirement. Their advantages increase if the owner prefers to keep the dew point below 45°F [7.2°C]. They have also been popular for installations where their deep-drying capability provides insurance against unexpected high moisture loads, such as those typical of hot and humid climates or in buildings with substantial air leaks.

Ice arena dehumidification is a special challenge because it requires removing moisture at low temperatures and low dew points, but also requires moisture removal from incoming ventilation air, which has high temperatures and moisture levels. Frequently, equipment suppliers will take advantage of the best characteristics of both types of dehumidifiers, with the mechanical technology removing the bulk of the moisture from the ventilation air, and the desiccant equipment removing the rest of the moisture and delivering air at very low dew points. These combined systems are often more economical than using only one or the other technology.

Each technology and each combination of technologies has advantages, and the merits are vigorously debated by the respective manufacturers. The designer can always clarify competing claims by asking to speak with owners of equipment in buildings similar to the project in question, and which are located in a similar climate zone.

Demand-controlled ventilation saves enormous costs

It's clear from figure 28.5 that ventilation air is by far the largest load component. The purpose of ventilation air is to ensure good indoor air quality. But if that goal can be achieved with less outdoor air, then most local building codes will be satisfied. Many ice arena owners choose to ventilate only when a carbon dioxide sensor indicates that the CO_2 level is excessive.

The CO_2 concentration is not a concern in itself, but it acts as a convenient indicator of the concentration of indoor air pollutants that are more hazardous and/or difficult to measure. For example, the gas of most concern to rink operators is carbon *monoxide*, produced by gas-powered ice resurfacing machines. Their engines also produce CO_2, and there is a relationship between the CO and CO_2 concentrations that can be used to trigger extra ventilation based on CO_2 rather than by measuring the CO directly. However, this relationship varies widely by type of equipment, and must be established through *careful measurements of both gases at the site* before the CO_2 sensor can be trusted as an indicator of CO concentration. Clearly, measuring CO directly is the preferred method.

Regardless of which sensor triggers the change, the ventilation system can be set up to bring in a very minimal amount of air on a continuous basis—perhaps 15 to 20% of the amount suggested by ASHRAE Std 62. Then the full amount is brought in when the sensor indicates the CO concentration is too high. The arrangement saves the huge cost of drying ventilation air at other times.

Load reducers can reduce construction costs

In many areas, building codes require the installation of enough equipment to handle the entire ventilation load continuously under full-load conditions, rather than allowing for time-weighted averaging of peak loads, or for demand-controlled ventilation.

In those locations, the designer can cut the construction cost of the dehumidification system by as much as 30 to 50% by using load reducers such as passive desiccant wheels. These enthalpy recovery devices use the dryness of the exhaust air to pre-dry the incoming ventilation air. They are especially useful in ice rinks because the air flow is large and the moisture difference between indoor and outdoor air is so great.

For example, if the outdoor design condition in Atlanta is 83°F, 133 gr/lb and the indoor control condition is 60°F, 30 gr/lb, then *66 lbs/h* of water must be removed from every 1,000 cfm of ventilation air brought into the building. With a passive desiccant wheel, that incoming load reduces to about *20 lb/h*.

[If the outdoor design condition in Atlanta is 28.3°C°, 19 g/kg and the indoor control condition is 15.6°F, 4.3 g/kg, then *30 kg/h* of water must be removed from every 470 l/s of ventilation air brought into the building. With a passive desiccant wheel, that incoming load reduces to about *9 kg/h*.]

For highly-ventilated ice arenas, passive desiccant wheels are nearly an economic necessity. The only reason *not* to use a passive desiccant wheel might be the cost of the ductwork to return the exhaust air to the wheel. In large buildings like municipal auditoriums, return ducting costs may be greater than the equipment cost saving because air is often exhausted from many remote locations. More information about load reducers is contained in Chapter 15.

Fig. 28.11
Ventilate In Proportion To The Need
Based on a 24-hour average, the occupancy in an ice arena is low, and the enclosed volume is very large. Ventilation can be minimal until a large number of spectators or ice resurfacing generates IAQ pollutants. Then the ventilation air quantity must be increased substantially.

Fig. 28.12 Load Reducer
Passive desiccant wheels can use the drying capacity of the building's exhaust air to partly dehumidify incoming ventilation air. This equipment can greatly reduce the ventilation moisture load.

Fig. 28.13
Consequences Of High Dew Points
The photo shows the thick ice that forms outside the rink near the refrigeration headers when the dew point is allowed to drift above 45°F [7.2°C]. Dehumidification prevents this hazardous and wasteful accumulation.

Be cautious with cold nightime air temperatures

In most ice arenas, the temperature is allowed to drop while the rink is not in use. It costs quite a bit of money to keep a large building warm when it contains an immense sheet of cold ice. But when the air cools down, surfaces get colder and become more likely to condense moisture. So don't shut the dehumidifiers off at night.

Install a make-up air system for lockers

Air from locker rooms and showers must be exhausted from the building. The designer may be tempted to "save money" by installing only an exhaust fan for these areas, and let the make-up air be pulled from the arena itself. But that's a very poor and expensive practice.

Make-up air brought through the arena is expensive because it must be dried and cooled deeply to avoid damaging the ice. Air for lockers and shower rooms just needs a bit of heat in the winter, and a little cooling in the summer. So it's much more economical to add a separate make-up air system to serve the lockers, showers and bathrooms than to add that burden to the main dehumidification system.

In either case, the HVAC designer should arrange the controls so the exhaust fan only runs while the lockers and toilets are in use. For example, an occupancy sensor with a time-delay relay can bring the system on when needed, and keep it running for a while after the rooms are vacant to make sure all humid and stale air is exhausted.

Use the heat from refrigeration condensers

Ice refrigeration continuously releases a large amount of heat though the refrigerant condensers. That heat can be used to pre-heat water for the building's boiler, or for heating the ground under the ice floor. This saves heating costs 24 hours a day, 7 days a week.

Use the jacket heat from engine refrigeration

Engine-driven refrigeration is sometimes used in ice rinks, especially where natural gas is available at low cost or where electrical power costs are exceptionally high. The engine releases a great deal of waste heat at a high temperature, and it operates continuously.

That heat can be used to pre-warm the reactivation air for a desiccant dehumidifier, which saves a considerable portion of the annual operating cost for that type of equipment. Or when mechanical dehumidifiers are installed, the engine heat can be used to reheat the supply air before it is delivered to the arena.

References

ASHRAE Handbook—Refrigeration 1998. Chapter 34 - "Ice Rinks."

Banks, Nancy J. 1990. "Desiccant dehumidifiers in ice arenas" ASHRAE Transactions, 1990, vol.96. pt.1.

Bradley, Todd J, Hamstra, S, Cooley, D., 1994. "Operating an ice rink year-round by using a desiccant dehumidifier to remove humidity." ASHRAE Transactions, Vol. 100, pt.1, 1994.

Cambron, Charles. 1995. "Arena system design" *Engineered Systems* magazine. March, 1995. Business News. Troy, MI.

"Drying wet ice rinks: A special report," *Contracting Business* magazine. May, 1999. Penton Publishing, Cleveland, OH.

EUN. 1999. "Commercial Building Award Winner: Skatetown Ice Arena, Roseville, CA." *Energy User News* magazine. December 1999. p. 20-21.

Kittler, Reinhold 2001. *Design guidelines for ice rink environmental control*. Dectron Internationale. Atlanta, GA.

Lenko, Brendan. 1999. "Engineered solutions for ice rinks" *Rink Magazine*, Volume 2, no.3, May, 1999. Liebert Publishing Ltd. Larchmont, NY.

Moore, D. 1995. "Ice making: the significance of water quality" *On Ice* newsletter. May-June 1995. University of Michigan Sports Facility Research Lab. 1000 S. State St. Ann Arbor, MI 48109-2201.

NHL. 1999. *Building conditions for games* National Hockey league instructions to clubs, April 19th, 1999, Colin Campbell, Senior VP and Director of Hockey Operations. NHL, New York, NY.

Overall, John. 1999. "Installation Monitoring Report: Development of information on the operating characteristics of desiccant dehumidification in an ice arena; Chesswood, Ontario" November, 1999. Caneta Research, 7145 West Credit Avenue, Suite 102, Building 2. Mississauga, Ontario, L5N 6J7.

Sutherland, Arthur. 2000. "Energy efficiency in the ice rink" *Air Conditioning, Heating & Refrigeration News*. August 7th, 2000. p.21. Business News Publishing. Troy, MI.

Vivian, Jack. 1995. "Humidity: it's more than a nuisance" *On Ice* newsletter. January-February 1995. University of Michigan Sprots Facility Research Lab. 1000 S. State St. Ann Arbor, MI 48109-2201.

Yang, Chunxin, Demotriku, P, Chen, Q, Spengler, J, Parson, A. 2000. "Ventilation and air quality in indoor skating arenas" ASHRAE Transactions, 2000. Vol. 106. pt.2.

Special Thanks

The authors are grateful to many experts who shared their experience and provided useful advice concerning ice rink humidity control, in particular, Todd Bradley and Brian Simkins, pioneers in the use of desiccant dehumidifiers in hot and humid climates, and also Jack Vivian of the Sports Research Laboratory as well as Claude L'Oiselle, Director of Operations of the National Hockey League.

Photos

Fig. 28.1 - © Steve Babineau Sports Photography

Fig. 28.2 - © Munters Dehumidification Division

Fig. 28.4 - © Munters Dehumidification Division

Fig. 28.11 - © Munters Dehumidification Division

Fig. 28.12 - © AAON, Inc.

Chapter 29

Dry Air Storage

By Lew Harriman

Dry Air Storage
Costs are low when the building is tight

Purposes for Projects

- Reduce heating costs
- Create usable space
- Protect unoccupied buildings
- Protect inactive machinery
- Protect bulk materials in storage

Fig. 29.1 Humidity Damage In Storage

The photo above shows the corrosion that can occur when power generation turbines are idle during maintenance. Storage in dry air protects these and a diverse range of similarly costly materials such as archives, agricultural products, electronic equipment and military hardware. The chief advantages of dry air over other storage alternatives are simplicity, low cost and speed of completion.

Fig. 29.2
Reducing Heating Costs in Warehouses
For products that do not need heat, dehumidified storage saves costs by reducing or eliminating the cost of heating or air conditioning the warehouse. Also note the lack of insulation in this Scandinavian installation. Dehumidification is fully effective without insulation.

Fig. 29.3
Creating Usable Space In Old Buildings
Lightweight plastic sheeting over a wooden frame contains dry air effectively. So poor-quality space in existing buildings—such as damp basements—can be upgraded to safe storage space very quickly at very low cost.

Dry Air Storage

Disposable baby diapers, military ammunition, photographic film, power plant boilers and Leonardo daVinci's mural "The Last Supper" all share one characteristic—they last longer in dry air.

When one examines the worldwide sales of commercial dehumidifiers over the last 50 years, it quickly becomes apparent that storage has been the largest single application for the equipment. In fact, even residential dehumidifiers are usually purchased to make the air in damp basements safer for storage.

Designers in other parts of the world are sometimes more familiar with dry air storage technology. Perhaps this is because in the past, energy costs in North America were so low that the operating cost advantage of dehumidified storage over heated and air conditioned storage was less compelling than in other countries. In recent years, using dehumidifiers in commercial and institutional storage in North America has become more common as the cost of heating and air conditioning has risen.

Purpose of Humidity Control

The reasons for using dry air storage are as varied as the objects being stored. There are five circumstances where dehumidifiers are likely to be useful in storage areas. Each application has a different purpose for controlling humidity.

1. Reducing warehouse heating costs in very hot or cold climates

The relative humidity of air can be changed by changing its temperature. So in small spaces, it is possible and often cost-effective to control relative humidity by controlling air temperature. But that method becomes less practical as the volume of the space increases, or as the climate becomes more extreme.

As sensible loads rise, heating and cooling equipment becomes expensive compared to a dehumidifier, and the cost of *insulating* the structure rises as well. (Insulation has no effect on the absolute moisture content of the air, so it is not needed when dehumidifiers are used to control humidity.) At some combination of size, required humidity level and outdoor climate, it becomes less expensive and simpler to control the relative humidity with a dehumidifier than with a heater. Figure 29.2 shows such an installation. In general warehousing, dehumidification has advantages over heating for:

- Very hot and humid climates, where further heating is impractical or it creates an unsafe condition for people or products.
- Very cold climates, where the annual cost of heating is excessive.
- Large structures with little or no insulation, where heating and cooling equipment and better insulation would be very costly.
- Products which lose value when they get hot.
- Objects or commodities with a high mass, which do not change temperature quickly, so that heaters would be enormous or ineffective, or both.

In all cases, the principal purpose of the humidity control project is to preserve the value of the stored objects. So control levels will depend on the sensitivity of the stored objects. But for commercial products, the optimal control level is likely to be between 40 and 60% rh. Objects that can tolerate humidity above 60% are seldom stored in dehumidified spaces, and materials that need humidity kept below 40% are usually stored in special industrial buildings.

2. Creating usable storage space in existing buildings

In many buildings, space is available but high humidity damages stored materials. Examples include museum basements, or unheated buildings that leak large amounts of air. In these situations, owners build interior humidity-controlled rooms. The surrounding building acts as the structure, as seen in figure 29.3. The interior room simply serves as a container for dry air. Where fire codes permit, the room walls can be as simple as steel or wood studs covered with flexible plastic sheeting, provided that all seams and joints are sealed with tape to form a reasonably air-tight enclosure.

Interior dry rooms can be erected very quickly, since they have no heavy structural requirement, and because little or no foundation work is necessary. This approach provides a high-quality storage environment at very low cost on a tight schedule.

3. Protecting historic structures and unoccupied buildings

In some cases, buildings themselves need protection. Dehumidifiers are often used as an alternative to heating or cooling a building when it is unoccupied, as shown in figure 29.4. Such museum buildings will deteriorate from high humidity when they are closed during the off-season.

When closed buildings are unoccupied and unheated in cold climates, the relative humidity indoors can exceed 75% for months. During those months, building materials absorb moisture, creating expansion and mold problems. Then during spring, rising temperatures adds energy that accelerates fungal growth. That's why vacation homes in cold climates smell moldy after they have been unoccupied for the winter season.

Dehumidifiers are also used in hot and humid climates, to protect buildings when there is no need to cool them. Examples include second homes, boats, churches or winter-season houses that remain vacant for much of the year.

In both cold and hot climates, the purpose of the project is to keep the structure intact, and free from mold and mildew while avoiding heating or cooling costs whenever there is no need for human comfort. 50% rh is usually the design goal for such buildings. The challenge is to ensure that air circulates throughout the building, so that no pockets of high or low humidity develop as the indoor temperature swings in response to the outdoor environment.

4. Protecting valuable inactive machinery

Military equipment provides the classic example of protecting machinery. Such equipment is designed to be temperature-tolerant and durable, but it has metal surfaces and electronics that are subject to corrosion when the equipment is not in use. Keeping military equipment dry greatly reduces its cost of maintenance, and increases both its mission-availability and its mean time between equipment failures. Savings have been rather dramatic.

For example, the Swedish Army studied the effect of dry air storage of active equipment for a period of 8 years, beginning in 1978. That study established the fact that their widespread use of dry air storage had reduced the entire army's annual cost of maintenance by about 20%, based on the costs typical of the previous 10 years (Swedish Army, 1997). Similarly, the United States Navy Audit Service determined that where dry air storage had been used for military aircraft, the mean time between failures was increased by 33%, the maintenance hours were reduced by 29% and the mission-capable rates were increased by 5% over rates for aircraft that were not protected with dry air. In other words, one extra aircraft was mission-ready for every 20 in the unit. (U.S. Navy, 1994)

Another classic example is the protection of power plants during maintenance outages as shown in figure 29.5. By flooding the key components with dry air, corrosion is eliminated. Also, the insulation of motors and generators cannot absorb the moisture that allows the current leakage that reduces generating capacity. In this application, dry air allows the plant to come back on line more quickly than either 'wet layup'' or protection with corrosion inhibitors. At periods of peak electrical demand, the value of such quickly-generated power can be measured in the millions of dollars per day.

In machinery-protection applications, the principal goal is to prevent corrosion, so there is seldom a lower limit on humidity. Drier is usually better. On the other hand, there is no great benefit to drying below 40% relative humidity, so there is seldom any advantage to investing additional funds for drying below that level. And in many cases, 45 to 50% rh is also tolerable, and the higher control level costs less to maintain.

5. Protecting bulk materials or metal parts in storage

In developing countries, product losses in agricultural storage can exceed 25% of the annual harvest. And in bulk storage of sugar

Fig. 29.4
Protecting Unoccupied Buildings
Seldom-occupied vacation homes and historic structures are often protected with dehumidification alone. This avoids the cost and energy waste of heating and cooling when the only needs are to avoid damage from mold or from humidity-related expansion and contraction of paintings or woodwork.

Fig. 29.5
Protecting Inactive Machinery
By flowing dry air through inactive generating equipment, the machinery and wire insulation can be fully protected, but also reactivated in a matter of hours rather than days or weeks, and without any need for the usual toxic protective chemicals.

Fig. 29.6 Protecting Metal Parts
Dry air protection costs less than either heating or air conditioning for metal parts or for bulk commodities in storage.

and fertilizer, excess humidity clogs machinery, slowing or halting production altogether. Finally, in auto factories, corrosion of metal parts while in production leads to high warrantee costs and to production interruptions that may cost millions of dollars. Under all these circumstances, there is a compelling economic reason to control humidity, but no need to control temperature. So dehumidifiers have been widely applied to reduce the cost of humidity-related damage. (Volvo, 1983, Renault, 1987)

Figure 29.6 shows a smaller installation where dry air protects machined steel. But often, these applications involve very large warehouse-style buildings, with frequent openings of large doorways. Open doors make it difficult to keep the humidity down to low levels when the weather is rainy and windy. When necessary, pre-drying the building to very low levels during night time provides a large "reserve" of dry air that helps keep humidity low during periods of peak door activity.

Also, the purpose of such projects is usually is to reduce the magnitude of the problems, since factors other than humidity also influence losses in storage. So the precision of the control level for commodities such as corn, rice, sugar, fertilizer and metal stampings is not as critical as when storing something more sensitive, such as lithium metal, which generates explosive hydrogen gas when relative humidity rises above 6% rh. Usually, keeping relative humidity below 50% provides adequate protection for bulk commodities, but every product is different, so the needs of any stored product must be defined by its owner.

Control Levels and Tolerances

The nature of the product and the nature of the potential damage will determine the appropriate humidity control level. Usually, the level will be between 40 and 60% rh, but the designer has several information sources to help define the optimal range more precisely.

First and most important is information from the owner. The owner lives with the problem, so he is usually in the best position to understand what humidity level will prevent it. Next, in the case of stored materials such as photographs, grain, sugar and fertilizer, the producer of the material itself has probably determined the humidity levels that are best for long-term storage. Absent that source, the technical associations for such products can assist the designer with general industry guidelines. Finally, Chapters 4 through 9 of this book provide the designer with some information about the effects of humidity on different materials.

In the end, of course, the HVAC designer is not usually responsible for solving the whole problem. He or she is simply charged with making sure the HVAC system keeps humidity within a defined range. So the owner—not the designer—must determine what control level will achieve the desired result.

Maximum rh vs. constant rh

In storage applications, it is especially important for the owner to define whether the relative humidity must *remain constant*, or whether it should simply stay *below some maximum*. The answer to this question will probably not effect the size of the dehumidifier, but it will determine the need for additional equipment and the need for additional air circulation throughout the space.

Consider two cases: museum storage for antique carriages in a large barn during winter, compared to factory storage for steel auto body panels that are not yet painted.

The problem for the auto body parts is that they rust. Drier air is better, so there is no need to keep humidity within a defined range—just keep the air as dry as possible. The dehumidifier can simply turn on when the humidity in the storage racks exceeds some value determined by the owner, perhaps 45% rh. After the unit turns on, the humidity near the dehumidifier falls way below that level well before the rh changes near the humidistat—but so much the better. Drier air means less corrosion.

On the other hand, the carriages in the barn are sensitive to low humidity as well as to high humidity, and are also sensitive to fast changes. Fast humidity changes create enormous stress on panels and joints. Thin wooden pieces will crack, damaging the carriages beyond repair. So in the storage barn, the designer will need to circulate and mix the warmer and cooler air constantly, so the *relative humidity stays more uniform* throughout the building. And if the barn is heated to allow comfort for visitors during winter months, humidifiers will be needed, and the building will need both a vapor barrier and insulation.

In other words, controlling humidity for the carriage barn will be far more complex and costly than in the factory. Wooden objects need relative humidity kept above some minimum as well as below some maximum, and the designer must also prevent fast changes within the allowable rh range.

That's why the owner must define the need as either keeping the humidity *below some maximum, or* keeping the humidity *constant* within some range. The answer to this question has large cost implications, so the designer and owner need to be clear on this point at an early stage in the project.

Moisture Loads

In most commercial buildings, ventilation air creates the largest part of the moisture load. Storage applications are very different. There is rarely any requirement to ventilate, since the space is seldom occupied. In storage, much more load comes from door activity, especially when the storage enclosure is outdoors.

In the load estimate shown by figure 29.8, we assume a fabric-covered structure, similar to that shown in figure 29.7. Our hypothetical structure measures roughly 150 ft long, by 50 ft wide, by 25 ft high, for a total enclosed volume of 187,500 ft^3. [45.7m x 15.2m x 7.6m, or 5,280 m^3] We assume it is located in Atlanta, Georgia. The climate in Atlanta is mostly warm and humid, but the city still has several months of cold weather. Outdoor design conditions are 83°F, 133 gr/lb in the summer, and 20°F, 16 gr/lb during the winter. [27.8°C with 19.0 g/kg during the summer and -6.6°C with 2.2 g/kg during the winter.]

The load estimate assumes the owner has asked the designer to keep humidity below 40% rh all year long. Since there is no heating or cooling, we take 40% rh at the outdoor design temperatures as the summer and winter control points indoors—maximums of 67 gr/lb and 6 gr/lb respectively. [9.6 and 0.9 g/kg]

Fig. 29.7
Storage In Warm, Humid Climates
Fabric structures are popular for dehumidified storage because they are relatively airtight, low-cost and can be installed quickly. The moisture load estimate in figure 29.8 assumes such a structure.

Fig. 29.8 Moisture Loads
These estimates highlight the fact that moisture loads exist during winter as well as summer. Also, the load increases tremendously when the door must be opened, even when that door's open area is reduced by installing a curtain of overlapping plastic strips.

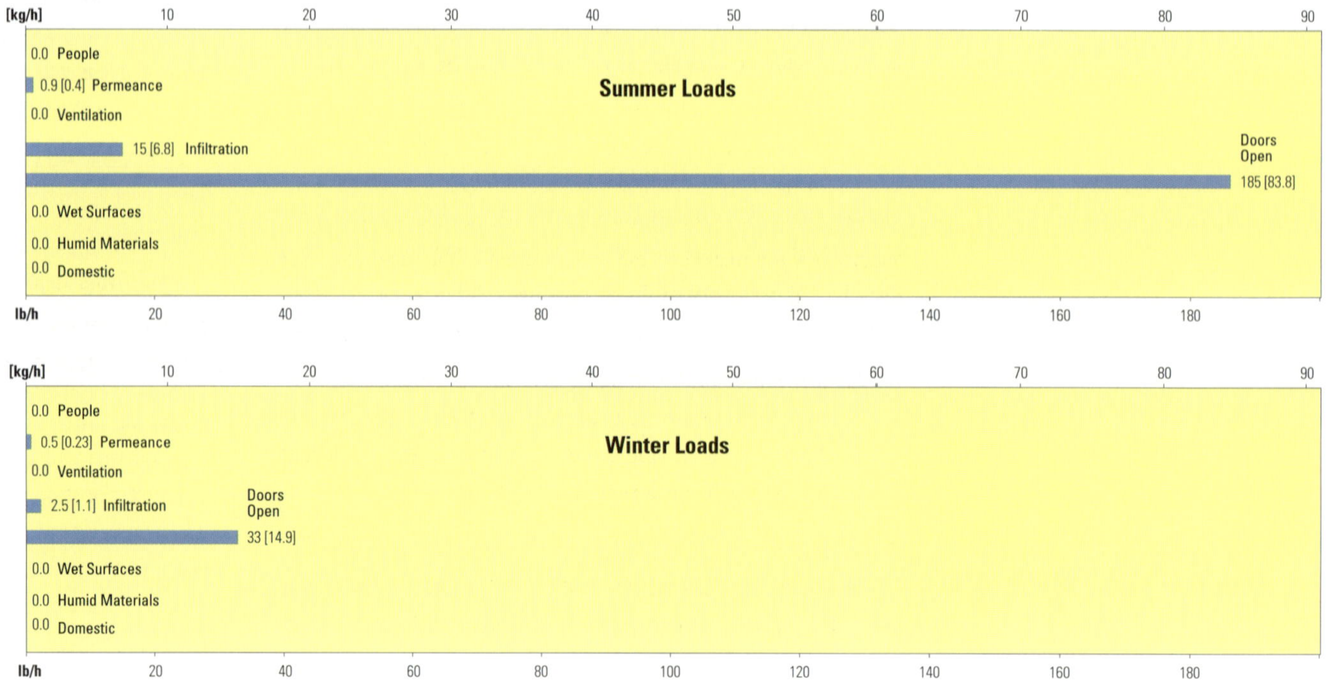

Infiltration

This type of building is very airtight compared to buildings with windows built with either metal siding or masonry walls. In fabric structures, there are no joints around windows, nor any joints between walls and roof. The exterior membrane is continuous, except for where the walls meet the floor slab, and around the entry door. Those joints are not likely to be completely sealed. So the load estimate assumes a leakage similar to "tight" commercial-grade construction for the building as a whole. The leakage will probably be much less than this estimate allows, provided that the owner takes the trouble to make a reasonable seal between walls and floor.

Doors

Doors either leak nothing at all, or a tremendous amount—depending on whether they stay closed or must be opened to move product. Door activity is by far the most important load element for storage applications. This load estimate assumes that wind will be pushing rainy air through that open doorway at a speed of 175 fpm (2 mph, or 0.9 m/s), but that there will be a set of overlapping plastic strips in the doorway to reduce the net open area.

The owner can choose between a dehumidifier with enough capacity to handle the large load, or one which can only accommodate the load when doors are closed. The smaller unit will cost less, but humidity is likely to rise above set point when the doors are open, especially if the weather is humid and windy at the same time. The designer can be guided by the owner's needs: what would be lost or damaged if humidity were above set point for a few hours per year?

Another point worth emphasizing is the winter moisture load. The cold weather moisture load in Atlanta is still considerable. The dehumidifier needs to be able to remove that load as well as the load that occurs during the peak summer design conditions.

Permeance

In storage applications, the loads from people and ventilation are usually so low that permeance becomes a slightly more significant part of the total load. But note how much larger is the infiltration load. Once again, the designer's time is better spent making sure the structure does not leak air rather than trying to find a less permeable material.

That said, it is still important to avoid having very porous materials for walls, floors and ceilings. For example, any unfinished masonry such as bricks, concrete block or poured concrete floors should be sealed with two coats of a high-quality, vapor-retardant paint. The exact perm rating of the paint is not especially important—any paint sold as a vapor retarder will be adequate to cut vapor permeation by more than 80% compared to unfinished masonry. Two coats of good paint costs very little, but greatly reduces the load on the dehumidifier. A continuous coat of paint and floor sealer also helps prevent humid spots from developing underneath stored material, or behind stacks of product that are pushed against masonry walls.

People and ventilation

There are no people in this structure unless they are moving material in or out. The amount of moisture they release during a 5-minute visit is not enough to matter in the load estimate. More importantly, there is no need to provide the structure with ventilation air. Without continuous ventilation, the moisture load is quite low.

Suggestions & Cautions

Dry air storage is considerably different than most commercial HVAC projects. Storage systems are simpler, and they cost much less. But there are several suggestions and cautions that are may not be obvious to the first-time designer.

Define the lowest temperature at which dehumidifiers must perform

Both desiccant and mechanical (cooling-type) dehumidifiers have been used for storage applications. But in one respect, the two equipment types are very different: desiccant units continue to dry the air at low temperatures, where cooling-type units do not. Therefore it is very important to define the lowest temperature at which the unit needs to operate.

This imperative is not usually obvious, because the peak moisture load is likely to occur at a fairly high outdoor temperature, so that design extreme will be used to size the equipment. But often, dehumidification must continue through cold months. To be certain the installation will be successful, the designer should define *both* operating extremes: the moisture removal requirement during the warm weather conditions that generate the peak moisture load, and *also* the moisture removal required when the outdoor temperature is at its lowest point. Based on those extremes, the manufacturers will be able to configure the equipment for moisture removal under all relevant weather conditions. Figure 29.9 shows how different locations can have very different climates.

Usually, cooling-type units have economic advantages where the weather stays warm and humid all year long, because in those conditions their operating cost is quite low. Desiccant-type units will have advantages in colder climates, because they remain functional in cold weather. Defining the performance required during both operating extremes will ensure that the owner ends up with optimal equipment for his particular circumstances.

Equipment size depends on wind pressure

When all other factors are equal, the cost of the installation depends on how hard the wind blows against the outside walls during rainy weather. Air infiltration, sometimes called unintended ventilation, is the largest component of the moisture load. And infiltration rises with wind pressure. Consequently, the site and any surrounding structures or vegetation will influence the cost of the installation. A site visit is very useful before design begins. Any building located in an open field or near the edge of a lake will have higher wind pressure than one located in the middle of a forest, or one sheltered from wind by nearby buildings.

This also explains why the costs are so low for keeping an *inside* room dry; there is little or no wind pressure to drive humid air into the storage enclosure.

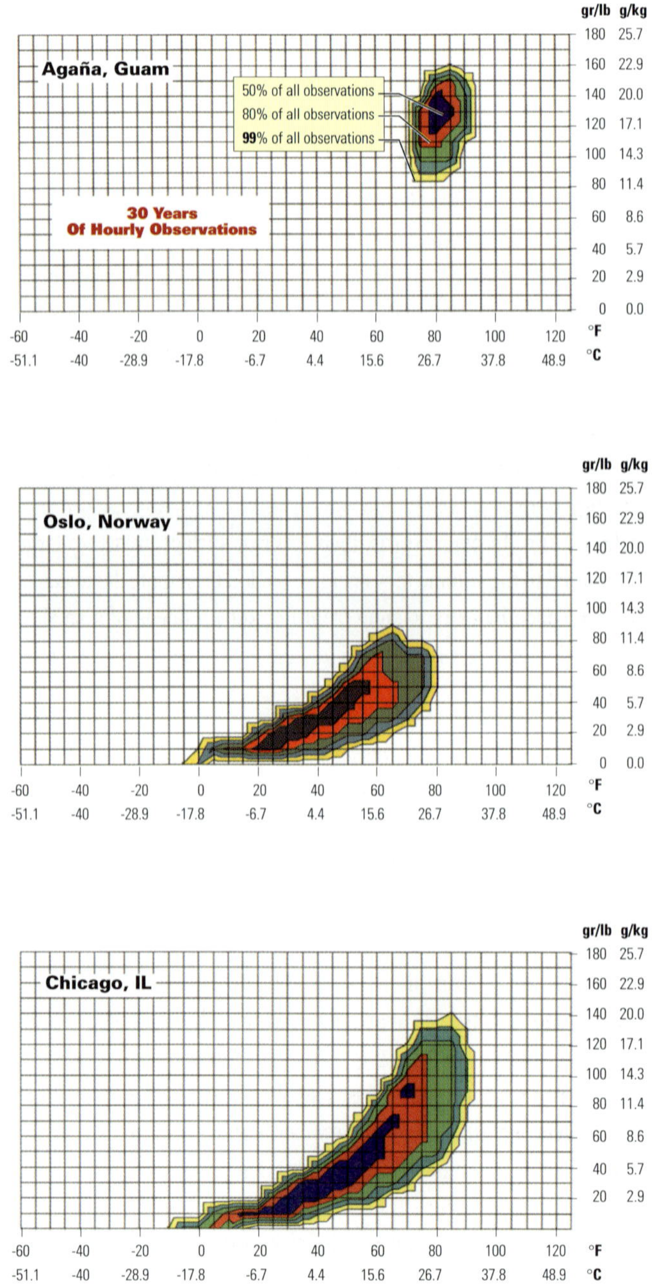

Fig. 29.9

Climates Influence Equipment

For unheated dry air storage in Guam, mechanical dehumidifiers will work well all year long at high capacity and high efficiency. But in Oslo, the cold temperatures heavily favor the use of desiccant-type dehumidifiers. In Chicago, the mixed climate provides each technology with different advantages.

Mechanical-type units have a much lower operating cost during summer, but desiccant-type units can dry the air during all seasons.

These 30-year Psychrometric Climate Summaries are published in the US DOD Engineering Weather Data.

Fig. 29.10

Dehumidifier Size Depends Mostly On The Exterior Surface Area Of The Enclosure

These two small dehumidifiers (1125 cfm or 529 l/s each), can easily maintain 50% rh in the large aircraft hanger shown above. Its exterior surface area is small compared to the cubic volume it encloses, and the fabric makes the structure nearly airtight.

Smaller, leakier structures will need proportionately larger dehumidifiers, because moisture load is largely dependent on exposed surface and on air infiltration rather than on interior volume.

Equipment size depends more on the storage structure's exposed surface area than on its internal volume

When dehumidifier manufacturers help designers estimate the required capacity of the equipment, the first question is likely to be: what is the internal volume of the controlled space? Surprisingly, the larger the volume, the smaller the dehumidifier will need to be in relation to that volume.

For example, given equal "leakiness" and the same humidity control level, a small space may need a dehumidifier air flow capacity that allows the entire internal volume of the building to be dried once every hour. But a very large building may only need to have enough dehumidifier air flow to dry the air once every 12 hours.

This is because the source of most of the load is air leakage through the exterior wall, and small spaces have a larger surface to volume ratio. In other words, their exterior surface areas are large compared to the volume they enclose. So don't estimate the needs of a large space by extrapolating from the known needs of a smaller space. The larger structure will be much *less* costly to dry per cubic foot of enclosed space. Photos in figure 29.10 show an example.

Combining heating and dehumidification

In cold climates, both dehumidification and heating can be used to control relative humidity. Usually, the technologies are used in combination. It is seldom practical to install no heating at all in commercial buildings, even when humidity is controlled primarily by a dehumidifier.

But the owner can minimize costs by only heating the building to the minimum temperature needed to protect the products, to prevent water pipes from freezing, and to provide an acceptable working environment for employees. Usually that temperature is somewhere between 40 and 50°F [between 5 and 10°C]. This practice keeps heating equipment cost to a minimum. And in many climates, keeping the temperature low does away with any need for insulation, allowing considerable construction cost savings.

Also, when comparing the annual cost of operating heaters to the cost of operating dehumidifiers, it may be helpful to know that in the past, owners report that in most installations, dehumidifiers operate for 20 to 50% of the hours in a year. Where units operate for only 20% of the time the units are larger, and have extra capacity for humidity peaks. Operating hours of 40 to 50% of the year probably indicate smaller and less-costly units are installed. Each building is different, and it is nearly impossible to predict the operating hours of any given unit with precision, but historical records kept by owners in hundreds of military and civilian installations show average run-hours between 20 and 50% of the hours in a year.

Locate humidistats with some care

In large spaces, there may be a considerable time lag between when the dehumidifier turns on versus when that change is sensed by the controlling humidistat. This has several implications.

In spaces that must be kept within a defined range of humidity instead of simply below some maximum, it will be important to circulate the air continuously in all parts of the space, so that big differences in relative humidity cannot develop.

Also, the humidistat should be located near the objects under protection, so that as much as possible the needs of the objects are controlling the humidity level. For example, if the humidistat is located near the *door* of a large warehouse instead of near product storage racks, the dehumidifier might operate needlessly. Similarly, if the humidistat is located *high up* in a tall warehouse, the higher temperature air at that level could "fool" the device into keeping the dehumidifier off, even through the relative humidity at the cooler lower levels remains excessive.

Finally, the humidistat must be located well away from the supply air discharge from the dehumidifier. Otherwise, the control will switch the unit on and off quickly, and the humidity in the rest of the building will not be controlled at all. In summary, always try to locate the sensor where the relative humidity matters the most.

Prevent condensation on cold roofs and floors

When NATO armed forces first implemented large-scale dry storage for military equipment, the organization built lightweight steel buildings to contain the dry air. Some of the stored equipment developed obvious rust stains, in spite of the fact that data logs showed that indoor moisture levels never exceeded 50% rh. Similar problems occurred in grain storage sheds in nonmilitary installations. Some of the grain developed mold, even though records showed the relative humidity was always below 40%.

Both problems were eventually explained by the phenomenon of roofs cooling by radiation. During cold, clear nights, roofs lose heat by radiation to the cold night sky. But at the same time, air inside the building was relatively warm, so the humidistat sensed (and the dataloggers recorded) a low relative humidity. Unfortunately, the air inside the buildings was still at a dew point high enough to condense moisture onto the inside surface of the cold roofs. Moisture dripped down from the roof to generate rust on the military equipment, and to feed mold in the grain storage buildings.

The solution was to install a surface temperature sensor on the coldest surface in the building—in these cases, on the underside of the roof. Then the control system was adapted to turn on the dehumidifiers in *either* of two circumstances: whenever the relative humidity is too high, *or* whenever the air's dew point rises to within 5°F [3°C] of the inside surface temperature of the cold roof.

Similar problems can also occur on floors and in basements in cold climates during the springtime. The winter-chilled ground remains cold, so the concrete floor stays cold. At the same time the dry bulb and dew point temperatures rise when spring rain humidifies the air and sunshine heats it. The indoor relative humidity may still be low, because the air is gaining heat. But since its dew point is also rising, condensation forms on the cold floors.

In that situation, moisture appears to be coming up through the concrete from below the ground, and the problem is blamed on a lack of a vapor barrier under the concrete. But the real problem is the rising dew point of air inside the building. As with roof drips, the problem can be avoided by using a dehumidifier to keep the air's dew point lower than the surface temperature of the floor.

Consider multiple units for large enclosures

In small rooms, and where the humidity must simply be kept below some maximum, there may be no need for air distribution ductwork. In larger spaces, the air must be mixed and distributed around the room or the building to avoid areas of high and low humidity. But even in such large spaces, multiple small dehumidifiers can be installed at less cost than a single large unit with extensive ductwork.

Multiple units provide better reliability. If one unit goes down for some reason, another unit can continue to operate and remove the load by operating for more hours than usual. Also, in large spaces the moisture loads will be quite different in each part of the building. So operating a single large unit to remove a small load in one isolated part of the building is less efficient than placing small units around the building to remove local loads at lower cost.

No benefit in keeping empty space dry

Figures 29.5 and 29.11 show a useful approach to minimizing the cost of dry air protection. In both cases the equipment is already inside a weather-tight building. Also, the equipment is only sensitive to high humidity *inside the machinery*—its exterior resists any negative effects of moisture.

In those situations, the owners simply connected dry air to the equipment itself instead of dehumidifying the entire volume of the enclosing building. The equipment is smaller and less expensive.

A similar approach is used in power plant installations. High-value spare valves and motors are stored in a small dry room built from simple studs and plastic sheeting inside the much larger space of the generating hall. A very small dehumidifier is installed in that dry room, avoiding the need to dry out the entire generator hall (which leaks outdoor air freely).

Fig. 29.11
Dry The Air In The Object Rather Than The Whole Building
When the space is much larger than the protected materials, sometimes it saves money to just inject dry air into the objects, or to cover the objects with a tarp into with dry air can be injected. This avoids the need to buy large dehumidifiers, or to invest extra money to tighten up a large building that leaks air freely.

Fig. 29.12
Shipboard Dry Air Storage
Marine environments present a special challenge, because the equipment must be able to dehumidify properly in any climate the ship is likely to encounter—from tropical to arctic weather. Consequently, desiccants are nearly always used in these applications.

Avoid dead spots with low-cost circulators

In spaces of all dimensions, it is useful to keep the humidity reasonably similar throughout the space, even when uniformity is not an absolute requirement. When the space is fairly open, air can be mixed at low cost by freestanding circulation fans, avoiding the higher cost of installing air distribution ductwork.

Such fans can operate intermittently, perhaps on a time clock set to ensure operation in late evening and early morning. That's when outdoor temperature changes can make the relative humidity indoors less uniform in any building that is neither heated nor cooled.

References

ASTM. 1989, *Air Change Rate and Air Tightness in Buildings*, ASTM STP 1067, (ISBN 0-8031-1451-6) American Society of Mechanical Engineers, 1916 Race St, Philadelphia, PA 19103 www.astm.org

Brault, Pascal. 1987. "Renault dry air storage, Douai, France." AB Carl Munters, Sollentuna, Sweden.

Harriman, Lewis G. III, *1985*. "Preventing metallic corrosion in storage areas." Proceedings of the International Symposium on Moisture & Humidity, Washington, DC, April 1985. (pp: 949-964) Instrument Society of America, Research Triangle, NC

Jeansson, Jan. 1982. "Comparison between different stores for pressed steel parts." AB Volvo, Göteborg, Sweden.

Naval Audit # 94-0016. 1994. *Dehumidification of in-service aircraft*. US Naval Audit Service, Pensacola, Fl. USA.

Sharp, Sandy. 1990. "Protection of control equipment from atmospheric corrosion." *Materials Performance* magazine, December, 1990. pp: 43-48. National Association of Corrosion Engineers (NACE International. Houston, TX. USA.

Swedish Army Material Command, 1997. "Evaluation of costs and savings for dry air storage." Swedish Defence, Material Administration, Stockholm Sweden.

Treschel, Heinz and Lagus, Peter. 1986. *Measured Air leakage of Buildings*. (ASTM STP 904) American Society of Mechanical Engineers, 1916 Race St, Philadelphia, PA 19103 www.astm.org. ISBN 0-8031-0469-3

U.S. Department of Defense, 2000. *Engineering Weather Data CD-ROM. (Worldwide)* National Climatic Data Center, Asheville, NC USA (http://www.ncdc.noaa.gov/

Special Thanks

The author is grateful to those who provided an excellent foundation for this chapter based on their thousands of dry air installations over 25 years, in particular: Gerard Mojet, Hansi Krüger, Inge Nodfors, Bo Ostrom, Leo Djursner and Chip Crotty.

Donated Photos & Diagrams

Fig. 29.1 - © Munters USA

Fig. 29.2 - © A. B. Carl Munters

Fig. 29.3 - © A. B. Carl Munters

Fig. 29.4 - © A. B. Carl Munters

Fig. 29.5 - © Pro-Tec Inc.

Fig. 29.6 - © A. B. Carl Munters

Fig. 29.7 - U.S. Marine Corps

Fig 29.9 - U.S. Department of Defense

Fig. 29.10 - U.S. Marine Corps

Fig. 29.11 - © Gerard Mojet

Fig. 29.12 - © Pro-Tec Inc.

Chapter 30
Laboratories

By Lew Harriman

Fig. 30.1 Lab Exhaust Hoods Demand Make Up Air
Laboratories use more outdoor air than any other type of commercial building. The key to controlling humidity in a laboratory is to control the dew point of the makeup air.

Laboratories

A laboratory HVAC system must keep people and animals safe from hazardous vapors and airborne particles. The designer must do nothing in relation to controlling temperature, humidity, or operating costs that might interfere with that purpose.

The HVAC system helps minimize risks from particles and gases by keeping air moving *from* people *towards* the exhaust, so that contamination is less likely to "swim upstream," to enter the occupants' breathing zones or to settle on their clothing. Air always moves from areas at high pressure towards areas with lower pressure. So the designer is principally concerned with defining and managing air pressures in the building, and slightly less concerned about maintaining precise levels of temperature and humidity. With that hierarchy of concerns in mind, the designer can proceed to the issue of how best to control humidity.

The principal task is to control the amount of moisture in the incoming make-up air. The outdoor air volume is so very large that any change in its moisture level makes an immediate change in the humidity level in the labs.

Purposes of Humidity Control & Control Levels

In laboratories, the value and importance of humidity control varies widely. For example, during 1999 in the United States, one pair of genetically-modified laboratory mice were sold for $750,000. The figure is especially startling because several thousand mice can be housed in a very small lab. It would be unfortunate if any HVAC system allowed those creatures to develop a skin rash, or allowed unhealthy fungus to grow in the bedding of their cages. Extremely high or low humidity would lead to those costly problems. In contrast, a teaching laboratory in a high school seldom has any need for humidity control, apart from the comfort of its infrequent occupants.

From such examples, it becomes clear that owners are likely to differ widely in their purposes for humidity control. Also, the need for humidity control may vary within the same building. It is very costly to assume that because one lab in a new building does not need humidity control, the lab in the next room is similar. To avoid expensive surprises, the HVAC designer should always seek guidance from the users of the space before finalizing system design concepts.

Finally, in most research buildings the work is eclectic. It is difficult to predict future activity or the future need for humidity control. The information below is only intended to serve as a starting point for discussions with the owner, who must define the purpose for humidity control, and who must also define the humidity levels that will achieve that purpose.

Biological laboratories

Often, the purpose for controlling humidity in biological labs is to simplify the task of maintaining a stable temperature. Temperature stability is usually more critical than humidity control, because the speed of biological reactions can depend in part on how much energy is received from the air. Fluctuating temperature can change the rates of reactions, and therefore change the research result. Specifications for the allowable temperature gradient (the difference between air temperatures at any two points in the same room) can be as tight as 1 or 2°F [0.5° to 1.0°C] (Harris, 1992)

Consider a typical problem: temperature in the lab begins to fluctuate in springtime or fall, as outdoor humidity changes with warm rain and cold nights. The indoor temperature fluctuates because the cooling coil performance varies just a bit as the moisture load rises and falls. With more condensation, the coil must remove more heat. With less condensation, the supply air temperature falls. Both situations tend to destabilize the indoor air temperature.

This problem can be solved by dehumidifying the incoming air to a dew point *below the summer supply air temperature*. Dry incoming air will eliminate nearly all condensation on the main supply air cooling coils, making it easier to maintain stable temperatures.

Biological research seldom needs whole-room humidification beyond the human comfort range. Most projects in biological labs can be satisfied by humidifying incoming air to a dew point of 35°F [1.7°C] during dry seasons.

Labs developing manufacturing processes for pharmaceutical or biological products often need a much tighter control of humidity, and usually at a lower level. For example, a steady dew point requirement as low as -20°F [-28.9°C] is common in rooms containing freeze-dry machinery (lyophilizer rooms) or rooms working with surgical sutures. Requirements for 20% rh as a maximum are common for manufacturing tablets, and medical devices are often assembled in low-humidity air. The list of humidity control levels is as varied as the products in question.

For pharmaceutical pre-production work, the HVAC designer can rely on the process engineers to define the required air temperature and humidity. Without that definition, the food and drug regulatory authorities could not "validate" the manufacturing process. In most countries, manufacturing of drugs and medical devices must take place under temperature and humidity levels which are identical to those used during product and process development.

Animal research

In a vivarium (a "mouse house") or a primate research facility, the usual purpose of humidity control is to keep the animals healthy.

Most laboratory animals remain healthy as long as humidity is held somewhere between 30 and 70% rh (ILAR, 1996). At typical lab temperatures, that means the dew point should stay roughly within the human comfort range—between dew points of 35 and 57° [between dew points of 1.7 and 13.9°C]. Below that range, skin problems and eye irritation are more likely, and above that range, heat stress is higher, and more infectious organisms can survive and grow in the air and in the bedding under the animals.

Chemical research

In most chemical research labs, any specification for humidity control is usually intended to improve human comfort. The low limit is likely to be a dew point of 35°F during the dry season. During humid weather, the upper limit is likely to be a dew point of 57°F. [a dew point of at least 1.7°C during dry weather, but less than 14.0°C during humid weather].

Occasionally, researchers ask for a specific control level for humidity. The usual purpose is either to prevent the buildup of electrostatic charges during dry seasons, or to prevent hygroscopic materials from absorbing moisture during humid months. In those situations, the researchers are likely to know the humidity range which will prevent problems that affect the research.

Physics & electronic research

The usual purpose of controlling humidity for physics or electrical experiments is to reduce the number of random electrostatic discharges, and to prevent hygroscopic materials from absorbing moisture. Keeping moisture levels within the human comfort range is often adequate for these purposes. But the designer should confirm the requirement with the lab users, since very high voltages and very hygroscopic materials are occasionally the subjects of research. In those circumstances, the user may need a "room-within-a-room" to accommodate the humidity sensitivity of the work without driving up the cost of the entire building.

Moisture Loads

For the example shown in figure 30.2, we have assumed a medium-sized chemistry lab building for a university. The building contains three stories and 21 classrooms, and is located in Atlanta, Georgia. Atlanta has a hot and humid climate for most of the year, but has cool and dry weather for a few months during the winter season. We have assumed the building will be humidified to the lower end of the human comfort range in winter, and dehumidified to a condition of 50% rh at 75°F [23.9°C] during the humid months. The summer control condition represents a dew point of 55°F, or a humidity ratio of 65 gr/lb [dew point of 12.8°C, or a humidity ratio of 9.3 g/kg].

No commercial or institutional building uses more ventilation air than a laboratory. The moisture load from that ventilation *far*

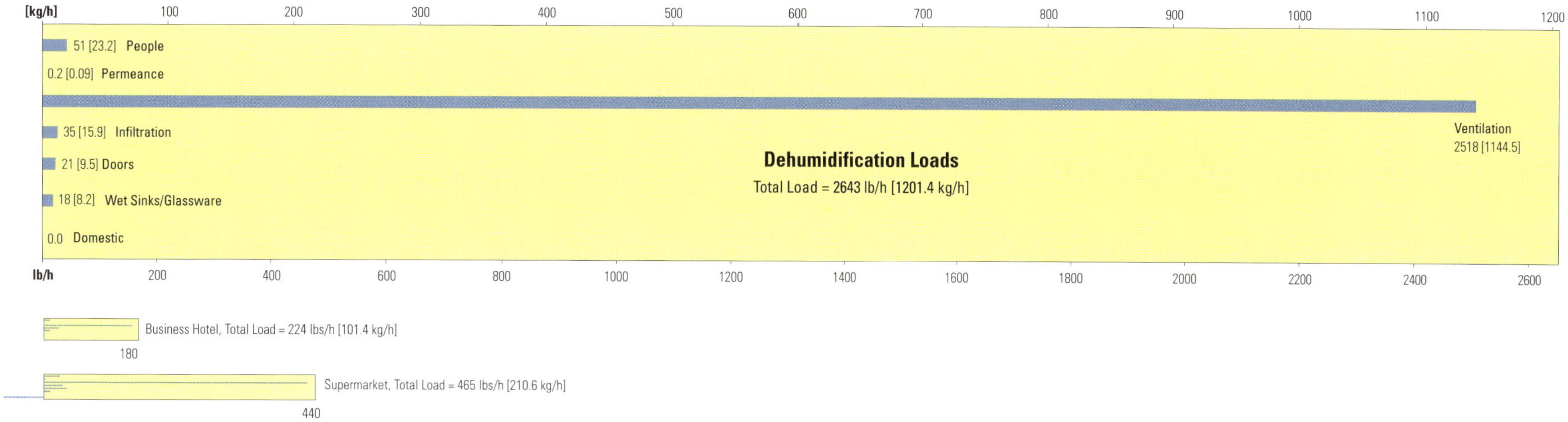

Fig. 30.2 Dehumidification Loads

Dehumidification loads in laboratories are truly immense. The diagrams above compare the loads in labs to those in supermarkets and hotels using the same scale for the horizontal axis.

Labs have much larger loads because they use much more ventilation air than nearly any other commercial structure. Only hospitals have similarly large moisture loads.

Fig. 30.3 Humidification Loads

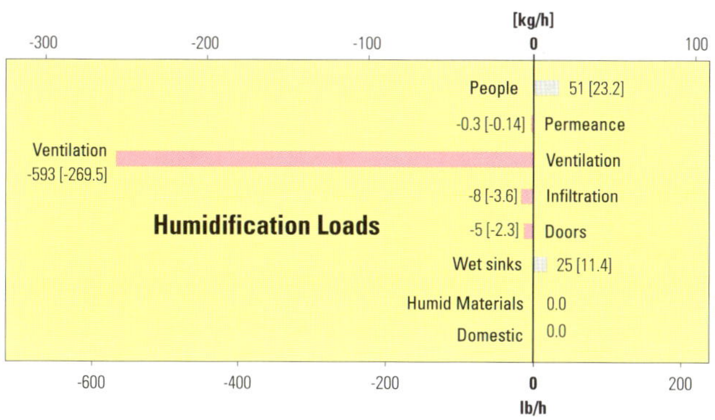

outweighs the combined total from all other load elements, which can be seen clearly in figure 30.2. To achieve humidity control inside the building, the designer must control the dew point of the incoming ventilation air.

Ventilation air

Successful humidity control begins by quantifying the amount of ventilation air required for the building. The total flow is often dictated either by policy regarding the number of air changes per hour in the labs, or by the amount of air needed to maintain an effective capture velocity across the face of the fume hoods.

To contain vapors and particulates, the exhaust fans pull air from the classrooms into each hood at an average velocity of 100 fpm [0.508 m/s]. But the total amount of air exhausted from the building varies, because the sash height of the fume hood (therefore the open area and the amount of air needed to maintain the capture velocity) depends on the activity inside the hood. Also, the number of hoods in operation varies, because in a teaching environment, most experiments are short-term and many hoods can be shut off entirely.

The designer must estimate the maximum amount of make-up air needed for the building under peak load conditions. The best way to ensure humidity control is to treat that make-up air through a separate system which can dehumidify, humidify and also modulate the total make-up air quantity in response to the changing needs of the hood exhaust system.

The make-up air system should dehumidify the incoming air to slightly *below* the control condition in the summer, so that incoming air can absorb any excess moisture generated inside the space. During winter, the incoming air is humidified to slightly *above* the control condition, to compensate for moisture losses caused by the infiltration of dry air.

To illustrate this point, consider that the example building needs about 57,600 cfm of make-up air. If that incoming air is dehumidified to only 5 gr/b *below* the control condition, it will have the capacity to remove 185 lb/h of excess moisture from the building. So all the other moisture loads in the building can easily be removed by this method. However, this large air flow can also create problems if the system *fails* to dry the incoming air completely. Leaving an extra

5 gr/lb in the air stream means the incoming air will flood the building with 185 lb/h of *excess moisture*, which will force the indoor humidity to rise above the control level.

[To illustrate this point, consider that the example building needs about 27,181 l/s of make-up air. If that incoming air is dehumidified to only 0.7 g/kg *below* the control condition, it will have the capacity to remove 84.1 kg/h of excess moisture from the building. So all the other moisture loads in the building can easily be removed by this method. However, this large air flow can also create problems if the system *fails* to dry the incoming air completely. Leaving an extra 0.7 g/kg in the air stream means the incoming air will flood the building with 84.1 kg/h of *excess moisture*, which will force the indoor humidity to rise above the control level.]

Keeping the humidity constant in the make-up air is rather tricky, because throughout the year its volume varies widely, and so does its moisture content. The system must be able to modulate smoothly through an unusually broad range of loads. The authors encourage the designer to size the humidification, dehumidification and fan capacities at peak load conditions as usual, but also to pick several weather conditions *between* the extremes, and to calculate system behavior in those moderate ranges that occur most frequently.

In laboratories, a common reason for poor humidity control is that the system has too much capacity for smooth modulation in the mid-range weather conditions. Components designed for peak loads do not modulate smoothly when only 10 or 20% of their capacity is in-use. Humidity (and temperature) becomes unstable as equipment turns on and off entirely during moderate weather. Installing several dehumidifiers in parallel simplifies the task of modulating capacity smoothly over a wide range of load conditions. Some units can be turned off entirely when the load is low. For the same reason, the multiple-unit design is also useful for humidifiers.

People

In most laboratories, the moisture load from people's respiration and perspiration is very small because there are so few people in the building. But in a university setting, a large number of students can occupy each room, and they often enter the building dressed in clothing that has become soaked with rain. So in this example, the load from people is significant.

On the other hand, if the make-up air is dried very slightly below the control dew point in summer, the system will have plenty of ca-

Fig. 30.4 Ventilation Pre-Treatment

Because the lab uses so much outdoor air, humidity control becomes largely a matter of controlling the dew point of the incoming air. That can be accomplished most effectively by using a separate, dedicated system.

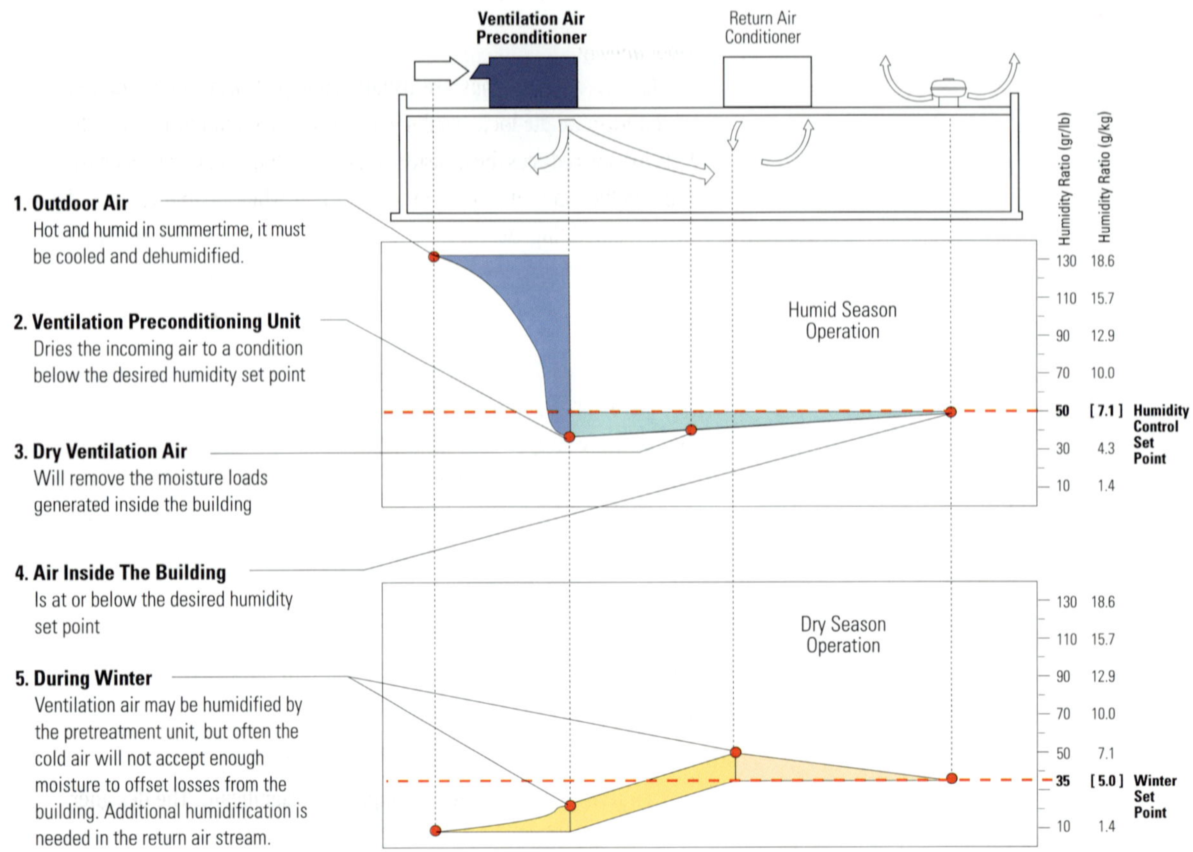

pacity to absorb water vapor from people and clothing without allowing the indoor dew point to rise above the set point.

Infiltration

Usually, laboratory buildings are very well-built in comparison to other commercial or institutional buildings. On the other hand, laboratories are also usually under a negative air pressure to make sure contaminants do not escape the building's exhaust air treatment system. So even though our building would probably be built much tighter than other commercial buildings, it will still pull in some untreated outdoor air. As with other internal moisture loads, the infiltration load can be easily removed by the make-up air, provided that air is dried low enough before it enters the building.

Door activity

Laboratory buildings are usually equipped with double doors, which form an air lock vestibule for access from outdoors. A vestibule greatly reduces the moisture load from door openings, because it limits the amount of incoming humid air when people enter and leave the building. Assuming our example building is equipped with such vestibules, the moisture load from door activity is very low, even when classes change and doors stay open for several minutes.

Wet sinks

In most labs, there is very little moisture load from wet surfaces. But in this example, we assumed that students will be washing up after class, and therefore the wetted surface will be significant. On the other hand, one must keep this load element in perspective. Even if all sinks are being flooded with warm water at the same moment (as we have estimated here), the evaporation rate is still only $^7/_{10}$ of one percent (0.007) of the load from ventilation air. Drying the ventilation air air remains the key to controlling the indoor humidity.

Permeation

At less than 1 lb/h [0.45 kg/h] for the entire building, the moisture load from permeation is completely insignificant. The designer should waste no time trying to minimize the load from permeation. Any extra effort would be better spent making certain of the volume of make-up air, and making certain that the entire mass of ventilation air is brought to the correct dew point under all weather conditions.

Humidification loads

The example building was placed in Atlanta, Georgia, which has a rather forgiving winter design condition. The extreme winter dew point is not low, especially when compared to conditions at locations closer to the pole, such as Northern Canada, Scandinavia or Argentina. In those climates, the humidification load is very large, and the winter months pose the greatest challenge to controlling humidity in a laboratory.

Once again, the greatest load is created by the make-up air, which has a large moisture deficit. This air must be humidified before it enters the building. In winter, that's difficult to accomplish without causing condensation in the duct work. Also, humidifiers sized for the dryness of outdoor air under extreme winter conditions are difficult to control during more moderate spring and fall weather, when the load is a small fraction of the maximum.

That's why it is very important to have excellent controls for the humidifiers. Also, most designers prefer to install several units, rather than one single, large unit. That way, capacity can be turned on and off and modulated smoothly in stages, avoiding the condensation problem that can occur when the outdoor dew point rises suddenly with rain or other weather changes.

Suggestions & Cautions

Our suggestions and cautions are divided into two sections, so they can be passed to the appropriate member of the design team. Many of the most important cautions relate to architectural aspects of the building.

Architectural considerations

Avoid placing fume hoods or biosafety cabinets near doors

While not strictly related to humidity control, this caution bears repeating because of its importance for safety. When doors open and close, they change the nearby air pressures. If a fume hood or biosafety cabinet is nearby, an opening door can easily pull contamination *out* of the protective enclosure and into the room, creating a potential hazard. Neither HVAC engineering nor fume hood design can overcome the safety risk that results from placing this equipment near doors. Don't do it.

Expect large floor-to-floor height

As the complexity of lab operations increases, so does the height needed between floors. While the ceiling may be at the 8 ft level, the mechanical systems above the ceiling may need the same height, yielding a floor-to-floor distance as high as 16 ft. [Ceiling at 2.4 m, but distance between floors of 4.8 m]

This space is needed to accommodate supply and exhaust ductwork, humidifiers, dehumidifiers, air flow controls and air filters. If the owner or architectural designer decides to "save money" by reducing the space above the ceiling, the owner will have to live with:

- Constant noise, because ducts are too small and air velocities too high.
- Increased safety risk, because air cannot flow smoothly through the room, which reduces the air's ability to control hazards inside the lab.
- High maintenance cost, because there may be no access to key components without disassembling others.
- More lab downtime and risk to experiments, because HVAC technicians will have to access the system by working inside the lab itself.
- Poor system performance, because lab managers will be reluctant to interrupt research to allow HVAC technicians to adjust the system.

In other words, although the cost of long floor-to-floor distances can be economically and aesthetically painful, the costs of too little vertical space are more serious. The architectural designer who may be new to designing lab buildings can save the owner quite a bit of trouble by keeping this fact in mind during the early design phases of the project.

Commissioning will be necessary

The complexity of a lab mechanical system demands an experienced and energetic commissioning authority. Most owners are eager to save large amounts of money, and also reduce the time between building completion and building occupancy by engaging the services of a commissioning agent. The architectural designer and HVAC system designer can also enjoy savings when the commissioning authority participates in the early design meetings. The alternative to early participation is costly redesign, or extra months of involvement while the commissioning agent overcomes design weaknesses during startup.

The design team should encourage the owner to engage the commissioning authority at the earliest possible time, so that overall project costs are minimized through that firm's valuable contribution of experience and by their training of the owner's operating staff.

Consider turnkey rooms for special requirements

In many cases, a lab building may have relatively simple requirements for temperature and humidity in all but one or two rooms, where the work demands special conditions. To keep the overall HVAC system simple and more economical, many architectural designers prefer to exempt such rooms from the mechanical contract, and to ask turnkey room manufacturers to create the mechanical systems for those special rooms. This reduces cost and simplifies responsibility. On the other hand, architectural designers new to special environments are often less familiar than HVAC designers with the cost implications of similar-sounding specifications for temperature and moisture.

For example, in the USA during 2001, a turnkey installation of a single small cold/clean room used for protein purification might cost

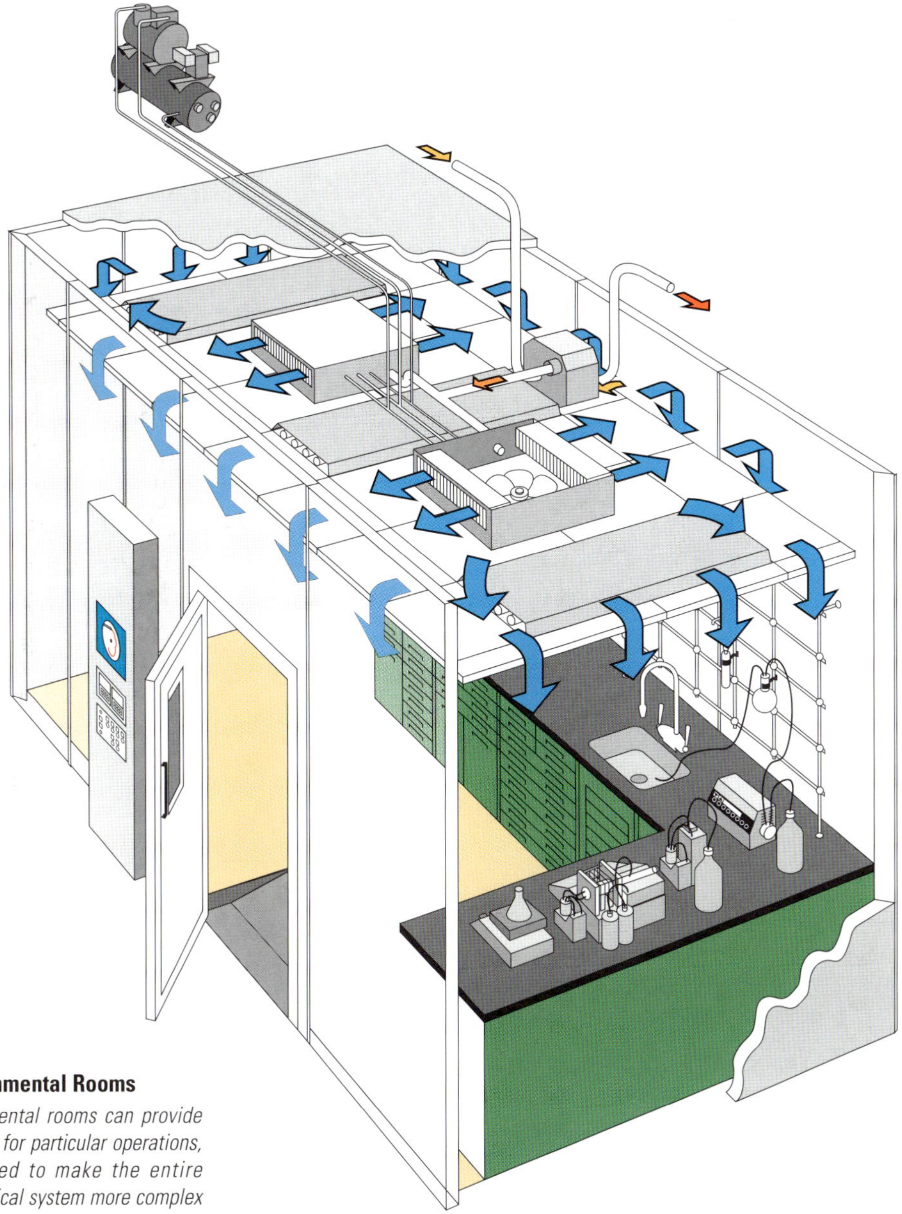

Fig. 30.5 Environmental Rooms

Turnkey environmental rooms can provide special conditions for particular operations, avoiding the need to make the entire building's mechanical system more complex and costly.

about \$75,000 if the temperature gradient is specified to be ±1°C throughout the room. But if the gradient must actually be ±1°***F*** instead of ±1°***C***, the task is much more difficult, because the allowable variation has been cut in half. With that easy-to-overlook change, the room cost may increase by 25% because of the need for much more air, tighter controls, larger cooling and heating components and a more elaborate air distribution arrangement. (Harris, 1992).

So when the architectural designer does not have the assistance of a mechanical engineer while specifying the special environment rooms, nasty surprises can be avoided by extensive consultation with the users of the rooms, to be certain of their exact requirements, and consultation with the turnkey vendors, to be certain of the cost implications. Over or under-specifying a few special environments can have an astonishingly large impact on the project budget.

HVAC considerations

Keep supply air diffusers away from fume hoods and biosafety cabinets

Earlier, we included a caution for the architectural designer against locating fume hoods and biosafety cabinets near doors. To preclude similar problems, the HVAC designer *must not* locate high-velocity supply air diffusers near these devices.

If supply air is blown towards or across the face of a biological safety cabinet, the air pressure can force contaminated air into its protected work area, or pull experimental contaminants out of the cabinet into the room, or both. The most common reason that biological safety cabinets fail to perform is because doors or supply air diffusers are too close to the face of the unit.

In some cases, the HVAC designer can be excused from total responsibility because the owner or architectural designer decided to move the fume hoods and biosafety cabinets late in the project, or he moves them during a renovation that does not involve the main supply and return systems. But that does not change the fact that the room becomes hazardous. To serve the owner properly, the designer of both new and renovated lab systems needs to know the entire room equipment layout before finalizing locations for the supply and return air connections to each room.

Use peak dew point, not peak dry bulb for summer ventilation design

In most commercial buildings, the sensible heat loads dominate the design of the cooling systems. But in a laboratory, the huge amount of excess moisture in the ventilation air should increase the attention given by the designer to the latent loads. In all but desert and high altitude climates, the annual latent load in ventilation air outweighs the sensible load by 4:1 or even 6:1 (Harriman et al, 1997).

The practical implication of this fact is that the total cooling load on the ventilation system is almost certainly *larger at the peak dew point condition* rather than at the peak dry bulb condition. For example, consider the cooling and dehumidification design conditions for Atlanta, Georgia, which are shown in figure 30.6. The maximum enthalpy of the outdoor air is 40.5 Btu/lb [79.1 kJ/kg], which occurs at the peak dew point condition, not the peak dry bulb temperature. If the ventilation system were designed for the peak sensible condition, it would be grossly short of overall capacity and especially short of dehumidification capacity.

This caution is useful in designing all ventilation systems, but is especially important in laboratories and hospitals, where so much outdoor air is needed.

Consider load reducers in extreme climates

Lab ventilation systems use a tremendous amount of air and therefore a very large amount of energy. Load reducers, such as passive desiccant wheels, heat pipes and plate-type heat exchangers and coil-to-coil glycol loops, have been applied to laboratory exhaust systems to use the energy contained in the exhaust air stream to reduce the load from the make-up air.

Chapter 15 of this book contains more guidance in the use of load reducers. But with respect to laboratories, here are some points the designer might wish to consider:

1. **Cross-contamination.** Heat pipes, plate-type heat exchangers and coil-to-coil glycol loops are more common in lab applications than rotary heat exchangers, because they are perceived to have less potential for cross contamination between the exhaust and supply air streams. While this is certainly a legitimate concern, the much greater efficiency of passive desiccant (enthalpy) wheels has encouraged many designers to include them in lab HVAC systems as well. The designer should consult the manufacturer of the devices in question for guidelines for safe use in laboratory exhaust air.
2. **Climate & Operating hours.** Operating economics for load reducers are improved when the climate is extreme, and when the system operates for all 8760 hours in a year. In other words, a high school lab with few operating hours, located in the mild climate of San Diego, California, would probably not be a good candidate for a load reducer. But an animal research building, with a continuous and filtered exhaust, located in Chicago, Illinois, which has extreme temperatures in both summer and winter, would be an excellent candidate for such a device.

Consider multiple small units in place of one large one

As mentioned in the load calculation section, it is very difficult to control humidity in the lab ventilation air stream because its volume changes constantly, and so does its temperature and moisture content. Very large dehumidification and humidification equipment is needed to remove peak loads.

However, by definition the peak load occurs for less than 0.4% of the life of the building. For the *other 99.6%* of the building's life, the equipment will be removing much smaller loads. That's when lab humidity can be unstable. The load is often so small that large equip-

Atlanta Design Conditions
2001 ASHRAE Handbook—Fundamentals

I-P Units	Cooling 0.4%	Dehumidification 0.4%
Dry Bulb Temp.	93° F	82° F
Wet Bulb Temp.	75° F	77° F
Dew Point Temp.	67.7°F	75.3°F
Humidity Ratio	102 gr/lb	**133** gr/lb
Enthalpy	38.4 Btu/lb	**40.5** Btu/lb
SI Units	**Cooling 0.4%**	**Dehumidification 0.4%**
Dry Bulb Temp.	34° C	28.0° C
Wet Bulb Temp.	24° C	25.6° C
Dew Point Temp.	20.0°C	24.9°C
Humidity Ratio	14.7 g/kg	**20.0** g/kg
Enthalpy	71.9 kJ/kg	**79.1** kJ/kg

Fig. 30.6
Comparing Moisture At Peak Cooling vs. Peak Dehumidification
Note that in Atlanta, when the dew point peaks, so does the enthalpy. This pattern is true in most climates. The designer should check ***both*** *design points—to be certain of the real extreme when sizing equipment for treating ventilation air.*

ment wants to turn off entirely, and yet there is still a load large enough to radically change indoor humidity quickly, because the air change rate may be as high as 10 times per hour (air in the building would change completely in 6 minutes).

The designer can avoid this problem by selecting several small units rather than a single large one. With multiple smaller units, one or two can be modulating fully, while the others are brought on or off-line when the load rises or falls by larger amounts. The modulating units provide the "fine control," and the non-modulating units provide the heavy capacity to meet less frequent peak loads. (Also of course all of the units must respond to a single controller. Otherwise the results are chaotic.)

This approach also provides partial capacity in case any single unit needs maintenance. But most importantly, it provides much smoother modulation, which results in more stable humidity.

References

Blazewicz, Bernard. 2000. "Designing for pharmaceutical research facilities." *ASHRAE Journal*, October, 2000. pp:45-50. This paper provides a very succinct overview of design issues in pharmaceutical research and production, together with excellent, simple suggestions for the first-time designer.

Chang, Stephen K.W, 1994. "Air velocity profiles around a person standing in front of exhaust hoods." ASHRAE Transactions, Vol. 100, Pt.2, 1994. This paper shows visually and quantitatively how drafts can affect the concentration of hazardous particles in the breathing zone of a lab worker, and how different types of exhaust hoods protect those individuals.

Chapter 13 - Laboratories. 1999. *ASHRAE Handbook—Applications*. This is a fundamental reference for any HVAC designer working on lab systems. The notes at the end of the chapter are very comprehensive, providing the designer with most of the North American regulatory references he or she will need to ensure a safe design.

Harriman, Plager & Kosar. 1997. "Dehumidification and cooling loads from ventilation air." *ASHRAE Journal*, November 1997. pp: 37-45. This paper quantifies the annual sensible and latent loads contained in ventilation air at 239 US locations. It demonstrates the overwhelming predominance of the latent component in all climates other than deserts and mountains, where the sensible load is larger.

Harris guide to environmental room design. 1992. Harris Environmental Systems, 11 Connector Rd., Andover, MA. USA. The guide explains the meaning of such terms as uniformity, gradient and recorder span, which help the architectural or HVAC designer understand the economic implications of the requirements in special environments.

Rake, Barbara. 1978. "Influence of crossdrafts on the performance of a biological safety cabinet." *Applied and Environmental Microbiology*, Vol. 36, no.2, August, 1978, pp.278-283. American Society for Microbiology. The research quantifies the risk to product protection and containment in a typical biosafety cabinet when it is exposed to crossdrafts, such as those from nearby door openings or supply air diffusers.

Wilkins, Christopher K. 1999. "Cooling loads in laboratories." ASHRAE Transactions, Vol. 105, pt. 1. 1999. The author has actually measured and recorded the sensible cooling loads from many different types of lab equipment while it is operating, as well as when it is in standby mode. These values are not published by the equipment manufacturers, so this paper is exceptionally useful for the designer of cooling systems for laboratories.

Animals

Besch, Emerson L. 1992. "Animal facility ventilation air quality and quantity." ASHRAE Transactions, Vol. 98, pt.2 1992. The author outlines the broad issues for designing animal ventilation systems, and includes descriptions of key references from regulatory authorities in North America.

ILAR. 1996. *Guide for the care and use of laboratory animals* Institute of Laboratory Animal Resources, Commission on Life Sciences, National Research Council, National Institutes of Health, Washington, DC. http://oacu.od.nih.gov.regs/guide/

In the following three papers, the authors provide field measurements of temperature, humidity and ammonia concentration in a total of 55 different buildings which contain lab animals. The papers are useful for understanding what actually occurs in a lab animal room when special caging is not included in the project.

Y. Zhang, Christianson, Riskowski, B. Zhang, Taylor, Gonyou & Harrison.1992. "A survey of laboratory rat environments." ASHRAE Transactions, Vol. 98, pt.2

Y. Zhang, Christianson & Riskowski. 1992. "Evaluation of ventilation rates through four types of rat cages." ASHRAE Transactions Vol. 98. pt.2.

Maghirang, Ronaldo, Riskowski, Christianson & Harrison. 1995. "Development of ventilation rates and design information for laboratory animal facilities—Part 1: Field study." ASHRAE Transactions Vol. 101, pt.2.

Commissioning & installation case histories

Laboratory HVAC Systems: Design, Validation, Commissioning. 1994. ASHRAE Technical Bulletin Vol.10, no.1.

This ASHRAE special publication is a collection of technical papers from the proceedings of the ASHRAE meeting held in new Orleans, LA in January, 1994. It provides an excellent overview of design and commissioning issues, based on real-world, personally-observed experience.

Special Thanks

The author is grateful for the advice and information provided by those with far greater experience in laboratory ventilation design and commissioning, in particular: Rob Smith, Carl Lawson, Robert Cox, Dan Frasier, Dan Ghidoni and Bob Jones.

Donated Diagram

Fig. 30.5 - © Harris Environmental Systems

Chapter 31
Weather Data For Design

Introduction By Lew Harriman

Data By Don Colliver and Rich Gates

Fig. 31.1 Design Data vs. Hourly Observations

Data in this chapter are used for estimating moisture loads under peak design conditions. By definition, these extreme conditions only occur for a small fraction of time.

Hourly observations can be used to simulate system behavior during normal operation, when loads are much smaller.

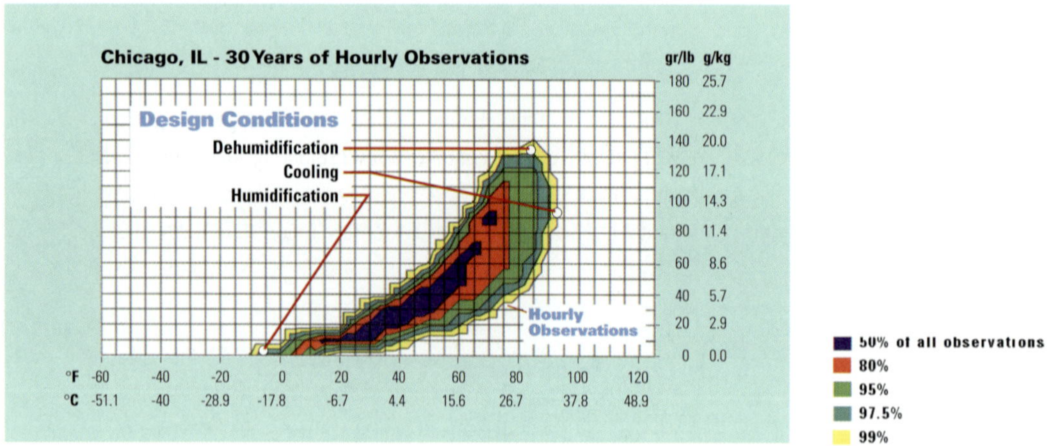

Fig. 31.2 Comparing Climates

Department of Defense Engineering Weather Data helps a designer understand a new climate, by visual psychrometric comparisons to climates in more familiar locations. These figures display 30 years of hourly observations. The DOD weather information also includes other visual displays and quantitative data tables. (DOD, 1999)

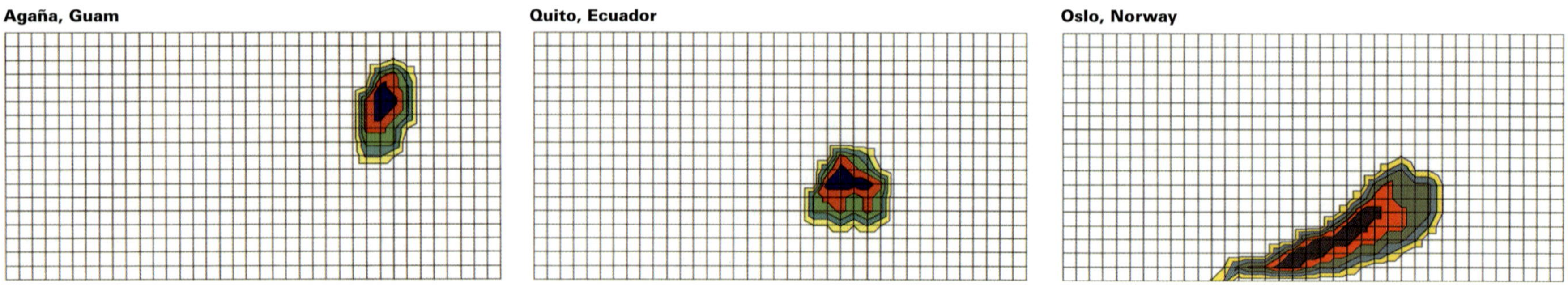

Engineering Weather Data For Humidity Control

For HVAC design, there are two main categories of engineering weather data: hourly observations and design conditions.

Hourly observations are used to simulate system operation, which helps estimate how the system will respond to the real world of varying loads. That sort of data is not displayed here, because it is very voluminous—8760 hourly observations of perhaps 10 variables for each geographic location. Such data does not lend itself to publication on paper. Electronic files are more useful. Sources for hourly weather are described in the reference section.

The tables in this chapter provide the second type of engineering weather data: design conditions for estimating the system's peak loads. With this data, the designer can use the formulas in chapters 11 and 12 to estimate the capacity of humidifiers and dehumidifiers that will be needed to satisfy the building owner's humidity control requirement.

There's a key point that many HVAC designers overlook. In the final analysis, the owner must determine which weather conditions to use for designing the system, because the basic question is: how many hours in a typical year is it acceptable to risk having the humidity drift outside of the desired range? The system's capacity will depend on what weather conditions are chosen for estimating the maximum load. When more extreme conditions are chosen, the system will cost more, but there will be fewer hours when the outdoor conditions will generate loads "greater than design."

0.4%, 1%, 2%, 99%, 99.6%, - What the percentages mean

The tables in this chapter describe the high and low extremes of humidity at 1400 North American and international locations. The extremes are arranged in columns according to their frequency of occurrence, and that frequency is defined by how many hours in a typical year are likely to be *above* or *below* the value in the column.

For example, for sizing dehumidifiers, the 0.4% humidity ratio is the highest outdoor moisture value. It's the amount of moisture that is *likely to be exceeded for only 0.4% of the hours in a typical year*. In other words, the outdoor humidity is only likely to be higher than the 0.4% value for about 35 of the 8760 hours in the year. The 1% humidity ratio is likely to be exceeded for roughly 88 hours and the 2% humidity ratio will be exceeded for about 175 hours in that typical year.

The tables allow the owner to choose between 35, 88 or 175 hours beyond the design load from outdoor humidity. As the reader will have noted, the humidity in the ventilation air (outdoor air) is the governing factor in moisture load estimates for *all* commercial buildings other than swimming pool enclosures. So the choice of the outdoor design condition is the single most influential decision in determining the initial cost of the system's equipment. That's why the client must decide: how many hours will it be acceptable for the building to have greater-than-design loads from the weather?

Similar values are provided for humidification design. The 99.6% value is the lowest humidity shown in the table. The outdoor moisture level is likely to be *less* than that value for only about 35 hours each year. The 99% value is slightly less extreme, and outdoor humidity is likely to be drier than the 99% value for about 88 hours in a typical year.

Note that we have been using the expressions "likely to exceed" and "typical year." Nature is more variable than engineers would prefer. In any year, the extremes and their occurrences will vary from the values in these tables. But over the long term, the values will be consistent. To construct these tables, the research team at the University of Kentucky examined between 8 and 30 years of hourly observations at each location (Colliver et al, 1998) The values presented in these tables refer to the occurrences within the period of record examined during the research, not the observations of one single year.

The period of record varies somewhat between stations, because the raw data is not always complete for all stations. For example, the researchers examined the data from 1982 through 1993 for Anniston, Alabama, but used all the observations from 1961 through 1993 for Key West, Florida. The weather records for Key West were simply more complete. The percentages described in the tables refer to the entire period of record selected. Those stations with longer weather records will have values closer to the average year than stations with shorter records. In other words, the values for Anniston are likely to be very slightly more extreme than the values for Key West. Those who need to understand more about the derivation of these values can consult the Climatic Design Information Chapter in the *ASHRAE Handbook—Fundamentals*, and the technical paper written by the researchers who calculated the values (Colliver et al, 1998).

Design for dehumidification vs. design for cooling

Designers often ask: "Should I size my system's equipment using the peak *sensible* design values—or the peak *dew point* values?" The best answer depends on the purpose of the system.

You will notice that the dry bulb temperatures in these tables are much lower than the values used for sizing cooling equipment. That's because the temperature is generally lower when the humidity is at its peak. Humidity usually peaks in the early morning or after rainstorms, when water evaporates from grass, plants, buildings and streets into the air. The dry bulb temperature of the air stays fairly low as the moisture evaporates, because the moisture uses some of the sensible heat of the air for evaporation.

For ventilation systems, the designer should calculate loads using both sets of extremes, because the peak dew point condition often generates a higher enthalpy (higher total heat) than the peak dry bulb condition. On the other hand, when ventilation loads are small, and when the client does not need humidity control during the summer, then the sensible values alone are probably adequate for design, since the principal purpose of the system is cooling.

When the client needs humidity control during summer months, the decision is easy. The peak dew point values must be used to calculate moisture loads if the system is to have adequate dehumidification capacity.

Failures caused by oversizing equipment

In typical commercial HVAC design, the natural tendency is to install plenty of equipment capacity, and to use the "next-size-larger" unit—one which has a capacity greater than the estimated load. The designer reasons that there will always be days when the loads exceed the estimate, and besides, the next-size unit really does not cost much more. That reasoning seems comfortably conservative, but it does not reflect how equipment behaves in operation. Conservative HVAC design is not like structural design; oversized mechanical equipment nearly always results in poor system performance.

When cooling equipment is oversized, it operates infrequently and for short periods. It shuts on and off so quickly that temperature fluctuates widely, and the cooling coil never really gets cold enough to dehumidify the air. Also, the rapid on-off cycles reduce the life of the equipment.

When dehumidifiers are oversized, the same thing happens to the humidity level—it fluctuates widely as the equipment turns on and off. Excess dehumidification capacity, especially in a single machine, often makes it impossible to actually control humidity in spite of the heavy investment in dehumidification equipment.

When humidifiers are oversized, they can oversaturate the air, which produces condensation, feeding the growth of mold and mildew and leading to water dripping out of the ductwork.

All of these problems come from oversizing large, single pieces of equipment, combined with a failure to invest in adequate controls to modulate the excess capacity during normal operating load conditions.

This discussion is relevant here because equipment can be oversized if the designer chooses an excessively extreme condition for calculating the loads. It is often better to risk a few more hours short of capacity than to overestimate the load and end up with equipment so large and unwieldy that the system fails to control humidity when the loads are at normal levels.

Said another way: don't risk failure during 99.9% of the system's life just to have extra capacity during 0.1% of its operating hours.

The Weather Data Tables

The data in this chapter were produced for the *ASHRAE Handbook—Fundamentals,* and were first published in the 1997 edition of that book. The work was performed by Professors Donald Colliver and Richard Gates of the University of Kentucky along with a team of researchers at that institution, with funding from ASHRAE through research project 890-RP. The data contained in the *Handbook—Fundamentals* is more complete than what is included here. The *Fundamentals* adds information for cooling and evaporative cooling design, as well as more wind speed data. It also describes the years used as the period of record for each station.

But otherwise the data in this chapter are the same as the data in the 1997 and later editions of the *Handbook*. Please note: editions before 1997 *do not* contain data that accurately described humidity extremes. (Harriman el al, 1998, 1999)

Humidification design

ASHRAE research project 890 did not calculate the lower extremes of humidity. So for sizing humidification systems, this chapter uses the lowest dry bulb temperatures, and makes the assumption that the incoming air is saturated at that dry bulb temperature.

In cold climates, this assumption is probably close enough to the true value, because the absolute moisture content of very cold air is nearly zero in any case. In more moderate climates, that assumption is less conservative, because the outdoor moisture content can be below saturation for more of the time.

Designers of humidification systems sometimes assume the outdoor air is completely dry, and size the equipment accordingly. This can result in the oversized equipment problem described above. As of the publication date of this book, the Society does not have better information than what is printed here. The best outdoor moisture content to use for humidification design is probably somewhere between zero and the saturation moisture content at the low dry bulb temperature for the location. Keep in mind that the most common problem with humidification systems in commercial buildings is *too much* rather than too little capacity.

1a and 1c - 99.6% Design - Dry Bulb Temperature

These data represent the dry bulb temperature that is lower than 99.6% of all the hourly observations during the years of the period of record. Units are provided in whole degrees Fahrenheit and 1/10th of a degree Celsius.

1b and 1d - 99.6% Humidity Ratio

These data are calculated. It is not contained in the *Handbook—Fundamentals*. It represents the humidity ratio at saturation for the 99.6% dry bulb temperature, adjusted for normal atmospheric pressure at the station's elevation.

The humidity ratio is the weight of water in an air sample, divided by the weight of the entire sample. In this case, the units are grains of water vapor per lb. of air, and grams of water vapor per kilogram of air.

1e and 1f - 99.6% Mean Coincident Wind Speed

These data define the average wind speed that was observed when the air was at the extreme low dry bulb temperature described by columns 1a and 1c. In other words, the values are not the average wind speed during the entire cold season—just its average speed when the air is very cold.

The information helps when estimating the wind pressure, which drives air infiltration through open doors and through cracks in the exterior wall. Units are in miles per hour and meters per second.

1g - 99.6% Prevailing Wind Direction

Wind can hit the building from several points of the compass at the same time. These data describe the prevailing wind direction in an open field when the air's dry bulb temperature was at the value described in columns 1a and 1c.

The prevailing direction is that which was "most frequently observed" during the hours when the temperature was at its peak value. The units are degrees true: 360 indicates wind coming from directly north, 90 indicates wind from due east, etc.

2a and 2c - 99% Dry Bulb Temperature

These data represent the dry bulb temperature that is lower than

99% of all the hourly observations during the years of the period of record. Units are provided in whole degrees Fahrenheit and 1/10th of a degree Celsius.

2b and 2d - 99% Humidity Ratio

These data are calculated. It represents the humidity ratio at saturation for the 99% dry bulb temperature, adjusted for normal atmospheric pressure at the station's elevation. See the notes under "Humidification Design" for cautions regarding sizing humidifiers.

The humidity ratio is the weight of water in an air sample, divided by the weight of the entire sample. In this case, the units are grains of water vapor per pound of air, and grams of water vapor per kilogram of air.

Location description

3a - Location

The locations described here are the same as those contained in the *Handbook —Fundamentals*. For many locations, the weather station is at an airport rather than in the middle of a dense urban area.

3b - World Meteorological Station Number

The world meteorological number uniquely identifies the location where the measurement was taken, although the precise location can change over time if the measurement equipment is moved when an observation point is relocated.

3c - Latitude

The latitude of the measurement station is given in degrees and minutes.

3d - Longitude

The longitude of the measurement station is also given in degrees and minutes.

3e and 3f - Station Altitude

The altitude of the measurement station is given in feet and meters above sea level.

3g and 3h - Station Pressure

This data represents the atmospheric pressure at standard conditions, adjusted for the altitude of the station. The pressure is given in pounds per square inch (absolute) and kiloPascals.

Dehumidification design

4a and 4c - 0.4% Humidity Ratio

These data represent the humidity ratio that is higher than all but 0.4% of the observations during the years of the period of record. It is the same as the peak dew point values contained in the *Handbook —Fundamentals*, beginning with the 1997 edition. When a year is typical of the long-term record, the humidity is not likely to be higher than this value, except for about 35 hours out of the years' total of 8760 hours.

The humidity ratio is the weight of water in an air sample, divided by the weight of the entire sample. In this case, the units are grains of water vapor per pound of air, and grams of water vapor per kilogram of air.

4b and 4d - 0.4% Mean Coincident Dry Bulb Temperature

This value is the average of all the dry bulb temperatures that were observed over the long-term record whenever the humidity ratio was at the 0.4% value. Units are provided in whole degrees Fahrenheit and 1/10th of a degree Celsius.

5a and 5c - 1% Humidity Ratio

This data represents the humidity ratio that is higher than all but 1% of the observations during the years of the period of record. When a year is typical of the long-term record, the humidity is not likely to be higher than this value, except for about 88 hours out of the years' total of 8760 hours.

The humidity ratio is the weight of water in an air sample, divided by the weight of the entire sample. In this case, the units are grains of water vapor per pound of air, and grams of water vapor per kilogram of air.

5b and 5d - 0.4% Mean Coincident Dry Bulb Temperature

This value is the average of all the dry bulb temperatures that were observed over the long-term record whenever the humidity ratio was at the 1% value. Units are provided in whole degrees Fahrenheit and 1/10th of a degree Celsius.

6a and 6c - 2% Humidity Ratio

These data represent the humidity ratio that is higher than all but 2% of the observations during the years of the period of record. When a year is typical of the long-term record, the humidity is not likely to be higher than this value, except for about 175 hours out of the years' total of 8760 hours.

The humidity ratio is the weight of water in an air sample, divided by the weight of the entire sample. In this case, the units are grains of water vapor per pound of air, and grams of water vapor per kilogram of air.

6b and 6d - 2% Mean Coincident Dry Bulb Temperature

This value is the average of all the dry bulb temperatures that were observed over the long-term record whenever the humidity ratio was at the 2% value. Units are provided in whole degrees Fahrenheit and 1/10th of a degree Celsius.

7a and 7b - Summer Wind Speed

This value is the average speed of the wind whenever the *dry bulb temperature* was observed to be at the 0.4% value listed in the handbook of Fundamentals, beginning with the 1997 edition. It's a useful "summer wind" value to use when estimating the outdoor wind pressure for use in calculating infiltration through doors and cracks in the exterior envelope. Units are given in miles per hour and meters per second.

7c - Prevailing Wind Direction

Wind can hit the building from several points of the compass at the same time. These data describes the prevailing wind direction in an open field when the air's dry bulb temperature was at the 0.4% value described in the *Handbook—Fundamentals*.

The prevailing direction is that which was "most frequently observed" during the hours when the temperature was at its peak value. The units are degrees true: 360 indicates wind coming from directly north, 90 indicates wind from due east, etc.

References

Harriman, L.G. III, 1998. "New peak moisture design data in the 1997 ASHRAE Handbook—Fundamentals." *Proceedings of the Symposium on Improving Building Systems in Hot and Humid Climates*. Energy Systems Laboratory, Texas A&M University, College Station, TX 77843 (409) 847-8950 drosen@esl.tamu.edu

Harriman, Lewis G., III, Colliver, D.G., and Hart, K.Q., 1999. "New Engineering Weather Data for Design and Energy Calculations." *ASHRAE Journal*, March, 1999. pp: 31-48

Weather data for design load calculations

Climatic Design Information. 2001 *ASHRAE Handbook—Fundamentals, Chapter 27*, American Society of Heating Refrigerating and Air Conditioning Engineers, 1791 Tullie Circle NE Atlanta, GA (404) 636-8400 www.ashrae.org

Colliver, D.G., Gates, R.S, Zhang, T.F. Burkes and Priddy, T.K. 1998. *Updating the Tables of Design Weather Conditions in the ASHRAE Handbook - Fundamentals*, Final Report. of ASHRAE Research Project 890. ASHRAE, Atlanta, GA

Department of Defense Engineering Weather Data 1999. USAF Handbook 32-1163 (806 US and international locations). HQ AFCESA/CESA 139 Barnes Dr., Suite 1 Tyndall AFB, FL 32403-5319 Distributed to the public as "Engineering Weather Data" CD-ROM by the US National Climatic Data Center: http://www.ncdc.noaa.gov/

Hourly weather data for system simulation

TMY-2 Typical Meteorological Years (239 U.S. cities) *BinMaker CD-ROM*. Published by InterEnergy Software, Distributed by Linric Co. Bedford, NH. http://www.linric.com

WYEC-2 Weather Year for Energy Calculations (76 U.S. cities) American Society of Heating Refrigerating and Air Conditioning Engineers, 1791 Tullie Circle NE Atlanta, GA (404) 636-8400 www.ashrae.org

Canadian Weather Energy & Engineering Data Sets (CWEEDS) 1953-1995. Environment Canada, Atmospheric Environment Service 4905 Dufferin Street Downsview, ON M3H 5T4 (416) 739-4361

The CIBSE Example Weather Year. Weather Data and Its Applications - A Symposium for Building Service Engineers. 9 March 1988, Charted Institution of Building Services Engineers, Delta House, 222 Balham High Road, London SW12 9BS, Great Britain. Phone (44) 181-675-5211 fax (44) 181-675-5449 www.cibse.org

Test Reference Years TRY, Weather Data Sets for Computer Simulations of Solar Energy Systems and Energy Consumption in Buildings. Commission of the European Communities, Directorate General XII for Science, Research and Development. Communication Unit (XII-AP4) European Commission, SDME 2/85 Wetstraat 200 / Rue de la Loi, 200 B-1049 Brussel / Bruxelles BELGIÀ / BELGIQUE Phone (32-2) 295.25.59 http://europa.eu.int/comm/dg12/helpline.html

Special Thanks

In 1981, the Munters Dehumidification Division conducted research to generate the first-ever engineering weather data for dehumidification design. The results were published in 1982 in the first edition of their *Dehumidification Handbook*. Initially received with great skepticism, their data showed the HVAC industry that the then-current ASHRAE climatic design data were not adequate for dehumidification design.

Stimulated by that work, ASHRAE Technical Committees 3.5 (Desiccant and Sorption Technologies) and 4.2 (Weather Data), combined their efforts to sponsor a research project to create worldwide design data for dehumidification. The resulting project, ASHRAE 890-RP, was performed for the society by Donald Colliver and Richard Gates of the University of Kentucky. Their work produced the design data in the *ASHRAE Handbook—Fundamentals*, beginning with the 1997 edition. Those data are also the basis of this chapter.

The author is deeply grateful to all of these organizations and individuals for the fact that finally, after more than 100 years of ASHRAE history, designers throughout the world have the data needed to design dehumidification systems with the same degree of confidence as cooling and heating systems.

United States of America

HUMIDIFICATION 99.6% (35 hours below...)							99% (88 hours below...)												DEHUMIDIFICATION 0.4% (35 hours above...)				1% (88 hours above...)				2% (175 hours above...)				Summer Wind		
°F 1a	gr/lb 1b	°C 1c	g/kg 1d	mph 1e	m/s 1f	PWD 1g	°F 2a	gr/lb 2b	°C 2c	g/kg 2d	Location 3a	WMO# 3b	Lat. 3c	Long. 3d	ft. 3e	m 3f	psia 3g	kPa 3h	gr/lb 4a	°F 4b	g/kg 4c	°C 4d	gr/lb 5a	°F 5b	g/kg 5c	°C 5d	gr/lb 6a	°F 6b	g/kg 6c	°C 6d	mph 7a	m/s 7b	PWD 7c
											ALABAMA																						
19	15	-7.1	2.1	6	2.5	300	24	19	-4.5	2.7	Anniston	722287	33.58	85.85	610	186	14.37	99.1	143	84	20.4	29.1	137	82	19.6	28.0	133	81	19.0	27.2	7	3.3	240
18	14	-7.8	2.0	7	3.3	340	23	17	-5.2	2.5	Birmingham	722280	33.57	86.75	630	192	14.36	99.0	135	83	19.3	28.6	131	82	18.7	27.9	127	81	18.1	27.4	9	4.0	320
28	22	-2.4	3.2	9	4.0	320	32	27	-0.2	3.8	Dothan	722268	31.32	85.45	400	122	14.48	99.9	144	83	20.6	28.4	139	82	19.9	27.9	136	82	19.4	27.5	8	3.4	320
15	12	-9.4	1.7	10	4.6	340	20	16	-6.6	2.2	Huntsville	723230	34.65	86.77	643	196	14.36	99.0	135	83	19.3	28.6	130	82	18.6	27.9	126	81	18.0	27.3	10	4.3	270
26	21	-3.1	2.9	10	4.3	360	30	25	-1.0	3.5	Mobile	722230	30.68	88.25	220	67	14.58	100.5	142	83	20.3	28.5	139	83	19.8	28.1	135	82	19.3	27.8	9	3.9	320
24	18	-4.7	2.6	7	3.1	360	27	21	-2.8	3.0	Montgomery	722260	32.30	86.40	203	62	14.59	100.6	139	85	19.8	29.3	134	84	19.2	28.7	130	83	18.6	28.3	8	3.5	270
16	13	-8.8	1.8	9	3.9	360	21	16	-6.0	2.3	Muscle Shoals	723235	34.75	87.62	551	168	14.41	99.3	137	82	19.6	27.9	133	82	19.0	27.6	130	81	18.6	27.3	7	3.1	290
28	22	-2.5	3.2	5	2.4	340	31	26	-0.5	3.7	Ozark, Fort Rucker	722269	31.28	85.72	299	91	14.54	100.2	146	85	20.9	29.4	142	84	20.3	28.7	138	83	19.7	28.3	5	2.2	300
20	15	-6.7	2.2	5	2.1	360	24	18	-4.2	2.6	Tuscaloosa	722286	33.22	87.62	171	52	14.61	100.7	142	84	20.3	28.8	137	83	19.6	28.2	134	82	19.1	27.9	7	3.3	240
											ALASKA																						
19	14	-7.0	2.1	4	2.0	210	23	17	-5.1	2.5	Adak, NAS	704540	51.88	176.65	13	4	14.69	101.3	59	58	8.4	14.6	55	56	7.8	13.1	51	53	7.3	11.9	10	4.6	170
-13	3	-25.1	0.4	3	1.3	50	-8	4	-22.0	0.5	Anch. (Elmendorf AFB)	702720	61.25	149.80	213	65	14.58	100.5	69	62	9.9	16.5	65	61	9.3	16.2	61	60	8.7	15.6	7	3.1	260
-19	2	-28.2	0.3	3	1.4	50	-13	3	-25.0	0.4	Anch. (Ft. Richardson)	702700	61.27	149.65	377	115	14.50	100.0	69	64	9.8	18.0	63	62	9.0	16.5	61	61	8.7	16.2	5	2.2	270
-14	3	-25.6	0.4	4	1.7	10	-9	4	-22.5	0.5	Anchorage, Airport	702730	61.17	150.02	131	40	14.63	100.8	68	62	9.7	16.8	64	61	9.2	16.3	61	60	8.7	15.6	8	3.7	290
13	11	-10.8	1.5	10	4.2	40	17	13	-8.4	1.9	Annette	703980	55.03	131.57	112	34	14.64	100.9	71	65	10.2	18.3	68	63	9.7	17.0	65	61	9.3	15.9	8	3.8	320
-41	1	-40.3	0.1	7	3.3	140	-36	1	-38.0	0.1	Barrow	700260	71.30	156.78	13	4	14.69	101.3	53	54	7.5	12.1	46	51	6.6	10.5	42	48	6.0	8.8	12	5.3	90
-28	1	-33.3	0.2	13	5.6	20	-24	1	-31.1	0.2	Bethel	702190	60.78	161.80	151	46	14.62	100.8	68	62	9.7	16.4	64	60	9.1	15.3	60	58	8.6	14.7	12	5.1	360
-49	0	-45.2	0.0	2	0.8	340	-44	0	-42.3	0.1	Bettles	701740	66.92	151.52	643	196	14.36	99.0	72	66	10.3	18.9	67	64	9.6	17.9	63	63	9.0	17.1	8	3.6	190
-45	0	-42.8	0.1	3	1.4	180	-39	1	-39.5	0.1	Big Delta, Ft.Greely	702670	64.00	145.73	1283	391	14.03	96.7	70	65	10.0	18.6	65	63	9.3	17.1	61	61	8.7	16.2	9	3.8	180
6	8	-14.3	1.1	15	6.6	340	10	9	-12.1	1.3	Cold Bay	703160	55.20	162.73	102	31	14.64	101.0	62	56	8.9	13.6	59	55	8.4	12.8	56	54	8.0	12.2	16	7.1	140
-4	4	-20.2	0.6	1	0.4	340	1	6	-17.4	0.8	Cordova	702960	60.50	145.50	43	13	14.67	101.2	67	63	9.6	17.2	62	60	8.9	15.5	60	59	8.5	14.9	8	3.7	240
-36	1	-37.6	0.1	12	5.3	240	-34	1	-36.7	0.1	Deadhorse	700637	70.20	148.47	56	17	14.67	101.1	61	62	8.7	16.6	56	59	8.0	15.2	51	56	7.3	13.6	12	5.4	60
-20	2	-28.9	0.3	5	2.3	40	-13	3	-25.1	0.4	Dillingham	703210	59.05	158.52	95	29	14.64	101.0	67	62	9.6	16.5	61	59	8.7	15.2	57	57	8.2	14.1	10	4.3	180
-33	1	-36.1	0.1	0	0.1	150	-31	1	-34.8	0.1	Fairbanks, Eielson AFB	702650	64.67	147.10	548	167	14.41	99.3	74	66	10.5	19.1	69	66	9.9	18.9	63	66	9.0	18.9	5	2.1	290
-47	0	-44.0	0.1	2	0.7	10	-41	1	-40.7	0.1	Fairbanks, Airport	702610	64.82	147.87	453	138	14.46	99.7	72	65	10.3	18.2	68	64	9.7	17.7	64	63	9.1	17.3	8	3.7	220
-33	1	-36.3	0.1	0	0.2	270	-31	1	-35.2	0.1	Galena	702220	64.73	156.93	151	46	14.62	100.8	73	66	10.4	19.0	69	65	9.8	18.2	63	64	9.0	17.7	5	2.2	320
-44	0	-42.4	0.1	3	1.3	360	-39	1	-39.3	0.1	Gulkana	702710	62.15	145.45	1578	481	13.88	95.7	63	62	9.0	16.8	60	60	8.5	15.7	55	59	7.9	15.1	7	3.2	180
0	6	-17.7	0.8	9	3.8	30	4	7	-15.5	1.0	Homer	703410	59.63	151.50	72	22	14.66	101.1	64	59	9.1	14.9	60	58	8.6	14.7	57	57	8.2	14.1	10	4.3	270
4	7	-15.8	1.0	5	2.1	360	7	8	-13.7	1.1	Juneau	703810	58.37	134.58	23	7	14.68	101.2	70	63	10.0	17.3	67	61	9.6	16.1	64	60	9.2	15.4	9	4.1	230
-22	2	-29.8	0.2	2	1.1	30	-14	3	-25.5	0.4	Kenai	702590	60.57	151.25	95	29	14.64	101.0	64	59	9.2	15.0	60	58	8.6	14.4	58	57	8.3	14.0	9	4.2	270
13	11	-10.3	1.5	5	2.4	280	20	15	-6.7	2.2	Ketchikan	703950	55.35	131.70	95	29	14.64	101.0	74	64	10.5	17.5	71	62	10.1	16.6	68	61	9.7	16.0	11	4.8	320
-24	1	-31.1	0.2	7	3.0	360	-19	2	-28.4	0.3	King Salmon	703260	58.68	156.65	49	15	14.67	101.1	66	61	9.4	16.3	62	60	8.8	15.3	58	58	8.3	14.5	12	5.4	270
7	8	-13.8	1.1	18	8.1	300	12	10	-11.4	1.4	Kodiak, USCG Base	703500	57.75	152.50	112	34	14.64	100.9	67	61	9.6	16.3	64	59	9.1	15.2	61	57	8.7	14.1	11	4.8	320
-36	1	-37.6	0.1	7	3.0	70	-31	1	-35.2	0.1	Kotzebue	701330	66.87	162.63	16	5	14.69	101.3	70	64	10.0	17.6	65	61	9.3	16.3	62	59	8.8	15.2	12	5.1	300
-47	0	-43.8	0.1	1	0.6	310	-42	1	-40.8	0.1	Mcgrath	702310	62.97	155.62	338	103	14.52	100.1	69	63	9.9	17.3	65	62	9.3	16.7	62	61	8.8	15.9	7	3.3	340
18	14	-7.6	2.0	18	8.2	330	21	16	-6.0	2.3	Middleton Island	703430	59.43	146.33	46	14	14.67	101.2	57	56	8.2	13.3	56	56	8.0	13.1	54	55	7.7	12.9	8	3.7	260
-51	0	-46.2	0.0	2	1.1	250	-44	0	-42.5	0.1	Nenana	702600	64.55	149.08	361	110	14.50	100.0	69	65	9.9	18.3	65	65	9.3	18.2	60	63	8.6	17.0	7	3.0	60
-31	1	-34.8	0.1	4	1.6	20	-26	1	-32.3	0.2	Nome	702000	64.50	165.43	23	7	14.68	101.2	64	61	9.1	15.9	60	59	8.6	14.8	56	57	8.0	14.0	12	5.2	260
-34	1	-36.7	0.1	0	0.1	300	-32	1	-35.5	0.1	Northway	702910	62.97	141.93	1722	525	13.80	95.2	66	62	9.4	16.4	62	61	8.9	15.9	60	59	8.5	15.1	7	3.2	290
-6	4	-21.1	0.6	17	7.4	60	-2	5	-18.8	0.7	Port Heiden	703330	56.95	158.62	95	29	14.64	101.0	57	59	8.1	14.8	54	57	7.7	14.0	51	55	7.3	12.7	15	6.9	160
-2	5	-18.7	0.7	19	8.3	350	3	6	-16.1	0.9	Saint Paul Island	703080	57.15	170.22	30	9	14.68	101.2	55	52	7.8	11.1	52	51	7.4	10.4	50	50	7.2	9.8	14	6.4	240

United States of America

HUMIDIFICATION 99.6% (35 hours below...)							99% (88 hours below...)												DEHUMIDIFICATION 0.4% (35 hours above...)				1% (88 hours above...)				2% (175 hours above...)				Summer Wind		
°F 1a	gr/lb 1b	°C 1c	g/kg 1d	mph 1e	m/s 1f	PWD 1g	°F 2a	gr/lb 2b	°C 2c	g/kg 2d	Location 3a	WMO# 3b	Lat. 3c	Long. 3d	ft. 3e	m 3f	psia 3g	kPa 3h	gr/lb 4a	°F 4b	g/kg 4c	°C 4d	gr/lb 5a	°F 5b	g/kg 5c	°C 5d	gr/lb 6a	°F 6b	g/kg 6c	°C 6d	mph 7a	m/s 7b	PWD 7c
4	7	-15.5	1.0	15	6.6	70	7	8	-13.8	1.1	Valdez	702750	61.13	146.35	33	10	14.68	101.2	60	59	8.6	14.9	59	57	8.4	14.1	57	56	8.2	13.3	10	4.3	240
-3	5	-19.6	0.7	2	0.8	100	2	6	-16.4	0.9	Yakutat	703610	59.52	139.67	30	9	14.68	101.2	67	58	9.6	14.6	64	57	9.2	14.1	62	57	8.9	13.7	9	3.8	320
											ARIZONA																						
1	8	-17.0	1.1	3	1.5	20	8	11	-13.5	1.5	Flagstaff	723755	35.13	111.67	7011	2137	11.33	78.2	93	65	13.3	18.1	88	64	12.5	17.7	83	63	11.8	17.3	9	4.2	220
22	19	-5.7	2.7	5	2.4	90	27	24	-3.0	3.4	Kingman	723700	35.27	113.95	3389	1033	12.98	89.5	112	77	16.0	24.9	92	75	13.2	23.8	85	76	12.1	24.3	13	5.8	240
20	18	-6.7	2.5	4	2.0	300	24	21	-4.7	3.0	Page	723710	36.93	111.45	4278	1304	12.56	86.6	92	74	13.1	23.5	85	74	12.2	23.2	80	74	11.4	23.4	7	3.3	360
34	30	1.2	4.3	5	2.4	90	37	34	3.0	4.9	Phoenix, Int'l Airport	722780	33.43	112.02	1106	337	14.12	97.3	118	82	16.8	27.8	111	84	15.9	28.6	104	85	14.9	29.6	9	4.2	270
35	31	1.6	4.4	4	1.9	340	38	35	3.3	5.0	Phoenix, Luke AFB	722785	33.53	112.38	1089	332	14.13	97.4	130	85	18.6	29.6	118	85	16.9	29.7	111	86	15.9	30.2	9	3.8	210
15	14	-9.3	2.1	6	2.9	190	20	18	-6.9	2.5	Prescott	723723	34.65	112.42	5043	1537	12.21	84.2	104	71	14.9	21.6	98	71	14.0	21.4	93	70	13.3	21.3	11	4.8	230
21	18	-6.3	2.5	4	1.6	110	26	22	-3.1	3.2	Safford, Agri Center	722747	32.82	109.68	3117	950	13.11	90.4	111	77	15.8	24.9	106	76	15.2	24.6	102	77	14.5	25.0	7	3.0	310
31	28	-0.6	4.0	7	3.1	140	34	31	1.1	4.5	Tucson	722740	32.12	110.93	2556	779	13.39	92.3	116	76	16.5	24.3	111	76	15.8	24.5	106	77	15.1	24.9	12	5.2	300
10	11	-12.4	1.6	5	2.3	140	14	13	-9.8	1.9	Winslow	723740	35.02	110.73	4882	1488	12.28	84.7	95	71	13.6	21.5	91	69	13.0	20.8	85	69	12.2	20.7	9	4.2	250
40	37	4.7	5.3	4	2.0	30	44	43	6.4	6.1	Yuma	722800	32.65	114.60	207	63	14.59	100.6	136	87	19.4	30.7	127	88	18.2	31.2	117	89	16.7	31.5	7	3.0	280
											ARKANSAS																						
12	10	-11.1	1.5	10	4.3	10	18	14	-7.8	2.0	Blytheville, Eaker AFB	723408	35.97	89.95	256	78	14.56	100.4	149	88	21.3	31.2	142	86	20.3	30.2	135	85	19.3	29.2	6	2.9	240
6	8	-14.4	1.1	9	3.9	350	13	11	-10.6	1.6	Fayetteville	723445	36.00	94.17	1250	381	14.04	96.8	136	85	19.4	29.4	132	84	18.8	28.7	124	81	17.7	27.4	10	4.6	190
13	11	-10.6	1.6	9	4.0	320	19	14	-7.5	2.0	Fort Smith	723440	35.33	94.37	463	141	14.45	99.6	134	85	19.2	29.5	130	84	18.6	28.8	126	83	18.0	28.3	9	4.2	270
16	13	-8.9	1.8	9	4.2	360	21	16	-6.2	2.3	Little Rock, AFB	723405	34.92	92.15	312	95	14.53	100.2	141	86	20.2	29.8	137	85	19.6	29.3	133	84	19.0	28.8	9	4.0	200
20	15	-6.5	2.2	9	3.9	50	25	19	-4.0	2.8	Texarkana	723418	33.45	93.98	390	119	14.49	99.9	143	85	20.4	29.7	139	85	19.8	29.2	135	84	19.3	28.8	9	3.8	190
											CALIFORNIA																						
40	37	4.4	5.2	6	2.6	120	42	39	5.4	5.6	Alameda, NAS	745060	37.78	122.32	13	4	14.69	101.3	85	70	12.1	20.9	80	69	11.5	20.3	78	67	11.2	19.7	8	3.6	300
30	24	-1.1	3.5	5	2.2	90	32	27	0.1	3.9	Arcata/Eureka	725945	40.98	124.10	217	66	14.58	100.5	78	64	11.2	17.8	75	63	10.7	17.1	71	62	10.2	16.4	10	4.3	320
32	27	0.2	3.9	5	2.1	90	35	30	1.6	4.3	Bakersfield	723840	35.43	119.05	492	150	14.44	99.5	92	84	13.1	28.9	85	83	12.1	28.6	79	83	11.3	28.4	12	5.4	310
28	24	-2.1	3.4	6	2.6	270	32	28	-0.2	4.0	Barstow/Daggett	723815	34.85	116.78	1926	587	13.70	94.5	103	81	14.7	27.4	91	85	13.0	29.6	81	85	11.5	29.5	12	5.2	290
21	19	-6.0	2.7	5	2.3	70	24	22	-4.7	3.2	Blue Canyon	725845	39.28	120.72	5285	1611	12.10	83.4	74	70	10.6	21.1	69	70	9.9	21.2	64	68	9.2	20.1	6	2.7	290
39	36	4.0	5.2	2	0.9	330	41	39	5.1	5.6	Burbank/Glendale	722880	34.20	118.35	774	236	14.29	98.5	108	80	15.4	26.6	103	78	14.7	25.7	98	77	14.0	24.8	8	3.5	180
31	25	-0.3	3.6	4	1.6	20	34	29	1.0	4.1	Fairfield, Travis AFB	745160	38.27	121.93	62	19	14.66	101.1	85	76	12.1	24.7	79	74	11.3	23.5	75	73	10.7	22.9	9	4.2	240
30	24	-1.2	3.5	4	1.7	90	32	27	0.1	3.8	Fresno	723890	36.77	119.72	328	100	14.52	100.1	92	85	13.1	29.7	85	84	12.1	28.6	80	82	11.4	27.7	9	3.9	290
22	18	-5.7	2.6	2	1.0	260	24	20	-4.3	2.8	Lancaster/Palmdale	723816	34.73	118.22	2346	715	13.49	93.0	92	80	13.1	26.5	84	81	12.0	27.2	78	81	11.2	27.2	14	6.4	240
30	24	-1.2	3.5	4	1.6	150	32	27	0.2	3.8	Lemoore, Reeves NAS	747020	36.33	119.95	236	72	14.57	100.5	101	89	14.4	31.7	94	87	13.4	30.6	85	86	12.1	30.0	7	3.0	360
40	37	4.5	5.2	4	1.6	300	43	40	5.8	5.8	Long Beach	722970	33.82	118.15	39	12	14.67	101.2	101	76	14.4	24.4	96	75	13.7	24.1	92	75	13.2	23.7	10	4.6	270
43	41	6.2	5.9	6	2.8	70	45	45	7.4	6.4	Los Angeles	722950	33.93	118.40	105	32	14.64	100.9	99	75	14.2	23.6	95	73	13.6	22.8	92	72	13.1	22.1	10	4.4	250
31	25	-0.4	3.6	3	1.3	20	34	29	0.9	4.1	Marysville (Beale AFB)	724837	39.13	121.43	112	34	14.64	100.9	86	85	12.3	29.7	80	82	11.5	27.8	76	81	10.9	27.0	5	2.2	200
30	24	-1.0	3.5	2	0.9	110	32	27	0.1	3.8	Merced, Castle Afb	724810	37.38	120.57	187	57	14.60	100.6	90	81	12.8	27.2	82	84	11.7	29.1	78	81	11.2	27.4	9	3.9	320
16	14	-8.9	2.0	4	1.9	60	21	18	-6.3	2.6	Mount Shasta	725957	41.32	122.32	3543	1080	12.91	89.0	76	74	10.9	23.4	69	73	9.8	22.9	64	71	9.1	21.5	4	2.0	180
36	31	2.2	4.5	1	0.3	140	39	35	3.8	5.0	Mt. View, Moffet NAS	745090	37.42	122.05	39	12	14.67	101.2	83	74	11.9	23.6	80	73	11.4	23.0	76	72	10.9	22.2	9	3.8	330
35	31	1.4	4.4	4	2.0	10	38	35	3.4	5.0	Ontario	722865	34.05	117.60	942	287	14.20	97.9	113	80	16.1	26.5	106	80	15.1	26.8	101	78	14.4	25.6	13	5.8	240
39	35	3.8	5.0	5	2.3	20	41	38	5.0	5.4	Oxnard, Pt. Mugu NAWS	723910	34.12	119.12	7	2	14.69	101.3	103	74	14.7	23.4	97	73	13.9	22.8	93	72	13.3	22.2	12	5.2	50
26	21	-3.2	3.0	3	1.3	110	29	24	-1.8	3.4	Paso Robles	723965	35.67	120.63	837	255	14.26	98.3	81	76	11.6	24.2	75	73	10.7	22.8	71	71	10.2	21.6	11	4.9	300
29	23	-1.5	3.3	6	2.7	340	32	27	-0.1	3.8	Red Bluff	725910	40.15	122.25	354	108	14.51	100.0	94	82	13.4	27.6	85	80	12.2	26.4	80	78	11.5	25.3	9	4.2	160
34	30	1.0	4.3	1	0.6	210	36	33	2.2	4.7	Riverside, March Afb	722860	33.88	117.27	1539	469	13.90	95.8	104	79	14.9	26.1	97	80	13.9	26.7	90	79	12.8	26.3	9	4.1	300
30	24	-1.3	3.5	2	0.7	120	32	27	0.1	3.8	Sacmnto., Mather AFB	724835	38.55	121.30	95	29	14.64	101.0	80	79	11.5	26.1	77	77	11.0	25.1	74	76	10.5	24.2	6	2.5	310

United States of America

HUMIDIFICATION 99.6% (35 hours below...)							HUMIDIFICATION 99% (88 hours below...)				United States of America								DEHUMIDIFICATION 0.4% (35 hours above...)				1% (88 hours above...)				2% (175 hours above...)				Summer Wind		
°F	gr/lb	°C	g/kg	mph	m/s	PWD	°F	gr/lb	°C	g/kg	Location	WMO#	Lat.	Long.	ft.	m	psia	kPa	gr/lb	°F	g/kg	°C	gr/lb	°F	g/kg	°C	gr/lb	°F	g/kg	°C	mph	m/s	PWD
1a	1b	1c	1d	1e	1f	1g	2a	2b	2c	2d	3a	3b	3c	3d	3e	3f	3g	3h	4a	4b	4c	4d	5a	5b	5c	5d	6a	6b	6c	6d	7a	7b	7c
31	25	-0.4	3.6	2	1.0	340	34	29	1.0	4.1	Sacmnto., McCl. AFB	724836	38.67	121.40	75	23	14.66	101.0	85	84	12.2	28.7	80	81	11.5	27.0	77	79	11.0	26.2	5	2.4	220
31	25	-0.8	3.6	3	1.3	340	33	28	0.6	3.9	Sacramento, Metro	724839	38.70	121.58	23	7	14.68	101.2	84	82	12.0	27.9	79	80	11.3	26.4	76	78	10.8	25.4	8	3.8	220
33	28	0.7	4.0	6	2.9	130	35	30	1.9	4.3	Salinas	724917	36.67	121.60	85	26	14.65	101.0	82	69	11.7	20.7	78	68	11.2	19.9	76	67	10.8	19.2	11	4.7	310
34	30	1.0	4.3	2	0.7	50	36	32	2.3	4.6	San Bernadino (Geo.)	722866	34.10	117.23	1158	353	14.09	97.2	107	83	15.3	28.5	101	83	14.4	28.4	95	82	13.6	28.0	8	3.4	250
44	43	6.7	6.1	3	1.5	70	46	46	7.9	6.6	San Diego, Airport	722900	32.73	117.17	30	9	14.68	101.2	111	77	15.8	25.1	104	76	14.9	24.3	99	74	14.2	23.5	10	4.6	310
39	36	4.0	5.1	3	1.2	90	42	40	5.3	5.7	San Diego, Miramar	722930	32.85	117.12	420	128	14.47	99.8	104	78	14.8	25.6	99	77	14.2	24.8	96	75	13.7	23.9	6	2.8	310
37	32	2.7	4.6	5	2.4	160	39	35	3.9	5.0	San Francisco	724940	37.62	122.38	16	5	14.69	101.3	76	67	10.8	19.4	73	66	10.4	18.9	71	65	10.1	18.5	13	5.6	300
35	30	1.6	4.3	1	0.4	160	38	34	3.1	4.8	San Jose Int'l Airport	724945	37.37	121.93	56	17	14.67	101.1	85	77	12.2	24.8	81	76	11.6	24.2	78	74	11.2	23.3	10	4.4	320
34	29	1.2	4.1	1	0.6	40	37	32	2.8	4.6	Santa Barbara	723925	34.43	119.83	10	3	14.69	101.3	96	74	13.7	23.1	91	71	13.0	21.9	85	70	12.2	21.2	10	4.4	260
32	27	0.1	3.8	4	1.9	110	35	30	1.4	4.2	Santa Maria	723940	34.90	120.45	240	73	14.57	100.5	80	70	11.4	21.0	77	69	11.0	20.7	74	68	10.5	19.9	11	4.7	300
30	24	-1.1	3.5	4	1.8	110	32	27	0.1	3.8	Stockton	724920	37.90	121.25	26	8	14.68	101.2	83	78	11.8	25.3	78	78	11.2	25.3	75	77	10.7	24.8	11	4.7	280
27	23	-2.7	3.3	3	1.4	160	30	27	-1.1	3.8	Victorville, George	723825	34.58	117.38	2874	876	13.23	91.2	102	78	14.5	25.5	90	79	12.8	26.0	83	78	11.8	25.4	9	4.0	180
											COLORADO																						
-17	3	-27.4	0.4	3	1.5	190	-11	4	-24.0	0.6	Alamosa	724620	37.45	105.87	7543	2299	11.11	76.6	87	62	12.4	16.7	81	62	11.6	16.5	77	62	11.0	16.9	12	5.4	240
-2	6	-18.7	0.9	7	3.2	20	4	9	-15.3	1.2	Colorado Springs	724660	38.82	104.72	6171	1881	11.70	80.7	92	66	13.2	19.1	88	66	12.5	18.7	83	65	11.9	18.5	12	5.4	160
-20	2	-28.8	0.3	2	1.1	270	-12	4	-24.7	0.5	Craig	725700	40.50	107.53	6283	1915	11.65	80.3	77	66	11.0	19.1	72	65	10.3	18.6	68	64	9.7	17.9	9	3.9	250
-3	6	-19.7	0.8	6	2.7	180	3	8	-16.1	1.1	Denver	724699	39.75	104.87	5331	1625	12.08		96	69	13.7	20.4	90	68	12.8	20.1	85	68	12.1	20.1	9	4.1	160
-13	3	-25.0	0.5	3	1.2	90	-7	5	-21.7	0.7	Eagle	724675	39.65	106.92	6539	1993	11.54	79.6	88	64	12.5	17.8	82	65	11.7	18.1	76	65	10.8	18.3	11	5.0	230
2	7	-16.8	1.0	5	2.2	70	7	10	-13.8	1.4	Grand Junction	724760	39.12	108.53	4839	1475	12.30	84.8	93	70	13.3	20.9	87	71	12.4	21.7	79	71	11.3	21.9	11	4.9	290
-6	5	-21.2	0.7	9	3.9	160	1	7	-17.0	1.0	Limon	724665	39.27	103.67	5364	1635	12.06	83.2	96	67	13.7	19.2	92	66	13.1	18.7	88	66	12.6	18.7	12	5.4	200
-1	6	-18.5	0.9	5	2.3	270	5	8	-15.0	1.2	Pueblo	724640	38.28	104.52	4721	1439	12.36	85.2	104	71	14.8	21.7	98	71	14.0	21.7	92	71	13.2	21.6	12	5.4	140
-2	6	-18.8	0.9	5	2.3	290	6	9	-14.7	1.3	Trinidad	724645	37.27	104.33	5761	1756	11.88	81.9	96	71	13.7	21.4	91	69	13.0	20.7	86	69	12.3	20.8	10	4.6	210
											CONNECTICUT																						
8	8	-13.3	1.2	14	6.1	320	12	10	-10.9	1.5	Bridgeport	725040	41.17	73.13	16	5	14.69	101.3	126	79	18.0	26.2	120	78	17.1	25.6	115	77	16.4	24.9	14	6.0	230
2	6	-16.9	0.9	7	3.3	320	6	8	-14.3	1.1	Hartford, Brainard	725087	41.73	72.65	20	6	14.68	101.3	121	81	32.6	21.8	116	80	30.9	20.9	110	78	29.5	-17.8	11	4.9	250
3	6	-15.9	0.9	7	3.3	360	8	8	-13.2	1.2	Windsor L., Bradley	725080	41.93	72.68	180	55	14.60	100.7	119	81	17.0	27.2	114	79	16.3	26.1	109	77	15.6	25.2	11	4.7	240
											DELAWARE																						
14	11	-10.2	1.6	8	3.6	340	18	14	-7.5	2.0	Dover, AFB	724088	39.13	75.47	30	9	14.68	101.2	141	84	20.1	28.8	135	82	19.3	27.8	129	81	18.4	27.3	9	3.9	240
10	9	-12.4	1.3	11	5.1	290	14	11	-9.9	1.6	Wilmington	724089	39.68	75.60	79	24	14.65	101.0	132	82	18.8	27.9	125	81	17.9	27.1	120	80	17.2	26.6	11	4.7	240
											FLORIDA																						
31	25	-0.5	3.6	6	2.5	360	35	30	1.7	4.3	Apalachicola	722200	29.73	85.03	20	6	14.68	101.3	148	85	21.2	29.3	145	84	20.7	29.1	141	84	20.2	28.7	9	4.1	220
38	34	3.1	4.8	8	3.6	320	42	39	5.3	5.6	Cape Canaveral	747946	28.62	80.72	10	3	14.69	101.3	145	84	20.7	29.1	141	84	20.2	28.7	138	83	19.7	28.2	8	3.5	220
34	28	0.9	4.0	7	3.2	310	37	33	3.0	4.7	Daytona Beach	722056	29.18	81.05	36	11	14.68	101.2	141	84	20.1	29.1	137	84	19.6	28.7	134	83	19.1	28.3	11	4.8	240
46	46	7.9	6.6	9	4.0	330	50	54	10.2	7.7	Fort Lauderdale	722025	26.07	80.15	23	7	14.68	101.2	147	85	21.0	29.2	145	84	20.7	29.0	141	84	20.2	28.8	11	4.8	120
42	39	5.7	5.6	6	2.8	30	47	48	8.2	6.8	Fort Myers	722106	26.58	81.87	16	5	14.69	101.3	147	84	21.0	28.7	144	83	20.5	28.5	140	83	20.0	28.2	9	4.0	70
30	24	-1.0	3.5	4	1.8	300	33	28	0.8	4.0	Gainesville	722146	29.68	82.27	151	46	14.62	100.8	143	84	20.4	28.9	139	83	19.8	28.4	136	82	19.5	28.0	9	3.8	270
48	50	8.8	7.1	6	2.6	360	52	58	10.9	8.3	Homestead, AFB	722026	25.48	80.38	7	2	14.69	101.3	150	87	21.4	30.3	145	86	20.7	29.8	141	85	20.2	29.6	7	3.2	120
31	25	-0.8	3.6	3	1.5	290	34	29	1.0	4.1	Jacksonville, Cecil	722067	30.22	81.88	82	25	14.65	101.0	138	84	19.7	28.8	134	83	19.1	28.2	130	82	18.6	27.9	7	3.0	270
29	23	-1.7	3.3	6	2.7	310	32	27	0.2	3.8	Jacksonville, Airport	722060	30.50	81.70	30	9	14.68	101.2	142	85	20.3	29.2	138	84	19.7	28.7	134	83	19.2	28.3	9	3.9	230
34	29	1.2	4.1	6	2.7	310	39	35	3.8	5.0	Jacksonville, Mayport	722066	30.40	81.42	16	5	14.69	101.3	147	86	21.0	29.9	142	85	20.3	29.6	139	85	19.8	29.2	7	3.3	270
55	64	12.6	9.1	12	5.5	50	58	72	14.4	10.3	Key West	722010	24.55	81.75	20	6	14.68	101.3	149	85	21.3	29.6	146	85	20.8	29.4	143	85	20.4	29.2	9	4.2	140
38	34	3.5	4.8	9	4.1	320	43	41	5.9	5.9	Melbourne	722040	28.10	80.65	36	11	14.68	101.2	155	86	22.2	30.0	150	85	21.5	29.7	146	85	20.9	29.4	11	4.9	120
46	45	7.6	6.5	10	4.5	340	50	53	9.8	7.5	Miami, Int'l Airport	722020	25.82	80.28	13	4	14.69	101.3	144	83	20.6	28.3	141	83	20.1	28.1	138	82	19.7	28.0	11	5.0	150

United States of America

HUMIDIFICATION 99.6% (35 hours below...)							99% (88 hours below...)												DEHUMIDIFICATION 0.4% (35 hours above...)				1% (88 hours above...)				2% (175 hours above...)				Summer Wind		
°F 1a	gr/lb 1b	°C 1c	g/kg 1d	mph 1e	m/s 1f	PWD 1g	°F 2a	gr/lb 2b	°C 2c	g/kg 2d	Location 3a	WMO# 3b	Lat. 3c	Long. 3d	ft. 3e	m 3f	psia 3g	kPa 3h	gr/lb 4a	°F 4b	g/kg 4c	°C 4d	gr/lb 5a	°F 5b	g/kg 5c	°C 5d	gr/lb 6a	°F 6b	g/kg 6c	°C 6d	mph 7a	m/s 7b	PWD 7c
45	44	7.1	6.3	8	3.6	360	49	52	9.6	7.4	Miami/Tamiami Airport	722029	25.65	80.43	10	3	14.69	101.3	145	83	20.7	28.5	141	83	20.2	28.3	138	83	19.7	28.1	11	5.0	130
28	22	-2.3	3.2	6	2.9	340	31	26	-0.3	3.6	Milton, Whiting NAS	722226	30.72	87.02	200	61	14.59	100.6	148	86	21.1	29.9	143	85	20.4	29.2	138	84	19.7	28.7	6	2.7	330
37	33	2.9	4.6	8	3.6	330	42	40	5.3	5.7	Orlando	722050	28.43	81.32	105	32	14.64	100.9	142	83	20.3	28.2	139	82	19.8	27.8	136	81	19.4	27.4	9	3.9	290
33	28	0.5	3.9	9	3.8	360	37	32	2.9	4.6	Panama City, Tyndall	747750	30.07	85.58	16	5	14.69	101.3	160	86	22.9	30.1	154	85	22.0	29.4	150	84	21.4	29.1	7	3.3	240
28	22	-2.0	3.2	9	4.1	360	32	27	-0.1	3.8	Pensacola, Sherman	722225	30.35	87.32	30	9	14.68	101.2	150	85	21.4	29.6	144	85	20.5	29.4	138	85	19.7	29.3	10	4.5	200
43	41	6.1	5.9	11	4.8	10	47	48	8.5	6.8	Saint Petersburg	722116	27.92	82.68	10	3	14.69	101.3	156	86	22.3	29.9	153	85	21.9	29.6	150	85	21.4	29.5	10	4.4	230
39	35	3.8	5.0	5	2.1	40	43	41	6.0	5.9	Sarasota/Bradenton	722115	27.40	82.55	30	9	14.68	101.2	153	87	21.8	30.4	148	86	21.2	30.2	146	86	20.9	29.8	9	4.1	270
25	19	-4.0	2.7	3	1.2	350	28	22	-2.1	3.2	Tallahassee	722140	30.38	84.37	69	21	14.66	101.1	142	83	20.3	28.3	138	82	19.7	27.8	135	82	19.3	27.5	8	3.6	360
36	31	2.2	4.5	8	3.5	20	40	36	4.2	5.1	Tampa, Int'l Airport	722110	27.97	82.53	10	3	14.69	101.3	144	85	20.5	29.2	140	84	20.0	28.7	137	83	19.5	28.5	10	4.4	270
30	24	-1.3	3.5	6	2.6	360	33	28	0.6	4.0	Valparaiso, Eglin AFB	722210	30.48	86.53	85	26	14.65	101.0	149	85	21.3	29.4	144	84	20.6	28.8	141	83	20.1	28.4	7	3.3	210
39	35	4.0	5.0	8	3.6	310	43	41	6.1	5.9	Vero Beach	722045	27.65	80.42	26	8	14.68	101.2	141	85	20.2	29.2	139	84	19.9	28.9	137	84	19.6	28.8	11	4.7	240
43	40	5.9	5.8	9	4.2	320	47	47	8.1	6.7	West Palm Beach	722030	26.68	80.12	20	6	14.68	101.3	143	84	20.4	28.9	139	84	19.9	28.7	137	83	19.5	28.5	12	5.4	110
											GEORGIA																						
27	21	-3.0	3.0	4	1.6	360	30	24	-1.2	3.5	Albany	722160	31.53	84.18	194	59	14.59	100.6	141	83	20.1	28.2	136	82	19.5	27.9	133	81	19.0	27.4	9	3.8	250
20	16	-6.6	2.2	10	4.2	290	25	19	-4.2	2.7	Athens	723110	33.95	83.32	810	247	14.27	98.4	133	82	19.0	27.8	129	81	18.4	27.4	125	80	17.9	26.9	9	3.9	270
18	14	-7.9	2.0	12	5.5	320	23	18	-4.9	2.6	Atlanta	722190	33.65	84.42	1033	315	14.15	97.6	133	82	19.0	27.8	128	81	18.3	27.1	124	80	17.7	26.6	9	4.1	300
21	16	-6.1	2.3	5	2.4	290	25	19	-4.1	2.7	Augusta	722180	33.37	81.97	148	45	14.62	100.8	135	84	19.3	28.6	130	83	18.6	28.1	127	82	18.1	27.7	9	4.1	250
30	24	-0.9	3.5	8	3.5	350	34	29	0.9	4.1	Brunswick	722137	31.15	81.38	20	6	14.68	101.3	147	86	21.0	29.9	144	85	20.6	29.4	141	84	20.1	28.9	10	4.5	250
23	18	-4.8	2.5	3	1.4	320	27	21	-2.6	3.0	Columbus, Ft Benning	722250	32.33	85.00	233	71	14.57	100.5	142	85	20.3	29.2	136	83	19.5	28.5	133	82	19.0	27.9	5	2.2	240
23	18	-4.8	2.6	7	3.0	310	27	22	-2.6	3.1	Columbus, Airport	722255	32.52	84.93	397	121	14.49	99.9	139	82	19.8	27.9	134	82	19.2	27.5	130	81	18.6	27.2	9	3.8	310
23	18	-5.1	2.5	7	3.1	320	27	21	-2.8	3.0	Macon	722170	32.70	83.65	361	110	14.50	100.0	136	83	19.4	28.4	132	82	18.9	27.9	129	82	18.4	27.5	9	4.2	270
21	16	-6.2	2.3	9	4.0	340	26	21	-3.2	3.0	Marietta, Dobbins AFB	722270	33.92	84.52	1070	326	14.14	97.5	134	82	19.2	27.6	130	81	18.5	27.1	123	79	17.6	26.1	6	2.5	300
15	12	-9.4	1.7	5	2.3	340	21	16	-6.0	2.3	Rome	723200	34.35	85.17	643	196	14.36	99.0	134	83	19.2	28.6	130	83	18.6	28.1	127	83	18.1	28.1	6	2.6	270
26	20	-3.5	2.8	7	3.0	270	29	23	-1.6	3.3	Savannah	722070	32.13	81.20	49	15	14.67	101.1	139	84	19.8	28.7	135	83	19.3	28.1	132	82	18.8	27.8	9	4.2	270
30	24	-0.9	3.5	4	1.7	360	34	29	1.0	4.1	Valdosta, Moody AFB	747810	30.97	83.20	233	71	14.57	100.5	142	85	20.3	29.5	139	84	19.8	28.8	135	83	19.3	28.5	5	2.4	300
28	22	-2.5	3.2	4	1.7	340	31	26	-0.8	3.6	Valdosta, Airport	722166	30.78	83.28	203	62	14.59	100.6	144	83	20.5	28.6	139	82	19.9	28.0	136	82	19.4	27.6	8	3.4	300
29	23	-1.7	3.3	4	1.6	250	32	27	-0.1	3.8	Waycross	722130	31.25	82.40	151	46	14.62	100.8	134	84	19.1	28.9	130	83	18.6	28.5	127	83	18.2	28.2	7	3.3	240
											HAWAII																						
59	75	14.9	10.7	5	2.2	40	61	81	16.2	11.5	Ewa, Barbers Pt NAS	911780	21.32	158.07	49	15	14.67	101.1	126	83	18.0	28.1	118	82	16.8	27.8	113	82	16.2	27.5	11	4.8	60
61	82	16.3	11.7	7	3.2	230	63	86	17.1	12.2	Hilo	912850	19.72	155.07	36	11	14.68	101.2	130	79	18.6	26.3	127	79	18.1	25.9	123	78	17.6	25.7	12	5.4	110
61	80	16.0	11.4	5	2.3	320	63	86	17.2	12.3	Honolulu	911820	21.35	157.93	16	5	14.69	101.3	125	80	17.8	26.6	120	80	17.2	26.4	116	79	16.6	26.2	15	6.6	60
59	75	14.9	10.7	6	2.5	160	61	80	15.9	11.4	Kahului	911900	20.90	156.43	66	20	14.66	101.1	127	80	18.1	26.9	122	80	17.4	26.8	118	80	16.9	26.7	19	8.3	50
67	100	19.3	14.2	7	3.3	190	68	103	19.9	14.8	Kaneohe, MCAS	911760	21.45	157.77	10	3	14.69	101.3	138	81	19.7	27.1	133	81	19.0	27.0	128	80	18.3	26.9	10	4.5	70
60	78	15.5	11.1	8	3.6	270	62	83	16.6	11.9	Lihue	911650	21.98	159.35	148	45	14.62	100.8	132	80	18.9	26.9	128	80	18.3	26.6	125	79	17.8	26.3	14	6.0	60
60	79	15.3	11.3	4	1.7	70	61	82	16.3	11.7	Molokai	911860	21.15	157.10	449	137	14.46	99.7	128	80	18.3	26.4	124	79	17.7	26.3	118	79	16.8	26.0	13	5.9	60
											IDAHO																						
2	7	-16.8	1.0	6	2.5	130	9	10	-12.6	1.4	Boise	726810	43.57	116.22	2867	874	13.24	91.3	79	72	11.3	22.3	72	71	10.3	21.8	67	71	9.5	21.8	11	4.7	320
-5	5	-20.4	0.7	7	3.0	60	2	7	-16.7	1.0	Burley	725867	42.55	113.77	4150	1265	12.62	87.0	90	75	12.9	23.9	84	72	12.0	22.2	78	72	11.1	22.1	8	3.6	280
-12	3	-24.4	0.5	7	3.3	360	-6	5	-20.9	0.7	Idaho Falls	725785	43.52	112.07	4741	1445	12.35	85.1	88	71	12.5	21.4	81	69	11.6	20.6	73	68	10.4	20.1	12	5.2	180
6	8	-14.2	1.1	5	2.2	280	15	12	-9.7	1.8	Lewiston	727830	46.38	117.02	1437	438	13.95	96.2	76	72	10.9	22.4	71	71	10.2	21.5	65	71	9.3	21.5	7	3.1	310
0	6	-17.9	0.9	2	1.0	90	5	8	-14.8	1.1	Mountain Home, AFB	726815	43.05	115.87	2995	913	13.17	90.8	79	71	11.3	21.4	70	69	10.0	20.5	64	71	9.2	21.6	8	3.5	350
-1	6	-18.5	0.8	2	0.8	10	7	9	-13.8	1.3	Mullan	727836	47.47	115.80	3317	1011	13.02	89.8	86	69	12.3	20.4	80	68	11.4	19.9	75	66	10.7	19.1	4	1.8	10

United States of America

HUMIDIFICATION 99.6% (35 hours below...) °F 1a	gr/lb 1b	°C 1c	g/kg 1d	mph 1e	m/s 1f	PWD 1g	99% (88 hours below...) °F 2a	gr/lb 2b	°C 2c	g/kg 2d	Location 3a	WMO# 3b	Lat. 3c	Long. 3d	ft. 3e	m 3f	psia 3g	kPa 3h	DEHUMIDIFICATION 0.4% (35 hours above...) gr/lb 4a	°F 4b	g/kg 4c	°C 4d	1% (88 hours above...) gr/lb 5a	°F 5b	g/kg 5c	°C 5d	2% (175 hours above...) gr/lb 6a	°F 6b	g/kg 6c	°C 6d	Summer Wind mph 7a	m/s 7b	PWD 7c
-7	4	-21.6	0.6	6	2.5	50	0	7	-17.7	0.9	Pocatello	725780	42.92	112.60	4478	1365	12.47	86.0	83	70	11.9	20.8	76	70	10.9	21.1	70	69	10.0	20.4	11	5.0	250
											ILLINOIS																						
3	7	-16.2	0.9	7	3.2	360	10	9	-12.4	1.3	Belleville, Scott AFB	724338	38.55	89.85	453	138	14.46	99.7	141	87	20.2	30.6	136	85	19.4	29.7	131	84	18.7	29.0	7	3.1	190
-4	5	-20.0	0.7	12	5.4	240	3	7	-16.1	0.9	Chicago, Meigs Field	725340	41.78	87.75	623	190	14.37	99.1	132	84	18.9	28.7	121	80	17.3	26.9	115	80	16.4	26.7	13	5.6	220
-6	4	-21.2	0.6	10	4.6	270	-1	5	-18.1	0.8	Chicago, O'Hare Int'l Airport	725300	41.98	87.90	673	205	14.34	98.9	130	84	18.6	28.9	123	82	17.5	27.8	115	80	16.4	26.6	12	5.4	230
-2	5	-19.0	0.7	13	5.7	310	3	7	-15.9	0.9	Decatur	725316	39.83	88.87	682	208	14.34	98.9	140	86	20.0	30.1	133	84	19.0	29.0	127	83	18.1	28.2	12	5.2	210
-3	5	-19.7	0.7	11	4.7	250	4	7	-15.7	1.0	Glenview, NAS	725306	42.08	87.82	653	199	14.35	99.0	130	85	18.6	29.5	120	82	17.2	28.0	113	81	16.2	27.3	10	4.3	240
-5	4	-20.4	0.6	12	5.2	290	1	6	-17.3	0.9	Marseilles	744600	41.37	88.68	738	225	14.31	98.7	135	85	19.3	29.4	126	82	18.0	27.6	117	81	16.7	27.2	10	4.5	250
-8	4	-22.4	0.5	9	4.2	290	-3	5	-19.3	0.7	Moline/Davenport IA	725440	41.45	90.52	594	181	14.38	99.2	134	85	19.2	29.5	127	83	18.1	28.3	120	82	17.2	27.5	12	5.2	200
-6	4	-21.1	0.6	9	4.1	290	-1	5	-18.1	0.8	Peoria	725320	40.67	89.68	663	202	14.35	98.9	137	85	19.5	29.2	130	83	18.5	28.3	123	81	17.6	27.2	11	4.9	180
-4	5	-19.9	0.7	12	5.4	330	2	6	-16.6	0.9	Quincy	724396	39.95	91.20	768	234	14.29	98.5	138	84	19.7	28.8	132	82	18.8	28.0	126	82	18.0	27.6	12	5.2	210
-10	3	-23.1	0.5	9	3.9	290	-4	5	-20.1	0.6	Rockford	725430	42.20	89.10	741	226	14.31	98.6	132	84	18.8	28.8	124	81	17.7	27.3	116	79	16.6	26.2	13	5.6	200
-4	5	-19.7	0.7	10	4.5	270	2	6	-16.6	0.9	Springfield	724390	39.85	89.67	614	187	14.37	99.1	139	86	19.9	29.8	132	84	18.9	28.7	125	82	17.9	27.6	12	5.3	230
-7	4	-21.6	0.6	11	5.1	290	0	6	-18.0	0.8	West Chicago	725305	41.92	88.25	758	231	14.30	98.6	138	85	19.7	29.3	130	83	18.6	28.1	119	80	17.0	26.8	11	4.8	240
											INDIANA																						
3	6	-16.2	0.9	7	3.1	320	9	9	-12.8	1.3	Evansville	724320	38.05	87.53	387	118	14.49	99.9	137	86	19.6	29.8	132	84	18.8	28.9	126	83	18.0	28.1	9	4.2	240
-4	5	-19.9	0.7	10	4.4	250	2	6	-16.9	0.9	Fort Wayne	725330	41.00	85.20	827	252	14.26	98.3	131	83	18.7	28.2	124	81	17.7	27.3	117	79	16.7	26.2	12	5.2	230
-3	5	-19.2	0.7	8	3.7	230	3	7	-16.1	1.0	Indianapolis	724380	39.73	86.27	807	246	14.27	98.4	137	84	19.6	28.7	131	82	18.7	28.0	125	81	17.8	26.9	11	4.8	230
-5	4	-20.5	0.6	9	4.2	270	3	7	-16.3	0.9	Lafayette, Purdue.	724386	40.42	86.93	607	185	14.38	99.1	139	85	19.8	29.5	132	83	18.8	28.3	125	82	17.9	27.5	12	5.3	220
-3	5	-19.2	0.7	11	4.9	270	4	7	-15.3	1.0	Peru, Grissom AFB	725335	40.65	86.15	810	247	14.27	98.4	142	85	20.3	29.7	134	83	19.1	28.3	127	81	18.2	27.3	9	4.1	210
-2	5	-18.9	0.7	13	5.7	230	3	7	-16.2	0.9	South Bend	725350	41.70	86.32	774	236	14.29	98.5	130	83	18.5	28.1	123	80	17.5	26.8	116	78	16.6	25.7	12	5.3	230
-3	5	-19.2	0.7	8	3.5	150	5	7	-15.1	1.0	Terre Haute	724373	39.45	87.32	584	178	14.39	99.2	144	86	20.5	29.9	136	84	19.5	28.9	131	82	18.7	27.9	11	4.7	230
											IOWA																						
-4	5	-20.0	0.7	9	3.9	310	1	6	-17.4	0.9	Burlington	725455	40.78	91.13	699	213	14.33	98.8	136	85	19.5	29.3	131	83	18.7	28.6	124	82	17.7	27.8	11	4.9	200
-11	3	-24.0	0.4	10	4.6	300	-5	4	-20.6	0.6	Cedar Rapids	725450	41.88	91.70	869	265	14.24	98.2	136	84	19.4	28.8	129	83	18.4	28.3	120	80	17.1	26.8	11	4.7	180
-9	4	-22.7	0.5	11	4.9	320	-4	5	-19.9	0.7	Des Moines	725460	41.53	93.65	965	294	14.19	97.8	133	85	19.0	29.4	126	83	18.0	28.4	120	81	17.1	27.4	12	5.5	180
-13	3	-25.0	0.4	11	4.7	340	-7	4	-21.9	0.6	Fort Dodge	725490	42.55	94.18	1165	355	14.09	97.1	133	84	19.0	29.0	123	82	17.5	27.9	116	79	16.6	26.3	11	5.0	190
-6	4	-21.1	0.6	7	3.2	320	0	6	-17.7	0.8	Lamoni	725466	40.62	93.95	1122	342	14.11	97.3	134	83	19.1	28.3	127	82	18.2	27.6	120	80	17.1	26.8	9	3.9	210
-15	3	-26.2	0.4	12	5.3	300	-10	3	-23.5	0.5	Mason City	725485	43.15	93.33	1214	370	14.06	97.0	135	84	19.3	28.6	126	82	18.0	27.5	118	80	16.9	26.4	14	6.3	200
-5	4	-20.6	0.6	13	5.9	320	0	6	-17.9	0.8	Ottumwa	725465	41.10	92.45	846	258	14.25	98.3	136	84	19.4	28.9	130	83	18.5	28.4	121	81	17.3	27.3	15	6.5	200
-11	3	-23.8	0.5	11	4.7	320	-6	4	-21.1	0.6	Sioux City	725570	42.40	96.38	1102	336	14.12	97.4	135	86	19.3	29.7	127	84	18.2	28.8	120	82	17.1	27.7	14	6.4	180
-16	2	-26.5	0.3	10	4.6	300	-11	3	-23.8	0.5	Spencer	726500	43.17	95.15	1339	408	14.00	96.5	134	84	19.1	29.0	123	82	17.6	27.5	117	79	16.7	25.9	12	5.4	180
-14	3	-25.7	0.4	9	4.1	300	-9	3	-22.9	0.5	Waterloo	725480	42.55	92.40	879	268	14.23	98.1	132	84	18.9	28.6	124	82	17.7	27.6	117	80	16.7	26.5	13	5.9	180
											KANSAS																						
-4	5	-19.9	0.7	13	6.0	360	3	7	-16.0	1.0	Concordia	724580	39.55	97.65	1483	452	13.92	96.0	133	84	19.0	28.9	123	82	17.6	27.5	118	81	16.9	27.1	16	7.2	10
0	6	-17.7	0.9	13	5.7	10	6	8	-14.7	1.2	Dodge City	724510	37.77	99.97	2592	790	13.37	92.2	120	79	17.1	26.1	114	78	16.3	25.4	109	77	15.6	25.2	17	7.7	200
-2	5	-19.0	0.7	5	2.2	350	5	7	-14.9	1.1	Ft Riley, Marshall AAF	724550	39.05	96.77	1066	325	14.14	97.5	136	86	19.5	29.8	130	83	18.5	28.6	120	82	17.2	28.0	9	4.2	180
-3	5	-19.2	0.7	12	5.3	360	4	8	-15.5	1.1	Garden City	724515	37.93	100.72	2890	881	13.22	91.2	118	79	16.8	26.1	113	78	16.1	25.6	108	77	15.5	25.1	16	7.2	190
-3	5	-19.7	0.8	12	5.4	270	2	7	-16.5	1.0	Goodland	724650	39.37	101.70	3688	1124	12.84	88.5	111	74	15.9	23.5	106	73	15.1	23.0	100	73	14.3	22.8	13	5.9	180
-4	5	-20.1	0.7	11	5.1	10	3	7	-16.0	1.0	Russell	724585	38.87	98.82	1864	568	13.73	94.7	126	83	18.0	28.3	120	82	17.2	27.5	116	80	16.5	26.6	16	7.3	190
-3	5	-19.4	0.7	11	4.7	360	4	7	-15.4	1.0	Salina	724586	38.80	97.65	1273	388	14.03	96.7	132	85	18.8	29.3	123	83	17.6	28.4	118	82	16.9	27.6	15	6.7	180
-2	5	-18.8	0.7	9	4.1	320	4	7	-15.6	1.0	Topeka	724560	39.07	95.62	886	270	14.23	98.1	139	87	19.8	30.3	132	85	18.9	29.3	126	83	18.0	28.5	12	5.5	180

United States of America

HUMIDIFICATION 99.6% (35 hours below...)							99% (88 hours below...)												DEHUMIDIFICATION 0.4% (35 hours above...)				1% (88 hours above...)				2% (175 hours above...)				Summer Wind		
°F 1a	gr/lb 1b	°C 1c	g/kg 1d	mph 1e	m/s 1f	PWD 1g	°F 2a	gr/lb 2b	°C 2c	g/kg 2d	Location 3a	WMO# 3b	Lat. 3c	Long. 3d	ft. 3e	m 3f	psia 3g	kPa 3h	gr/lb 4a	°F 4b	g/kg 4c	°C 4d	gr/lb 5a	°F 5b	g/kg 5c	°C 5d	gr/lb 6a	°F 6b	g/kg 6c	°C 6d	mph 7a	m/s 7b	PWD 7c
2	6	-16.6	0.9	13	5.7	360	8	9	-13.3	1.2	Wichita, Airport	724500	37.65	97.43	1339	408	14.00	96.5	129	83	18.4	28.4	123	82	17.6	27.7	118	81	16.9	27.2	16	7.2	200
2	6	-16.5	0.9	11	5.1	360	10	10	-12.3	1.4	Wichita, Mcconnell	724505	37.62	97.27	1371	418	13.98	96.4	133	84	19.0	28.8	124	83	17.7	28.5	119	82	17.0	27.7	12	5.4	190
											KENTUCKY																						
7	8	-14.1	1.2	6	2.6	220	14	11	-10.2	1.6	Bowling Green	746716	36.97	86.42	548	167	14.41	99.3	136	84	19.5	28.7	132	82	18.9	28.0	127	81	18.2	27.4	9	4.1	230
1	6	-17.5	0.8	9	4.2	250	7	8	-14.1	1.1	Covington/Cincinnati	724210	39.05	84.67	876	267	14.24	98.2	132	84	18.9	28.6	126	81	18.0	27.4	120	80	17.2	26.4	10	4.6	230
9	9	-12.6	1.3	4	1.8	330	15	12	-9.4	1.7	Fort Campbell, AAF	746710	36.67	87.50	571	174	14.39	99.3	143	85	20.4	29.4	136	84	19.5	29.0	132	83	18.8	28.4	6	2.6	240
9	9	-12.8	1.3	4	1.7	290	15	12	-9.7	1.7	Fort Knox, Godman	724240	37.90	85.97	755	230	14.30	98.6	138	85	19.7	29.4	132	83	18.8	28.6	126	82	18.0	27.9	6	2.5	270
7	8	-14.0	1.2	7	3.2	230	14	12	-10.1	1.7	Jackson	724236	37.60	83.32	1381	421	13.98	96.4	135	83	19.3	28.6	130	81	18.5	27.3	122	79	17.4	26.1	6	2.8	230
4	7	-15.8	1.0	8	3.8	270	10	9	-12.4	1.3	Lexington	724220	38.03	84.60	988	301	14.18	97.8	130	83	18.5	28.1	124	81	17.7	27.3	120	80	17.1	26.5	9	4.0	240
6	8	-14.5	1.1	10	4.3	290	12	10	-11.4	1.4	Louisville	724230	38.18	85.73	489	149	14.44	99.5	134	85	19.2	29.5	129	84	18.4	28.7	125	82	17.8	27.8	10	4.5	250
7	8	-13.9	1.1	8	3.6	40	13	11	-10.4	1.6	Paducah	724350	37.07	88.77	413	126	14.48	99.8	143	86	20.4	30.0	138	85	19.7	29.2	132	83	18.9	28.6	9	3.9	180
											LOUISIANA																						
27	21	-3.0	3.0	7	3.0	360	30	24	-1.1	3.5	Alexandria, England	747540	31.33	92.55	89	27	14.65	101.0	147	86	21.0	29.8	142	85	20.3	29.6	138	85	19.7	29.3	3	1.5	180
27	21	-3.0	3.0	8	3.4	360	30	25	-0.9	3.5	Baton Rouge	722317	30.53	91.15	69	21	14.66	101.1	145	84	20.7	28.9	141	84	20.1	28.7	137	83	19.6	28.3	8	3.6	270
22	17	-5.5	2.4	7	3.2	360	27	21	-2.8	3.0	Bossier Cty, Barksdale	722485	32.50	93.67	167	51	14.61	100.7	144	84	20.5	29.1	139	83	19.8	28.5	134	83	19.2	28.2	5	2.2	180
28	22	-2.4	3.2	9	4.1	10	32	27	-0.2	3.8	Lafayette	722405	30.20	91.98	43	13	14.67	101.2	146	84	20.9	29.1	143	83	20.4	28.6	140	83	20.0	28.3	8	3.5	200
29	23	-1.8	3.3	10	4.2	20	32	27	0.1	3.8	Lake Charles	722400	30.12	93.22	33	10	14.68	101.2	148	84	21.1	28.9	145	84	20.7	28.7	141	83	20.2	28.4	8	3.8	230
27	21	-2.9	3.0	4	2.0	20	30	24	-0.9	3.5	Leesville, Fort Polk	722390	31.05	93.20	328	100	14.52	100.1	144	83	20.6	28.4	140	82	20.0	28.0	136	82	19.4	27.6	4	1.7	180
22	17	-5.7	2.4	9	3.8	10	27	21	-2.9	3.0	Monroe	722486	32.52	92.03	79	24	14.65	101.0	147	86	21.0	30.2	143	85	20.4	29.6	139	84	19.9	29.1	7	3.2	230
30	24	-1.3	3.4	7	3.3	340	34	28	0.8	4.0	New Orleans, Intl. Apt.	722310	29.98	90.25	30	9	14.68	101.2	151	86	21.5	30.2	146	85	20.9	29.4	142	84	20.3	29.0	8	3.5	360
35	30	1.8	4.3	14	6.4	360	39	35	3.9	5.0	New Orleans, Lakefnt.	722315	30.05	90.03	10	3	14.69	101.3	150	85	21.4	29.3	145	84	20.7	28.8	141	83	20.2	28.5	9	3.8	300
22	17	-5.5	2.4	9	3.9	360	26	20	-3.2	2.9	Shreveport	722480	32.47	93.82	259	79	14.56	100.4	139	84	19.9	28.8	135	83	19.3	28.4	132	83	18.9	28.2	8	3.8	180
											MAINE																						
-3	5	-19.7	0.7	10	4.6	320	1	6	-17.2	0.8	Augusta	726185	44.32	69.80	351	107	14.51	100.0	113	77	16.1	24.9	106	75	15.1	23.9	100	74	14.3	23.1	11	5.1	210
-7	4	-21.5	0.5	6	2.9	300	-2	5	-19.0	0.7	Bangor	726088	44.80	68.83	194	59	14.59	100.6	111	78	15.9	25.6	104	75	14.9	24.1	99	73	14.1	22.9	10	4.6	240
-2	5	-19.1	0.7	4	1.6	340	2	6	-16.4	0.9	Brunswick, NAS	743920	43.88	69.93	75	23	14.66	101.0	111	78	15.9	25.4	105	76	15.0	24.3	100	74	14.3	23.4	9	3.8	190
-14	3	-25.8	0.4	10	4.4	270	-10	3	-23.2	0.5	Caribou	727120	46.87	68.02	623	190	14.37	99.1	112	76	16.0	24.6	104	75	14.8	23.7	97	72	13.8	22.3	13	5.6	250
-13	3	-25.1	0.4	7	3.1	300	-9	3	-22.7	0.5	Limestone, Loring AFB	727125	46.95	67.88	745	227	14.30	98.6	107	75	15.3	23.7	101	72	14.4	22.5	94	71	13.5	21.9	9	3.9	260
-3	5	-19.6	0.7	7	3.1	320	2	6	-16.7	0.9	Portland	726060	43.65	70.32	62	19	14.66	101.1	114	79	16.3	26.1	107	76	15.3	24.6	101	74	14.4	23.3	12	5.3	270
											MARYLAND																						
13	11	-10.3	1.5	7	3.2	350	18	14	-7.6	2.0	Andrews AFB	745940	38.82	76.87	282	86	14.55	100.3	134	83	19.2	28.4	129	82	18.4	27.5	124	80	17.7	26.8	9	3.8	230
11	10	-11.6	1.4	10	4.5	290	15	12	-9.2	1.7	Baltimore, BWI Airport	724060	39.18	76.67	154	47	14.61	100.8	132	83	18.8	28.1	125	81	17.9	27.3	120	80	17.2	26.6	11	4.9	280
16	12	-9.0	1.8	9	3.8	340	21	16	-6.2	2.3	Patuxent River NAS	724040	38.28	76.40	39	12	14.67	101.2	136	84	19.5	29.0	131	83	18.7	28.1	125	82	17.9	27.5	9	3.9	270
13	11	-10.5	1.5	6	2.8	10	18	14	-7.7	2.0	Salisbury	724045	38.33	75.52	52	16	14.67	101.1	144	84	20.6	28.8	137	82	19.6	28.0	132	81	18.9	27.2	9	4.2	240
											MASSACHUSETTS																						
7	8	-13.7	1.1	17	7.5	320	12	10	-11.3	1.4	Boston	725090	42.37	71.03	30	9	14.68	101.2	119	80	17.0	26.7	113	79	16.2	25.9	108	78	15.4	25.4	14	6.2	270
11	10	-11.9	1.4	9	4.2	300	14	11	-9.8	1.6	East Falmouth, Otis	725060	41.65	70.52	131	40	14.63	100.8	125	78	17.9	25.3	118	76	16.9	24.7	113	75	16.1	23.7	10	4.5	240
6	8	-14.6	1.1	7	3.0	320	11	10	-11.8	1.4	South Weymouth NAS	725097	42.15	70.93	161	49	14.61	100.7	129	82	18.4	27.6	118	79	16.9	26.2	111	78	15.8	25.3	9	3.9	260
0	6	-17.7	0.8	14	6.1	270	5	7	-15.1	1.0	Worcester	725095	42.27	71.88	1010	308	14.17	97.7	119	78	17.0	25.7	112	76	16.0	24.6	105	75	15.0	23.7	10	4.3	270
											MICHIGAN																						
-7	4	-21.6	0.6	5	2.3	270	-1	5	-18.6	0.7	Alpena	726390	45.07	83.57	692	211	14.33	98.8	116	79	16.5	26.1	107	76	15.3	24.7	100	74	14.3	23.6	11	5.0	240
0	6	-17.8	0.8	11	5.0	240	5	7	-15.1	1.0	Detroit, Metro	725370	42.23	83.33	663	202	14.35	98.9	125	83	17.8	28.2	118	80	16.8	26.5	111	78	15.9	25.7	13	5.6	230

United States of America

HUMIDIFICATION																			DEHUMIDIFICATION														
99.6% (35 hours below...)							**99%** (88 hours below...)												**0.4%** (35 hours above...)				**1%** (88 hours above...)				**2%** (175 hours above...)				Summer Wind		
°F 1a	gr/lb 1b	°C 1c	g/kg 1d	mph 1e	m/s 1f	PWD 1g	°F 2a	gr/lb 2b	°C 2c	g/kg 2d	Location 3a	WMO# 3b	Lat. 3c	Long. 3d	ft. 3e	m 3f	psia 3g	kPa 3h	gr/lb 4a	°F 4b	g/kg 4c	°C 4d	gr/lb 5a	°F 5b	g/kg 5c	°C 5d	gr/lb 6a	°F 6b	g/kg 6c	°C 6d	mph 7a	m/s 7b	PWD 7c
-2	5	-18.9	0.7	8	3.8	230	3	6	-16.3	0.9	Flint	726370	42.97	83.75	764	233	14.29	98.6	125	81	17.8	27.2	116	78	16.6	25.7	110	77	15.7	24.9	13	5.6	230
0	6	-17.8	0.8	8	3.7	180	5	7	-15.2	1.0	Grand Rapids	726350	42.88	85.52	804	245	14.27	98.4	126	81	18.0	27.4	118	79	16.9	26.3	112	77	16.0	25.2	13	5.8	240
-9	4	-22.6	0.5	8	3.7	270	-4	5	-19.8	0.7	Hancock	727440	47.17	88.50	1079	329	14.13	97.4	116	79	16.6	25.9	109	76	15.6	24.6	103	74	14.7	23.6	10	4.2	250
9	9	-13.0	1.3	10	4.6	220	12	10	-11.1	1.5	Harbor Beach	725386	44.02	82.80	600	183	14.38	99.1	113	82	16.1	28.0	106	80	15.1	26.6	100	78	14.3	25.3	9	3.9	230
-3	5	-19.4	0.7	9	3.9	240	4	7	-15.7	1.0	Jackson	725395	42.27	84.47	1001	305	14.17	97.7	134	83	19.1	28.2	123	81	17.5	27.0	117	78	16.7	25.6	11	5.1	210
-3	5	-19.6	0.7	8	3.6	290	2	6	-16.8	0.9	Lansing	725390	42.77	84.60	873	266	14.24	98.2	127	81	18.2	27.3	120	79	17.2	26.2	114	78	16.3	25.3	13	5.7	250
-11	3	-24.0	0.5	6	2.5	280	-6	4	-21.3	0.6	Marquette, K. I. S. AFB	727435	46.35	87.40	1220	372	14.06	96.9	113	77	16.1	25.1	106	74	15.1	23.4	99	73	14.2	22.8	10	4.5	210
-13	3	-24.9	0.4	8	3.5	270	-8	4	-22.1	0.5	Marquette/Airport	727430	46.53	87.55	1424	434	13.95	96.2	111	77	15.9	25.0	104	75	14.9	23.8	98	72	14.0	22.5	11	4.9	230
3	7	-15.9	0.9	7	3.3	280	7	8	-14.0	1.2	Mt Clemens, Selfridge	725377	42.62	82.83	581	177	14.39	99.2	131	83	18.7	28.4	120	81	17.1	27.0	113	78	16.2	25.7	9	4.0	230
3	7	-16.2	0.9	10	4.6	290	7	8	-13.9	1.2	Muskegon	726360	43.17	86.25	633	193	14.36	99.0	122	80	17.4	26.4	115	77	16.4	25.1	109	76	15.5	24.2	12	5.4	200
0	6	-17.9	0.8	6	2.5	220	3	7	-16.1	0.9	Oscoda, Wurtsmith	726395	44.45	83.40	633	193	14.36	99.0	120	80	17.2	26.6	112	79	16.0	25.9	106	77	15.1	24.8	11	5.1	200
-9	3	-22.7	0.5	4	1.6	300	-3	5	-19.5	0.7	Pellston	727347	45.57	84.80	719	219	14.32	98.7	115	78	16.4	25.3	108	76	15.5	24.5	103	75	14.7	24.0	14	6.2	250
0	6	-17.5	0.8	10	4.6	260	4	7	-15.6	1.0	Saginaw	726379	43.53	84.08	669	204	14.34	98.9	132	83	18.8	28.1	120	80	17.1	26.5	112	78	16.0	25.5	13	5.8	240
-12	3	-24.4	0.4	7	2.9	90	-7	4	-21.8	0.5	Sault Ste. Marie	727340	46.47	84.37	725	221	14.31	98.7	111	76	15.8	24.7	103	74	14.7	23.4	95	72	13.6	22.1	10	4.4	230
0	6	-17.6	0.8	9	4.2	300	4	7	-15.8	1.0	Seul Choix Point	726399	45.92	85.92	591	180	14.38	99.2	106	74	15.2	23.2	101	72	14.4	22.3	94	70	13.4	21.2	8	3.7	200
-3	5	-19.6	0.7	7	2.9	180	2	6	-16.8	0.9	Traverse City	726387	44.73	85.58	623	190	14.37	99.1	117	80	16.7	26.6	109	78	15.6	25.4	103	76	14.7	24.3	13	5.7	230
											MINNESOTA																						
-20	2	-29.0	0.3	10	4.4	300	-15	3	-26.0	0.4	Alexandria	726557	45.87	95.40	1424	434	13.95	96.2	123	82	17.6	27.7	116	79	16.5	25.9	109	77	15.6	24.9	14	6.2	180
-24	2	-31.1	0.2	3	1.4	320	-17	2	-27.3	0.3	Brainerd	727500	46.60	94.32	1280	390	14.03	96.7	108	81	15.5	27.1	102	77	14.6	24.9	96	75	13.7	23.7	5	2.2	190
-21	2	-29.2	0.3	10	4.5	310	-16	2	-26.6	0.3	Duluth	727450	46.83	92.18	1417	432	13.96	96.2	110	77	15.7	25.2	102	75	14.5	23.7	94	72	13.4	22.1	12	5.2	230
-25	1	-31.8	0.2	6	2.8	330	-20	2	-28.8	0.3	Hibbing	727455	47.38	92.83	1352	412	13.99	96.5	116	78	16.5	25.7	108	76	15.4	24.4	101	73	14.4	22.6	11	5.0	200
-29	1	-33.6	0.2	6	2.6	270	-23	2	-30.8	0.2	International Falls	727470	48.57	93.38	1184	361	14.08	97.1	112	78	16.0	25.4	103	75	14.7	23.9	96	73	13.7	22.7	11	5.0	180
-16	2	-26.5	0.3	9	4.2	300	-11	3	-23.7	0.5	Minneapolis-St. Paul	726580	44.88	93.22	837	255	14.26	98.3	124	83	17.7	28.5	116	81	16.6	27.2	109	79	15.6	26.0	14	6.3	180
-17	2	-27.1	0.3	11	4.7	280	-12	3	-24.4	0.4	Redwood Falls	726556	44.55	95.08	1024	312	14.16	97.6	135	83	19.3	28.6	123	81	17.5	27.4	116	80	16.5	26.4	14	6.1	180
-17	2	-27.3	0.3	13	5.8	300	-12	3	-24.4	0.4	Rochester	726440	43.92	92.50	1319	402	14.01	96.6	128	81	18.3	27.4	120	79	17.1	26.3	111	77	15.9	24.9	15	6.8	200
-20	2	-28.7	0.3	8	3.6	300	-14	3	-25.8	0.4	Saint Cloud	726550	45.55	94.07	1024	312	14.16	97.6	125	83	17.8	28.2	116	80	16.5	26.9	109	78	15.5	25.7	12	5.4	200
-10	3	-23.1	0.5	8	3.7	260	-6	4	-21.0	0.6	Tofte	727554	47.58	90.83	791	241	14.28	98.5	92	70	13.1	20.9	83	69	11.8	20.5	77	68	11.0	19.8	8	3.4	330
											MISSISSIPPI																						
31	25	-0.7	3.6	8	3.4	360	35	30	1.6	4.3	Biloxi, Keesler AFB	747686	30.42	88.92	33	10	14.68	101.2	151	86	21.6	29.8	147	85	21.0	29.4	144	84	20.5	29.1	7	3.0	210
20	15	-6.5	2.2	6	2.5	360	25	19	-4.1	2.8	Columbus, AFB	723306	33.65	88.45	220	67	14.58	100.5	141	85	20.2	29.3	136	83	19.4	28.5	132	82	18.9	27.9	6	2.7	240
20	15	-6.9	2.2	6	2.9	360	24	18	-4.5	2.6	Greenwood	722359	33.50	90.08	154	47	14.61	100.8	148	86	21.1	29.8	143	85	20.4	29.5	139	84	19.8	29.1	6	2.9	180
21	16	-6.3	2.2	7	3.2	340	25	19	-4.1	2.7	Jackson	722350	32.32	90.08	331	101	14.52	100.1	142	84	20.3	28.7	138	83	19.7	28.4	134	82	19.2	27.9	8	3.6	270
23	18	-4.9	2.5	6	2.5	350	28	22	-2.1	3.2	Mccomb	722358	31.18	90.47	413	126	14.48	99.8	141	83	20.1	28.3	138	82	19.7	27.8	135	81	19.3	27.3	7	3.3	230
21	16	-6.2	2.3	6	2.7	360	25	19	-3.9	2.8	Meridian	722340	32.33	88.75	308	94	14.53	100.2	139	84	19.9	28.9	134	83	19.2	28.6	130	83	18.6	28.1	8	3.6	360
18	14	-7.9	2.0	7	3.2	10	22	17	-5.3	2.4	Tupelo	723320	34.27	88.77	361	110	14.50	100.0	137	83	19.6	28.4	134	83	19.1	28.1	131	82	18.7	27.5	7	3.3	260
											MISSOURI																						
6	8	-14.4	1.1	9	4.2	360	13	11	-10.7	1.5	Cape Girardeau	723489	37.23	89.57	341	104	14.52	100.1	141	86	20.1	30.2	136	85	19.4	29.4	132	83	18.8	28.6	10	4.3	200
-1	6	-18.1	0.8	11	4.8	310	5	7	-14.9	1.1	Columbia	724450	38.82	92.22	899	274	14.22	98.1	137	85	19.5	29.4	130	83	18.5	28.3	124	82	17.7	27.5	11	4.8	200
3	7	-15.9	1.0	10	4.4	10	11	10	-11.9	1.4	Joplin	723495	37.15	94.50	981	299	14.18	97.8	137	85	19.6	29.4	132	85	18.9	29.2	125	83	17.8	28.1	11	4.9	220
-1	5	-18.6	0.8	10	4.2	320	4	7	-15.4	1.0	Kansas City	724460	39.32	94.72	1024	312	14.16	97.6	137	86	19.5	29.7	130	84	18.6	29.1	125	83	17.9	28.1	13	5.7	190
8	8	-13.2	1.2	7	3.0	360	13	11	-10.3	1.6	Poplar Bluff	723300	36.77	90.47	479	146	14.44	99.6	144	85	20.6	29.3	138	83	19.7	28.6	133	82	19.0	28.0	7	3.0	200
1	6	-17.4	0.9	8	3.5	360	6	8	-14.6	1.1	Spickard/Trenton	725400	40.25	93.72	886	270	14.23	98.1	139	83	19.8	28.6	128	83	18.3	28.2	118	81	16.8	27.2	11	5.0	200

United States of America

HUMIDIFICATION 99.6% (35 hours below...)							99% (88 hours below...)												DEHUMIDIFICATION 0.4% (35 hours above...)				1% (88 hours above...)				2% (175 hours above...)				Summer Wind		
°F 1a	gr/lb 1b	°C 1c	g/kg 1d	mph 1e	m/s 1f	PWD 1g	°F 2a	gr/lb 2b	°C 2c	g/kg 2d	Location 3a	WMO# 3b	Lat. 3c	Long. 3d	ft. 3e	m 3f	psia 3g	kPa 3h	gr/lb 4a	°F 4b	g/kg 4c	°C 4d	gr/lb 5a	°F 5b	g/kg 5c	°C 5d	gr/lb 6a	°F 6b	g/kg 6c	°C 6d	mph 7a	m/s 7b	PWD 7c
3	7	-16.2	1.0	10	4.6	340	9	9	-12.8	1.3	Springfield	724400	37.23	93.38	1270	387	14.03	96.8	134	84	19.2	29.1	128	83	18.3	28.2	124	81	17.7	27.4	10	4.4	230
2	6	-16.8	0.9	12	5.4	290	8	8	-13.6	1.2	St. Louis, Int'l Airport	724340	38.75	90.37	564	172	14.40	99.3	138	85	19.7	29.3	132	83	18.9	28.5	127	82	18.1	27.8	11	4.8	240
1	6	-17.3	0.9	9	4.0	360	7	8	-13.8	1.2	Warrensburg/Sedalia	724467	38.73	93.55	869	265	14.24	98.2	139	86	19.9	30.2	134	85	19.1	29.3	128	83	18.3	28.6	9	4.0	190
											MONTANA																						
-13	3	-25.1	0.4	10	4.2	230	-7	4	-21.8	0.6	Billings	726770	45.80	108.53	3570	1088	12.90	88.9	83	71	11.9	21.8	78	71	11.1	21.4	74	70	10.5	20.8	10	4.4	240
-20	2	-29.0	0.3	4	1.9	140	-12	3	-24.7	0.5	Bozeman	726797	45.78	111.15	4475	1364	12.47	86.0	83	69	11.9	20.5	78	67	11.1	19.4	71	66	10.1	19.0	9	4.0	360
-22	2	-30.1	0.3	4	1.6	150	-14	3	-25.8	0.5	Butte	726785	45.95	112.50	5545	1690	11.98	82.6	76	61	10.9	16.2	70	63	10.0	17.0	66	62	9.4	16.5	13	5.6	120
-21	2	-29.2	0.3	7	3.2	320	-16	3	-26.4	0.4	Cut Bank	727796	48.60	112.37	3855	1175	12.76	88.0	77	67	11.0	19.5	70	65	10.0	18.2	65	63	9.3	17.4	13	5.8	270
-22	2	-30.2	0.3	8	3.6	330	-17	2	-27.1	0.3	Glasgow	727680	48.22	106.62	2297	700	13.52	93.2	91	74	13.0	23.2	83	71	11.9	21.4	77	69	11.0	20.7	13	5.9	160
-19	2	-28.1	0.3	7	3.3	240	-13	3	-24.8	0.5	Gt.Falls, Int'l Apt	727750	47.48	111.37	3658	1115	12.85	88.6	81	69	11.5	20.6	74	67	10.6	19.6	69	66	9.9	19.1	12	5.4	230
-17	2	-27.4	0.4	4	1.8	240	-11	3	-23.9	0.5	Gt.Falls, Malmstrom	727755	47.50	111.18	3527	1075	12.92	89.1	84	71	12.0	21.5	78	69	11.1	20.5	71	68	10.2	20.2	8	3.5	260
-25	2	-31.6	0.2	6	2.5	240	-19	2	-28.4	0.3	Havre	727770	48.55	109.77	2598	792	13.37	92.2	84	72	12.0	22.0	78	69	11.1	20.4	74	68	10.5	19.9	9	3.9	270
-18	2	-27.5	0.3	5	2.0	290	-10	4	-23.5	0.5	Helena	727720	46.60	112.00	3898	1188	12.74	87.8	80	68	11.4	19.9	73	66	10.4	19.1	68	66	9.7	19.1	12	5.4	280
-12	3	-24.2	0.5	7	2.9	20	-3	5	-19.7	0.7	Kalispell	727790	48.30	114.27	2972	906	13.18	90.9	82	69	11.7	20.7	76	67	10.9	19.6	71	67	10.2	19.4	9	4.1	170
-18	2	-27.9	0.3	7	3.3	250	-12	3	-24.2	0.5	Lewistown	726776	47.05	109.47	4167	1270	12.61	87.0	85	71	12.2	21.5	79	69	11.3	20.5	74	67	10.5	19.4	11	4.7	90
-19	2	-28.2	0.3	8	3.7	290	-13	3	-24.7	0.4	Miles City	742300	46.43	105.87	2628	801	13.35	92.1	95	76	13.6	24.6	88	75	12.6	23.7	82	73	11.7	22.8	11	4.9	140
-9	4	-22.8	0.5	7	2.9	120	-1	6	-18.3	0.8	Missoula	727730	46.92	114.08	3189	972	13.08	90.2	82	68	11.7	20.2	76	68	10.9	20.2	71	66	10.2	18.9	10	4.6	290
											NEBRASKA																						
-5	4	-20.7	0.6	8	3.4	330	1	6	-17.2	0.9	Bellevue, Offutt AFB	725540	41.12	95.92	1047	319	14.15	97.6	141	85	20.1	29.6	134	83	19.2	28.6	127	82	18.2	27.7	10	4.4	190
-8	4	-22.2	0.5	11	4.8	270	-2	5	-19.1	0.7	Grand Island	725520	40.97	98.32	1857	566	13.74	94.7	127	82	18.1	27.7	120	81	17.1	26.9	113	79	16.1	26.2	15	6.7	180
-7	4	-21.9	0.6	9	4.0	350	-2	5	-18.8	0.7	Lincoln	725510	40.85	96.75	1188	362	14.08	97.1	136	84	19.5	29.1	130	83	18.5	28.4	121	82	17.3	27.7	15	6.8	180
-11	3	-23.7	0.5	11	4.7	340	-5	4	-20.7	0.6	Norfolk	725560	41.98	97.43	1552	473	13.89	95.8	129	83	18.4	28.6	121	82	17.3	27.7	115	81	16.4	27.1	15	6.6	190
-10	4	-23.1	0.5	7	3.3	320	-4	5	-19.8	0.7	North Platte	725620	41.13	100.68	2785	849	13.28	91.5	118	80	16.9	26.8	111	78	15.9	25.8	105	77	15.0	25.1	12	5.4	180
-7	4	-21.8	0.6	10	4.6	340	-2	5	-18.9	0.7	Omaha, Eppley	725500	41.30	95.90	981	299	14.18	97.8	136	85	19.4	29.7	128	84	18.3	29.0	121	82	17.3	28.0	12	5.5	180
-8	4	-22.1	0.5	10	4.4	310	-2	5	-18.8	0.7	Omaha, WSO	725530	41.37	96.02	1332	406	14.00	96.5	134	84	19.2	28.9	126	83	18.0	28.1	120	82	17.1	27.6	11	4.7	170
-11	4	-23.6	0.5	8	3.5	300	-3	5	-19.4	0.8	Scottsbluff	725660	41.87	103.60	3957	1206	12.71	87.7	102	76	14.6	24.2	97	74	13.8	23.2	91	73	13.0	22.9	11	5.0	300
-8	4	-22.5	0.6	9	4.2	290	-1	6	-18.2	0.9	Sidney	725610	41.10	102.98	4304	1312	12.55	86.5	97	73	13.9	22.6	91	72	13.0	22.4	86	71	12.3	21.8	12	5.4	160
-16	3	-26.5	0.4	9	3.8	250	-8	4	-22.0	0.6	Valentine	725670	42.87	100.55	2598	792	13.37	92.2	110	79	15.7	26.2	103	78	14.7	25.4	94	77	13.4	24.9	15	6.6	180
											NEVADA																						
-5	5	-20.6	0.7	4	1.6	70	1	7	-17.1	1.0	Elko	725825	40.83	115.78	5135	1565	12.17	83.9	84	68	12.0	19.9	75	66	10.7	19.1	67	67	9.5	19.4	10	4.5	230
-6	5	-21.2	0.7	11	4.9	190	0	7	-17.8	1.0	Ely 724860	39.28	114.85	6263	1909	11.66	80.4	82	64	11.7	17.9	75	6410.7		17.8	68	65	9.7	18.2	13	5.7	230	
27	23	-2.7	3.3	7	3.3	250	30	27	-0.9	3.8	Las Vegas, Int'l Airport	723860	36.08	115.17	2178	664	13.57	93.6	102	79	14.5	26.2	92	81	13.2	27.4	84	85	12.0	29.4	12	5.5	230
24	21	-4.5	2.9	8	3.4	50	28	25	-2.2	3.6	Mercury	723870	36.62	116.02	3310	1009	13.02	89.8	102	72	14.5	22.2	89	77	12.7	24.8	80	80	11.4	26.5	12	5.5	230
28	24	-2.4	3.4	2	0.7	20	31	27	-0.7	3.9	N. Las Vegas, Nellis	723865	36.23	115.03	1870	570	13.73	94.7	106	79	15.2	26.1	97	82	13.8	27.9	86	84	12.3	29.1	9	4.0	210
8	10	-13.4	1.4	3	1.4	160	13	13	-10.6	1.8	Reno	724880	39.50	119.78	4400	1341	12.50	86.2	77	69	11.0	20.4	69	69	9.9	20.3	63	68	9.0	20.1	10	4.6	290
7	10	-13.7	1.4	9	4.2	340	13	13	-10.8	1.8	Tonopah	724855	38.05	117.08	5427	1654	12.03	83.0	83	67	11.8	19.3	74	68	10.6	20.2	67	69	9.5	20.6	12	5.2	180
1	7	-17.1	1.0	5	2.2	160	7	9	-13.7	1.3	Winnemucca	725830	40.90	117.80	4314	1315	12.54	86.5	79	68	11.3	20.1	69	67	9.9	19.6	62	68	8.8	19.7	11	4.9	250
											NEW HAMPSHIRE																						
-8	4	-21.9	0.5	4	2.0	320	-2	5	-18.9	0.7	Concord	726050	43.20	71.50	344	105	14.51	100.1	118	79	16.8	25.9	111	77	15.9	24.9	105	76	15.0	24.3	10	4.6	230
-7	4	-21.8	0.6	2	1.1	360	-3	5	-19.3	0.7	Lebanon	726116	43.63	72.30	597	182	14.38	99.2	113	79	16.2	26.2	108	77	15.5	24.8	103	75	14.7	23.7	9	4.1	220
-23	2	-30.8	0.3	73	32.4	280	-19	2	-28.2	0.4	Mount Washington	726130	44.27	71.30	6266	1910	11.66	80.4	90	58	12.8	14.7	84	57	12.0	13.8	78	55	11.1	12.6	21	9.3	270
4	7	-15.5	1.0	8	3.4	280	9	9	-12.9	1.3	Portsmouth	726055	43.08	70.82	102	31	14.64	101.0	123	85	17.6	29.5	113	77	16.1	25.2	106	76	15.2	24.4	8	3.7	270

United States of America

HUMIDIFICATION 99.6% (35 hours below...)							99% (88 hours below...)												DEHUMIDIFICATION 0.4% (35 hours above...)				1% (88 hours above...)				2% (175 hours above...)				Summer Wind		
°F	gr/lb	°C	g/kg	mph	m/s	PWD	°F	gr/lb	°C	g/kg	Location	WMO#	Lat.	Long.	ft.	m	psia	kPa	gr/lb	°F	g/kg	°C	gr/lb	°F	g/kg	°C	gr/lb	°F	g/kg	°C	mph	m/s	PWD
1a	1b	1c	1d	1e	1f	1g	2a	2b	2c	2d	3a	3b	3c	3d	3e	3f	3g	3h	4a	4b	4c	4d	5a	5b	5c	5d	6a	6b	6c	6d	7a	7b	7c
											NEW JERSEY																						
8	8	-13.4	1.2	9	3.9	310	13	10	-10.8	1.5	Atlantic City	724070	39.45	74.57	66	20	14.66	101.1	131	81	18.7	27.3	125	80	17.9	26.6	120	79	17.2	26.0	11	5.1	250
10	9	-12.2	1.3	7	3.0	300	15	12	-9.5	1.7	Millville	724075	39.37	75.07	82	25	14.65	101.0	134	81	19.1	27.4	129	80	18.4	26.8	125	80	17.8	26.5	11	4.9	240
10	9	-12.3	1.3	13	5.9	260	14	11	-10.0	1.6	Newark	725020	40.70	74.17	30	9	14.68	101.2	127	81	18.1	27.4	121	80	17.3	26.9	116	80	16.5	26.5	13	5.7	230
10	9	-12.0	1.3	11	4.8	280	14	11	-9.8	1.6	Teterboro	725025	40.85	74.07	10	3	14.69	101.3	134	84	19.2	28.7	128	82	18.3	27.5	119	81	17.0	27.1	12	5.2	240
11	10	-11.8	1.4	8	3.6	330	15	12	-9.6	1.7	Trenton/Mcguire AFB	724096	40.02	74.60	135	41	14.62	100.8	132	83	18.9	28.6	127	82	18.1	27.5	118	80	16.9	26.8	8	3.4	240
											NEW MEXICO																						
20	18	-6.8	2.5	3	1.2	10	23	20	-5.2	2.9	Alamogordo, Holloman	747320	32.85	106.10	4094	1248	12.65	87.2	106	72	15.2	22.3	98	72	14.0	22.3	92	73	13.2	22.6	8	3.5	250
13	13	-10.4	1.9	8	3.6	360	18	17	-7.6	2.4	Albuquerque	723650	35.05	106.62	5315	1620	12.08	83.3	98	68	14.0	20.0	93	69	13.3	20.3	89	69	12.7	20.7	10	4.5	240
19	16	-7.4	2.3	8	3.5	340	23	20	-5.2	2.8	Carlsbad	722687	32.33	104.27	3294	1004	13.03	89.8	121	76	17.3	24.4	116	76	16.6	24.4	111	75	15.9	23.7	12	5.5	150
1	7	-17.1	1.0	10	4.3	40	9	11	-12.8	1.5	Clayton	723600	36.45	103.15	4970	1515	12.24	84.4	98	72	14.0	22.0	94	71	13.4	21.5	90	70	12.9	21.0	13	5.9	200
10	11	-12.0	1.5	8	3.4	50	15	14	-9.2	2.0	Clovis, Cannon AFB	722686	34.38	103.32	4295	1309	12.55	86.6	114	75	16.3	24.0	109	74	15.6	23.3	105	73	15.0	22.9	11	5.0	220
8	10	-13.3	1.5	6	2.5	60	13	13	-10.8	1.9	Farmington	723658	36.75	108.23	5502	1677	12.00	82.7	94	69	13.5	20.4	90	68	12.8	20.2	85	69	12.2	20.4	10	4.6	240
-1	7	-18.1	0.9	1	0.6	140	5	9	-15.1	1.3	Gallup	723627	35.52	108.78	6470	1972	11.57	79.8	94	65	13.4	18.1	90	64	12.8	17.7	85	64	12.2	17.9	11	5.0	270
14	13	-9.8	1.8	8	3.4	360	20	17	-6.8	2.5	Roswell	722680	33.30	104.53	3668	1118	12.85	88.6	111	73	15.9	23.0	108	73	15.4	22.9	104	73	14.8	22.8	11	4.9	140
22	20	-5.6	2.8	8	3.5	350	26	24	-3.2	3.4	Truth Or Conseq.	722710	33.23	107.27	4859	1481	12.29	84.8	94	71	13.5	21.7	90	71	12.9	21.7	87	72	12.4	22.1	10	4.4	170
9	10	-12.6	1.5	8	3.3	50	15	14	-9.5	1.9	Tucumcari	723676	35.18	103.60	4065	1239	12.66	87.3	109	73	15.6	22.7	104	73	14.9	22.6	100	72	14.3	22.4	12	5.2	230
											NEW YORK																						
-7	4	-21.9	0.5	5	2.1	300	-2	5	-18.8	0.7	Albany	725180	42.75	73.80	292	89	14.54	100.3	118	79	16.8	25.9	111	77	15.9	24.9	106	76	15.1	24.3	10	4.6	230
-2	5	-18.9	0.8	13	5.6	270	2	7	-16.6	0.9	Binghamton	725150	42.22	75.98	1631	497	13.85	95.5	118	77	16.8	25.2	111	75	15.9	24.1	106	74	15.1	23.2	11	4.8	220
2	6	-16.8	0.9	12	5.2	270	5	7	-14.8	1.1	Buffalo	725280	42.93	78.73	705	215	14.32	98.8	118	78	16.9	25.8	113	77	16.1	24.9	106	75	15.2	24.1	13	5.8	240
11	10	-11.7	1.4	10	4.6	340	15	12	-9.6	1.7	Central Islip	725035	40.80	73.10	98	30	14.64	101.0	129	79	18.4	26.3	124	78	17.7	25.4	116	77	16.6	24.8	11	4.7	210
-2	5	-18.9	0.7	5	2.4	240	3	7	-15.9	1.0	Elmira/Corning	725156	42.17	76.90	955	291	14.20	97.9	122	81	17.4	27.0	116	78	16.5	25.6	110	76	15.7	24.3	11	5.0	210
-10	3	-23.1	0.5	2	1.1	350	-4	5	-19.8	0.6	Glens Falls	725185	43.33	73.62	328	100	14.52	100.1	127	81	18.1	27.1	116	79	16.5	25.9	108	76	15.5	24.6	10	4.4	190
-15	2	-26.3	0.3	4	1.6	270	-10	3	-23.2	0.5	Massena	726223	44.93	74.85	213	65	14.58	100.5	118	80	16.9	26.8	111	78	15.9	25.4	105	76	15.0	24.5	10	4.6	230
11	10	-11.4	1.4	17	7.6	320	15	12	-9.2	1.7	New York, JFK	744860	40.65	73.78	23	7	14.68	101.2	125	80	17.8	26.8	120	80	17.1	26.5	114	79	16.3	25.9	13	5.9	230
13	11	-10.7	1.5	18	7.9	310	17	13	-8.1	1.9	New York, La Guardia	725030	40.77	73.90	30	9	14.68	101.2	129	81	18.4	27.0	125	80	17.8	26.5	116	80	16.6	26.5	12	5.5	280
6	8	-14.7	1.1	8	3.5	260	10	9	-12.0	1.3	Newburgh	725038	41.50	74.10	492	150	14.44	99.5	130	82	18.6	27.6	119	80	17.0	26.6	111	78	15.9	25.3	10	4.4	230
4	7	-15.7	1.0	11	5.1	240	7	8	-13.9	1.2	Niagara Falls	725287	43.10	78.95	591	180	14.38	99.2	125	81	17.9	27.2	116	78	16.6	25.4	111	76	15.8	24.5	13	5.8	230
-9	3	-22.6	0.5	2	1.0	350	-4	4	-20.2	0.6	Plattsburgh, AFB	726225	44.65	73.47	236	72	14.57	100.5	115	79	16.4	26.1	108	76	15.4	24.7	102	75	14.5	23.8	8	3.5	260
2	6	-16.7	0.9	3	1.2	250	6	8	-14.2	1.1	Poughkeepsie	725036	41.63	73.88	167	51	14.61	100.7	126	82	18.0	27.8	116	80	16.6	26.6	111	78	15.9	25.8	9	3.9	250
1	6	-17.2	0.9	10	4.4	230	5	7	-15.0	1.0	Rochester	725290	43.12	77.67	554	169	14.40	99.3	123	81	17.6	27.2	116	79	16.5	25.9	109	77	15.6	24.8	12	5.4	250
-5	4	-20.6	0.6	3	1.4	330	1	6	-17.4	0.8	Rome, Griffiss AFB	725196	43.23	75.40	505	154	14.43	99.5	117	80	16.7	26.7	111	78	15.9	25.5	105	76	15.0	24.2	8	3.4	260
-3	5	-19.6	0.7	7	3.3	90	2	6	-16.4	0.9	Syracuse	725190	43.12	76.12	407	124	14.48	99.8	120	80	17.2	26.9	113	78	16.2	25.7	107	77	15.3	24.8	11	4.8	250
-12	3	-24.7	0.4	5	2.2	80	-6	4	-20.9	0.6	Watertown	726227	44.00	76.02	325	99	14.52	100.1	118	78	16.8	25.7	111	77	15.9	25.0	106	75	15.2	24.1	11	5.1	240
7	8	-13.8	1.1	13	5.7	310	12	10	-11.1	1.5	White Plains	725037	41.07	73.70	440	134	14.46	99.7	128	80	18.3	26.9	120	80	17.2	26.4	114	78	16.3	25.3	9	3.9	260
											NORTH CAROLINA																						
11	11	-11.4	1.5	11	5.0	340	16	14	-8.8	1.9	Asheville	723150	35.43	82.55	2169	661	13.58	93.6	128	79	18.3	25.8	123	78	17.5	25.3	118	76	16.9	24.6	9	3.9	340
26	20	-3.6	2.8	11	4.9	340	29	23	-1.9	3.2	Cape Hatteras	723040	35.27	75.55	10	3	14.69	101.3	147	83	21.0	28.5	142	83	20.3	28.1	138	82	19.7	27.5	11	4.8	230
18	14	-7.6	2.0	6	2.8	50	23	18	-5.1	2.5	Charlotte	723140	35.22	80.93	768	234	14.29	98.5	130	82	18.5	27.5	125	80	17.9	26.9	122	80	17.4	26.4	9	3.8	240
24	18	-4.7	2.6	5	2.3	10	28	22	-2.2	3.2	Cherry Point, MCAS	723090	34.90	76.88	30	9	14.68	101.2	146	87	20.9	30.5	141	86	20.1	30.0	136	85	19.5	29.2	7	3.3	240
22	17	-5.5	2.4	4	1.9	10	27	21	-3.0	3.0	Fayetteville, Fort Bragg	746930	35.13	78.93	243	74	14.57	100.4	139	84	19.9	29.1	135	83	19.3	28.6	131	83	18.7	28.1	6	2.7	240

United States of America

HUMIDIFICATION 99.6% (35 hours below...)							99% (88 hours below...)												DEHUMIDIFICATION 0.4% (35 hours above...)				1% (88 hours above...)				2% (175 hours above...)				Summer Wind		
°F	gr/lb	°C	g/kg	mph	m/s	PWD	°F	gr/lb	°C	g/kg	Location	WMO#	Lat.	Long.	ft.	m	psia	kPa	gr/lb	°F	g/kg	°C	gr/lb	°F	g/kg	°C	gr/lb	°F	g/kg	°C	mph	m/s	PWD
1a	1b	1c	1d	1e	1f	1g	2a	2b	2c	2d	3a	3b	3c	3d	3e	3f	3g	3h	4a	4b	4c	4d	5a	5b	5c	5d	6a	6b	6c	6d	7a	7b	7c
22	17	-5.6	2.4	4	2.0	270	27	21	-3.0	3.0	Goldsboro, S. J. AFB	723066	35.33	77.97	108	33	14.64	100.9	139	84	19.8	29.0	135	83	19.3	28.4	132	82	18.8	27.8	8	3.4	260
15	12	-9.7	1.7	7	3.2	290	19	15	-7.2	2.1	Greensboro	723170	36.08	79.95	886	270	14.23	98.1	132	82	18.9	27.6	127	81	18.2	27.0	123	80	17.6	26.4	8	3.8	230
18	14	-7.7	2.0	4	1.9	320	23	18	-5.2	2.6	Hickory	723145	35.73	81.38	1188	362	14.08	97.1	133	80	19.0	26.7	128	80	18.3	26.6	120	78	17.2	25.5	9	3.9	240
23	17	-5.1	2.5	5	2.4	350	27	21	-2.6	3.0	Jacksonville, New Riv.	723096	34.72	77.45	26	8	14.68	101.2	145	86	20.7	30.0	140	85	20.0	29.4	136	84	19.4	28.7	7	3.3	240
22	17	-5.4	2.4	6	2.5	10	27	21	-2.8	3.0	New Bern	723095	35.07	77.05	20	6	14.68	101.3	144	86	20.6	29.9	139	84	19.9	29.0	134	83	19.2	28.2	8	3.6	240
16	12	-9.1	1.8	8	3.4	360	20	16	-6.4	2.2	Raleigh/Durham	723060	35.87	78.78	440	134	14.46	99.7	134	82	19.2	27.9	130	81	18.5	27.3	125	80	17.9	26.7	9	4.1	240
23	18	-4.9	2.5	7	3.3	320	27	21	-2.9	3.0	Wilmington	723013	34.27	77.90	33	10	14.68	101.2	146	85	20.8	29.2	141	83	20.1	28.6	137	83	19.5	28.1	10	4.6	220
18	14	-8.0	2.0	7	3.3	290	23	18	-5.2	2.6	Winston-Salem	723193	36.13	80.22	971	296	14.19	97.8	134	81	19.1	27.3	129	80	18.4	26.9	121	79	17.3	26.2	8	3.6	240
											NORTH DAKOTA																						
-21	2	-29.6									Bismarck	727640	46.77	100.75	1660	506	13.83	95.4	109	79	15.5	26.3	100	77	14.3	24.8	92	75	13.2	23.7	13	5.9	180
-23	2	-30.6	0.2	9	4.0	300	-19	2	-28.1	0.3	Devils Lake	727580	48.10	98.87	1453	443	13.94	96.1	108	78	15.5	25.6	100	77	14.3	25.1	90	75	12.8	24.0	11	5.0	10
-22	2	-29.7	0.2	8	3.6	180	-17	2	-27.3	0.3	Fargo	727530	46.90	96.80	899	274	14.22	98.1	122	82	17.4	27.6	112	80	16.0	26.7	104	77	14.8	24.9	14	6.4	160
-20	2	-29.1	0.3	7	3.3	290	-16	2	-26.6	0.3	Grand Forks, AFB	727575	47.97	97.40	912	278	14.22	98.0	118	81	16.9	27.3	109	78	15.6	25.6	101	76	14.4	24.3	13	5.6	180
-21	2	-29.2	0.3	10	4.5	310	-16	2	-26.5	0.4	Minot, AFB	727675	48.42	101.35	1667	508	13.83	95.4	109	80	15.6	26.4	100	78	14.3	25.3	90	75	12.9	23.9	12	5.3	150
-20	2	-28.9	0.3	12	5.3	290	-16	2	-26.6	0.4	Minot, Int'l Airport	727676	48.27	101.28	1716	523	13.81	95.2	106	78	15.1	25.4	96	75	13.7	23.9	89	73	12.7	22.7	13	5.8	200
-24	2	-31.0	0.2	8	3.6	220	-18	2	-27.7	0.3	Williston	727670	48.18	103.63	1906	581	13.71	94.5	103	78	14.7	25.6	92	76	13.1	24.2	85	73	12.2	23.0	14	6.4	150
											OHIO																						
0	6	-17.9	0.8	11	5.0	270	5	8	-14.9	1.1	Akron/Canton	725210	40.92	81.43	1237	377	14.05	96.9	125	80	17.8	26.7	118	78	16.8	25.7	113	77	16.1	24.9	10	4.6	230
5	7	-14.9	1.0	9	3.8	260	12	10	-11.3	1.5	Cincinnati, Lunken	724297	39.10	84.42	482	147	14.44	99.6	132	82	18.9	27.8	128	81	18.3	27.1	120	80	17.1	26.9	10	4.3	210
1	6	-17.4	0.8	12	5.5	230	6	8	-14.7	1.1	Cleveland	725240	41.42	81.87	804	245	14.27	98.4	125	82	17.8	27.6	118	80	16.9	26.4	112	78	16.0	25.4	12	5.3	230
1	6	-17.4	0.8	9	4.1	270	6	8	-14.3	1.1	Columbus, Int'l Airport	724280	40.00	82.88	817	249	14.27	98.4	128	82	18.3	27.8	123	81	17.5	27.1	117	79	16.7	26.2	11	4.7	270
3	7	-16.3	0.9	7	3.2	210	10	9	-12.5	1.4	Columbus, Rickenb.	724285	39.82	82.93	745	227	14.30	98.6	130	83	18.6	28.4	120	82	17.2	27.7	115	79	16.4	26.3	8	3.7	270
-1	6	-18.2	0.8	11	4.8	270	5	7	-15.2	1.0	Dayton, Int'l Airport	724290	39.90	84.20	1004	306	14.17	97.7	129	82	18.4	28.0	123	80	17.5	26.9	117	79	16.7	26.1	11	4.8	240
1	6	-17.1	0.9	7	3.1	270	8	9	-13.2	1.2	Dayton, Wright-Pat.	745700	39.83	84.05	823	251	14.26	98.3	136	84	19.4	28.8	127	83	18.2	28.1	119	81	17.0	27.0	9	3.8	240
-2	5	-18.9	0.7	11	4.8	250	4	7	-15.3	1.0	Findlay	725366	41.02	83.67	810	247	14.27	98.4	132	81	18.9	27.4	121	80	17.3	26.6	116	78	16.5	25.5	12	5.3	210
-1	5	-18.6	0.8	13	5.6	240	4	7	-15.8	1.0	Mansfield	725246	40.82	82.52	1296	395	14.02	96.7	128	81	18.3	27.1	122	79	17.4	26.3	116	78	16.6	25.4	12	5.3	240
-2	5	-19.0	0.7	10	4.4	250	3	7	-16.2	0.9	Toledo	725360	41.60	83.80	692	211	14.33	98.8	129	82	18.4	28.0	122	80	17.4	26.7	115	78	16.4	25.7	11	5.0	230
-1	6	-18.1	0.8	10	4.6	230	4	7	-15.4	1.0	Youngstown	725250	41.27	80.67	1184	361	14.08	97.1	122	80	17.4	26.4	116	78	16.5	25.4	110	76	15.7	24.6	10	4.6	230
2	6	-16.8	0.9	7	3.1	240	9	9	-12.8	1.3	Zanesville	724286	39.95	81.90	899	274	14.22	98.1	130	82	18.5	27.5	120	80	17.2	26.7	116	78	16.6	25.8	9	4.2	220
											OKLAHOMA																						
13	11	-10.6	1.6	9	3.9	20	19	15	-7.1	2.2	Altus, AFB	723520	34.67	99.27	1378	420	13.98	96.4	132	84	18.9	28.7	124	83	17.7	28.6	119	82	17.0	28.0	10	4.4	190
5	7	-14.8	1.1	12	5.4	10	12	11	-11.0	1.5	Enid, Vance Afb	723535	36.33	97.92	1306	398	14.02	96.6	130	85	18.5	29.4	121	83	17.3	28.5	116	82	16.6	28.0	11	4.9	190
12	11	-11.3	1.5	11	4.9	10	19	15	-7.3	2.1	Lawton, Fort Sil	723550	34.65	98.40	1188	362	14.08	97.1	135	83	19.3	28.3	129	82	18.4	27.7	121	81	17.3	27.0	11	4.8	170
10	9	-12.2	1.4	9	3.8	360	17	13	-8.1	1.9	Mcalester	723566	34.88	95.78	771	235	14.29	98.5	141	85	20.2	29.5	137	83	19.6	28.6	133	83	19.0	28.2	9	4.1	190
10	10	-12.1	1.4	10	4.6	10	17	14	-8.5	2.0	Ok. City, Tinker AFB	723540	35.42	97.38	1293	394	14.02	96.7	138	87	19.7	30.6	132	85	18.9	29.4	123	83	17.6	28.5	11	4.8	190
9	9	-12.6	1.3	15	6.9	360	15	12	-9.6	1.8	Ok. City, W. R. Apt.	723530	35.40	97.60	1302	397	14.02	96.6	129	83	18.4	28.4	125	82	17.8	27.8	120	81	17.2	27.4	13	5.9	180
9	9	-12.9	1.3	11	4.9	360	14	11	-10.1	1.6	Tulsa	723560	36.20	95.90	676	206	14.34	98.9	137	87	19.6	30.6	132	85	18.8	29.6	127	84	18.1	29.1	12	5.4	180
											OREGON																						
25	19	-3.8	2.8	8	3.6	90	29	23	-1.8	3.3	Astoria	727910	46.15	123.88	23	7	14.68	101.2	81	69	11.5	20.6	76	66	10.9	19.1	74	65	10.5	18.2	12	5.2	320
21	16	-6.0	2.3	8	3.5	360	26	20	-3.6	2.9	Eugene	726930	44.12	123.22	374	114	14.50	100.0	83	74	11.9	23.3	78	73	11.2	22.5	74	71	10.6	21.4	12	5.2	360
19	14	-7.0	2.1	8	3.6	60	24	18	-4.2	2.6	Hillsboro	726986	45.53	122.95	203	62	14.59	100.6	90	79	12.9	26.3	82	75	11.7	23.9	78	72	11.1	22.3	9	3.9	360
4	8	-15.8	1.1	6	2.5	320	10	11	-12.5	1.5	Klamath Falls	725895	42.15	121.73	4091	1247	12.65	87.2	85	74	12.2	23.4	80	73	11.5	22.8	75	71	10.7	21.6	9	3.9	320

United States of America

HUMIDIFICATION 99.6% (35 hours below...)							HUMIDIFICATION 99% (88 hours below...)												DEHUMIDIFICATION 0.4% (35 hours above...)				1% (88 hours above...)				2% (175 hours above...)				Summer Wind		
°F	gr/lb	°C	g/kg	mph	m/s	PWD	°F	gr/lb	°C	g/kg	Location	WMO#	Lat.	Long.	ft.	m	psia	kPa	gr/lb	°F	g/kg	°C	gr/lb	°F	g/kg	°C	gr/lb	°F	g/kg	°C	mph	m/s	PWD
1a	1b	1c	1d	1e	1f	1g	2a	2b	2c	2d	3a	3b	3c	3d	3e	3f	3g	3h	4a	4b	4c	4d	5a	5b	5c	5d	6a	6b	6c	6d	7a	7b	7c
-9	4	-23.0	0.6	1	0.6	130	0	6	-17.7	0.9	Meacham	726885	45.52	118.40	4055	1236	12.67	87.3	67	66	9.6	19.1	63	64	9.0	18.0	59	64	8.4	17.8	5	2.3	360
21	16	-6.3	2.3	3	1.2	130	24	19	-4.5	2.7	Medford	725970	42.37	122.87	1329	405	14.00	96.6	81	75	11.6	23.7	76	74	10.8	23.2	71	73	10.2	22.7	9	4.2	290
30	24	-1.3	3.4	7	2.9	140	32	27	0.2	3.9	North Bend	726917	43.42	124.25	13	4	14.69	101.3	76	65	10.9	18.2	73	64	10.4	17.5	70	63	10.0	17.0	14	6.2	340
3	7	-15.9	1.0	6	2.6	140	11	10	-11.7	1.5	Pendleton	726880	45.68	118.85	1496	456	13.92	96.0	74	71	10.5	21.5	68	69	9.7	20.8	62	70	8.9	21.0	9	4.1	310
22	16	-5.8	2.3	13	5.7	120	27	21	-2.9	3.0	Portland	726980	45.60	122.60	39	12	14.67	101.2	83	75	11.8	23.7	78	72	11.2	22.3	75	71	10.7	21.6	11	5.0	340
1	7	-17.2	0.9	6	2.5	320	9	10	-12.7	1.4	Redmond	726835	44.25	121.15	3077	938	13.13	90.6	71	68	10.2	19.9	65	67	9.3	19.4	60	66	8.6	19.1	11	4.7	340
20	15	-6.6	2.2	6	2.6	350	25	19	-4.1	2.7	Salem	726940	44.92	123.00	200	61	14.59	100.6	81	75	11.5	24.1	76	73	10.8	22.9	72	71	10.3	21.6	10	4.4	360
21	18	-6.1	2.6	9	4.1	120	24	21	-4.4	3.0	Sexton Summit	725975	42.62	123.37	3842	1171	12.77	88.0	76	70	10.8	21.3	69	68	9.9	20.2	66	66	9.4	18.9	6	2.9	340
											PENNSYLVANIA																						
5	7	-14.9	1.0	9	4.2	270	10	9	-12.3	1.3	Allentown	725170	40.65	75.43	384	117	14.49	99.9	123	81	17.5	27.0	117	79	16.7	26.1	111	78	15.9	25.6	11	4.9	240
5	8	-15.0	1.1	9	4.0	270	10	10	-12.3	1.4	Altoona	725126	40.30	78.32	1503	458	13.91	95.9	119	79	17.0	26.0	113	77	16.2	25.2	109	76	15.6	24.3	8	3.4	250
-6	4	-21.2	0.6	7	3.1	270	-1	6	-18.3	0.8	Bradford	725266	41.80	78.63	2142	653	13.59	93.7	116	75	16.6	24.1	111	73	15.8	23.0	105	72	15.0	22.4	9	4.2	240
0	6	-17.8	0.8	11	5.0	280	5	8	-15.0	1.1	Du Bois	725125	41.18	78.90	1818	554	13.76	94.8	116	76	16.6	24.2	112	74	16.0	23.6	108	73	15.4	22.9	10	4.3	270
2	6	-16.6	0.9	14	6.1	200	7	8	-14.1	1.1	Erie	725260	42.08	80.18	738	225	14.31	98.7	122	79	17.4	25.9	115	77	16.4	25.2	109	76	15.6	24.4	12	5.4	250
9	9	-13.0	1.2	8	3.7	270	13	11	-10.7	1.5	Harrisburg	725115	40.20	76.77	308	94	14.53	100.2	130	82	18.6	27.8	123	80	17.6	26.8	118	79	16.8	26.1	10	4.2	250
11	10	-11.9	1.4	12	5.2	290	15	12	-9.7	1.7	Philadelphia, Int'l Apt.	724080	39.88	75.25	30	9	14.68	101.2	132	83	18.8	28.2	126	81	18.0	27.3	121	80	17.3	26.6	11	4.8	230
11	10	-11.7	1.4	10	4.5	300	15	12	-9.5	1.7	Philadelphia, N.E. Apt.	724085	40.08	75.02	121	37	14.63	100.9	135	83	19.3	28.6	129	82	18.4	27.7	121	82	17.3	27.5	10	4.4	260
10	9	-12.0	1.3	5	2.4	300	14	11	-9.8	1.6	Philadelphia, Willow G.	724086	40.20	75.15	361	110	14.50	100.0	131	83	18.7	28.5	125	82	17.9	27.7	116	81	16.6	27.3	6	2.8	250
4	7	-15.3	1.0	11	5.0	250	11	10	-11.9	1.4	Pittsburgh, Allegheny	725205	40.35	79.93	1253	382	14.04	96.8	122	79	17.4	26.3	117	78	16.7	25.6	113	77	16.1	24.9	11	4.8	240
2	6	-16.9	0.9	10	4.5	260	7	8	-14.1	1.2	Pittsburgh, Int'l Apt.	725200	40.50	80.22	1224	373	14.06	96.9	121	80	17.3	26.4	115	78	16.4	25.7	109	77	15.6	24.8	11	4.9	230
2	6	-16.7	0.9	8	3.8	230	7	8	-14.2	1.1	Wilkes-Barre/Scranton	725130	41.33	75.73	948	289	14.20	97.9	120	79	17.1	25.9	115	77	16.4	25.2	109	76	15.6	24.5	11	4.8	220
2	6	-16.6	0.9	8	3.4	270	7	8	-13.7	1.2	Williamsport	725140	41.25	76.92	525	160	14.42	99.4	125	80	17.8	26.6	118	78	16.9	25.7	113	77	16.2	24.8	10	4.6	250
											PUERTO RICO																						
68	103	20.2	14.8	2	0.8	330	70	111	21.1	15.9	Cieba, Roosevelt Rds.	785350	18.25	65.63	39	12	14.68	101.2	150	85	21.5	29.4	143	84	20.4	28.9	141	84	20.1	28.9	11	4.8	80
69	107	20.3	15.3	3	1.3	190	69	107	20.8	15.3	San Juan	785260	18.43	66.00	62	19	14.67	101.1	147	85	21.0	29.3	146	84	20.8	29.1	143	84	20.4	28.8	12	5.4	170
											RHODE ISLAND																						
5	7	-14.8	1.0	12	5.1	340	10	9	-12.3	1.3	Providence	725070	41.73	71.43	62	19	14.66	101.1	124	80	17.7	26.5	118	78	16.8	25.6	112	77	16.0	25.1	13	5.7	230
											SOUTH CAROLINA																						
28	22	-2.3	3.2	4	2.0	300	31	25	-0.4	3.6	Beaufort, MCAS	722085	32.48	80.72	39	12	14.67	101.2	145	85	20.7	29.4	141	85	20.1	29.2	137	84	19.6	28.7	7	3.2	270
25	19	-3.9	2.7	7	3.3	20	28	22	-2.0	3.2	Charleston	722080	32.90	80.03	49	15	14.67	101.1	145	84	20.7	29.0	139	83	19.8	28.6	134	83	19.2	28.1	10	4.5	230
21	16	-6.3	2.2	5	2.1	220	24	19	-4.2	2.7	Columbia	723100	33.95	81.12	226	69	14.58	100.5	134	82	19.2	27.8	130	81	18.6	27.3	127	81	18.1	26.9	9	4.1	240
23	17	-5.1	2.5	7	3.0	360	27	21	-2.7	3.0	Florence	723106	34.18	79.72	148	45	14.62	100.8	142	85	20.3	29.2	136	83	19.5	28.3	132	82	18.9	27.7	10	4.3	240
19	15	-7.1	2.1	6	2.7	50	23	18	-4.8	2.6	Greer/Greenville	723120	34.90	82.22	971	296	14.19	97.8	130	81	18.5	27.3	126	80	18.0	26.8	122	80	17.4	26.4	9	3.9	230
25	19	-3.8	2.7	4	1.6	360	29	23	-1.4	3.3	Myrtle Beach, AFB	747910	33.68	78.93	26	8	14.68	101.2	150	87	21.4	30.8	144	86	20.6	29.9	140	84	20.0	29.1	7	3.1	290
24	18	-4.2	2.6	5	2.2	10	29	23	-1.9	3.3	Sumter, Shaw AFB	747900	33.97	80.47	243	74	14.57	100.4	136	83	19.5	28.2	132	82	18.9	27.6	129	81	18.4	27.0	8	3.5	240
											SOUTH DAKOTA																						
-13	3	-25.0	0.4	11	5.0	270	-7	4	-21.8	0.6	Chamberlain	726530	43.80	99.32	1739	530	13.80	95.1	124	84	17.7	29.0	116	82	16.6	27.7	109	80	15.6	26.9	13	6.0	190
-17	2	-27.1	0.3	9	3.9	290	-12	3	-24.4	0.4	Huron	726540	44.38	98.22	1289	393	14.02	96.7	126	84	18.0	28.7	117	81	16.7	27.2	110	79	15.7	26.1	14	6.4	180
-14	3	-25.4	0.4	11	4.8	320	-9	4	-22.5	0.5	Pierre	726686	44.38	100.28	1742	531	13.79	95.1	116	81	16.5	27.1	109	80	15.5	26.4	102	78	14.5	25.6	14	6.3	180
-11	3	-23.8	0.5	9	4.2	350	-5	5	-20.6	0.7	Rapid City	726620	44.05	103.07	3169	966	13.09	90.2	104	76	14.9	24.6	98	75	14.0	23.8	92	73	13.1	22.9	13	5.8	160
-16	2	-26.5	0.4	8	3.5	310	-11	3	-23.7	0.5	Sioux Falls	726510	43.58	96.73	1427	435	13.95	96.2	127	84	18.1	28.7	119	82	17.0	27.7	112	80	16.0	26.5	15	6.7	180

United States of America

HUMIDIFICATION 99.6% (35 hours below...)							99% (88 hours below...)												DEHUMIDIFICATION 0.4% (35 hours above...)				1% (88 hours above...)				2% (175 hours above...)				Summer Wind		
°F	gr/lb	°C	g/kg	mph	m/s	PWD	°F	gr/lb	°C	g/kg	Location	WMO#	Lat.	Long.	ft.	m	psia	kPa	gr/lb	°F	g/kg	°C	gr/lb	°F	g/kg	°C	gr/lb	°F	g/kg	°C	mph	m/s	PWD
1a	1b	1c	1d	1e	1f	1g	2a	2b	2c	2d	3a	3b	3c	3d	3e	3f	3g	3h	4a	4b	4c	4d	5a	5b	5c	5d	6a	6b	6c	6d	7a	7b	7c
											TENNESSEE																						
9	9	-12.8	1.3	6	2.5	270	14	12	-9.8	1.7	Bristol	723183	36.48	82.40	1519	463	13.91	95.9	125	81	17.9	27.1	120	79	17.2	25.8	116	77	16.6	25.2	8	3.5	250
15	12	-9.7	1.7	7	3.2	360	20	15	-6.9	2.2	Chattanooga	723240	35.03	85.20	689	210	14.33	98.8	134	82	19.2	27.8	130	82	18.5	27.5	125	81	17.9	26.9	8	3.6	280
7	8	-13.7	1.2	4	1.7	310	15	13	-9.7	1.8	Crossville	723265	35.95	85.08	1880	573	13.72	94.6	134	80	19.2	26.7	125	79	17.9	26.1	121	78	17.3	25.5	8	3.4	270
12	10	-11.0	1.5	9	3.8	360	18	14	-7.5	2.0	Jackson	723346	35.60	88.92	433	132	14.47	99.7	140	85	20.0	29.7	135	85	19.3	29.2	132	84	18.8	28.7	8	3.5	240
13	11	-10.5	1.6	7	3.2	50	19	15	-7.5	2.1	Knoxville	723260	35.82	83.98	981	299	14.18	97.8	131	82	18.7	27.8	127	81	18.1	27.2	123	80	17.5	26.5	8	3.4	250
16	12	-8.9	1.8	10	4.4	20	21	16	-6.3	2.2	Memphis	723340	35.05	90.00	285	87	14.54	100.3	143	87	20.4	30.3	137	86	19.6	29.8	133	84	19.0	29.0	9	4.1	240
10	9	-12.2	1.4	8	3.7	340	16	13	-9.0	1.8	Nashville	723270	36.13	86.68	591	180	14.38	99.2	134	83	19.2	28.3	130	82	18.5	27.7	126	81	18.0	27.2	9	4.2	230
											TEXAS																						
16	13	-8.9	1.9	12	5.2	0	22	18	-5.7	2.5	Abilene	722660	32.42	99.68	1791	546	13.77	94.9	123	81	17.6	27.1	119	80	17.0	26.6	115	79	16.4	26.2	11	5.0	140
6	9	-14.4	1.2	14	6.3	20	12	11	-11.3	1.6	Amarillo	723630	35.23	101.70	3606	1099	12.88	88.8	112	76	16.0	24.2	107	75	15.3	23.8	104	74	14.8	23.3	15	6.7	200
25	20	-3.7	2.8	12	5.3	10	30	24	-1.3	3.5	Austin	722540	30.30	97.70	620	189	14.37	99.1	137	81	19.6	26.9	134	80	19.1	26.7	130	80	18.6	26.6	10	4.3	180
29	23	-1.8	3.3	10	4.6	340	32	27	0.2	3.9	Beaumont/Port Arthur	722410	29.95	94.02	23	7	14.68	101.2	152	86	21.7	29.7	148	85	21.2	29.3	145	84	20.7	28.9	9	4.2	200
28	22	-2.1	3.2	13	5.7	350	33	28	0.8	4.0	Beeville, Chase NAS	722556	28.37	97.67	190	58	14.59	100.6	155	86	22.2	29.9	148	85	21.2	29.2	144	84	20.6	28.8	9	4.1	150
36	31	2.1	4.4	13	5.9	330	40	36	4.2	5.1	Brownsville	722500	25.90	97.43	20	6	14.68	101.3	146	83	20.8	28.3	142	83	20.3	28.2	140	82	20.0	28.0	16	7.0	160
22	17	-5.8	2.4	12	5.5	350	29	23	-1.9	3.3	College Station/Bryan	722445	30.58	96.37	322	98	14.53	100.2	141	82	20.1	27.6	138	81	19.7	27.4	134	81	19.2	27.0	9	4.2	170
32	27	0.1	3.8	13	5.7	360	36	31	2.1	4.4	Corpus Christi	722510	27.77	97.50	43	13	14.67	101.2	148	84	21.2	28.9	146	83	20.9	28.6	143	83	20.4	28.3	15	6.8	140
17	13	-8.1	1.9	13	5.6	350	24	19	-4.4	2.7	Dallas/Fort Worth Apt.	722590	32.90	97.03	597	182	14.38	99.2	132	82	18.9	28.0	130	82	18.5	27.5	126	81	18.0	27.3	10	4.6	170
28	23	-2.3	3.3	7	3.0	10	32	28	0.1	3.9	Del Rio, Laughlin AFB	722615	29.37	100.78	1083	330	14.13	97.4	136	82	19.4	27.8	131	82	18.7	27.7	124	81	17.7	27.2	9	4.1	140
21	18	-6.2	2.6	5	2.1	20	25	22	-3.9	3.1	El Paso	722700	31.80	106.40	3917	1194	12.73	87.8	114	73	16.3	22.9	109	73	15.5	22.7	103	74	14.7	23.1	8	3.8	180
18	14	-7.9	2.0	11	5.0	10	24	19	-4.2	2.7	Fort Worth, Carswell	722595	32.77	97.45	650	198	14.35	99.0	141	85	20.1	29.3	135	84	19.3	29.0	130	84	18.6	28.7	8	3.6	10
19	15	-7.2	2.1	13	5.8	350	24	19	-4.2	2.7	Fort Worth, Meacham	722596	32.82	97.37	709	216	14.32	98.8	135	83	19.3	28.4	131	82	18.7	27.9	127	82	18.1	27.6	10	4.6	180
13	13	-10.6	1.9	19	8.7	70	19	18	-7.2	2.5	Guadalupe Pass	722620	31.83	104.80	5453	1662	12.02	82.9	102	71	14.6	21.7	96	71	13.7	21.7	91	71	13.0	21.4	13	5.9	250
29	23	-1.8	3.3	13	5.9	350	34	29	1.0	4.1	Houston, Hobby	722435	29.65	95.28	46	14	14.67	101.2	147	84	21.0	28.7	144	83	20.5	28.2	141	82	20.1	27.9	7	3.3	190
27	22	-2.6	3.1	8	3.6	340	31	26	-0.4	3.7	Houston, Intcntl. Apt.	722430	29.97	95.35	108	33	14.64	100.9	144	83	20.5	28.4	141	83	20.1	28.3	137	83	19.6	28.2	10	4.2	180
19	15	-7.2	2.2	6	2.8	360	23	19	-4.8	2.6	Junction	747400	30.50	99.77	1713	522	13.81	95.2	130	80	18.5	26.6	121	79	17.3	26.2	118	79	16.9	26.0	9	4.1	150
20	16	-6.6	2.2	11	4.8	360	27	22	-2.8	3.1	Killeen, Fort Hood	722576	31.07	97.83	1014	309	14.16	97.7	137	81	19.6	27.2	132	81	18.9	27.1	128	80	18.3	26.7	9	3.9	160
31	25	-0.8	3.6	11	4.9	360	36	31	2.0	4.5	Kingsville, NFB	722516	27.50	97.82	49	15	14.67	101.1	149	85	21.3	29.5	144	84	20.6	29.0	141	84	20.1	28.9	11	4.9	150
32	27	-0.1	3.9	9	3.9	320	36	32	2.2	4.5	Laredo	722520	27.55	99.47	509	155	14.43	99.5	138	82	19.7	27.9	136	81	19.4	27.4	132	81	18.9	27.2	13	5.8	140
11	11	-11.7	1.6	12	5.3	0	17	15	-8.3	2.1	Lubbock, Int'l Arpt	722670	33.65	101.82	3241	988	13.05	90.0	120	77	17.1	25.0	115	76	16.4	24.6	111	76	15.8	24.2	14	6.2	160
11	11	-11.4	1.6	10	4.6	20	18	15	-7.8	2.2	Lubbock, Reese AFB	722675	33.60	102.05	3337	1017	13.01	89.7	122	78	17.4	25.8	115	77	16.4	25.2	110	77	15.7	24.9	11	4.9	170
23	18	-4.9	2.5	6	2.7	330	27	22	-2.6	3.1	Lufkin	722446	31.23	94.75	289	88	14.54	100.3	143	83	20.4	28.6	139	83	19.8	28.3	134	82	19.2	28.0	8	3.5	230
15	14	-9.6	2.0	5	2.3	360	19	17	-7.3	2.5	Marfa	722640	30.37	104.02	4859	1481	12.29	84.8	110	72	15.7	22.2	103	71	14.7	21.7	98	71	14.0	21.4	9	4.1	220
34	29	1.3	4.1	11	4.8	350	40	37	4.2	5.2	Mcallen	722506	26.18	98.23	108	33	14.64	100.9	146	83	20.9	28.5	143	82	20.4	28.0	140	82	20.0	27.7	14	6.3	130
17	15	-8.2	2.1	9	4.2	20	22	19	-5.4	2.7	Midland/Odessa	722650	31.95	102.18	2861	872	13.24	91.3	120	76	17.2	24.5	115	75	16.4	24.0	111	75	15.8	24.0	13	5.7	180
20	16	-6.9	2.3	10	4.3	20	24	20	-4.4	2.8	San Angelo	722630	31.37	100.50	1909	582	13.71	94.5	123	80	17.5	26.4	118	79	16.9	25.9	116	78	16.5	25.7	11	4.9	160
26	21	-3.3	3.0	10	4.4	350	30	25	-1.0	3.6	San Antonio, Int'l Apt.	722530	29.53	98.47	794	242	14.28	98.5	139	81	19.9	27.3	135	81	19.3	26.9	132	80	18.8	26.8	10	4.4	160
27	22	-2.7	3.1	8	3.7	360	32	27	-0.2	3.9	San Antonio, Kelly AFB	722535	29.38	98.58	689	210	14.33	98.8	145	83	20.7	28.3	140	82	20.0	27.6	136	81	19.4	27.0	8	3.5	160
27	22	-2.9	3.1	7	3.2	340	31	26	-0.6	3.7	San Antonio, Randolph	722536	29.53	98.28	761	232	14.30	98.6	138	82	19.7	27.6	134	81	19.2	27.1	132	81	18.8	27.0	7	3.3	150
23	19	-4.9	2.8	6	2.6	360	28	24	-2.5	3.5	Sanderson	747300	30.17	102.42	2838	865	13.25	91.4	123	79	17.6	26.3	119	78	17.0	25.5	114	77	16.3	25.2	7	3.3	120
29	23	-1.7	3.3	12	5.3	360	33	28	0.7	4.0	Victoria	722550	28.85	96.92	118	36	14.63	100.9	145	83	20.7	28.1	141	82	20.2	27.8	139	82	19.8	27.6	12	5.1	180
22	17	-5.7	2.4	13	5.9	360	26	21	-3.2	3.0	Waco	722560	31.62	97.22	509	155	14.43	99.5	135	83	19.3	28.2	131	82	18.7	27.9	127	82	18.2	27.7	12	5.1	180
14	12	-9.9	1.7	12	5.3	360	19	15	-7.1	2.2	Wichita Fls., Sheppard	723510	33.98	98.50	1030	314	14.16	97.6	129	82	18.4	27.7	124	82	17.7	27.6	120	81	17.2	27.3	13	5.6	180

United States of America

HUMIDIFICATION 99.6% (35 hours below...)							99% (88 hours below...)												DEHUMIDIFICATION 0.4% (35 hours above...)				1% (88 hours above...)				2% (175 hours above...)				Summer Wind		
°F 1a	gr/lb 1b	°C 1c	g/kg 1d	mph 1e	m/s 1f	PWD 1g	°F 2a	gr/lb 2b	°C 2c	g/kg 2d	Location 3a	WMO# 3b	Lat. 3c	Long. 3d	ft. 3e	m 3f	psia 3g	kPa 3h	gr/lb 4a	°F 4b	g/kg 4c	°C 4d	gr/lb 5a	°F 5b	g/kg 5c	°C 5d	gr/lb 6a	°F 6b	g/kg 6c	°C 6d	mph 7a	m/s 7b	PWD 7c
											UTAH																						
2	8	-16.5	1.1	4	1.7	140	8	10	-13.3	1.5	Cedar City	724755	37.70	113.10	5623	1714	11.94	82.4	93	68	13.3	19.7	85	68	12.2	19.8	78	68	11.2	20.2	12	5.3	200
6	9	-14.7	1.3	9	4.2	110	11	12	-11.6	1.6	Ogden, Hill AFB	725755	41.12	111.97	4787	1459	12.33	85.0	91	72	13.0	22.2	83	73	11.9	22.6	77	73	11.0	22.9	6	2.7	190
6	9	-14.7	1.2	7	2.9	160	11	11	-11.7	1.6	Salt Lake City	725720	40.78	111.97	4226	1288	12.59	86.8	92	73	13.1	22.5	84	73	12.0	22.5	77	73	11.0	22.8	11	5.0	340
											VERMONT																						
-11	3	-23.9	0.4	6	2.9	70	-6	4	-21.2	0.6	Burlington	726170	44.47	73.15	341	104	14.52	100.1	115	79	16.4	26.2	109	77	15.5	25.1	102	75	14.6	24.0	11	4.9	180
-10	3	-23.1	0.5	4	1.6	320	-6	4	-21.0	0.6	Montpelier/Barre	726145	44.20	72.57	1165	355	14.09	97.1	111	78	15.9	25.5	106	75	15.1	24.0	99	73	14.2	22.9	9	4.0	220
											VIRGINIA																						
12	10	-11.0	1.5	2	0.9	320	18	14	-7.8	2.0	Fort Belvoir	724037	38.72	77.18	69	21	14.66	101.1	139	86	19.9	29.9	133	85	19.0	29.4	127	83	18.2	28.5	6	2.5	160
21	16	-6.3	2.3	10	4.3	330	24	18	-4.2	2.6	Hampton, Langley AFB	745980	37.08	76.37	10	3	14.69	101.3	141	85	20.2	29.6	136	84	19.5	28.9	132	83	18.9	28.4	9	4.1	240
12	11	-11.1	1.5	8	3.3	360	17	14	-8.2	1.9	Lynchburg	724100	37.33	79.20	938	286	14.20	97.9	129	81	18.4	27.3	125	80	17.8	26.7	120	79	17.2	26.3	9	3.8	230
18	14	-7.5	2.0	8	3.5	350	22	17	-5.3	2.4	Newport News	723086	37.13	76.50	43	13	14.67	101.2	139	84	19.9	29.1	135	83	19.3	28.5	132	82	18.8	27.9	10	4.5	220
20	15	-6.7	2.2	12	5.4	340	24	18	-4.7	2.6	Norfolk	723080	36.90	76.20	30	9	14.68	101.2	135	83	19.3	28.6	130	82	18.6	27.8	126	81	18.0	27.4	12	5.2	230
22	17	-5.8	2.4	8	3.4	310	25	19	-3.9	2.7	Oceana, NAS	723075	36.82	76.03	23	7	14.68	101.2	139	85	19.9	29.2	134	83	19.1	28.5	129	82	18.4	27.9	9	4.0	220
16	12	-9.0	1.8	6	2.8	340	21	16	-6.3	2.3	Quantico, MCAS	724035	38.50	77.30	13	4	14.69	101.3	136	87	19.4	30.5	130	85	18.5	29.6	125	83	17.8	28.3	5	2.4	230
14	11	-10.1	1.6	7	3.1	340	18	14	-7.6	2.0	Richmond	724010	37.50	77.33	177	54	14.60	100.7	137	84	19.5	28.9	131	82	18.7	28.0	126	81	18.0	27.2	10	4.4	230
12	11	-11.1	1.5	10	4.6	320	17	13	-8.6	1.9	Roanoke	724110	37.32	79.97	1175	358	14.08	97.1	123	80	17.6	26.7	118	79	16.9	26.1	115	78	16.4	25.6	10	4.4	290
9	9	-12.8	1.3	7	2.9	340	14	11	-10.2	1.6	Sterling	724030	38.95	77.45	322	98	14.53	100.2	130	83	18.6	28.2	125	81	17.8	27.4	120	80	17.1	26.6	10	4.2	250
15	12	-9.3	1.7	11	5.0	340	20	15	-6.5	2.2	Washington, National	724050	38.85	77.03	66	20	14.66	101.1	137	83	19.6	28.5	132	83	18.9	28.2	127	81	18.2	27.4	11	4.8	170
											WASHINGTON																						
15	12	-9.2	1.7	17	7.4	40	21	16	-6.0	2.3	Bellingham	727976	48.80	122.53	157	48	14.61	100.7	81	73	11.6	22.8	78	70	11.1	21.2	74	67	10.6	19.5	9	4.0	290
5	7	-15.1	1.0	6	2.7	20	12	10	-11.1	1.5	Hanford	727840	46.57	119.60	732	223	14.31	98.7	73	72	10.4	22.4	68	75	9.7	23.7	62	74	8.8	23.6	8	3.4	20
18	14	-7.8	2.0	5	2.1	180	23	18	-4.8	2.5	Olympia	727920	46.97	122.90	200	61	14.59	100.6	81	73	11.5	22.7	76	71	10.9	21.9	73	69	10.4	20.8	8	3.8	50
23	17	-5.1	2.5	7	2.9	60	27	21	-2.8	3.0	Quillayute	727970	47.95	124.55	203	62	14.59	100.6	76	65	10.9	18.3	74	63	10.5	17.4	71	62	10.1	16.7	9	3.8	240
23	18	-4.8	2.6	10	4.4	10	28	22	-2.2	3.2	Seattle, Int'l Airport	727930	47.45	122.30	449	137	14.46	99.7	78	71	11.1	21.6	74	69	10.6	20.5	71	68	10.2	19.8	10	4.5	350
1	6	-17.4	0.9	7	3.1	50	7	9	-13.8	1.2	Spokane, Fairchild AFB	727855	47.62	117.65	2461	750	13.43	92.6	77	68	11.0	20.2	71	68	10.1	19.9	67	67	9.5	19.4	9	4.0	240
3	7	-16.1	1.1	13	5.9	90	10	11	-12.4	1.5	Stampede Pass	727815	47.28	121.33	3967	1209	12.71	87.6	70	63	10.0	17.3	65	61	9.3	16.0	62	58	8.8	14.4	7	3.3	100
18	14	-7.5	2.0	2	1.0	180	24	18	-4.7	2.6	Tacoma, McChord AFB	742060	47.13	122.48	322	98	14.53	100.2	79	71	11.3	21.8	76	70	10.8	20.9	72	68	10.3	20.2	7	3.1	20
4	7	-15.6	1.0	6	2.7	180	12	11	-11.1	1.5	Walla Walla	727846	46.10	118.28	1204	367	14.07	97.0	82	74	11.7	23.3	76	72	10.8	22.5	71	72	10.2	22.4	9	4.1	300
3	7	-15.9	1.0	3	1.4	100	9	9	-12.6	1.3	Wenatchee	727825	47.40	120.20	1243	379	14.05	96.9	78	75	11.1	23.8	73	75	10.4	23.8	68	74	9.7	23.3	9	3.9	280
4	7	-15.4	1.0	7	2.9	250	11	10	-11.8	1.4	Yakima	727810	46.57	120.53	1066	325	14.14	97.5	78	75	11.2	23.8	71	74	10.2	23.2	67	72	9.5	22.4	7	3.3	90
											WEST VIRGINIA																						
5	8	-14.9	1.1	6	2.9	270	12	11	-11.3	1.6	Bluefield	724125	37.30	81.20	2858	871	13.24	91.3	120	75	17.2	24.1	116	75	16.5	23.7	111	73	15.9	23.0	6	2.6	290
6	8	-14.4	1.1	7	2.9	250	11	10	-11.5	1.5	Charleston	724140	38.37	81.60	981	299	14.18	97.8	129	81	18.4	27.3	123	80	17.6	26.6	118	78	16.9	25.8	8	3.5	240
-2	5	-18.8	0.8	4	1.6	280	5	8	-15.0	1.1	Elkins	724170	38.88	79.85	1998	609	13.66	94.2	121	78	17.3	25.7	116	77	16.6	24.8	111	75	15.9	23.8	8	3.7	290
6	8	-14.6	1.1	8	3.4	270	11	10	-11.5	1.4	Huntington	724250	38.37	82.55	837	255	14.26	98.3	132	82	18.9	27.9	127	81	18.1	26.9	121	79	17.3	26.3	8	3.6	270
8	8	-13.1	1.2	7	3.3	270	14	11	-10.2	1.6	Martinsburg	724177	39.40	77.98	558	170	14.40	99.3	130	81	18.6	27.3	120	80	17.2	26.9	116	79	16.5	26.0	9	4.0	290
4	7	-15.3	1.0	6	2.8	210	11	10	-11.7	1.4	Morgantown	724176	39.65	79.92	1247	380	14.05	96.8	124	79	17.7	25.9	119	78	17.0	25.4	115	76	16.4	24.6	8	3.6	240
4	7	-15.4	1.0	7	3.1	240	11	10	-11.5	1.4	Parkersburg	724273	39.35	81.43	860	262	14.24	98.2	132	82	18.8	27.6	122	80	17.4	26.9	118	78	16.9	25.6	8	3.7	270
											WISCONSIN																						
-18	2	-27.7	0.3	7	3.1	250	-13	3	-24.9	0.4	Eau Claire	726435	44.87	91.48	906	276	14.22	98.1	125	82	17.8	27.8	116	80	16.6	26.4	109	78	15.6	25.3	13	5.8	220
-13	3	-24.8	0.4	10	4.3	270	-8	4	-21.9	0.5	Green Bay	726450	44.48	88.13	702	214	14.33	98.8	124	82	17.7	27.5	116	79	16.6	26.3	109	77	15.5	25.1	12	5.5	200

United States of America

HUMIDIFICATION 99.6% (35 hours below...)							99% (88 hours below...)												DEHUMIDIFICATION 0.4% (35 hours above...)				1% (88 hours above...)				2% (175 hours above...)				Summer Wind		
°F	gr/lb	°C	g/kg	mph	m/s	PWD	°F	gr/lb	°C	g/kg	Location	WMO#	Lat.	Long.	ft.	m	psia	kPa	gr/lb	°F	g/kg	°C	gr/lb	°F	g/kg	°C	gr/lb	°F	g/kg	°C	mph	m/s	PWD
1a	1b	1c	1d	1e	1f	1g	2a	2b	2c	2d	3a	3b	3c	3d	3e	3f	3g	3h	4a	4b	4c	4d	5a	5b	5c	5d	6a	6b	6c	6d	7a	7b	7c
-14	3	-25.6	0.4	7	3.0	310	-8	4	-22.4	0.5	La Crosse	726430	43.87	91.25	663	202	14.35	98.9	132	83	18.9	28.2	125	81	17.8	27.0	117	78	16.7	25.8	12	5.3	180
-11	3	-24.1	0.4	8	3.6	300	-6	4	-21.2	0.6	Madison	726410	43.13	89.33	866	264	14.24	98.2	126	83	18.0	28.1	118	80	16.8	26.7	111	78	15.8	25.6	12	5.4	230
-7	4	-21.7	0.6	13	5.8	290	-2	5	-18.7	0.7	Milwaukee	726400	42.95	87.90	692	211	14.33	98.8	127	83	18.2	28.1	119	80	17.0	26.7	111	78	15.9	25.5	15	6.5	220
-15	3	-26.2	0.4	7	3.2	300	-9	4	-22.7	0.5	Wausau	726463	44.93	89.63	1201	366	14.07	97.0	120	79	17.1	25.9	113	77	16.2	25.0	108	75	15.4	23.8	10	4.6	200
											WYOMING																						
-22	2	-29.9	0.3	3	1.5	60	-15	3	-25.9	0.5	Big Piney	726710	42.57	110.10	6969	2124	11.35	78.3	69	60	9.9	15.8	64	60	9.1	15.3	57	59	8.1	14.9	11	5.1	260
-13	3	-24.7	0.5	9	4.1	260	-5	5	-20.3	0.8	Casper	725690	42.92	106.47	5289	1612	12.10	83.4	85	66	12.2	18.7	78	66	11.2	19.0	73	65	10.4	18.5	13	5.8	240
-7	5	-21.5	0.7	10	4.4	290	0	7	-17.6	1.0	Cheyenne, Warren AFB	725640	41.15	104.82	6142	1872	11.71	80.8	90	66	12.9	18.7	85	65	12.1	18.3	80	64	11.4	17.9	13	5.7	290
-14	3	-25.5	0.4	6	2.5	40	-7	5	-21.8	0.7	Cody	726700	44.52	109.02	5095	1553	12.18	84.0	76	70	10.9	21.0	69	66	9.8	18.7	64	65	9.1	18.4	11	4.7	70
-16	3	-26.5	0.4	8	3.4	260	-7	4	-21.9	0.6	Gillette	726650	44.35	105.53	4035	1230	12.68	87.4	88	73	12.5	22.6	80	69	11.4	20.7	73	68	10.4	20.1	11	5.0	140
-14	3	-25.6	0.5	3	1.3	120	-7	5	-21.6	0.7	Lander	725760	42.82	108.73	5558	1694	11.97	82.6	81	69	11.5	20.4	74	68	10.6	19.8	69	67	9.8	19.6	10	4.6	270
-9	4	-22.6	0.6	7	3.0	70	-2	6	-18.7	0.9	Rock Springs	725744	41.60	109.07	6759	2060	11.44	78.9	78	62	11.2	16.4	71	61	10.2	16.3	66	61	9.4	15.9	13	5.6	280
-14	3	-25.7	0.4	5	2.1	280	-8	4	-21.9	0.6	Sheridan	726660	44.77	106.97	3967	1209	12.71	87.6	88	71	12.6	21.9	82	71	11.7	21.7	76	69	10.9	20.8	9	4.2	120
-22	2	-30.1	0.3	3	1.4	210	-13	3	-25.0	0.5	Worland	726665	43.97	107.95	4245	1294	12.58	86.7	94	75	13.4	24.0	86	75	12.3	23.8	80	73	11.5	22.7	9	4.1	220

Canada

HUMIDIFICATION 99.6% (35 hours below...)							99% (88 hours below...)												DEHUMIDIFICATION 0.4% (35 hours above...)				1% (88 hours above...)				2% (175 hours above...)				Summer Wind		
°F 1a	gr/lb 1b	°C 1c	g/kg 1d	mph 1e	m/s 1f	PWD 1g	°F 2a	gr/lb 2b	°C 2c	g/kg 2d	Location 3a	WMO# 3b	Lat. 3c	Long. 3d	ft. 3e	m 3f	psia 3g	kPa 3h	gr/lb 4a	°F 4b	g/kg 4c	°C 4d	gr/lb 5a	°F 5b	g/kg 5c	°C 5d	gr/lb 6a	°F 6b	g/kg 6c	°C 6d	mph 7a	m/s 7b	PWD 7c
											ALBERTA																						
-22	2	-30.0	0.3	7	2.9	0	-17	3	-27.1	0.4	Calgary Int'l. A	718770	51.12	114.02	3556	1084	12.90	89.0	79	67	11.3	19.6	74	65	10.6	18.4	69	64	9.8	17.8	11	4.7	160
-31	1	-35.1	0.1	3	1.2	270	-26	1	-32.2	0.2	Cold Lake A	711200	54.42	110.28	1785	544	13.77	95.0	88	71	12.5	21.9	82	69	11.7	20.5	76	67	10.9	19.2	9	4.2	180
-27	1	-33.0	0.2	9	3.8	320	-23	2	-30.3	0.3	Coronation	718730	52.07	111.45	2595	791	13.37	92.2	84	69	12.0	20.7	78	67	11.1	19.5	73	66	10.4	18.7	11	4.9	160
-28	1	-33.4	0.2	6	2.5	180	-23	2	-30.5	0.2	Edmonton Int'l. A	711230	53.30	113.58	2372	723	13.48	92.9	89	73	12.7	22.5	83	69	11.8	20.7	77	66	11.0	19.1	9	3.8	180
-32	1	-35.8	0.1	3	1.2	90	-29	1	-33.9	0.2	Fort Mcmurray A	719320	56.65	111.22	1211	369	14.06	97.0	84	71	12.0	21.5	78	68	11.1	20.1	74	66	10.5	19.0	9	3.8	250
-32	1	-35.4	0.1	3	1.3	320	-27	1	-32.6	0.2	Grande Prairie A	719400	55.18	118.88	2195	669	13.57	93.5	81	68	11.6	20.1	76	65	10.8	18.4	71	63	10.1	17.4	8	3.5	270
-22	2	-29.9	0.3	5	2.4	250	-16	3	-26.9	0.4	Lethbridge A	718740	49.63	112.80	3048	929	13.15	90.7	84	70	12.0	21.0	78	69	11.2	20.4	73	67	10.4	19.2	13	5.7	270
-24	2	-31.2	0.2	5	2.1	230	-19	2	-28.4	0.3	Medicine Hat A	718720	50.02	110.72	2349	716	13.49	93.0	83	70	11.9	21.3	78	69	11.1	20.6	72	68	10.3	20.0	11	4.7	220
-32	1	-35.3	0.1	4	1.9	0	-27	1	-32.9	0.2	Peace River A	710680	56.23	117.43	1873	571	13.73	94.7	82	69	11.7	20.8	76	67	10.9	19.4	71	65	10.2	18.3	9	4.1	270
-27	1	-32.8	0.2	6	2.8	200	-21	2	-29.6	0.3	Red Deer A	718780	52.18	113.90	2969	905	13.19	90.9	85	71	12.2	21.8	80	69	11.4	20.3	74	66	10.6	19.0	10	4.5	180
-25	2	-31.7	0.2	3	1.4	340	-20	2	-29.1	0.3	Rocky Mtn. House	719280	52.43	114.92	3245	989	13.05	90.0	87	70	12.4	20.9	81	68	11.6	19.9	76	66	10.9	18.9	8	3.7	160
-30	1	-34.3	0.2	4	1.9	270	-24	2	-31.2	0.2	Whitecourt	719300	54.15	115.78	2566	782	13.38	92.3	86	69	12.3	20.7	80	67	11.4	19.3	75	64	10.7	18.0	7	3.3	90
											BRITISH COLUMBIA																						
15	12	-9.3	1.7	12	5.5	90	20	15	-6.5	2.2	Abbotsford	711080	49.03	122.37	190	58	14.59	100.6	85	77	12.1	24.7	80	73	11.4	23.0	76	70	10.8	21.0	7	3.3	220
25	19	-4.0	2.7	22	9.8	50	29	23	-1.7	3.3	Cape St. James	710310	51.93	131.02	302	92	14.54	100.2	74	61	10.6	15.9	71	59	10.2	15.2	69	59	9.9	14.7	11	5.0	300
5	8	-15.0	1.1	8	3.5	0	9	9	-12.9	1.3	Castlegar	718840	49.30	117.63	1624	495	13.85	95.5	83	71	11.8	21.9	78	70	11.1	21.0	74	68	10.5	19.7	7	3.3	180
21	16	-6.2	2.2	7	3.0	290	25	19	-4.0	2.7	Comox	718930	49.72	124.90	79	24	14.65	101.0	78	68	11.2	19.9	75	66	10.7	19.1	71	65	10.2	18.3	7	3.0	340
-15	3	-25.9	0.4	2	1.0	200	-8	4	-22.2	0.6	Cranbrook	718800	49.60	115.78	3081	939	13.13	90.5	77	66	11.0	18.6	72	66	10.3	18.6	67	65	9.5	18.5	10	4.6	210
-33	1	-36.3	0.1	1	0.5	220	-30	1	-34.4	0.2	Fort Nelson	719450	58.83	122.58	1253	382	14.04	96.8	80	68	11.4	19.8	75	66	10.7	18.8	70	64	10.0	18.0	5	2.4	120
-30	1	-34.4	0.2	7	2.9	0	-25	1	-31.8	0.2	Fort St. John	719430	56.23	120.73	2280	695	13.52	93.2	80	67	11.4	19.3	74	65	10.6	18.2	69	63	9.8	17.3	9	3.8	230
-8	4	-22.1	0.5	4	1.9	90	-1	5	-18.4	0.8	Kamloops	718870	50.70	120.45	1135	346	14.10	97.2	78	70	11.2	20.9	74	69	10.5	20.3	69	68	9.9	20.2	8	3.4	270
5	7	-15.0	1.1	8	3.7	340	10	9	-12.4	1.3	Penticton	718890	49.47	119.60	1129	344	14.11	97.3	81	73	11.6	22.9	76	72	10.9	22.0	72	71	10.3	21.7	9	3.8	180
22	17	-5.5	2.4	8	3.4	110	26	20	-3.5	2.8	Port Hardy	711090	50.68	127.37	72	22	14.66	101.1	72	62	10.3	16.9	69	61	9.9	16.0	67	60	9.5	15.4	9	4.1	340
-25	1	-31.9	0.2	2	0.9	0	-18	2	-27.8	0.3	Prince George	718960	53.88	122.68	2267	691	13.53	93.3	77	66	11.0	18.6	73	64	10.4	17.8	67	62	9.6	16.9	6	2.7	180
7	8	-13.7	1.2	6	2.5	70	13	11	-10.5	1.5	Prince Rupert	718980	54.30	130.43	112	34	14.64	100.9	71	61	10.2	16.2	69	60	9.8	15.5	66	59	9.4	14.9	8	3.6	270
-22	2	-29.8	0.3	1	0.3	340	-14	3	-25.6	0.4	Quesnel	711030	53.03	122.52	1788	545	13.77	94.9	81	67	11.5	19.2	75	65	10.7	18.4	71	63	10.1	17.3	5	2.4	340
21	16	-6.3	2.2	18	8.0	320	25	19	-4.0	2.7	Sandspit	711010	53.25	131.82	20	6	14.68	101.3	75	62	10.7	16.9	71	61	10.2	16.3	69	60	9.8	15.8	9	4.2	270
-19	2	-28.2	0.3	3	1.3	140	-12	3	-24.2	0.4	Smithers	719500	54.82	127.18	1716	523	13.81	95.2	74	65	10.6	18.3	70	64	10.0	17.6	65	62	9.3	16.8	6	2.8	320
29	23	-1.7	3.3	6	2.6	50	31	26	-0.5	3.7	Spring Island	714790	50.12	127.93	322	98	14.53	100.2	76	62	10.8	16.7	74	61	10.5	16.1	71	60	10.2	15.4	6	2.8	320
-2	5	-19.1	0.7	19	8.6	0	2	6	-16.7	0.9	Terrace	719510	54.47	128.58	712	217	14.32	98.7	73	66	10.4	18.8	69	64	9.8	17.8	65	64	9.3	17.7	8	3.7	270
25	19	-3.7	2.8	5	2.1	70	29	23	-1.9	3.2	Tofino	711060	49.08	125.77	79	24	14.65	101.0	76	63	10.8	17.3	73	62	10.4	16.6	71	60	10.1	15.8	7	3.1	290
18	14	-7.8	2.0	6	2.7	90	24	18	-4.7	2.5	Vancouver Int'l.	718920	49.18	123.17	7	2	14.69	101.3	83	71	11.8	21.7	79	70	11.3	20.9	76	68	10.9	19.9	7	3.3	290
23	17	-5.3	2.4	10	4.5	50	26	20	-3.2	2.9	Victoria Int'l. A	717990	48.65	123.43	62	19	14.66	101.1	75	69	10.7	20.3	71	67	10.2	19.3	69	65	9.8	18.3	6	2.8	90
-20	2	-29.0	0.3	3	1.2	320	-14	3	-25.5	0.4	Williams Lake	711040	52.18	122.05	3084	940	13.13	90.5	73	62	10.4	16.9	67	61	9.5	16.2	62	60	8.9	15.8	6	2.8	140
											MANITOBA																						
-29	1	-33.7	0.2	9	4.1	270	-24	2	-31.3	0.2	Brandon	711400	49.92	99.95	1342	409	14.00	96.5	105	77	15.0	24.7	97	74	13.8	23.3	88	72	12.6	22.3	12	5.5	160
-36	1	-37.7	0.1	15	6.5	270	-33	1	-36.2	0.1	Churchill	719130	58.75	94.07	95	29	14.64	101.0	76	68	10.8	20.1	68	65	9.7	18.3	60	62	8.6	16.9	13	5.8	230
-28	1	-33.5	0.2	9	4.0	250	-23	2	-30.8	0.2	Dauphin	718550	51.10	100.05	1001	305	14.17	97.7	102	77	14.5	25.0	94	74	13.4	23.2	86	72	12.3	22.1	13	5.8	200
-25	1	-31.7	0.2	9	3.8	250	-21	2	-29.6	0.3	Portage La Prairie	718510	49.90	98.27	883	269	14.23	98.1	108	78	15.4	25.6	99	75	14.1	23.8	92	73	13.2	22.8	12	5.2	180
-32	1	-35.3	0.1	6	2.6	290	-28	1	-33.3	0.2	The Pas	718670	53.97	101.10	889	271	14.23	98.1	93	73	13.3	22.6	85	70	12.2	21.3	80	68	11.4	20.2	11	4.8	160
-38	1	-38.9	0.1	3	1.4	270	-34	1	-36.8	0.1	Thompson	710790	55.80	97.87	715	218	14.32	98.7	85	71	12.1	21.9	78	69	11.1	20.4	73	66	10.4	19.1	10	4.5	180
-27	1	-32.8	0.2	7	3.2	320	-23	2	-30.6	0.2	Winnipeg Int'l A	718520	49.90	97.23	784	239	14.28	98.5	107	78	15.3	25.4	99	75	14.1	23.9	92	73	13.2	22.9	13	5.7	180

Canada

HUMIDIFICATION 99.6% (35 hours below...)							HUMIDIFICATION 99% (88 hours below...)												DEHUMIDIFICATION 0.4% (35 hours above...)				1% (88 hours above...)				2% (175 hours above...)				Summer Wind		
°F 1a	gr/lb 1b	°C 1c	g/kg 1d	mph 1e	m/s 1f	PWD 1g	°F 2a	gr/lb 2b	°C 2c	g/kg 2d	Location 3a	WMO# 3b	Lat. 3c	Long. 3d	ft. 3e	m 3f	psia 3g	kPa 3h	gr/lb 4a	°F 4b	g/kg 4c	°C 4d	gr/lb 5a	°F 5b	g/kg 5c	°C 5d	gr/lb 6a	°F 6b	g/kg 6c	°C 6d	mph 7a	m/s 7b	PWD 7c
											NEW BRUNSWICK																						
-14	3	-25.5	0.4	11	5.1	250	-10	3	-23.5	0.5	Charlo	717110	47.98	66.33	125	38	14.63	100.9	102	74	14.6	23.3	97	72	13.8	22.2	90	70	12.9	21.1	11	4.8	250
-12	3	-24.3	0.4	7	3.2	270	-7	4	-21.7	0.5	Chatham	717170	47.00	65.45	102	31	14.64	101.0	104	75	14.8	24.1	98	73	14.0	22.9	92	71	13.2	21.9	11	5.1	230
-12	3	-24.3	0.4	5	2.3	270	-7	4	-21.5	0.6	Fredericton	717000	45.87	66.53	66	20	14.66	101.1	104	77	14.9	25.0	99	75	14.1	23.6	93	72	13.3	22.3	11	5.0	230
-10	3	-23.3	0.5	13	5.6	270	-5	4	-20.8	0.6	Moncton	717050	46.12	64.68	233	71	14.57	100.5	104	75	14.9	24.1	99	73	14.1	22.9	93	71	13.3	21.8	13	5.8	250
-9	3	-22.7	0.5	9	4.0	340	-4	4	-20.2	0.6	Saint John	716090	45.32	65.88	358	109	14.51	100.0	96	71	13.7	21.6	90	69	12.8	20.3	84	66	12.0	19.0	11	5.1	230
											NEWFOUNDLAND																						
-14	3	-25.4	0.4	18	8.1	270	-10	3	-23.2	0.5	Battle Harbour	718170	52.30	55.83	26	8	14.68	101.2	69	61	9.9	16.2	62	58	8.8	14.5	57	55	8.2	13.0	17	7.6	230
3	6	-16.1	0.9	24	10.8	280	7	8	-14.0	1.1	Bonavista	711960	48.67	53.12	89	27	14.65	101.0	92	70	13.1	21.1	85	68	12.2	19.8	80	66	11.4	18.6	17	7.6	230
-18	2	-28.0	0.3	12	5.5	220	-15	3	-25.9	0.4	Cartwright	718180	53.70	57.03	46	14	14.67	101.2	76	67	10.8	19.2	70	65	10.0	18.2	64	63	9.2	17.0	12	5.2	210
-7	4	-21.6	0.5	12	5.4	270	-3	5	-19.3	0.7	Daniels Harbour	711850	50.23	57.58	62	19	14.66	101.1	90	67	12.8	19.2	83	65	11.9	18.1	78	63	11.2	17.3	15	6.6	230
-13	3	-25.1	0.4	3	1.1	240	-7	4	-21.7	0.5	Deer Lake	718090	49.22	57.40	72	22	14.66	101.1	95	73	13.6	22.5	88	71	12.6	21.8	81	69	11.6	20.3	14	6.1	220
-4	5	-20.0	0.6	16	7.1	270	0	6	-17.8	0.8	Gander Int'l. A	718030	48.95	54.57	495	151	14.43	99.5	95	71	13.5	21.9	88	70	12.6	20.9	83	68	11.8	19.7	13	5.7	230
-23	2	-30.5	0.2	11	4.9	250	-20	2	-28.7	0.3	Goose	718160	53.32	60.37	160	49	14.36	99.0	82	70	11.7	21.0	76	68	10.8	19.8	71	65	10.1	18.4	12	5.4	250
-21	2	-29.2	0.3	12	5.4	250	-18	2	-27.5	0.3	Hopedale	719000	55.45	60.23	26	8	14.68	101.2	69	64	9.9	17.6	62	61	8.9	16.1	57	59	8.1	14.9	13	6.0	250
3	7	-16.0	0.9	17	7.6	290	7	8	-14.1	1.1	St. John's	718010	47.62	52.73	459	140	14.45	99.7	98	71	14.0	21.7	92	69	13.2	20.6	85	67	12.2	19.5	17	7.6	250
-2	5	-18.9	0.7	11	4.9	50	3	6	-16.3	0.9	Stephenville	718150	48.53	58.55	85	26	14.65	101.0	95	70	13.5	20.9	88	68	12.6	20.2	83	66	11.8	19.1	9	3.9	250
-33	1	-36.2	0.1	5	2.1	270	-30	1	-34.4	0.2	Wabush Lake	718250	52.93	66.87	1808	551	13.76	94.9	83	66	11.8	19.0	76	63	10.9	17.4	70	62	10.0	16.4	12	5.2	240
											NORTHWEST TERRITORIES																						
-41	1	-40.4	0.1	12	5.4	0	-39	1	-39.3	0.1	Baker Lake	719260	64.30	96.08	59	18	14.66	101.1	63	64	9.0	17.5	57	59	8.2	14.8	53	57	7.6	13.7	10	4.3	270
-38	1	-39.0	0.1	9	4.0	320	-35	1	-37.4	0.1	Cambridge Bay	719250	69.10	105.12	89	27	14.65	101.0	55	56	7.8	13.5	50	53	7.1	11.9	46	51	6.5	10.7	11	5.1	140
-34	1	-36.7	0.1	7	3.3	270	-33	1	-36.1	0.1	Cape Parry	719480	70.17	124.68	56	17	14.67	101.1	53	56	7.5	13.1	48	52	6.9	11.2	43	50	6.2	9.7	9	3.8	110
-35	1	-37.0	0.1	14	6.4	320	-34	1	-36.6	0.1	Chesterfield	719164	63.33	90.72	36	11	14.68	101.2	53	60	7.6	15.4	49	55	7.0	12.5	46	52	6.5	11.2	13	5.7	320
-40	1	-40.2	0.1	9	3.8	340	-38	1	-38.9	0.1	Coral Harbour	719150	64.20	83.37	210	64	14.58	100.6	54	58	7.7	14.5	50	55	7.1	12.5	46	52	6.6	11.2	13	5.8	270
-34	1	-36.6	0.1	3	1.1	150	-32	1	-35.4	0.1	Fort Smith	719340	60.02	111.95	666	203	14.35	98.9	81	69	11.6	20.6	76	67	10.8	19.5	71	66	10.1	18.7	10	4.4	180
-42	1	-40.9	0.1	10	4.3	320	-38	1	-39.0	0.1	Hall Beach	710810	68.78	81.25	26	8	14.68	101.2	48	53	6.8	11.7	42	50	6.0	10.0	39	47	5.6	8.5	11	4.7	180
-43	0	-41.4	0.1	1	0.6	70	-40	1	-39.9	0.1	Inuvik Ua	719570	68.30	133.48	223	68	14.58	100.5	67	67	9.6	19.5	61	65	8.7	18.3	57	64	8.2	17.6	8	3.5	180
-39	1	-39.6	0.1	4	1.7	320	-36	1	-37.9	0.1	Iqualuit A (Frobisher)	719090	63.75	68.55	108	33	14.64	100.9	46	53	6.6	11.9	43	52	6.1	10.9	40	50	5.7	9.8	12	5.2	320
-40	1	-39.7	0.1	2	0.8	170	-36	1	-37.9	0.1	Norman Wells	710430	65.28	126.80	243	74	14.57	100.4	75	68	10.7	19.7	71	66	10.1	19.0	67	65	9.5	18.4	8	3.7	140
-42	1	-40.9	0.1	9	4.2	320	-40	1	-39.7	0.1	Resolute	719240	74.72	94.98	220	67	14.58	100.5	39	47	5.6	8.4	36	45	5.2	7.4	34	43	4.9	6.2	12	5.5	110
-39	1	-39.5	0.1	5	2.2	50	-36	1	-37.9	0.1	Yellowknife	719360	62.47	114.45	676	206	14.34	98.9	73	66	10.4	18.9	68	65	9.7	18.1	62	63	8.9	17.3	10	4.4	160
											NOVA SCOTIA																						
-2	5	-18.8	0.7	7	3.1	300	3	6	-16.2	0.9	Greenwood	713970	44.98	64.92	92	28	14.65	101.0	108	76	15.4	24.2	102	74	14.6	23.4	96	72	13.7	22.4	15	6.5	250
-2	5	-19.0	0.7	11	5.1	320	2	6	-16.7	0.9	Halifax Int'l. Apt.	713950	44.88	63.50	476	145	14.44	99.6	106	73	15.1	22.7	99	71	14.2	21.6	95	69	13.5	20.5	12	5.2	200
14	11	-9.8	1.6	24	10.6	290	17	13	-8.1	1.9	Sable Island	716000	43.93	60.02	13	4	14.69	101.3	101	69	14.4	20.5	97	68	13.8	19.7	92	66	13.2	19.1	13	5.7	200
1	6	-17.0	0.9	12	5.2	340	5	7	-15.0	1.0	Shearwater	716010	44.63	63.50	167	51	14.61	100.7	102	71	14.5	21.9	96	69	13.7	20.6	92	68	13.1	19.7	10	4.5	230
-1	5	-18.4	0.7	13	5.9	270	3	6	-16.1	0.9	Sydney	717070	46.17	60.05	203	62	14.59	100.6	105	74	15.0	23.5	99	72	14.1	22.2	93	70	13.3	21.1	14	6.2	230
-9	3	-22.8	0.5	5	2.4	0	-4	5	-19.8	0.7	Truro	713980	45.37	63.27	131	40	14.63	100.8	106	75	15.2	23.7	102	73	14.5	22.5	97	70	13.9	21.3	11	4.7	270
7	8	-14.0	1.1	12	5.3	320	10	9	-12.1	1.3	Yarmouth	716030	43.83	66.08	141	43	14.62	100.8	98	69	14.0	20.8	93	68	13.3	19.8	86	66	12.3	18.9	10	4.4	190

Canada

HUMIDIFICATION — DEHUMIDIFICATION

99.6% (35 hours below...)							99% (88 hours below...)												0.4% (35 hours above...)				1% (88 hours above...)				2% (175 hours above...)				Summer Wind		
°F 1a	gr/lb 1b	°C 1c	g/kg 1d	mph 1e	m/s 1f	PWD 1g	°F 2a	gr/lb 2b	°C 2c	g/kg 2d	Location 3a	WMO# 3b	Lat. 3c	Long. 3d	ft. 3e	m 3f	psia 3g	kPa 3h	gr/lb 4a	°F 4b	g/kg 4c	°C 4d	gr/lb 5a	°F 5b	g/kg 5c	°C 5d	gr/lb 6a	°F 6b	g/kg 6c	°C 6d	mph 7a	m/s 7b	PWD 7c
											ONTARIO																						
-33	1	-36.3	0.1	3	1.2	270	-30	1	-34.4	0.2	Armstrong	718410	50.30	89.03	1152	351	14.09	97.2	102	73	14.6	22.6	94	71	13.4	21.8	85	69	12.1	20.5	12	5.5	0
-31	1	-35.2	0.1	1	0.3	270	-27	1	-32.8	0.2	Atikokan	717480	48.75	91.62	1289	393	14.02	96.7	109	75	15.5	24.1	102	73	14.5	23.0	95	71	13.5	21.8	8	3.4	230
-32	1	-35.8	0.1	6	2.5	290	-30	1	-34.4	0.2	Big Trout Lake	718480	53.83	89.87	735	224	14.31	98.7	91	70	13.0	21.3	84	68	12.0	20.1	78	67	11.2	19.3	9	4.0	200
-27	1	-32.6	0.2	4	1.8	320	-21	2	-29.6	0.3	Earlton	717350	47.70	79.85	797	243	14.28	98.4	108	76	15.4	24.6	101	74	14.4	23.2	95	72	13.5	22.0	12	5.2	200
-32	1	-35.7	0.1	2	0.7	270	-28	1	-33.5	0.2	Geraldton	718340	49.78	86.93	1152	351	14.09	97.2	103	74	14.7	23.1	97	72	13.8	22.1	87	69	12.4	20.6	11	4.7	0
-12	3	-24.3	0.4	7	3.1	0	-7	4	-21.4	0.6	Gore Bay	717330	45.88	82.57	633	193	14.36	99.0	108	74	15.4	23.3	102	72	14.6	22.3	97	71	13.8	21.6	11	4.8	180
-30	1	-34.4	0.2	4	1.8	270	-25	1	-31.7	0.2	Kapuskasing	718310	49.42	82.47	745	227	14.30	98.6	102	75	14.5	23.8	95	72	13.6	22.4	87	70	12.4	21.2	10	4.6	230
-27	1	-32.7	0.2	8	3.6	320	-22	2	-30.2	0.2	Kenora	718500	49.78	94.37	1348	411	13.99	96.5	106	75	15.2	24.0	99	73	14.1	22.9	92	72	13.1	22.0	11	4.9	180
-3	5	-19.3	0.7	9	3.8	250	2	6	-16.7	0.9	London	716230	43.03	81.15	912	278	14.22	98.0	119	79	17.0	26.1	113	77	16.2	24.9	108	75	15.4	23.8	11	5.1	250
-7	4	-21.5	0.6	6	2.7	90	-3	5	-19.3	0.7	Mount Forest	716310	43.98	80.75	1362	415	13.99	96.4	115	77	16.4	24.8	109	74	15.5	23.5	103	73	14.7	22.5	10	4.4	250
-17	2	-27.0	0.3	7	2.9	320	-11	3	-24.0	0.4	Muskoka	716300	44.97	79.30	925	282	14.21	98.0	113	76	16.1	24.5	106	74	15.2	23.3	100	72	14.3	22.1	9	4.1	270
-18	2	-27.9	0.3	6	2.8	0	-13	3	-25.1	0.4	North Bay	717310	46.35	79.43	1217	371	14.06	96.9	108	74	15.4	23.2	102	72	14.5	22.1	96	70	13.7	21.2	10	4.4	230
-13	3	-24.8	0.4	9	3.9	290	-8	4	-22.2	0.5	Ottawa Int'l	716280	45.32	75.67	374	114	14.50	100.0	112	78	16.0	25.5	106	76	15.1	24.6	99	75	14.2	23.7	10	4.5	250
-13	3	-25.2	0.4	4	1.9	90	-8	4	-22.3	0.5	Sault Ste. Marie	712600	46.48	84.52	630	192	14.36	99.0	108	76	15.4	24.3	102	73	14.5	22.9	95	71	13.5	21.6	9	3.9	220
-2	5	-18.6	0.7	10	4.6	270	3	7	-16.2	0.9	Simcoe	715270	42.85	80.27	791	241	14.28	98.5	118	79	16.8	25.9	112	77	16.0	24.9	107	75	15.3	23.7	11	5.0	230
-30	1	-34.5	0.2	4	1.9	270	-25	1	-31.9	0.2	Sioux Lookout	718420	50.12	91.90	1280	390	14.03	96.7	104	74	14.8	23.2	97	71	13.8	21.9	90	70	12.8	21.1	9	4.0	200
-19	2	-28.4	0.3	10	4.6	0	-14	3	-25.6	0.4	Sudbury	717300	46.62	80.80	1142	348	14.10	97.2	105	74	15.0	23.4	99	72	14.1	22.4	92	71	13.2	21.4	14	6.2	230
-22	2	-30.2	0.2	9	3.8	250	-18	2	-27.6	0.3	Thunder Bay	717490	48.37	89.32	653	199	14.35	99.0	101	76	14.4	24.4	93	72	13.3	22.2	86	70	12.3	20.9	11	5.1	200
-28	1	-33.5	0.2	5	2.0	180	-23	2	-30.5	0.2	Timmins	717390	48.57	81.37	968	295	14.19	97.8	102	74	14.6	23.5	95	72	13.6	22.4	88	70	12.6	21.0	10	4.4	250
-4	5	-19.9	0.7	9	4.1	340	1	6	-17.2	0.8	Toronto Int'l A	716240	43.67	79.63	568	173	14.40	99.3	116	79	16.6	26.1	110	77	15.7	25.0	104	75	14.8	23.8	12	5.3	270
-8	4	-22.1	0.5	6	2.8	50	-3	5	-19.6	0.7	Trenton	716210	44.12	77.53	282	86	14.55	100.3	116	78	16.5	25.5	110	76	15.7	24.5	104	74	14.9	23.5	12	5.5	230
-5	4	-20.3	0.6	7	3.2	340	0	6	-17.6	0.8	Wiarton	716330	44.75	81.10	728	222	14.31	98.7	113	77	16.1	24.8	106	75	15.2	23.6	101	73	14.4	22.6	12	5.4	230
2	6	-16.9	0.9	12	5.3	230	6	8	-14.6	1.1	Windsor	715380	42.27	82.97	623	190	14.37	99.1	125	82	17.9	27.7	118	79	16.8	26.1	112	77	16.0	25.1	12	5.5	250
											PRINCE EDWARD ISLAND																						
-6	4	-21.0	0.6	13	5.7	270	-2	5	-18.9	0.7	Charlottetown Cda	717060	46.28	63.13	177	54	14.60	100.7	105	75	15.0	23.6	99	72	14.2	22.4	94	71	13.4	21.4	12	5.3	230
-5	4	-20.5	0.6	15	6.7	270	-1	5	-18.4	0.7	Summerside	717020	46.43	63.83	79	24	14.65	101.0	104	74	14.8	23.4	99	72	14.1	22.2	94	70	13.4	21.1	13	5.6	200
											QUEBEC																						
-23	2	-30.7	0.2	6	2.8	270	-19	2	-28.4	0.3	Bagotville	717270	48.33	71.00	522	159	14.42	99.4	99	74	14.1	23.1	93	72	13.3	22.1	85	70	12.2	21.0	10	4.6	270
-19	2	-28.4	0.3	11	4.7	270	-14	3	-25.7	0.4	Baie Comeau	711870	49.13	68.20	72	22	14.66	101.1	85	68	12.1	20.1	79	66	11.3	18.8	75	64	10.7	17.6	13	5.7	230
-33	1	-36.2	0.1	8	3.5	120	-30	1	-34.3	0.1	Kuujjuarapik	719050	55.28	77.77	39	12	14.67	101.2	76	67	10.9	19.5	69	64	9.9	18.0	62	62	8.8	16.4	13	6.0	180
-34	1	-36.5	0.1	5	2.2	230	-31	1	-34.8	0.1	Kuujuaq	719060	58.10	68.42	121	37	14.63	100.9	72	65	10.3	18.4	65	63	9.3	17.1	58	60	8.3	15.4	11	4.9	180
-33	1	-36.0	0.1	6	2.6	270	-30	1	-34.3	0.2	La Grande Riviere	718270	53.63	77.70	640	195	14.36	99.0	83	68	11.9	19.8	77	65	11.0	18.3	70	63	10.0	17.4	13	5.7	240
-31	1	-34.9	0.1	5	2.2	270	-27	1	-33.0	0.2	Lake Eon	714210	51.87	63.28	1841	561	13.74	94.8	83	65	11.8	18.5	78	63	11.1	17.4	74	62	10.5	16.7	10	4.3	230
-12	3	-24.5	0.4	15	6.5	290	-8	4	-22.3	0.5	Mont Joli	717180	48.60	68.22	171	52	14.61	100.7	97	75	13.9	23.8	90	72	12.9	22.3	83	70	11.9	21.0	14	6.3	230
-12	3	-24.4	0.4	7	3.2	250	-7	4	-21.8	0.5	Montreal Int'l. A	716270	45.47	73.75	118	36	14.63	100.9	112	79	16.0	26.1	106	77	15.1	24.9	101	75	14.4	23.8	11	5.0	230
-16	2	-26.9	0.3	6	2.6	240	-11	3	-24.0	0.4	Montreal Mirabel	716278	45.68	74.03	269	82	14.55	100.3	110	78	15.7	25.3	103	75	14.7	24.0	98	73	14.0	22.8	9	3.8	240
-16	2	-26.4	0.3	10	4.3	250	-11	3	-24.0	0.4	Quebec	717140	46.80	71.38	240	73	14.57	100.5	111	77	15.8	24.9	103	75	14.7	23.6	95	73	13.6	22.5	12	5.2	250
-13	3	-25.1	0.4	10	4.3	180	-10	3	-23.4	0.5	Riviere Du Loup	717150	47.80	69.55	486	148	14.44	99.6	106	74	15.2	23.3	99	72	14.2	22.4	93	70	13.3	21.3	11	5.1	230
-23	2	-30.7	0.2	7	3.1	270	-19	2	-28.5	0.3	Roberval	717280	48.52	72.27	587	179	14.39	99.2	105	76	15.0	24.3	98	74	14.0	23.1	92	72	13.1	22.0	12	5.3	220
-33	1	-36.3	0.1	7	3.2	320	-31	1	-34.7	0.2	Schefferville	718280	54.80	66.82	1709	521	13.81	95.2	76	64	10.9	17.7	71	62	10.1	16.8	64	60	9.1	15.7	12	5.5	270
-20	2	-28.6	0.3	8	3.7	300	-15	2	-26.3	0.3	Sept-Iles	718110	50.22	66.27	180	55	14.60	100.7	82	66	11.7	18.6	77	64	11.0	17.7	73	62	10.4	16.9	11	4.9	220

Canada

HUMIDIFICATION 99.6% (35 hours below...)							99% (88 hours below...)											
°F 1a	gr/lb 1b	°C 1c	g/kg 1d	mph 1e	m/s 1f	PWD 1g	°F 2a	gr/lb 2b	°C 2c	g/kg 2d	Location 3a	WMO# 3b	Lat. 3c	Long. 3d	ft. 3e	m 3f	psia 3g	kPa 3h
-20	2	-29.1	0.3	5	2.1	110	-14	3	-25.7	0.4	Sherbrooke	716100	45.43	71.68	791	241	14.28	98.5
-12	3	-24.4	0.4	7	3.1	20	-7	4	-21.7	0.5	St. Hubert	713710	45.52	73.42	89	27	14.65	101.0
-19	2	-28.5	0.3	4	1.9	290	-15	3	-25.9	0.4	Ste. Agathe Des Mon	717200	46.05	74.28	1296	395	14.02	96.7
-27	1	-33.0	0.2	5	2.1	310	-22	2	-30.2	0.2	Val D'or	717250	48.07	77.78	1106	337	14.12	97.3
											SASKATCHEWAN							
-30	1	-34.2	0.2	8	3.4	290	-25	1	-31.7	0.2	Broadview	718610	50.38	102.68	1975	602	13.68	94.3
-25	1	-31.8	0.2	10	4.5	290	-21	2	-29.3	0.3	Estevan	718620	49.22	102.97	1906	581	13.71	94.5
-26	1	-32.2	0.2	9	3.9	290	-21	2	-29.6	0.3	Moose Jaw	718640	50.33	105.55	1893	577	13.72	94.6
-31	1	-34.9	0.1	4	1.9	320	-26	1	-32.1	0.2	North Battleford	718760	52.77	108.25	1798	548	13.77	94.9
-34	1	-36.5	0.1	3	1.4	270	-29	1	-34.1	0.2	Prince Albert	718690	53.22	105.68	1404	428	13.96	96.3
-29	1	-33.9	0.2	9	3.8	270	-24	2	-31.1	0.2	Regina	718630	50.43	104.67	1893	577	13.72	94.6
-31	1	-34.8	0.1	7	3.3	290	-26	1	-32.1	0.2	Saskatoon	718660	52.17	106.68	1654	504	13.84	95.4
-25	1	-31.9	0.2	14	6.3	270	-21	2	-29.4	0.3	Swift Current	718700	50.28	107.68	2684	818	13.33	91.9
-38	1	-39.0	0.1	4	1.6	70	-35	1	-37.0	0.1	Uranium City	710760	59.57	108.48	1043	318	14.15	97.6
-29	1	-34.1	0.2	8	3.6	270	-25	1	-31.6	0.2	Wynyard	718650	51.77	104.20	1841	561	13.74	94.8
-30	1	-34.6	0.2	7	3.0	290	-26	1	-32.0	0.2	Yorkton	711380	51.27	102.47	1634	498	13.85	95.5
											YUKON TERRITORY							
-34	1	-36.6	0.1	0	0.2	290	-32	1	-35.3	0.1	Burwash	719670	61.37	139.05	2644	806	13.34	92.0
-34	1	-36.8	0.1	2	0.9	340	-31	1	-34.7	0.2	Whitehorse	719640	60.72	135.07	2306	703	13.51	93.2

	DEHUMIDIFICATION 0.4% (35 hours above...)				1% (88 hours above...)				2% (175 hours above...)				Summer Wind		
Location 3a	gr/lb 4a	°F 4b	g/kg 4c	°C 4d	gr/lb 5a	°F 5b	g/kg 5c	°C 5d	gr/lb 6a	°F 6b	g/kg 6c	°C 6d	mph 7a	m/s 7b	PWD 7c
Sherbrooke	109	77	15.6	25.0	103	75	14.7	23.6	97	73	13.8	22.5	9	4.1	250
St. Hubert	113	78	16.2	25.8	106	77	15.2	24.8	101	75	14.4	23.9	13	5.6	250
Ste. Agathe Des Mon	110	75	15.7	23.7	103	72	14.7	22.4	97	70	13.9	21.3	9	3.9	270
Val D'or	102	73	14.6	23.0	95	72	13.6	22.3	88	70	12.6	21.1	11	4.8	230
SASKATCHEWAN															
Broadview	97	74	13.9	23.5	89	71	12.7	21.9	83	70	11.8	20.9	12	5.3	160
Estevan	103	76	14.7	24.5	92	73	13.2	22.6	86	71	12.3	21.8	14	6.4	180
Moose Jaw	92	73	13.1	22.7	85	71	12.1	21.4	79	69	11.3	20.5	13	5.7	180
North Battleford	88	71	12.6	21.9	83	69	11.8	20.7	77	68	11.0	19.7	11	5.0	140
Prince Albert	90	74	12.9	23.1	84	71	12.0	21.6	78	69	11.1	20.4	11	5.1	180
Regina	92	75	13.1	23.6	85	71	12.1	21.7	79	69	11.3	20.7	14	6.2	180
Saskatoon	88	73	12.6	22.5	82	70	11.7	21.2	76	68	10.9	20.2	13	5.6	180
Swift Current	88	71	12.6	21.6	82	68	11.7	19.9	76	67	10.9	19.4	13	6.0	180
Uranium City	78	68	11.2	19.7	74	66	10.5	18.9	69	64	9.9	17.7	7	3.1	230
Wynyard	94	74	13.4	23.3	85	70	12.2	21.0	81	68	11.5	19.8	13	5.6	180
Yorkton	96	74	13.7	23.3	87	71	12.4	21.8	81	69	11.5	20.7	12	5.3	180
YUKON TERRITORY															
Burwash	62	60	8.9	15.5	58	58	8.3	14.6	55	57	7.9	14.0	9	4.1	110
Whitehorse	62	61	8.9	15.9	58	59	8.3	15.0	55	58	7.8	14.5	8	3.7	140

International

HUMIDIFICATION 99.6% (35 hours below...)							99% (88 hours below...)												DEHUMIDIFICATION 0.4% (35 hours above...)				1% (88 hours above...)				2% (175 hours above...)				Summer Wind		
°F 1a	gr/lb 1b	°C 1c	g/kg 1d	mph 1e	m/s 1f	PWD 1g	°F 2a	gr/lb 2b	°C 2c	g/kg 2d	Location 3a	WMO# 3b	Lat. 3c	Long. 3d	ft. 3e	m 3f	psia 3g	kPa 3h	gr/lb 4a	°F 4b	g/kg 4c	°C 4d	gr/lb 5a	°F 5b	g/kg 5c	°C 5d	gr/lb 6a	°F 6b	g/kg 6c	°C 6d	mph 7a	m/s 7b	PWD 7c
											ALGERIA																						
36	31	2.0	4.5	2	0.7	200	38	34	3.1	4.8	Algiers	603900	36.72	3.25	82	25	14.66	101.0	129	82	18.4	27.5	125	81	17.8	27.1	118	80	16.9	26.6	12	5.2	60
39	35	4.1	5.0	5	2.2	220	41	38	5.1	5.4	Annaba	603600	36.83	7.82	13	4	14.69	101.3	134	82	19.2	27.8	130	81	18.5	27.2	125	80	17.8	26.8	11	4.8	240
41	38	5.0	5.5	7	3.0	10	43	41	6.2	5.9	Biskra	605250	34.80	5.73	285	87	14.55	100.3	92	83	13.2	28.4	86	82	12.3	27.8	82	83	11.7	28.4	11	4.9	180
31	28	-0.6	3.9	2	0.8	320	33	30	0.4	4.3	Constantine	604190	36.28	6.62	2276	694	13.53	93.3	106	76	15.1	24.7	100	76	14.3	24.4	94	75	13.4	23.8	12	5.2	240
33	29	0.3	4.1	4	1.8	320	35	31	1.7	4.5	El Golea	605900	30.57	2.87	1302	397	14.02	96.6	87	82	12.4	28.0	80	84	11.5	29.0	76	85	10.8	29.2	8	3.7	180
35	30	1.9	4.3	2	0.9	200	38	34	3.2	4.9	Oran	604900	35.63	0.60	295	90	14.54	100.2	123	80	17.5	26.4	119	79	17.0	26.3	115	79	16.4	26.0	14	6.3	240
29	25	-1.9	3.6	1	0.6	280	31	28	-0.5	4.0	Tebessa	604750	35.48	8.13	2667	813	13.34	91.9	97	73	13.8	23.0	92	72	13.2	22.3	88	72	12.6	22.3	11	5.0	220
											ARGENTINA																						
31	25	-0.7	3.6	4	2.0	270	34	29	1.0	4.1	Buenos Aires	875760	34.82	58.53	65	20	14.67	101.1	126	80	18.0	26.9	120	79	17.1	26.2	115	78	16.4	25.3	11	4.9	270
29	23	-1.7	3.3	11	4.7	270	31	25	-0.4	3.6	Comodoro Rivadavia	878600	45.78	67.50	150	46	14.62	100.8	70	66	10.0	18.8	65	64	9.3	17.8	60	65	8.6	18.5	22	9.8	290
31	27	-0.4	3.8	3	1.4	270	34	30	1.3	4.3	Cordoba	873440	31.32	64.22	1555	474	13.89	95.8	136	84	19.4	28.8	127	82	18.2	27.5	120	80	17.2	26.5	16	7.2	20
30	24	-1.3	3.5	1	0.6	270	32	27	0.2	3.8	Junin Airport	875480	34.55	60.95	266	81	14.56	100.4	125	80	17.8	26.5	118	78	16.8	25.8	111	77	15.9	24.9	12	5.2	360
40	37	4.7	5.3	6	2.6	270	44	43	6.6	6.1	Formosa	871620	26.20	58.23	196	60	14.60	100.6	145	85	20.7	29.7	141	84	20.1	29.0	136	83	19.5	28.4	17	7.5	360
29	23	-1.5	3.3	3	1.2	50	32	27	0.2	3.8	Marcos Juarez	874670	32.70	62.15	374	114	14.50	100.0	136	85	19.5	29.2	129	82	18.4	27.7	123	80	17.6	26.7	11	4.8	360
30	26	-0.9	3.8	3	1.3	230	33	30	0.6	4.3	Mendoza	874180	32.83	68.78	2309	704	13.51	93.1	112	81	16.0	27.0	106	80	15.1	26.5	100	79	14.3	26.0	11	5.1	50
36	31	2.2	4.5	1	0.6	180	39	35	4.0	5.1	Paso De Los Libres	872890	29.68	57.15	229	70	14.58	100.5	137	84	19.6	28.8	132	83	18.8	28.3	126	82	18.0	27.5	13	5.9	360
39	36	4.1	5.1	6	2.5	180	43	42	5.9	5.9	Posadas	871780	27.37	55.97	410	125	14.48	99.8	142	87	20.3	30.6	136	85	19.4	29.7	132	84	18.9	29.1	11	5.1	360
37	33	2.8	4.7	2	0.7	200	40	37	4.4	5.2	Reconquista	872700	29.18	59.67	173	53	14.61	100.7	150	87	21.4	30.5	144	86	20.5	29.8	138	84	19.7	28.8	12	5.3	50
35	30	1.8	4.3	2	0.8	50	39	35	3.9	5.0	Resistencia	871550	27.45	59.05	170	52	14.61	100.7	144	86	20.5	29.9	140	85	20.0	29.3	134	83	19.2	28.5	11	4.7	20
17	13	-8.6	1.9	6	2.7	270	22	17	-5.8	2.4	Rio Gallegos	879250	51.62	69.28	62	19	14.67	101.1	60	63	8.5	17.0	56	61	8.0	16.0	52	60	7.5	15.4	20	9.1	320
30	24	-1.0	3.5	2	1.0	180	33	28	0.5	4.0	Rosario	874800	32.92	60.78	82	25	14.66	101.0	132	83	18.9	28.2	125	81	17.9	27.4	119	79	17.0	26.3	13	5.7	360
30	28	-0.9	4.0	2	0.9	270	33	32	0.6	4.6	Salta Airport	870470	24.85	65.48	3990	1216	12.70	87.5	123	76	17.6	24.5	119	76	17.0	24.2	115	74	16.4	23.5	10	4.6	70
28	24	-2.0	3.4	1	0.3	360	31	27	-0.5	3.9	San Juan	873110	31.57	68.42	1961	598	13.69	94.3	110	82	15.7	27.9	102	82	14.6	27.5	97	81	13.9	27.0	11	5.0	180
37	34	2.9	4.9	5	2.1	360	40	39	4.3	5.5	San Miguel De Tuc.	871210	26.85	65.10	1476	450	13.93	96.0	144	85	20.6	29.7	138	84	19.7	28.9	132	83	18.9	28.2	12	5.2	90
											ARMENIA																						
7	9	-14.1	1.3	1	0.4	180	11	11	-11.7	1.5	Yerevan	377890	40.13	44.47	2919	890	13.21	91.1	104	85	14.8	29.5	95	82	13.6	27.6	90	81	12.8	27.0	6	2.7	210
											ASCENSION ISLAND																						
69	108	20.8	15.4	16	7.3	90	70	112	21.3	16.0	Georgetown	619020	7.97	14.40	259	79	14.56	100.4	132	81	18.9	27.3	130	80	18.5	26.9	126	80	18.0	26.9	19	8.6	120
											AUSTRALIA																						
39	35	4.0	5.0	2	1.0	50	41	38	5.2	5.4	Adelaide	946720	34.93	138.52	13	4	14.69	101.3	97	74	13.8	23.4	89	73	12.7	22.7	81	72	11.6	22.3	13	5.7	310
34	31	1.0	4.4	2	0.9	270	36	33	2.3	4.7	Alice Springs	943260	23.80	133.90	1774	541	13.78	95.0	123	77	17.6	25.1	115	77	16.4	24.9	104	77	14.9	25.1	8	3.7	100
44	43	6.6	6.1	4	1.8	220	46	46	7.8	6.6	Brisbane	945780	27.38	153.10	16	5	14.69	101.3	133	81	19.0	27.3	126	80	18.0	26.5	121	79	17.3	26.1	11	5.0	20
56	67	13.2	9.6	8	3.4	170	59	75	14.9	10.7	Cairns	942870	16.88	145.75	22	7	14.69	101.2	150	85	21.5	29.3	146	84	20.9	28.9	141	83	20.2	28.4	8	3.6	120
26	21	-3.1	3.1	0	0.0	310	29	25	-1.8	3.5	Canberra	949260	35.30	149.18	1893	577	13.72	94.6	96	71	13.7	21.7	90	70	12.8	21.2	84	68	12.0	20.0	12	5.3	310
64	90	17.9	12.8	7	3.1	140	66	97	19.0	13.8	Darwin	941200	12.40	130.87	98	30	14.65	101.0	160	86	22.8	30.2	152	85	21.7	29.4	151	85	21.6	29.2	12	5.2	290
36	33	2.0	4.6	1	0.5	220	38	35	3.3	5.0	Kalgoorlie/Boulder	946370	30.77	121.45	1181	360	14.08	97.1	98	74	14.0	23.4	90	71	12.8	21.9	84	71	12.0	21.7	9	3.9	320
49	52	9.4	7.4	5	2.1	210	51	56	10.8	8.0	Learmouth	943020	22.23	114.08	19	6	14.69	101.3	140	84	20.0	28.7	134	83	19.1	28.3	127	83	18.2	28.6	14	6.1	210
41	38	4.8	5.4	1	0.3	50	43	41	6.1	5.9	Perth	946100	31.93	115.95	95	29	14.65	101.0	101	75	14.4	24.1	95	74	13.6	23.4	91	74	13.0	23.1	10	4.3	270
51	56	10.7	8.0	5	2.3	160	54	62	12.0	8.9	Port Hedland	943120	20.37	118.62	19	6	14.69	101.3	156	87	22.3	30.5	151	86	21.6	30.1	147	86	21.0	29.8	12	5.4	120
42	39	5.8	5.6	2	1.1	320	44	43	6.8	6.1	Sydney Intl Airport	947670	33.95	151.18	10	3	14.69	101.3	115	77	16.4	24.8	111	76	15.8	24.3	107	75	15.3	23.9	12	5.3	300
48	50	9.1	7.1	0	0.2	190	52	58	11.1	8.3	Townsville	942940	19.25	146.75	19	6	14.69	101.3	150	84	21.5	29.1	147	84	21.0	28.9	141	83	20.2	28.5	9	4.1	50

International

HUMIDIFICATION 99.6% (35 hours below...)							99% (88 hours below...)												DEHUMIDIFICATION 0.4% (35 hours above...)				1% (88 hours above...)				2% (175 hours above...)				Summer Wind		
°F 1a	gr/lb 1b	°C 1c	g/kg 1d	mph 1e	m/s 1f	PWD 1g	°F 2a	gr/lb 2b	°C 2c	g/kg 2d	Location 3a	WMO# 3b	Lat. 3c	Long. 3d	ft. 3e	m 3f	psia 3g	kPa 3h	gr/lb 4a	°F 4b	g/kg 4c	°C 4d	gr/lb 5a	°F 5b	g/kg 5c	°C 5d	gr/lb 6a	°F 6b	g/kg 6c	°C 6d	mph 7a	m/s 7b	PWD 7c
											AUSTRIA																						
2	7	-16.9	0.9	1	0.3	60	7	9	-14.1	1.2	Aigen/Ennstal (Mil)	111570	47.53	14.13	2129	649	13.60	93.8	93	74	13.3	23.4	88	72	12.5	22.1	83	69	11.8	20.7	7	3.1	60
5	7	-14.8	1.1	1	0.5	180	12	11	-11.2	1.5	Graz	112400	47.00	15.43	1138	347	14.11	97.2	105	75	15.0	23.8	99	73	14.2	22.7	94	71	13.4	21.8	7	3.2	140
10	10	-12.2	1.4	1	0.5	260	14	12	-10.2	1.7	Innsbruck	111200	47.27	11.35	1945	593	13.70	94.4	91	69	13.0	20.3	86	67	12.3	19.5	81	67	11.6	19.4	8	3.5	70
4	7	-15.8	1.0	3	1.3	310	9	9	-13.0	1.3	Klagenfurt	112310	46.65	14.33	1482	452	13.93	96.0	92	72	13.2	22.0	88	71	12.6	21.6	84	68	12.0	20.2	6	2.6	100
5	7	-14.9	1.1	6	2.7	90	12	11	-11.2	1.5	Linz	110100	48.23	14.20	1026	313	14.16	97.6	95	70	13.6	21.1	90	69	12.9	20.8	88	68	12.5	20.2	7	3.3	110
7	8	-14.1	1.2	3	1.4	130	12	11	-11.1	1.5	Salzburg	111500	47.80	13.00	1476	450	13.93	96.0	91	72	13.0	22.4	85	71	12.2	21.4	83	70	11.8	21.3	7	3.0	330
12	10	-11.1	1.5	6	2.8	240	17	13	-8.6	1.9	Vienna, Hohe Warte	110350	48.25	16.37	656	200	14.35	98.9	98	74	14.0	23.3	94	73	13.4	23.0	90	72	12.8	22.4	9	4.0	140
9	9	-12.6	1.3	6	2.5	320	14	11	-9.9	1.6	Vienna, Schwechat	110360	48.12	16.57	623	190	14.37	99.1	92	73	13.2	22.7	88	72	12.6	22.0	83	71	11.9	21.8	12	5.3	150
0	6	-17.7	0.9	1	0.3	250	5	8	-14.8	1.1	Zeltweg	111650	47.20	14.75	2237	682	13.55	93.4	94	74	13.4	23.3	89	72	12.7	22.0	85	69	12.1	20.8	7	3.0	190
											AZORES																						
46	46	8.0	6.6	3	1.3	300	48	50	9.1	7.1	Lajes	85090	38.77	27.10	180	55	14.60	100.7	115	75	16.4	24.0	111	75	15.8	23.7	106	74	15.1	23.3	8	3.7	250
											BAHAMAS																						
57	70	14.1	9.9	3	1.5	300	60	78	15.8	11.1	Nassau	780730	25.05	77.47	22	7	14.69	101.2	151	83	21.6	28.6	150	83	21.4	28.5	144	83	20.6	28.3	10	4.5	130
											BAHRAIN																						
52	58	11.0	8.3	13	5.8	290	54	62	12.2	8.9	Al-Manamah	411500	26.27	50.65	6	2	14.70	101.3	189	93	27.0	33.7	181	92	25.8	33.4	176	92	25.2	33.4	11	5.0	340
											BELARUS																						
-9	3	-23.0	0.5	4	1.6	210	-2	5	-19.1	0.7	Babruysk (Bobruysk)	269610	53.12	29.25	541	165	14.41	99.4	94	73	13.5	23.0	88	71	12.5	21.4	83	69	11.8	20.4	8	3.6	200
-6	4	-21.3	0.6	4	1.8	330	0	6	-17.9	0.8	Homyel (Gomel')	330410	52.45	31.00	416	127	14.48	99.8	95	72	13.6	22.5	89	71	12.7	21.8	84	70	12.0	21.2	7	3.3	150
-5	4	-20.3	0.6	5	2.2	270	1	6	-17.1	0.8	Hrodna (Grodno)	268250	53.68	23.83	442	135	14.47	99.7	92	72	13.2	22.5	87	71	12.4	21.4	81	69	11.6	20.3	9	3.9	180
-9	3	-22.9	0.5	7	3.0	30	-3	5	-19.5	0.7	Mahilyow (Mogilev)	268630	53.90	30.32	633	193	14.37	99.0	94	72	13.4	22.4	88	70	12.6	21.2	83	68	11.8	20.2	9	4.1	200
-5	4	-20.7	0.6	5	2.1	300	0	6	-17.6	0.8	Minsk	268500	53.87	27.53	767	234	14.30	98.5	91	71	13.0	21.7	85	69	12.2	20.7	80	69	11.5	20.4	9	4.0	70
-8	4	-22.3	0.5	3	1.5	30	-2	5	-18.7	0.7	Vitsyebsk (Vitebsk)	266660	55.17	30.13	577	176	14.40	99.2	94	72	13.4	22.2	88	70	12.6	21.1	83	68	11.8	20.0	8	3.4	210
											BELGIUM/LUXEMBOURG																						
16	12	-8.7	1.8	7	3.0	50	21	16	-6.0	2.3	Antwerp	64500	51.20	4.47	45	14	14.68	101.2	97	75	13.9	24.0	91	73	13.0	22.6	85	71	12.2	21.4	7	3.2	90
15	12	-9.3	1.7	7	3.1	50	21	16	-6.2	2.3	Brussels	64510	50.90	4.53	190	58	14.60	100.6	95	74	13.6	23.2	90	72	12.9	22.2	85	71	12.1	21.4	8	3.6	60
15	12	-9.3	1.7	10	4.4	50	21	16	-6.2	2.3	Charleroi	64490	50.47	4.45	629	192	14.37	99.0	94	74	13.5	23.1	89	71	12.7	21.8	85	69	12.1	20.6	8	3.5	50
13	11	-10.3	1.6	9	4.1	70	19	15	-7.1	2.1	Florennes	64560	50.23	4.65	980	299	14.19	97.8	97	73	13.8	22.8	90	70	12.8	21.3	85	68	12.1	20.1	8	3.6	170
17	13	-8.4	1.9	10	4.4	90	22	17	-5.8	2.4	Koksijde	64000	51.08	2.65	29	9	14.68	101.2	93	72	13.3	22.5	88	70	12.5	20.9	83	68	11.9	19.9	10	4.6	100
13	11	-10.5	1.6	11	4.9	50	18	14	-7.9	2.0	Luxembourg	65900	49.62	6.22	1243	379	14.05	96.9	92	71	13.1	21.8	87	68	12.4	20.2	82	67	11.7	19.7	9	3.8	80
18	14	-7.8	2.0	11	4.9	70	22	17	-5.4	2.4	Oostende	64070	51.20	2.87	16	5	14.69	101.3	92	71	13.1	21.4	88	69	12.5	20.7	83	67	11.9	19.6	10	4.6	100
11	10	-11.8	1.5	11	5.1	90	16	13	-8.7	1.9	Saint Hubert	64760	50.03	5.40	1827	557	13.75	94.8	91	69	13.0	20.4	85	67	12.2	19.2	80	65	11.5	18.5	8	3.5	30
											BENIN																						
71	115	21.7	16.4	5	2.1	20	72	119	22.3	17.0	Cotonou	653440	6.35	2.38	29	9	14.68	101.2	156	85	22.3	29.6	153	85	21.9	29.6	151	85	21.6	29.5	12	5.5	200
65	97	18.2	13.9	4	1.6	40	67	105	19.3	14.9	Parakou	653300	9.35	2.62	1289	393	14.03	96.7	138	84	19.7	29.1	135	84	19.3	28.8	132	83	18.9	28.4	5	2.4	40
											BERMUDA																						
55	65	12.9	9.2	16	7.2	310	57	70	13.9	9.9	Hamilton, Nas	780160	32.37	64.68	9	3	14.69	101.3	150	84	21.5	28.9	144	83	20.6	28.5	141	83	20.1	28.1	11	4.8	190
											BOLIVIA																						
35	41	1.8	5.8	0	0.0	180	37	44	2.9	6.3	Cochabamba	852230	17.45	66.10	8303	2531	10.79	74.4	97	64	13.8	18.0	91	62	13.0	16.7	90	62	12.8	16.6	6	2.8	360
25	32	-4.0	4.5	2	0.9	330	27	35	-3.0	5.0	La Paz	852010	16.52	68.18	13169	4014	8.93	61.5	73	51	10.4	10.6	71	50	10.2	10.0	68	49	9.7	9.4	8	3.4	60
											BOSNIA-HERZEGOVINA																						
10	9	-12.0	1.3	2	1.1	320	16	13	-8.9	1.8	Banja Luka	132420	44.78	17.22	511	156	14.43	99.5	106	79	15.1	26.1	99	76	14.1	24.3	94	73	13.4	23.0	4	2.0	360

International

HUMIDIFICATION 99.6% (35 hours below...)							99% (88 hours below...)												DEHUMIDIFICATION 0.4% (35 hours above...)				1% (88 hours above...)				2% (175 hours above...)				Summer Wind		
°F 1a	gr/lb 1b	°C 1c	g/kg 1d	mph 1e	m/s 1f	PWD 1g	°F 2a	gr/lb 2b	°C 2c	g/kg 2d	Location 3a	WMO# 3b	Lat. 3c	Long. 3d	ft. 3e	m 3f	psia 3g	kPa 3h	gr/lb 4a	°F 4b	g/kg 4c	°C 4d	gr/lb 5a	°F 5b	g/kg 5c	°C 5d	gr/lb 6a	°F 6b	g/kg 6c	°C 6d	mph 7a	m/s 7b	PWD 7c
											BRAZIL																						
73	124	22.9	17.7	7	3.1	130	74	128	23.3	18.3	Archip. De Fernan.	824000	3.85	32.42	183	56	14.60	100.7	147	84	21.0	28.7	144	83	20.5	28.5	141	83	20.2	28.4	10	4.6	150
72	119	22.4	17.0	2	0.7	90	73	123	22.8	17.6	Belem	821930	1.38	48.48	52	16	14.67	101.1	150	84	21.5	28.8	149	84	21.3	28.7	143	83	20.4	28.1	6	2.9	90
48	57	9.1	8.1	0	0.2	90	51	63	10.8	9.0	Brasilia	833780	15.87	47.93	3480	1061	12.94	89.2	120	73	17.2	22.6	118	72	16.9	22.3	116	72	16.6	22.1	7	3.2	90
47	52	8.2	7.4	8	3.7	150	50	58	9.9	8.3	Campinas	837210	23.00	47.13	2168	661	13.58	93.6	136	79	19.5	26.0	134	78	19.1	25.6	128	77	18.3	25.0	7	3.3	330
46	49	8.0	7.0	16	7.3	180	50	57	10.2	8.2	Campo Grande	836120	20.47	54.67	1824	556	13.76	94.8	141	80	20.1	26.8	134	80	19.2	26.4	132	79	18.8	26.3	11	4.8	360
61	80	16.3	11.5	1	0.4	240	63	87	17.3	12.4	Caravelas	834970	17.63	39.25	13	4	14.69	101.3	142	81	20.3	27.4	140	81	20.0	27.2	137	80	19.6	26.9	12	5.3	60
36	35	2.4	5.0	2	1.1	270	40	41	4.5	5.8	Curitiba	838400	25.52	49.17	2979	908	13.18	90.9	128	75	18.3	24.0	123	74	17.5	23.3	119	74	17.0	23.3	9	3.8	300
72	119	22.5	17.0	5	2.1	180	73	123	22.9	17.6	Fortaleza	823980	3.78	38.53	82	25	14.66	101.0	151	81	21.6	27.3	149	81	21.3	27.4	143	81	20.4	27.4	13	6.0	120
53	66	11.5	9.4	1	0.3	180	55	71	12.9	10.1	Goiania	834240	16.63	49.22	2450	747	13.44	92.7	129	78	18.4	25.5	125	77	17.9	25.0	123	76	17.5	24.5	4	1.8	360
67	101	19.2	14.4	0	0.2	180	68	105	19.8	15.0	Maceio	829930	9.52	35.78	377	115	14.50	100.0	141	81	20.2	27.2	136	80	19.4	26.6	134	80	19.1	26.4	10	4.5	100
70	111	21.1	15.8	1	0.3	120	71	115	21.8	16.4	Manaus, E. Gomes Int'l	821110	3.03	60.05	6	2	14.70	101.3	161	86	23.0	29.8	158	85	22.6	29.5	151	83	21.6	28.6	5	2.2	60
71	116	21.9	16.6	4	1.6	30	73	124	22.7	17.8	Manaus, Ponta Pelada	823320	3.15	59.98	275	84	14.55	100.3	153	84	21.8	29.1	151	84	21.6	28.9	146	83	20.8	28.5	10	4.3	90
70	111	21.0	15.9	6	2.9	150	71	115	21.6	16.5	Natal	825990	5.92	35.25	170	52	14.61	100.7	148	82	21.1	27.8	142	82	20.3	27.6	141	82	20.1	27.5	11	5.1	100
40	37	4.4	5.2	2	0.7	240	43	41	6.1	5.9	Porto Alegre	839710	30.00	51.18	9	3	14.69	101.3	136	83	19.5	28.5	132	82	18.8	27.6	126	80	18.0	26.8	7	3.0	290
70	111	21.1	15.9	5	2.3	240	71	115	21.8	16.4	Recife	828990	8.07	34.85	62	19	14.67	101.1	143	83	20.4	28.6	142	83	20.3	28.6	140	83	20.0	28.4	12	5.2	120
59	75	14.9	10.7	3	1.2	320	61	81	15.9	11.5	Rio De Janeiro, Galeao	837460	22.82	43.25	19	6	14.69	101.3	150	86	21.5	30.1	142	84	20.3	28.9	141	84	20.1	28.9	7	3.1	50
68	103	20.1	14.8	2	0.7	180	70	111	21.0	15.8	Salvador	832480	12.90	38.33	19	6	14.69	101.3	150	85	21.5	29.3	142	84	20.3	28.9	141	84	20.2	28.7	12	5.5	90
72	120	22.2	17.1	6	2.6	210	73	124	22.9	17.7	Santarem	822440	2.43	54.72	236	72	14.57	100.5	146	83	20.9	28.1	144	82	20.5	28.0	143	82	20.4	28.0	12	5.4	90
48	55	8.8	7.8	4	2.0	160	50	59	9.9	8.4	Sao Paulo	837800	23.62	46.65	2634	803	13.35	92.0	127	77	18.2	24.9	122	75	17.4	24.1	120	75	17.1	23.8	6	2.6	330
61	80	16.3	11.5	1	0.6	210	63	87	17.2	12.4	Vitoria	836490	20.27	40.28	13	4	14.69	101.3	148	83	21.1	28.3	141	81	20.1	27.2	137	82	19.6	27.6	11	5.1	30
											BULGARIA																						
-7	5	-21.5	0.7	30	13.4	320	-2	7	-19.0	0.9	Botevgrad	156270	42.67	24.83	7837	2389	10.99	75.7	72	53	10.3	11.8	68	52	9.7	11.0	64	50	9.2	10.2	9	3.9	270
17	13	-8.3	1.9	9	4.2	270	21	16	-6.2	2.3	Burgas	156550	42.48	27.48	91	28	14.65	101.0	118	79	16.8	26.2	111	77	15.8	25.2	104	76	14.9	24.4	11	4.9	110
14	11	-10.0	1.6	2	0.8	270	18	14	-7.6	2.0	Lom	155110	43.82	23.25	108	33	14.64	100.9	114	83	16.3	28.5	108	81	15.4	27.3	102	79	14.6	26.1	4	1.9	50
-10	5	-23.4	0.7	17	7.5	20	-5	6	-20.8	0.9	Musala	156150	42.18	23.58	9603	2927	10.27	70.8	62	48	8.8	9.0	58	47	8.3	8.2	55	46	7.9	7.6	7	3.3	320
14	11	-9.9	1.6	2	1.1	40	19	15	-7.1	2.1	Plovdiv	156250	42.13	24.75	606	185	14.38	99.1	105	83	15.0	28.3	99	80	14.2	26.8	94	78	13.5	25.3	6	2.6	90
12	10	-11.3	1.5	5	2.3	50	16	12	-8.8	1.8	Ruse	155350	43.85	25.95	147	45	14.62	100.8	107	85	15.3	29.2	100	80	14.3	26.9	95	78	13.6	25.6	8	3.5	270
10	10	-12.1	1.4	2	1.1	360	14	12	-9.9	1.7	Sofia	156140	42.65	23.38	1952	595	13.69	94.4	92	74	13.1	23.5	87	72	12.4	22.5	83	72	11.8	22.0	5	2.4	110
17	13	-8.5	1.9	15	6.7	360	20	15	-6.5	2.2	Varna	155520	43.20	27.92	141	43	14.63	100.8	120	80	17.1	26.5	112	78	16.0	25.6	106	77	15.2	24.8	9	4.0	90
											BRUNEI																						
71	115	21.4	16.4	0	0.0	220	72	119	22.0	17.0	Brunei Intl Airport	963150	4.93	114.93	49	15	14.67	101.1	158	86	22.6	30.0	155	85	22.1	29.7	152	85	21.7	29.4	8	3.7	320
											CANARY ISLANDS																						
55	65	13.0	9.3	4	1.8	320	57	70	13.9	10.0	Las Palmas	600300	27.93	15.38	82	25	14.66	101.0	132	79	18.9	25.9	125	77	17.8	25.2	118	76	16.8	24.6	19	8.7	30
57	70	13.8	10.0	8	3.4	360	57	70	14.1	10.0	Santa Cruz De Tenerife	600250	28.05	16.57	236	72	14.57	100.5	119	80	17.0	26.4	117	79	16.7	26.2	111	78	15.9	25.4	19	8.5	60
											CAPE VERDE																						
63	87	17.1	12.4	10	4.6	30	64	90	17.8	12.9	Sal Island	85940	16.73	22.95	180	55	14.60	100.7	141	81	20.2	27.0	136	80	19.5	26.7	133	80	19.0	26.5	15	6.5	60
											CHILE																						
50	54	10.2	7.8	6	2.7	90	52	59	11.0	8.4	Antofagasta	854420	23.43	70.43	393	120	14.49	99.9	100	73	14.3	22.9	97	73	13.9	22.6	92	71	13.2	21.6	15	6.7	190
52	58	11.2	8.3	4	2.0	90	55	65	12.6	9.3	Arica	854060	18.33	70.33	193	59	14.60	100.6	114	78	16.3	25.8	107	78	15.3	25.5	100	76	14.3	24.4	15	6.6	210
35	30	1.8	4.3	4	1.8	140	37	32	2.8	4.6	Concepcion	856820	36.77	73.05	52	16	14.67	101.1	85	67	12.2	19.7	80	66	11.5	18.9	78	65	11.2	18.6	19	8.5	240
54	63	12.1	9.0	6	2.9	160	55	65	12.9	9.3	Iquique	854180	20.53	70.18	170	52	14.61	100.7	95	75	13.6	24.0	92	75	13.1	23.7	89	74	12.7	23.3	14	6.4	210

International

HUMIDIFICATION 99.6% (35 hours below...)							99% (88 hours below...)				International								DEHUMIDIFICATION 0.4% (35 hours above...)				1% (88 hours above...)				2% (175 hours above...)				Summer Wind		
°F 1a	gr/lb 1b	°C 1c	g/kg 1d	mph 1e	m/s 1f	PWD 1g	°F 2a	gr/lb 2b	°C 2c	g/kg 2d	Location 3a	WMO# 3b	Lat. 3c	Long. 3d	ft. 3e	m 3f	psia 3g	kPa 3h	gr/lb 4a	°F 4b	g/kg 4c	°C 4d	gr/lb 5a	°F 5b	g/kg 5c	°C 5d	gr/lb 6a	°F 6b	g/kg 6c	°C 6d	mph 7a	m/s 7b	PWD 7c
42	40	5.8	5.7	6	2.8	120	44	43	6.6	6.2	La Serena	854880	29.92	71.20	479	146	14.45	99.6	84	67	12.0	19.7	81	67	11.6	19.4	78	66	11.2	18.7	13	5.8	270
28	22	-2.1	3.2	2	0.7	280	30	24	-0.9	3.5	Puerto Montt	857990	41.42	73.08	282	86	14.55	100.3	81	66	11.6	19.1	77	65	11.0	18.4	72	64	10.3	17.6	10	4.4	150
23	17	-5.0	2.5	6	2.9	290	25	19	-3.7	2.7	Punta Arenas	859340	53.00	70.85	121	37	14.64	100.9	59	57	8.4	14.1	55	56	7.9	13.6	52	55	7.5	12.9	18	8.2	270
29	24	-1.4	3.5	2	0.9	20	32	28	-0.1	4.0	Santiago	855740	33.38	70.78	1561	476	13.89	95.7	85	76	12.1	24.3	81	75	11.6	23.8	78	74	11.2	23.2	12	5.4	210
29	23	-1.9	3.3	2	1.0	70	31	26	-0.7	3.7	Temuco	857430	38.75	72.63	393	120	14.49	99.9	82	71	11.7	21.9	76	70	10.9	20.9	72	67	10.3	19.7	10	4.5	240
											CHINA																						
-19	2	-28.6	0.3	3	1.3	250	-16	2	-26.5	0.3	Anda	508540	46.38	125.32	492	150	14.44	99.5	122	79	17.4	26.2	115	77	16.4	25.2	108	75	15.5	23.9	12	5.3	250
-1	6	-18.1	0.9	1	0.6	200	2	7	-16.4	1.0	Andirlangar	518480	37.93	83.65	4146	1264	12.63	87.0	90	71	12.8	21.6	83	71	11.8	21.9	76	72	10.9	22.1	6	2.5	40
18	14	-7.9	2.0	1	0.5	290	21	16	-6.3	2.3	Anyang	538980	36.12	114.37	249	76	14.57	100.4	151	86	21.6	29.8	144	84	20.6	28.7	138	83	19.7	28.4	7	3.2	180
13	11	-10.4	1.5	2	0.7	220	16	12	-8.9	1.8	Baoding	546020	38.85	115.57	62	19	14.67	101.1	146	84	20.9	28.7	139	83	19.8	28.1	132	82	18.8	27.7	7	3.0	200
-5	5	-20.5	0.7	6	2.5	270	-1	6	-18.5	0.9	Bayan Mod	524950	40.75	104.50	4360	1329	12.53	86.3	88	69	12.5	20.7	80	68	11.4	19.8	74	69	10.5	20.4	8	3.4	180
13	11	-10.4	1.5	4	2.0	340	15	12	-9.2	1.7	Beijing	545110	39.93	116.28	180	55	14.60	100.7	142	83	20.3	28.4	134	82	19.2	27.6	130	81	18.5	27.1	7	3.1	200
22	17	-5.6	2.4	3	1.3	20	24	18	-4.2	2.6	Bengbu	582210	32.95	117.37	72	22	14.66	101.1	156	87	22.3	30.8	152	87	21.7	30.4	147	86	21.0	29.8	8	3.7	180
-12	3	-24.7	0.4	6	2.6	270	-9	4	-22.7	0.5	Changchun	541610	43.90	125.22	780	238	14.29	98.5	130	81	18.6	27.0	123	78	17.6	25.8	116	76	16.5	24.6	9	4.2	250
29	23	-1.5	3.3	3	1.5	320	31	26	-0.4	3.6	Changsha (576790)	576870	28.23	112.87	223	68	14.58	100.5	158	87	22.5	30.3	153	86	21.8	29.8	148	85	21.1	29.6	7	3.1	200
32	28	-0.1	4.0	0	0.2	20	34	31	1.2	4.4	Chengdu	562940	30.67	104.02	1666	508	13.84	95.4	155	84	22.2	29.0	148	82	21.2	27.8	144	81	20.5	27.2	4	1.9	200
10	9	-12.5	1.3	15	6.5	360	13	11	-10.7	1.5	Dalian	546620	38.90	121.63	318	97	14.53	100.2	135	80	19.3	26.6	132	78	18.8	25.8	126	78	18.0	25.5	9	4.1	180
1	6	-17.0	0.8	5	2.4	340	5	7	-14.8	1.0	Dandong	544970	40.05	124.33	45	14	14.68	101.2	135	80	19.3	26.4	131	78	18.7	25.6	126	77	18.0	25.0	6	2.5	220
-6	5	-20.9	0.7	5	2.4	360	-3	5	-19.2	0.8	Datong	534870	40.10	113.33	3507	1069	12.93	89.1	111	73	15.9	22.9	104	72	14.9	22.2	98	72	14.0	22.2	9	3.8	160
18	21	-7.7	3.0	2	1.1	320	20	23	-6.6	3.3	Deqen	564440	28.50	98.90	11443	3488	9.56	65.9	93	58	13.3	14.3	90	57	12.8	13.7	87	56	12.4	13.2	12	5.3	180
31	25	-0.7	3.6	7	3.3	340	33	28	0.5	4.0	Dinghai	584770	30.03	122.12	121	37	14.64	100.9	156	87	22.3	30.4	151	85	21.6	29.5	148	84	21.1	28.7	6	2.9	140
-20	2	-28.8	0.3	5	2.3	70	-16	3	-26.9	0.4	Erenhot	530680	43.65	112.00	3169	966	13.09	90.2	94	70	13.4	21.1	88	69	12.5	20.3	81	69	11.6	20.4	12	5.3	140
40	37	4.2	5.3	7	3.0	320	42	40	5.3	5.7	Fuzhou	588470	26.08	119.28	278	85	14.55	100.3	154	89	22.0	31.7	150	88	21.4	30.9	148	86	21.1	30.0	10	4.4	140
0	8	-17.6	1.1	3	1.4	250	4	10	-15.8	1.4	Golmud	528180	36.42	94.90	9215	2809	10.42	71.8	69	60	9.8	15.8	62	60	8.9	15.7	57	58	8.2	14.6	6	2.8	70
42	40	5.3	5.6	6	2.7	360	44	43	6.7	6.1	Guangzhou	592870	23.13	113.32	26	8	14.69	101.2	156	85	22.3	29.5	153	84	21.9	29.1	150	84	21.5	29.0	5	2.3	270
34	29	1.3	4.2	11	4.9	20	36	32	2.2	4.5	Guilin	579570	25.33	110.30	544	166	14.41	99.3	155	84	22.1	29.0	150	83	21.4	28.5	147	83	21.0	28.3	6	2.5	20
28	25	-2.2	3.6	4	2.0	40	30	28	-1.0	3.9	Guiyang	578160	26.58	106.72	3523	1074	12.92	89.1	130	77	18.6	25.2	126	77	18.0	24.9	123	76	17.5	24.5	8	3.4	160
0	6	-18.0	0.9	3	1.2	40	3	7	-16.0	1.0	Hami	522030	42.82	93.52	2424	739	13.46	92.8	95	80	13.6	26.4	87	77	12.4	25.1	80	75	11.4	24.1	2	1.1	250
28	22	-2.4	3.2	4	1.9	340	30	24	-1.1	3.5	Hangzhou	584570	30.23	120.17	141	43	14.63	100.8	160	89	22.8	31.4	152	86	21.7	30.0	149	85	21.3	29.7	8	3.6	160
-21	2	-29.2	0.3	2	1.1	200	-16	2	-26.8	0.3	Harbin	509530	45.75	126.77	469	143	14.45	99.6	127	80	18.1	26.7	119	78	17.0	25.6	111	76	15.9	24.4	11	5.0	180
24	18	-4.4	2.6	4	1.7	340	27	21	-2.8	3.0	Hefei	583210	31.87	117.23	118	36	14.64	100.9	157	88	22.4	30.9	152	87	21.7	30.5	148	86	21.1	29.9	8	3.6	180
-4	5	-20.2	0.7	2	0.7	360	-1	6	-18.6	0.8	Hohhot	534630	40.82	111.68	3494	1065	12.94	89.2	113	75	16.1	23.9	106	73	15.1	23.0	99	72	14.1	22.1	8	3.5	180
18	14	-8.0	2.0	5	2.2	70	20	15	-6.4	2.2	Jinan	548230	36.68	116.98	190	58	14.60	100.6	145	85	20.7	29.5	139	84	19.9	29.0	134	83	19.1	28.3	10	4.3	200
28	22	-2.2	3.2	4	1.7	40	31	26	-0.7	3.6	Jingdezhen	585270	29.30	117.20	196	60	14.60	100.6	153	85	21.8	29.7	149	85	21.3	29.5	145	85	20.7	29.3	7	3.2	250
2	6	-16.4	0.9	6	2.5	360	5	7	-14.8	1.0	Jinzhou	543370	41.13	121.12	229	70	14.58	100.5	136	81	19.4	27.0	130	79	18.5	26.1	124	78	17.7	25.6	9	3.9	200
-14	3	-25.6	0.4	6	2.7	250	-10	3	-23.3	0.5	Jixi	509780	45.28	130.95	767	234	14.30	98.5	121	79	17.3	26.0	113	77	16.2	25.2	106	76	15.2	24.3	8	3.4	270
8	10	-13.6	1.4	1	0.6	320	12	12	-11.1	1.7	Kashi	517090	39.47	75.98	4235	1291	12.59	86.7	108	78	15.4	25.6	99	76	14.1	24.4	92	75	13.1	24.1	4	1.8	180
6	8	-14.4	1.2	1	0.5	70	10	10	-12.0	1.5	Korla	516560	41.75	86.13	3061	933	13.15	90.6	97	81	13.9	27.2	89	78	12.7	25.4	83	76	11.9	24.7	4	2.0	40
48	50	9.0	7.1	8	3.5	330	51	56	10.8	8.0	Kowloon	450070	22.33	114.18	78	24	14.66	101.0	160	85	22.8	29.7	152	85	21.7	29.3	150	85	21.5	29.3	10	4.6	250
32	33	0.1	4.8	2	0.7	140	34	36	1.3	5.2	Kunming	567780	25.02	102.68	6207	1892	11.69	80.6	120	72	17.1	22.3	118	71	16.8	21.9	116	71	16.5	21.4	11	4.9	250
10	11	-12.2	1.6	0	0.2	90	13	13	-10.7	1.8	Lanzhou	528890	36.05	103.88	4980	1518	12.24	84.4	106	74	15.1	23.3	99	73	14.1	22.6	92	72	13.2	22.0	5	2.2	70
13	17	-10.3	2.4	4	1.9	90	16	20	-9.1	2.8	Lhasa	555910	29.67	91.13	11975	3650	9.36	64.5	89	59	12.7	15.1	85	58	12.2	14.6	83	58	11.9	14.5	6	2.6	290

International

HUMIDIFICATION 99.6% (35 hours below...) °F	gr/lb	°C	g/kg	mph	m/s	PWD	99% (88 hours below...) °F	gr/lb	°C	g/kg	Location	WMO#	Lat.	Long.	ft.	m	psia	kPa	DEHUMIDIFICATION 0.4% (35 hours above...) gr/lb	°F	g/kg	°C	1% (88 hours above...) gr/lb	°F	g/kg	°C	2% (175 hours above...) gr/lb	°F	g/kg	°C	Summer Wind mph	m/s	PWD
1a	1b	1c	1d	1e	1f	1g	2a	2b	2c	2d	3a	3b	3c	3d	3e	3f	3g	3h	4a	4b	4c	4d	5a	5b	5c	5d	6a	6b	6c	6d	7a	7b	7c
38	34	3.1	4.9	6	2.5	340	40	37	4.2	5.3	Liuzhou	590460	24.35	109.40	318	97	14.53	100.2	151	85	21.6	29.6	148	85	21.1	29.3	145	84	20.7	29.1	6	2.6	200
43	42	6.0	6.0	1	0.4	90	45	45	7.3	6.4	Longzhou	594170	22.37	106.75	423	129	14.48	99.8	158	86	22.6	30.2	155	86	22.1	29.8	152	85	21.7	29.6	5	2.1	200
44	43	6.9	6.1	13	6.0	360	47	48	8.3	6.9	Macau	450110	22.20	113.53	193	59	14.60	100.6	161	87	23.0	30.4	157	86	22.4	30.0	156	86	22.3	29.8	7	3.0	200
-15	2	-26.1	0.4	3	1.3	180	-12	3	-24.2	0.4	Mudanjiang	540940	44.57	129.60	242	74	14.57	100.4	120	80	17.2	26.8	113	78	16.2	25.8	106	76	15.2	24.4	7	3.2	200
30	24	-1.0	3.5	6	2.7	20	32	27	-0.1	3.8	Nanchang	586060	28.60	115.92	164	50	14.61	100.7	160	87	22.8	30.6	155	87	22.2	30.4	151	86	21.6	30.1	6	2.8	220
23	17	-5.2	2.5	2	0.8	340	26	20	-3.5	2.9	Nanjing	582380	32.00	118.80	39	12	14.68	101.2	160	87	22.9	30.8	154	87	22.0	30.3	150	86	21.4	29.8	9	4.0	220
42	40	5.4	5.7	2	0.8	90	44	43	6.4	6.1	Nanning	594310	22.82	108.35	239	73	14.57	100.5	158	86	22.5	29.9	155	85	22.1	29.6	152	85	21.7	29.3	4	1.8	160
-32	1	-35.4	0.1	2	0.7	20	-27	1	-32.6	0.2	Nenjiang	505570	49.17	125.23	797	243	14.28	98.4	113	76	16.2	24.7	106	74	15.2	23.3	100	73	14.3	22.6	13	5.8	160
19	14	-7.4	2.1	17	7.4	340	21	16	-6.1	2.3	Qingdao	548570	36.07	120.33	252	77	14.57	100.4	145	81	20.7	27.0	140	80	20.0	26.5	136	79	19.4	26.0	11	4.8	160
-18	2	-27.9	0.3	4	1.7	290	-14	3	-25.7	0.4	Qiqihar	507450	47.38	123.92	485	148	14.44	99.6	123	80	17.5	26.4	113	77	16.2	25.0	106	75	15.2	24.0	10	4.3	180
26	20	-3.1	2.9	6	2.5	290	29	23	-1.8	3.3	Shanghai	583670	31.17	121.43	22	7	14.69	101.2	161	88	23.0	31.0	157	87	22.4	30.6	150	85	21.5	29.6	8	3.5	200
44	43	6.7	6.1	7	3.0	20	47	48	8.1	6.8	Shantou	593160	23.40	116.68	9	3	14.69	101.3	156	85	22.3	29.6	153	85	21.8	29.2	150	84	21.5	28.9	8	3.6	200
36	31	2.5	4.5	2	1.1	360	39	35	3.8	5.1	Shaoguan	590820	24.80	113.58	223	68	14.58	100.5	149	85	21.3	29.3	146	84	20.9	29.1	144	84	20.5	28.8	5	2.2	180
-6	4	-21.0	0.6	3	1.5	70	-2	5	-18.7	0.7	Shenyang	543420	41.77	123.43	141	43	14.63	100.8	133	82	19.0	27.9	128	81	18.3	27.1	122	79	17.4	26.3	9	4.2	200
14	11	-9.8	1.6	2	1.0	160	17	13	-8.1	1.9	Shijiazhuang	536980	38.03	114.42	265	81	14.56	100.4	149	85	21.3	29.3	141	83	20.2	28.4	134	82	19.1	27.6	6	2.8	160
4	7	-15.4	1.1	2	0.8	90	8	9	-13.5	1.3	Taiyuan	537720	37.78	112.55	2555	779	13.39	92.3	132	79	18.9	26.3	125	78	17.8	25.6	117	77	16.7	24.9	6	2.9	180
7	8	-13.8	1.1	2	1.1	290	10	9	-12.1	1.3	Tangshan	545340	39.67	118.15	95	29	14.65	101.0	142	83	20.3	28.4	136	82	19.4	27.7	130	80	18.5	26.6	7	3.1	180
14	11	-10.0	1.6	4	1.9	340	17	13	-8.5	1.9	Tianjin	545270	39.10	117.17	16	5	14.69	101.3	145	84	20.7	29.0	139	83	19.8	28.2	133	82	19.0	27.7	6	2.7	250
-10	4	-23.4	0.5	2	0.8	180	-5	5	-20.6	0.7	Urumqi	514630	43.78	87.62	3015	919	13.17	90.8	79	66	11.3	19.1	74	67	10.6	19.5	70	68	10.0	19.9	8	3.6	320
12	10	-11.3	1.5	6	2.7	290	15	12	-9.6	1.7	Weifang	548430	36.70	119.08	167	51	14.61	100.7	150	85	21.4	29.3	142	83	20.3	28.3	136	82	19.5	27.5	10	4.5	200
33	28	0.8	3.9	4	1.9	290	36	31	2.0	4.5	Wenzhou	586590	28.02	120.67	22	7	14.69	101.2	160	88	22.9	31.2	156	87	22.3	30.3	153	85	21.8	29.6	7	3.2	90
27	21	-2.8	3.0	2	1.0	20	29	23	-1.5	3.3	Wuhan	574940	30.62	114.13	75	23	14.66	101.0	161	87	23.0	30.8	156	87	22.3	30.7	152	87	21.7	30.4	8	3.5	220
20	16	-6.4	2.3	1	0.4	40	23	18	-4.8	2.6	Xi'an	570360	34.30	108.93	1305	398	14.02	96.6	141	84	20.1	28.8	134	83	19.2	28.4	128	81	18.3	27.4	5	2.2	40
43	42	5.9	6.0	10	4.4	70	45	45	7.0	6.4	Xiamen	591340	24.48	118.08	456	139	14.46	99.7	153	85	21.9	29.5	150	84	21.5	28.9	148	83	21.2	28.5	8	3.7	160
4	9	-15.3	1.3	1	0.6	290	7	10	-13.7	1.5	Xining	528660	36.62	101.77	7421	2262	11.16	76.9	91	65	13.0	18.4	86	63	12.3	17.2	81	62	11.6	16.9	6	2.5	140
19	14	-7.0	2.1	2	1.1	270	22	17	-5.4	2.4	Xuzhou	580270	34.28	117.15	137	42	14.63	100.8	155	86	22.1	30.1	148	85	21.2	29.3	143	84	20.4	28.8	8	3.7	160
59	75	15.1	10.7	6	2.6	20	62	83	16.5	11.9	Yaxian	599480	18.23	109.52	22	7	14.69	101.2	160	87	22.9	30.3	157	86	22.4	30.0	155	86	22.2	29.9	8	3.4	180
30	25	-1.2	3.5	2	0.8	110	32	27	-0.1	3.8	Yichang	574610	30.70	111.30	439	134	14.47	99.7	158	88	22.5	30.9	152	87	21.7	30.3	148	85	21.1	29.6	7	3.3	140
-28	1	-33.3	0.2	1	0.6	220	-24	2	-30.9	0.2	Yichun	507740	47.72	128.90	761	232	14.30	98.6	113	78	16.2	25.4	106	75	15.2	24.1	101	73	14.4	23.0	7	3.2	220
-1	6	-18.2	0.9	2	1.0	20	4	8	-15.6	1.1	Yinchuan	536140	38.48	106.22	3648	1112	12.86	88.7	118	78	16.9	25.4	111	76	15.8	24.4	104	75	14.8	23.7	6	2.5	20
1	6	-17.3	0.8	6	2.6	20	4	7	-15.6	1.0	Yingkou	544710	40.67	122.20	13	4	14.69	101.3	136	82	19.4	27.7	130	80	18.5	26.9	125	80	17.9	26.4	10	4.4	200
-10	3	-23.1	0.5	1	0.3	70	-3	5	-19.2	0.7	Yining	514310	43.95	81.33	2175	663	13.58	93.6	94	76	13.5	24.6	89	74	12.7	23.2	85	73	12.1	22.8	3	1.3	270
30	24	-1.2	3.5	7	3.2	20	32	27	-0.1	3.8	Yueyang	575840	29.38	113.08	170	52	14.61	100.7	161	89	23.0	31.8	155	88	22.2	31.1	150	86	21.5	30.1	8	3.4	250
1	6	-17.0	0.9	7	3.1	340	4	7	-15.4	1.1	Zhangjiakou	544010	40.78	114.88	2381	726	13.48	92.9	119	78	17.0	25.3	111	76	15.8	24.6	104	75	14.9	24.1	6	2.5	140
46	46	7.6	6.6	8	3.6	340	48	50	8.9	7.1	Zhanjiang	596580	21.22	110.40	91	28	14.65	101.0	162	86	23.1	29.8	158	85	22.6	29.5	155	85	22.2	29.4	8	3.5	250
19	15	-7.4	2.1	3	1.2	180	21	16	-5.9	2.3	Zhengzhou	570830	34.72	113.65	364	111	14.51	100.0	154	86	22.0	30.1	147	84	21.0	28.9	141	83	20.1	28.4	8	3.4	160
											COLOMBIA																						
36	43	2.2	6.1	0	0.2	320	39	48	3.9	6.9	Bogota	802220	4.70	74.13	8359	2548	10.77	74.2	97	64	13.8	17.5	94	63	13.5	17.0	90	62	12.9	16.5	10	4.5	90
											COOK ISLANDS																						
62	83	16.8	11.9	1	0.5	150	64	90	17.8	12.8	Rarotonga Island	918430	21.20	159.82	22	7	14.69	101.2	148	82	21.2	28.0	146	82	20.9	27.8	142	82	20.3	27.6	11	4.9	80
											CROATIA																						
25	19	-4.1	2.8	7	3.3	20	27	21	-2.8	3.0	Pula	132090	44.90	13.92	206	63	14.59	100.6	117	79	16.7	26.0	109	77	15.6	25.1	103	75	14.7	24.1	6	2.8	270

International

HUMIDIFICATION 99.6% (35 hours below...)							99% (88 hours below...)												DEHUMIDIFICATION 0.4% (35 hours above...)				1% (88 hours above...)				2% (175 hours above...)				Summer Wind		
°F 1a	gr/lb 1b	°C 1c	g/kg 1d	mph 1e	m/s 1f	PWD 1g	°F 2a	gr/lb 2b	°C 2c	g/kg 2d	Location 3a	WMO# 3b	Lat. 3c	Long. 3d	ft. 3e	m 3f	psia 3g	kPa 3h	gr/lb 4a	°F 4b	g/kg 4c	°C 4d	gr/lb 5a	°F 5b	g/kg 5c	°C 5d	gr/lb 6a	°F 6b	g/kg 6c	°C 6d	mph 7a	m/s 7b	PWD 7c
29	23	-1.9	3.3	9	3.9	340	32	27	-0.1	3.8	Split	133330	43.53	16.30	68	21	14.66	101.1	104	79	14.8	25.9	97	78	13.9	25.3	92	76	13.1	24.7	8	3.7	230
8	8	-13.2	1.2	2	1.0	240	14	11	-10.0	1.6	Zagreb	131310	45.73	16.07	351	107	14.51	100.0	106	78	15.1	25.6	99	77	14.2	24.9	94	74	13.5	23.6	6	2.9	230
											CUBA																						
67	100	19.2	14.3	8	3.5	360	68	103	20.1	14.8	Guantanamo	783670	19.90	75.15	55	17	14.67	101.1	150	89	21.5	31.5	146	88	20.9	31.3	141	87	20.2	30.8	12	5.2	130
											CYPRUS																						
40	37	4.6	5.2	5	2.3	350	43	41	6.0	5.9	Akrotiri	176010	34.58	32.98	75	23	14.66	101.0	137	82	19.6	27.7	132	82	18.9	27.6	127	81	18.2	27.3	10	4.3	260
37	32	3.0	4.6	7	3.2	310	40	37	4.6	5.2	Larnaca	176090	34.88	33.63	6	2	14.70	101.3	134	83	19.1	28.5	131	83	18.7	28.2	125	82	17.9	27.7	12	5.5	200
39	35	4.0	5.0	9	3.9	30	42	40	5.4	5.6	Paphos	176000	34.72	32.48	26	8	14.69	101.2	141	85	20.2	29.2	137	84	19.6	28.8	132	83	18.9	28.2	9	4.0	280
											CZECH REPUBLIC																						
6	8	-14.4	1.1	8	3.4	60	12	10	-10.9	1.5	Brno	117230	49.15	16.70	807	246	14.28	98.4	92	74	13.1	23.2	88	71	12.5	21.9	83	70	11.9	21.3	10	4.5	180
4	7	-15.6	1.0	2	1.0	40	10	10	-12.4	1.4	Cheb	114060	50.08	12.40	1545	471	13.90	95.8	89	71	12.7	21.8	84	68	12.0	20.2	80	67	11.4	19.5	5	2.3	220
1	6	-17.1	0.9	5	2.3	20	9	9	-12.9	1.3	Ostrava	117820	49.68	18.12	839	256	14.26	98.3	93	73	13.3	22.6	88	71	12.6	21.5	84	70	12.0	21.2	10	4.6	190
2	6	-16.7	0.9	2	1.0	20	9	9	-12.8	1.3	Plzen	114480	49.65	13.27	1194	364	14.08	97.0	94	74	13.4	23.6	89	73	12.7	22.6	84	69	12.0	20.8	8	3.5	120
-2	6	-19.0	0.8	19	8.4	20	2	7	-16.4	1.0	Praded Mountain	117350	50.07	17.23	4895	1492	12.28	84.6	79	59	11.3	15.0	74	57	10.6	14.1	70	56	10.0	13.3	12	5.2	180
3	7	-16.1	1.0	4	1.9	10	10	10	-12.4	1.4	Prague	115180	50.10	14.28	1200	366	14.07	97.0	90	72	12.9	22.1	85	69	12.2	20.6	82	69	11.7	20.3	8	3.5	160
3	7	-16.2	1.0	5	2.1	360	9	9	-13.0	1.3	Pribyslav	116590	49.58	15.77	1758	536	13.79	95.0	90	71	12.8	21.7	85	69	12.1	20.6	80	67	11.5	19.7	9	3.9	130
											DENMARK																						
8	8	-13.1	1.2	6	2.6	220	15	12	-9.2	1.7	Alborg	60300	57.10	9.87	9	3	14.69	101.3	80	68	11.4	20.2	75	67	10.7	19.3	71	65	10.1	18.1	11	4.7	100
12	10	-11.1	1.5	11	5.1	360	18	14	-8.0	2.0	Copenhagen	61800	55.63	12.67	16	5	14.69	101.3	80	68	11.5	20.0	77	67	11.0	19.4	74	66	10.5	18.9	11	4.7	160
20	15	-6.7	2.2	20	8.9	70	22	17	-5.3	2.4	Hammerodde	61930	55.30	14.78	36	11	14.68	101.2	90	68	12.9	20.2	85	67	12.1	19.2	80	66	11.5	18.7	12	5.2	230
18	14	-8.0	2.0	14	6.2	320	22	17	-5.7	2.4	Mon Island	61790	54.95	12.55	49	15	14.67	101.1	90	69	12.8	20.8	84	68	12.0	20.0	80	66	11.4	19.1	9	4.0	70
14	11	-10.2	1.6	8	3.4	40	18	14	-7.7	2.0	Odense	61200	55.47	10.33	55	17	14.67	101.1	85	70	12.1	21.1	79	68	11.3	20.1	75	66	10.7	19.1	11	4.9	120
15	12	-9.3	1.7	17	7.4	40	20	15	-6.4	2.2	Skagen	60410	57.77	10.65	22	7	14.69	101.2	94	69	13.5	20.7	88	67	12.6	19.4	84	65	12.0	18.5	10	4.6	360
9	9	-13.0	1.3	5	2.4	20	16	12	-9.1	1.8	Tirstrup	60700	56.30	10.62	82	25	14.66	101.0	85	68	12.2	19.8	80	66	11.4	19.0	75	65	10.7	18.2	11	4.8	280
											ECUADOR																						
67	100	19.7	14.3	8	3.6	210	68	103	19.9	14.8	Guayaquil	842030	2.15	79.88	29	9	14.68	101.2	148	85	21.1	29.3	143	83	20.4	28.2	141	82	20.1	27.9	7	3.2	40
45	63	7.0	9.0	1	0.3	350	46	65	7.9	9.3	Quito	840710	0.15	78.48	9225	2812	10.42	71.8	93	61	13.3	16.1	88	59	12.6	15.1	88	59	12.5	14.8	9	4.1	150
											EGYPT																						
44	43	6.8	6.1	5	2.1	190	46	46	7.8	6.6	Alexandria	623180	31.20	29.95	22	7	14.69	101.2	128	82	18.3	27.9	125	82	17.8	27.5	120	81	17.2	27.4	10	4.3	340
45	45	7.0	6.4	6	2.6	210	46	46	8.0	6.6	Cairo	623660	30.13	31.40	242	74	14.57	100.4	119	79	17.0	26.0	117	78	16.7	25.8	112	78	16.0	25.6	13	5.6	350
40	37	4.4	5.3	2	0.9	180	42	40	5.7	5.7	Luxor	624050	25.67	32.70	288	88	14.55	100.3	97	92	13.8	33.4	90	91	12.9	32.6	85	90	12.2	32.5	6	2.6	330
											ESTONIA																						
5	7	-15.1	1.0	5	2.3	80	11	10	-11.9	1.4	Kopu/Cape Ristna	261150	58.92	22.07	29	9	14.68	101.2	88	69	12.6	20.4	83	67	11.9	19.4	78	65	11.2	18.4	6	2.7	70
-4	4	-19.8	0.6	6	2.9	140	3	6	-16.0	0.9	Tallinn	260380	59.35	24.80	144	44	14.62	100.8	88	69	12.5	20.8	82	68	11.7	20.0	76	66	10.8	18.7	8	3.6	40
											FAEROE ISLANDS																						
26	20	-3.2	2.9	13	5.8	320	28	22	-2.3	3.2	Torshavn	60110	62.02	6.77	127	39	14.63	100.9	64	56	9.1	13.3	62	55	8.8	12.7	60	54	8.5	12.2	11	4.8	210
											FIJI																						
61	81	16.0	11.5	4	1.6	120	63	87	17.1	12.4	Nadi	916800	17.75	177.45	59	18	14.67	101.1	146	83	20.9	28.5	143	83	20.4	28.2	140	82	20.0	28.0	13	5.8	350
62	83	16.9	11.9	1	0.3	320	64	90	17.9	12.8	Nausori	916830	18.05	178.57	22	7	14.69	101.2	148	83	21.2	28.2	146	82	20.9	28.0	142	82	20.3	27.7	11	4.7	60
											FINLAND																						
-11	3	-23.7	0.4	5	2.4	340	-3	5	-19.5	0.7	Helsinki	29740	60.32	24.97	183	56	14.60	100.7	85	67	12.2	19.6	80	67	11.4	19.2	75	65	10.7	18.4	11	4.8	210
-21	2	-29.2	0.3	2	0.7	330	-13	3	-24.8	0.4	Jyvaskyla	29350	62.40	25.68	475	145	14.45	99.6	84	69	12.0	20.5	78	65	11.1	18.5	72	64	10.3	17.6	9	3.8	180

International

HUMIDIFICATION 99.6% (35 hours below...)							99% (88 hours below...)												DEHUMIDIFICATION 0.4% (35 hours above...)				1% (88 hours above...)				2% (175 hours above...)				Summer Wind		
°F	gr/lb	°C	g/kg	mph	m/s	PWD	°F	gr/lb	°C	g/kg	Location	WMO#	Lat.	Long.	ft.	m	psia	kPa	gr/lb	°F	g/kg	°C	gr/lb	°F	g/kg	°C	gr/lb	°F	g/kg	°C	mph	m/s	PWD
1a	1b	1c	1d	1e	1f	1g	2a	2b	2c	2d	3a	3b	3c	3d	3e	3f	3g	3h	4a	4b	4c	4d	5a	5b	5c	5d	6a	6b	6c	6d	7a	7b	7c
-20	2	-29.0	0.3	2	0.8	80	-14	3	-25.6	0.4	Kauhava	29130	63.10	23.03	144	44	14.62	100.8	83	68	11.8	20.2	76	65	10.9	18.5	71	64	10.2	17.7	9	3.8	230
-21	2	-29.7	0.3	1	0.6	140	-14	3	-25.6	0.4	Kuopio	29170	63.02	27.80	334	102	14.52	100.1	84	69	12.0	20.7	79	66	11.3	19.1	74	65	10.5	18.1	7	3.3	170
-15	2	-26.3	0.4	1	0.6	350	-7	4	-21.9	0.5	Lahti	29650	60.97	25.63	275	84	14.55	100.3	86	70	12.3	21.2	79	68	11.3	20.0	75	66	10.7	18.8	6	2.7	150
-25	1	-31.4	0.2	1	0.4	300	-20	2	-29.1	0.3	Pello	28440	66.80	24.00	275	84	14.55	100.3	78	66	11.1	19.1	72	64	10.3	17.9	67	62	9.6	16.8	7	3.0	340
-12	3	-24.3	0.4	5	2.3	90	-4	4	-20.2	0.6	Pori	29520	61.47	21.80	55	17	14.67	101.1	83	67	11.8	19.7	76	65	10.9	18.5	72	64	10.3	17.5	11	5.0	140
-21	2	-29.7	0.3	1	0.5	360	-17	2	-27.2	0.3	Suomussalmi	28790	64.90	29.02	734	224	14.31	98.7	78	66	11.2	19.0	73	64	10.4	17.7	69	62	9.8	16.8	7	3.1	270
-15	2	-26.2	0.4	2	0.8	10	-8	4	-22.2	0.5	Tampere	29440	61.42	23.58	367	112	14.51	100.0	83	67	11.8	19.6	76	66	10.9	18.9	71	64	10.2	17.6	9	4.0	10
-10	3	-23.1	0.5	6	2.7	40	-3	5	-19.6	0.7	Turku	29720	60.52	22.27	193	59	14.60	100.6	84	69	12.0	20.6	77	66	11.0	19.0	73	65	10.4	18.1	9	4.0	230
											FRANCE																						
22	17	-5.8	2.4	4	1.6	40	27	21	-3.0	3.0	Bordeaux	75100	44.83	0.70	200	61	14.59	100.6	111	79	15.9	25.9	104	75	14.8	24.0	98	74	14.0	23.1	7	3.3	80
16	13	-9.1	1.8	3	1.4	360	20	16	-6.8	2.2	Clermont-Ferrand	74600	45.78	3.17	1082	330	14.13	97.4	104	77	14.8	25.0	98	76	14.0	24.3	93	74	13.3	23.2	8	3.4	20
14	12	-9.8	1.6	7	3.1	20	20	15	-6.8	2.2	Dijon	72800	47.27	5.08	744	227	14.31	98.6	106	77	15.1	25.1	99	74	14.1	23.3	93	72	13.3	22.2	10	4.6	170
27	21	-2.8	3.0	8	3.4	120	30	24	-1.0	3.5	Brest	71100	48.45	4.42	337	103	14.52	100.1	93	69	13.3	20.6	89	67	12.7	19.5	85	66	12.2	18.9	9	4.0	40
17	13	-8.5	1.9	3	1.4	20	23	18	-5.1	2.6	Lyon	74810	45.73	5.08	787	240	14.29	98.5	105	78	15.0	25.7	99	75	14.1	24.1	93	73	13.3	22.9	12	5.2	180
25	19	-3.9	2.7	9	3.8	360	28	22	-2.0	3.2	Marseille	76500	43.45	5.23	118	36	14.64	100.9	118	79	16.8	26.2	111	78	15.8	25.8	104	77	14.9	24.9	13	5.6	280
25	19	-3.8	2.7	7	3.3	340	29	23	-1.8	3.3	Montpellier	76430	43.58	3.97	19	6	14.69	101.3	131	79	18.7	26.0	123	78	17.5	25.3	116	76	16.5	24.7	13	5.6	180
14	12	-10.2	1.6	7	3.2	60	17	13	-8.1	1.9	Nancy	71800	48.68	6.22	711	217	14.33	98.7	100	76	14.3	24.6	94	73	13.5	22.9	90	71	12.8	21.4	8	3.7	220
23	17	-5.2	2.5	7	3.1	60	27	21	-2.8	3.0	Nantes	72220	47.17	1.60	88	27	14.65	101.0	104	76	14.8	24.4	97	73	13.9	22.9	92	70	13.1	21.3	9	4.1	60
35	30	1.6	4.3	11	4.7	340	37	32	2.9	4.6	Nice	76900	43.65	7.20	32	10	14.68	101.2	136	81	19.4	27.0	131	80	18.7	26.6	123	79	17.5	25.9	8	3.6	160
26	20	-3.3	2.9	10	4.3	20	30	24	-1.1	3.5	Nimes	76450	43.87	4.40	203	62	14.59	100.6	113	77	16.2	24.9	107	76	15.3	24.6	100	75	14.3	23.8	9	4.1	40
17	13	-8.4	1.9	8	3.6	60	22	17	-5.3	2.4	Orleans	72490	47.98	1.75	410	125	14.48	99.8	102	78	14.5	25.6	93	73	13.3	22.7	88	71	12.6	21.8	9	3.8	80
18	14	-7.8	2.0	10	4.6	60	23	18	-5.0	2.5	Paris, De Gaulle	71570	49.02	2.53	357	109	14.51	100.0	104	76	14.9	24.6	97	74	13.9	23.3	91	72	13.0	22.2	9	4.1	60
19	15	-7.1	2.1	8	3.7	20	23	18	-4.8	2.5	Paris, Orly	71490	48.73	2.40	314	96	14.53	100.2	99	75	14.1	23.7	93	73	13.3	22.9	88	71	12.6	21.9	8	3.4	100
17	13	-8.2	1.9	11	5.0	60	22	17	-5.6	2.4	St.-Quentin	70610	49.82	3.20	331	101	14.52	100.1	102	75	14.6	23.9	94	72	13.4	22.1	88	70	12.6	21.2	9	3.8	120
12	10	-11.0	1.5	6	2.9	340	17	13	-8.2	1.9	Strasbourg	71900	48.55	7.63	505	154	14.43	99.5	104	76	14.9	24.7	98	75	14.0	23.8	92	72	13.2	22.5	8	3.4	20
22	17	-5.8	2.4	5	2.2	280	27	21	-3.0	3.1	Toulouse	76300	43.63	1.37	501	153	14.44	99.5	111	79	15.8	26.3	104	77	14.9	25.2	98	75	14.0	23.8	7	3.3	140
											FRENCH POLYNESIA																						
67	100	19.6	14.2	16	7.0	140	68	103	20.2	14.8	Moruroa Island	919520	21.82	138.80	9	3	14.69	101.3	151	83	21.6	28.2	147	82	21.0	27.9	144	82	20.6	27.8	11	4.7	60
68	103	19.8	14.8	3	1.5	120	69	107	20.6	15.3	Papeete, Tahiti	919380	17.55	149.62	6	2	14.70	101.3	148	86	21.1	30.0	145	85	20.7	29.6	141	85	20.2	29.3	7	3.0	260
											GERMANY																						
14	12	-10.1	1.6	4	1.9	50	19	15	-7.2	2.1	Aachen	105010	50.78	6.10	672	205	14.35	98.9	93	73	13.3	22.6	88	71	12.6	21.7	83	70	11.9	20.9	6	2.5	210
11	10	-11.8	1.4	7	3.0	90	16	12	-9.0	1.8	Ahlhorn (Ger-Afb)	102180	52.88	8.23	184	56	14.60	100.7	90	71	12.9	21.9	85	70	12.1	21.0	80	69	11.4	20.4	9	4.0	10
11	10	-11.8	1.4	8	3.4	80	15	12	-9.2	1.7	Berlin	103840	52.47	13.40	160	49	14.61	100.7	90	72	12.9	22.3	85	70	12.1	21.2	80	69	11.4	20.8	8	3.7	150
12	11	-10.9	1.5	11	4.7	60	18	14	-8.0	2.0	Bitburg	106100	49.95	6.57	1227	374	14.06	96.9	94	72	13.5	22.5	89	71	12.7	21.9	83	70	11.9	21.0	7	3.1	10
12	10	-11.3	1.5	8	3.7	70	16	12	-8.7	1.8	Bremen	102240	53.05	8.80	16	5	14.69	101.3	91	71	13.0	21.8	85	71	12.2	21.4	80	68	11.4	20.2	10	4.5	100
15	12	-9.4	1.7	8	3.6	60	19	14	-7.0	2.1	Bremerhaven	101290	53.53	8.58	36	11	14.68	101.2	94	71	13.4	21.6	88	70	12.5	21.3	83	69	11.8	20.3	10	4.3	130
8	9	-13.3	1.2	4	1.9	320	13	11	-10.3	1.6	Dresden	104880	51.13	13.78	741	226	14.31	98.6	91	72	13.0	22.4	86	71	12.3	21.6	81	69	11.6	20.5	7	3.0	990
14	11	-9.9	1.6	6	2.8	60	20	15	-6.9	2.2	Dusseldorf	104000	51.28	6.78	144	44	14.62	100.8	92	73	13.2	22.9	88	72	12.5	22.0	83	71	11.8	21.4	9	3.8	130
11	10	-11.9	1.4	7	3.2	30	17	13	-8.6	1.9	Eggebek (Ger-Navy)	100340	54.63	9.35	72	22	14.66	101.1	85	70	12.2	21.3	80	68	11.4	20.2	75	67	10.7	19.3	11	5.0	90
5	8	-14.8	1.1	15	6.9	100	10	10	-12.1	1.5	Ehrenberg	105440	50.50	9.95	3034	925	13.16	90.7	87	68	12.4	20.0	81	65	11.6	18.3	77	63	11.0	17.1	11	5.1	190
12	10	-11.0	1.5	7	3.3	30	17	13	-8.2	1.9	Frankfurt Am Main	106370	50.05	8.60	370	113	14.50	100.0	93	73	13.3	22.6	88	71	12.6	21.5	83	70	11.9	20.9	9	3.9	40
-2	5	-18.9	0.7	1	0.6	10	5	8	-14.8	1.1	Grafenwohr	106870	49.70	11.95	1361	415	13.99	96.4	90	69	12.9	20.3	88	71	12.6	21.4	83	69	11.9	20.7	5	2.1	10

International

HUMIDIFICATION 99.6% (35 hours below...)							99% (88 hours below...)												DEHUMIDIFICATION 0.4% (35 hours above...)				1% (88 hours above...)				2% (175 hours above...)				Summer Wind		
°F 1a	gr/lb 1b	°C 1c	g/kg 1d	mph 1e	m/s 1f	PWD 1g	°F 2a	gr/lb 2b	°C 2c	g/kg 2d	Location 3a	WMO# 3b	Lat. 3c	Long. 3d	ft. 3e	m 3f	psia 3g	kPa 3h	gr/lb 4a	°F 4b	g/kg 4c	°C 4d	gr/lb 5a	°F 5b	g/kg 5c	°C 5d	gr/lb 6a	°F 6b	g/kg 6c	°C 6d	mph 7a	m/s 7b	PWD 7c
9	9	-12.8	1.3	4	1.6	250	15	12	-9.4	1.7	Greifswald	101840	54.10	13.40	19	6	14.69	101.3	91	72	13.0	22.1	85	70	12.1	20.9	80	68	11.4	20.0	8	3.6	50
11	10	-11.6	1.4	6	2.5	60	16	12	-8.9	1.8	Hamburg	101470	53.63	10.00	52	16	14.67	101.1	90	72	12.8	22.1	85	70	12.1	21.0	80	68	11.4	20.1	11	4.7	90
9	9	-12.7	1.3	6	2.5	80	14	11	-9.8	1.6	Hannover	103380	52.47	9.70	177	54	14.61	100.7	93	72	13.3	22.5	87	71	12.4	21.4	82	69	11.7	20.8	9	4.2	110
14	11	-10.0	1.6	4	1.9	170	19	15	-7.1	2.1	Heidelberg	107340	49.40	8.65	357	109	14.51	100.0	98	77	14.0	25.0	92	75	13.1	23.8	87	71	12.4	21.9	6	2.8	10
3	7	-16.0	1.0	6	2.5	140	9	9	-13.0	1.3	Hof	106850	50.32	11.88	1863	568	13.74	94.7	89	70	12.7	21.3	83	67	11.9	19.7	79	66	11.3	18.7	7	3.3	150
12	10	-11.2	1.5	9	3.9	50	17	13	-8.2	1.9	Husum (Ger-Afb)	100260	54.52	9.15	121	37	14.64	100.9	90	70	12.8	21.3	84	68	12.0	19.8	79	66	11.3	19.0	9	4.2	90
17	13	-8.1	1.9	15	6.7	360	21	16	-5.9	2.3	Kap Arkona	100910	54.68	13.43	134	41	14.63	100.8	90	70	12.8	20.9	85	68	12.1	19.9	80	67	11.5	19.2	12	5.4	70
14	11	-9.9	1.6	9	4.0	40	19	14	-7.2	2.1	Kiel/Holtenau(Gnvy)	100460	54.38	10.15	102	31	14.65	101.0	84	71	12.0	21.5	79	68	11.3	20.2	75	67	10.7	19.2	9	3.8	160
11	10	-11.4	1.4	4	1.9	110	17	13	-8.1	1.9	Koln	105130	50.87	7.17	324	99	14.53	100.1	92	73	13.2	22.7	87	71	12.4	21.5	83	69	11.8	20.8	8	3.5	130
11	10	-11.4	1.4	4	1.7	20	17	13	-8.2	1.9	Lahr	108050	48.37	7.83	511	156	14.43	99.5	99	77	14.2	25.1	94	74	13.4	23.3	91	73	13.0	22.9	5	2.4	120
5	8	-14.9	1.1	3	1.5	70	10	10	-12.1	1.4	Landsberg (Ger-Afb)	108570	48.07	10.90	2060	628	13.64	94.0	90	72	12.9	22.0	85	70	12.2	21.3	80	69	11.5	20.6	6	2.9	260
11	10	-11.5	1.4	5	2.3	80	17	13	-8.2	1.9	Leck (Ger-Afb)	100220	54.80	8.95	43	13	14.68	101.2	85	71	12.1	21.8	80	68	11.4	20.2	75	67	10.7	19.3	10	4.5	110
8	8	-13.4	1.2	6	2.8	70	13	11	-10.4	1.6	Leipzig	104690	51.42	12.23	436	133	14.47	99.7	91	73	13.0	22.9	86	71	12.3	21.7	81	70	11.6	21.1	9	4.0	190
5	8	-14.8	1.1	5	2.2	50	10	10	-12.0	1.4	Memmingen (Ger-Afb)	109470	47.98	10.23	2113	644	13.61	93.8	88	73	12.5	22.9	82	70	11.7	20.9	80	69	11.4	20.7	9	3.8	220
4	7	-15.4	1.0	4	1.6	80	10	10	-12.5	1.4	Munich	108660	48.13	11.70	1735	529	13.80	95.1	91	72	13.0	22.2	87	71	12.4	21.4	83	69	11.9	20.7	8	3.6	30
3	7	-15.9	1.0	4	1.6	60	10	10	-12.5	1.4	Neuburg (Ger-Afb)	108530	48.72	11.22	1270	387	14.04	96.8	90	73	12.8	22.9	85	71	12.1	21.8	79	69	11.3	20.7	4	1.9	200
12	10	-10.9	1.5	9	3.8	80	17	13	-8.2	1.9	Nordholz (Ger-Navy)	101360	53.77	8.67	102	31	14.65	101.0	88	70	12.6	21.0	83	69	11.9	20.3	79	67	11.3	19.5	11	5.0	120
11	10	-11.8	1.4	2	0.9	10	16	13	-9.0	1.8	Ramstein (Usafb)	106140	49.43	7.60	780	238	14.29	98.5	94	72	13.5	22.1	88	72	12.6	22.4	83	71	11.9	21.6	5	2.4	240
13	11	-10.8	1.6	6	2.6	30	17	13	-8.1	1.9	Sollingen (Can-Afb)	107220	48.77	8.10	419	128	14.48	99.8	99	75	14.2	23.9	97	74	13.8	23.4	92	72	13.1	22.5	6	2.5	10
9	9	-12.7	1.3	4	1.9	90	14	12	-10.0	1.7	Stuttgart	107380	48.68	9.22	1374	419	13.98	96.4	91	74	13.0	23.1	87	71	12.4	21.9	83	70	11.8	21.1	7	3.0	90
											GEORGIA																						
29	23	-1.7	3.3	13	6.0	130	32	27	-0.2	3.8	Batumi	374840	41.65	41.63	19	6	14.69	101.3	123	79	17.5	26.1	118	78	16.8	25.3	113	76	16.1	24.5	9	4.2	300
28	22	-2.4	3.2	9	3.9	90	30	25	-1.1	3.5	K'ut'aisi (Kutaisi)	373950	42.27	42.63	380	116	14.50	99.9	125	79	17.9	26.3	120	78	17.1	25.4	116	77	16.5	25.0	22	9.6	90
29	23	-1.5	3.3	6	2.7	50	31	25	-0.5	3.6	Sokhumi (Sukhumi)	372600	42.87	41.13	42	13	14.68	101.2	127	79	18.2	26.2	122	78	17.4	25.5	118	77	16.9	25.0	9	4.0	320
21	17	-6.0	2.4	6	2.6	320	24	19	-4.7	2.8	Tbilisi	375490	41.68	44.95	1532	467	13.90	95.8	111	81	15.8	27.2	105	79	15.0	26.2	99	78	14.2	25.6	10	4.5	180
											GIBRALTAR																						
46	46	7.8	6.6	9	4.0	270	48	50	8.9	7.1	North Front	84950	36.15	5.35	16	5	14.69	101.3	120	76	17.2	24.6	117	76	16.7	24.6	113	75	16.1	23.7	14	6.2	200
											GREECE																						
32	27	-0.2	3.8	2	0.9	130	34	29	1.1	4.1	Andravida	166820	37.92	21.28	39	12	14.68	101.2	124	80	17.7	26.4	117	79	16.7	26.2	111	78	15.8	25.5	12	5.3	350
34	29	1.2	4.1	8	3.5	360	37	32	2.9	4.6	Athens	167160	37.90	23.73	49	15	14.67	101.1	116	83	16.6	28.2	108	82	15.5	27.5	102	80	14.5	26.7	14	6.1	30
33	28	0.8	4.0	4	2.0	360	36	31	2.1	4.5	Elefsis (Hel-Afb)	167180	38.07	23.55	102	31	14.65	101.0	110	83	15.7	28.6	103	82	14.7	27.6	96	80	13.7	26.7	12	5.5	10
41	38	5.0	5.4	14	6.2	340	44	43	6.8	6.1	Iraklion	167540	35.33	25.18	127	39	14.63	100.9	116	80	16.5	26.7	110	79	15.7	26.3	104	79	14.9	25.9	11	4.7	320
23	18	-5.1	2.5	1	0.4	360	26	20	-3.5	2.9	Larisa	166480	39.63	22.42	242	74	14.57	100.4	98	74	14.0	23.6	92	74	13.2	23.3	89	74	12.7	23.1	7	3.2	270
36	31	2.3	4.5	10	4.3	40	38	34	3.5	4.8	Preveza	166430	38.95	20.77	13	4	14.69	101.3	126	80	18.0	26.6	122	79	17.4	26.1	118	79	16.8	25.9	10	4.3	250
41	38	5.0	5.4	13	6.0	360	44	43	6.6	6.1	Rodhos	167490	36.40	28.08	36	11	14.68	101.2	126	79	18.0	26.1	122	79	17.4	26.1	117	79	16.7	25.9	14	6.4	270
39	36	4.0	5.1	10	4.4	20	41	39	5.2	5.5	Soudha	167460	35.48	24.12	495	151	14.44	99.5	111	75	15.8	24.1	105	75	15.0	24.0	99	75	14.2	23.9	10	4.3	300
25	19	-3.8	2.7	8	3.6	110	28	22	-2.0	3.1	Thessaloniki	166220	40.52	22.97	13	4	14.69	101.3	108	81	15.5	27.0	102	79	14.6	26.3	97	78	13.8	25.4	10	4.4	180
											GREENLAND																						
-35	1	-37.1	0.1	7	3.0	80	-33	1	-35.9	0.1	Dundas, Thule Ab	42020	76.53	68.50	193	59	14.60	100.6	36	47	5.1	8.1	33	45	4.7	7.1	31	44	4.5	6.4	9	4.0	310
-12	3	-24.5	0.4	13	5.9	350	-8	4	-22.4	0.5	Godthab	42500	64.17	51.75	229	70	14.58	100.5	49	50	7.0	10.2	46	49	6.5	9.3	43	47	6.2	8.6	10	4.6	20
-42	0	-41.0	0.1	9	3.8	60	-38	1	-39.0	0.1	Kangerlussuaq	42310	67.00	50.80	173	53	14.61	100.7	45	52	6.4	11.3	43	52	6.1	11.3	41	52	5.8	11.1	11	4.9	70
-18	2	-27.8	0.3	2	0.9	60	-12	3	-24.7	0.4	Narsarsuaq	42700	61.18	45.42	85	26	14.65	101.0	52	51	7.4	10.6	49	50	7.0	10.1	46	49	6.6	9.7	17	7.5	70

International

HUMIDIFICATION 99.6% (35 hours below...)							HUMIDIFICATION 99% (88 hours below...)												DEHUMIDIFICATION 0.4% (35 hours above...)				1% (88 hours above...)				2% (175 hours above...)				Summer Wind		
°F 1a	gr/lb 1b	°C 1c	g/kg 1d	mph 1e	m/s 1f	PWD 1g	°F 2a	gr/lb 2b	°C 2c	g/kg 2d	Location 3a	WMO# 3b	Lat. 3c	Long. 3d	ft. 3e	m 3f	psia 3g	kPa 3h	gr/lb 4a	°F 4b	g/kg 4c	°C 4d	gr/lb 5a	°F 5b	g/kg 5c	°C 5d	gr/lb 6a	°F 6b	g/kg 6c	°C 6d	mph 7a	m/s 7b	PWD 7c
											GUAM																						
74	130	23.3	18.6	9	3.8	70	75	135	23.7	19.3	Andersen Afb	912180	13.58	144.93	606	185	14.38	99.1	158	84	22.5	29.0	154	83	22.0	28.5	150	83	21.4	28.2	9	3.8	90
											HUNGARY																						
8	9	-13.2	1.2	2	0.9	170	14	11	-10.2	1.6	Budapest	128390	47.43	19.27	606	185	14.38	99.1	94	71	13.4	21.6	92	74	13.1	23.4	87	74	12.4	23.5	10	4.5	200
6	8	-14.6	1.1	4	1.6	50	11	10	-11.6	1.4	Debrecen	128820	47.48	21.63	367	112	14.51	100.0	102	80	14.6	26.5	97	78	13.8	25.5	92	76	13.1	24.2	6	2.5	90
8	8	-13.5	1.2	5	2.1	360	14	11	-10.2	1.6	Nagykanizsa	129250	46.45	16.98	462	141	14.46	99.6	104	78	14.9	25.3	97	75	13.9	23.9	92	73	13.2	23.0	6	2.7	230
12	10	-11.2	1.5	6	2.6	320	16	13	-9.0	1.8	Pecs	129420	46.00	18.23	666	203	14.35	98.9	102	80	14.5	26.9	96	79	13.7	26.3	90	76	12.9	24.7	7	3.2	270
12	10	-11.3	1.5	4	2.0	320	17	13	-8.5	1.9	Siofok	129350	46.92	18.03	354	108	14.51	100.0	110	81	15.7	27.4	103	79	14.7	26.0	97	76	13.8	24.7	5	2.2	270
10	9	-12.0	1.4	8	3.4	270	15	12	-9.5	1.7	Szombathely	128120	47.27	16.63	725	221	14.32	98.7	103	76	14.7	24.7	97	75	13.8	23.7	92	72	13.1	22.5	7	3.2	180
											ICELAND																						
8	8	-13.4	1.2	7	3.0	160	11	10	-11.5	1.4	Akureyri	40630	65.68	18.08	88	27	14.65	101.0	62	61	8.8	16.0	57	58	8.1	14.7	52	56	7.5	13.3	11	4.9	180
17	13	-8.1	1.9	14	6.3	20	20	15	-6.9	2.2	Keflavik	40180	63.97	22.60	177	54	14.61	100.7	58	54	8.3	12.3	57	53	8.1	11.8	54	52	7.7	11.3	11	5.0	350
10	9	-12.2	1.3	14	6.2	230	13	11	-10.4	1.5	Raufarhofn	40770	66.45	15.95	32	10	14.68	101.2	55	57	7.9	13.9	52	54	7.4	12.0	50	52	7.1	10.9	14	6.2	320
14	11	-9.8	1.6	10	4.3	90	17	13	-8.1	1.9	Reykjavik	40300	64.13	21.90	200	61	14.59	100.6	59	56	8.4	13.4	56	54	8.0	12.5	54	53	7.7	11.7	12	5.5	360
											INDIA																						
52	58	11.3	8.3	2	1.0	360	55	65	12.8	9.3	Ahmadabad	426470	23.07	72.63	180	55	14.60	100.7	166	88	23.7	31.1	161	87	23.0	30.5	157	86	22.4	30.1	7	3.0	270
59	84	14.9	12.0	3	1.4	90	60	87	15.7	12.4	Bangalore	432950	12.97	77.58	3021	921	13.16	90.7	132	77	18.9	25.1	127	76	18.2	24.6	125	76	17.8	24.3	4	1.7	90
62	84	16.5	11.9	0	0.2	360	64	90	17.6	12.8	Bombay	430030	19.12	72.85	45	14	14.68	101.2	156	86	22.3	30.2	153	86	21.9	29.9	150	85	21.5	29.6	7	3.1	320
54	62	12.0	8.9	0	0.2	360	56	67	13.1	9.6	Calcutta	428090	22.65	88.45	19	6	14.69	101.3	171	90	24.4	32.2	167	89	23.8	31.6	164	88	23.4	31.1	4	2.0	180
68	103	19.9	14.8	2	0.7	320	69	107	20.7	15.3	Cuddalore	433290	11.77	79.77	39	12	14.68	101.2	166	88	23.7	31.3	162	88	23.1	31.0	159	87	22.7	30.8	6	2.6	250
67	100	19.6	14.3	5	2.2	50	69	108	20.3	15.4	Goa/Panaji	431920	15.48	73.82	196	60	14.60	100.6	163	87	23.3	30.5	158	86	22.5	29.8	153	85	21.9	29.3	6	2.7	320
58	77	14.5	11.0	1	0.4	50	60	83	15.8	11.8	Hyderabad	431280	17.45	78.47	1788	545	13.77	94.9	139	81	19.8	27.3	135	80	19.3	26.8	133	80	19.0	26.4	8	3.7	320
44	45	6.8	6.4	0	0.2	90	47	50	8.2	7.2	Jaipur	423480	26.82	75.80	1279	390	14.03	96.7	157	84	22.4	28.7	153	83	21.9	28.3	149	82	21.3	28.0	9	3.8	320
68	103	19.9	14.8	2	0.9	290	69	107	20.5	15.3	Madras	432790	13.00	80.18	52	16	14.67	101.1	162	87	23.2	30.6	159	86	22.7	30.2	155	86	22.2	29.8	8	3.7	270
53	62	11.8	8.9	2	1.0	360	55	67	13.0	9.6	Nagpur	428670	21.10	79.05	1017	310	14.17	97.7	151	83	21.6	28.6	148	83	21.1	28.2	145	82	20.7	27.9	6	2.8	320
44	44	6.6	6.3	2	0.7	270	46	47	7.6	6.8	New Delhi	421820	28.58	77.20	708	216	14.33	98.8	162	87	23.2	30.4	158	86	22.6	30.0	155	86	22.1	29.8	7	3.3	320
50	57	9.8	8.2	0	0.0	70	52	62	10.9	8.8	Poona	430630	18.53	73.85	1833	559	13.75	94.8	134	80	19.1	26.6	131	79	18.7	25.9	127	78	18.2	25.5	3	1.5	270
61	85	16.2	12.2	1	0.5	90	63	92	17.2	13.1	Sholapur	431170	17.67	75.90	1571	479	13.88	95.7	148	86	21.2	30.0	141	83	20.2	28.6	136	82	19.4	28.0	2	0.9	320
72	120	22.0	17.1	2	1.1	360	73	124	22.6	17.7	Trivandrum	433710	8.48	76.95	209	64	14.59	100.6	153	85	21.8	29.2	150	84	21.5	29.0	148	84	21.1	28.8	6	2.5	320
											INDIAN OCEAN ISLANDS																						
73	123	23.0	17.6	11	4.8	110	75	132	23.8	18.8	Diego Garcia Isl.	619670	7.30	72.40	9	3	14.69	101.3	161	85	23.0	29.2	157	84	22.4	29.1	151	84	21.6	28.8	7	3.1	90
											IRELAND																						
30	24	-1.2	3.5	8	3.6	90	32	27	0.2	3.8	Belmullet	39760	54.23	10.00	32	10	14.68	101.2	83	65	11.9	18.6	79	64	11.3	17.8	76	62	10.8	16.9	11	4.7	180
24	18	-4.2	2.6	1	0.6	90	28	22	-2.5	3.2	Birr	39650	53.08	7.88	236	72	14.57	100.5	87	69	12.4	20.3	81	66	11.6	18.8	78	65	11.1	18.6	7	3.0	150
26	20	-3.6	2.9	6	2.7	70	28	22	-2.2	3.2	Claremorris	39700	53.72	8.98	226	69	14.58	100.5	86	67	12.3	19.6	82	65	11.7	18.6	78	64	11.1	17.9	8	3.6	90
25	19	-3.7	2.8	4	1.8	60	28	22	-2.1	3.2	Clones	39740	54.18	7.23	291	89	14.55	100.3	85	67	12.2	19.7	80	66	11.5	18.8	76	65	10.9	18.3	7	3.1	120
29	24	-1.4	3.4	13	5.7	40	32	27	-0.2	3.9	Cork	39550	51.85	8.48	531	162	14.42	99.4	86	65	12.3	18.4	82	64	11.7	17.8	78	63	11.2	17.1	9	4.2	330
29	23	-1.6	3.3	9	4.2	250	31	26	-0.4	3.7	Dublin	39690	53.43	6.25	278	85	14.55	100.3	85	67	12.1	19.5	80	65	11.4	18.5	76	64	10.8	17.6	11	4.9	230
25	19	-3.7	2.8	2	1.0	360	28	22	-2.3	3.2	Kilkenny	39600	52.67	7.27	209	64	14.59	100.6	87	68	12.4	20.2	82	66	11.7	19.1	78	65	11.1	18.3	7	3.1	180
31	25	-0.3	3.6	14	6.2	170	33	28	0.8	4.0	Malin	39800	55.37	7.33	82	25	14.66	101.0	79	64	11.3	17.5	76	62	10.8	16.8	72	61	10.3	16.2	16	7.2	200
25	19	-3.7	2.8	3	1.4	70	28	22	-2.4	3.2	Mullingar	39710	53.53	7.37	341	104	14.52	100.1	86	67	12.3	19.6	81	66	11.6	18.7	78	64	11.1	18.0	9	3.9	100
32	27	0.2	3.8	15	6.5	90	34	29	1.2	4.1	Rosslare	39570	52.25	6.33	82	25	14.66	101.0	83	65	11.8	18.3	79	64	11.3	17.5	76	62	10.8	16.8	11	5.1	220

International

HUMIDIFICATION 99.6% (35 hours below...)							99% (88 hours below...)												DEHUMIDIFICATION 0.4% (35 hours above...)				1% (88 hours above...)				2% (175 hours above...)				Summer Wind		
°F 1a	gr/lb 1b	°C 1c	g/kg 1d	mph 1e	m/s 1f	PWD 1g	°F 2a	gr/lb 2b	°C 2c	g/kg 2d	Location 3a	WMO# 3b	Lat. 3c	Long. 3d	ft. 3e	m 3f	psia 3g	kPa 3h	gr/lb 4a	°F 4b	g/kg 4c	°C 4d	gr/lb 5a	°F 5b	g/kg 5c	°C 5d	gr/lb 6a	°F 6b	g/kg 6c	°C 6d	mph 7a	m/s 7b	PWD 7c
28	22	-2.0	3.2	6	2.9	70	31	25	-0.6	3.6	Shannon	39620	52.70	8.92	65	20	14.67	101.1	85	68	12.2	20.0	81	67	11.6	19.4	78	65	11.1	18.6	9	4.1	110
31	25	-0.6	3.6	7	3.0	60	33	28	0.8	4.0	Valentia Observatory	39530	51.93	10.25	45	14	14.68	101.2	88	66	12.5	18.9	83	65	11.9	18.5	80	64	11.4	17.8	10	4.6	270
											ISRAEL																						
33	30	0.6	4.3	6	2.5	270	35	33	1.6	4.7	Jerusalem	401840	31.78	35.22	2473	754	13.43	92.6	110	74	15.7	23.6	104	72	14.8	22.4	100	71	14.3	21.9	10	4.5	290
40	37	4.2	5.2	4	1.7	150	42	40	5.6	5.7	Lod	401800	32.00	34.90	160	49	14.61	100.7	132	83	18.9	28.6	126	82	18.0	28.0	122	82	17.4	27.5	11	4.8	320
36	33	2.3	4.7	5	2.2	210	39	37	3.7	5.3	Ovda (Isr-Afb/Civ)	401980	30.00	34.83	1417	432	13.96	96.2	109	80	15.6	26.8	102	78	14.5	25.8	95	76	13.6	24.5	10	4.3	40
44	43	6.4	6.1	7	3.1	120	46	46	7.6	6.6	Tel Aviv-Yafo	401760	32.10	34.78	13	4	14.69	101.3	137	83	19.6	28.2	132	82	18.9	27.9	127	82	18.2	27.6	9	4.1	310
											ITALY																						
22	17	-5.5	2.4	1	0.5	220	25	19	-3.9	2.7	Bologna/Borgo (Afb)	161400	44.53	11.30	138	42	14.63	100.8	125	83	17.8	28.2	118	81	16.9	27.4	111	81	15.9	27.0	5	2.4	80
36	31	2.1	4.5	9	3.9	360	39	35	3.9	5.0	Brindisi	163200	40.65	17.95	32	10	14.68	101.2	149	83	21.3	28.6	141	82	20.1	28.0	133	81	19.0	27.2	10	4.5	180
35	30	1.8	4.3	5	2.3	230	37	32	3.0	4.6	Catania	164600	37.47	15.05	55	17	14.67	101.1	141	82	20.2	27.9	133	81	19.0	27.3	126	80	18.0	26.9	10	4.6	90
32	27	0.1	3.8	15	6.8	40	36	31	2.0	4.5	Genova	161200	44.42	8.85	9	3	14.69	101.3	129	82	18.4	27.6	123	81	17.6	27.0	117	79	16.7	26.3	7	3.3	50
43	41	6.1	5.9	4	2.0	310	45	45	7.3	6.4	Messina	164200	38.20	15.55	167	51	14.61	100.7	142	83	20.3	28.1	136	83	19.5	28.1	132	82	18.8	27.6	6	2.8	60
21	16	-6.0	2.3	1	0.4	90	25	19	-4.1	2.8	Milan, Linate	160800	45.43	9.28	337	103	14.52	100.1	123	81	17.6	27.4	117	80	16.7	26.4	111	78	15.9	25.8	5	2.2	220
15	12	-9.6	1.7	0	0.1	360	18	14	-7.9	2.0	Milan, Malpensa	160660	45.62	8.73	692	211	14.34	98.8	131	83	18.7	28.6	121	81	17.3	27.1	115	79	16.4	26.3	4	1.8	210
32	27	0.0	3.8	4	1.8	340	34	29	1.2	4.1	Naples	162890	40.85	14.30	236	72	14.57	100.5	142	83	20.3	28.6	133	82	19.0	27.5	125	80	17.9	26.7	8	3.4	200
44	43	6.9	6.1	11	4.8	210	46	46	7.9	6.6	Palermo	164050	38.18	13.10	111	34	14.64	100.9	149	85	21.3	29.2	142	83	20.3	28.5	137	82	19.6	27.9	11	5.1	250
24	19	-4.4	2.7	1	0.5	360	27	22	-2.9	3.1	Perugia	161810	43.08	12.50	672	205	14.35	98.9	110	79	15.7	26.0	104	77	14.9	25.1	99	76	14.1	24.3	5	2.3	270
26	20	-3.1	2.9	4	1.6	90	29	23	-1.9	3.3	Pisa	161580	43.68	10.38	3	1	14.70	101.3	125	80	17.9	26.7	118	79	16.9	26.2	113	78	16.1	25.5	9	4.0	270
30	24	-0.9	3.5	9	3.8	60	32	27	0.1	3.8	Rome	162420	41.80	12.23	9	3	14.69	101.3	142	83	20.3	28.1	136	81	19.5	27.3	131	80	18.7	26.8	12	5.2	270
21	16	-5.9	2.3	2	1.1	10	25	19	-4.1	2.7	Ronchi Legionari Ab	161080	45.82	13.48	39	12	14.68	101.2	125	80	17.9	26.9	116	79	16.6	26.0	109	78	15.6	25.6	5	2.3	220
20	16	-6.6	2.2	0	0.0	260	23	18	-5.0	2.6	Torino	160590	45.22	7.65	941	287	14.21	97.9	125	78	17.8	25.6	120	79	17.1	25.9	113	77	16.1	25.2	2	0.7	70
23	17	-4.9	2.5	4	1.8	60	26	20	-3.1	2.9	Venice	161050	45.50	12.33	19	6	14.69	101.3	132	81	18.9	27.4	123	80	17.6	26.8	116	78	16.6	25.8	6	2.7	160
											JAMAICA																						
71	115	21.9	16.4	6	2.5	330	72	119	22.3	17.0	Kingston	783970	17.93	76.78	29	9	14.68	101.2	150	85	21.4	29.5	143	85	20.4	29.3	141	85	20.1	29.2	25	11.2	110
70	111	21.3	15.8	5	2.3	140	71	115	21.9	16.4	Montego Bay	783880	18.50	77.92	9	3	14.69	101.3	148	86	21.1	29.8	142	85	20.3	29.7	141	86	20.1	29.8	21	9.5	70
											JAPAN																						
18	14	-7.5	2.0	8	3.7	230	20	15	-6.4	2.2	Aomori	475750	40.82	140.77	9	3	14.69	101.3	125	80	17.9	26.6	120	80	17.1	26.4	115	78	16.4	25.5	10	4.6	220
-2	5	-19.1	0.7	2	0.7	80	2	6	-16.4	0.9	Asahikawa	474070	43.77	142.37	380	116	14.50	99.9	118	82	16.9	27.5	112	78	16.0	25.6	106	76	15.1	24.6	6	2.9	270
29	23	-1.9	3.3	4	2.0	360	30	24	-0.9	3.5	Atsugi	476790	35.45	139.45	213	65	14.59	100.5	141	82	20.2	27.8	135	83	19.3	28.1	132	81	18.9	27.4	10	4.6	180
30	24	-1.0	3.5	8	3.4	10	32	27	0.0	3.8	Fukuoka	478080	33.58	130.45	39	12	14.68	101.2	141	83	20.2	28.5	134	82	19.1	27.6	133	82	19.0	27.9	11	5.0	10
13	11	-10.3	1.5	5	2.3	290	16	12	-8.8	1.8	Hakodate	474300	41.82	140.75	118	36	14.64	100.9	123	79	17.6	25.9	117	77	16.7	24.9	111	76	15.8	24.3	7	3.3	220
30	24	-1.2	3.5	8	3.7	10	32	27	-0.2	3.8	Hamamatsu	476810	34.75	137.70	157	48	14.62	100.7	148	81	21.2	27.4	142	81	20.3	27.0	140	81	20.0	27.1	13	5.6	270
30	24	-1.3	3.5	6	2.8	20	31	25	-0.3	3.6	Hiroshima	477650	34.40	132.47	173	53	14.61	100.7	142	84	20.3	29.0	138	83	19.7	28.3	134	82	19.2	27.9	10	4.4	220
19	14	-7.0	2.1	2	1.1	250	23	17	-5.2	2.5	Hyakuri (Jasdf)	477150	36.18	140.42	114	35	14.64	100.9	142	83	20.3	28.1	139	82	19.9	27.9	134	81	19.1	27.0	11	4.8	130
50	54	10.1	7.7	10	4.3	70	52	58	11.2	8.3	Kadena	479310	26.35	127.77	147	45	14.62	100.8	162	87	23.1	30.7	161	87	23.0	30.4	159	86	22.7	30.1	12	5.2	240
33	28	0.3	3.9	6	2.5	300	35	30	1.4	4.3	Kagoshima	478270	31.57	130.55	16	5	14.69	101.3	146	84	20.9	28.7	142	83	20.3	28.5	140	83	20.0	28.4	9	4.0	270
28	22	-2.1	3.2	3	1.4	360	30	24	-0.9	3.5	Kumamoto	478190	32.82	130.72	127	39	14.63	100.9	146	83	20.9	28.5	142	82	20.3	28.0	139	82	19.8	27.8	8	3.6	240
26	20	-3.4	2.9	7	3.3	330	28	22	-2.2	3.2	Maebashi	476240	36.40	139.07	370	113	14.50	100.0	134	82	19.1	28.0	130	83	18.6	28.1	125	82	17.9	27.6	8	3.6	110
28	22	-2.2	3.2	4	1.8	240	30	24	-1.3	3.5	Maizuru	477500	35.45	135.32	72	22	14.66	101.1	137	83	19.6	28.5	132	82	18.9	27.8	130	81	18.5	27.4	7	3.1	20
16	13	-8.9	1.9	3	1.2	30	18	15	-7.5	2.1	Matsumoto	476180	36.25	137.97	2004	611	13.67	94.2	123	79	17.6	26.0	118	78	16.9	25.4	113	77	16.2	25.1	9	4.0	170
31	25	-0.6	3.6	4	2.0	110	33	28	0.4	4.0	Matsuyama	478870	33.83	132.78	111	34	14.64	100.9	136	82	19.4	27.8	133	82	19.0	27.6	129	82	18.4	27.6	8	3.4	270

International

HUMIDIFICATION 99.6% (35 hours below...)							99% (88 hours below...)												DEHUMIDIFICATION 0.4% (35 hours above...)				1% (88 hours above...)				2% (175 hours above...)				Summer Wind		
°F 1a	gr/lb 1b	°C 1c	g/kg 1d	mph 1e	m/s 1f	PWD 1g	°F 2a	gr/lb 2b	°C 2c	g/kg 2d	Location 3a	WMO# 3b	Lat. 3c	Long. 3d	ft. 3e	m 3f	psia 3g	kPa 3h	gr/lb 4a	°F 4b	g/kg 4c	°C 4d	gr/lb 5a	°F 5b	g/kg 5c	°C 5d	gr/lb 6a	°F 6b	g/kg 6c	°C 6d	mph 7a	m/s 7b	PWD 7c
30	24	-1.2	3.5	14	6.2	260	31	25	-0.8	3.6	Miho (Civ/Jasdf)	477430	35.48	133.25	29	9	14.68	101.2	141	84	20.2	28.7	134	82	19.1	27.9	132	81	18.9	27.4	13	5.9	260
54	63	12.3	8.9	13	5.6	360	56	67	13.2	9.6	Miyako Jima Island	479270	24.78	125.28	134	41	14.63	100.8	160	85	22.8	29.5	156	84	22.3	28.9	152	84	21.7	28.8	13	5.8	210
14	11	-9.8	1.6	4	1.9	140	18	14	-8.0	2.0	Morioka	475840	39.70	141.17	515	157	14.43	99.5	130	80	18.5	26.6	125	79	17.9	26.2	120	78	17.1	25.7	9	4.2	190
33	28	0.4	4.0	5	2.4	300	35	30	1.4	4.3	Nagasaki	478170	32.73	129.87	114	35	14.64	100.9	150	84	21.4	28.8	146	83	20.8	28.1	141	82	20.2	27.7	6	2.9	230
27	21	-3.0	3.0	3	1.4	350	29	23	-1.9	3.3	Nagoya	476350	35.25	136.93	55	17	14.67	101.1	141	82	20.1	28.0	133	81	19.0	27.4	131	82	18.7	27.5	9	3.9	10
53	60	11.8	8.6	16	7.1	10	55	65	12.8	9.2	Naha	479300	26.18	127.65	26	8	14.69	101.2	159	86	22.7	29.9	151	85	21.6	29.3	151	85	21.6	29.2	13	5.8	200
49	52	9.2	7.4	7	3.1	190	50	54	10.1	7.7	Naze	479090	28.38	129.50	22	7	14.69	101.2	150	85	21.5	29.6	147	85	21.0	29.4	145	84	20.7	29.1	7	3.3	210
23	17	-5.1	2.5	5	2.1	330	25	19	-3.9	2.7	New Tokyo Intl Arpt	476860	35.77	140.38	144	44	14.62	100.8	143	82	20.4	28.0	141	82	20.1	27.8	134	81	19.1	27.2	11	4.7	10
27	21	-2.7	3.0	11	5.0	200	29	23	-1.7	3.3	Niigata	476040	37.92	139.05	22	7	14.69	101.2	135	83	19.3	28.5	130	82	18.5	27.8	125	82	17.9	27.6	10	4.5	140
28	22	-2.0	3.2	7	3.0	270	31	26	-0.8	3.7	Nyutabaru (Jasdf)	478540	32.08	131.45	269	82	14.56	100.3	149	82	21.3	27.7	143	82	20.4	27.5	141	82	20.2	27.5	12	5.5	230
30	24	-1.0	3.5	6	2.9	170	32	27	0.0	3.8	Oita	478150	33.23	131.62	42	13	14.68	101.2	141	83	20.1	28.4	136	82	19.5	28.0	133	82	19.0	27.7	9	3.9	10
28	22	-2.0	3.2	5	2.1	10	30	24	-1.0	3.5	Osaka	477710	34.78	135.45	49	15	14.67	101.1	141	83	20.1	28.3	134	81	19.2	27.3	132	82	18.9	28.0	9	4.0	10
30	24	-1.0	3.5	3	1.3	280	32	27	0.1	3.8	Owase	476630	34.07	136.20	88	27	14.65	101.0	139	83	19.8	28.1	134	82	19.2	27.6	130	81	18.6	27.1	9	4.1	70
12	10	-11.0	1.5	3	1.4	130	15	12	-9.5	1.7	Sapporo	474120	43.05	141.33	62	19	14.67	101.1	118	80	16.8	26.6	111	78	15.9	25.8	105	76	15.0	24.5	7	3.2	150
24	18	-4.6	2.6	6	2.9	350	26	20	-3.5	2.9	Sendai	475900	38.27	140.90	141	43	14.63	100.8	134	81	19.1	27.1	129	80	18.4	26.4	124	78	17.7	25.8	9	3.9	130
33	28	0.8	4.0	10	4.6	330	35	30	1.8	4.3	Shimonoseki	477620	33.95	130.93	62	19	14.67	101.1	139	83	19.9	28.3	136	82	19.4	28.0	133	82	19.0	27.7	9	4.1	100
30	24	-1.0	3.5	9	4.0	10	32	27	0.0	3.8	Shizuhama (Jasdf)	476580	34.82	138.30	32	10	14.68	101.2	150	83	21.4	28.5	143	83	20.4	28.1	140	82	20.0	28.0	13	5.7	260
31	25	-0.8	3.6	7	3.3	280	32	27	0.2	3.8	Tokyo, Intl Airport	476710	35.55	139.78	26	8	14.69	101.2	143	83	20.4	28.6	141	83	20.1	28.6	134	82	19.1	27.8	14	6.4	10
34	29	1.2	4.1	8	3.4	350	36	31	2.5	4.5	Tosashimizu	478980	32.72	133.02	108	33	14.64	100.9	151	83	21.6	28.4	147	83	21.0	28.1	145	82	20.7	27.8	8	3.4	250
11	10	-11.7	1.4	9	4.0	190	13	11	-10.3	1.5	Wakkanai	474010	45.42	141.68	36	11	14.68	101.2	113	75	16.1	23.9	106	73	15.2	22.9	100	72	14.3	22.0	10	4.5	240
											JORDAN																						
33	30	0.8	4.3	6	2.7	90	35	33	1.8	4.7	Amman	402700	31.98	35.98	2536	773	13.40	92.4	115	76	16.4	24.4	106	74	15.2	23.2	100	72	14.3	22.5	8	3.6	290
											KAZAKHSTAN																						
-3	5	-19.5	0.7	1	0.6	360	3	7	-16.1	1.0	Almaty (Alma Ata)	368700	43.23	76.93	2778	847	13.28	91.6	92	76	13.1	24.7	84	73	12.0	23.0	79	73	11.3	22.9	3	1.4	330
-21	2	-29.4	0.3	5	2.3	270	-17	2	-27.0	0.3	Aqmola (Tselinograd)	351880	51.13	71.37	1141	348	14.10	97.2	89	71	12.7	21.5	83	69	11.9	20.5	77	69	11.0	20.4	6	2.8	230
-19	2	-28.6	0.3	2	1.1	190	-14	3	-25.6	0.4	Aqtobe (Aktyubinsk)	352290	50.30	57.23	744	227	14.31	98.6	90	73	12.9	22.9	85	72	12.2	22.2	80	72	11.5	22.3	8	3.4	60
-7	4	-21.9	0.5	4	1.7	90	-3	5	-19.3	0.7	Atyrau (Gur'yev)	357000	47.02	51.85	-49	-15	14.73	101.5	104	78	14.8	25.7	96	76	13.7	24.2	89	76	12.7	24.2	10	4.4	140
-18	2	-27.6	0.3	6	2.5	360	-13	3	-25.1	0.4	Oral (Ural'sk)	351080	51.25	51.40	118	36	14.64	100.9	94	75	13.4	23.7	88	74	12.6	23.2	83	72	11.9	22.1	11	4.9	140
-24	1	-31.1	0.2	5	2.4	60	-19	2	-28.5	0.3	Pavlodar	360030	52.28	76.95	403	123	14.49	99.9	92	72	13.2	22.5	86	71	12.3	21.8	80	71	11.5	21.5	7	3.2	230
-18	2	-28.0	0.3	5	2.2	120	-13	3	-25.2	0.4	Qaraghandy (Karag.)	353940	49.80	73.13	1820	555	13.76	94.8	85	68	12.2	20.0	78	67	11.2	19.5	74	66	10.5	19.1	9	3.9	20
-21	2	-29.6	0.3	6	2.6	180	-17	2	-27.1	0.3	Qostanay (Kustanay)	289520	53.22	63.62	511	156	14.43	99.5	93	73	13.3	22.9	87	72	12.4	22.5	83	71	11.8	21.9	9	4.2	90
-25	1	-31.4	0.2	1	0.4	90	-20	2	-28.8	0.3	Semey (Semipalatinsk)	361770	50.35	80.25	643	196	14.36	99.0	90	72	12.9	22.5	86	72	12.3	22.2	81	71	11.6	21.7	6	2.8	90
-5	5	-20.6	0.7	2	0.8	180	1	6	-17.0	0.9	Zhambyl (Dzhambul)	383410	42.85	71.38	2142	653	13.60	93.7	83	72	11.8	22.2	78	72	11.1	22.2	73	72	10.4	22.3	9	3.9	20
											KENYA																						
70	113	21.1	16.1	7	3.1	180	71	117	21.8	16.7	Arissa	637230	0.47	39.63	482	147	14.45	99.6	139	83	19.8	28.1	135	81	19.3	27.3	132	80	18.9	26.9	8	3.6	180
60	89	15.8	12.7	4	1.6	90	62	96	16.4	13.7	Kisumu	637080	0.10	34.75	3759	1146	12.81	88.3	123	76	17.5	24.7	119	76	17.0	24.2	116	75	16.6	23.8	13	5.7	230
69	114	20.8	16.3	4	1.7	270	71	122	21.8	17.5	Lodwar	636120	3.12	35.62	1689	515	13.82	95.3	129	79	18.4	26.3	125	79	17.8	26.1	120	79	17.2	26.1	13	5.8	90
49	63	9.5	9.0	6	2.6	240	51	68	10.8	9.7	Nairobi	637400	1.32	36.92	5328	1624	12.08	83.3	106	67	15.2	19.6	104	66	14.9	19.0	102	66	14.6	18.7	14	6.3	60
47	60	8.1	8.6	2	0.8	350	48	63	9.0	9.0	Nakuru	637140	0.27	36.10	6236	1901	11.68	80.5	101	66	14.4	18.9	97	65	13.9	18.6	95	65	13.6	18.3	9	3.8	360
											KOREA, NORTH																						
-1	5	-18.1	0.7	3	1.2	140	4	7	-15.7	1.0	Anju	470500	39.62	125.65	88	27	14.65	101.0	136	80	19.4	26.9	130	79	18.6	26.2	125	78	17.8	25.7	5	2.4	230
7	8	-14.0	1.1	6	2.9	320	10	9	-12.0	1.3	Ch'ongjin	470080	41.78	129.82	141	43	14.63	100.8	123	77	17.5	25.2	116	75	16.5	24.1	110	74	15.7	23.2	3	1.3	90

International

HUMIDIFICATION 99.6% (35 hours below...) °F 1a	gr/lb 1b	°C 1c	g/kg 1d	mph 1e	m/s 1f	PWD 1g	99% (88 hours below...) °F 2a	gr/lb 2b	°C 2c	g/kg 2d	Location 3a	WMO# 3b	Lat. 3c	Long. 3d	ft. 3e	m 3f	psia 3g	kPa 3h	DEHUMIDIFICATION 0.4% (35 hours above...) gr/lb 4a	°F 4b	g/kg 4c	°C 4d	1% (88 hours above...) gr/lb 5a	°F 5b	g/kg 5c	°C 5d	2% (175 hours above...) gr/lb 6a	°F 6b	g/kg 6c	°C 6d	Summer Wind mph 7a	m/s 7b	PWD 7c
-19	2	-28.2	0.3	1	0.5	320	-15	3	-26.1	0.4	Changjin	470310	40.37	127.25	3546	1081	12.91	89.0	115	71	16.4	21.9	108	70	15.5	20.9	102	68	14.6	20.1	6	2.7	320
10	9	-12.2	1.3	7	3.2	320	13	11	-10.7	1.5	Haeju	470690	38.03	125.70	265	81	14.56	100.4	136	80	19.5	26.7	132	79	18.8	26.2	128	78	18.3	25.3	7	3.0	180
7	8	-13.9	1.1	7	3.2	360	10	9	-12.1	1.3	Hamhung	470410	39.93	127.55	124	38	14.63	100.9	130	81	18.5	27.2	124	79	17.7	26.1	119	77	17.0	25.2	9	4.0	230
8	8	-13.2	1.2	7	3.1	320	11	10	-11.5	1.4	Namp'o	470600	38.72	125.37	154	47	14.62	100.8	139	80	19.8	26.9	134	79	19.1	26.3	130	78	18.5	25.8	6	2.8	270
3	6	-16.2	0.9	2	1.1	110	7	8	-14.1	1.1	P'yongyang	470580	39.03	125.78	124	38	14.63	100.9	135	81	19.3	27.0	130	80	18.6	26.5	125	79	17.9	26.0	4	1.8	270
3	6	-16.0	0.9	6	2.6	50	7	8	-14.1	1.1	Sinuiju	470350	40.10	124.38	22	7	14.69	101.2	135	80	19.3	26.7	130	78	18.5	25.8	125	78	17.9	25.3	5	2.4	230
12	10	-11.0	1.5	6	2.8	270	15	12	-9.2	1.7	Wonsan	470550	39.18	127.43	118	36	14.64	100.9	134	82	19.1	27.5	126	80	18.0	26.7	120	79	17.1	26.1	4	1.6	250
											KOREA, SOUTH																						
30	24	-1.1	3.5	14	6.2	40	32	27	-0.1	3.8	Cheju	471820	33.50	126.55	87	27	14.65	101.0	160	84	22.8	29.1	151	83	21.6	28.4	148	83	21.2	28.2	14	6.3	230
12	10	-11.2	1.5	11	4.7	320	15	12	-9.5	1.7	Inch'on	471120	37.48	126.63	229	70	14.58	100.5	136	81	19.4	27.2	131	80	18.7	26.4	127	79	18.1	26.0	7	3.1	230
16	12	-8.7	1.8	12	5.2	250	20	15	-6.9	2.2	Kangnung	471050	37.75	128.90	88	27	14.65	101.0	132	82	18.8	27.9	126	82	18.0	27.6	122	80	17.4	26.7	6	2.8	90
19	14	-7.1	2.1	4	1.9	20	22	17	-5.8	2.4	Kwangju	471560	35.13	126.92	236	72	14.57	100.5	144	83	20.6	28.4	139	82	19.9	27.7	135	81	19.3	27.4	7	3.3	250
7	8	-13.9	1.1	3	1.3	10	11	10	-11.8	1.4	Osan	471220	37.08	127.03	39	12	14.68	101.2	148	86	21.1	29.8	139	83	19.9	28.4	132	82	18.9	27.6	6	2.6	10
7	8	-14.1	1.1	3	1.2	10	10	9	-12.1	1.3	Seoul	471100	37.55	126.80	62	19	14.67	101.1	150	82	21.4	28.0	142	81	20.3	27.1	134	79	19.2	26.3	9	4.2	160
17	13	-8.2	1.9	8	3.6	290	20	15	-6.7	2.2	Taegu	471430	35.88	128.62	200	61	14.59	100.6	141	85	20.1	29.3	135	83	19.3	28.3	130	82	18.6	27.9	9	3.9	270
12	10	-11.0	1.5	1	0.3	110	15	12	-9.3	1.7	Taejon	471330	36.30	127.40	255	78	14.56	100.4	141	82	20.2	27.7	136	81	19.4	27.1	131	80	18.7	26.7	6	2.8	270
20	15	-6.8	2.2	7	3.0	320	22	17	-5.4	2.4	Ulsan	471520	35.55	129.32	108	33	14.64	100.9	142	84	20.3	28.8	138	83	19.7	28.4	134	82	19.2	27.9	8	3.6	140
											KUWAIT																						
38	34	3.2	4.9	4	1.7	300	41	38	5.0	5.5	Kuwait	405820	29.22	47.98	180	55	14.60	100.7	153	92	21.8	33.3	132	87	18.8	30.5	112	85	16.0	29.3	14	6.1	340
											KYRGYZSTAN																						
-8	4	-22.4	0.6	3	1.2	150	-2	5	-18.8	0.8	Bishkek (Frunze)	383530	42.85	74.53	2083	635	13.63	93.9	92	77	13.2	25.1	87	74	12.4	23.4	83	74	11.8	23.1	8	3.4	220
-27	2	-32.6	0.3	1	0.3	360	-23	2	-30.8	0.3	Tianshan (Mtn Stn)	369820	41.92	78.23	11856	3614	9.40	64.8	57	46	8.2	7.6	53	44	7.6	6.6	50	43	7.1	5.9	11	4.7	210
											LATVIA																						
1	6	-17.1	0.8	7	3.3	30	9	9	-12.9	1.3	Liepaja	264060	56.55	21.02	26	8	14.69	101.2	90	70	12.9	21.1	85	68	12.1	19.8	80	66	11.4	18.8	9	3.8	120
-3	5	-19.6	0.7	4	2.0	40	4	7	-15.5	1.0	Riga	264220	56.97	24.07	9	3	14.69	101.3	91	71	13.0	21.8	85	69	12.2	20.5	80	68	11.4	19.9	9	4.1	150
											LIBYA																						
44	43	6.7	6.2	5	2.3	90	46	47	7.5	6.7	Banghazi	620530	32.08	20.27	433	132	14.47	99.7	134	83	19.2	28.1	127	81	18.1	27.0	123	80	17.5	26.5	15	6.6	350
39	35	4.1	5.1	4	1.7	240	41	38	5.1	5.5	Tripoli	620100	32.67	13.15	265	81	14.56	100.4	139	88	19.9	30.9	130	84	18.6	29.1	123	83	17.6	28.3	13	5.6	60
											LIECHTENSTEIN																						
12	11	-11.1	1.5	3	1.2	180	17	14	-8.6	2.0	Vaduz	69900	47.13	9.53	1519	463	13.91	95.9	94	74	13.4	23.4	90	72	12.8	22.2	85	71	12.2	21.4	10	4.5	320
											LITHUANIA																						
-4	5	-19.9	0.6	6	2.5	70	3	7	-15.9	0.9	Kaunas	266290	54.88	23.88	246	75	14.57	100.4	94	73	13.4	23.0	88	71	12.6	21.9	83	68	11.8	20.2	8	3.7	180
1	6	-17.4	0.8	8	3.4	70	8	8	-13.3	1.2	Klaipeda	265090	55.70	21.15	32	10	14.68	101.2	91	71	13.0	21.8	85	69	12.2	20.3	80	67	11.5	19.3	8	3.7	140
-5	4	-20.4	0.6	5	2.2	70	2	6	-16.7	0.9	Vilnius	267300	54.63	25.28	511	156	14.43	99.5	91	71	13.0	21.9	85	70	12.2	21.0	80	68	11.5	19.9	11	4.7	140
											MACEDONIA																						
10	9	-12.4	1.4	1	0.4	50	15	12	-9.3	1.7	Skopje	135860	41.97	21.65	784	239	14.29	98.5	94	78	13.4	25.5	88	76	12.6	24.4	86	75	12.3	24.0	4	2.0	270
											MADEIRA ISLANDS																						
53	60	11.9	8.6	8	3.6	310	55	65	12.8	9.3	Funchal	85210	32.68	16.77	180	55	14.60	100.7	111	76	15.8	24.2	105	75	15.0	23.8	102	74	14.6	23.6	11	4.9	30
											MALAYSIA																						
73	123	22.8	17.6	2	1.1	350	73	123	22.9	17.6	George Town	486010	5.30	100.27	13	4	14.69	101.3	158	85	22.6	29.6	151	84	21.6	29.1	150	84	21.5	28.9	8	3.7	270
71	115	21.6	16.4	1	0.5	190	72	119	22.2	17.0	Kota Baharu	486150	6.17	102.28	16	5	14.69	101.3	150	85	21.5	29.3	147	84	21.0	29.1	145	84	20.7	28.9	9	4.0	90
71	115	21.6	16.4	1	0.5	340	72	119	22.0	17.0	Kuala Lumpur	486470	3.12	101.55	72	22	14.66	101.1	152	85	21.7	29.4	149	84	21.3	29.0	146	84	20.8	28.7	8	3.4	270

International

HUMIDIFICATION 99.6% (35 hours below...)							99% (88 hours below...)												DEHUMIDIFICATION 0.4% (35 hours above...)				1% (88 hours above...)				2% (175 hours above...)				Summer Wind		
°F 1a	gr/lb 1b	°C 1c	g/kg 1d	mph 1e	m/s 1f	PWD 1g	°F 2a	gr/lb 2b	°C 2c	g/kg 2d	Location 3a	WMO# 3b	Lat. 3c	Long. 3d	ft. 3e	m 3f	psia 3g	kPa 3h	gr/lb 4a	°F 4b	g/kg 4c	°C 4d	gr/lb 5a	°F 5b	g/kg 5c	°C 5d	gr/lb 6a	°F 6b	g/kg 6c	°C 6d	mph 7a	m/s 7b	PWD 7c
70	111	21.1	15.9	5	2.1	350	71	115	21.5	16.4	Kuantan	486570	3.78	103.22	52	16	14.67	101.1	150	85	21.5	29.2	147	84	21.0	28.9	146	84	20.9	28.7	8	3.5	230
72	119	22.0	17.0	3	1.3	10	72	119	22.4	17.0	Malacca	486650	2.27	102.25	29	9	14.68	101.2	151	84	21.6	28.9	150	84	21.4	28.8	147	83	21.0	28.6	8	3.5	20
71	115	21.8	16.4	1	0.6	60	72	119	22.3	17.0	Sitiawan	486200	4.22	100.70	26	8	14.69	101.2	151	85	21.6	29.7	150	85	21.4	29.6	147	85	21.0	29.2	7	3.3	180
71	115	21.8	16.4	2	0.9	260	72	119	22.0	17.0	Kuching	964130	1.48	110.33	88	27	14.65	101.0	151	86	21.6	30.0	148	85	21.2	29.4	143	83	20.4	28.5	5	2.2	360
72	119	22.4	17.0	2	1.1	120	73	123	22.8	17.6	Miri	964490	4.33	113.98	59	18	14.67	101.1	155	86	22.2	29.8	152	85	21.7	29.3	150	84	21.5	29.1	9	3.9	270
											MALI																						
59	78	15.1	11.2	7	3.0	40	62	87	16.8	12.5	Bamako	612910	12.53	7.95	1250	381	14.05	96.8	147	83	21.0	28.6	140	82	20.0	27.9	139	82	19.8	27.7	9	4.0	80
											MALTA																						
44	43	6.8	6.2	6	2.6	270	46	47	7.8	6.7	Luqa	165970	35.85	14.48	298	91	14.54	100.2	134	80	19.2	26.8	127	80	18.2	26.4	125	79	17.8	26.2	9	4.1	310
											MARSHALL ISLANDS																						
76	137	24.4	19.5	12	5.5	70	77	141	24.8	20.2	Kwajalein Atoll	913660	8.73	167.73	26	8	14.69	101.2	150	86	21.4	29.8	147	85	21.0	29.5	146	85	20.9	29.3	11	4.9	70
											MAURITANIA																						
55	65	12.9	9.2	14	6.3	360	57	70	13.9	9.9	Nouadhibou	614150	20.93	17.03	9	3	14.69	101.3	126	79	18.0	26.1	118	78	16.8	25.8	112	77	16.0	24.8	14	6.3	20
55	65	12.8	9.2	9	3.8	60	57	70	13.9	9.9	Nouakchott	614420	18.10	15.95	9	3	14.69	101.3	150	84	21.5	29.1	148	84	21.1	28.8	141	83	20.2	28.5	14	6.3	80
											MEXICO																						
68	103	20.0	14.8	2	1.0	320	70	111	20.9	15.8	Acapulco	768056	16.77	99.75	16	5	14.69	101.3	151	87	21.6	30.4	150	86	21.5	30.2	150	86	21.4	29.8	17	7.4	200
57	70	13.9	9.9	4	2.0	90	60	78	15.8	11.1	Merida	766440	20.98	89.65	32	10	14.68	101.2	147	86	21.0	29.9	143	84	20.4	28.9	141	83	20.1	28.3	15	6.6	140
39	46	4.0	6.6	5	2.1	90	42	52	5.4	7.4	Mexico City	766790	19.43	99.08	7329	2234	11.20	77.2	98	65	14.0	18.4	92	64	13.2	17.8	92	63	13.1	17.3	11	4.8	360
59	75	14.8	10.7	0	0.2	10	60	78	15.6	11.1	Puerto Vallarta	766014	20.68	105.25	19	6	14.69	101.3	160	88	22.9	30.9	153	87	21.8	30.3	150	86	21.5	30.1	17	7.5	330
50	54	9.9	7.7	9	3.8	270	53	60	11.8	8.6	Tampico	765494	22.28	97.87	78	24	14.66	101.0	168	88	24.0	30.9	158	86	22.6	30.1	152	84	21.7	29.1	11	4.9	90
57	70	14.0	10.0	4	2.0	330	59	75	15.2	10.7	Veracruz	766910	19.20	96.13	45	14	14.68	101.2	152	85	21.7	29.7	150	85	21.5	29.6	148	85	21.1	29.4	22	9.6	90
											MICRONESIA																						
75	132	24.0	18.8	9	3.9	100	76	136	24.4	19.5	Truk Intl/Moen Isl	913340	7.47	151.85	7	2	14.70	101.3	151	86	21.6	29.8	148	85	21.2	29.5	147	85	21.0	29.3	9	3.9	40
											MIDWAY ISLAND																						
59	75	14.8	10.7	10	4.6	360	60	78	15.4	11.1	Midway Island NAS	910660	28.22	177.37	13	4	14.69	101.3	132	82	18.9	27.7	129	82	18.4	27.5	126	82	18.0	27.5	9	4.2	110
											MOLDOVA																						
6	8	-14.2	1.1	5	2.1	300	10	9	-12.0	1.3	Chisinau (Kishinev)	338150	47.02	28.87	590	180	14.39	99.2	98	76	14.0	24.2	92	74	13.1	23.1	87	72	12.4	22.1	6	2.8	200
											MONGOLIA																						
-23	2	-30.3	0.3	2	0.8	320	-19	2	-28.6	0.3	Ulaanbataar	442920	47.93	106.98	4317	1316	12.55	86.5	86	67	12.3	19.2	80	66	11.4	18.8	75	64	10.7	17.9	8	3.7	270
-40	1	-40.2	0.1	1	0.6	180	-37	1	-38.4	0.1	Ulaangom	442120	49.97	92.08	3070	936	13.14	90.6	80	68	11.5	20.2	75	68	10.7	20.2	70	66	10.0	19.1	5	2.1	50
											MOROCCO																						
44	43	6.9	6.1	3	1.3	180	46	46	7.8	6.6	Al Hoceima	601070	35.18	3.85	45	14	14.68	101.2	134	80	19.1	26.9	130	79	18.5	26.3	125	78	17.8	25.6	12	5.2	360
42	40	5.7	5.7	5	2.3	180	44	43	6.7	6.1	Casablanca	601550	33.57	7.67	203	62	14.59	100.6	126	78	18.0	25.3	122	77	17.4	24.8	118	76	16.8	24.3	8	3.4	360
38	35	3.1	4.9	2	0.7	160	40	37	4.2	5.3	Casablanca/Nouasser	601560	33.37	7.58	675	206	14.34	98.9	114	80	16.3	26.6	108	77	15.5	25.0	106	77	15.1	24.8	13	5.9	340
29	28	-1.7	4.0	6	2.7	260	31	30	-0.6	4.3	Midelt	601950	32.68	4.73	4970	1515	12.24	84.4	81	66	11.6	19.1	76	67	10.9	19.7	72	67	10.3	19.3	9	4.1	200
33	32	0.4	4.5	2	0.8	320	35	34	1.5	4.9	Ouarzazate	602650	30.93	6.90	3740	1140	12.82	88.4	84	72	12.0	22.1	77	71	11.0	21.5	71	71	10.1	21.8	12	5.4	240
34	30	1.0	4.3	4	1.6	240	36	33	2.2	4.7	Oujda	601150	34.78	1.93	1541	470	13.90	95.8	116	80	16.6	26.5	111	78	15.8	25.6	106	77	15.2	24.9	15	6.6	360
42	40	5.4	5.7	7	3.1	60	44	43	6.4	6.1	Safi	601850	32.28	9.23	147	45	14.62	100.8	116	78	16.6	25.8	111	77	15.9	24.8	107	75	15.3	24.1	12	5.2	20
41	38	4.8	5.4	4	1.8	100	43	41	5.9	5.9	Tanger	601010	35.73	5.90	68	21	14.66	101.1	114	79	16.3	26.1	110	78	15.7	25.3	104	77	14.9	24.9	24	10.6	80
											NETHERLANDS																						
17	13	-8.3	1.9	11	5.0	70	21	16	-6.0	2.3	Amsterdam	62400	52.30	4.77	-6	-2	14.70	101.3	94	72	13.5	22.2	90	69	12.8	20.8	85	68	12.1	19.8	11	4.9	70
14	11	-10.0	1.6	10	4.6	60	19	15	-7.0	2.1	Beek	63800	50.92	5.78	380	116	14.50	99.9	97	74	13.9	23.2	92	71	13.1	21.8	87	70	12.4	20.9	8	3.7	40

International

HUMIDIFICATION 99.6% (35 hours below...)							99% (88 hours below...)												DEHUMIDIFICATION 0.4% (35 hours above...)				1% (88 hours above...)				2% (175 hours above...)				Summer Wind		
°F 1a	gr/lb 1b	°C 1c	g/kg 1d	mph 1e	m/s 1f	PWD 1g	°F 2a	gr/lb 2b	°C 2c	g/kg 2d	Location 3a	WMO# 3b	Lat. 3c	Long. 3d	ft. 3e	m 3f	psia 3g	kPa 3h	gr/lb 4a	°F 4b	g/kg 4c	°C 4d	gr/lb 5a	°F 5b	g/kg 5c	°C 5d	gr/lb 6a	°F 6b	g/kg 6c	°C 6d	mph 7a	m/s 7b	PWD 7c
16	12	-9.1	1.8	7	3.1	50	20	15	-6.5	2.2	De Bilt	62600	52.10	5.18	13	4	14.69	101.3	92	73	13.2	22.8	88	70	12.5	21.3	83	69	11.9	20.6	8	3.7	70
16	12	-9.0	1.8	7	3.3	40	21	16	-6.2	2.3	Eindhoven	63700	51.45	5.42	72	22	14.66	101.1	92	73	13.1	22.7	87	70	12.4	21.0	83	69	11.8	20.3	9	4.1	50
15	12	-9.7	1.7	9	3.8	20	20	15	-6.9	2.2	Gilze/Rijen	63500	51.57	4.93	42	13	14.68	101.2	92	72	13.1	22.4	87	69	12.4	20.8	83	68	11.8	20.2	10	4.3	70
14	11	-10.1	1.6	7	3.0	50	18	14	-7.6	2.0	Groningen	62800	53.13	6.58	13	4	14.69	101.3	95	74	13.6	23.1	90	70	12.9	21.1	85	68	12.1	20.0	9	3.9	100
16	12	-8.8	1.8	8	3.4	80	20	15	-6.7	2.2	Leeuwarden	62700	53.22	5.75	6	2	14.70	101.3	90	70	12.9	21.3	85	68	12.2	20.2	81	67	11.6	19.3	11	4.7	80
17	13	-8.3	1.9	9	4.1	50	21	16	-5.9	2.3	Rotterdam	63440	51.95	4.45	-13	-4	14.71	101.4	96	73	13.7	22.7	90	71	12.9	21.5	85	68	12.2	20.2	10	4.4	90
											NETHERLANDS ANTILLES																						
74	128	23.3	18.3	10	4.6	100	75	133	23.9	19.0	Willemstad	789880	12.20	68.97	219	67	14.58	100.5	153	86	21.9	30.1	153	86	21.8	30.0	150	86	21.5	29.9	18	7.9	80
											NEW CALEDONIA																						
61	81	16.1	11.6	6	2.9	60	62	84	16.7	12.0	Noumea	915920	22.27	166.45	236	72	14.57	100.5	143	83	20.4	28.1	139	81	19.8	27.3	134	80	19.2	26.9	12	5.3	80
											NEW ZEALAND																						
35	30	1.8	4.3	10	4.6	240	37	32	2.8	4.6	Auckland	931190	37.02	174.80	19	6	14.69	101.3	104	72	14.9	22.3	99	71	14.2	21.8	94	70	13.5	21.1	13	5.9	20
28	22	-2.2	3.2	1	0.6	280	30	24	-1.2	3.5	Christchurch	937800	43.48	172.55	111	34	14.64	100.9	83	67	11.8	19.4	78	67	11.2	19.3	74	65	10.6	18.4	16	7.0	300
38	34	3.1	4.9	19	8.3	240	39	35	3.7	5.1	Taiaroa Head	938960	45.77	170.73	249	76	14.57	100.4	76	62	10.9	16.7	74	61	10.5	16.2	71	60	10.1	15.8	17	7.8	320
35	30	1.8	4.3	14	6.3	10	36	31	2.0	4.5	Wellington	934360	41.33	174.80	22	7	14.69	101.2	90	69	12.9	20.3	86	67	12.3	19.7	83	67	11.8	19.3	17	7.7	360
											NIGER																						
51	59	10.3	8.4	7	3.1	100	53	64	11.7	9.1	Agadez	610240	16.97	7.98	1646	502	13.85	95.4	122	81	17.4	27.4	116	82	16.6	27.5	111	82	15.8	27.9	11	4.8	120
60	80	15.5	11.4	6	2.8	40	62	86	16.8	12.2	Niamey	610520	13.48	2.17	744	227	14.31	98.6	142	85	20.3	29.3	138	84	19.7	28.9	135	84	19.3	28.9	8	3.7	40
											NORWAY																						
16	12	-9.0	1.8	3	1.5	60	20	15	-6.8	2.2	Bergen	13110	60.30	5.22	164	50	14.61	100.7	74	61	10.6	16.0	70	60	10.0	15.4	66	59	9.5	14.8	8	3.5	240
9	9	-12.8	1.3	19	8.6	80	13	11	-10.8	1.5	Bodo	11520	67.27	14.37	42	13	14.68	101.2	69	61	9.9	16.0	65	59	9.3	15.2	62	58	8.8	14.2	11	5.0	100
0	6	-18.0	0.8	2	0.7	360	5	7	-14.9	1.0	Oslo/Fornebu	14880	59.90	10.62	55	17	14.67	101.1	83	67	11.8	19.5	78	66	11.2	18.8	74	65	10.5	18.1	7	3.3	180
-8	4	-22.0	0.5	2	1.1	30	-2	5	-18.9	0.7	Oslo/Gardermoen	13840	60.20	11.08	669	204	14.35	98.9	76	63	10.8	17.1	71	62	10.1	16.8	66	61	9.4	15.9	7	3.3	180
13	11	-10.3	1.5	3	1.5	150	18	14	-7.9	2.0	Stavanger	14150	58.88	5.63	29	9	14.68	101.2	77	64	11.0	17.5	74	62	10.5	16.8	69	61	9.9	16.0	12	5.2	320
28	22	-2.5	3.2	14	6.2	140	29	23	-1.4	3.3	Svinoy (Lgt-H)	12050	62.33	5.27	134	41	14.63	100.8	71	60	10.2	15.6	68	59	9.7	15.0	65	58	9.3	14.4	13	5.6	150
6	8	-14.2	1.1	2	1.1	170	10	9	-12.5	1.3	Tromso	10250	69.68	18.92	32	10	14.68	101.2	66	61	9.4	16.0	62	59	8.8	14.8	57	57	8.2	13.8	8	3.4	180
-1	5	-18.1	0.7	9	3.8	120	6	8	-14.2	1.1	Trondheim	12710	63.47	10.93	55	17	14.67	101.1	80	66	11.4	18.7	75	64	10.7	17.5	70	61	10.0	16.3	10	4.4	260
23	18	-4.8	2.5	14	6.1	90	27	21	-2.8	3.0	Utsira	14030	59.30	4.88	183	56	14.60	100.7	76	62	10.8	16.5	73	60	10.4	15.6	70	59	10.0	14.9	11	5.0	120
											OMAN																						
61	81	16.1	11.5	5	2.1	200	63	87	17.0	12.4	Masqat	412560	23.58	58.28	49	15	14.67	101.1	181	91	25.8	32.8	174	91	24.9	32.6	169	90	24.2	32.4	11	5.0	340
63	87	17.4	12.4	10	4.4	20	65	93	18.3	13.3	Salalah	413160	17.03	54.08	65	20	14.67	101.1	160	86	22.9	30.1	158	86	22.5	29.9	152	85	21.7	29.5	12	5.2	200
48	52	9.0	7.5	7	3.1	160	51	59	10.8	8.4	Thamarit	413140	17.67	54.03	1459	445	13.94	96.1	142	86	20.3	29.8	134	84	19.2	28.8	130	82	18.6	28.0	11	4.8	340
63	87	17.1	12.4	14	6.3	300	65	93	18.4	13.3	Tur'at Masirah	412880	20.67	58.90	62	19	14.67	101.1	168	88	24.0	31.0	160	86	22.9	30.1	156	86	22.3	29.8	13	5.7	210
											PANAMA																						
73	123	22.8	17.6	2	0.8	10	73	123	22.9	17.6	Panama	788060	8.92	79.60	52	16	14.67	101.1	158	86	22.6	30.1	152	85	21.7	29.4	150	85	21.5	29.6	12	5.2	10
68	103	19.8	14.8	0	0.2	300	68	103	20.2	14.8	Tocumen	787920	9.05	79.37	36	11	14.68	101.2	150	85	21.5	29.6	148	85	21.1	29.5	143	84	20.4	28.8	10	4.3	30
											PARAGUAY																						
41	38	4.9	5.5	2	1.0	180	44	43	6.9	6.2	Asuncion	862180	25.27	57.63	331	101	14.52	100.1	144	86	20.5	30.0	136	83	19.4	28.6	134	83	19.2	28.5	14	6.2	360
											PERU																						
42	54	5.3	7.7	6	2.8	30	43	56	6.1	8.0	Arequipa	847520	16.32	71.55	8267	2520	10.81	74.5	90	64	12.8	17.6	85	62	12.1	16.8	83	61	11.8	16.2	14	6.3	240
32	40	-0.2	5.7	0	0.0	90	34	43	0.9	6.1	Cuzco	846860	13.55	71.98	10659	3249	9.85	67.9	80	61	11.5	16.1	76	60	10.9	15.3	74	59	10.6	15.2	4	2.0	330
66	98	19.0	14.0	2	1.1	170	68	105	20.1	15.0	Iquitos	843770	3.75	73.25	413	126	14.48	99.8	153	87	21.8	30.5	151	87	21.6	30.5	147	87	21.0	30.4	4	1.7	330

HUMIDIFICATION 99.6% (35 hours below...)							99% (88 hours below...)				International								DEHUMIDIFICATION 0.4% (35 hours above...)				1% (88 hours above...)				2% (175 hours above...)				Summer Wind		
°F 1a	gr/lb 1b	°C 1c	g/kg 1d	mph 1e	m/s 1f	PWD 1g	°F 2a	gr/lb 2b	°C 2c	g/kg 2d	Location 3a	WMO# 3b	Lat. 3c	Long. 3d	ft. 3e	m 3f	psia 3g	kPa 3h	gr/lb 4a	°F 4b	g/kg 4c	°C 4d	gr/lb 5a	°F 5b	g/kg 5c	°C 5d	gr/lb 6a	°F 6b	g/kg 6c	°C 6d	mph 7a	m/s 7b	PWD 7c
57	70	13.9	10.0	4	1.9	170	58	72	14.5	10.3	Lima	846280	12.00	77.12	42	13	14.68	101.2	126	80	18.0	26.8	124	80	17.7	26.8	118	79	16.8	26.3	13	5.9	170
53	60	11.9	8.6	1	0.3	90	55	65	12.8	9.2	Pisco	846910	13.75	76.28	22	7	14.69	101.2	123	83	17.5	28.2	117	81	16.7	27.0	111	79	15.9	26.1	11	5.0	210
60	78	15.8	11.2	22	9.8	150	61	81	16.0	11.6	Talara	843900	4.57	81.25	295	90	14.54	100.2	143	82	20.4	28.0	139	82	19.9	27.5	134	81	19.2	27.2	15	6.6	190
											PHILLIPPINES																						
68	106	19.8	15.1	5	2.2	10	69	110	20.8	15.6	Angeles, Clark Afb	983270	15.18	120.55	643	196	14.36	99.0	164	86	23.4	30.2	161	86	23.0	30.0	154	85	22.0	29.3	7	3.1	990
52	69	11.3	9.9	3	1.3	90	54	75	12.3	10.7	Baguio	983280	16.42	120.60	4924	1501	12.27	84.5	145	77	20.7	25.1	135	75	19.3	24.1	130	74	18.5	23.6	4	1.6	140
73	123	22.9	17.6	5	2.3	40	74	128	23.4	18.2	Cebu/Mandaue	986460	10.30	123.97	78	24	14.66	101.0	153	87	21.9	30.6	152	87	21.7	30.5	150	87	21.5	30.4	9	4.0	40
70	111	21.0	15.9	4	1.7	70	71	115	21.9	16.4	Olongapo	984260	14.80	120.27	56	17	14.67	101.1	160	88	22.9	30.9	152	86	21.7	30.0	150	86	21.5	29.8	11	4.7	70
69	107	20.4	15.3	2	0.7	90	71	115	21.5	16.4	Manila, Aquino Apt.	984290	14.52	121.00	68	21	14.66	101.1	161	89	23.0	31.5	158	88	22.5	31.1	152	87	21.7	30.4	21	9.5	90
											POLAND																						
-4	5	-20.1	0.6	2	0.9	310	3	7	-16.2	0.9	Bialystok	122950	53.10	23.17	495	151	14.44	99.5	97	74	13.9	23.3	90	71	12.9	21.6	84	69	12.0	20.4	6	2.8	180
1	6	-17.2	0.8	2	0.8	130	9	9	-12.9	1.3	Gdansk	121500	54.38	18.47	452	138	14.46	99.7	91	70	13.0	21.0	82	68	11.7	20.1	77	67	11.0	19.3	11	5.1	10
3	7	-15.9	1.0	2	1.1	20	9	9	-12.6	1.3	Katowice	125600	50.23	19.03	931	284	14.21	98.0	94	72	13.4	22.1	88	71	12.6	21.4	85	69	12.1	20.8	8	3.4	250
-1	5	-18.5	0.8	4	1.9	60	6	8	-14.7	1.1	Kielce	125700	50.82	20.70	856	261	14.25	98.2	94	73	13.4	22.7	90	71	12.8	21.5	84	69	12.0	20.6	7	3.2	190
10	9	-12.2	1.3	5	2.3	110	16	12	-8.7	1.8	Kolobrzeg	121000	54.18	15.58	16	5	14.69	101.3	90	70	12.9	21.1	85	68	12.2	20.1	80	67	11.5	19.5	6	2.5	140
-1	5	-18.2	0.8	3	1.3	60	6	8	-14.4	1.1	Krakow	125660	50.08	19.80	777	237	14.29	98.5	99	76	14.1	24.2	93	73	13.3	22.8	88	71	12.6	21.6	6	2.5	240
2	6	-16.8	0.9	5	2.3	90	9	9	-13.0	1.3	Lodz	124650	51.73	19.40	616	188	14.38	99.1	92	72	13.1	22.1	86	71	12.3	21.8	82	70	11.7	20.9	8	3.6	130
-2	5	-18.8	0.7	5	2.2	180	5	7	-14.9	1.0	Lublin	124950	51.22	22.40	787	240	14.29	98.5	96	74	13.7	23.5	90	72	12.9	22.0	85	70	12.2	21.1	5	2.3	220
3	7	-15.9	0.9	3	1.3	90	11	10	-11.7	1.4	Poznan	123300	52.42	16.83	301	92	14.54	100.2	92	72	13.1	22.1	86	71	12.3	21.6	80	69	11.5	20.3	8	3.4	220
1	6	-17.2	0.9	5	2.4	270	7	8	-13.8	1.2	Przemysl	126950	49.80	22.77	918	280	14.22	98.0	98	74	14.0	23.6	92	72	13.2	22.0	88	70	12.5	21.2	6	2.7	250
-3	6	-19.5	0.8	36	16.1	340	2	7	-16.6	1.1	Snezka	125100	50.73	15.73	5291	1613	12.10	83.4	79	58	11.3	14.7	74	57	10.5	13.7	69	55	9.9	12.8	13	5.6	200
-5	4	-20.7	0.6	3	1.5	20	2	6	-16.8	0.9	Suwalki	121950	54.13	22.95	610	186	14.38	99.1	94	73	13.5	23.0	88	70	12.6	21.3	83	68	11.8	20.2	7	3.3	300
7	8	-13.7	1.1	4	2.0	40	13	11	-10.3	1.5	Szczecin	122050	53.40	14.62	9	3	14.69	101.3	96	74	13.7	23.6	90	72	12.8	22.2	84	70	12.0	21.1	9	3.9	220
1	6	-17.0	0.8	4	1.9	30	9	9	-13.0	1.3	Torun	122500	53.03	18.58	236	72	14.57	100.5	92	72	13.2	22.0	86	71	12.3	21.4	80	70	11.5	20.9	7	3.1	110
0	6	-17.5	0.8	5	2.1	90	8	8	-13.4	1.2	Warsaw	123750	52.17	20.97	351	107	14.51	100.0	97	76	13.9	24.2	91	72	13.0	22.5	86	71	12.3	21.4	9	3.9	150
2	6	-16.7	0.9	4	1.8	110	10	9	-12.4	1.3	Wroclaw	124240	51.10	16.88	396	121	14.49	99.9	94	73	13.4	22.9	88	71	12.5	21.8	83	71	11.9	21.5	8	3.5	170
											PORTUGAL																						
36	32	2.1	4.6	7	3.3	90	38	35	3.4	5.0	Beja	85620	38.02	7.87	810	247	14.27	98.4	94	74	13.4	23.2	89	74	12.7	23.1	85	73	12.2	22.7	10	4.4	180
26	22	-3.6	3.1	2	1.1	180	28	24	-2.4	3.4	Braganca	85750	41.80	6.73	2270	692	13.53	93.3	85	72	12.2	22.2	81	70	11.6	21.1	78	69	11.1	20.8	8	3.5	240
35	30	1.9	4.3	4	1.8	180	38	34	3.2	4.9	Coimbra	85490	40.20	8.42	459	140	14.46	99.7	100	79	14.3	26.3	96	76	13.7	24.4	92	74	13.2	23.3	6	2.8	310
37	34	2.7	4.8	11	4.7	320	39	36	4.0	5.2	Evora	85570	38.57	7.90	1053	321	14.15	97.5	93	70	13.3	21.3	89	71	12.7	21.6	85	69	12.1	20.8	8	3.4	300
41	38	4.8	5.4	5	2.3	20	43	41	5.9	5.9	Faro	85540	37.02	7.97	13	4	14.69	101.3	113	77	16.1	25.1	108	77	15.5	24.8	103	76	14.7	24.2	11	4.8	110
39	36	4.0	5.1	4	2.0	50	41	39	5.1	5.5	Lisbon	85360	38.78	9.13	403	123	14.49	99.9	106	76	15.1	24.4	103	76	14.7	24.5	97	75	13.9	23.7	11	4.9	330
34	31	1.3	4.4	11	4.7	290	37	35	2.7	5.0	Portalegre	85710	39.28	7.42	1935	590	13.70	94.4	90	69	12.8	20.8	85	69	12.1	20.8	81	69	11.6	20.4	8	3.7	240
35	30	1.8	4.3	7	3.2	90	37	33	2.9	4.7	Porto	85450	41.23	8.68	239	73	14.57	100.5	98	72	14.0	22.0	93	70	13.3	20.9	92	69	13.1	20.6	9	4.0	330
33	28	0.4	4.0	1	0.5	50	35	30	1.5	4.3	Viana Do Castelo	85430	41.70	8.80	59	18	14.67	101.1	100	76	14.3	24.6	96	74	13.7	23.1	92	72	13.2	22.3	6	2.6	160
											QATAR																						
51	56	10.3	8.0	7	3.2	290	53	60	11.6	8.6	Ad Dawhah	411700	25.25	51.57	32	10	14.68	101.2	184	92	26.3	33.2	180	91	25.7	33.0	171	92	24.4	33.1	15	6.9	350
											ROMANIA																						
8	8	-13.5	1.2	3	1.3	250	14	11	-10.2	1.6	Bucharest	154200	44.50	26.13	298	91	14.54	100.2	118	78	16.8	25.7	109	77	15.6	24.8	104	75	14.8	23.7	5	2.2	230
4	7	-15.6	1.0	2	0.9	270	8	9	-13.5	1.2	Cluj-Napoca	151200	46.78	23.57	1354	413	13.99	96.5	107	75	15.3	24.1	99	73	14.1	22.6	93	71	13.3	21.5	6	2.6	140
15	12	-9.7	1.7	12	5.2	360	19	14	-7.1	2.1	Constanta	154800	44.22	28.63	45	14	14.68	101.2	126	78	18.0	25.6	118	77	16.9	25.0	113	76	16.1	24.2	8	3.7	180

International

HUMIDIFICATION 99.6% (35 hours below...)							99% (88 hours below...)												DEHUMIDIFICATION 0.4% (35 hours above...)				1% (88 hours above...)				2% (175 hours above...)				Summer Wind		
°F 1a	gr/lb 1b	°C 1c	g/kg 1d	mph 1e	m/s 1f	PWD 1g	°F 2a	gr/lb 2b	°C 2c	g/kg 2d	Location 3a	WMO# 3b	Lat. 3c	Long. 3d	ft. 3e	m 3f	psia 3g	kPa 3h	gr/lb 4a	°F 4b	g/kg 4c	°C 4d	gr/lb 5a	°F 5b	g/kg 5c	°C 5d	gr/lb 6a	°F 6b	g/kg 6c	°C 6d	mph 7a	m/s 7b	PWD 7c
10	9	-12.3	1.3	5	2.2	270	15	12	-9.6	1.7	Craiova	154500	44.23	23.87	639	195	14.36	99.0	127	83	18.1	28.6	118	80	16.8	26.8	110	79	15.7	25.9	4	1.9	180
7	8	-13.7	1.1	9	3.9	20	12	10	-10.9	1.5	Galati	153100	45.50	28.02	236	72	14.57	100.5	116	80	16.5	26.6	108	77	15.4	24.9	102	75	14.5	24.0	8	3.7	230
-13	4	-25.0	0.5	33	14.8	230	-7	5	-21.7	0.7	Omul Mountain	152800	45.45	25.45	8231	2509	10.82	74.6	74	54	10.6	12.0	69	52	9.9	11.0	64	49	9.2	9.7	6	2.8	230
0	6	-17.8	0.8	2	0.9	90	6	8	-14.4	1.1	Satu Mare	150100	47.78	22.88	406	124	14.49	99.8	106	78	15.1	25.6	100	76	14.3	24.6	94	74	13.5	23.4	5	2.4	230
9	9	-12.7	1.3	4	1.7	360	14	11	-9.9	1.6	Timisoara	152470	45.77	21.25	288	88	14.55	100.3	100	75	14.3	23.8	96	74	13.7	23.2	91	73	13.0	22.7	6	2.5	200
											RUSSIA																						
-29	1	-33.9	0.2	1	0.3	350	-24	2	-31.1	0.2	Abakan	298650	53.75	91.40	803	245	14.28	98.4	90	71	12.8	21.6	85	70	12.1	21.0	80	69	11.4	20.4	5	2.2	50
-41	1	-40.6	0.1	2	0.7	200	-37	1	-38.1	0.1	Aldan	310040	58.62	125.37	2237	682	13.55	93.4	83	67	11.8	19.7	77	66	11.0	18.9	72	65	10.3	18.6	5	2.3	180
-17	2	-27.2	0.3	7	3.0	130	-13	3	-25.1	0.4	Aleksandrovsk-Sahal	320610	50.90	142.17	102	31	14.65	101.0	92	69	13.1	20.8	85	67	12.2	19.5	80	65	11.5	18.4	12	5.5	220
-37	1	-38.5	0.1	14	6.1	320	-34	1	-36.7	0.1	Anadyr'	255630	64.78	177.57	203	62	14.59	100.6	62	60	8.9	15.8	58	58	8.3	14.7	55	56	7.8	13.3	9	4.2	140
-18	2	-27.7	0.3	15	6.5	60	-14	3	-25.8	0.4	Apuka	259560	60.45	169.58	26	8	14.69	101.2	62	57	8.9	13.9	60	56	8.5	13.3	56	54	8.0	12.5	12	5.4	270
-29	1	-34.1	0.2	2	0.9	130	-23	2	-30.3	0.2	Arkhangel'sk	225500	64.53	40.47	42	13	14.68	101.2	90	72	12.9	22.3	83	69	11.9	20.5	77	67	11.0	19.6	6	2.8	140
4	7	-15.3	1.0	2	1.0	140	10	9	-12.5	1.3	Armavir	370310	44.98	41.12	524	160	14.42	99.4	111	79	15.9	26.3	104	77	14.8	24.8	98	75	14.0	23.8	6	2.5	90
-1	5	-18.4	0.7	7	3.0	270	4	7	-15.7	1.0	Astrakhan'	348800	46.27	48.03	59	18	14.67	101.1	113	79	16.1	26.1	105	78	15.0	25.4	99	76	14.1	24.5	11	4.9	90
-21	2	-29.6	0.3	7	3.1	170	-16	2	-26.7	0.3	Barnaul	298380	53.40	83.70	826	252	14.27	98.3	93	73	13.3	22.7	88	72	12.6	22.0	83	70	11.9	21.0	9	4.1	60
-27	1	-32.8	0.2	2	0.7	310	-23	2	-30.5	0.2	Blagoveshchensk	315100	50.25	127.50	449	137	14.46	99.7	116	77	16.6	25.1	109	76	15.6	24.2	104	74	14.8	23.2	7	3.1	180
-37	1	-38.1	0.1	3	1.4	80	-32	1	-35.7	0.1	Borzya	309650	50.38	116.52	2244	684	13.55	93.4	97	72	13.9	22.4	92	70	13.1	21.3	87	69	12.4	20.3	10	4.3	150
-32	1	-35.4	0.1	3	1.3	280	-27	1	-32.9	0.2	Bratsk	303090	56.07	101.83	1604	489	13.87	95.6	90	70	12.8	21.1	84	69	12.0	20.4	78	67	11.2	19.7	7	3.3	130
-8	4	-22.1	0.5	7	3.0	110	-3	5	-19.3	0.7	Bryansk	268980	53.33	34.23	711	217	14.33	98.7	97	72	13.8	22.4	90	71	12.9	21.6	85	69	12.2	20.4	10	4.5	220
-19	2	-28.3	0.3	7	3.1	340	-15	3	-26.2	0.4	Chelyabinsk	286420	55.30	61.53	744	227	14.31	98.6	94	74	13.5	23.1	90	72	12.8	22.0	85	70	12.1	21.3	12	5.2	190
-25	1	-31.5	0.2	2	0.9	20	-17	2	-27.4	0.3	Cherepovets	271130	59.12	37.93	429	131	14.47	99.8	97	74	13.9	23.1	90	70	12.9	21.2	83	68	11.9	20.0	6	2.9	170
-33	1	-35.9	0.1	0	0.0	310	-29	1	-33.7	0.2	Chita	307580	52.02	113.33	2247	685	13.54	93.4	98	74	14.0	23.1	91	70	13.0	21.2	85	68	12.2	19.8	8	3.5	210
-50	0	-45.4	0.0	6	2.5	100	-44	0	-42.5	0.1	Dudinka	230740	69.40	86.17	62	19	14.67	101.1	77	69	11.0	20.6	69	66	9.9	19.1	64	63	9.1	17.0	10	4.4	60
-34	1	-36.6	0.1	2	0.7	190	-29	1	-33.7	0.2	Egvekinot	253780	66.35	179.12	85	26	14.65	101.0	56	57	8.0	13.9	53	54	7.6	12.2	50	52	7.2	11.2	8	3.6	160
5	7	-14.8	1.0	3	1.3	270	10	9	-12.4	1.3	Groznyy	372350	43.35	45.68	531	162	14.42	99.4	108	79	15.4	25.9	102	77	14.6	25.0	99	75	14.1	24.1	9	4.2	90
-22	2	-29.8	0.2	4	1.7	200	-18	2	-28.0	0.3	Habarovsk/Novy	317350	48.52	135.17	236	72	14.57	100.5	115	76	16.4	24.7	108	76	15.5	24.2	102	73	14.5	23.0	9	3.8	250
-29	1	-33.7	0.2	4	1.7	80	-24	2	-31.1	0.2	Irkutsk	307100	52.27	104.35	1683	513	13.83	95.3	92	70	13.1	21.1	86	68	12.3	19.9	80	67	11.5	19.2	8	3.6	190
-21	2	-29.7	0.3	5	2.2	100	-16	2	-26.6	0.3	Izhevsk	284110	56.82	53.27	518	158	14.43	99.4	94	74	13.5	23.6	89	72	12.7	22.1	84	70	12.0	21.3	11	4.8	160
10	9	-12.1	1.3	13	6.0	320	13	11	-10.6	1.5	Juzno-Kurilsk	321650	44.02	145.87	131	40	14.63	100.8	94	67	13.5	19.3	90	65	12.8	18.4	86	64	12.3	17.7	8	3.5	320
-11	3	-24.0	0.4	3	1.5	360	-8	4	-22.2	0.5	Juzno-Sahalinsk	321500	46.92	142.73	102	31	14.65	101.0	106	73	15.1	22.9	99	71	14.1	21.5	91	69	13.0	20.4	7	3.3	180
-3	5	-19.2	0.7	4	1.8	350	6	8	-14.4	1.1	Kaliningrad	267020	54.70	20.62	88	27	14.65	101.0	91	72	13.0	22.4	85	70	12.1	21.2	79	68	11.3	20.0	5	2.4	120
-12	3	-24.5	0.4	3	1.5	340	-7	4	-21.6	0.6	Kaluga	277030	54.57	36.37	659	201	14.35	98.9	97	74	13.8	23.6	91	72	13.0	22.1	85	69	12.2	20.6	9	3.8	130
-18	2	-27.6	0.3	9	4.2	330	-12	3	-24.7	0.4	Kazan'	275950	55.78	49.18	380	116	14.50	99.9	98	75	14.0	23.8	91	73	13.0	22.9	85	71	12.2	21.6	10	4.4	170
-27	1	-32.7	0.2	7	3.0	250	-18	2	-27.9	0.3	Kirov	271960	58.65	49.62	482	147	14.45	99.6	97	75	13.9	24.0	90	72	12.9	22.2	85	70	12.1	21.0	9	3.9	90
-37	1	-38.5	0.1	3	1.4	290	-30	1	-34.5	0.1	Kolpashevo	292310	58.30	82.88	249	76	14.57	100.4	97	73	13.8	23.0	90	72	12.9	22.0	84	69	12.0	20.8	7	3.2	160
3	6	-15.9	0.9	7	3.0	50	9	9	-13.0	1.3	Krasnodar	349290	45.03	39.15	108	33	14.64	100.9	111	79	15.9	26.2	106	78	15.1	25.4	100	77	14.3	24.8	8	3.6	90
-29	1	-33.8	0.2	3	1.5	290	-24	2	-31.2	0.2	Krasnoyarsk	295740	56.00	92.88	908	277	14.22	98.0	93	72	13.3	22.1	88	70	12.5	21.1	82	69	11.7	20.3	6	2.9	50
-25	1	-31.6	0.2	5	2.4	30	-19	2	-28.6	0.3	Kurgan	286610	55.47	65.40	259	79	14.56	100.4	96	75	13.7	23.8	90	73	12.9	22.8	85	72	12.2	22.1	10	4.3	200
-9	3	-22.7	0.5	5	2.3	360	-3	5	-19.6	0.7	Kursk	340090	51.73	36.27	688	210	14.34	98.8	99	74	14.1	23.4	92	72	13.1	22.4	87	71	12.4	21.4	9	4.2	90
-21	2	-29.4	0.3	0	0.1	150	-18	2	-27.9	0.3	Kyakhta	309250	50.37	106.45	2627	801	13.36	92.1	89	69	12.7	20.7	83	68	11.8	20.0	78	67	11.1	19.2	6	2.5	180
-20	2	-28.7	0.3	1	0.6	50	-15	2	-26.0	0.4	Magadan	259130	59.58	150.78	387	118	14.50	99.9	64	58	9.1	14.2	61	56	8.7	13.4	58	55	8.3	13.0	9	3.9	270
-19	2	-28.5	0.3	1	0.6	30	-15	3	-25.9	0.4	Magnitogorsk	288380	53.35	59.08	1253	382	14.05	96.8	92	73	13.2	23.0	86	72	12.3	22.0	80	70	11.5	21.0	9	4.2	170

International

HUMIDIFICATION 99.6% (35 hours below...)							99% (88 hours below...)												DEHUMIDIFICATION 0.4% (35 hours above...)				1% (88 hours above...)				2% (175 hours above...)				Summer Wind		
°F 1a	gr/lb 1b	°C 1c	g/kg 1d	mph 1e	m/s 1f	PWD 1g	°F 2a	gr/lb 2b	°C 2c	g/kg 2d	Location 3a	WMO# 3b	Lat. 3c	Long. 3d	ft. 3e	m 3f	psia 3g	kPa 3h	gr/lb 4a	°F 4b	g/kg 4c	°C 4d	gr/lb 5a	°F 5b	g/kg 5c	°C 5d	gr/lb 6a	°F 6b	g/kg 6c	°C 6d	mph 7a	m/s 7b	PWD 7c
-54	0	-47.5	0.0	0	0.2	200	-49	0	-45.2	0.0	Markovo	255510	64.68	170.42	108	33	14.64	100.9	71	67	10.1	19.2	66	63	9.4	17.4	62	62	8.8	16.5	5	2.2	180
-10	3	-23.1	0.5	3	1.5	20	-4	5	-20.1	0.6	Moscow	276120	55.75	37.63	511	156	14.43	99.5	96	73	13.7	22.9	90	72	12.9	22.1	85	70	12.2	21.2	4	2.0	210
-11	3	-24.0	0.4	6	2.6	10	-6	4	-21.1	0.6	Moscow, Vnukovo	275185	55.65	37.27	666	203	14.35	98.9	94	74	13.4	23.2	88	69	12.6	20.6	83	68	11.8	20.2	9	4.1	240
-20	2	-28.7	0.3	10	4.6	210	-12	3	-24.4	0.4	Murmansk	221130	68.97	33.05	167	51	14.61	100.7	70	64	10.0	18.0	64	63	9.1	17.1	60	60	8.6	15.7	10	4.3	150
-28	1	-33.1	0.2	4	1.6	290	-23	2	-30.8	0.2	Nikolayevsk	313690	53.15	140.70	223	68	14.58	100.5	95	73	13.6	22.6	90	71	12.9	21.5	85	68	12.1	20.2	11	4.7	120
12	10	-11.0	1.5	14	6.1	360	15	12	-9.4	1.7	Nikolskoe/Beringa	326180	55.20	165.98	20	6	14.69	101.3	62	56	8.9	13.3	59	54	8.4	12.3	56	53	8.0	11.8	10	4.6	170
-17	2	-27.2	0.3	4	1.8	350	-11	3	-23.8	0.4	Nizhniy Novgorod	275530	56.22	43.82	269	82	14.56	100.3	98	75	14.0	23.8	92	72	13.1	22.2	85	71	12.2	21.6	9	4.1	150
-26	1	-32.2	0.2	5	2.4	20	-20	2	-28.9	0.3	Nizhniy Tagil	282400	57.88	60.07	846	258	14.26	98.3	93	72	13.3	22.2	88	71	12.6	21.4	83	68	11.8	20.0	8	3.6	250
-23	2	-30.8	0.2	4	1.8	340	-17	2	-27.4	0.3	Novokuznetsk	298460	53.73	87.18	1010	308	14.17	97.7	93	72	13.3	22.5	88	71	12.5	21.7	83	70	11.8	20.9	7	3.1	150
-25	1	-31.7	0.2	7	3.0	220	-19	2	-28.4	0.3	Novosibirsk	296340	55.03	82.90	580	177	14.39	99.2	97	72	13.9	22.3	91	71	13.0	21.4	85	70	12.2	21.0	10	4.5	180
-63	0	-52.9	0.0	2	0.8	20	-58	0	-50.1	0.0	Nyurba	246390	63.28	118.33	423	129	14.48	99.8	92	73	13.1	22.8	84	71	12.0	21.4	76	68	10.9	19.9	6	2.9	20
-54	0	-47.7	0.0	2	1.0	90	-48	0	-44.7	0.0	Olekminsk	249440	60.40	120.42	741	226	14.31	98.6	88	71	12.5	21.8	82	72	11.7	22.1	76	68	10.9	20.2	6	2.8	90
-24	1	-31.3	0.2	7	3.0	210	-18	2	-27.9	0.3	Omsk	286980	54.93	73.40	403	123	14.49	99.9	90	74	12.9	23.1	85	72	12.2	22.3	80	70	11.5	21.3	10	4.5	120
-10	3	-23.5	0.5	7	3.0	360	-4	5	-19.9	0.7	Orel	279060	53.00	36.03	666	203	14.35	98.9	96	73	13.7	22.8	90	72	12.8	22.1	85	70	12.1	21.2	10	4.5	280
-18	2	-27.7	0.3	5	2.2	260	-13	3	-24.8	0.4	Orenburg	351210	51.78	55.22	357	109	14.51	100.0	90	72	12.9	22.5	85	72	12.1	22.0	80	71	11.4	21.5	11	4.8	160
4	7	-15.3	1.0	11	5.0	80	7	8	-13.7	1.1	Ozernaja	325940	51.48	156.48	95	29	14.65	101.0	66	58	9.5	14.7	62	56	8.9	13.4	59	55	8.4	12.7	16	7.1	90
-14	3	-25.5	0.4	6	2.5	220	-9	3	-22.8	0.5	Penza	279620	53.13	45.02	570	174	14.40	99.3	94	74	13.5	23.3	90	72	12.8	22.1	85	70	12.1	21.1	11	4.9	130
-24	1	-30.9	0.2	6	2.5	280	-18	2	-27.6	0.3	Perm'	282250	58.02	56.30	564	172	14.40	99.3	95	75	13.6	24.1	89	72	12.7	22.4	84	70	12.0	21.0	10	4.5	200
5	7	-14.8	1.0	8	3.7	360	8	8	-13.2	1.2	Petropavlovsk-Kamca	325400	52.97	158.75	79	24	14.66	101.0	71	64	10.1	18.0	65	61	9.3	15.9	62	59	8.8	14.8	8	3.7	330
-19	2	-28.3	0.3	4	1.7	90	-11	3	-24.1	0.4	Petrozavodsk	228200	61.82	34.27	367	112	14.51	100.0	90	70	12.8	21.1	82	67	11.7	19.7	76	65	10.9	18.6	6	2.7	290
-13	3	-24.9	0.4	2	1.1	150	-4	4	-20.1	0.6	Pskov	262580	57.80	28.42	137	42	14.63	100.8	94	73	13.5	22.6	88	70	12.5	21.0	81	69	11.6	20.4	6	2.8	130
2	6	-16.9	0.9	11	4.8	80	5	7	-14.9	1.0	Rostov-Na-Donu	347310	47.25	39.82	252	77	14.57	100.4	104	78	14.9	25.3	99	76	14.2	24.4	95	75	13.6	24.0	11	5.1	110
-26	1	-32.2	0.2	4	1.8	360	-20	2	-29.1	0.3	Rubtsovsk	360340	51.50	81.22	705	215	14.33	98.8	98	73	14.0	22.7	92	71	13.2	21.8	87	71	12.4	21.4	8	3.6	30
-10	3	-23.5	0.5	6	2.8	310	-6	4	-20.9	0.6	Ryazan'	277310	54.62	39.72	557	170	14.41	99.3	97	74	13.9	23.6	92	72	13.1	22.2	87	71	12.4	21.5	7	3.3	140
-19	2	-28.4	0.3	5	2.3	320	-12	3	-24.7	0.4	Rybinsk	272250	58.00	38.83	374	114	14.50	100.0	96	73	13.7	23.0	90	72	12.9	22.0	86	69	12.3	20.8	8	3.5	210
-17	2	-27.1	0.3	2	1.0	320	-12	3	-24.7	0.4	Samara (Kuybyshev)	289000	53.25	50.45	144	44	14.62	100.8	101	77	14.4	25.0	94	76	13.4	24.5	88	73	12.5	22.9	8	3.4	110
-8	4	-22.4	0.5	13	5.8	300	-5	4	-20.3	0.6	Saratov	341720	51.57	46.03	511	156	14.43	99.5	94	74	13.4	23.6	89	73	12.7	22.9	84	72	12.0	22.5	11	4.7	70
-9	4	-22.7	0.5	6	2.6	280	-4	5	-19.8	0.7	Smolensk	267810	54.75	32.07	790	241	14.28	98.5	95	72	13.6	22.5	90	71	12.8	21.4	84	69	12.0	20.3	6	2.8	230
28	22	-2.5	3.2	8	3.4	70	30	24	-1.2	3.5	Sochi	371710	43.45	39.90	52	16	14.67	101.1	126	80	18.0	26.4	120	78	17.2	25.6	116	77	16.5	25.2	8	3.4	270
-9	3	-22.6	0.5	3	1.2	40	-2	5	-18.9	0.7	St Petersburg	260630	59.97	30.30	13	4	14.69	101.3	90	72	12.9	22.1	84	69	12.0	20.8	79	68	11.3	19.8	6	2.6	180
-35	1	-37.3	0.1	1	0.6	290	-30	1	-34.7	0.1	Svobodnyy	314450	51.45	128.12	646	197	14.36	99.0	109	75	15.6	24.1	103	73	14.7	23.0	97	72	13.9	22.3	7	3.2	180
-32	1	-35.5	0.1	4	1.6	10	-25	1	-31.6	0.2	Syktyvkar	238040	61.72	50.83	390	119	14.49	99.9	94	74	13.4	23.5	88	71	12.5	21.9	82	68	11.7	20.2	7	3.0	140
-13	3	-25.1	0.4	5	2.1	360	-8	4	-22.3	0.5	Tambov	279470	52.73	41.47	456	139	14.46	99.7	97	75	13.8	23.9	92	73	13.1	22.6	86	71	12.3	21.6	9	4.2	140
-34	1	-36.9	0.1	2	0.8	330	-29	1	-33.7	0.2	Tayshet	295940	55.95	98.00	990	302	14.18	97.7	92	72	13.1	22.4	86	70	12.3	21.0	81	68	11.6	20.2	7	3.1	40
-24	1	-31.2	0.2	3	1.5	140	-18	2	-27.7	0.3	Ufa	287220	54.75	56.00	344	105	14.52	100.1	98	76	14.0	24.5	91	73	13.0	22.9	86	72	12.3	22.0	8	3.7	50
-33	1	-35.9	0.1	0	0.1	50	-29	1	-33.8	0.2	Ulan Ude	308230	51.80	107.43	1673	510	13.83	95.3	90	72	12.8	22.1	84	70	12.0	21.2	79	68	11.3	19.8	8	3.6	20
12	10	-11.1	1.5	24	10.5	320	14	11	-10.0	1.6	Urup Island	321860	46.20	150.50	230	70	14.58	100.5	78	61	11.1	16.2	71	58	10.1	14.6	65	56	9.3	13.4	11	4.8	180
-40	1	-40.1	0.1	9	4.0	300	-35	1	-37.4	0.1	Ust'ilimsk	301170	58.03	102.73	1318	402	14.01	96.6	90	70	12.9	21.1	83	69	11.9	20.3	77	67	11.0	19.2	7	3.1	170
-23	2	-30.6	0.2	6	2.5	360	-18	2	-27.7	0.3	Ust-Kamcatsk	324080	56.22	162.47	89	27	14.65	101.0	66	60	9.5	15.8	63	59	9.0	14.8	60	57	8.6	13.9	9	3.9	180
-16	2	-26.6	0.3	7	3.0	10	-10	3	-23.3	0.5	Vladimir	275320	56.13	40.38	557	170	14.41	99.3	97	74	13.8	23.5	91	71	13.0	21.8	85	70	12.2	21.0	9	4.1	210
-8	4	-22.1	0.5	21	9.4	360	-4	5	-20.2	0.7	Vladivostok	319600	43.12	131.90	604	184	14.38	99.1	116	73	16.6	23.0	110	72	15.7	22.0	104	70	14.8	21.0	10	4.4	240
-6	4	-21.2	0.6	9	4.1	340	-2	5	-18.9	0.7	Volgograd	345600	48.68	44.35	475	145	14.45	99.6	93	73	13.3	23.0	89	72	12.7	22.5	84	72	12.0	22.5	15	6.6	110

HUMIDIFICATION 99.6% (35 hours below...)							99% (88 hours below...)				**International**							DEHUMIDIFICATION 0.4% (35 hours above...)				1% (88 hours above...)				2% (175 hours above...)				Summer Wind		
°F 1a	gr/lb 1b	°C 1c	g/kg 1d	mph 1e	m/s 1f	PWD 1g	°F 2a	gr/lb 2b	°C 2c	g/kg 2d	Location 3a	WMO# 3b	Lat. 3c	Long. 3d	ft. 3e	m 3f	psia 3g	kPa 3h	gr/lb 4a	°F 4b	g/kg 4c	°C 4d	gr/lb 5a	°F 5b	g/kg 5c	°C 5d	gr/lb 6a	°F 6b	g/kg 6c	°C 6d	mph 7a	m/s 7b	PWD 7c
											SWITZERLAND																						
18	14	-8.0	2.1	6	2.9	230	23	18	-5.2	2.6	Geneva	67000	46.25	6.13	1364	416	13.99	96.4	93	72	13.3	22.2	89	71	12.7	21.6	84	70	12.0	21.1	8	3.4	210
15	13	-9.7	1.8	6	2.6	190	19	15	-7.3	2.2	Interlaken	67340	46.67	7.88	1902	580	13.72	94.5	91	72	13.0	22.2	87	70	12.4	20.9	83	69	11.8	20.3	8	3.4	280
-15	4	-26.3	0.5	16	7.2	310	-11	5	-23.9	0.7	Jungfrau Mountain	67300	46.55	7.98	11732	3576	9.45	65.1	47	38	6.7	3.1	43	36	6.2	2.4	41	35	5.9	1.9	11	5.1	140
6	8	-14.4	1.2	3	1.2	230	12	11	-11.3	1.6	La Chaux-De-Fonds	66120	47.08	6.80	3343	1019	13.01	89.7	85	67	12.1	19.5	80	66	11.5	18.9	76	65	10.9	18.1	7	3.2	250
21	16	-6.3	2.3	4	1.8	90	24	19	-4.7	2.7	Locarno	67620	46.17	8.88	649	198	14.36	99.0	108	78	15.4	25.4	102	76	14.6	24.6	97	74	13.9	23.5	6	2.7	240
25	20	-3.7	2.8	3	1.5	360	28	23	-2.0	3.3	Lugano	67700	46.00	8.97	905	276	14.23	98.1	112	79	16.0	25.9	106	77	15.1	25.2	100	76	14.3	24.3	6	2.8	190
14	12	-10.2	1.7	4	1.7	40	19	15	-7.2	2.2	Payerne	66100	46.82	6.95	1610	491	13.86	95.6	95	74	13.6	23.5	91	73	13.0	22.6	86	70	12.3	21.3	7	3.2	230
-3	6	-19.5	0.9	14	6.1	260	1	8	-17.1	1.1	Saentis (Aut)	66800	47.25	9.35	8202	2500	10.83	74.7	67	51	9.6	10.5	63	49	9.0	9.6	60	48	8.5	9.0	6	2.9	210
6	9	-14.4	1.3	4	1.7	310	10	11	-12.2	1.6	San Bernardino	67830	46.47	9.18	5374	1638	12.06	83.1	81	62	11.6	16.5	76	60	10.9	15.6	71	58	10.2	14.5	7	3.2	120
13	11	-10.4	1.6	6	2.6	60	18	15	-7.8	2.1	Zurich	66600	47.38	8.57	1866	569	13.74	94.7	94	71	13.4	21.8	90	70	12.8	21.1	86	69	12.3	20.5	5	2.2	230
											SYRIA																						
25	21	-4.1	2.9	3	1.3	30	28	24	-2.2	3.4	Damascus	400800	33.42	36.52	1984	605	13.68	94.3	103	72	14.7	22.3	97	72	13.8	22.0	91	71	13.0	21.6	7	3.3	210
											TAIWAN																						
47	48	8.6	6.9	9	4.1	40	50	54	9.8	7.7	Hsinchu	467570	24.83	120.00	88	27	14.65	101.0	158	89	22.5	31.6	154	88	22.0	31.2	150	87	21.4	30.7	9	3.9	270
53	60	11.7	8.6	6	2.8	250	55	65	12.8	9.3	Hualien	466990	23.98	121.60	62	19	14.67	101.1	156	87	22.3	30.4	153	86	21.8	30.2	149	86	21.3	30.0	8	3.5	140
52	58	11.2	8.3	8	3.4	360	55	65	12.6	9.2	Kaohsiung	467400	22.58	120.35	29	9	14.68	101.2	158	85	22.6	29.3	151	84	21.6	29.1	150	84	21.4	29.1	13	5.8	280
46	47	7.5	6.7	5	2.4	30	48	50	8.9	7.2	T'aichung	467510	24.18	120.65	367	112	14.51	100.0	169	90	24.2	32.4	162	89	23.2	31.9	159	89	22.7	31.5	10	4.3	240
51	56	10.4	8.0	11	4.7	20	53	60	11.6	8.6	T'ainan	467410	23.00	120.22	45	14	14.68	101.2	162	86	23.2	30.0	158	85	22.6	29.7	155	85	22.1	29.6	11	4.9	200
48	50	8.8	7.1	4	1.8	110	50	54	10.0	7.7	Taipei	466960	25.07	121.55	19	6	14.69	101.3	153	86	21.9	30.1	150	86	21.5	29.9	148	86	21.2	29.8	11	5.1	290
48	50	8.8	7.1	16	7.3	40	50	54	9.8	7.7	Taipei Intl Airport	466860	25.08	121.22	108	33	14.64	100.9	159	88	22.7	30.9	152	86	21.7	30.1	149	86	21.3	30.0	16	7.1	260
											TAJIKISTAN																						
19	16	-7.3	2.3	2	1.1	60	23	19	-5.1	2.7	Dushanbe	388360	38.55	68.78	2634	803	13.35	92.0	98	83	14.0	28.6	91	81	13.0	27.4	85	80	12.1	26.5	3	1.5	270
17	14	-8.4	2.0	14	6.1	240	21	17	-6.3	2.4	Khujand (Leninabad)	385990	40.22	69.73	1404	428	13.97	96.3	90	79	12.8	26.2	84	79	12.0	26.2	79	78	11.3	25.5	11	5.0	240
											THAILAND																						
65	93	18.4	13.3	4	1.9	40	68	103	19.9	14.8	Bangkok	484560	13.92	100.60	39	12	14.68	101.2	164	88	23.4	31.3	160	87	22.9	30.6	156	87	22.3	30.3	11	4.8	180
53	62	11.9	8.9	1	0.4	360	56	70	13.1	9.9	Chiang Mai	483270	18.78	98.98	1030	314	14.16	97.6	146	83	20.9	28.1	139	82	19.9	27.8	138	81	19.7	27.4	7	3.1	190
49	54	9.6	7.7	0	0.0	20	52	61	11.0	8.7	Chiang Rai	483030	19.92	99.83	1295	395	14.02	96.7	147	84	21.0	28.7	144	83	20.6	28.4	141	82	20.2	28.0	5	2.1	180
66	96	19.1	13.8	0	0.0	30	68	103	20.2	14.8	Chumphon	485170	10.48	99.18	16	5	14.69	101.3	150	87	21.5	30.8	147	87	21.0	30.3	145	86	20.7	30.1	9	4.0	120
70	111	20.9	15.9	0	0.2	330	71	115	21.8	16.5	Hat Yai	485690	6.92	100.43	114	35	14.64	100.9	149	83	21.3	28.5	144	82	20.5	27.9	143	82	20.4	27.8	7	3.3	240
56	68	13.6	9.7	0	0.1	360	59	76	15.2	10.8	Phetchabun	483790	16.43	101.15	380	116	14.50	99.9	153	87	21.9	30.8	150	87	21.5	30.3	148	86	21.1	29.9	6	2.6	180
55	66	12.9	9.4	0	0.2	30	58	74	14.3	10.5	Phrae	483300	18.17	100.17	531	162	14.42	99.4	153	86	21.8	30.0	149	85	21.3	29.7	146	85	20.9	29.4	5	2.2	240
57	71	14.0	10.1	0	0.0	270	60	79	15.5	11.3	Tak	483760	16.88	99.15	406	124	14.49	99.8	146	83	20.9	28.6	144	83	20.5	28.5	140	83	20.0	28.2	5	2.1	270
											TRINIDAD																						
68	103	20.1	14.8	0	0.1	90	70	111	20.9	15.9	Port Of Spain	789700	10.62	61.35	49	15	14.67	101.1	147	83	21.0	28.5	143	82	20.4	27.9	141	82	20.2	27.8	12	5.2	90
											TUNIS IA																						
38	34	3.3	4.8	1	0.4	320	40	37	4.6	5.2	Bizerte	607140	37.25	9.80	9	3	14.69	101.3	129	81	18.4	27.1	123	81	17.6	27.0	118	80	16.9	26.4	12	5.3	100
42	39	5.8	5.6	4	1.9	230	44	43	6.8	6.1	Gabes	607650	33.88	10.10	16	5	14.69	101.3	142	85	20.3	29.5	136	85	19.5	29.2	130	84	18.5	28.7	10	4.6	250
36	32	2.1	4.6	5	2.4	60	38	35	3.5	5.0	Gafsa	607450	34.42	8.82	1030	314	14.16	97.6	110	79	15.7	26.0	104	78	14.8	25.7	99	78	14.1	25.4	9	3.8	240
42	40	5.7	5.7	5	2.4	300	44	43	6.6	6.1	Kelibia	607200	36.85	11.08	98	30	14.65	101.0	138	81	19.7	27.3	133	81	19.0	27.1	127	80	18.1	26.5	9	3.8	300
40	37	4.4	5.3	2	1.0	240	42	40	5.5	5.7	Qairouan (Kairouan)	607350	35.67	10.10	223	68	14.58	100.5	125	81	17.9	27.1	118	80	16.8	26.7	112	80	16.0	26.5	7	3.0	180
41	38	4.9	5.4	6	2.8	240	43	41	5.9	5.9	Tunis	607150	36.83	10.23	13	4	14.69	101.3	134	82	19.1	28.0	128	82	18.3	27.5	123	81	17.5	27.2	11	4.9	180

International

HUMIDIFICATION 99.6% (35 hours below...)							99% (88 hours below...)				International								DEHUMIDIFICATION 0.4% (35 hours above...)				1% (88 hours above...)				2% (175 hours above...)				Summer Wind		
°F 1a	gr/lb 1b	°C 1c	g/kg 1d	mph 1e	m/s 1f	PWD 1g	°F 2a	gr/lb 2b	°C 2c	g/kg 2d	Location 3a	WMO# 3b	Lat. 3c	Long. 3d	ft. 3e	m 3f	psia 3g	kPa 3h	gr/lb 4a	°F 4b	g/kg 4c	°C 4d	gr/lb 5a	°F 5b	g/kg 5c	°C 5d	gr/lb 6a	°F 6b	g/kg 6c	°C 6d	mph 7a	m/s 7b	PWD 7c
											TURKEY																						
32	27	-0.2	3.8	6	2.5	30	34	29	1.1	4.1	Adana	173500	37.00	35.42	216	66	14.59	100.5	137	82	19.6	27.9	133	82	19.0	28.0	128	82	18.3	27.7	8	3.5	210
2	7	-16.9	1.0	1	0.4	20	8	9	-13.1	1.3	Ankara	171280	40.12	32.98	3113	949	13.12	90.4	83	73	11.8	23.0	78	72	11.1	22.0	74	70	10.5	21.3	8	3.5	270
-23	2	-30.7	0.3	0	0.0	310	-17	3	-27.4	0.4	Erzurum	170960	39.92	41.27	5767	1758	11.88	81.9	88	74	12.5	23.4	82	72	11.7	22.4	76	70	10.9	21.2	10	4.3	90
12	11	-11.2	1.6	3	1.4	120	16	14	-9.1	1.9	Eskisehir	171240	39.78	30.57	2575	785	13.38	92.2	105	80	15.0	26.7	97	78	13.9	25.3	92	75	13.1	24.1	9	3.9	320
26	20	-3.2	2.9	14	6.2	360	29	23	-1.8	3.3	Istanbul	170600	40.97	28.82	121	37	14.64	100.9	117	78	16.7	25.3	111	76	15.8	24.7	105	76	15.0	24.4	13	5.9	60
28	22	-2.0	3.1	6	2.7	360	31	25	-0.8	3.6	Izmir/Cigli(Cv/Afb)	172180	38.50	27.02	16	5	14.69	101.3	104	83	14.8	28.3	98	82	14.0	27.5	95	81	13.6	27.0	14	6.2	350
10	10	-12.1	1.5	4	1.6	210	16	14	-9.1	2.0	Malatya	172000	38.43	38.08	2785	849	13.28	91.5	88	87	12.5	30.7	82	85	11.7	29.3	77	83	11.0	28.1	7	3.2	60
6	9	-14.7	1.3	4	1.9	90	9	11	-13.0	1.5	Van	171700	38.45	43.32	5449	1661	12.03	82.9	116	80	16.6	26.6	108	78	15.5	25.8	101	77	14.4	25.1	3	1.4	300
											TURKMENISTAN																						
20	15	-6.9	2.2	4	1.8	110	23	18	-5.0	2.5	Ashgabat (Ashkhabad)	388800	37.97	58.33	688	210	14.34	98.8	99	85	14.1	29.7	92	85	13.2	29.3	87	84	12.4	28.9	10	4.3	90
5	7	-14.9	1.0	6	2.7	200	10	9	-12.1	1.3	Dashhowuz (Tashauz)	383920	41.83	59.98	288	88	14.55	100.3	112	92	16.0	33.5	105	90	15.0	32.3	98	88	14.0	31.1	10	4.5	360
											UNITED KINGDOM & NORTHERN IRELAND																						
22	17	-5.7	2.4	3	1.4	360	27	21	-3.0	3.0	Aberdeen/Dyce	30910	57.20	2.22	213	65	14.59	100.5	79	67	11.3	19.6	75	64	10.7	18.0	70	62	10.0	16.9	11	4.7	170
26	20	-3.1	2.9	14	6.4	90	29	23	-1.5	3.4	Aberporth	35020	52.13	4.57	439	134	14.47	99.7	84	65	12.0	18.6	80	64	11.5	17.8	77	62	11.0	16.9	13	5.6	130
26	20	-3.5	2.9	8	3.7	130	28	22	-2.1	3.2	Aughton	33220	53.55	2.92	183	56	14.60	100.7	84	68	12.0	19.8	80	66	11.4	18.9	76	64	10.9	17.9	9	3.9	130
15	12	-9.4	1.7	1	0.5	360	21	16	-6.3	2.3	Aviemore	30630	57.20	3.83	721	220	14.32	98.7	76	65	10.9	18.6	71	65	10.2	18.1	67	62	9.6	16.9	9	4.0	200
27	21	-2.8	3.0	4	1.8	180	29	23	-1.4	3.3	Belfast	39170	54.65	6.22	265	81	14.56	100.4	82	66	11.7	19.0	78	65	11.1	18.1	74	64	10.5	17.5	9	3.9	110
21	16	-6.2	2.3	4	1.9	70	24	18	-4.2	2.6	Birmingham	35340	52.45	1.73	324	99	14.53	100.1	84	68	12.0	20.1	80	67	11.4	19.3	76	65	10.8	18.5	9	3.9	100
22	17	-5.4	2.4	4	1.9	20	25	19	-3.8	2.7	Bournemouth	38620	50.78	1.83	36	11	14.68	101.2	86	69	12.3	20.5	83	67	11.8	19.4	80	66	11.4	18.9	10	4.4	90
26	20	-3.5	2.9	8	3.4	70	29	23	-1.7	3.3	Bristol	37260	51.47	2.60	36	11	14.68	101.2	85	71	12.2	21.4	82	68	11.7	20.0	78	66	11.2	19.1	9	3.9	90
30	24	-1.3	3.5	13	5.6	50	32	27	0.2	3.8	Camborne	38080	50.37	5.53	288	88	14.55	100.3	86	65	12.3	18.4	83	64	11.8	17.7	79	63	11.3	17.1	13	5.8	100
25	19	-4.0	2.8	14	6.2	60	28	22	-2.2	3.2	Cardiff	37150	51.40	3.35	219	67	14.58	100.5	87	69	12.4	20.3	83	66	11.8	19.1	79	65	11.3	18.4	9	4.0	60
21	16	-5.9	2.3	2	0.7	250	25	19	-3.8	2.7	Edinburgh	31600	55.95	3.35	134	41	14.63	100.8	78	66	11.2	18.8	74	64	10.6	17.6	71	63	10.1	17.1	9	4.2	250
24	18	-4.3	2.6	7	3.0	40	27	21	-2.7	3.0	Exeter	38390	50.73	3.42	98	30	14.65	101.0	88	70	12.6	21.2	84	68	12.0	20.0	81	67	11.6	19.3	10	4.4	150
23	17	-5.1	2.5	4	1.8	170	26	20	-3.3	2.9	Finningley	33600	53.48	1.00	55	17	14.67	101.1	83	69	11.8	20.4	78	67	11.2	19.5	75	65	10.7	18.5	10	4.4	150
21	16	-6.2	2.3	2	1.1	270	24	18	-4.2	2.6	Glasgow	31400	55.87	4.43	26	8	14.69	101.2	79	67	11.3	19.5	75	65	10.7	18.4	71	63	10.1	17.3	9	4.2	230
27	21	-3.0	3.0	15	6.5	70	29	23	-1.6	3.3	Hemsby	34960	52.68	1.68	45	14	14.68	101.2	87	69	12.4	20.3	83	67	11.8	19.2	79	65	11.3	18.4	12	5.4	230
24	18	-4.5	2.6	9	3.9	40	27	21	-2.7	3.0	Herstmonceux	38840	50.87	0.33	55	17	14.67	101.1	88	70	12.5	20.9	83	67	11.9	19.5	80	66	11.5	19.0	9	4.0	130
28	22	-2.5	3.2	13	5.6	100	31	26	-0.7	3.7	Jersey/Channel Islands	38950	49.22	2.20	275	84	14.55	100.3	87	67	12.4	19.5	83	65	11.9	18.6	81	65	11.6	18.2	10	4.6	70
29	23	-1.7	3.3	11	5.1	270	31	25	-0.5	3.6	Kirkwall	30170	58.95	2.90	68	21	14.66	101.1	73	62	10.4	16.4	69	59	9.8	15.2	66	58	9.4	14.3	11	5.0	120
28	22	-2.2	3.2	14	6.3	350	30	24	-1.2	3.5	Lerwick	30050	60.13	1.18	275	84	14.55	100.3	69	58	9.8	14.5	66	57	9.4	13.8	64	56	9.1	13.1	12	5.3	180
24	18	-4.7	2.6	5	2.2	250	27	21	-2.9	3.0	Leuchars	31710	56.38	2.87	39	12	14.68	101.2	76	65	10.9	18.3	73	63	10.4	17.2	69	62	9.9	16.6	13	5.8	240
22	17	-5.6	2.4	3	1.5	80	25	19	-3.7	2.8	London, Gatwick Arpt	37760	51.15	0.18	203	62	14.59	100.6	88	70	12.5	21.0	83	68	11.8	19.9	79	67	11.3	19.2	8	3.7	70
25	19	-4.0	2.7	6	2.7	20	28	22	-2.3	3.2	London, Heathrow	37720	51.48	0.45	78	24	14.66	101.0	88	70	12.5	21.3	83	69	11.9	20.7	80	68	11.4	20.0	10	4.5	90
22	17	-5.6	2.4	11	5.1	30	26	20	-3.6	2.9	Lyneham	37400	51.50	1.98	511	156	14.43	99.5	85	68	12.1	20.2	81	66	11.6	18.7	77	65	11.0	18.3	10	4.4	70
28	22	-2.3	3.2	16	7.1	190	31	25	-0.7	3.6	Lynemouth	32620	55.02	1.42	98	30	14.65	101.0	80	64	11.4	17.8	76	63	10.8	17.1	73	62	10.4	16.7	15	6.5	260
24	18	-4.2	2.6	6	2.5	90	27	21	-2.7	3.0	Manchester	33340	53.35	2.27	255	78	14.56	100.4	83	68	11.9	20.0	79	67	11.3	19.2	74	65	10.6	18.3	9	3.9	130
23	18	-4.9	2.5	8	3.5	20	26	20	-3.2	2.9	Nottingham	33540	53.00	1.25	383	117	14.50	99.9	87	70	12.4	21.3	81	68	11.6	19.8	77	65	11.0	18.6	8	3.6	210
26	20	-3.2	2.9	3	1.4	180	29	23	-1.5	3.3	Oban	31140	56.42	5.47	13	4	14.69	101.3	77	65	11.0	18.3	73	64	10.4	17.5	70	62	10.0	16.8	6	2.9	180
29	23	-1.7	3.3	8	3.7	80	31	25	-0.3	3.6	Plymouth	38270	50.35	4.12	88	27	14.65	101.0	85	67	12.2	19.5	83	66	11.9	18.7	80	64	11.4	18.0	10	4.5	80
23	18	-5.2	2.5	7	3.2	30	26	20	-3.5	2.9	Stansted Airport	36830	51.88	0.23	347	106	14.52	100.1	87	70	12.4	21.0	82	67	11.7	19.5	78	66	11.1	18.8	10	4.5	130

International

HUMIDIFICATION 99.6% (35 hours below...)							99% (88 hours below...)												DEHUMIDIFICATION 0.4% (35 hours above...)				1% (88 hours above...)				2% (175 hours above...)				Summer Wind		
°F 1a	gr/lb 1b	°C 1c	g/kg 1d	mph 1e	m/s 1f	PWD 1g	°F 2a	gr/lb 2b	°C 2c	g/kg 2d	Location 3a	WMO# 3b	Lat. 3c	Long. 3d	ft. 3e	m 3f	psia 3g	kPa 3h	gr/lb 4a	°F 4b	g/kg 4c	°C 4d	gr/lb 5a	°F 5b	g/kg 5c	°C 5d	gr/lb 6a	°F 6b	g/kg 6c	°C 6d	mph 7a	m/s 7b	PWD 7c
29	23	-1.8	3.3	6	2.8	300	31	25	-0.4	3.6	Stornoway	30260	58.22	6.32	42	13	14.68	101.2	74	62	10.5	16.4	71	60	10.1	15.5	67	59	9.6	14.8	10	4.4	160
27	21	-2.7	3.0	10	4.5	80	30	24	-1.1	3.5	Valley	33020	53.25	4.53	36	11	14.68	101.2	82	66	11.7	18.9	78	64	11.2	17.8	75	63	10.7	17.1	9	4.2	50
22	17	-5.3	2.4	8	3.4	40	26	20	-3.6	2.9	Wyton Raf	35660	52.35	0.12	134	41	14.63	100.8	87	70	12.4	21.2	82	68	11.7	19.8	78	66	11.1	18.9	10	4.4	100
											UKRANIAN																						
-7	4	-21.6	0.6	5	2.4	110	-1	5	-18.3	0.8	Chernihiv (Chernigov)	331350	51.48	31.28	449	137	14.46	99.7	99	75	14.1	23.9	93	73	13.3	22.8	89	71	12.7	21.9	9	4.1	160
2	6	-16.6	0.9	8	3.7	320	7	8	-14.1	1.2	Chernivtsi (Chernov.)	336580	48.27	25.97	787	240	14.29	98.5	98	74	14.0	23.6	92	73	13.1	22.6	88	72	12.5	22.0	8	3.7	110
0	6	-17.6	0.8	9	4.1	50	4	7	-15.3	1.0	Dnipropetrovs'k	345040	48.37	35.08	465	142	14.45	99.6	99	75	14.2	23.8	94	74	13.4	23.1	89	72	12.7	22.3	11	4.9	90
-1	5	-18.6	0.8	6	2.9	70	3	7	-16.3	0.9	Donets'k	345190	48.07	37.77	741	226	14.31	98.6	97	74	13.8	23.6	92	73	13.2	22.8	88	72	12.5	22.1	11	4.8	100
11	10	-11.6	1.4	11	5.1	30	15	12	-9.5	1.7	Kerch	339830	45.40	36.42	160	49	14.61	100.7	111	78	15.9	25.7	105	77	15.0	24.8	99	75	14.2	24.0	11	5.0	40
-3	5	-19.2	0.7	6	2.8	30	2	6	-16.8	0.9	Kharkiv (Khar'kov)	343000	49.87	36.13	498	152	14.44	99.5	98	74	14.0	23.1	93	73	13.3	22.8	88	71	12.6	21.8	11	4.8	110
4	7	-15.8	1.0	7	3.3	270	8	8	-13.2	1.2	Kherson	339020	46.67	32.62	157	48	14.62	100.7	102	75	14.6	23.9	97	75	13.8	23.7	92	73	13.1	22.6	8	3.4	80
-2	5	-19.0	0.7	5	2.1	310	2	6	-16.5	0.9	Kirovohrad	337110	48.48	32.25	564	172	14.40	99.3	96	73	13.7	22.8	90	72	12.9	22.0	85	70	12.1	20.9	11	4.8	100
0	6	-18.0	0.8	9	4.0	50	4	7	-15.6	1.0	Kryvyy Rih (Krivoy Rog)	337910	47.93	33.33	410	125	14.48	99.8	98	75	14.0	23.8	92	73	13.1	23.0	87	72	12.4	22.0	10	4.3	90
-2	5	-19.0	0.7	7	3.0	270	3	7	-15.9	0.9	Kyyiv (Kiev)	333450	50.40	30.45	551	168	14.41	99.3	99	74	14.1	23.5	93	73	13.3	22.6	88	71	12.5	21.6	6	2.9	180
-4	4	-20.1	0.6	6	2.9	90	0	6	-17.7	0.8	Luhans'k (Voroshil.)	345230	48.60	39.27	203	62	14.59	100.6	96	76	13.7	24.3	90	72	12.9	22.5	86	72	12.3	22.1	10	4.5	90
4	7	-15.5	1.0	12	5.3	70	8	8	-13.4	1.2	Mariupol' (Zdanov)	347120	47.07	37.50	229	70	14.58	100.5	118	78	16.8	25.4	109	77	15.6	24.8	103	75	14.7	24.0	10	4.4	90
7	8	-14.1	1.1	11	5.0	360	12	10	-11.2	1.5	Odesa	338370	46.48	30.63	114	35	14.64	100.9	109	76	15.6	24.5	102	74	14.5	23.6	96	73	13.7	22.6	10	4.4	180
-3	5	-19.4	0.7	6	2.9	360	2	6	-16.6	0.9	Poltava	335060	49.60	34.55	521	159	14.42	99.4	98	75	14.0	23.7	92	73	13.1	22.6	87	71	12.4	21.7	7	3.1	70
-3	5	-19.5	0.7	6	2.8	270	3	7	-16.0	0.9	Rivne (Rovno)	333010	50.58	26.13	767	234	14.30	98.5	97	75	13.9	23.7	92	72	13.1	22.0	86	70	12.3	20.9	9	4.2	130
8	9	-13.1	1.2	9	3.8	50	13	11	-10.6	1.6	Simferopol'	339460	45.02	33.98	593	181	14.39	99.2	100	74	14.3	23.2	94	73	13.5	22.8	89	72	12.7	22.1	11	4.7	50
-7	4	-21.8	0.6	6	2.5	330	-1	5	-18.5	0.8	Sumy	332750	50.88	34.78	570	174	14.40	99.3	99	74	14.1	23.4	92	72	13.2	22.1	88	71	12.5	21.5	8	3.4	130
6	8	-14.7	1.1	3	1.3	100	10	9	-12.0	1.3	Uzhhorod (Uzhgorod)	336310	48.63	22.27	387	118	14.50	99.9	97	77	13.9	25.2	92	75	13.1	24.0	88	73	12.5	22.9	7	3.0	170
-2	5	-19.1	0.7	8	3.5	340	3	7	-16.3	1.0	Vinnytsya (Vinnitsa)	335620	49.23	28.47	977	298	14.19	97.8	97	73	13.8	22.9	91	71	13.0	21.9	86	70	12.3	20.9	9	4.2	180
0	6	-17.7	0.8	8	3.7	360	5	7	-15.1	1.0	Zaporizhzhya	346010	47.80	35.25	282	86	14.55	100.3	100	76	14.3	24.3	94	74	13.4	23.4	89	72	12.7	22.2	9	4.2	220
-4	5	-19.8	0.7	6	2.9	90	2	6	-16.4	0.9	Zhytomyr (Zhitomir)	333250	50.27	28.63	744	227	14.31	98.6	96	74	13.7	23.2	91	71	13.0	21.9	86	70	12.3	20.9	9	3.8	190
											UNITED ARAB EMIRATES																						
52	58	10.9	8.3	4	2.0	200	54	62	12.0	8.9	Abu Dhabi	412170	24.43	54.65	88	27	14.65	101.0	182	91	26.0	32.9	178	91	25.4	32.8	169	90	24.2	32.5	9	4.2	320
54	62	12.0	8.9	4	1.8	170	55	65	13.0	9.2	Dubai	411940	25.25	55.33	16	5	14.69	101.3	181	91	25.9	33.0	177	91	25.3	32.8	169	91	24.1	32.7	11	4.9	270
49	52	9.7	7.4	2	1.0	210	52	58	11.0	8.3	Ra's Al Khaymah	411840	25.62	55.93	101	31	14.65	101.0	174	94	24.8	34.2	169	93	24.1	34.0	160	92	22.9	33.6	10	4.6	320
49	52	9.2	7.4	5	2.1	120	51	56	10.7	8.0	Sharjah	411960	25.33	55.52	108	33	14.64	100.9	178	92	25.4	33.5	169	91	24.2	33.0	160	91	22.9	32.7	10	4.5	270
											URUGUAY																						
39	35	3.8	5.0	9	4.0	50	41	38	5.0	5.4	Colonia Del Sacra.	865600	34.45	57.83	75	23	14.66	101.0	127	82	18.2	27.7	122	80	17.4	26.7	117	79	16.7	25.9	9	3.9	360
35	30	1.8	4.3	8	3.5	330	38	34	3.1	4.8	Montevideo	865800	34.83	56.00	104	32	14.64	100.9	125	79	17.9	26.2	118	77	16.9	25.0	116	76	16.5	24.7	14	6.4	360
34	29	1.1	4.1	2	1.0	280	36	31	2.5	4.5	Paso De Los Toros	864600	32.80	56.52	246	75	14.57	100.4	127	82	18.1	27.5	122	80	17.4	26.7	117	78	16.7	25.8	10	4.3	330
34	29	0.9	4.1	1	0.6	310	36	31	2.3	4.5	Rocha	865650	34.48	54.30	59	18	14.67	101.1	130	80	18.5	26.8	123	78	17.6	25.8	118	77	16.8	24.9	9	4.1	360
34	29	1.3	4.1	1	0.6	120	37	33	2.9	4.7	Salto	863600	31.38	57.95	111	34	14.64	100.9	134	85	19.2	29.7	129	83	18.4	28.1	123	81	17.6	27.3	9	3.9	20
33	28	0.4	4.0	2	0.9	270	35	30	1.8	4.3	Treinta Y Tres	865000	33.22	54.38	151	46	14.62	100.8	130	80	18.5	26.9	123	79	17.6	26.2	118	78	16.9	25.5	6	2.7	290
											UZBEKISTAN																						
12	11	-11.1	1.6	8	3.6	140	17	14	-8.2	2.0	Samarqamd	386960	39.70	67.00	2375	724	13.48	92.9	91	81	13.0	27.1	85	77	12.2	25.0	80	76	11.4	24.6	10	4.5	50
13	11	-10.3	1.6	2	0.9	90	18	14	-7.8	2.1	Tashkent	384570	41.27	69.27	1604	489	13.87	95.6	111	89	15.8	31.5	99	84	14.2	28.7	91	82	13.0	28.0	4	1.7	300
											VANUATU																						
66	97	18.8	13.8	2	0.7	290	68	104	19.9	14.8	Luganville	915540	15.52	167.22	144	44	14.62	100.8	148	83	21.1	28.3	146	82	20.8	28.0	143	82	20.4	27.7	9	4.0	100

International

HUMIDIFICATION 99.6% (35 hours below...)							99% (88 hours below...)												DEHUMIDIFICATION 0.4% (35 hours above...)				1% (88 hours above...)				2% (175 hours above...)				Summer Wind		
°F 1a	gr/lb 1b	°C 1c	g/kg 1d	mph 1e	m/s 1f	PWD 1g	°F 2a	gr/lb 2b	°C 2c	g/kg 2d	Location 3a	WMO# 3b	Lat. 3c	Long. 3d	ft. 3e	m 3f	psia 3g	kPa 3h	gr/lb 4a	°F 4b	g/kg 4c	°C 4d	gr/lb 5a	°F 5b	g/kg 5c	°C 5d	gr/lb 6a	°F 6b	g/kg 6c	°C 6d	mph 7a	m/s 7b	PWD 7c
											VENEZUELA																						
70	111	20.9	15.9	1	0.3	140	71	115	21.6	16.5	Caracas	804150	10.60	66.98	157	48	14.62	100.7	190	89	27.2	31.7	181	88	25.9	31.1	177	87	25.3	30.6	6	2.7	340
											VIETNAM																						
68	104	20.0	14.8	4	1.9	360	70	111	21.0	15.9	Ho Chi Minh City	489000	10.82	106.67	62	19	14.67	101.1	150	86	21.5	29.8	147	85	21.0	29.3	143	84	20.4	28.7	9	4.1	160
											WAKE ISLAND																						
71	115	21.8	16.4	15	6.5	40	72	119	22.4	17.0	Wake Island	912450	19.28	166.65	13	4	14.69	101.3	155	85	22.2	29.2	150	84	21.4	28.9	146	84	20.8	28.7	14	6.2	80
											WALLIS & FUTUNA ISLAND																						
72	119	22.1	17.0	2	1.1	160	73	123	22.8	17.6	Wallis Islands	917530	13.23	176.17	88	27	14.65	101.0	157	85	22.4	29.5	153	85	21.8	29.3	151	84	21.6	29.0	9	4.2	100
											YUGOSLAVIA																						
11	10	-11.5	1.4	6	2.5	10	16	13	-8.9	1.8	Belgrade	132720	44.82	20.28	324	99	14.53	100.1	105	80	15.0	26.8	99	78	14.1	25.4	94	76	13.4	24.4	6	2.7	120
10	9	-12.5	1.3	4	1.9	50	15	12	-9.6	1.7	Palic	130670	46.10	19.77	344	105	14.52	100.1	99	77	14.1	25.2	94	76	13.4	24.3	90	75	12.8	23.8	5	2.4	180
25	19	-4.1	2.7	7	3.1	360	27	21	-2.8	3.0	Podgorica	134620	42.37	19.25	108	33	14.64	100.9	104	79	14.9	25.9	99	80	14.2	26.4	96	78	13.7	25.8	7	3.3	180
											ZIMBABWE																						
45	53	7.0	7.6	6	2.5	120	46	55	8.0	7.9	Harare	677750	17.92	31.13	4931	1503	12.26	84.5	116	70	16.5	21.0	112	69	16.0	20.6	110	69	15.7	20.5	10	4.5	60

Chapter 32
Reference Material

Psychrometric Chart

Sea Level (Barometric Pressure = 29.921 in.hg.)

This chart was donated by James Judge, P.E. of Linric Company, Bedford, NH. The chart was constructed using the algorithms contained in the ASHRAE Handbook of Fundamentals 1997.

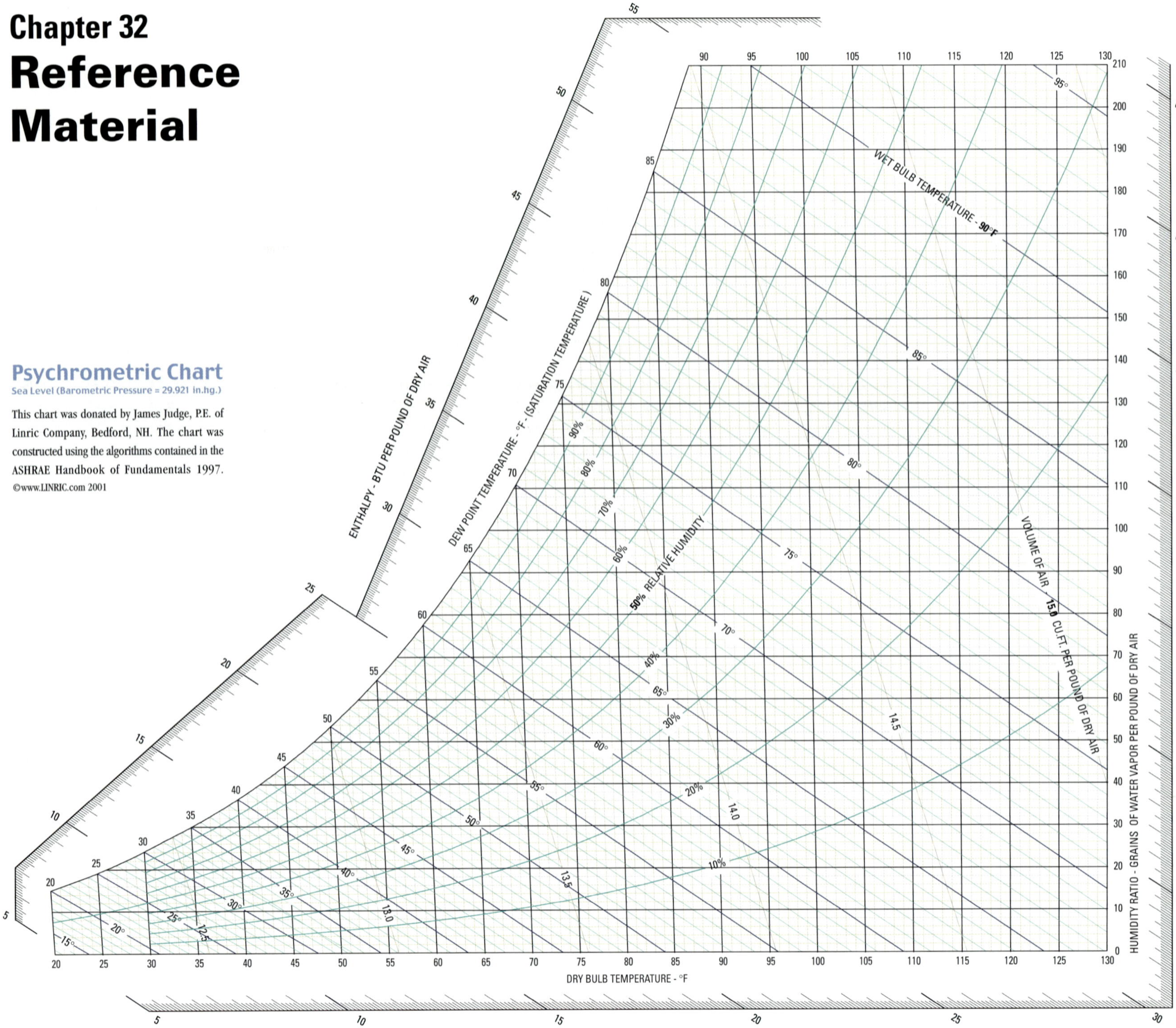

Table 32.1
Dew Point Temperatures With:
• Humidity Ratio At Saturation
• Vapor Pressure At Saturation

°F	gr/lb	in.hg.	°C	g/kg	kPa
160	2096	9.665	71.11	299.70	32.67
159	2023	9.436	70.56	289.34	31.89
158	1954	9.213	70.00	279.40	31.14
157	1887	8.994	69.44	269.87	30.40
156	1823	8.779	68.89	260.73	29.67
155	1762	8.569	68.33	251.95	28.96
154	1703	8.362	67.78	243.52	28.27
153	1646	8.161	67.22	235.42	27.58
152	1592	7.963	66.67	227.62	26.91
151	1539	7.769	66.11	220.13	26.26
150	1489	7.580	65.56	212.91	25.62
149	1440	7.394	65.00	205.97	24.99
148	1394	7.212	64.44	199.27	24.38
147	1348	7.034	63.89	192.83	23.78
146	1305	6.860	63.33	186.61	23.19
145	1263	6.689	62.78	180.62	22.61
144	1223	6.522	62.22	174.84	22.05
143	1184	6.359	61.67	169.27	21.49
142	1146	6.199	61.11	163.89	20.95
141	1110	6.042	60.56	158.69	20.42
140	1075	5.889	60.00	153.67	19.91
139	1041	5.740	59.44	148.82	19.40
138	1008	5.593	58.89	144.14	18.90
137	976	5.450	58.33	139.61	18.42
136	946	5.310	57.78	135.24	17.95
135	916	5.172	57.22	131.01	17.48
134	888	5.038	56.67	126.91	17.03
133	860	4.907	56.11	122.96	16.59
132	833	4.779	55.56	119.13	16.15
131	807	4.654	55.00	115.42	15.73
130	782	4.531	54.44	111.83	15.32
129	758	4.412	53.89	108.36	14.91
128	734	4.295	53.33	105.00	14.52
127	711	4.180	52.78	101.74	14.13
126	689	4.069	52.22	98.59	13.75
125	668	3.959	51.67	95.54	13.38
124	647	3.853	51.11	92.58	13.02
123	627	3.749	50.56	89.71	12.67
122	608	3.647	50.00	86.93	12.33
121	589	3.548	49.44	84.24	11.99
120	571	3.451	48.89	81.63	11.66
119	553	3.356	48.33	79.10	11.34
118	536	3.263	47.78	76.65	11.03
117	519	3.173	47.22	74.27	10.72
116	503	3.085	46.67	71.97	10.43
115	488	2.999	46.11	69.74	10.14
114	473	2.915	45.56	67.57	9.85
113	458	2.833	45.00	65.47	9.58
112	444	2.753	44.44	63.43	9.30
111	430	2.675	43.89	61.45	9.04
110	416	2.599	43.33	59.54	8.78
109	403	2.525	42.78	57.68	8.53
108	391	2.452	42.22	55.87	8.29
107	378	2.382	41.67	54.12	8.05
106	367	2.313	41.11	52.43	7.82
105	355	2.246	40.56	50.78	7.59
104	344	2.180	40.00	49.18	7.37
103	333	2.117	39.44	47.63	7.15
102	323	2.054	38.89	46.13	6.94
101	312	1.994	38.33	44.67	6.74
100	302	1.935	37.78	43.26	6.54
99	293	1.877	37.22	41.88	6.35
98	284	1.821	36.67	40.55	6.16
97	275	1.767	36.11	39.26	5.97
96	266	1.714	35.56	38.00	5.79
95	257	1.662	35.00	36.79	5.62
94	249	1.612	34.44	35.61	5.45
93	241	1.562	33.89	34.46	5.28
92	233	1.515	33.33	33.35	5.12
91	226	1.468	32.78	32.27	4.96
90	218	1.423	32.22	31.23	4.81
89	211	1.379	31.67	30.22	4.66
88	204	1.336	31.11	29.23	4.52
87	198	1.294	30.56	28.28	4.38
86	191	1.254	30.00	27.35	4.24
85	185	1.214	29.44	26.46	4.10

Table 32.1 (continued)
Dew Point Temperatures With:
- **Humidity Ratio At Saturation**
- **Vapor Pressure At Saturation**

°F	gr/lb	in.hg.	°C	g/kg	kPa
84	179	1.176	28.89	25.58	3.98
83	173	1.139	28.33	24.74	3.85
82	167	1.103	27.78	23.92	3.73
81	162	1.067	27.22	23.13	3.61
80	156	1.033	26.67	22.36	3.49
79	151	1.000	26.11	21.61	3.38
78	146	0.967	25.56	20.89	3.27
77	141	0.936	25.00	20.19	3.16
76	136	0.905	24.44	19.51	3.06
75	132	0.876	23.89	18.85	2.96
74	127	0.847	23.33	18.21	2.86
73	123	0.819	22.78	17.59	2.77
72	119	0.792	22.22	16.99	2.68
71	115	0.765	21.67	16.41	2.59
70	111	0.740	21.11	15.85	2.50
69	107	0.715	20.56	15.30	2.42
68	103	0.691	20.00	14.77	2.33
67	100	0.667	19.44	14.26	2.26
66	96	0.645	18.89	13.76	2.18
65	93	0.622	18.33	13.28	2.10
64	90	0.601	17.78	12.82	2.03
63	86	0.580	17.22	12.37	1.96
62	83	0.560	16.67	11.93	1.89
61	80	0.541	16.11	11.51	1.83
60	78	0.522	15.56	11.10	1.76
59	75	0.504	15.00	10.70	1.70
58	72	0.486	14.44	10.32	1.64
57	70	0.469	13.89	9.95	1.58
56	67	0.452	13.33	9.59	1.53
55	65	0.436	12.78	9.24	1.47
54	62	0.420	12.22	8.90	1.42
53	60	0.405	11.67	8.58	1.37
52	58	0.391	11.11	8.27	1.32
51	56	0.376	10.56	7.96	1.27
50	54	0.363	10.00	7.67	1.23
49	52	0.349	9.44	7.38	1.18
48	50	0.337	8.89	7.11	1.14
47	48	0.324	8.33	6.84	1.10

°F	gr/lb	in.hg.	°C	g/kg	kPa
46	46	0.312	7.78	6.59	1.05
45	44	0.300	7.22	6.34	1.02
44	43	0.289	6.67	6.10	0.98
43	41	0.278	6.11	5.87	0.94
42	39	0.268	5.56	5.64	0.91
41	38	0.258	5.00	5.43	0.87
40	37	0.248	4.44	5.22	0.84
39	35	0.238	3.89	5.02	0.81
38	34	0.229	3.33	4.82	0.77
37	32	0.220	2.78	4.64	0.74
36	31	0.212	2.22	4.46	0.72
35	30	0.204	1.67	4.28	0.69
34	29	0.196	1.11	4.11	0.66
33	28	0.188	0.56	3.95	0.64
32	26.5	0.1805	0.00	3.79	0.610
31	25.3	0.1724	-0.56	3.62	0.583
30	24.2	0.1646	-1.11	3.46	0.556
29	23.1	0.1572	-1.67	3.30	0.531
28	22.0	0.1501	-2.22	3.15	0.507
27	21.0	0.1432	-2.78	3.01	0.484
26	20.1	0.1367	-3.33	2.87	0.462
25	19.1	0.1304	-3.89	2.74	0.441
24	18.3	0.1244	-4.44	2.61	0.420
23	17.4	0.1186	-5.00	2.49	0.401
22	16.6	0.1131	-5.56	2.37	0.382
21	15.8	0.1078	-6.11	2.26	0.365
20	15.1	0.1028	-6.67	2.16	0.347
19	14.4	0.0980	-7.22	2.05	0.331
18	13.7	0.0933	-7.78	1.96	0.315
17	13.0	0.0889	-8.33	1.86	0.300
16	12.4	0.0847	-8.89	1.77	0.286
15	11.8	0.0806	-9.44	1.69	0.272
14	11.2	0.0767	-10.00	1.61	0.259
13	10.7	0.0730	-10.56	1.53	0.247
12	10.2	0.0695	-11.11	1.46	0.235
11	9.7	0.0661	-11.67	1.38	0.224
10	9.2	0.0629	-12.22	1.32	0.213
9	8.8	0.0598	-12.78	1.25	0.202

Table 32.1 (continued)
Dew Point Temperatures With:
• Humidity Ratio At Saturation
• Vapor Pressure At Saturation

°F	gr/lb	in.hg.	°C	g/kg	kPa
8	8.3	0.0569	-13.33	1.19	0.192
7	7.9	0.0541	-13.89	1.13	0.183
6	7.5	0.0514	-14.44	1.08	0.174
5	7.1	0.0488	-15.00	1.02	0.165
4	6.8	0.0464	-15.56	0.97	0.157
3	6.4	0.0440	-16.11	0.92	0.149
2	6.1	0.0418	-16.67	0.88	0.141
1	5.8	0.0397	-17.22	0.83	0.134
0	5.5	0.0377	-17.78	0.79	0.127
-1	5.2	0.0357	-18.33	0.75	0.121
-2	5.0	0.0339	-18.89	0.71	0.115
-3	4.7	0.0322	-19.44	0.67	0.109
-4	4.5	0.0305	-20.00	0.64	0.103
-5	4.2	0.0289	-20.56	0.60	0.098
-6	4.0	0.0274	-21.11	0.57	0.093
-7	3.8	0.0260	-21.67	0.54	0.088
-8	3.6	0.0246	-22.22	0.51	0.083
-9	3.4	0.0233	-22.78	0.49	0.079
-10	3.2	0.0220	-23.33	0.46	0.075
-11	3.1	0.0209	-23.89	0.44	0.071
-12	2.9	0.0198	-24.44	0.41	0.067
-13	2.7	0.0187	-25.00	0.39	0.063
-14	2.6	0.0177	-25.56	0.37	0.060
-15	2.4	0.0167	-26.11	0.35	0.057
-16	2.3	0.0158	-26.67	0.33	0.053
-17	2.2	0.0149	-27.22	0.31	0.050
-18	2.1	0.0141	-27.78	0.30	0.048
-19	2.0	0.0133	-28.33	0.28	0.045

°F	gr/lb	in.hg.	°C	g/kg	kPa
-20	1.84	0.012595	-28.89	0.263	0.0426
-21	1.74	0.011892	-29.44	0.249	0.0402
-22	1.64	0.011226	-30.00	0.235	0.0379
-23	1.55	0.010594	-30.56	0.222	0.0358
-24	1.46	0.009995	-31.11	0.209	0.0338
-25	1.38	0.009428	-31.67	0.197	0.0319
-26	1.30	0.008890	-32.22	0.186	0.0300
-27	1.23	0.008380	-32.78	0.175	0.0283
-28	1.16	0.007898	-33.33	0.165	0.0267
-29	1.09	0.007441	-33.89	0.156	0.0252
-30	1.03	0.007009	-34.44	0.147	0.0237
-31	0.97	0.006600	-35.00	0.138	0.0223
-32	0.91	0.006213	-35.56	0.130	0.0210
-33	0.86	0.005848	-36.11	0.122	0.0198
-34	0.80	0.005502	-36.67	0.115	0.0186
-35	0.76	0.005175	-37.22	0.108	0.0175
-36	0.71	0.004866	-37.78	0.102	0.0164
-37	0.67	0.004574	-38.33	0.096	0.0155
-38	0.63	0.004299	-38.89	0.090	0.0145
-39	0.59	0.004039	-39.44	0.084	0.0137
-40	0.55	0.003793	-40.00	0.079	0.0128

Knowns:

Q (supply air flow rate - scfm or l/s)
w_1 (humidity ratio of air in the controlled space - gr/lb or g/kg)
w_2 (humidity ratio of the supply air - gr/lb or g/kg)

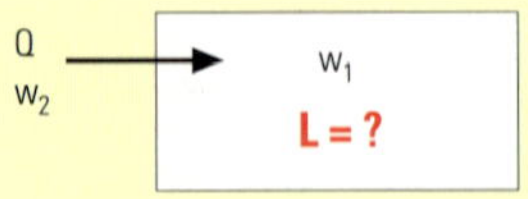

$$\textbf{lbs/h} = \frac{Q \times (w_1 - w_2) \times 4.5}{7000}$$

$$\textbf{kg/h} = \frac{Q \times (w_1 - w_2) \times 4.32}{1000}$$

Equation 32.2
Dehumidification Capacity Of Air
What is the dehumidification capacity of the supply air? Said another way; how much moisture can a known amount of air with a known humidity ratio remove from the conditioned space every hour?

Knowns: Q (supply air flow rate - scfm or l/s)
w_1 (humidity ratio of air in the controlled space - gr/lb or g/kg)
L (moisture load in the controlled space - lb/h or kg/h)

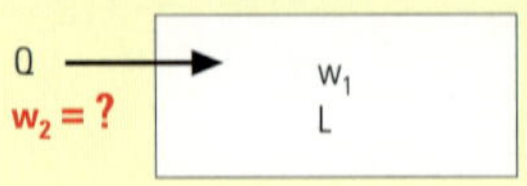

$$\textbf{gr/lb} = \left(\frac{7000 \times L}{Q \times 4.5}\right) - w_1$$

$$\textbf{g/kg} = \left(\frac{1000 \times L}{Q \times 0.003}\right) - w_1$$

Equation 32.3
Required Supply Air Humidity Ratio
If the supply air flow rate and the indoor humidity ratio are fixed, how dry does the supply air have to be in order to remove a known internal moisture load?

Knowns: w_1 (humidity ratio of air in the controlled space - gr/lb or g/kg)
w_2 (humidity ratio of the supply air - gr/lb or g/kg)
L (moisture load in the controlled space - lb/h or kg/h)

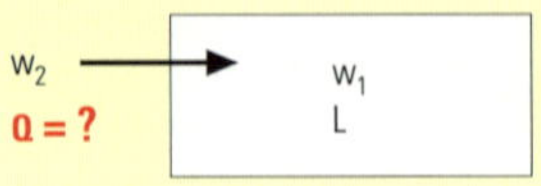

$$\textbf{scfm} = \frac{7000 \times L}{(w_1 - w_2) \times 4.5}$$

$$\textbf{l/s} = \frac{1000 \times L}{(w_1 - w_2) \times 0.003}$$

Equation 32.4
Required Supply Air Flow Rate
If the supply air humidity ratio and the indoor humidity ratio are both fixed, how much supply air is needed to remove a known internal moisture load?

Knowns: Q_1 (Air flow rate of the first air stream - scfm or l/s)
Q_2 (Air flow rate of the second air stream - scfm or l/s)
w_1 (humidity ratio of air in the controlled space - gr/lb or g/kg)
w_2 (humidity ratio of the supply air - gr/lb or g/kg)

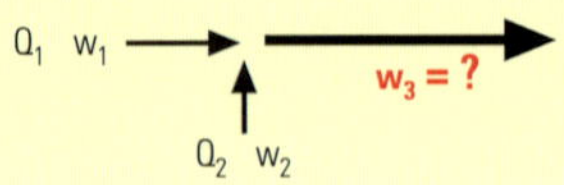

$$\textbf{gr/lb} = \frac{(Q_1 \times w_1) + (Q_2 \times w_2)}{Q_1 - Q_2}$$

$$\textbf{g/kg} = \frac{(Q_1 \times w_1) + (Q_2 \times w_2)}{Q_1 - Q_2}$$

Equation 32.5
Mixed-air Humidity Ratio
When two air flows with different humidity ratios are blended, what will be the humidity ratio of the resulting mixture?

Equation 32.6
Dehumidified Air vs. Bypass Air

Usually, a dehumidifier only needs to dry ***part*** *of the supply air rather than its entire volume. How much of the supply air should be dried by the dehumidifier in order to bring the entire supply air flow to the required humidity ratio?*

Knowns: Q_3 (Supply air flow rate - scfm or l/s)
w_1 (Humidity ratio of air leaving the dehumidifier - gr/lb or g/kg)
w_3 (Required humidity ratio of the supply air - gr/lb or g/kg)

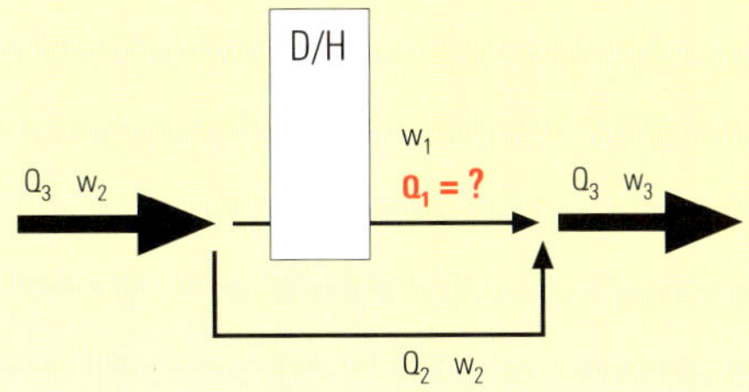

$$\textbf{scfm} = \frac{(Q_3 \times w_3) - (Q_3 \times w_2)}{w_1 - w_2}$$

$$\textbf{l/s} = \frac{(Q_3 \times w_3) - (Q_3 \times w_2)}{w_1 - w_2}$$

Table 32.7 Conversion Factors

To Convert	Into	Multiply By
Btu	joules	1055
Btu/h	watts	0.293
Btu/lb	kJ/kg	2.326
celsius	fahrenheit	(°C x 1.8) + 32
cubic feet/min (cfm)	litres/sec	0.4720
fahrenheit	celsius	(°F -32) ÷ 1.8
feet	metres	0.3048
feet/min (fpm)	metres/sec (m/s)	5.08×10^{-3}
gallons	pounds	8.34
gallons	ft^3	0.1337
gallons	litres	3.785
gallons/hr	millitres/sec	1.05
gallons/min	litres/sec	0.06308
grains/lb	grams/kg	0.143
grains	grams	15.43
grams	pounds	2.2046×10^{-3}
inches	millimetres	25.4
inches of water column	inches of mercury	0.07355
inches of water column	pascals	249
joules	Btu	9.48×10^{-4}

To Convert	Into	Multiply By
kilograms	pounds	2.205
kilowatts	Btu/h	3414
litres/sec	cfm	2.19
metres	feet	3.281
metres/sec	miles/hr	2.237
metres/sec	ft/min	88
millimetres	inches	0.03937
pascals	inches of water	4.016×10^{-3}
pascals	inches of mercury	2.959×10^{-4}
pints	pounds	1
pints/day	pounds/hr	0.0416
pounds (mass)	grains	7000
pounds (mass)	kilograms	0.4536
pounds of water	ft^3	0.01603
pounds of water	gallons	0.1199
pounds	pints	1
pounds/hr	pints/day	24
tons, refrigeration	Btu/h	12000
tons, refrigeration	kW	3.52
watts	Btu/h	3.414

Book Production Notes

This book was written, designed and illustrated by Lew Harriman of Mason-Grant in Portsmouth, New Hampshire. He also produced the press-ready electronic files, using a Macintosh G3/233 computer with 256MB of RAM and a Sony 19" monitor.

The page layout program was Adobe PageMaker 6.0. Illustrations were produced using Macromedia Freehand 8.0. Photos were edited by Lynne Fountain of Fountain Graphics in Kittery, Maine using Adobe Photoshop 6.0 for the Mac.

All type faces are PostScript. The text is Adobe Garamond Book Condensed. Headings and diagram text are Adobe Univers. The book title and all page numbers are Adobe Stone Sans.